**Plant Physiology,
Biochemistry and
Molecular Biology**

Plant Physiology, Biochemistry and Molecular Biology

edited by
DAVID T. DENNIS
and
DAVID H. TURPIN

*Department of Biology,
Queen's University,
Kingston,
Canada*

Longman Scientific & Technical
Longman Group UK Limited
Longman House, Burnt Mill, Harlow
Essex CM20 2JE, England
and Associated Companies throughout the world.

© Longman Group UK Limited 1990

All rights reserved; no part of this publication may be reproduced, stored in a retrieval system, or transmitted in any form or by any means, electronic, mechanical, photocopying, recording, or otherwise, without the prior written permission of the Publishers, or a licence permitting restricted copying in the United Kingdom issued by the Copyright Licensing Agency Ltd, 33 – 34 Alfred Place, London, WC1E 7DP

First published in 1990

British Library Cataloguing in Publication Data
Plant physiology, biochemistry and molecular biology.
 1. Plants
 I. Dennis, David T.   II. Turpin, David H.
 581

ISBN 0-582-46052-2 CSD
ISBN 0-582-06114-8 PPR

53316

Set in 9/11½ pt Times
Produced by Longman Singapore Publishers (Pte) Ltd.
Printed in Singapore

Contents

Preface vii
List of contributors ix

I The Control of Metabolism 1

1 Fundamentals of gene structure and control 3
   D. D. Lefebvre
2 Regulation of gene expression during development 16
   R. Casey
3 Biochemical regulation 28
   W. Plaxton
4 Regulation by compartmentation 45
   D. T. Dennis and M. J. Emes

II Cytosolic Carbon Metabolism 57

5 Carbohydrate synthesis and degradation 59
   N. J. Kruger
6 Glycolysis, the oxidative pentose phosphate pathway and anaerobic respiration 77
   J. A. Miernyk

III Mitochondrial Metabolism 101

7 Mitochondrial structure 103
   W. Newcomb
8 Carbon metabolism in mitochondria 106
   T. ap Rees
9 Oxidation of mitochondrial NADH and the synthesis of ATP 124
   H. Lambers

IV Mitochondrion–Cytosol Interaction 145

10 The mitochondrial genome and its expression 147
    M. Gray
11 Protein import into the mitochondrion 160
    K. Freeman, S. Reichling and A. Balogh
12 Metabolite exchange between the mitochondrion and the cytosol 173
    R. Douce and M. Neuburger

V Photosynthesis 191

13 Plastid structure and development 193
    W. Newcomb
14 Molecular biology of photosynthesis in higher plants 198
    J. Mullet
15 The formation of ATP and reducing power in the light 212
    N. Nelson and B. Prezelin
16 Ribulose 1,5-bisphosphate carboxylase/oxygenase: mechanisms, activation and regulation 224
    R. G. Jensen
17 The reductive pentose phosphate pathway and its regulation 249
    F. D. Macdonald and B. B. Buchanan
18 Photorespiration and CO_2-concentrating mechanisms 263
    D. T. Canvin
19 The flux of metabolites in C4 and CAM plants 284
    R. C. Leegood and C. B. Osmond

VI Chloroplast–Cytosol Interactions 309

20 Transport of proteins into chloroplasts 311
 K. Keegstra
21 The flux of carbon between the chloroplast and the cytosol 319
 M. Stitt

VII The Formation and Breakdown of Lipids 339

22 The structure and formation of microbodies 341
 C. Halpin and M. Lord
23 Fatty acid and lipid biosynthesis and degradation 351
 J. E. Andrews and J. Ohlrogge
24 Terpene biosynthesis and metabolism 363
 C. A. West

VIII Nitrogen Metabolism 381

25 The molecular biology of N metabolism 383
 C. P. Vance and S. M. Griffith
26 N_2 fixation, NO_3^- reduction and NH_4^+ assimilation 399
 D. B. Layzell
27 Amino acid and ureide biosynthesis 417
 R. Ireland
28 Interactions between nitrogen assimilation, photosynthesis and respiration 430
 D. H. Turpin and H. G. Weger
29 Long-distance transport of carbon and nitrogen from sources to sinks in higher plants 442
 M. Peoples and R. Gifford
30 Protein turnover 456
 P. M. Hatfield and R. D. Vierstra
31 Protein storage and utilization in seeds 464
 J. D. Bewley and J. S. Greenwood

IX Prospects for Plant Improvement 479

32 Fundamentals of gene transfer in plants 481
 B. Miki and V. N. Iyer
33 The biochemical basis for plant improvement 498
 C. R. Somerville

Index 511

Preface

The last decade has seen a rapid expansion in our knowledge of plant metabolism. Our motivation for preparing this text was our own need for a plant physiology, biochemistry and molecular biology text which provided both beginning and advanced students with an overview of carbon and nitrogen metabolism while at the same time allowing them the opportunity of examining the subject with some vigor.

The purpose of this text is to examine the assimilation and metabolism of carbon and nitrogen. Our approach is to deal with these processes in an integrative fashion assessing the physiology, biochemistry and molecular biology of each topic being discussed. The book is divided into nine sections. Section I examines the principles of metabolic control from the fundamentals of gene structure and the regulation of expression, through biochemical regulation and the role of compartmentation. Section II deals with cytosolic carbon metabolism, specifically the metabolism of hexoses and polysaccharides, and the integration and control of glycolysis, the oxidative pentose phosphate pathway and anaerobic respiration. Section III examines the use of glycolytic carbon in mitochondrial metabolism. This section emphasizes that mitochondrial carbon metabolism involves both anabolic and catabolic processes and that these reactions are closely integrated with the oxidation of NADH and the formation of ATP. Organelles cannot be viewed in isolation and Section IV explores the interactions which occur between the mitochondrion and cytosol at the genomic, protein and metabolite levels.

In Section V, the focus shifts to the chloroplast and the process of photosynthesis. These processes are examined first at the molecular level followed by an assessment of the processes for ATP and NADPH formation in the light. The control of carboxylation and oxygenation and the regulation of both the photosynthetic carbon reduction and oxidation cycles are examined. Finally, strategies by which plants have suppressed photorespiration by CO_2-concentrating mechanisms are examined. These include CO_2 and HCO_3^- pumps, CAM and C4 photosynthesis. Section VI examines chloroplast interactions with the cytosol through protein and metabolite exchange. Section VII explores the function and structure of microbodies and the process of fatty acid, lipid and terpene biosynthesis.

Section VIII provides a synthesis of the processes involved in nitrogen metabolism at the molecular, biochemical and physiological levels. It deals with genetic and molecular control of N assimilation and its biochemistry. We have also endeavored to illustrate the important interactions that occur between photosynthesis, respiration and nitrogen assimilation. This section also provides a whole plant component which deals with long distance transport of carbon and nitrogen between sources and sinks and the role of protein turnover, storage and utilization in seeds.

Finally, Section IX explores the fundamentals of gene transfer in plants and the potential to exploit these techniques in a rational approach to plant improvement.

Our hope is that this book will provide students of plant physiology with a strong awareness of primary processes in plant metabolism and an appreciation for the integration and control of carbon and nitrogen metabolism.

Contributors

J. E. Andrews
Agrigenetics, 5649 E. Buckeye Road, Madison, WI 53716 U.S.A.

T. ap Rees
Botany School, University of Cambridge, Cambridge, CB2 3EA, U.K.

A. Balogh
Department of Biochemistry, McMaster University, Hamilton, Ontario L8N 3Z5, Canada

J. D. Bewley
Department of Botany, University of Guelph, Guelph, Ontario, N1G 2W1, Canada

B. B. Buchanan
Division of Molecular Plant Biology, University of California, Berkeley, CA 94720, U.S.A.

D. T. Canvin
Biology Department, Queen's University, Kingston, Ontario, K7L 3N6, Canada

R. Casey
AFRC Institute of Plant Science Research, Department of Applied Genetics, John Innes Institute, Colney Lane, Norwich NR4 7UH, U.K.

D. T. Dennis
Department of Biology, Queen's University, Kingston, Ontario, K7L 3N6, Canada

R. Douce
Laboratoire de Physiologie Cellulaire Végétale, Unité Associeé au CNRS No. 521, Department Recherche Fondamentale, Centre d'Etudes Nucleaires, Université Joseph Fournier, 85X 38041 Grenoble, Cedex, France

M. J. Emes
Department of Cell & Structural Biology, University of Manchester, School of Biological Sciences, Williamson Building, Oxford Rd., Manchester, M13 9PL, U.K.

K. Freeman
Department of Biochemistry, McMaster University, Hamilton, Ontario L8N 3Z5, Canada

R. Gifford
CSIRO, Division of Plant Industry, G.P.O. Box 1600, Canberra, ACT, Australia 2601

M. W Gray
Department of Biochemistry, Dalhousie University, Halifax, N.S. B3H 4H7, Canada

J. S. Greenwood
Department of Botany, University of Guelph, Guelph, Ontario N1G 2W1, Canada

S. M. Griffith
United States Department of Agriculture, Agricultural Research Service and Department of Agronomy & Plant Genetics, The University of Minnesota, St. Paul, MN 55108, U.S.A.

C. Halpin
Department of Biological Sciences,

University of Warwick, Coventry, CV4 7AL, U.K.

P. M. Hatfield
Department of Horticulture,
University of Wisconsin-Madison, Madison, WI 53706, U.S.A.

R. Ireland
Biology Department, Mount Allison University, Sackville, New Brunswick E0A 3C0, Canada

B. Iyer
Plant Research Centre, Agriculture Canada, Ottawa, Ontario, K1A 0C6, Canada

R. G. Jensen
Biochemistry Department,
University of Arizona, Tucson, AZ 85721, U.S.A.

K. Keegstra
Department of Botany, University of Wisconsin, Madison, WI 53706, U.S.A.

N. J. Kruger
Department of Biochemistry, AFRC Institute of Arable Crop Research, Rothamsted Experimental Station, Harpenden, Herts AL5 2JQ, U.K.

H. Lambers
Department of Plant Ecology and Evolutionary Biology, University of Utrecht,
Lange Nieuwstraat 106, Utrecht 3512 PN, The Netherlands

D. B. Layzell
Department of Biology, Queen's University, Kingston, Ontario, K7L 3N6, Canada

R. C. Leegood
Robert Hill Institute and Department of Plant Sciences, University of Sheffield, Sheffield, S10 2TN, U.K.

D. D. Lefebvre
Department of Biology, Queen's University, Kingston, Ontario, K7L 3N6, Canada

M. Lord
Department of Biological Sciences,
University of Warwick, Coventry, CV4 7AL, U.K.

F. D. Macdonald
Division of Molecular Plant Biology,
University of California, Berkeley, CA 94720, U.S.A.

J. A. Miernyk
Northern Regional Research Center, USDA, ARS, 1815 N. University St., Peoria, IL 61604, U.S.A.

B. Miki
Plant Research Centre, Agriculture Canada, Ottawa, Ontario, K1A 0C6, Canada

J. Mullet
Department of Biochemistry & Biophysics, Texas A & M University, College Station, TX 77843-2128, U.S.A.

N. Nelson
Department of Biological Sciences,
University of California, Santa Barbara, CA 93106, U.S.A.

M. Neuburger
Laboratoire de Physiologie Cellulaire Vegétale, Unité Associeé au CNRS No. 521,
Department Recherche Fondamentale,
Centre d'Etudes Nucleaires,
Université Joseph Fournier, 85X 38041 Grenoble, Cedex, France

W. Newcomb
Department of Biology, Queen's University, Kingston, Ontario, K7L 3N6, Canada

J. B. Ohlrogge
Department of Botany and Plant Pathology, Michigan State University, East Lansing, MI 48824-112, U.S.A.

C. B. Osmond
Department of Botany, Duke University, Durham, NC 27706, U.S.A.

M. Peoples
CSIRO, Division of Plant Industry,
G.P.O. Box 1600, Canberra, ACT, Australia 2601

W. Plaxton
Department of Biology, Queen's University, Kingston, Ontario, K7L 3N6, Canada

B. B. Prézelin
Department of Biological Sciences, University of
California, Santa Barbara, CA 93106, U.S.A.

S. Reichling
Department of Biochemistry,
McMaster University, Hamilton, Ontario,
L8N 3Z5, Canada

C. R. Somerville
MSU/DOE Plant Research Laboratory,
Michigan State University, East Lansing,
MI 48824, U.S.A.

M. Stitt
Universität Bayreuth,
Lehrstuhl für Pflanzenphysiologie,
Universitätsstrasse 30, D-8580 Bayreuth,
W. Germany

D. H. Turpin
Department of Biology, Queen's University,
Kingston, Ontario, K7L 3N6, Canada

C. Vance
United States Department of Agriculture,
Agricultural Research Service and Department
of Agronomy and Plant Genetics, The
University of Minnesota, St. Paul, MN 55108,
U.S.A.

R. D. Vierstra
Department of Horticulture,
University of Wisconsin-Madison, Madison,
WI 53706, U.S.A.

H. G. Weger
Department of Biology, Queen's University,
Kingston, Ontario, K7L 3N6, Canada

C. A. West
Department of Chemistry and Biochemistry,
University of California, Los Angeles,
Los Angeles, CA 90024, U.S.A.

I

The Control
of Metabolism

1 Fundamentals of gene structure and control
Daniel D. Lefebvre

Introduction

The purpose of this chapter is to familiarize the reader with the basic features of gene structure and regulation. It is not intended to deal in any way with recombinant DNA methodology, but to present an outline of current knowledge so that the more detailed chapters on plant genes and their expression can be more easily understood. In addition, since the examples employed in this chapter will often be of organisms other than plants, the reader will be presented with a perspective from which to judge the present state of understanding of gene regulation in plants.

The field of molecular genetics is enormous and extremely dynamic. Areas of most relevance to plant molecular genetics will therefore be emphasized. These include the composition of prokaryotic and eukaryotic genes, transcription and translation, production and characterization of messenger RNA. Some details will be given on the organization of genes, particularly on the identification of consensus sequences of regulatory regions, and the post-transcriptional control of gene expression. Where possible plant genes will be examined and compared with those of bacteria or animals.

A detailed elementary account of the basic processes of protein synthesis can be found in textbooks such as Goodenough (1984).

The gene is the basic hereditary unit in all living organisms. It is defined as a segment of deoxyribonucleic acid (DNA) involved in the production of a polypeptide chain. Each gene is composed of a coding region and the regions preceding and following it. Genes may also contain intervening sequences (non-coding regions) which are situated within the coding region. The coding region provides the information for the appropriate amino acid sequence of a protein and is also referred to as the open reading frame (ORF). Regulatory sequences usually precede the ORF and are responsible for the control and rate of production of ribonucleic acid (RNA) synthesized from the DNA template. Their characteristics are a property of a number of specific types of DNA sequences which act as recognition sites for various proteins. Transcription is the process by which messenger RNAs (mRNA) are produced. This mRNA acts in turn as the direct template for protein synthesis. The triplet base code of the nucleic acid sequence (genetic code) in the form of codons governs the sequential joining of amino acids to make an enzyme or structural protein. The translation of mRNAs into their corresponding proteins occurs on ribosomes situated either in the cytoplasm or organelle; the latter occurs when the gene is within the mitochondrial or chloroplast genome.

Transcription and translation are the two major genetic events where the control of gene expression can be asserted. Post-translational control also occurs and this aspect of metabolic regulation is dealt with in Chapters 3 and 30.

Prokaryotic genes

The operon

The genes of prokaryotic organisms and their regulation were the first to be studied. In bacteria, genes for proteins catalyzing related activities are often arranged in tandem so that they are transcribed on a single polycistronic mRNA in which each cistronic component is the messenger template for a single protein. Such a collection of genes is referred to as an operon.

Bacteria must be able to adapt to rapid environmental changes in a similar manner as plants; hence their genes are constructed to enable them to function at highly variable rates. When required mRNA molecules can be produced at high levels, but only when their genes receive appropriate signals. The compounds responsible for these signals are called inducers.

The classical study of Jacob and Monod in 1961 on the lactose (*lac*) operon in *Escherichia coli* paved the way to an understanding of the regulatory components of prokaryotic genes (Fig. 1.1). The *lac* operon contains the structural genes for three proteins: β-galactosidase which hydrolyzes lactose into glucose and galactose; lactose permease which is the transporter for the uptake of lactose into the cell; and thiogalactosidase transacetylase, an enzyme that transfers an acetyl group from acetyl-CoA to β-galactosides. The last enzyme appears to be involved in the detoxification of non-metabolizable β-galactosides which can then be secreted from the cell.

Under normal conditions lactose is not available to cells as a source of carbon skeletons and glucose is employed for this purpose. Hence, in the absence of lactose the *lac* operon is dormant or is in a repressed state. A specific regulator gene is involved in this repression which is accomplished through the production of a repressor protein. There are two domains on this repressor: one has an affinity for the *lac* operator which is the DNA sequence of the operon situated just before (upstream of) the open reading frames; the other binds lactose, the inducer. Lactose-free repressor binds to the operator and prevents transcription by RNA polymerase. When lactose forms a complex with the repressor it no longer binds to the operator and the mRNA is transcribed.

The promoter

The promoter is the site where RNA polymerase physically binds to genes. Since in the *lac* operon this is only slightly upstream of the operator, the repressor appears to sterically interfere with this binding, thereby preventing transcription.

The sequence of bases in promoters determine the maximum possible rate of gene transcription. Between different genes this rate can vary by over three orders of magnitude. In *E. coli* promoters there are two separate sets of nucleotide sequences which are highly conserved. If the transcriptional start site is designated base position +1, at −35 there is the RNA polymerase recognition and binding site. At −10 the double-stranded DNA is thought to open for transcription of the mRNA from the complementary minus strand template of DNA. Transcription is in the 5′ to 3′ direction based on the sugar–phosphate bonds of the DNA coding or plus strand, that which contains the protein coding sequence.

Consensus sequences

The consensus sequence of a particular functional component of a gene is a DNA sequence that

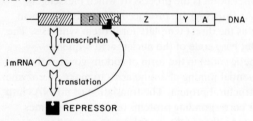
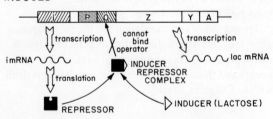

Fig. 1.1 Lactose operon. In the absence of lactose inducer the repressor binds to the operator (O) preventing transcription of the structural genes. When lactose is present it binds to form an inducer–repressor complex which cannot bind the operator and transcription proceeds to produce a single messenger encoding β-galactosidase (Z), β-galactoside permease (Y) and β-galactoside transacetylase (A).

appears to be conserved with only minor changes in a group of closely related genes. Over 100 promoters have been sequenced in *E. coli* and only the −35 and −10 regions are consensus sequences. The RNA polymerase binding site at −35 has the consensus sequence TTGACA, and that at −10 TATAAT which is also referred to as the Pribnow box. Conservation of the base at each position varies from 45 to 100% and alterations can mildly or severely affect the rate of transcription. The distance between these two consensus sequences is 16–18 base pairs (bp) in 90% of promoters, with limits appearing to be as low as 15 or as high as 20. This distance is probably critical with respect to the geometry of RNA polymerase.

The site for initiation of RNA polymerase activity – the start base for transcription – is usually a purine (A or G). Commonly it is the center of the sequence CAT, although this is certainly not an essential signal.

In a few bacterial promoters one of the consensus sequences is missing and ancillary proteins are required for RNA polymerase to initiate transcription. However, these are rare exceptions and typically the −35 and −10 boxes are present.

Although consensus sequences are useful and attractive means of identifying essential DNA regions, they tend to over-simplify the signal concept. Promoter efficiency cannot be predicted entirely from the degree of homology with the consensus. Almost all bacterial promoters vary from the consensus and, in addition, adjacent base partners within the consensus can have dramatic positive or negative effects on transcription rates.

The *lac* operon is a prime example of negative genetic control. Bacteria also employ positive regulation of gene transcription. The best understood system is that of cyclic AMP (cAMP). In the absence of glucose the concentration of cAMP builds up and binds to CAP, a DNA-binding protein. This complex in turn binds to promoters which activate operons involved in the utilization of alternative energy sources to glucose.

Cis- and trans-acting factors

In both prokaryotes and eukaryotes, the control of transcription involves DNA sequences and factors which can act in two ways: (1) directly, by being associated with the gene and recognized by a controlling factor (*cis*-regulating) or (2) indirectly, by producing a product that recognizes a target gene or genes (*trans*-regulating). *Cis*-regulating elements include promoters and closely associated DNA regions such as operators which exert their effects as a consequence of their DNA sequences being associated with the DNA of the gene. *Trans*-acting factors need not be in close proximity with the target gene. The concept of *cis*- and *trans*-acting factors applies equally well to eukaryotic genes. In eukaryotes the genes for *trans*-acting factors are often located on a different chromosome from the the target gene. These genes specify diffusible products which can act on one or a number of separate target sites simultaneously.

Mutations which result in an inability to transcribe a gene can be *cis*- or *trans*-acting. For example, a *cis* defect in *lac P* of the *lac* operon (Fig. 1.1) prevents the binding of RNA polymerase to this DNA region effectively preventing transcription of the gene. This is genotypically distinct from a *trans*-acting *lac I* mutation which alters this gene's product, the repressor, in such a way that it is no longer able to bind lactose. These mutant cells, however, are phenotypically identical.

Transcriptional attenuation

The process of transcription is controlled not only at initiation as described above but also by the process of transcriptional attenuation which has been described in many of the operons encoding enzymes involved in amino acid biosynthesis in bacteria. You will recall that prokaryotic translation can begin well before transcription is complete. Attenuation involves the cessation of transcription after the translation of only a small leader region of 14 to 32 residues in length. This region is usually rich in the amino acid that the respective operon is involved in synthesizing. Two interesting features of attenuation are that less than 300 bp of DNA contain all the genetic information needed for very effective control and that RNA secondary structures known as hairpins are directly involved (Fig. 1.2).

Transcription of the entire operon is dependent on the translation rate of the leader peptide. When the

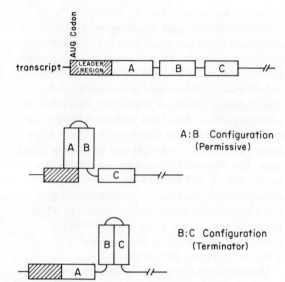

Fig. 1.2 Transcription attenuator in operons of amino acid biosynthesis. Translation of the messenger RNA begins shortly after the mRNA starts to be transcribed from the DNA. When the amino acid of the operon's biosynthetic pathway is at a deficient level, translation stops on the leader sequence of the protein because this region is rich in this amino acid. Stalling allows the A:B secondary structure to form and prevents the B:C configuration. This A:B secondary structure enhances the production of the messenger. If translation of the leader region proceeds without stalling because of sufficient levels of the amino acid, the B:C terminator forms instead and this prevents further transcription.

amino acid for which the operon is responsible is at a low level in the cell, stalling of translation occurs within the leader whereas under sufficient amino acid nutrition, translation proceeds quickly to the stop codon of the leader peptide. Therefore, the codon content (i.e. the number of codons of the specific amino acid) in the initial translation region acts as a fine tuning parameter for this process which in turn controls transcription in the following manner. When the amino acid is plentiful the nascent RNA still closely associated with the DNA adopts a termination structure which is a hairpin configuration that prevents further transcription. Alternatively, if translation is delayed during the leader peptide because the amino acid is scarce, this terminator does not form and transcription of all major structural genes proceeds unimpeded, thereby leading to the synthesis of the enzymes responsible for producing the respective amino acid.

Translational control

Control of gene expression can also occur post-transcriptionally. Initiation of transcription requires a purine-rich nucleotide region of six to eight bases just upstream from the AUG start codon. Shine and Dalgarno identified this as a ribosome-binding site in 1974. They found that there was homology between this RNA region and a section of the 16S ribosomal RNA. Base pairing between the mRNA and the 16S RNA could implement initiation. mRNA is translated most efficiently if this sequence is exactly eight bases upstream of the start codon.

In certain circumstances proteins bind to specific mRNA Shine–Dalgarno sequences, thereby preventing the initiation of translation. For example, this is how ribosomal (r) protein synthesis is kept strictly in parallel with the production of ribosomal RNA. Specific r-proteins of their respective operons are able to bind rRNA in a very strong association while they bind their own mRNA at the Shine–Dalgarno sequences less strongly. When rRNA production is below that of r-proteins the latter bind their own mRNAs, thus preventing translation.

Secondary structure loops in mRNA can also modify translation. These override the presence of an otherwise effective initiation region for translation by physically preventing the binding of the Shine–Dalgarno region to the 16S rRNA.

Another possible means of regulating translation is codon usage frequency. The levels of transfer RNA (tRNA) species within cells can be different and result in different translation rates depending on whether the messenger requires tRNAs with prevalent or rare anticodons.

Regulation by antisense RNA

Antisense RNA is a transcript that has a high degree of complementation with a target mRNA so that it can hybridize with the mRNA *in vivo*. Hence, the antisense RNA acts as a repressor of the function of the target RNA. Although not always the case, the locus of the target and regulatory RNA are often the same; the DNA strands are transcribed in opposite directions.

This type of regulation occurs at translation, an illustration of which is the prokaryotic transposable element Tn10. This is regulated by a small antisense RNA transcript (pOUT) which contains a 35 base stretch exactly complementary to a region of the transposable gene (pIN) which includes the Shine–Dalgarno sequence and the translation-initiation codon. Regulation is effected by complementation interference. These antisense RNAs are also referred to as mRNA interfering-complementary RNA or micRNA.

Inhibition by micRNA may be due to several mechanisms (Harland and Weintraub, 1985). Perhaps the double-stranded RNA cannot be transported across the nuclear envelope into the cytoplasm. Decreases in stability of mRNA–micRNA hybrids has also been proposed since it is known that *in vitro* synthesized antisense RNA injected into the frog *Xenopus* oocytes inhibits translation of mRNAs already bound to polyribosomes in the cytoplasm.

In theory, the mechanism of antisense RNA regulation should function equally well in prokaryotic and eukaryotic cells, and in fact, this is the case. Functional micRNA inhibition has been demonstrated in a number of higher organisms including mammals and higher plants. Ecker and Davis (1986) introduced into carrot cells synthetic gene constructs composed of promoters associated with a reporter gene (chloramphenicol acetyltransferase) in correct and reverse orientation (Fig. 1.3). Protein synthesis was effectively inhibited when the nonsense orientation was present.

Structure of cyanobacterial genes

Although the available knowledge of cyanobacterial gene organization is not as extensive as that for bacterial genes, there is reason to believe that they are very similar and probably regulated identically.

The genes of the nitrogenase complex of the bacterium *Klebsiella* are sufficiently homologous to those from the cyanobacterium, *Anabaena*, that they can be used as high stringency probes to detect these

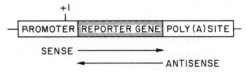

Fig. 1.3 Gene constructs employed in the study of plant antisense RNA by Ecker and Davis. The promoters employed in this construct were from nopaline synthase, cauliflower mosaic virus 35S RNA, and phenylalanine ammonia lyase. The reporter gene was chloramphenicol acetyltransferase (CAT). The nopaline synthase polyadenylation signal was used in all constructs. Experiments were performed by the electroporation of sense and antisense genes in different proportions to one another into carrot protoplasts followed by the measurement of transient CAT expression.

genes in the blue–green alga (Haselkorn, 1986). In the bacterium the genes are arranged in the *nifHDK* operon which contains three open reading frames, each for a polypeptide component of the complex. This is also the case in heterocysts in *Anabaena*. Heterocysts are nitrogen-fixing structures which differentiate from normal vegetative cells. During this differentiation an interesting gene rearrangement occurs which results in a novel means of gene regulation (Fig. 1.4). In vegetative cells the *nifD* ORF is interrupted by 11 kilobase pairs (kbp) of DNA, effectively preventing any possibility of

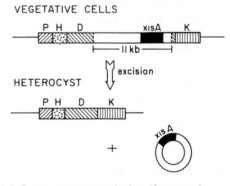

Fig. 1.4 Gene rearrangement in the *nif* operon of *Anabaena*. In vegetative cells which do not fix nitrogen, *nif* D is interrupted by an 11 kbp insert, which is excised during cellular differentiation to heterocysts. Excision is performed by the gene product of *xisA*, itself located within the insert. Excision results in a functional operon. P, promoter; H, D, and K are open reading frames.

transcription. Removal of the insertion creates the mature functional operon in the heterocyst. The gene for excision, *xisA*, is actually located on the 11 kbp sequence which circularizes into a stable plasmid during the differentiation.

Anabaena genes sometimes possess more than one promoter. This is the case with *glnA*, a gene involved in ammonium assimilation (Tumer *et al.*, 1983). When ammonia is plentiful, promoter sequences similar to those of *E. coli* are regulatory. However, under conditions where anaerobic nitrogen fixation occurs and involves the *nif* gene products, a promoter region similar to those of the *nif* genes controls transcription. In this way the synthesis of related enzymes is maintained in the same proportion.

Gene arrangement in the cauliflower mosaic virus

The cauliflower mosaic virus (CaMV) genome is a circular double-stranded DNA molecule approximately 8 kbp in length. In the host two transcripts are produced. One is from the entire viral genome (35S) and the other (19S) is the messenger for the portion of the genome responsible for the inclusion body protein involved in viral assembly. Although CaMV has promoters and a polyadenylation signal for these transcripts that are characteristics of eukaryotic genes (see next section of this chapter), the 35S transcript is polycistronic containing seven ORFs and, therefore, prokaryotic in nature (Hull and Covey, 1985). A 'relay race' model involving the movement of ribosomes directly from one ORF to the next has been proposed for the translation of the first four genes of this large transcript. Foreign genes inserted in this region are also strongly expressed (Lefebvre *et al.*, 1987). There is some evidence, however, that the other three genes of this polycistronic region of the 35S transcript are translated from smaller RNAs, derived post-transcriptionally from the 35S species, possibly by the splicing processes available in the eukaryotic host cells.

Eukaryotic genes

Eukaryotic genes differ in a number of ways from those of lower organisms. For example, they are not arranged in operons and, therefore, each coding region is associated with its own regulatory sequence. The coding regions are often interrupted, requiring post-transcriptional splicing in order to form mature mRNAs that are functional. Transcription of mRNA is performed by RNA polymerase II, one of three eukaryotic RNA polymerases. RNA polymerase I and III are responsible for the synthesis of ribosomal and transfer RNA, respectively. In prokaryotes all these functions are performed by a single RNA polymerase.

The messenger

Eukaryotic transcription initially results in the formation of heterogeneous nuclear RNA (hnRNA) in the nucleoplasm which is the precursor of mRNA. The average hnRNA molecule is four to five times longer than that of the average mRNA (1800–2000 bases). After transcription hnRNA is rapidly 5' capped with 7-methylguanosine residues linked to the RNA by a triphosphate bridge (m^7Gppp). Poly(A)polymerase uses ATP as a substrate to add a poly(A)-tail to the 3' end of certain mRNAs. The 5' caps and 3' poly(A)-tails remain in the resultant mRNA species, although a few bases may be removed from the 3' end during transport from the nucleus to the cytoplasm.

Processing of hnRNA into mature mRNA involves the removal of the non-translated regions, introns or intervening sequences, situated within the coding regions. Exons are RNA segments found both in the precursor hnRNA and in the mature RNA. The presence of introns may have resulted from genetic recombination bringing together domains from originally separate genes. Perhaps introns, which can be very long, act by ensuring that coding sequences are kept intact during genetic crossing-over events. They would also be expected to increase the likelihood of DNA crossing-over events which would be expected, in turn, to have had important evolutionary effects at the level of individual alleles.

Specific base sequences delineate exon–intron

boundaries, but RNA sequence complementation between the two boundaries is not involved in the bringing together of the appropriate sites for splicing. Instead, adapter RNAs which are attached to the splicing enzymes have been implicated in this process. The two consensus sequences in all higher eukaryotes for the left and right intron junctions, the donor and acceptor sites respectively, are

$$\text{Left Junction} \downarrow$$
$$5' \ldots \text{exon} \ldots \text{AG} \quad \text{GTAAGT} \ldots \text{intron} \ldots$$
$$\text{Right Junction} \downarrow$$
$$\ldots \text{intron} \ldots \text{Py}_{10} \text{CAG} \ldots \text{exon} \ldots 3'$$

where Py is a pyrimidine (C or T). Nuclear splicing requires ATP and occurs in two stages (Fig. 1.5). Initially the left junction is cut and the free 5' end of the intron forms a 5'–2' bond with an A which is always present at position 6 in a consensus sequence called the TACTAAC box situated 18–40 nucleotides upstream of the right junction. This circular DNA, a lariat, is released and linearized when the right junction is cut and spliced to the left

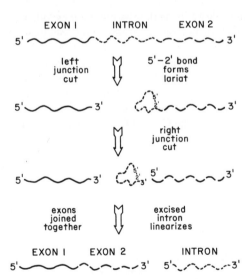

Fig. 1.5 Removal of introns produces mature messenger RNA. *In vitro* studies have shown that the left exon–intron junction is cleaved and the intron forms a lariat with a 5' to 2' bond. The right junction is then cut and the exons are covalently linked together. The intron assumes a linear form which is probably rapidly degraded *in vivo*.

junction in stage two of the process. Except for this invariant A the TACTAAC box sequence is not particularly well conserved between genes and this is essentially only a convenient label.

Most eukaryotic genes contain introns. Two notable plant examples which are intron free are the genes for zein storage proteins (Heidecker and Messing, 1986) and the chlorophyll a/b binding protein of wheat (Kuhlemeier *et al.*, 1987).

All eukaryotic mRNAs are monocistronic, but each messenger contains many more nucleotides than are necessary for encoding its respective protein. At the 5' end the non-translated leader is usually less than 100 bases, although exceptions that are much longer do occur. The 3' non-translated trailer can be very long – for example up to 1000 bases; however, it has no known function in translation. The 5' methylated cap is important in ribosome binding. This involves the 40S small subunit of the ribosome and the so-called cap binding proteins which are numerous and not particularly well understood. One of their functions may be to remove secondary structure from the mRNA leader region ensuring that the appropriate sites are available in single-stranded form. The cap is necessary for successful complexing of the initiation codon even in the rare messengers where they are as much as 1000 bases apart.

The 40S subunit recognizes the 5' cap and travels along the mRNA until the AUG initiation codon is encountered. It appears that the AUG triplet must be in the right environment which may be determined by two or three nearby bases. As a result of these requirements not all apparent start codons are recognized for translation initiation, although in the majority of cases it is the AUG triplet closest to the 5' end (Kozak, 1984). There is usually a purine at position −3 (A:75%, G:20%). In the rare mRNAs that have AUG triplets upstream of the actual initiation site, pyrimidines are present three bases upstream from this site. The true start codon also often has a G at position +4. These generalities, initially deduced from animal sequences, have since been shown to be also true for plant genes. The consensus sequence for animal and plant translational start sites respectively are

CCACCAUGGC and NNANNAUGGCU

where N is any base (Heidecker and Messing, 1986).

It has been proposed that the animal sequence may function by being complementary to the 18S rRNA sequence 3'GGUGG, but this clearly cannot be the case with plant transcripts. At initiation the 60S large subunit associates with the mRNA–40S complex to begin translation.

In contrast to the 5' cap, there is no evidence that polyadenylation has a role in translation. In mammals up to one-third of total mRNA may not be polyadenylated. One-third of this poly(A)$^-$ messenger is histone mRNA which is never polyadenylated. Generally speaking, poly(A)$^+$ and poly(A)$^-$ mRNA is similar in size, stability, translation rates, and transfer rate from the nucleus to the cytoplasm. There is also considerable overlap between the two mRNA groups with respect to their coding for the same proteins.

Polyadenylation signals have the consensus sequence AAUAAA in plants and animals, although plant genes may deviate from this, particularly at the last position. These signals precede the end of the mature messages by 15 to 23 nucleotides. Plant poly(A)$^+$ mRNA usually contains more than one signal, but only one is functional. It is not yet clear what determines this since there is no correlation between the actual sequences of the preferred sites.

The presence of poly(A) can be exploited experimentally for the separation of mRNA from rRNA. Oligo(dT) sepharose is used to bind the poly(A)-tails. In addition, small oligo(dT) fragments can be used to prime reverse transcriptase on the poly(A)-tail for first strand synthesis of complementary DNA. Although this is a useful characteristic of eukaryotic messengers, researchers must bear in mind that not all mRNAs are polyadenylated.

Transcriptional regulation

Cis-regulatory elements

Promoters Eukaryotic promoters consist of three important regions: (1) the transcriptional start (designated position 0), (2) a sequence 20–30 bp upstream of this start point and (3) a region further upstream at about −75 (Fig. 1.6).

The Hogness or TATA box is located approximately 25 bp upstream and has the consensus sequence

$$\text{TATA}^A_T{}^A_T{}^A.$$

It is surrounded by GC-rich sequences which may play a role in its function. Except for its position at −25 it could pass for the bacterial Pribnow box located at 10 bp upstream of the prokaryotic transcriptional start. In some genes such as those for zein (the maize storage proteins) two TATA boxes may be found as close as 10 bp apart.

The function of the TATA sequence is to direct RNA polymerase II to the correct transcriptional start site. This site is like the prokaryotic CAT start sequence since the first base of eukaryotic mRNA is almost always A flanked by pyrimidines; otherwise it shows little homology between different eukaryotic genes.

Centered around −75 are one or more elements responsible for regulating the frequency of initiation. These upstream elements often show homologies between genes. Two of the most common of these elements are the GC and CAAT boxes which have the respective animal consensus sequences

$$\text{GGGCGG} \quad \text{and} \quad \text{GG}^C_T\text{CAATCT}.$$

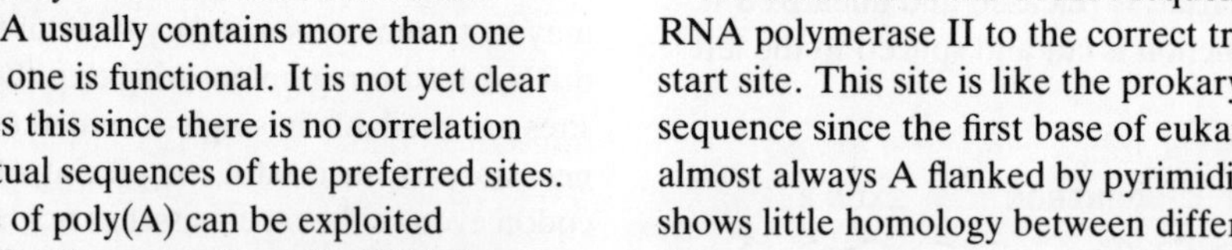

Fig. 1.6 *Cis*-regulatory DNA elements. The promoter region includes the TATA box and upstream elements. Enhancer elements may be situated further upstream or downstream of the protein coding region.

These elements may be present in the same gene in either orientation and in single or numerous copies.

CAAT boxes are often absent from plant genes. Cereal genes have a different consensus sequence

$$CCATCTCNA\underline{CC}$$

at position -90 called the CATC box which may serve as a substitute for CAAT (Kreis et al., 1986). Certain ribulose bisphosphate carboxylase small subunit genes (rbcS) may have no requirements for CAAT or functionally equivalent sequences (Morelli et al., 1985; Kuhlemeier et al., 1987). The regions between the TATA box and the upstream elements, when present, are not involved in promoter function and the distance between them can be artificially lengthened to some degree without impairing promoter function.

Enhancers In several eukaryotic genes, enhancer elements are also involved in the regulation of transcription. They are 100–200 bp in length and can act in either orientation and at a considerable distance from the target gene's promoter. They are linked to the target genes (on the same chromosome) and, as such, are cis-acting DNA sequences. They may be 5', or 3' to the transcriptional start and usually act by stimulating the nearest promoter(s). These elements commonly contain repeated sequences which individually possess a small stimulatory ability and no DNA consensus sequences are common to all enhancers. One sequence which occurs in several animal genes has the consensus

$$NTGTGG^{AA}_{TT}.$$

In plants many rbcS genes from dicots contain a closely related sequence to the above animal enhancer

$$GTGTGG^{TT}_{CC}A^{A}_{T}TA^{T}_{A}G$$

at -140 from the transcriptional start (Kuhlemeier et al., 1987). A similar sequence is also present at -215 indicating that redundancy of information is an advantage for at least some plant enhancers.

A number of other enhancer regions have been identified in specific plant genes by deletion analyses of upstream elements in transgenic plants (for methods see Chapter 32). In the α-subunit β-conglycinin gene there are five 6 bp repeats with the consensus

$$AGCCC^{A}_{C}.$$

One is located at position -149 relative to the site of transcriptional initiation. The other four repeats occur at -255, -204, -182, and -171. The pair of repeats at -204 and -182 and their intervening sequence form a larger imperfect direct repeat of 28 nucleotides with the -171 and -149 repeats and their intervening sequence. An additional characteristic of this 5' region is that the -149 and -171 repeats as well as the -182 and -204 repeats are 16 bp apart. This separation of 16 nucleotides places them on the same face of the DNA helix. This could be important in a regulatory role since both would be accessible in concert to trans-acting factors (see next section). Petunia plants which were transformed with a truncated form of this gene containing only the 6 bp repeat at -149 had only 5% of the expression level found in plants transformed with the complete gene. Similar repeats are present in the genes for β-subunit β-conglycinin and phaseolin (Chen et al., 1986).

Maize alcohol dehydrogenase genes are transcribed under anaerobic conditions (Walker et al., 1987). Adh1 contains 16 bp anaerobic regulatory elements at two positions

-140 to -125, CTGCAGCCCCGGTTTC,
and -99 to -114, CCGTGGTTTGCTTGCC,

which are responsible for expression of the gene at low oxygen levels. Adh2 also has similar sequences at -150 to -135 and -104 to -119.

The Ti-plasmid of *Agrobacterium tumefaciens* has evolved the ability to transfer functional genes into plants. The promoters of these genes are eukaryotic in nature and one that has been studied extensively is responsible for the synthesis of the amino acid, octopine. Between positions -193 to -178 an element was found to possess perfect dyad symmetry

5'..ACGTAAGC GCTTACGT..3'
3'..TGCATTCG CGAATGCA..5'

which can be referred to as a palindrome where the sequences to each side of a central axis are the same. This 16 bp palindrome possesses all the attributes of a conventional enhancer sequence (Ellis et al., 1987). Its enhancing effect diminishes with increasing distance from the promoter and it has been shown to be functional when placed at the 3' end of genes. As would be expected from a palindrome reversing its orientation has no effect. Similar properties of dyad sequences have been found in *Drosophila* and yeast genes.

One plant gene is known to require a 3' sequence for expression. Potato proteinase inhibitor II does not function in transgenic tobacco when a 200 bp downstream region is not included in chimaeric gene constructs (Sanchez-Serrano *et al.*, 1987).

Trans-*acting factors*

Trans-acting factors are proteins which act by binding to *cis*-regulatory DNA sequences.

Affecting upstream elements The Sp1 factor of simian virus (SV40) binds specifically to a 21 bp repeat found in the promoter of the viral 'early gene'. It also interacts with Herpes thymidine kinase gene and with monkey genome promoters. Sp1 associates with the GGGCGG upstream element (Dynan and Tjian, 1985), but it is not known how this factor acts to stimulate transcription. No *trans*-acting factors affecting the upstream elements of plant genes have been identified.

Affecting enhancers The SV40 enhancer has two domains, A and B, both of which are required for full activity (Zenke *et al.*, 1986). Both A and B interact with a single *trans*-acting factor. This factor is also known to associate with other animal enhancers. The mechanisms of positive control factors, therefore, may be conserved to some degree dependent on their specific regulatory roles. Since eukaryotic genes are not arranged in operons, these factors probably coordinate the expression of one to several genes.

Two *trans*-acting factors have been identified in plants. The first factor was isolated from a nuclear fraction from the endosperm of maize. It binds a 15 bp conserved region in zein storage protein genes (Maier *et al.*, 1987). This sequence

CACATGTGTAAAGGT

is similar to the enhancer consensus sequence of animal genes (see above) which are involved in positive gene regulation. Similar sequences are found in the 5' regions of other cereal storage protein genes. Zein genes are only expressed in the endosperm tissue of seeds. Since a number of different zein-deficient mutants have been isolated it is likely that there are several *trans*-acting regulatory genes that control the transcription of these genes. It is possible that the nuclear factor is the product of one of these regulatory genes and its binding to the zein gene may elicit the tissue specific response.

Another nuclear protein factor, GT-1, binds GT-rich regions in the ribulose bisphosphate carboxylase gene *rbcS-3A* (Green *et al.*, 1987). These GT-rich DNA sequences, of which there are four, act as both positive and negative light-responsive elements and once again share homology with animal enhancers.

Negative regulation of transcription

Only recently have attempts been made to identify *cis*-acting DNA elements and *trans*-acting factors that repress gene transcription in eukaryotes. Here again SV40 provides the best characterized example. A protein isolated from cells infected with SV40 and encoded on the viral genome has been termed the T antigen. It regulates its own production and that of other virally encoded proteins by reducing the synthesis of their respective mRNAs. T antigen binds to sites near the viral replication origin. The *cis*-acting factors involved in negative control are referred to as blockers or silencers.

The gene for pea chlorophyll *a/b* binding protein and *rbcS-3A* have upstream sequences which inhibit transcription. Normally, constitutive photosynthetic genes have their expression greatly reduced in the dark and in non-green tissue as a result of the activity of these elements (Simpson *et al.*, 1986; Nagy *et al.*, 1988). Interestingly, both these silencer elements appear to have enhancer properties in the light.

It is generally felt that *trans*-acting factors which stimulate or retard transcription act on virtually all eukaryotic genes. How initiation of transcription by RNA polymerase II is affected must somehow involve stereospecific hindrance of interacting complexes. Interactions with other DNA-associated proteins must also be implicated in these processes.

Messenger RNA turnover

Two features of eukaryotes imply that post-transcriptional regulation of gene expression must be an integral component of the overall control of genes in these organisms. First, transcription and translation occur in different compartments, the nucleus and cytoplasm. Secondly, there is the presence of non-translated intervening sequences in their genes. Several processes, therefore, are involved in the production of functional mRNA. The subsequent stability and availability for translation of the mRNA must also be a factor. However, of all these possible steps where control mechanisms could be predicted, so far only RNA turnover has been extensively investigated.

The degradation of mRNA may very well be the most important of all these processes. The half-lives of individual species may vary from 30 min to days. The mammalian oncogenes, *c-myc* and *c-fos*, for example, produce very short-lived messengers. Three areas of interest in mRNA turnover are: (1) signals that elicit degradation, (2) the structural composition of mRNA, and (3) enzymes and other *trans*-acting factors responsible for modulation of mRNA breakdown.

Endogenous and exogenous signals

'Housekeeping' genes are responsible for the production of enzymes and proteins of essential life processes and their rates of transcription are maintained under diverse conditions. Changes in cell cycle do, however, affect their respective mRNA turnover rates. Included in these genes are those encoding histones, thymidylate synthase, thymidine kinase and dihydrofolate reductase of mammalian systems. The steady-state level of the mRNA for these proteins may vary several fold (Graves *et al.*, 1987).

In animals type I procollagen mRNA stability is affected by exogenous glucocorticoids and growth factors, respectively increasing and decreasing breakdown.

Evidence suggests that these triggers act via proteinaceous intermediates since cycloheximide treatment attenuates both positive and negative effects.

mRNA structural features

Very little is known about the stabilizing influence of different mRNA sequences. The viral and cytoplasmic oncogenes, *v-fos* and *c-fos* respectively, differ by a unique 67 nucleotide sequence in the 3' untranslated region of *c-fos* which is absent in the viral gene. When this segment is fused at a similar site in *v-fos* its stability is decreased to that of the *c-fos* transcript (Miller *et al.*, 1984).

Shaw and Kamen (1986) identified a 50–60 nucleotide AT-rich sequence in the 3' non-coding region of mRNAs of transiently expressed inflammatory genes. Inserting this sequence into the 3' end of the normally very stable globin messenger resulted in a 10–30 fold increase in degradation rates. Comparisons among a number of these rapidly degraded mRNA sequences did not reveal a conserved region despite their richness in A–T dimers.

Translation and mRNA degradation

The degradation of histone messengers can be prevented by translation inhibitors which cause either mRNA immobilization on or complete mRNA dissociation from ribosomes. Therefore, the process relies on events which occur during active translation to or beyond a certain point on the transcript. In fact, histone mRNA breakdown depends on a conserved 3' stem–loop (hairpin) structure. Artificially changing this hairpin's distance from the translation termination codon severely affects the rate of degradation. As such, it appears that messenger movement along the ribosome acts to bring nucleolytic activities to the appropriate site, in the process possibly removing the stabilizing secondary structure.

Ross and Kobs (1986) demonstrated with H4 histone transcripts in cell-free systems that degradation always commenced at the 3' terminus and proceeded in the 5' direction by exonuclease activity. The presence of discrete intermediate mRNA breakdown products, however, suggests that there is also some endonuclease involvement.

The importance of this facet of the regulation of gene expression has now been clearly demonstrated and studies are now being aimed at isolating the responsible degradative nucleases and identifying the factors involved in messenger turnover. Even though no examples are available at present, it is highly likely that this area of gene regulation is extremely important in plants.

Organelle genes

Mitochondrial and plastid genomes code for some, but not all, proteins found in these organelles. They produce their own ribosomal and transfer RNAs, and a small number of mRNAs (i.e. around ninety in chloroplasts and eight or more in mitochondria). The nuclear genome codes for most of the organellar proteins which are imported from the cytoplasm. In some cases separate protein subunits are encoded in the different genomes (e.g. nuclear *rbcS* and chloroplast *rbcL* genes of ribulose bisphosphate carboxylase), which poses interesting questions regarding regulatory interactions.

Organelle genes are essentially prokaryotic in nature. They possess Shine–Dalgarno promoter sequences which have been shown to be recognized by *E. coli* RNA polymerase. They are usually polycistronic, but they do occasionally contain introns, a characteristic very rarely seen in prokaryotic genes. In at least one case, that of the ribosomal protein S12 of tobacco chloroplasts, mature mRNA is formed by the *trans*-splicing of two pre-mRNAs. In this case, exon 1 is encoded by a separate gene than exons 2 and 3 (Hildebrand *et al.*, 1988). The lack of a membrane separating the genome from the ribosomes suggests that there is some process other than physical compartmentation involved in determining which RNAs are translated into protein: recall that the nuclear membrane separates the eukaryotic hnRNA from the mature mRNA found in the cytoplasm.

The regulation of gene expression in organelles and the intimate relationships of organellar and nuclear genes are the subjects of other chapters in this volume.

Concluding remarks

The regulation of gene expression both during and after transcription involves numerous and diverse mechanisms, some of which are becoming well understood. The identification and characterization of *cis*-regulatory elements in plants is proceeding at a considerable pace, and attempts to purify *trans*-acting factors are now underway. Still, even in the most advanced systems, the genes for these factors remain to be identified. Their isolation will open the way to an understanding of how primary signals (e.g. low O_2 for alcohol dehydrogenase genes) elicit differential gene expression.

Further reading

Fluhr, R., Kuhlemeier, C., Nagy, F. and Chua, N.-H. (1986). Organ-specific and light-induced expression of plant genes. *Science* **232**, 1106–12.

Goldberg, R. B. (1986). Regulation of plant gene expression. *Phil. Trans. R. Soc. Lond. B* **14**, 343–53.

Miller, J. H. and Reznikoff, W. S. (1980). *The Operon*, 2nd edn, Cold Spring Harbor, New York.

Pines, O. and Inouye, M. (1986). Antisense RNA regulation in prokaryotes. *Trends Genet.* **1**, 284–7.

Raghow, R. (1987). Regulation of messenger RNA turnover in eukaryotes. *Trends Biochem. Sci.* **12**, 358–60.

Sassone-Corsi, P. and Borrelli, E. (1986). Transcriptional regulation by *trans*-acting factors. *Trends Genet.* **1**, 215–9.

Yanofsky, C. (1987). Operon-specific control by transcription attenuation. *Trends Genet.* **3**, 356–60.

Willmitzer, L. (1988). The use of transgenic plants to study plant gene expression. *Trends Genet.* **4**, 13–18.

References

Chen, Z.-L., Schuler, M. A. and Beachy, R. N. (1986). Functional analysis of regulatory elements in a plant embryo-specific gene. *Proc. Natl Acad. Sci. USA* **83**, 8560–4.

Dynan, W. S. and Tjian, R. (1985). Control of eukaryotic messenger RNA synthesis by sequence-specific DNA-binding proteins. *Nature* **316**, 774–8.

Ecker, J. R. and Davis, R. W. (1986). Inhibition of gene expression in plant cells by expression of antisense RNA. *Proc. Natl Acad. Sci. USA* **83**, 5372–6.

Ellis, J. G., Llewellyn, D. J., Walker, J. C., Dennis, E. S. and Peacock, W. J. (1987). The *ocs* element: a 16 base pair palindrome essential for activity of the octopine synthase enhancer. *EMBO J.* **6**, 3203–8.

Goodenough, U. (1984). *Genetics*, 3rd edn, Saunders College Publishing, Philadelphia.

Graves, R. A., Pandey, N. B., Chodchoy, N. and Marzluff, W. F. (1987). Translation is required for regulation of histone mRNA degradation. *Cell* **48**, 615–26.

Green, P. L., Kay, S. A. and Chua, N.-H. (1987). Sequence-specific interactions of a pea nuclear factor with light-responsive elements upstream of the *rbcS-3A* gene. *EMBO J.* **6**, 2543–9.

Harland, R. and Weintraub, H. (1985). Translation of mRNA injected into *Xenopus* oocytes is specifically inhibited by antisense RNA. *J. Cell Biol.* **101**, 1094–9.

Haselkorn, R. (1986). Organization of the genes for nitrogen fixation in photosynthetic bacteria and cyanobacteria. *Ann. Rev. Microbiol.* **40**, 525–47.

Heidecker, G. and Messing, J. (1986). Structural analysis of plant genes. *Ann. Rev. Plant Physiol.* **37**, 439–66.

Hildebrand, M., Hallick, R. B., Passavant, C. W. and Bourque, D. P. (1988). *Trans*-splicing in chloroplasts: The *rps12* loci of *Nicotiana tabacum*. *Proc. Natl Acad. Sci. USA* **85**, 372–6.

Hull, R. and Covey, S. N. (1985). Cauliflower mosaic virus: pathways of infection. *BioEssays* **3**, 160–3.

Kozak, M. (1984). Compilation and analysis of sequences upstream from the translational start site in eukaryotic mRNAs. *Nucl. Acid. Res.* **12**, 857–73.

Kreis, M., Williamson, M. S., Forde, J., Schmutz, D., Clark, J., Buxton, B., Pywell, J., Marris, C., Hendersen, J., Harris, N., Shewry, P. R., Forde, B. G. and Miflin, B. J. (1986). Differential gene expression in the developing barley endosperm. *Phil. Trans. R. Soc. Lond. B.* **314**, 355–65.

Kuhlemeier, C., Green, P. J. and Chua, N.-H. (1987). Regulation of gene expression in higher plants. *Ann. Rev. Plant Physiol.* **38**, 221–57.

Lefebvre, D. D., Miki, B. L. and Laliberte, J.-F. (1987). Mammalian metallothionein functions in plants. *Bio/Technology* **5**, 1053–6.

Maier, U.-G., Brown, J. W. S., Toloczyki, C. and Feix, G. (1987). Binding of a nuclear factor to a consensus sequence in the 5′ flanking region of the zein genes from maize. *EMBO J.* **6**, 17–22.

Miller, A. D., Curran, T. and Verma, I. M. (1984). *c-fos* protein can induce cellular transformation: a novel mechanism of activation of a cellular oncogene. *Cell* **36**, 51–60.

Morelli, G., Nagy, F., Fraley, R. T., Rogers, S. G. and Chua, N.-H. (1985). A short conserved sequence is involved in the light-inducibility of a gene encoding ribulose bisphosphate carboxylase small subunit of pea. *Nature* **315**, 200–4.

Nagy, F., Kay, S. A. and Chua N.-H. (1988). Gene regulation by phytochrome. *Trends Genet.* **4**, 37–42.

Ross, J. and Kobs, G. (1986). H4 histone messenger RNA decay in cell-free extracts initiates at or near the 3′ terminus and proceeds 3′ to 5′. *J. Mol. Biol.* **188**, 579–93.

Sanchez-Serrano, J., Keil, M., O'Conner, A., Schell, J. and Willmitzer, L. (1987). Wound-induced expression of a potato proteinase inhibitor II gene in transgenic tobacco plants. *EMBO J.* **6**, 303–6.

Shaw, G. and Kamen, R. (1986). A conserved AU sequence from the 3′ untranslated region of GM-CSF mRNA mediates selective mRNA degradation. *Cell* **46**, 659–67.

Shine, J. and Dalgarno, L. (1974). The 3′-terminal sequence of *Escherichia coli* 16S ribosomal RNA: Complementarity to nonsense triplets and ribosome binding sites. *Proc. Natl. Acad. Sci. USA* **71**, 1342–6.

Simpson, J., Schell, J., Van Montagu, M. and Herrera-Estrella, L. (1986). Light-inducible and tissue-specific pea *lhcp* gene expression involves an upstream combining enhancer- and silencer-like properties. *Nature* **323**, 551–4.

Tumer, N. E., Robinson, S. J. and Haselkorn, R. (1983). Different promoters for the *Anabaena* glutamine synthetase gene during growth using molecular or fixed nitrogen. *Nature* **306**, 337–42.

Walker, J. C., Howard, E. A., Dennis, E. S. and Peacock, W. J. (1987). DNA sequences required for the anaerobic expression of maize alcohol dehydrogenase 1 gene. *Proc. Natl Acad. Sci. USA* **84**, 6624–8.

Zenke, M., Grundstrom, T., Matthes, H., Wintzerith, M., Schatz, C., Wildeman, A. and Chambon, P. (1986). Multiple motifs are involved in SV40 enhancer function. *EMBO J.* **5**, 387–97.

2 Regulation of gene expression during development
Rod Casey

Introduction

Development is characterized by the differential synthesis of gene products (transcripts) in time and space. While these transcripts are produced in response to developmental signals, some may themselves influence the developmental process. Few plant genes have been identified whose expression directly influences development, and consideration of the developmental regulation of gene expression is therefore usually limited to those genes whose expression is regulated during development. Such regulation has been investigated using several systems. Some important developmental processes, such as root development, or the transition from vegetative to floral morphology, have received relatively little attention, whereas others have been the subject of intense investigation. Considerable advances have been made in the study of light-induced changes in gene expression (Kuhlemeier et al., 1987; Chapter 14, this book), which provide the basis for several aspects of plant development; other aspects of plant development that have been studied at the level of gene expression include fruit ripening, seed germination, leaf and flower development, cell wall formation and pigmentation (see Verma and Goldberg, 1988).

Seed development has proved an especially attractive system for the study of gene expression. This is in part due to the availability of large amounts of material, the ability to determine a fixed starting point for the process (fertilization) and the recognition that many seed products are absent from all other parts of the plant. Furthermore, embryogenesis and seed development, in cereal and grain legumes in particular, is one of the most significant processes in plants with respect to their usefulness to man. This chapter describes the regulation of the expression of storage protein and other seed protein genes during the development of legume, cereal and some other seeds. The nature of these proteins and their deposition in seeds is discussed in detail in Chapter 31 (see also Casey et al., 1986).

As in Chapter 31, the 11S and 7S globulins of legume seeds and the prolamins of cereals are used here as the main examples of products of genes whose expression is regulated during seed development.

Seed development

It is not necessary for the purposes of this chapter to have a detailed knowledge of the process of seed development. There are a number of reviews of the subject; that by Dure (1975) is particularly informative and useful. In both monocotyledonous (monocot) and dicotyledonous (dicot) plants, storage proteins are synthesized during seed development and are finally deposited in membrane-bound organelles known as protein bodies, the location of which differs between monocots and dicots. In most cereals, protein bodies are found in mature endosperm cells, whereas in most dicots, including legumes and cotton, the protein bodies accumulate in cotyledonary parenchyma cells of the embryo; the endosperm of dicot seeds is totally consumed during embryogenesis.

One of the striking features of seed development is the large increase in amounts of DNA during the course of embryogenesis. Pea cotyledon cells, for instance, can achieve values in excess of 64C (copies), and DNA in maize endosperm nuclei can increase to 200C. Early speculation that DNA endoreduplication provided a means of selective amplification and consequent high levels of expression of storage protein genes proved unfounded (Millerd, 1975; Goldberg et al., 1981a), and the function of such large DNA amounts during embryogenesis/seed development is still unclear.

The arrangement of seed storage protein genes

The storage proteins of legume, cotton and cereal seeds are all the products of small gene families. Details of the structures of members of the various gene families can be found in several reviews (Kreis et al., 1985; Casey et al., 1986; Heidecker and Messing 1986; Casey and Domoney, 1987). The numbers of members within the families vary from about ten (pea 11S globulins (legumins) for example; see Casey et al., 1986) to approximately 100 (zein genes in maize; see Heidecker and Messing, 1986; Shotwell and Larkins, 1989). The members of a family can be sub-classified in a number of ways; zein genes, for instance, can be grouped either according to DNA sequence homologies or on the basis of the properties of the polypeptides that they encode (Heidecker and Messing, 1986; Shotwell and Larkins, 1989). The 11S globulins from pea, soybean and broad bean can be arranged into two groups according to their cDNA and gene sequences (see Casey et al., 1986). Similarly, the genes encoding phaseolin, the major storage protein of French bean seeds, can be grouped into α- and β-types on the basis of cDNA sequences (Hall et al., 1983).

It is of interest to establish the genetic map positions of storage protein genes because, as will be seen later, this permits an analysis of the developmental regulation of gene expression in relation to genomic location. Such map positions may be determined by classical linkage analysis, using either variation in polypeptide gel electrophoresis patterns or variation in the sizes of restriction fragments of genomic DNA that hybridize to storage protein probes (restriction fragment length polymorphism; RFLP). Table 2.1 summarizes the present state of knowledge with respect to the map positions of storage protein genes in wheat, barley, maize and pea.

Table 2.1 Genetic locations of seed storage protein genes

Protein	Species	Chromosome	Locus
Legumin (11S globulin)	Pea	1 7	near a (2 loci) near r (Lg-1)
Vicilin (7S globulin)	Pea	7	$Vc1$ $Vc2$ $Vc3$
		?	$Vc4$ $Vc5$
Convicilin (7S globulin)	Pea	2	near k (Cvc)
Zein (19–22 000 M_r)	Maize	4 (both arms) 7 (short arm) 10 (long arm)	
Zein (15 000 M_r)	Maize	6 (long arm)	
HMW glutenins	Wheat[a]	1A, 1B, 1D (long arms)	Glu-1
α/β gliadins γ gliadins Ω gliadins LMW glutenins	Wheat[a]	1A, 1B, 1D (short arms) 6A, 6B, 6D (short arms)	Gli-1; Gli-2; Glu-B2
B and C hordeins	Barley	5 (short arm)	Hor-1, Hor-2
D hordein	Barley	5 (long arm)	Hor-3

For details of the nomenclature of seed storage proteins see Chapter 31; Kreis et al., 1985; Casey et al., 1986; Shotwell and Larkins, 1989.

[a]Wheat is an allohexaploid with three genomes designated A, B and D, each of which contributes seven pairs of chromosomes to the total genome. Chromosomes from the A, B and D genomes are related, but do not pair (homoeologous) and are given similar numbers (e.g. 1A, 1B, 1D, etc.).

The general structure of storage protein genes in relation to gene expression

The overall features of plant gene structure have been discussed in Chapter 1 and the details of storage protein genes have been described elsewhere (Heidecker and Messing, 1986; Casey and Domoney, 1987). The general characteristics of storage protein genes are similar to those of other plant genes. They are flanked by 5'- and 3'-non-coding sequences, interrupted by small introns (although cereal prolamin genes seem to be an exception to this; see Shotwell and Larkins, 1988) and have defined eukaryotic promoter elements, such as TATA and CAAT or AGGA, and sequences that direct polyadenylation of mRNA precursors. Most plant storage protein genes have more than one signal sequence for polyadenylation; the significance of this is unclear, but different transcripts, perhaps with different stabilities yet containing a common coding sequence, may offer a means of developmental regulation.

Conserved elements in 5'-non-coding (upstream) sequences have been noted in storage protein genes. A sequence about 300 bp upstream of the transcription start point in maize, barley and wheat prolamin genes (the *−300 element*) has been reported to be conserved (Kreis *et al.*, 1985) and thereby implicated in the regulation of prolamin gene expression. Further substantiation of the role of this region in gene regulation has come from the analysis of the interaction of putative regulatory factors with prolamin gene promoters (see below). The upstream regions of 11S globulin genes contain a highly conserved region referred to as the *legumin box* (Bäumlein *et al.*, 1986) in which 25 of 28 nucleotides are identical in genes from broad bean, pea and soybean. Such conservation in genes whose coding sequences are only 50–70% identical (Casey *et al.*, 1986) is suggestive of a role for the legumin box in developmental regulation of 11S globulin gene expression (Gatehouse *et al.*, 1986).

While almost exclusive attention has been paid to 5'-flanking promoter elements in the regulation of gene expression during plant development (see Kuhlemeier *et al.*, 1987), 3'-non-coding sequences and introns can modulate the expression of the gene in which they reside. Little is known about the role of 3'-flanking elements in the regulation of plant storage protein gene expression (but see Willmitzer, 1988), although a possible role for secondary structures in stabilizing zein mRNA has been proposed (Heidecker and Messing, 1986). Evidence is accumulating from *in vitro* assay systems, however, that plant gene introns can enhance the levels of gene activity.

Mutations in seed protein structural genes

There are certain to be many mutations at structural loci that can affect the expression of a seed protein gene. Many of these will be undetected because they occur in a single member of a gene family and synthesis from the normal members of the family obscures the effect of the mutation. There are, however, a few such mutants that have been characterized.

Risø 56 mutants of barley

The Risø 56 mutant, which was generated by mutagenesis of barley variety Carlsberg II, has normal amounts of D hordein, double the normal amount of C hordein and a considerable reduction in B hordein. The mutation, which seems to correspond to a deletion of 80–90 kb of DNA at the *Hor*-2 locus, removes about 13 copies of B hordein genes and results in a virtual absence of B hordein mRNA from developing seeds; C hordein mRNAs are increased. It is not surprising that a deletion in B hordein genes leads to reduction in the amounts of B hordein mRNA; this is, however, in contrast to naturally-occurring variants of pea, where there seems to be no variation in storage protein gene number between lines that have a wide range of legumin/vicilin ratios (Domoney and Casey, 1985).

Stop codons in zein genes

Zein genes have a high proportion of codons that

can mutate into a stop codon in a single event. This is predominantly a consequence of the high glutamine content of zein; glutamine is encoded by CAA and CAG and the C–T transition is the most frequently occurring mutation.

A mutation of a codon for an amino acid to a stop (nonsense) codon should not prevent the formation of mRNA, but the mRNA would be expected to be translated into a shorter polypeptide in the absence of suppression of the nonsense codon. Evidence for such truncated polypeptides does exist. Efficient suppression of a nonsense mutation, however, is not a general phenomenon, since almost all zein cDNA (mRNA copy) clones obtained to date do not carry nonsense mutations. Several of the known zein genomic sequences, however, have one or more stop codons, suggesting that the mRNAs of these mutant genes are rapidly degraded. Interestingly, none of the genes that contain nonsense codons appears to have accumulated point mutations, deletions or insertions, as is often the case for inactive pseudogenes. Heidecker and Messing (1986) have suggested that a system of unequal crossing over and gene conversion (homogenization) maintains a relatively constant ratio between active and inactive genes and eliminates mutational drift in inactive zein genes.

Null-alleles for Gy_4 glycinin genes in soybean

The Gy_4 gene in soybean is the structural gene for the $A_5A_4B_3$ subunit of glycinin (11S globulin). The soybean cultivar 'Raiden' lacks this subunit and sequencing of the Gy_4 gene from cv. 'Raiden' revealed a point mutation of the translation initiation codon ATG to ATA. This is probably the cause of the null phenotype; the mutant gene is transcribed in developing seed, but the mRNAs do not become associated with polysomes. Although ribosomes may bind to other AUG codons that are 3' to the mutated initiation codon, such initiation events would not necessarily be 'in frame' (Scallon et al., 1987).

Null-allele for the α'-polypeptide of β-conglycinin in soybean

The soybean cultivar 'Keburi' lacks the α'-polypeptides of β-conglycinin (7S globulin) but has a normal complement of other β-conglycinin polypeptides; cell-free translation of mRNA from developing 'Keburi' seeds indicated the absence of α'-polypeptide mRNA. The basis of this seems to be a deletion that begins immediately 5' to the α'-polypeptide gene and extends through most of the coding sequence (Ladin et al., 1984).

Defects in the expression of soybean lectin genes

Investigation of a mutant line of soybean that produces no detectable seed lectin (Le⁻ line) showed nuclear and cytoplasmic lectin mRNA levels in Le⁻ embryos to be reduced to approximately 0.01% of those in Le⁺ embryos. Structural analysis of the seed lectin gene in Le⁻ plants showed that a 3.4 kb DNA segment, consisting of repeated sequences, had been inserted into the lectin gene; this insertion seems to result in dramatically reduced transcription of the mutant gene. The insertion element has features suggestive of a transposon (Goldberg et al., 1983).

Temporal changes in gene expression during seed development

During seed development many different mRNAs can be detected by RNA–DNA hybridization techniques. For example, at any given stage of cotton or soybean seed development, 15 000–30 000 genes are expressed (Goldberg, 1986), and mRNA complexity appears to increase slightly as embryos develop (Dure, 1985). More striking than this small increase in complexity, however, is the dramatic change in the frequency distribution of mRNA classes. By mid-maturation of soybean embryos, a small number (7–10) of superabundant mRNAs (those encoding seed storage proteins) comprise approximately 50% of the total mRNA mass (Goldberg *et al.*, 1981b). These mRNA molecules, which number > 10 000 per cotyledon cell, are developmentally regulated in as much as they accumulate during early embryogenesis and then decay prior to seed desiccation. They are also tightly regulated in their location, being undetectable (<1 molecule per 20 cells) in leaf tissues. In contrast, lectin and Kunitz trypsin inhibitor genes, which produce abundant, highly regulated mRNAs during soybean seed development, are also expressed at low levels in root (lectin) or leaf, stem and root (Kunitz trypsin inhibitor) tissue (Goldberg, 1986).

The temporal pattern of gene expression during seed development is similar for all seed protein genes, in the sense that mRNAs for relatively abundant species appear at early to mid development, accumulate over a period of time and finally decay before seed maturation and desiccation; the specific timing of this process may, however, vary both between and within gene classes (see Higgins, 1984).

Temporal changes in expression within gene groups

Legumin in peas

Three groups of legumin (11S globulin) gene can be recognized in pea by DNA–DNA hybridization studies; two of these have been mapped to genetic loci on chromosome 1 and the other to a locus on chromosome 7. Analyses of these three loci show that sequences of functional genes at a locus are extremely similar, but between loci there are appreciable differences. The genes at the two loci on chromosome 1 are more similar to each other than to those on chromosome 7. Examination of the amounts of mRNA produced from the various classes of gene during the course of seed development shows that the genes on chromosome 7 provide most of the legumin mRNA found throughout most of seed development. The genes on chromosome 1, however, are responsible for most of the legumin mRNA during the very *early* stages of embryogenesis; there is subsequently a developmental transition, after which the chromosome 1-located legumin genes contribute

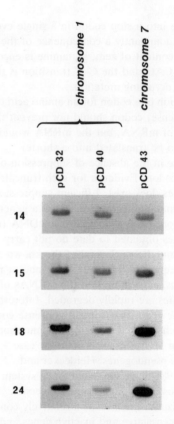

Fig. 2.1 The relative hybridization of ^{32}P-labelled RNA (prepared from pea embryos at 14, 15, 18 and 24 days after fertilization) to nitrocellulose filters containing equal amounts of insert from the three legumin cDNA plasmids pCD32, pCD40 and pCD43. The first two are derived from mRNAs that correspond to genetic loci on chromosome 1 and the latter from mRNA from the legumin locus on chromosome 7. The intensity of hybridization is proportional to the amount of the appropriate mRNA species in the total mRNA at a given developmental stage. At 14 days after fertilization, the mRNA corresponding to the genes on chromosome 7 is the least abundant, but by 18 days after fertilization it is the major species. From Domoney and Casey (1987), with permission.

less to the total mRNA than those on chromosome 7 (Domoney and Casey, 1987; see Fig. 2.1).

Thus, different groups of legumin gene are expressed at different times and to different extents during pea seed development. It should, however, be noted that the number of genes at each locus is different, varying from one or two at one of the loci on chromosome 1, to four to six (probably five including one pseudogene) at the locus on chromosome 7. Measurements such as those described above normally include *all* of the mRNA species produced by the various members of a gene group and will not discriminate between mRNAs from individual genes within a locus. The measurements are of mRNA amounts at a particular developmental stage, rather than of gene activity, and thus reflect the balance between mRNA production and degradation under any given set of conditions. A clearer indication of gene expression may possibly be obtained by measuring the 'run-on' of nascent transcripts in isolated nuclei from developing seeds.

7S proteins in soybean and pea

β-Conglycinin, the 7S major soybean storage protein of soybean seeds, contains α, α' and β-polypeptides that can associate to form a range of trimeric (7S) species comprising different complements of the three types of polypeptide. The α and α'-polypeptides accumulate earlier in seed development than β-polypeptides. The α and α' sequences are more similar to each other than either is to the β-polypeptide, and one of the major differences between α/α' and β-polypeptides is the presence of an insertion, of 20 000–30 000 M_r, near the N-termini of the α and α'-polypeptides relative to the β-polypeptides. An analogous situation exists with the polypeptides of vicilin and convicilin, the 7S storage proteins from pea seeds; the major polypeptides of vicilin are ~ 20 000 M_r smaller than those of convicilin, to which they show sequence homology. In contrast to the soybean situation, the larger (convicilin) polypeptides are synthesized later in pea seed development than those of vicilin, a phenomenon that reflects temporal changes in amounts of mRNA (Higgins, 1984). Although vicilin and convicilin polypeptides, like the β-conglycinin α/α' and β-polypeptides, probably can form mixed trimers, the genetic loci encoding vicilin and convicilin are dispersed throughout the pea genome on at least two linkage groups (see Casey *et al.*, 1986).

Prolamins in maize and barley

Zein, the alcohol-soluble fraction of the seed proteins of maize, comprises several groups of polypeptide that include two major groups of M_r 20 000 and 21 000–22 000; it accounts for more than 50% of the total seed proteins. Examination of the accumulation of the two major zein classes during seed development shows that the smaller size class ($M_r \sim$ 20 000) polypeptides accumulate at about three times the rate of the larger size group. Analogous alcohol-soluble proteins exist in the grains of other cereals, including wheat, sorghum and barley; in the last they are termed *hordeins*. Hordeins are conventionally separated into groups by SDS-polyacrylamide gel electrophoresis, the B and C hordeins accounting for 95% of the total hordein. Quantitative examination of the accumulation of hordein fractions during barley seed development shows that the various polypeptides of the B and C hordein make different contributions to the total accumulation of hordein at different developmental stages and that the kinetics of accumulation of individual polypeptides varies within the B hordein as a function of developmental stage (see Kreis *et al.*, 1985). Thus, even within a class or subgroup of proteins/polypeptides, individual members are differentially regulated with respect both to timing of synthesis and to the amount of product.

Temporal changes in expression between gene groups

It has been known for some time that there are differences in the appearance of major protein classes in developing seeds; one of the early examples was the appearance of vicilin before legumin in pea and broad bean seeds (see Millerd, 1975; Müntz, 1977). As discussed earlier, however, some minor legumin species appear early in development, whereas convicilin, which is closely related to vicilin, is synthesized late in development, at about the same time as legumin. Similarly, the majority of β-conglycinin polypeptides appear earlier in soybean seed development than the 11S globulin (glycinin), but the β-polypeptides of β-conglycinin are produced late in development. Thus, serological measurements of native 7S and 11S globulins suggest that the former are synthesized before the latter, but more detailed analysis of polypeptides indicates greater complexity (see above). Measurements of the amounts of mRNA corresponding to vicilin and legumin throughout pea seed development show that this differential accumulation of proteins and polypeptides reflects the appearance of the various vicilin and legumin mRNAs (see Higgins, 1984).

As a general rule, the synthesis of any particular storage protein reflects the amounts of mRNA encoding it and in this sense storage protein gene expression is regulated primarily at the level of transcription (see Gatehouse et al., 1986; Goldberg, 1986). Small differences in transcription rates in response to genotype, environment or developmental stage will, however, generally be missed by the methods currently employed to examine gene regulation. Although storage protein synthesis is controlled largely through differential mRNA synthesis, modulation of mRNA amounts through differential stability may well play a role in gene regulation during seed development. Post-transcriptional events clearly do play a significant part in the generalized developmental regulation of gene expression, because seed and non-seed protein genes that encode mRNAs which differ in prevalence by four orders of magnitude in soybean are transcribed at relatively similar rates (Goldberg, 1986).

The onset of gene activity, its spatial regulation, the relationship between gene activity and protein accumulation, and the complexity of mRNA populations during embryogenesis have all been studied in some detail using several plant species, especially pea, soybean, French bean, broad bean, wheat, barley, maize, cotton and rape. These considerations in turn prompt questions concerning the determination of organ-specificity and the initiation and promotion (or repression) of gene activity. Experiments with transgenic plants (see below) have given strong indications that elements within the 5'-flanking sequences of storage protein genes play a major role in determining site, timing and extent of expression.

Regulation of gene expression in transgenic plants

Plant genetic engineering – the introduction of alien or modified genes into plant cells and subsequent regeneration of *transgenic* plants – has proved to be a useful tool in the understanding of the regulation of gene expression (see Kuhlemeier et al., 1987; Willmitzer, 1988; and Chapter 32).

The first storage protein gene to be successfully, stably introduced into the genome of another species was that for the French bean 7S protein β-phaseolin, which was transformed into tobacco (Sengupta-Gopalan et al., 1985). The DNA fragment contained 863 bp of 5'-, and 1226 bp of 3'-flanking sequence in addition to the transcribed sequence of 1700 bp. The gene was fully developmentally regulated in the transgenic plant, being expressed specifically in seeds at the corresponding time in seed development. This suggested that all the necessary signals for developmentally regulated expression of the phaseolin gene reside in the 2089 bp of sequence flanking the transcribed sequence. Once this had been established, a number of systems were exploited to examine the role of specific DNA sequences in developmental regulation, special attention being paid to upstream (5'-flanking) sequences. These elements are often referred to as '*cis-acting*' sequences, to distinguish them from so-called '*trans-acting*' factors that operate through some interaction with *cis*-acting elements.

Using transgenic plants, such *cis*-acting elements have been observed in the upstream regions of genes that are photoregulated, or that respond to heat shock, UV light, or anaerobiosis (see Kuhlemeier et al., 1987), as well as those that are expressed during embryo development. Chen et al. (1986) made a series of deletions in the 5'-flanking sequence of the gene encoding an α'-polypeptide of β-conglycinin and introduced the normal and mutant genes into petunia plants. The unaltered gene was developmentally regulated in a normal fashion in petunia seeds. Analysis of the deletion mutants showed that if the α'-polypeptide gene was flanked by at least 257 bp 5' to the transcription initiation site, a high level of expression was obtained. Removal of approximately 100 bp more, leaving a gene flanked by 159 bp 5' to the transcription site, reduced expression by at least 20-fold. Analysis of the sequence between −159 and −257 revealed four repeats of a 6 bp sequence (AGCCCA); the deletion mutant in which only 159 bp remained upstream contained a single AACCCA sequence, which led Chen *et al.* (1986) to suggest that these repeats play a critical role in determining the level of expression of the soybean β-conglycinin α'-polypeptide gene in transgenic petunia plants. (The β-phaseolin gene and the gene for the β-polypeptide of β-conglycinin also contain AACCCA or CACCCA sequences in their upstream regions.) While this established that certain sequences were required for high levels of expression, it still remained to demonstrate that these, or other, sequences determined that seed protein genes were expressed only in seeds.

Experiments to show this have been carried out using transgenic tobacco plants. Instead of using whole seed protein genes, parts of the upstream sequences from wheat (Colot *et al.*, 1987), soybean (Chen *et al.*, 1988) or barley (Marris *et al.*, 1988) seed storage protein genes have been fused to the coding region of the bacterial chloramphenicol acetyltransferase (CAT) gene; these constructs were then introduced into tobacco plants. Thus, the measurement of CAT activity (the CAT sequence acting as a reporter gene) in various tobacco organs throughout their development gave an indication of the way in which the various upstream sequences of the storage protein genes were directing the expression of the genes to which they were directly coupled. In this way, it could be shown that the sequences present upstream of the transcription start point of wheat and barley seed storage protein genes were necessary to confer endosperm-specific CAT activity. Similarly, the soybean β-conglycinin gene element that was identified as necessary for optimal expression (see above) was also shown to direct the temporally regulated expression of CAT activity only in the seeds of transgenic tobacco plants; the element had no significant effect on the expression of CAT in roots, stems, or leaves, but increased CAT activity 25–40 fold in developing embryos. Furthermore, this increase was independent of the orientation of the element, fulfilling the criterion for recognition of the sequence as an organ-specific *enhancer* (see Chapter 1).

A third approach to studying the regulation of plant genes has been to transform a large (17.1 kbp) soybean DNA fragment containing a seed lectin gene and at least four non-seed protein genes into tobacco (Okamuro *et al.*, 1986). The fact that the five genes were physically linked prior to transformation meant that, provided no DNA rearrangement accompanied the integration of the soybean DNA into the tobacco genome, the possibility could be eliminated that differential regulation of the five genes in tobacco was a consequence of different sites of integration. As in soybean plants, the soybean lectin mRNA was present in tobacco seeds, accumulated and decayed during tobacco seed development, and was translated into protein. Each soybean non-seed protein mRNA was present in various tobacco organs in amounts similar to those found in soybean plants. Thus, it was concluded that a cluster of differentially regulated soybean genes are expressed in transformed tobacco plants as in soybean and that sequences controlling their expression are recognized by regulatory factors present in tobacco cells. Attention has recently turned to the nature of these factors and the regions of the gene that are recognized by them (see Goldberg, 1986).

'*Trans*-acting' factors and gene regulation

Interest in the action of possible *trans*-acting regulatory proteins has focused on interactions of nuclear proteins with the 5'-flanking (promoter) regions of developmentally regulated plant genes. Several different methods, all of which are based on the interaction of nuclear extracts from organs at defined developmental stages with DNA fragments of interest, have been used to detect and identify factors that recognize the promoter regions of developmentally regulated seed protein genes from, for example, maize (Maier et al., 1987) and soybean (Jofuku et al., 1987). In maize, a nuclear factor binds specifically to the highly conserved −300 element, discussed earlier, that is conserved in all zein genes and that contains a sequence (TGTGTAAAG) which is virtually identical to a viral/animal enhancer core sequence.

The regulation of the expression of seed storage protein genes in time and space is likely to be very complex, involving multiple interactions between the flanking regions of the genes and nuclear factors and between different DNA–protein complexes. It is also likely that these regulatory proteins themselves are the products of developmentally regulated genes.

Up to this point, the consideration of developmental regulation has been confined to studies made under standard conditions. Examination of gene expression in response to changes in light regime, temperature or nutrient status, or to salt and osmotic stress or anaerobiosis, however, makes it clear that genes can respond to changes in the environment. Further, the expression of specific genes can vary depending on the genotype under examination.

Environmental effects on gene expression

The effect of light on gene expression is discussed elsewhere in this book (Chapter 14; see also Kuhlemeier et al., 1987). Increasing the temperature of plant growth has only subtle effects on seed protein gene expression (Higgins, 1984), in contrast to the dramatic response of heat shock genes. The most profound effects on seed protein gene transcription and translation in whole plants result from deficiencies in nutrient (sulfur, potassium or phosphorus) supply.

Nutrient deficiencies

Alterations in nutritional factors can substantially affect the total amount and composition of seed proteins in peas, soybean, wheat, barley, rape and lupins (see Higgins, 1984). The relative proportion of 11S globulin is significantly increased in phosphorus- and potassium-deficient peas, whereas sulfur deficiency reduces the amount of 11S globulin to barely detectable levels. Sulfur deficiency also has striking effects in wheat, barley, soybean and lupins.

In plants grown without adequate sulfur supply, total protein is usually reduced, but the effect of the deficiency is confined largely to a reduction in synthesis of the sulfur-rich polypeptides. In barley, for instance, the sulfur-rich hordeins are reduced whereas the sulfur-poor C hordein is essentially unaffected. In soybeans, sulfur deficiency results in an over-production of a sulfur-poor isomer of conglycinin and, similarly, sulfur-deficient peas have relatively elevated amounts of vicilin. The other response in sulfur-deficient peas is the reduction of the (sulfur-rich) total albumin fraction and virtual elimination of legumin. This last effect is a direct consequence of reduced synthesis resulting from a reduced level of legumin mRNA in sulfur-deficient seeds; the reduction in legumin mRNA is a consequence of an interplay between reduced transcription and reduced stability of legumin mRNA. The mechanism whereby sulfur deficiency causes these effects is not known.

Developmental mutants and seed protein gene expression

A potentially fruitful area of research into the regulation of seed protein genes during development is the use of developmental mutants in which seed protein gene expression is perturbed. There are, however, few such mutants, variation at the r locus in peas and the opaque mutants in maize being examples.

Opaque mutants in maize

The opaque-2 (O_2) mutants of maize have a floury endosperm; they have a specific effect on zein synthesis, resulting in an approximately 50% reduction in the amount of zein. Although almost all groups of zein appear to be reduced by *opaque*-2 mutations, some alleles almost completely suppress the synthesis of the 22 000 M_r zeins.

Genetic linkage analysis has placed the O_2 locus on the short arm of maize chromosome 7, and although a few zein structural genes have been mapped to the same chromosome arm, they appear to be unaffected by O_2 mutations. Those genes that *are* affected by O_2 mutations map to chromosomes 4 and 10, which suggests that the O_2 gene produces a *trans*-acting regulator of the expression of some zein genes.

Other non-allelic *opaque* mutants also result in reduced zein synthesis. In *opaque*-6 all zein groups are approximately equally affected, whereas in *opaque*-7 the reduction in the M_r 20 000 zeins is proportionally greater than in other zeins.

There appear to be no major changes in zein gene number or organization in *opaque* mutants compared with wild types; zein mRNA levels are reduced during seed development relative to normal genotypes, but little is known about the mechanism of these effects.

The *r* locus in peas

The difference between round-seeded (RR or Rr) and wrinkled-seeded (rr) peas has been noted for over 100 years, but it was appreciated only recently that the nature of the allele at the *r* locus has an effect on seed protein synthesis. The synthesis of vicilin is essentially unaffected by the allele at the *r* locus, but round-seeded pea lines tend to have appreciably more legumin than corresponding wrinkled-seeded lines. The numbers and arrangements of legumin genes are the same in near-isogenic RR or rr types, but the amount of legumin mRNA is reduced in the rr line, apparently as a consequence of relatively reduced mRNA stability. Thus, the RR/rr mutation has a specific regulatory effect on legumin synthesis *via* mRNA stability.

Conclusions

The use of variants gives useful clues to the regulation of gene expression, but it will usually focus attention on rather narrow, detailed examples of gene regulation, although an accumulation of such information will provide a wider picture of the lesions that affect seed protein gene expression. Other insights into the developmental regulation of seed protein gene expression are likely to come from the use of transgenic plants (see Kuhlemeier *et al.*, 1987; Willmitzer, 1988, for instance) in the analysis of mutants, especially in promoter and other possible controlling elements; this will have broader implications for plant gene expression. The use of transgenic plants in this way necessarily assumes that the factors that regulate gene expression are common to many plants; experiments with seed protein genes in tobacco and petunia suggest that this may be so and that the seed storage protein genes respond to developmental signals which appear to be common to several plant species. It may be, however, that such experiments miss some aspects of regulation that could only be appreciated using a homologous transformation system. There is, for instance, an interesting difference between the expression of the genes for phaseolin and for the α-polypeptide of β-conglycinin in transgenic tobacco and petunia seed (see above). Although seed-specific expression was observed in both cases, the phaseolin gene did not adapt to the timing of tobacco seed protein gene expression, but retained the timing of expression typical of its behavior in French bean seed. The soybean β-conglycinin gene, however, followed the expression pattern of petunia seed proteins. Thus, in one case the host seems to control the temporal expression of the introduced gene, whereas in the other case the transferred gene and/or its product seems to determine the timing of its expression.

In the short term, much attention will focus on the nature and role of '*trans*-acting' factors that bind to upstream elements in seed protein genes. In one instance (see above) such a factor has been shown to be a protein that itself is developmentally regulated; its amount varies during development in parallel with the activity of the seed protein gene to which it binds. It will be interesting to establish the complexity and role of such factors, but attention in developmental regulation will then be focused on the *trans*-acting factors as much as the genes affected by their action.

We still understand little of the process of seed protein gene regulation during development; what activates and 'up-regulates' genes? Are they activated or de-repressed? What determines differences in quantity and timing of expression? What 'down-regulates' gene expression prior to dormancy and desiccation? What is the basis of spatial regulation of expression? It is to be expected that the use of transgenic plants, an analysis of regulatory factors, and a judicious study of variants and genetics will all play a role in the further analysis of the regulation of gene expression during seed development.

Acknowledgments

I am indebted to Claire Domoney, Roy Davies and Noel Ellis for their helpful criticisms and discussion of this chapter. I am very grateful to Tarn Dalzell for typing the manuscript. This work was supported by the Agricultural and Food Research Council via a grant-in-aid to the John Innes Institute.

References

Bäumlein, H., Wobus, U., Pustell, J. and Kafatos, F. C. (1986). The legumin gene family: structure of a B type gene of *Vicia faba* and a possible legumin gene specific regulatory element. *Nucl. Acids Res.* **14**, 2707–20.

Casey, R. and Domoney, C. (1987). The structure of plant storage protein genes. *Plant Mol. Biol. Reporter* **5**, 261–81.

Casey, R., Domoney, C. and Ellis, N. (1986). Legume storage proteins and their genes. *Oxford Surv. Plant Mol. Cell Biol.* **3**, 1–95.

Chen, Z-L., Schuler, M. A. and Beachy, R. N. (1986). Functional analysis of regulatory elements in a plant embryo-specific gene. *Proc. Natl Acad. Sci. USA* **83**, 8560–4.

Chen, Z-L., Pan, N-S. and Beachy, R. N. (1988). A DNA sequence element that confers seed-specific enhancement to a constitutive promoter. *EMBO J.* **7**, 297–302.

Colot, V., Robert, L. S., Kavanagh, T. A., Bevan, M. W. and Thompson, R. D. (1987). Localization of sequences in wheat endosperm protein genes which confer tissue-specific expression in tobacco. *EMBO J.* **6**, 3559–64.

Domoney, C. and Casey, R. (1985). Measurement of gene number for seed storage proteins in *Pisum. Nucl. Acids Res.* **13**, 687–99.

Domoney, C. and Casey, R. (1987). Changes in legumin messenger RNAs throughout seed development in *Pisum sativum* L. *Planta* **170**, 562–66.

Dure, L. S. (1975). Seed formation. *Ann. Rev. Plant Physiol.* **26**, 259–78.

Dure, L. (1985). Embryogenesis and gene expression during seed formation. *Oxford Surv. Plant Mol. Cell Biol.* **2**, 179–97.

Gatehouse, J. A., Evans, I. M., Croy, R. R. D. and Boulter, D. (1986). Differential expression of genes during legume seed development. *Phil. Trans. R. Soc. Lond. B* **314**, 367–84.

Goldberg, R. B. (1986). Regulation of plant gene expression. *Phil. Trans. R. Soc. Lond. B* **314**, 343–53.

Goldberg, R. B., Hoschek, G., Ditta, G. S. and Breidenbach, R. W. (1981a). Developmental regulation of cloned superabundant embryo mRNAs in soybean. *Develop. Biol.* **83**, 218–31.

Goldberg, R. B., Hoschek, G., Tam, S. H., Ditta, G. S. and Breidenbach, R. W. (1981b). Abundance, diversity, and regulation of mRNA sequence sets in soybean embryogenesis. *Develop. Biol.* **83**, 201–17.

Goldberg, R. B., Hoschek, G. and Vodkin, L. O. (1983). An insertion sequence blocks the expression of a soybean lectin gene. *Cell* **33**, 465–75.

Hall, T. C., Slightom, J. L., Ersland, D. R., Murray, M. G., Hoffman, L. M., Adang, M. J., Brown, J. W. S., Ma, Y., Matthews, J. A., Cramer, J. H., Barker, R. F., Sutton, D. W. and Kemp, J. D. (1983). Phaseolin: nucleotide sequence explains molecular weight and charge heterogeneity of a small multigene family and also assists vector construction for gene expression in alien tissue. In *Structure and Function of Plant Genomes*, eds O. Ciref and L. S. Dure, Plenum Press, New York, pp. 123–42.

Heidecker, G. and Messing, J. (1986). Structural analysis of plant genes. *Ann. Rev. Plant Physiol.* **37**, 439–66.

Higgins, T. J. V. (1984). Synthesis and regulation of major proteins in seeds. *Ann. Rev. Plant Physiol.* **35**, 191–221.

Jofuku, K. D., Okamuro, J. K. and Goldberg, R. B. (1987). Interaction of an embryo DNA binding protein with a soybean lectin gene upstream region. *Nature* **328**, 734–7.

Kreis, M., Shewry, P. R., Forde, B. G., Forde, J. and Miflin, B. J. (1985). Structure and evolution of seed storage proteins and their genes with particular reference to those of wheat, barley and rye. *Oxford Surv. Plant Mol. Cell Biol.* **2**, 253–317.

Kuhlemeier, C., Green, P. J. and Chua, N.-H. (1987). Regulation of gene expression in higher plants. *Ann. Rev. Plant Physiol.* **38**, 221–57.

Ladin, B. F., Doyle, J. J. and Beachy, R. N. (1984). Molecular characterization of a deletion mutation affecting the α'-subunit of β-conglycinin of soybean. *J. Mol. Appl. Genetics* **2**, 372–80.

Maier, U.-G., Brown, J. W. S., Toloczyki, C. and Feix, G. (1987). Binding of a nuclear factor to a consensus sequence in the 5' flanking region of zein genes from maize. *EMBO J.* **6**, 17–22.

Marris, C., Gallois, P., Copley, J. and Kreis, M. (1988). The 5' flanking region of a barley B hordein gene controls tissue and developmental specific CAT expression in tobacco plants. *Plant Mol. Biol.* **10**, 359–66

Millerd, A. (1975). Biochemistry of legume seed proteins. *Ann. Rev. Plant Physiol.* **26**, 53–72.

Müntz, K. (1977). Cell specialization processes during biosynthesis and storage of proteins in plant seeds. In *Proceedings of the International Conference on Regulation of Developmental Processes in Plants*, Halle, GDR, pp. 70–97.

Okamuro, J. K., Jofuku, K. D. and Goldberg, R. B. (1986). Soybean seed lectin gene and flanking nonseed protein genes are developmentally regulated in transformed tobacco plants. *Proc. Natl Acad. Sci. USA* **83**, 8240–4.

Scallon, B. J., Dickinson, C. D. and Nielsen, N. C. (1987). Characterization of a null-allele for the Gy_4 glycinin gene from soybean. *Mol. Gen. Genet.* **208**, 107–13.

Sengupta-Gopalan, C., Reichert, N. A,. Barker, R. F., Hall, T. C. and Kemp, J. D. (1985). Developmentally regulated expression of the bean β-phaseolin gene in tobacco seed. *Proc. Natl Acad. Sci. USA* **82**, 3320–4.

Shotwell, M. A. and Larkins, B. A. (1989). The biochemistry and molecular biology of seed storage proteins. In *The Biochemistry of Plants Vol. 15 (Molecular Biology)*, ed. A. Marcus, Academic Press, New York, pp. 297–345.

Verma, D. P. S. and Goldberg, R. B. (eds) (1988). *Plant Gene Research Vol. 5; Temporal and Spatial Regulation of Plant Genes*. Springer-Verlag, New York.

Willmitzer, L. (1988). The use of transgenic plants to study plant gene expression. *Trends Genet.* **4**, 13–8.

3 Biochemical regulation
William C. Plaxton

'Regulation is not a late development superposed on metabolism after catalysis had become well-established . . . Regulation is the most fundamental difference between living and nonliving systems, and it must have coevolved with other properties of life from the beginning' (Atkinson, 1977).

Key concepts

The ability to regulate the rates of metabolic processes in response to changes in the internal and/or external environment is a fundamental feature which is inherent in all organisms. This adaptability is necessary for conserving the stability of the intracellular environment (homeostasis) which is essential for maintaining an efficient functional state in the organism.

Complexity of metabolism and concept of biochemical unity

Even the most primitive of species is metabolically complex. For example, although the unicellular bacterium *Escherichia coli* is approximately 500-fold smaller than a typical eukaryotic cell, each *E. coli* cell contains over 2000 different types of proteins, most of which are enzymes. The metabolic complexity of even the smallest prokaryote is a result of these many separate enzymatic reactions which make up the metabolic pathways which constitute metabolism.

The substantial metabolic complexity of green plants is reflected by their autotrophic nature, extensive biosynthetic capabilities, and the presence of metabolic compartments not found in bacteria or animals, namely the plastid and cell vacuole. The sessile nature of most plants further complicates their metabolic needs. In order to survive many plants must possess extensive anatomical, physiological, and biochemical adaptations to environmental stresses such as heat, cold, drought, salinity stress, nutrient limitation, and anoxia.

Despite its complexity, a general understanding of metabolism, applicable to all the phyla, has been achieved. This is because common evolutionary 'solutions' to the problem of 'biochemical design' occur in a wide variety of organisms. Thus, the types of substrates, cosubstrates, coenzymes, fuels and metabolic pathways used (as well as the genetic code) are in many cases ubiquitous to all life forms. This is the concept of *biochemical unity*.

In general, biochemical unity still applies to metabolic regulation. Comparative biochemistry has shown that the design of regulation (i.e. the types of regulatory mechanisms found in metabolic pathways) is similar from one organism to the next. However, it is the implementation of these designs – the regulatory details – which not only can differ widely from species to species, but can differ widely for similar metabolic pathways in different cell types of a single organism, or even within different organelles of a single cell.

For example, the enzyme citrate synthase, which catalyzes the following reaction:

acetylCoA + oxaloacetate → citrate,

is regulated in different manners in different organisms. In respiring aerobic eukaryotic cells, a major function for this enzyme is in the TCA cycle which is operating in the mitochondria to produce ATP via oxidative phosphorylation. Here, the overall endproduct, ATP, feedback inhibits citrate synthase. This is logical since at high levels of ATP, the ATP-generating TCA cycle will be slowed down, but will speed up again as the ATP levels fall. *E. coli*, which is primarily anaerobic, generates ATP mainly through glycolytic fermentation of glucose. The main role of the TCA cycle in *E. coli* is in the production of biosynthetic precursors and reducing power (NADH). In this organism, citrate synthase is unaffected by ATP, but is inhibited by the ultimate endproduct, NADH. In contrast, germinating oil seeds contain a glyoxysomal citrate synthase, which is not inhibited by ATP or NADH. Here the enzyme functions as part of the glyoxylate cycle, a key component in the metabolic conversion of storage triglycerides to sucrose.

Thus, the concept of biochemical unity tends to break down when individual metabolic controls are compared. Although the structure and products of a metabolic pathway may be identical in various organisms, the *environment* and *function* of that pathway may not be the same. Nonetheless, all metabolic controls do have a common basis, and certain regulatory strategies are ubiquitous.

The basis of metabolic control

'Pacemaker' enzymes

It is self evident that the flux, or rate of movement, of metabolites through any given pathway must be closely coordinated with the needs of the cell, tissue, or organism for the final endproduct(s) of the pathway. A major way of accomplishing such regulation is through altering the activity of at least one *rate-limiting* enzyme of the pathway (Cohen, 1983; Crabtree and Newsholme, 1985). Therefore, a major focus of metabolic control has been on these key enzyme reactions – the *pacemakers* or *rate-determining steps* – which control metabolite flow, *in vivo*. Normally, the rate-determining step(s) of a pathway is essentially irreversible (i.e. has a high negative free energy change, *in vivo*), has a low activity overall, and frequently occurs at the first committed step of a pathway, directly before major branch points, and at the last step of a 'multi-input' pathway. Much effort has been expended in identifying not only the rate-determining step(s) of metabolic pathways, but also the regulatory mechanisms which are used to modulate the activity of these key enzymes.

Control analysis

In pursuing the 'pacemaker' theory, one attempts to formulate a theory of metabolic control by performing accurate and detailed analyses on the *in vitro* kinetic and regulatory properties of purified 'key' enzymes. A potential problem with this reductionist approach is that biological systems may display regulatory properties which are not possessed by their isolated components (i.e. the properties of biological systems are usually greater than the sum of the properties of their isolated parts). Thus, another important approach to examining metabolic control is to analyze the whole system. The control analysis theory developed by Kacser in 1973 attempts to provide a quantifiable mechanism for probing intact biological systems and interprets resultant data without resorting to preconceived notions as to which enzymes in a pathway are 'rate-determining steps' or 'pacemakers'. In fact, an important prediction of the control analysis theory is that metabolic control is shared among many if not all steps in a pathway. Kacser has established the concept of the *flux control coefficient* (C_E^J) whose value specifies the change in metabolic flux (ΔJ) to small changes in the activity of any enzyme (ΔE) in the metabolic system as follows: $C_E^J = (\Delta J/\Delta E)$. If this value is very small (i.e. 0.01), it means that any change in the activity of the chosen enzyme will have little effect on pathway flux, and that this enzyme therefore exercises very little control on overall pathway flux. If the flux control coefficient is determined to be close to 1.0, it shows that the change in pathway flux responds almost proportionally to a small change in the enzyme's activity. Such an enzyme would clearly be important to regulating the overall rate of metabolite movement (flux) through the pathway. In other words, flux through the pathway is peculiarly sensitive to small perturbations in the activity of that particular enzyme. Experimental determination of the magnitude of appropriate flux control coefficients apparently yields an unambiguous evaluation of the existing quantitative allotment of control among the various steps in a pathway, under specified conditions. As the magnitude of any one flux control coefficient is not a property of the enzyme *per se* but depends upon the concurrent activities of all the other enzymes in the system, individual flux coefficients must be determined experimentally from the intact system. Flux control coefficients are therefore determined by measuring how pathway flux changes following alteration of the activity of a specific enzyme *in situ*. Recent advances in molecular biology now allow for direct manipulations of *in vivo* enzyme activities and promise to yield exciting information on the control of metabolism. Readers should refer to recent reviews (Kacser, 1987; Kacser and Porteous, 1987) for further details and insights concerning the 'control analysis' approach to examining metabolic regulation.

Types of metabolic control

Basic mechanisms

The magnitude of metabolite flux through any metabolic pathway will depend upon the activities of the individual enzymes involved. Two basic mechanisms can potentially be used by the cell to vary the reaction velocity of a particular enzyme. These are 'coarse' and 'fine' metabolic control.

Coarse metabolic control

Coarse metabolic control is a long-term (i.e. hours to days), energetically expensive, response which is achieved through changes in the total cellular population of enzyme molecules. Coarse control can be applied to one or all of the enzymes in a particular pathway, and most frequently comes into play during tissue differentiation or long-term environmental (adaptive) changes. The total amount of a given enzyme is dependent on the rates of its biosynthesis versus degradation. Thus, any alteration in the rates of transcription, translation, mRNA processing or degradation, or proteolysis can be construed as coarse metabolic control. The regulation of gene expression, protein synthesis and protein turnover have been covered in some detail in Sections I and VIII of this book, and will therefore not be discussed here.

Fine metabolic control

Fine metabolic controls are generally fast (i.e. seconds to minutes), energetically inexpensive, regulatory devices which modulate the activity of the pre-existing enzyme molecule. Operating mainly on a pathway's regulatory, or rate-limiting, enzyme(s), fine controls allow the cell to prevent metabolic chaos. Fine controls can be thought of as 'metabolic transducers' which 'sense' the momentary metabolic requirements of the cell, and modulate the rate of metabolite flux through the various pathways accordingly.

The enormous advances which have been made in our understanding of plant biochemistry over the past 10 years are providing a clear indication that there is a comparable range and sophistication of fine metabolic controls operating in both plant and animal systems. It is important to be aware that the fine controls to be discussed in detail below are not mutually exclusive, but often interact with, or may actually be dependent upon, one another.

Fine control 1. Alteration in substrate or cosubstrate concentration

The rate of an enzyme-catalyzed reaction is dependent on its substrate concentration when that substrate is subsaturating. Substrate concentrations for most enzymes are indeed subsaturating *in vivo*. In other words, enzymes do not normally operate at their V_{max}. Often the *in vivo* substrate concentration is less than or equal to the K_m or

$S_{0.5}$ value of the enzyme for that particular substrate (K_m or $S_{0.5}$ = substrate concentration yielding $0.5V_{max}$ for enzymes which show hyperbolic or sigmoidal substrate saturation kinetics, respectively). Hence, an important question arises as to whether or not the rate of pathway flux could be controlled by changes in the substrate concentration for any of the enzymes which comprise a pathway.

Following activation of a metabolic pathway, the concentration of its constituent metabolites may increase by about two-fold. However, the rate of pathway flux may increase up to 10-fold during the same period. In order to obtain a 10-fold change in activity of an enzyme which shows hyperbolic substrate saturation (i.e. Michaelis–Menten) kinetics, the increase in its substrate concentration would have to be about 80-fold (Fig. 3.1A). Thus, variation in substrate concentration cannot be the sole determinant of the activity of enzymes which show hyperbolic substrate saturation kinetics.

Not all enzymes show simple Michaelis–Menten substrate kinetics, however. Multisubunit, polymeric enzymes often have more than one substrate binding site which can interact via conformational change to yield a sigmoidal plot of activity versus substrate concentration (i.e. show cooperative binding of substrate) (Fig. 3.1B). Sigmoidal enzymes can increase their activity 10-fold with only a two to five-fold increase in substrate concentration (Fig. 3.1B). The actual increase in substrate concentration which would be required is dependent upon the degree of cooperativity with which the enzyme binds its substrate. With increased cooperativity (i.e. higher values for n_H, the Hill coefficient) smaller-fold increases in substrate concentration are required to effect the same relative change in enzyme activity.

In summary, changes in substrate concentrations which normally occur *in vivo*, could alter the rate of pathway flux, but most significantly for enzymes which show sigmoidal substrate saturation kinetics. Enzymes of this nature have been found in bacteria, animals and plants. Invariably these enzymes have been identified as regulatory, or rate-limiting. A possible evolutionary 'advantage' of sigmoidal substrate saturation kinetics is that it allows a much more sensitive control of reaction rate by substrate concentration.

Fine control 2. Variation in pH

Most enzymes have a characteristic pH at which their activity is maximal, i.e. the pH optimum. Above or below this pH the activity normally declines, although to varying degrees depending on the particular enzyme. Thus, enzymes can typically show pH activity profiles ranging in shape from very broad to very narrow. The pH optimum of an enzyme is not always the same as the pH of its intracellular surroundings. This suggests that the pH dependence of enzyme activity may be a factor which determines its overall activity in the cell. As cells contain thousands of enzymes, all

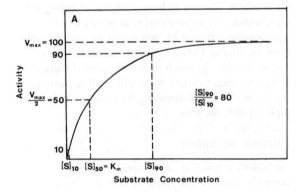
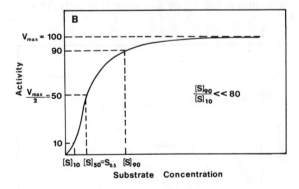

Fig. 3.1 (A) Relationship between substrate concentration and reaction rate for enzymes which show hyperbolic substrate saturation kinetics; (B) relationship between substrate concentration and reaction rate for enzymes which show sigmoidal substrate saturation kinetics.

differentially responsive to pH, the intracellular pH may represent an important element of fine metabolic control.

The light-dependent activation of several of the enzymes of the reductive pentose phosphate pathway (Calvin cycle) provides a well characterized example of how changes in pH can contribute to the overall regulation of plant enzymes. As discussed in Section V of this book, photosynthetic electron transport has been linked to H^+ uptake into the thylakoid lumen. This establishes a proton gradient between lumen and stroma which is believed to drive photophosphorylation. The transport of H^+ ions into the lumen results in a light-dependent increase in stromal pH from about 7.0 to 8.0. In the absence of light, H^+ ions leak back into the stroma and its pH falls from 8.0 to 7.0. As several of the reductive pentose phosphate pathway enzymes have a relatively sharp alkaline pH optima of between 7.8 and 8.2, the increase in stromal pH which is contingent upon photosynthetic electron transport ensures that these enzymes, and hence the overall cycle, will only be fully operative in the light.

Fine metabolic control of plant enzymes brought about by alterations in intracellular pH is not restricted to the chloroplast. For example, stimulation of the terminal glycolytic enzyme alcohol dehydrogenase during anaerobiosis of plant tissues may be partially dependent on an anoxia induced reduction in cytosolic pH (see Section II: Anaerobic Respiration).

Fine control 3. Allosteric regulation

Multisubunit regulatory enzymes often have allosteric sites where specific inhibiting or activating metabolites can reversibly bind. By varying the concentration of these non-substrate effector molecules it is possible to alter the velocity of a rate-determining step and, therefore, change the flux of the entire metabolic pathway. These allosteric effectors alter enzyme activity by binding to an allosteric site and eliciting a precise change in the enzyme's conformation. This conformational change ultimately affects substrate interactions with the enzyme in such a way that some or all of the kinetic constants (i.e. V_{max}, K_m or $S_{0.5}$, and n_H) are significantly altered. This regulatory strategy allows for metabolites which are remote from a specific reaction to act as feedback or feedforward control signals. The key regulatory significance of this interaction is that, at the low substrate concentrations that are found *in vivo*, the rate of a reaction, and therefore the pathway flux, can be dramatically increased or decreased by the binding of an allosteric effector. This is illustrated graphically in Fig. 3.2 which shows that with the addition of an activator or inhibitor it is possible to vary an enzyme's reaction rate from $0.1 V_{max}$ to $0.9 V_{max}$ with little or no change in substrate concentration.

Essentially all rate-determining enzymes are at least partially controlled by allosteric effectors. For an allosteric effector to have meaningful regulatory significance, the concentration at which it significantly activates or inhibits an enzyme *in vitro* must be similar to the effectors' actual *in vivo* concentration. As much of the material discussed in various pages of this book will testify, allosteric regulation is used extensively by plants to regulate pathway flux, *in vivo*.

Enzyme activation

Activators interact at an allosteric site and normally increase the reaction rate at subsaturating substrate concentrations by causing a reduction in the K_m or $S_{0.5}$ value (Fig. 3.2). This

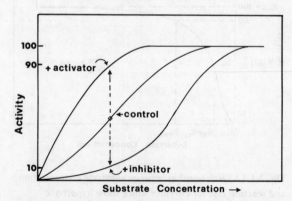

Fig. 3.2 Typical effect of the addition of an activator or inhibitor on substrate saturation kinetics for an allosteric enzyme.

may or may not be accompanied by a corresponding increase in V_{max}.

One of the first reports of physiologically meaningful allosteric activation of a plant enzyme was provided by Ghosh and Preiss (1966) for spinach leaf ADP-glucose synthetase. ADP-glucose synthetase, which is found only in plastids, is the rate-limiting enzyme in the pathway of starch biosynthesis from hexose phosphates (Preiss, 1984; see also Section VI). This enzyme catalyzes the formation of pyrophosphate and ADP-glucose from ATP and glucose 1-phosphate and shows significant feedforward activation by low concentrations of 3-phosphoglyceric acid (3-PGA), the immediate product of photosynthetic CO_2 fixation by ribulose 1,5-bisphosphate carboxylase. In the case of ADP-glucose synthetase the activator 3-PGA causes a very marked enhancement in V_{max} as well as a significant reduction in the K_m values for the substrates ATP and glucose 1-phosphate. The amount of activator required for half-maximal activation (i.e. $A_{0.5}$) is less than 0.05 mM. This is in the range of 3-PGA concentrations found in the cell and suggests this activating effect is relevant, *in vivo*.

Enzyme inhibition

Competitive inhibitors compete with the substrate for a common substrate binding site (i.e. the active site). Competitive inhibition causes an increase in the K_m or $S_{0.5}$ values, does not affect V_{max}, and is reversed by increasing the substrate concentration. Competitive inhibitors are not true allosteric effectors because they do not bind to an allosteric site, a site which is distinct from the active site. Nonetheless, one occasionally sees reports in the literature incorrectly describing 'allosteric enzymes' as being 'allosterically inhibited' by competitive inhibitors.

Non-competitive enzyme inhibition is a much rarer form of regulation in which the inhibitor combines with the enzyme, or enzyme–substrate complex, at a true allosteric site. In this instance the affinity of the enzyme for its substrate is unaffected whereas the V_{max} is reduced. Occasionally, an inhibitor can interact both at the active site, as in competitive inhibition, and at a separate allosteric site, as in non-competitive inhibition. This gives rise to mixed type inhibition. The inhibition of plant ADP-glucose synthetase by P_i is a typical example of mixed type inhibition (Preiss, 1984).

Interacting effectors

The activity of an allosteric enzyme, *in vivo*, is dependent on the relative concentrations of its activators versus inhibitors. Often, the presence of an activator can override, or cancel, inhibitory signals. This is shown clearly for the aforementioned ADP-glucose synthetase, in which the activator 3-PGA negates the inhibitory effect of P_i (Preiss, 1984). Thus, it is the ratio of plastidic $[3\text{-PGA}]/[P_i]$ which is believed to be a major factor in determining the rate of ADP-glucose synthetase, and thus starch biosynthetic activity, *in vivo*. This hypothesis, formulated in 1966 on the basis of ADP-glucose synthetase allosteric properties which were observed *in vitro*, has been subsequently confirmed by various *in vivo* evidence (Preiss, 1984).

The adenine nucleotides as allosteric regulators

Energy transduction and energy storage, involving the adenine nucleotides (ATP, ADP, and AMP) are a fundamental feature of metabolism. Thus, it is of no surprise that the most extensive system of allosteric regulators are the adenine nucleotides which operate on a large number of reactions to adjust the rate of formation of ATP to its utilization (Atkinson, 1977; Crabtree and Newsholme, 1985). In many plant and animal tissues, the adenine nucleotides are maintained in equilibrium by the enzyme adenylate kinase which catalyzes the reaction, $ATP + AMP \longleftrightarrow 2ADP$. Thus, a decrease in the concentration of ATP will occur following an increase in that of AMP, and vice versa. The widespread importance of the adenine nucleotides in metabolic regulation has led Atkinson (1977) to introduce the concept of 'energy charge' (energy charge = 'the state of the adenylates'), which he defined as:

$$\text{energy charge} = \frac{[\text{ATP}] + \tfrac{1}{2}[\text{ADP}]}{([\text{ATP}] + [\text{ADP}] + [\text{AMP}])}.$$

Energy charge can theoretically vary between 0 (i.e. all the adenylate is present as AMP) and 1 (i.e. all of the adenylate is present as ATP). As this parameter does fluctuate to a certain extent *in vivo*, it provides a basis for metabolic control by energy charge. Regulatory enzymes that occur in pathways in which ATP is consumed (i.e. anabolic pathways) respond to changes in energy charge in the general way shown by curve 2 of Fig. 3.3.

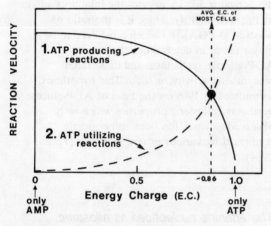

Fig. 3.3 Responses to the cellular energy charge of regulatory enzymes in metabolic pathways in which ATP is produced (curve 1) and in which it is utilized (curve 2) (adapted from Atkinson, 1977).

Regulatory enzymes which participate in pathways in which ATP is regenerated (i.e. catabolic pathways) respond in the general way shown by curve 2, Fig. 3.3. It is clear from Fig. 3.3 that any tendency for the energy charge to fall would be prevented by the consequential increase in metabolic flux through ATP-regenerating pathways (curve 1) and by the decrease in flux through ATP consuming pathways (curve 2). In healthy cells, energy charge is normally controlled at a value of between 0.7 and 0.9. This provides a constant environment for the application of all the other mechanisms of metabolic regulation.

Fine control 4. Covalent modification

This method of fine control functions cooperatively with allosteric regulation. The general model is that an enzyme is interconverted between a less active and more active form. This interconversion is not due to a relatively freely reversible equilibrium, as encountered with simple allosteric regulation, but is governed by two thermodynamically favorable enzyme catalyzed reactions which result in the formation of new stable covalent bonds. Interconversion in either direction can be very fast (i.e. minutes) and very complete (up to 100% conversion). A change in enzyme conformation which is induced by covalent modification leads to an alteration in normal enzyme–substrate interactions such that kinetic parameters such as V_{max}, $S_{0.5}$ and n_H are significantly elevated or reduced. Either interconversion can be used to override or cancel existing allosteric effector signals. This energetically inexpensive control mechanism can quickly provide the cell with an essentially 'new' enzyme form whose kinetic properties are well geared to the cell's momentary metabolic needs.

A key facet of enzyme regulation by reversible covalent modification is that it is the major mechanism in higher eukaryotes whereby extracellular stimuli, such as hormones or light, coordinate the regulation of key enzymes of intracellular pathways (Cohen, 1983; Poovaiah and Reddy, 1987; Ranjeva and Boudet, 1987). Although some 150 types of post-translational modifications of proteins have been reported *in vivo*, very few appear to be important in enzyme regulation (Cohen, 1983). Dithiol–disulfide interconversions and phosphorylation–dephosphorylation are the most important mechanisms of reversible covalent modification used in higher eukaryote enzyme regulation.

Dithiol–disulfide interconversion

Covalent modification by dithiol–disulfide exchange involves reactions which are chemically similar to those which occur in the formation of disulfide bonds which stabilize the tertiary structure of proteins. The only difference from the latter is that in the native enzyme, disulfides participating in regulation must be accessible to reduction by external thiols since regulation by this mechanism in response to cellular redox state can occur only if the reaction is freely reversible (Ziegler, 1985).

Thioredoxin is a 12 kD heat-stable protein which in its reduced SH form can act as a protein-disulfide reductase (Cohen, 1983). Distributed throughout the phyla, thioredoxin has been found in all cell types in which it has been sought (Holmgren, 1985). This protein appears to play a variety of roles in different cell types. In bacterial and animal systems it may act as a general protein-disulfide reductant, or may participate in the reduction of specific enzymes. In these organisms reduced thioredoxin is regenerated by thioredoxin reductase, at the expense of NADPH (Holmgren, 1985).

Dithiol–disulfide enzyme interconversion appears to have a much greater regulatory significance in photosynthetic organisms. As discussed in further detail in Section V and elsewhere (Cohen, 1983; Buchanan, 1984; Ziegler, 1985), this type of reversible covalent modification is extremely important in linking photosynthetic electron flow to light regulation of several key chloroplastic enzymes involved in photosynthetic CO_2 fixation and related processes. Reducing equivalents (electrons) are ultimately generated from H_2O by the photosynthetic electron transport chain. The proteins, ferredoxin, ferredoxin–thioredoxin reductase, and thioredoxin constitute the machinery which shuttles the reducing equivalents from the electron transport chain to selected target enzymes (Fig. 3.4A). The recent finding of a plant cytosolic thioredoxin (Florencio et al., 1988) will possibly extend this type of regulation to enzymes localized outside the chloroplast.

By mechanisms not yet fully understood, enzymes which are reduced by thioredoxin in the light are oxidized to disulfide forms in the dark. This might be catalyzed by compounds such as oxidized glutathione (Cohen, 1983; Buchanan, 1984).

Phosphorylation–dephosphorylation

Enzyme modification by the reversible covalent incorporation of phosphate is a widespread

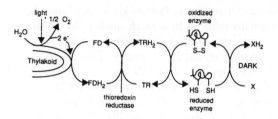

(a)

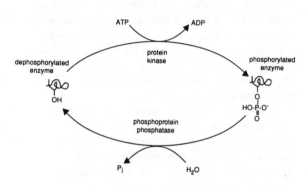

(b)

Fig. 3.4 The regulation of enzyme activity via reversible covalent modification. For details refer to the text.
(a) Dithiol–disulfide interconversion. FD and FDH_2, oxidized and reduced ferredoxin, respectively; TR and TRH_2, oxidized and reduced thioredoxin, respectively; X, some oxidant normally kept reduced in the light;
(b) Phosphorylation–dephosphorylation.

phenomenon with important consequences for metabolic control in vivo. As outlined in Fig. 3.4B, the phosphorylation reaction is catalyzed by a protein kinase, usually in an ATP-dependent manner, whereas the reverse reaction is catalyzed by a phosphoprotein phosphatase. Both protein kinase and phosphoprotein phosphatases can have wide or narrowly defined substrate specificities. Protein kinase and phosphoprotein phosphatase are always subject to their own fine controls. Each class of protein kinase is allosterically stimulated by a specific 'signal metabolite' (i.e. cyclic nucleotides (cAMP or cGMP), diacylglycerol, Ca^{2+}) which makes enzyme phosphorylation responsive to extracellular signals. Protein kinase activation is normally initiated by a much smaller change in effector concentration than with respect to allosteric controls. As the smallest increase in effector concentration yields a maximal effect, a complete on/off control can be obtained. An increase in signal amplification can be made possible by designing an 'amplification cascade' with more enzymes involved (Cohen, 1983, 1985). Animal phosphoprotein phosphatases are frequently controlled by an inhibitor protein which complexes with the enzyme under conditions when the corresponding kinase is activated. Although phosphoprotein phosphatases have been reported to occur in plant tissues, very little is known about their distribution and regulation in plants (Poovaiah and Reddy, 1987).

It has been definitively shown that in animal systems enzyme phosphorylation serves to coordinate the relative activities of competing metabolic pathways in multifunctional tissues (Cohen, 1983, 1985). At least 50 rate-limiting enzymes found in all the major metabolic pathways of higher animals, are now known to be regulated by reversible protein kinase mediated phosphorylation.

At present there is compelling evidence that plants utilize reversible enzyme phosphorylation to regulate the rate of various metabolic pathways. The light harvesting chlorophyll–protein complex has been shown to undergo phosphorylation–dephosphorylation and that this probably plays a role in regulating photosynthetic electron flow (Ranjeva & Boudet, 1987). The first plant enzyme which was determined to be regulated by phosphorylation–dephosphorylation control is the mitochondrial pyruvate dehydrogenase complex (PDC) from broccoli buds (reviewed by Ranjeva and Boudet, 1987). This enzyme complex can apparently be interconverted between an activated dephosphorylated form and an inactivated phosphorylated form, in a process very similar to that known for animal PDC. This interconversion is believed to play a key role in regulating carbon flow between glycolysis, the TCA cycle and fatty acid metabolism. The description of this important mechanism for the regulation of higher plant PDC extended our knowledge of this type of metabolic control into the higher plant kingdom and raised the possibility that other plant enzymes could be controlled in a similar fashion. Not surprisingly, there is now evidence that at least eight other plant enzymes may be controlled *in vivo* by the reversible incorporation of covalently bound phosphate (Poovaiah and Reddy, 1987; Budde and Chollet, 1988; Yang *et al.*, 1988). These findings, along with the existence in plants of several types of protein kinase and at least one type of phosphoprotein phosphatase (Hepler and Wayne, 1985; Poovaiah and Reddy, 1987), raises the likelihood that phosphorylation–dephosphorylation will prove to be a common and important control mechanism in mediating plant cellular responses to external stimuli.

Fine control 5. Subunit association–disassociation

Usually, multimeric enzymes become less active or inactive when the constituent subunits are dissociated. This property has been exploited to some degree by various organisms to help regulate the activities of certain rate-determining enzymes. Aggregation or dissociation of subunits is normally induced by the binding of a small molecular weight effector molecule, but can also be instigated by reversible covalent modification.

A plant enzyme that is probably regulated in this fashion, *in vivo*, is the pyrophosphate-dependent phosphofructokinase (PFP; see Section II). PFP can be interconverted between an active, approximately 260 kD tetramer, and a less active 130 kD dimer (Black *et al.*, 1987; Dennis and Greyson, 1987). The state of aggregation of this enzyme is dependent on the concentrations of its allosteric activator, fructose 2,6-bisphosphate (Fru-2, 6-P_2), and product of the forward reaction, PP_i. The tetrameric form of the enzyme, as well as its activity in the forward, or glycolytic direction, is enhanced by the presence of Fru-2, 6-P_2. The dimeric form of PFP, as well as its activity in the reverse, or gluconeogenic direction, is promoted by PP_i (Dennis and Greyson, 1987). Thus, the reversible association of PFP subunits, as determined by the relative concentrations of Fru-2, 6-P_2, and PP_i, may represent a glycolytic/gluconeogenic regulatory mechanism.

Fine control 6. Reversible associations of metabolically sequential enzymes

Recent studies are providing convincing evidence for subcellular structuring of metabolic pathways which were once thought of as being entirely soluble. For example, various workers have described associations between many of the so-called 'soluble' enzymes of glycolysis and membrane fractions or structural proteins in animal cells (Storey, 1985; Masters *et al.*, 1987; Srere, 1987). Of regulatory significance is that the extent of enzyme binding appears to be closely related to the rate of flux through the pathway. During mammalian muscle contraction, for example, a marked increase in the binding of key glycolytic enzymes to contractile proteins has been observed. The degree of binding was found to dissipate once the contractile activity, and hence the need for a high rate of glycolytic flux, ceased. Of equal significance are the findings that sequential glycolytic enzymes can also show specific interactions with each other – interactions which have mutual kinetic effects on the enzyme activities. Similar correlations between the degree of enzyme binding and rate of pathway flux have been observed elsewhere in animal systems. Enzymes which show a metabolic dependence in the degree to which they are associated with subcellular particulate structures have been termed *ambiquitous* (Masters *et al.*, 1987).

The micro-compartmentation of enzymes and metabolic pathways which can result from ambiquitous interactions could provide an effective means of metabolic control via (1) channeling of substrates between consecutive enzymes, and (2) altering enzyme kinetic properties due to conformational changes occurring during binding. It has been suggested, therefore, that metabolic activation of pathways such as glycolysis can occur not only by allosteric regulation and covalent modification of key enzymes, but also by altering the partitioning of enzymes from the soluble to the bound phase (Storey, 1985; Masters *et al.*, 1987).

The existence of known stable multienzyme complexes provides favorable evidence for the interaction of soluble enzymes that are metabolically sequential. Due to high intracellular protein concentrations, such interactions are far more likely *in vivo* than in dilute *in vitro* enzymological studies (Srere, 1987). The limited solvent capacity of the cell also supports interactions between sequential enzymes of a metabolic pathway as the need for a large pool of free intermediates would be eliminated (Atkinson, 1977).

The molecular mechanisms which control the extent of enzyme binding *in vivo* are not yet fully understood. Binding of animal glycolytic enzymes *in vitro* can be influenced by pH, concentrations of substrates, products and allosteric effectors, enzyme phosphorylation, as well as by changes in osmotic and/or ionic strength (Masters *et al.*, 1987; Srere, 1987).

Although there is now compelling evidence for complexing and/or binding of so-called 'soluble' enzymes and the direct transfer of substrates between sequential enzymes of various pathways in animal tissues, these possibilities have only recently been addressed in plant systems. For example, the chloroplastidic ferredoxin, ferredoxin–thioredoxin reductase, and thioredoxin system used for disulfide–dithiol enzyme interconversion may exist *in vivo* as an enzyme complex termed 'protein modulase' (Ford *et al.*, 1987). It has been suggested that ionic strength affects the interactions among these proteins and in part could determine the fate of the protein modulase complex *in vitro*. Likewise, recent studies indicate that there may be direct channeling of substrates between reductive pentose phosphate pathway enzymes in illuminated pea leaf chloroplasts (Marques *et al.*, 1987). Subsequently, a complex of five enzymes which catalyze five consecutive reactions of the reductive pentose phosphate pathway was purified and shown to be functionally active (Gontero *et al.*, 1988). These findings corroborate an earlier study which demonstrated that in the absence of added stromal protein, isolated membrane fractions prepared from spinach leaf chloroplasts could assimilate CO_2, yielding the intermediates of the reductive pentose phosphate pathway (Wah Kow and Gibbs, 1982). Similarly, the biosynthesis of flavonoids may take place in a multienzyme complex wholly or partially associated with the cytoplasmic face of the endoplasmic reticulum membrane (Hrazdina et al., 1987). Furthermore, the well characterized stimulation of respiration which accompanies aging of underground storage organ slices may result, in part, from an association of glycolytic enzymes with the particulate fraction of the cell promoting an elevated glycolytic rate (Moorhead and Plaxton, 1988). The above studies indicate that before an overall understanding of metabolic control in plants can be achieved we must determine not only how, where and when 'soluble', metabolically sequential plant enzymes might be aggregated into micro-compartments forming an organized multienzyme system (or 'metabolon'; Srere, 1987), but also how such associations may alter individual enzyme kinetic/regulatory properties. Equally significant will be the evaluation as to what extent the *reversible* formation and dissolution of metabolons contributes to the overall integration and control of plant metabolism.

The central role of calcium in plant metabolic regulation

The free Ca^{2+} ion is now recognized as having an important role in animals and plants as a 'second messenger' or 'signal molecule' which couples various extracellular stimuli such as hormones, light, gravity, or stress with intracellular metabolic events. The coupling of stimulus to response by Ca^{2+} implies that an external stimulus leads to a change in the rate at which Ca^{2+} is transported into or out of a specific subcellular compartment. The resulting alteration of the intracellular free Ca^{2+} concentration thus represents the 'signal' which, through its ability to regulate the activity of specific target enzymes, leads to the appropriate metabolic response.

If Ca^{2+} does act as a second messenger in plants it should be possible to demonstrate that its *in vivo* concentration changes in response to extracellular primary signals such as light or hormones. As yet, direct measurements have not been made. However, through manipulation of external or internal Ca^{2+} concentrations it has been possible to indirectly demonstrate a role for Ca^{2+} in stimulus–response coupling (Hepler and Wayne, 1985; Kauss, 1987; Poovaiah and Reddy, 1987).

Regulation of intracellular free calcium concentration

Although information is still limited, it is thought that animal and plant cytoplasmic free Ca^{2+} concentrations are normally maintained at <1 μM and that processes are activated by an increase in free Ca^{2+} concentration from 1 to 10 μM (Cohen, 1985; Kauss, 1987). This low Ca^{2+} concentration is maintained against high concentrations of Ca^{2+} both outside the cell (i.e. in the cell wall) and inside various subcellular compartments (i.e. cell vacuole, lumen of thylakoids and endoplasmic reticulum).

The existence of specific Ca^{2+} channels or pumps in plant membranes indicates that plant cells have the 'machinery' in place with which to maintain or alter Ca^{2+} concentration gradients. Two main types of Ca^{2+} transport mechanisms are thought to exist in higher plants. One is a Ca^{2+}-ATPase which may be partially dependent on calmodulin and is believed to be primarily localized in the endoplasmic reticulum and plasma membranes (Kauss, 1987; Poovaiah and Reddy, 1987). This system has a very high affinity for Ca^{2+} and may be the 'primary' pump which maintains the low free cytoplasmic Ca^{2+} concentration found in unstimulated cells. The second type of mechanism for Ca^{2+} transport is a 'secondary pump' which utilizes proton gradients as the driving force and has been characterized as a Ca^{2+}/H^+ antiport (Kauss, 1987). The Ca^{2+}/H^+ antiport is found in the tonoplast (with H^+ gradients generated by the tonoplast located H^+-ATPase) and possibly in the plasma membrane. Plastids and mitochondrial membranes are also believed to contain Ca^{2+} transport systems which are used to regulate free Ca^{2+} concentrations within these organelles (Kauss, 1987). Changes in the concentration of free Ca^{2+} probably plays an important role in metabolic regulation in all the subcellular compartments found in plants.

Calcium-modulated proteins

Many lines of evidence accrued since 1970 have demonstrated that a class of Ca^{2+} binding proteins referred to as 'Ca^{2+}-modulated proteins' are the targets or receptors for Ca^{2+} acting as a signal transducer in eukaryotic cells. These proteins have the ability to bind Ca^{2+} in a reversible manner with dissociation constants ranging from the nanomolar to micromolar range under physiological conditions (Roberts et al., 1986). Stimulation of the cell increases cellular free Ca^{2+} concentrations up to 10 μM, allowing formation of the Ca^{2+}-modulated protein : Ca^{2+} complex, which is the active species. The Ca^{2+}-modulated protein may be enzymatic, as in the case with Ca^{2+}-dependent protein kinase, or may be a regulatory protein, which in the presence of Ca^{2+} binds to, and thereby regulates the activity of, an appropriate enzyme. The enzyme whose activity is altered, in turn, regulates some cellular processes.

Calmodulin

The most important of all known Ca^{2+}-modulated proteins is calmodulin, a highly conserved, heat-stable, small molecular weight protein found in all eukaryotic cells (Roberts et al., 1986). Containing 4 Ca^{2+} binding sites per molecule, calmodulin is composed of a single subunit and 148 amino acids. The binding of Ca^{2+} induces a conformational change resulting in increased alpha helicity. This Ca^{2+}-induced conformational change allows the protein to interact with and alter the activity of calmodulin-dependent enzymes. The Ca^{2+}–calmodulin complex has a wide range of biochemical activities, including effects on both 'coarse' and 'fine' metabolic regulation (Hepler and Wayne, 1985; Roberts et al., 1986). Because of these characteristics, calmodulin has become the standard against which all other Ca^{2+}-modulated proteins have been compared.

Ca^{2+} and Ca^{2+}–calmodulin regulated plant enzymes

Plant enzymes which have been reported to be directly regulated by Ca^{2+} or Ca^{2+}–calmodulin include NAD-kinase, β-glucan synthase, Ca^{2+} and H^+-ATPase, quinate : NAD^+-oxidoreductase, and protein kinases (Girard and Maclachlan, 1987; Poovaiah and Reddy, 1987).

NAD kinase

NAD kinase is found in the plastid, cytosol and mitochondrion of plants and catalyzes the ATP-dependent conversion of NAD^+ to $NADP^+$ The pioneering work of Anderson and Cormier (1978) showed that the activity of NAD kinase was lost during its purification from pea leaves, but could be recovered by adding a Ca^{2+}-dependent, heat-stable, protein activator. This was the first report which demonstrated a probable link between a Ca^{2+}-dependent regulatory protein and a specific enzyme in higher plants. The protein activator was subsequently identified as calmodulin. Figure 3.5 shows a typical calmodulin dose–response curve for NAD kinase. It demonstrates that the activation of NAD kinase occurs at nanogram levels of calmodulin, which makes this stimulatory effect physiologically relevant. Addition of the Ca^{2+} chelator EGTA abolishes the effect of calmodulin (Fig. 3.5). The physiological roles for calmodulin activation of plant NAD kinase have yet to be completely elucidated (Roberts et al., 1986), but this regulation will obviously contribute to modulating

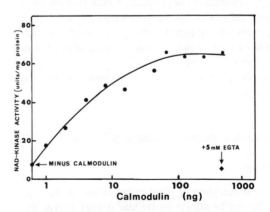

Fig. 3.5 Dose–response curve of the activation of plant NAD kinase by bovine brain calmodulin (figure reproduced with the permission of Prof. D. Marme and Elsevier Science Publishers B.V.).

the pools of NAD versus NADP in various subcellular compartments. A wide variety of metabolic processes would be affected, particularly reductive biosynthetic pathways which require NADPH.

Protein kinase

Many of the effects of Ca^{2+} on animal metabolism have been found to be mediated by Ca^{2+} acting alone, or by Ca^{2+} acting through calmodulin, to activate protein kinases which results in the phosphorylation of key regulatory enzymes. This has led to the concepts of signal 'amplification' and multienzyme 'cascades' (Cohen, 1983, 1985; Ranjeva and Boudet, 1987). There is now good evidence that plants also utilize Ca^{2+} to mediate protein phosphorylation, and this may be an important mechanism whereby extracellular stimuli are coupled with intracellular metabolic events (Elliot, 1985; Roberts et al., 1986; Pooviah and Reddy, 1987; Ranjeva and Boudet, 1987). However, although both membrane bound and soluble Ca^{2+} or Ca^{2+}–calmodulin dependent protein phosphorylating activities have been described in plants, very little is known about plant protein kinase structure or regulation. Even less is known about the endogenous protein substrates for these kinases, or how phosphorylation affects their activity. Thus, the purification and characterization of Ca^{2+} and Ca^{2+}–calmodulin dependent protein kinases and their substrates will be necessary before a comprehensive picture emerges as to the overall role of Ca^{2+} in plant metabolic regulation. Despite its presence in plants, cAMP has no known effect on phosphorylation of plant proteins (Ranjeva and Boudet, 1987).

Stimulus–response coupling by Ca^{2+}

Light

Photoregulation of plant metabolism can be achieved by effects on cellular energy charge or reducing power, rather than by a classical second messenger. The coupling of photosynthetic electron flow with the disulfide–dithiol interconversion of several chloroplastidic enzymes provides a good example of how light can directly regulate plant metabolic processes (see Chapter 17). However, animal photoreceptor cells have provided plant biochemists with a model whereby Ca^{2+} can act as a second messenger in coupling light with cellular function (Hepler and Wayne, 1985). Indeed, a growing body of evidence is becoming available in plant systems to indicate that light does play an important regulatory role through effects on Ca^{2+} fluxes within the plant cell. Light can stimulate Ca^{2+} uptake by the chloroplast as well as Ca^{2+} efflux from internal stores (Kauss, 1987; Poovaiah and Reddy, 1987; Ranjeva and Boudet, 1987). Of particular interest are the recent reports which suggest that at least some phytochrome-induced responses are mediated by Ca^{2+} and calmodulin (Roux et al., 1986).

Phytochrome is an important proteinaceous pigment which regulates many light-mediated growth and developmental responses in plants (Roux et al., 1986). This photoreceptor is found in all green plants and has the ability to interconvert between two spectrally different forms, a red absorbing P_r form and a far-red absorbing P_{fr} form. This interconversion allows phytochrome to initiate responses that are characteristically promoted by red light and inhibited or reversed by far-red illumination. That Ca^{2+} mediates at least some phytochrome initiated responses is suggested by the findings that: (1) P_{fr}, the active form of phytochrome, can affect many membrane properties including promoting the uptake of Ca^{2+} into plant cells; (2) several photomorphogenic effects of red light can be mimicked in darkness by the chemical induction of Ca^{2+} uptake into cells, and these same effects can be blocked by calmodulin antagonists; and (3) Ca^{2+}, calmodulin and P_{fr} all play a role in nuclear protein phosphorylation (Hepler and Wayne, 1985; Roux et al., 1986). These data have led to the hypothesis that phosphorylation of nuclear proteins may be an intermediate step in linking the photoactivation of phytochrome with the photoregulation of gene expression (Roux et al., 1986). Of interest is the discovery that, when prepared from etiolated oat seedlings, phytochrome itself is subject to reversible phosphorylation by a polycation-dependent protein kinase which copurifies with the photoreceptor (reviewed by Poovaiah and Reddy, 1987). The regulatory significance of a phytochrome protein kinase activity, as well as phytochrome phosphorylation, remains to be elucidated.

Plant hormones

Many hormones and neurological stimuli in animals regulate cell metabolism by inducing an alteration in the Ca^{2+} concentration in the cytoplasm. As with animal systems, the role of plant growth hormones (i.e. gibberellins, auxins, cytokinins, and abscisic acid) in manipulating cellular Ca^{2+} levels and/or protein phosphorylation has been firmly established. Consequently, a role for Ca^{2+} as a second messenger important to modulation of hormone-responsive plant systems has been proposed (Elliot, 1985; Hepler and Wayne, 1985; Poovaiah and Reddy, 1987; Owen, 1988).

Model of stimulus–response coupling by Ca^{2+}

The current models of how Ca^{2+} mediates transmembrane signaling in plants (Elliot, 1985; Roux et al., 1986; Pooviah and Reddy, 1987; Owen, 1988) are based upon similar models which have been proposed for animal systems. The initial event in this signaling is the interaction of a hormone with a plasma membrane receptor, or of light with the photoreceptor phytochrome (Fig. 3.6). These interactions are thought to alter Ca^{2+} transport across the plasma membrane or to activate phospholipase C, a membrane-bound enzyme which catalyzes the hydrolysis of a particular class of phospholipids, the phosphoinositides: phosphatidylinositol (PI), phosphatidylinositol 4-phosphate (PIP), and phosphatidylinositol 4,5-bisphosphate (PIP_2). Diacylglycerol (DAG) is a product of phospholipase C catalyzed phosphatidylinositide breakdown. Both Ca^{2+} and DAG are required to activate protein kinase C which in turn phosphorylates a distinct population of intracellular proteins, some of which may be involved in cellular replication (Elliot, 1985). Inositol trisphosphate (IP_3), another product of PIP_2 hydrolysis, causes Ca^{2+} release from intracellular Ca^{2+} storage areas (i.e. endoplasmic reticulum). The consequent rise in intracellular Ca^{2+} concentration then activates Ca^{2+} and Ca^{2+}–calmodulin dependent reactions, either by direct activation of an enzyme, or if that enzyme is a protein kinase, then indirectly by phosphorylation of a different subset of cellular proteins.

Although there are still several gaps in applying the model shown in Fig. 3.6 to plants, it is clear that much of the model agrees with the available information – namely the occurrence within plant cells of PIP and PIP_2, PI kinase, PIP kinase, phospholipase C, Ca^{2+} transport mechanisms, calmodulin, Ca^{2+}-dependent enzymes, various types of protein kinases (including protein kinase C), as well as plant growth hormone (or light) dependent effects on the transport and intracellular distribution of Ca^{2+}, and protein phosphorylation (Elliot, 1985; Poovaiah and Reddy, 1987; Owen, 1988; Sommarin and Sandelius, 1988).

Future prospects

As should be apparent from the preceding discussion remarkable advances have been made in the area of plant metabolism and its regulation. Equally apparent, however, is that much has yet to be learned before our understanding is complete. Future workers must continue to identify and purify the key regulatory enzymes of plant metabolic pathways and establish the mechanisms whereby their activities are controlled, *in vivo*. Only once these basic mechanisms are understood can geneticists and molecular biologists fully optimize conditions which will lead to increased crop yields.

Key areas for future research should include the occurrence, mechanism, sites and potential regulatory significance of reversible subcellular structuring, or micro-compartmentation, of so-called 'soluble' plant enzymes and metabolic pathways (i.e. '3D' metabolism). Studies concerning the relationship between enzyme structure and function have necessarily placed much emphasis on enzyme catalytic and allosteric

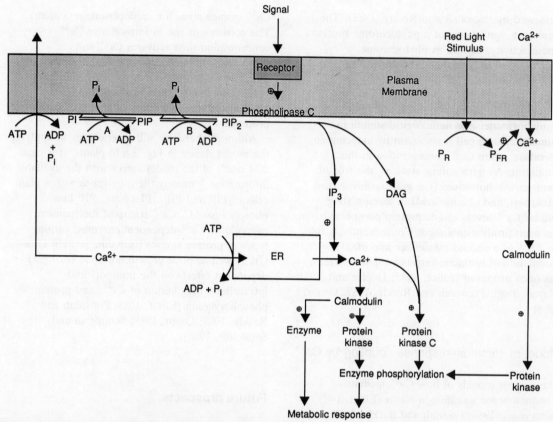

Fig. 3.6 Current model of how Ca^{2+} is believed to mediate transmembrane signaling in plants. For details refer to the text. PI, phosphatidylinositol: PIP, phosphatidylinositol 4-phosphate; PIP_2, phosphatidylinositol 4,5-bisphosphate; A, PI kinase; B, PIP kinase; ER, endoplasmic reticulum; IP_3, inositol 1,4,5 trisphosphate; DAG, diacylglycerol; P_R and P_{FR}, red and far-red light absorbing forms of phytochrome, respectively; +, stimulation of Ca^{2+} flux across the plasma and endoplasmic reticulum membranes, or activation of an enzyme (adapted from Poovaiah and Reddy, 1987 and Roux et al., 1986).

sites. However, as discussed by Srere (1987), if protein : protein interactions are important for normal cell functions, then the evolution and conservation of protein surface binding sites should also be examined.

Continued examination of the mechanisms whereby external environmental stimuli such as light, gravity, and stress are transduced into intracellular metabolic events is another important area for future research. In this regard more information on the methods of transmembrane signaling, phosphoinositide metabolism, as well as the concentration of Ca^{2+} in various compartments of living plant cells under changing environmental or physiological conditions is urgently needed. Also needed is the continued identification, purification, and characterization of plant enzymes which are under direct or indirect regulation by Ca^{2+} (i.e. calmodulin regulated enzymes, protein kinases, phosphoprotein phosphatases and their substrates). Of further interest will be the determination of the possible interaction between Ca^{2+}-dependent regulatory pathways and other regulatory pathways. For example, there is a probable interrelationship between calmodulin regulated pathways and those regulated by cAMP in animal systems (Cohen, 1985; Roberts et al., 1986). Although efforts by plant physiologists to establish a second messenger role for cAMP have been unsuccessful (Poovaiah and Reddy, 1987), its prominent involvement in extracellular signal transduction in animal systems, coupled with its occurrence in plant cells, indicates that cAMP may have an as yet unknown regulatory function(s) in plants. Undoubtedly, other compounds will be found which fulfill second messenger roles in plant cells. For example, changes in the intracellular concentration of the polyamines spermine and spermidine are known to occur following exposure of plants to environmental stress, changes in light intensity, as well as variations in growth hormones (Ranjeva and Boudet, 1987). Because polyamines can influence many different physiological and biochemical phenomena, including protein phosphorylation (Poovaiah and Reddy, 1987), additional studies should be directed at establishing their potential role as 'signal' metabolites in plants. Resolving these and other problems will contribute greatly to our appreciation as to how the metabolic flexibility of plants permits them to respond appropriately to their immediate environment and energy requirements, thereby providing for efficient maintenance of cellular homeostasis.

References

Anderson, J. M. and Cormier, M. J. (1978). Calcium-dependent regulator of NAD kinase in higher plants. *Biochem. Biophys. Res. Comm.* **84**, 595–602.

Atkinson, D. E. (1977) *Cellular Energy Metabolism and its Regulation*, Academic Press, New York.

Black, C. C., Mustardy, L., Sung, S. S., Kormanik, P. P., Xu, D.-P. and Paz, N. (1987). Regulation and roles for alternative pathways of hexose metabolism in plants. *Physiol. Plant.* **69**, 387–94.

Buchanan, B. (1984). The ferredoxin/thioredoxin system: A key element in the regulatory function of light in photosynthesis. *Bio-Science* **34**, 378–83.

Budde, R. J. A. and Chollet, R. (1988). Regulation of enzyme activity in plants by reversible phosphorylation. *Physiol. Plant.* **72**, 435–9.

Cohen, P. (1983). *Control of Enzyme Activity*, Chapman and Hall, London and New York.

Cohen, P. (1985). The role of protein phosphorylation in the hormonal control of enzyme activity. *Eur. J. Biochem.* **151**, 439–48.

Crabtree, B. and Newsholme, E. A. (1985). A quantitative approach to metabolic control. *Curr. Top. in Cell. Reg.* **25**, 21–76.

Dennis, D. T. and Greyson, M. F. (1987). Fructose 6-phosphate metabolism in plants. *Physiol. Plant.* **69**, 395–404.

Elliott, D. C. (1985) Calcium involvement in plant hormone action. In *Molecular and Cellular Aspects of Calcium in Plants*, ed. A. J. Trewavas, NATO ASI series, Series A, Life Sciences, Vol. 104, pp. 285–92.

Florencio, F. J., Yee, B. C. and Buchanan, B. B. (1988). An NADP/thioredoxin system in green leaves. *Plant Physiol.* **86**, S30.

Ford, D. M., Jablonski, P. P., Mohamed, A. H. and Anderson, L. E. (1987). Protein modulase appears to be a complex of ferredoxin, ferredoxin/thoredoxin reductase, and thioredoxin. *Plant Physiol.* **83**, 628–32.

Ghosh, H. P. and Preiss, J. (1966). Adenosine diphosphate glucose pyrophosphorylase: A regulatory enzyme in the biosynthesis of starch in spinach leaf chloroplasts. *J. Biol. Chem.* **241**, 4491–504.

Girard, V. and Maclachlan, G. (1987). Modulation of pea membrane β-glucan synthase activity by calcium, polycation, endogenous protease and protease inhibitor. *Plant Physiol.* **85**, 131–6.

Gontero, B., Cardenas, M. L. and Ricard, J. (1988). A functional five-enzyme complex of chloroplasts involved in the Calvin cycle. *Eur. J. Biochem.* **173**, 437–43.

Hepler, P. K. and Wayne, R. O. (1985). Calcium and plant development. *Ann. Rev. Plant Physiol.* **36**, 397–439.

Holmgren, A. (1985). Thioredoxin. *Ann. Rev. Biochem.* **54**, 237–71.

Hrazdina, G., Zobel, A. M, and Hoch, H. C. (1987). Biochemical, immunological, and immunocytochemical evidence for the association of chalcone synthase with endoplasmic reticulum membranes. *Proc. Natl Acad. Sci. USA* **84**, 8966–70.

Kacser, H. (1987). Control of metabolism. In *The Biochemistry of Plants*, Vol. 12, eds P. K. Stumpf and E. E. Conn, Academic Press, New York, pp. 39–67.

Kacser, H. and Porteous, J. W. (1987). Control of metabolism: what do we have to measure? *Trends Biochem. Sci.* **12**, 5–14.

Kauss, H. (1987). Some aspects of calcium-dependent regulation in plant metabolism. *Ann. Rev. Plant Physiol.* **38**, 47–72.

Marques, I. A., Ford, D. M, Muschinek, G. and Anderson, L. E. (1987). Photosynthetic carbon metabolism in isolated pea chloroplasts: metabolite levels and enzyme activities. *Arch. Biochem. Biophys.* **252**, 206–8.

Masters, C. J., Reid, S. and Don, M. (1987). Glycolysis – new concepts in an old pathway. *Mol. Cell. Biochem.* **76**, 3–14.

Moorhead, G. B. G. and Plaxton, W. C. (1988). Binding of glycolytic enzymes to a particulate fraction in carrot and sugar beet storage roots: dependence on metabolic state. *Plant Physiol.* **86**, 348–51.

Owen, J. H. (1988). Role of abscisic acid in a Ca^{2+} second messenger system. *Physiol. Plant.* **72**, 637–41.

Poovaiah, B. W. and Reddy, A. S. N. (1987). Calcium messenger system in plants. *CRC Crit. Rev. Plant Sci.* **6**, 47–103.

Preiss, J. (1984). Starch, sucrose biosynthesis and partition of carbon in plants are regulated by orthophosphate and triose-phosphates. *Trends Biochem. Sci.* **9**, 24–7.

Ranjeva, R. and Boudet, A. M. (1987). Phosphorylation of proteins in plants: Regulatory effects and potential involvement in stimulus/response coupling. *Ann. Rev. Plant Physiol.* **38**, 73–93.

Roberts, D. M., Lukas, T. J. and Watterson, D. M. (1986). Structure, function, and mechanism of action of calmodulin. *CRC Crit. Rev. Plant Sci.* **4**, 311–39.

Roux, S. J., Wayne, R. O. and Datta, N. (1986). Role of calcium ions in phytochrome response: an update. *Physiol. Plant.* **66**, 344–48.

Sommarin, M. and Sandelius, A. S. (1988). Phosphatidylinositol and phosphatidylinositolphosphate kinases in plant plasma membranes. *Biochim. Biophys. Acta* **958**, 268–78.

Srere, P. A. (1987). Complexes of sequential metabolic enzymes. *Ann. Rev. Biochem.* **56**, 89–124.

Storey, K. B. (1985). A re-evaluation of the Pasteur effect: New mechanisms in anaerobic metabolism. *Mol. Physiol.* **8**, 439–61.

Wah Kow, Y. and Gibbs, M. (1982). Characterization of a photosynthesizing reconstituted spinach chloroplast preparation. *Plant Physiol.* **69**, 179–86.

Yang, Y. P., Randall, D. D. and Trelease, R. N. (1988). Phosphorylation of glyoxysomal malate synthase from castor oil seeds *Ricinus communis* L. *FEBS Lett.* **234**, 275–9.

Ziegler, D. M. (1985). Role of reversible oxidation–reduction of enzyme thiols–disulfides in metabolic regulation. *Ann. Rev. Biochem.* **54**, 305–29.

4 Regulation by compartmentation
D. T. Dennis and M. J. Emes

The sequestration of metabolic pathways

Compartmentation of metabolism and the sequestration of metabolic pathways into membrane-bound organelles occurs in all eukaryotes but is most apparent in the cells of higher plants. Compartmentation is a means by which pathways can be regulated independently from one another, thereby providing a control of the flux through the pathways in response to the demands of cellular metabolism. The division of the cell into smaller volumes can serve to concentrate the intermediates of a pathway and may prevent futile cycling by opposing reactions. It may also provide specialized environments more favorable to certain reactions. Related pathways that interact through the supply and demand of substrates may also be kept in close proximity by being sequestered together. The major organelles found in plant cells are the nucleus, mitochondria, Golgi apparatus, microbodies, peroxisomes, glyoxysomes, vacuoles and plastids. In addition, it should not be forgotten that the cytoplasm is also a compartment in its own right. All these compartments are enclosed by the plasma membrane. Ultimately, the shape of the cell is supported by a rigid cellulose cell wall not found in animal cells. The cell wall space is not totally inert since it may contain enzymes (e.g. invertase) essential for metabolism.

The integration of cell metabolism requires controlled interaction between the pathways in the various compartments. This is achieved through a highly selective permeability of the organellar membranes so that a limited number of intermediates can be transported from one compartment to another. These transport systems are highly selective. There may also be a close physical association between the compartments.

In some cases, the sequence of reactions that comprises a metabolic pathway may be located in more than one organelle. An excellent example of this is photorespiration which requires the chloroplast, peroxisome, mitochondrion and cytosol (Chapter 18). Electron microscopy has indicated that these organelles may be closely associated, facilitating transfer of common intermediates.

Throughout this book, some examples of the compartmentation of plant metabolism will be presented. It is not the intention to attempt to describe all of them here. Examples of a few pathways will be presented as illustrations of the importance and complexity of compartmentation.

The unique role of plastids

The most fundamental difference between plants and other eukaryotic organisms is the presence of an additional compartment, the plastid. There is a wide variety of types of plastid, each with quite distinct metabolic function and capacity. The most commonly studied plastid is the chloroplast and, more specifically, its role in the photochemical generation of reducing power and ATP and their utilization in the reductive biosynthesis of carbohydrates from CO_2. Indeed, non-plant biochemists often regard the reductive pentose phosphate cycle and the photochemical reactions of the chloroplast as the only aspects of plant metabolism that seem to be worth investigating, since it is generally assumed that plants are otherwise biochemically more or less the same as other organisms. It should become apparent throughout this chapter and the rest of this book that this is a fallacy arising, at least to some extent, from the fact that many activities that are cytosolic or located elsewhere in other organisms are localized in plastids in plants.

The major part of the biosynthetic capacity of a plant cell is localized in plastids of one type or another, whether they are present as photosynthetically competent organelles or are completely achlorophyllous. This is also true for different tissues such as leaves, roots and seeds. The association of biosynthetic activity with the chloroplast is logical since this organelle is also the primary source of energy and primary carbon fixation, both of which are required for biosynthetic activity. In non-green plastids, carbon and energy must be imported for these pathways. The range of other activities found in all plastids include: fatty acid synthesis, the primary assimilation of nitrite into glutamate, the biosynthesis of a number of other amino acids (though not all), starch synthesis, and pigment biosynthesis (Dennis and Miernyk, 1982). In addition to these anabolic activities, plastids are also able to catabolize hexose phosphates via glycolysis and the oxidative pentose phosphate pathway (ap Rees, 1988a).

In chloroplasts the primary activity of the organelle is the generation of triose phosphates and hexose phosphates within the reductive pentose phosphate cycle which serve as the precursors of other pathways. This is not the case in non-photosynthetic plastids, where the predominating metabolic pathways differ depending on the tissue in which the organelle is found. These differences must be a result of differential gene expression occurring both as a function of the plant species and tissue location. For example, in the endosperm of developing castor oil seeds metabolism is directed predominantly toward fatty acid synthesis which occurs in the plastids. In contrast, the Gramineae store starch in the developing endosperm and this is also synthesized in the plastid. Nevertheless, within the leaves of both these plants, the activity of the plastid must be directed towards production of photosynthate that can be used in the synthesis of such essential intermediates as amino acids for growth and development.

All plastids contain DNA which in most cases consists of a circular molecule, from 120 to 150 kbp in length. Although plastids from various tissues in a plant can be very different both in morphology and biochemistry, the DNA they contain is identical (Dennis *et al.*, 1985). This suggests that all plastids arise from the same precursor plastid and differentiate in response to the development of the tissue. This differential response is undoubtedly under the control of the nucleus of the cell.

In chloroplasts, metabolic intermediates, reducing power and ATP for the above biosynthetic activities are generated substantially within the organelle, whereas in non-photosynthetic plastids they have to be imported from the cytoplasm either directly or generated from imported precursors. This difference is important, for it means that one cannot assume that the study of chloroplasts alone will provide an insight into the control of metabolism in all types of plastid. Many of the biosynthetic pathways in different plastids have common features but the regulation of these pathways and their integration with the rest of cellular metabolism must, inevitably, be different. For example, in chloroplasts, starch is synthesized in the light and degraded in both the light and dark (Preiss, 1982); in root plastids it is synthesized and degraded in the dark but turnover is negligible (Hargreaves and ap Rees, 1988). In contrast, in endosperm amyloplasts it is synthesized during development but broken down extracellularly during germination (Duffus, 1984). The only way to obtain information about the regulation of starch metabolism in these different plastids is to study each one separately, since extrapolation from any one case to another is bound to have limited usefulness. The contents of this book will deal extensively with metabolism in photosynthetic cells. Hence, the scope of this chapter will be confined to what is known of metabolism in non-photosynthetic plastids.

Metabolism of non-photosynthetic plastids

In this section, the metabolic capabilities of non-green plastids will be illustrated by specific examples where the organelles have been isolated and the pathways characterized. As has already been pointed out, the biosynthetic pathways in non-green plastids are dependent on a supply of precursors, ATP and reductant that are generated by the oxidation of carbohydrates. Initially, therefore, consideration is given to this aspect of cellular metabolism and its compartmentation.

Carbohydrate oxidation: glycolysis and the oxidative pentose phosphate pathway in plastids

In other organisms, the ubiquitous pathways of glycolysis and the oxidative pentose phosphate pathway are not associated with any membrane-bound organelle and it was assumed that plants would follow this pattern. With improved methods of cell fractionation, it became increasingly apparent that plant cells are unique in having part or all of these pathways present both in the cytosol and in the plastid. These pathways are shown in Fig. 4.1 where it is illustrated that they share some common intermediates. The organization and regulation of these pathways has recently been reviewed by Dennis and Miernyk (1982) and ap Rees (1988a, b) (see also Chapter 6). From studies of roots, stems, seeds and cell cultures it appears that non-photosynthetic plastids have the enzymic capacity to convert hexose and hexose phosphates at least as far as pyruvate, and in some cases there is evidence for the presence of a pyruvate dehydrogenase complex for its further conversion to acetyl-CoA. However, there are both qualitative and quantitative differences arising between plastids with respect to these pathways. For example, it appears that the first enzyme unique to the oxidative pentose phosphate pathway (OPPP), glucose 6-phosphate dehydrogenase, is absent from plastids of developing castor-bean endosperm (Simcox et al., 1977) but so far has been reported in the majority of non-green plastids from other plants.

In the case of glycolysis, it is generally agreed that the reactions for the conversion of glucose 6-phosphate to phosphoglycerate are present in all the non-photosynthetic plastids. However, the level of some of the enzymes for the formation of pyruvate from phosphoglycerate are at low levels in some plastid preparations. This is particularly the case with phosphoglyceromutase and enolase. There now seems good evidence, from studies of particulate and soluble isozymes, that these two enzymes are present

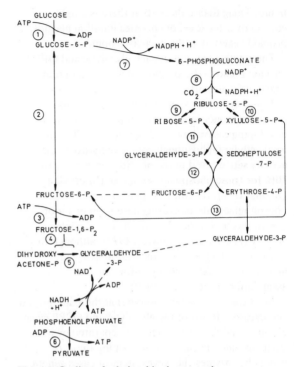

Fig. 4.1 Outline of relationships between the pentose phosphate pathway and glycolysis. Glycolysis: 1, hexokinase; 2, glucose 6-phosphate isomerase; 3, phosphofructokinase; 4, aldolase; 5, triose phosphate isomerase; 6, pyruvate kinase. Pentose phosphate pathway: 7, glucose 6-phosphate dehydrogenase; 8, 6-phosphogluconate dehydrogenase; 9, ribose 5-phosphate isomerase; 10, ribulose 5-phosphate 3-epimerase; 11, transketolase; 12, transaldolase; 13, transketolase.

in plastids of developing seeds (Miernyk and Dennis, 1984; Denyer and Smith, 1988), but whilst they have been reported as latent in plastids from other tissues such as stems, the amount and specific activity associated with the organelle is small and variable (Journet and Douce, 1985). When the plastid form of enolase was measured in many different tissues from the castor oil plant, it was found to vary from zero to 30% (Miernyk and Dennis, unpublished). Chloroplasts from mature leaves had none of the enzyme, whereas plastids from developing seeds had a large amount. Plastids from other tissues, such as young leaves and roots, had intermediate amounts, and the level appeared to correlate with the anticipated rate of fatty acid biosynthesis in the tissue. These data indicate that there is a tight control over the level of enzymes found in plastids, probably exerted at the nuclear genome.

The relative amount of each glycolytic and OPPP enzyme found in the plastid can also be different. The plastid proportion of each enzyme may also be different in various species and under various nutritional conditions. For example, plastids from developing castor-bean endosperm have 70% of the total cellular 6-phosphogluconate dehydrogenase activity sequestered in the plastid, compared with 10% for triose phosphate isomerase (Miernyk and Dennis, 1982). By contrast, plastid 6-phosphogluconate dehydrogenase from pea roots constitutes only about 30% of the total cellular activity (Ashihara and Fowler, 1979), although this percentage changes when the plastid isozyme increases in specific activity during nitrate assimilation (Emes *et al.*, 1979).

In all cases, it has been found that there are plastid and cytosolic forms of the glycolytic and OPPP cycle enzymes and they can usually be separated by ion exchange chromatography, suggesting they are probably isozymes. However, in most cases genetic analysis is not available to support this conclusion (Miernyk and Dennis, 1982). In some cases, such as enolase and phosphoglyceromutase, the isozymes have very similar kinetic and physical properties (Miernyk and Dennis, 1983; Botha and Dennis, 1986), whereas in others they are very different. The plastid isozyme of pyruvate kinase from developing castor-bean endosperm has a pH optimum of 8.0 compared with a value of 7.0 for its cytosolic counterpart. Further, the cytosolic enzyme has a much greater affinity for ADP (approximately 15-fold) than the plastid enzyme, is more stable and has a different reaction mechanism (Dennis and Miernyk, 1982). In the case of phosphofructokinase from the same source, the plastid isozyme shows cooperative binding of fructose 6-phosphate at pH 7.2 and is especially sensitive to inhibition by phosphoenolpyruvate. The cytoplasmic isozyme on the other hand, shows hyperbolic kinetics at this pH. Inhibition by phosphoenolpyruvate in both cases is relieved by inorganic phosphate (Dennis and Miernyk, 1982). In contrast to pyruvate kinase, the plastid isozyme is much more stable than its cytosolic counterpart.

Biosynthesis in non-photosynthetic plastids

Starch

Although starch represents the major staple food and calorific intake in the human diet, our understanding of the regulation of starch synthesis and its compartmentation in storage tissue is limited. It was originally postulated that it would be similar to starch storage in the chloroplast, but more recent studies of amyloplasts suggest otherwise. The primary substrate for starch synthesis in non-photosynthetic cells is sucrose imported from the leaf, the initial catabolism of which takes place in the cytoplasm. The key question to be addressed is whether starch is synthesized within the amyloplast from either hexose phosphates or triose phosphates imported from the cytoplasm. The two possible routes of starch synthesis are summarized in Fig. 4.2.

Comparison of amyloplasts with chloroplasts is not entirely valid since, in chloroplasts, substrates for starch synthesis are generated within the organelle during photosynthesis. Nonetheless, because chloroplasts possess a phosphate translocator capable of transporting triose phosphates into the organelle and all the enzymes necessary for their conversion to starch, the extrapolation has been made that this is also true for amyloplasts, particularly as hexose phosphates do not readily cross the chloroplast membrane. The findings that daffodil flower chromoplasts (Liedvogel and Kleinig, 1980) and root plastids of peas contain starch and also possess a phosphate translocator (Emes and Traska, 1987) are in agreement with this scheme. Amyloplasts from soybean suspension culture (Macdonald and ap Rees, 1983) and cauliflower bud plastids (Journet and Douce, 1985) have all the enzymes necessary for the conversion of triose phosphates to starch, including fructose 1,6-bisphosphatase. Maize kernel amyloplasts have been shown to possess a similar enzyme complement with an estimated 80% of the total cellular fructose 1,6-bisphosphatase sequestered within the organelle (Echeverria *et al.*, 1988). Of particular relevance in this study is the observation that dihydroxyacetone phosphate is a better substrate for starch synthesis than hexose phosphates

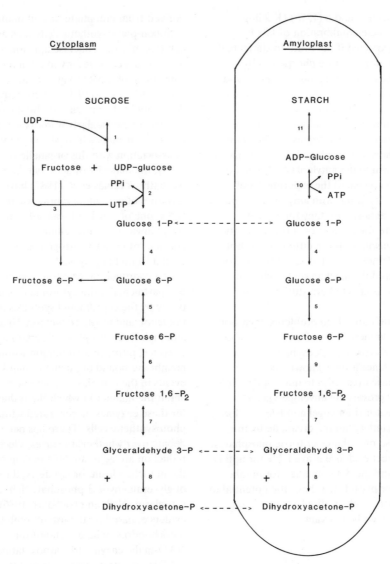

Fig. 4.2 Enzymes involved in the interconversion of sucrose to starch in non-photosynthetic cells. 1, Sucrose synthase; 2, UDP-glucose pyrophosphorylase; 3, hexokinase; 4, phosphoglucomutase; 5, hexose phosphate isomerase; 6, phosphofructokinase or pyrophosphate-dependent fructose 6-phosphate 1-phosphotransferase; 7, aldolase; 8, triose phosphate isomerase; 9, fructose 1,6-bisphosphatase; 10, ADP-glucose pyrophosphorylase; 11, starch synthase.

when supplied to the amyloplast preparations (of which approximately 50% were intact). It would seem from this discussion that the route from sucrose to starch is via triose phosphates which enter the plastid via the phosphate translocator.

However, two recent papers reporting studies of starch synthesis in developing wheat endosperm reach a conclusion that contradicts the above. When wheat grains are supplied with [1-^{13}C] or [6-^{13}C]-glucose, and the subsequent distribution of ^{13}C in the glycosyl moieties of starch determined, it is found that only partial redistribution of label occurs between carbon atoms 1 and 6 (15%) (Keeling et al., 1988). Complete equilibration of label might have been expected if the route to starch had involved the formation of triose phosphates by action of triose phosphate isomerase. Keeling and coworkers have suggested that this small amount of label redistribution is a result of recycling of substrates between hexose phosphates and triose phosphates solely within the cytoplasm. They have concluded that a hexose phosphate is the most likely candidate for entry into the amyloplast. When Tyson and ap Rees (1988) repeated the experiments of Echeverria et al. (1988), but with amyloplasts from developing wheat endosperm, they found that glucose l-phosphate was a superior substrate to triose phosphate. The reason for these differences is not clear but it does demonstrate that the pathway from sucrose to starch and the mechanism by which its compartmentation is effected are still far from understood.

In addition to the unresolved problems of carbon supply for starch synthesis in amyloplasts, the source of ATP for ADP-glucose pyrophosphorylase has yet to be established, since little is known about the permeability of these organelles to nucleotides. It is possible that ATP crosses the envelope via an adenylate translocator (Liedvogel and Kleinig, 1980) or that it is generated within the organelle by the oxidation of hexose phosphates or triose phosphates. At present, no direct evidence is available for uptake into amyloplasts and the data obtained for uptake into chloroplasts or plastids that have the potential to become chloroplasts, such as etioplasts or chromoplasts, may not be relevant.

Amino acids

The principal form of inorganic nitrogen available from the soil is nitrate, the assimilation of which gives rise to amino acids (see Chapter 26). Nitrate is reduced in a two-stage process catalyzed by nitrate reductase and nitrite reductase, both of which are inducible enzymes. The ammonia so generated then enters into organic compounds via the glutamine synthetase/glutamate synthase pathway with net production of one molecule of glutamate for each molecule of ammonia assimilated (Miflin and Lea, 1980). All other amino acids are subsequently derived from glutamate or glutamine (Chapter 26).

In non-photosynthetic cells such as roots and cell cultures, nitrate reductase appears to be cytosolic, whereas the enzymes involved in the subsequent conversion of nitrite to glutamate are plastidic (e.g. Emes and Fowler, 1979). There remains a lingering doubt about the amount of glutamine synthetase in the plastid, particularly in roots where some distribution studies have shown only a small degree of association with this organelle (Oaks and Hirel, 1985), the majority of activity being cytosolic. Recently, Vezinas et al. (1987) have separated two isozymes of glutamine synthetase in pea and alfalfa roots, one of which is localized in the plastid. Hence, the pathway and intracellular distribution of the enzymes of primary nitrogen assimilation in roots is similar to that in leaves.

The coordination of nitrate assimilation in the two compartments of the cytosol and plastid is achieved partly by the coordinated synthesis of nitrate reductase and nitrite reductase. This ensures that, although there is a physical separation of these enzymes, nitrogen assimilation within the membrane-bound organelle remains in step with events in the cytoplasm. There is less information about the manner in which the reductant, necessary for these enzymes, is generated within non-photosynthetic cells. There has been considerable debate over whether, in leaves, electrons for nitrate reduction are generated via malate oxidation within the mitochondrion, or are derived from the activity of glyceraldehyde 3-phosphate dehydrogenase in the cytoplasm (Naik and Nicholas, 1986). The weight of evidence, based on the use of inhibitors of mitochondrial malate metabolism, suggests that NAD-malic enzyme (decarboxylating) supplies NADH for nitrate reductase and that this is true for roots as well as leaves (Naik et al., 1982).

There seems to be a strong case for the role of the oxidative pentose phosphate pathway (OPPP) in meeting the needs of nitrite reductase for reductant within the non-photosynthetic plastid. Isozymes of the OPPP have been demonstrated within these organelles from a number of sources (see earlier) and it may serve a role in supplying reductant for nitrite reduction (Butt and Beevers, 1961). One problem with this suggestion is that root nitrite reductase, which like its photosynthetic counterpart can use reduced leaf ferredoxin *in vitro*, is unable to use NADPH as the immediate electron donor (Bowsher *et al.*, 1988). However, evidence for such a role for the plastid-located OPPP has come from work with isolated organelles which shows that glucose 6-phosphate is capable of supporting nitrite reduction (Oji *et al.*, 1985). More recently, it has been demonstrated that the release of CO_2 from carbon atom 1 of glucose 6-phosphate, supplied to intact pea root plastids, is stimulated several-fold when reduction of nitrite is taking place (Fig. 4.3, taken from Bowsher *et al.*, 1989). CO_2 is lost from this position as a result of the activity of 6-phosphogluconate dehydrogenase in the OPPP. No such stimulation in release was seen from carbon atoms 2, 3, 4 and 6 of glucose 6-phosphate, and was not observed when nitrite was supplied to plastids obtained from roots in which the enzymes of nitrate assimilation had not been induced.

These results suggest that, in addition to the enzymes of the OPPP, these plastids contain the other components necessary to mediate electron transport from NADPH to nitrite. Suzuki *et al.* (1985) have evidence of a ferredoxin-like compound and a pyridine nucleotide reductase in maize roots. However, whilst these two components, in the presence of NAD(P)H, could support ferredoxin-dependent glutamate synthase activity *in vitro*, they were unable to support ferredoxin-dependent nitrite reduction under similar conditions, suggesting that other additional components are necessary to mediate this transfer of electrons.

The data presented in Fig. 4.3 indicate a substantial release of CO_2 from carbon atoms 3 and 4 of glucose 6-phosphate supplied to intact pea root plastids. This CO_2 probably arises through the action of the pyruvate dehydrogenase complex. If this is confirmed, it implies that these organelles contain an active glycolytic sequence of reactions culminating in the decarboxylation of pyruvate. These preparations of plastids therefore afford the possibility of studying glycolytic flux and control in an isolated system, whereby the effect of available solutes in the cytoplasm and their impact on particulate glycolysis could be studied.

Non-photosynthetic plastids have been less well studied than chloroplasts with respect to the biosynthesis of other amino acids, but where comparison has been made, the intracellular distribution of enzymes seems similar.

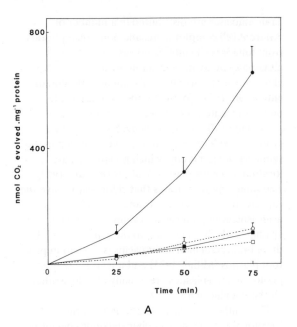

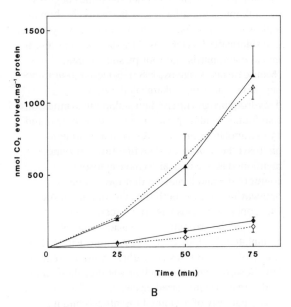

Fig. 4.3 Evolution of CO_2 from different carbon atoms after labeled glucose 6-phosphate had been supplied to intact root plastids in the presence (closed symbols) and absence (open symbols) of 1 mM sodium nitrite. (A) ○, ● [1-^{14}C]-glucose 6-phosphate; □, ■ [6-^{14}C]-glucose 6-phosphate. (B) △, ▲ [3,4-^{14}C]-glucose 6-phosphate; ◇, ◆ [2-^{14}C]-glucose 6-phosphate. Mean values ± SE of 3 determinations. (From Bowsher *et al.*, 1989.)

Transaminases are distributed in a number of different cell compartments and non-green plastids from tobacco cell cultures possess glutamate–pyruvate transaminase and glutamate–oxaloacetate transaminase (Washitani and Satoh, 1978). With regard to the aspartate family of amino acids (threonine, lysine and methionine) all the enzymes necessary for their synthesis, with the exception of the final step in the synthesis of methionine which is cytosolic, are localized in chloroplasts (Wallsgrove et al., 1983). The same report indicates that chloroplasts have the capacity to synthesize the branched-chain amino acids isoleucine, leucine and valine. Acetolactate synthase, which is common to the synthesis of all three branched-chain amino acids is present in pea root plastids (Miflin, 1974) and it is likely that non-photosynthetic plastids are equally able to synthesize all these amino acids.

The synthesis of the aromatic amino acids (tryptophan, tyrosine and phenylalanine) begins with the shikimic acid pathway. 5-Enolpyruvylshikimate 3-phosphate synthase, 3-dehydroquinate dehydratase and shikimate : $NADP^+$ oxidoreductase are present in leaf chloroplasts and root plastids of pea (Mousdale and Coggins, 1985). 3-Deoxy-D-arabino-heptulosonate 7-phosphate synthase, 3-dehydroquinate synthase and shikimate synthase are also found in chloroplast stromal preparations, and by extrapolation they are also probably in non-photosynthetic plastids. The first three enzymes mentioned above appear to have cytosolic counterparts and it may be that the shikimate pathway operates in both the cytosol and plastids. Chloroplast fractions have been reported to synthesize the aromatic amino acids from either assimilated $^{14}CO_2$ or from ^{14}C-shikimic acid (Bickel et al., 1978). The rates of synthesis were extremely low, however, and it is unclear whether these amino acids are also synthesized elsewhere in the cell.

The pathway of arginine biosynthesis and its compartmentation has been studied in soybean cell suspensions (Shargool et al., 1977). Carbamoyl phosphate synthetase and ornithine carbamoyltransferase were found associated with a fraction composed primarily of plastids.

Fatty acids

Details of fatty acid biosynthesis have been presented in Chapter 23 and will not be described in detail here. It was demonstrated by Zilkey and Canvin (1971) that fatty acids are synthesized in plastids in the developing endosperm of castor oil seed. Later, it was demonstrated by immunochemical techniques that acyl carrier protein, an essential component of the fatty acid synthetase complex, is found only in plastids and it is now generally accepted that fatty acid synthesis only occurs in these organelles (reviewed by Stumpf, 1984).

The precursor of fatty acid biosynthesis is acetyl-CoA which cannot cross the plastid membrane. An internal source of acetyl-CoA is, therefore, required. The plastids from the developing endosperm of the castor oil plant have all the enzymes of the glycolytic pathway for the conversion of hexose to pyruvate and they also contain a pyruvate dehydrogenase complex to generate acetyl-CoA (Dennis and Miernyk, 1982). In addition, these plastids on isolation will incorporate hexose, phosphoglycerate and pyruvate into fatty acids, indicating that the complete pathway is operative (Miernyk and Dennis, 1983a). The actual substrate that is imported from the cytosol in vivo is not known and it could be any of the above.

The supply of acetyl-CoA in other plastids has been in dispute and several different mechanisms have been proposed (Liedvogel, 1986). A pyruvate dehydrogenase complex is present in young active chloroplasts but may not be present in mature chloroplasts. In addition, mature chloroplasts may not have a complete glycolytic pathway as phosphoglyceromutase appears to be missing (Dennis and Miernyk, 1982). An incomplete glycolytic pathway in mature chloroplasts may ensure that metabolites are not removed from the reductive pentose phosphate cycle by the plastid glycolytic pathway.

Acetate will readily cross the membrane of the plastid and is rapidly incorporated into fatty acids. The conversion of acetate to acetyl-CoA is catalyzed by acetyl-CoA synthetase, an enzyme found in plastids. It has been suggested that hydrolysis of acetyl-CoA occurs in the mitochondrion and the free acetate that is formed diffuses to the plastid where it is converted to acetyl-CoA (Murphy and Stumpf, 1981). However, the level of free acetate appears to be low and is probably inadequate for normal fatty acid biosynthesis.

In plastids with a complete glycolytic pathway, some of the reducing power for fatty acid synthesis could be supplied by the glyceraldehyde 3-phosphate dehydrogenase reaction. The ATP required by the acetyl-CoA carboxylase could also be supplied by glycolysis. The NADPH is probably supplied by the oxidative pentose phosphate pathway in the organelle.

Conclusions

This chapter has summarized some of the major metabolic activities in plant cells that are sequestered within plastids. It has become clear that other pathways such as purine, ureide and terpene synthesis are also localized, at least to some extent, in these organelles. Within the various tissues of a plant, the morphology and biochemical capability of the plastid may be very different but they all appear to be differentiated from identical proplastids. Even within a single type of plastid such as chloroplasts, there may be very marked differences between the organelle in young and mature leaves or between different tissues, e.g. the bundle sheath and mesophyll cells of the C4 plants (Chapter 19). Since the principal biosynthetic activity of a plant cell resides in the plastids, it is clear that the overall metabolic activity is determined to a large extent by the type of plastid present in the cells of that tissue.

How the differentiation of the proplastid is controlled is not known. It is most likely a function of the nuclear genome interacting with the genome in the organelle. It is known that different concentrations of the various messengers transcribed from the organellar genome are found in different types of plastid, but this is most likely a result of differential stabilities of the different plastid messengers, rather than differential transcription. Without doubt, a knowledge of the control of plastid differentiation is of major importance if plant metabolism is to be fully understood.

Although there is considerable information about the enzyme complement of some plastids, the majority have not been examined in detail. There are still considerable gaps in our knowledge even of economically important plastids such as amyloplasts. This has resulted in a great deal of controversy in the literature which may have been a result of the different plastids that have been studied rather than differences in technique or competence of the investigators. A major problem in studying plastids from many tissues is that they are fragile and it is difficult to obtain the organelle intact. New methods of organelle isolation are now available and this should expedite these studies.

Probably the most important but least studied aspect of plastids is their interaction with the other compartments of the cell. The only transporters studied in detail are those found in chloroplasts from mature leaves and, in particular, the phosphate translocator. The role played by this translocator in other types of plastid is not known although it is certainly present. However, recent evidence suggests that carbon uptake into amyloplasts may be by a different route even though the endproduct, starch, is the same as in chloroplasts. In addition, little is known about the differences in specificity of the transporters in different plastids.

Clearly, the compartmentation of plant metabolism is complex. It is of great importance in controlling flux through the various metabolic pathways. It can be quite different in the various plant tissues and between the same tissues in different plant species. The transport of metabolites between the various compartments must be a site of control and again will vary between tissues. There is still a great deal to learn about these processes.

This chapter has concentrated on the importance of the plastid compartment and especially non-green plastids. Other chapters will deal in detail with other compartments such as the chloroplast and mitochondrion.

References

ap Rees, T. (1988a). Compartmentation of plant metabolism. In *The Biochemistry of Plants*, Vol. 12, ed. D. D. Davies, Academic Press, New York, pp. 87–115.

ap Rees, T. (1988b). Hexose phosphate metabolism by nonphotosynthetic tissues of higher plants. In *The Biochemistry of Plants*, Vol. 14, ed. J. Preiss, Academic Press, New York, pp. 1–33.

Ashihara, H. and Fowler, M. W. (1979). 6-phosphologluconate dehydrogenase species from roots of *Pisum sativum*. *Int. J. Biochem.* **10**, 675–81.

Bickel, H., Palme, L. and Schultz, G. (1978). Incorporation of shikimate and other precursors into aromatic amino acids and prenylquinones of isolated spinach chloroplasts. *Phytochemistry* **17**, 119–24.

Botha, F. C. and Dennis, D. T. (1986). Isozymes of phosphoglyceromutase from the developing endosperm of *Ricinus communis*: Isolation and kinetic properties. *Arch. Biochem. Biophys.* **245**, 96–103.

Bowsher, C. G., Emes, M. J., Cammack, R. and Hucklesby, D. P. (1988). Purification and properties of nitrite reductase from roots of pea (*Pisum sativum* cv. meteor). *Planta* **175**, 334–40.

Bowsher, C. G., Hucklesby, D. P. and Emes, M. J. (1989). Nitrite reduction and carbohydrate metabolism in plastids purified from roots of *Pisum sativum* L. *Planta* **177**, 359–66.

Butt, V. S. and Beevers H. (1961). The regulation of pathways of glucose catabolism in maize roots. *Biochem. J.* **80**, 21–7.

Dennis, D. T. and Miernyk, J. A. (1982). Compartmentation of nonphotosynthetic carbohydrate metabolism *Ann. Rev. Plant Physiol.* **33**, 27–50.

Dennis, D. T., Hekman, W. E., Thomson, A., Ireland, R. J., Botha, F. C. and Kruger, N. J. (1985). In *Regulation of Carbon Partitioning in Photosynthetic Tissue*, eds R. L. Heath and J. Preiss, Amer. Soc. Pl. Physiol. Symp., Riverside, California, pp. 127–46.

Denyer, K. and Smith, A. M. (1988). The capacity of plastids from developing pea cotyledons to synthesize acetyl CoA. *Planta* **173**, 172–82.

Duffus, C. M. (1984). Metabolism of reserve starch. In *Storage Carbohydrates in Vascular Plants*, ed. D. H. Lewis, Cambridge University Press, Cambridge, pp. 231–52.

Echeverria, E., Boyer, C. D., Thomas, P. A., Liu, K.-C. and Shannon, J. C. (1988). Enzyme activities associated with maize kernel amyloplasts *Pl. Physiol.* **86**, 786–92.

Emes, M. J. and Fowler, M. W. (1979). The intracellular location of the enzymes of nitrate assimilation in the apices of seedling pea roots. *Planta* **144**, 249–53.

Emes, M. J. and Traska, A. (1987). Uptake of inorganic phosphate by plastids purified from the roots of *Pisum sativum* L. *J. Exp. Bot.* **38**, 1781–88.

Emes, M. J., Ashihara, H. and Fowler, M. W. (1979). The influence of NO_3^- on particulate 6-phosphogluconate dehydrogenase activity in pea roots. *FEBS Lett.* **105**, 370–2.

Hargreaves, J. A. and ap Rees, T. (1988). Sucrose and hexose metabolism by clubs of *Typhonium giraldii* and roots of *Pisum sativum*. *Phytochemistry* **27**, 1621–25.

Journet, E.-P. and Douce, R. (1985). Enzymic capacities of purified cauliflower bud plastids for lipid synthesis and carbohydrate metabolism. *Plant Physiol.* **79**, 458–69.

Keeling, P. L., Wood, J. R., Tyson, R. H. and Bridges, I. G. (1988). Starch biosynthesis in developing wheat grain. *Plant Physiol.* **87**, 311–19.

Liedvogel, B. (1986). Acetyl coenzyme A and isopentenylpyrophosphate as lipid precursors in plant cells – biosynthesis and compartmentation. *J. Plant Physiol.* **124**, 211–22.

Liedvogel, B. and Kleinig, H. (1980). Phosphate translocator and adenylate translocator in chromoplast membranes. *Planta* **150**, 170–3.

Macdonald, F. D. and ap Rees, T. (1983). Enzymic properties of amyloplasts from suspension cultures of soybean. *Biochim. Biophys. Acta* **755**, 81–9.

Miernyk, J. A. and Dennis, D. T. (1982). Isozymes of glycolytic enzymes in endosperm from developing castor oil seeds. *Plant Physiol.* **69**, 825–8.

Miernyk, J. A. and Dennis, D. T. (1983). Mitochondrial, plastid, and cytosolic isozymes of hexokinase from developing endosperm of *Ricinus communis*. *Arch. Biochem. Biophys.* **226**, 458–68.

Miernyk, J. A. and Dennis, D. T. (1984). Enolase isozymes from *Ricinus communis*: Partial purification and characterization of the isozymes. *Arch. Biochem. Biophys.* **233**, 643–51.

Miflin, B. J. (1974). The location of nitrite reductase and other enzymes related to amino acid biosynthesis in the plastids of root and leaves. *Plant Physiol.* **54**, 550–5.

Miflin, B. J. and Lea, P. J. (1980). Ammonia assimilation. In *The Biochemistry of Plants*, Vol. 5, ed. B. J. Miflin, Academic Press, New York, pp. 169–202.

Mousdale, D. M. and Coggins, J. R. (1985). Subcellular localization of the common shikimate-pathway enzymes in *Pisum sativum* L. *Planta* **163**, 241–9.

Murphy, D. J. and Stumpf, P. K. (1981). The origin of chloroplast acetyl-CoA. *Arch. Biochem. Biophys.* **212**, 730–9.

Naik, M. S. and Nicholas, D. J. D. (1986). Malate metabolism and its relation to nitrate assimilation in plants. *Phytochemistry* **25**, 571–6.

Naik, M. S., Abrol, Y. P., Nair, T. V. R. and Ramarao, C. S. (1982). Nitrogen assimilation – its regulation and relationship to reduced nitrogen in higher plants. *Phytochemistry* **21**, 495–504.

Oaks, A. and Hirel, B. (1985). Nitrogen metabolism in roots. *Ann. Rev. Plant Physiol.* **36**, 345–65.

Oji, Y., Watanabe, M., Wakiuchi, N. and Okamoto, E. (1985). Nitrite reduction in barley-root plastids: Dependence on NADPH coupled with glucose-6-phosphate and 6-phosphogluconate dehydrogenases, and possible involvement of an electron carrier and diaphorase. *Planta* **165**, 85–90.

Preiss, J. (1982). Regulation of the biosynthesis and degradation of starch. *Ann. Rev. Plant Physiol.* **33**, 431–54.

Shargool, P. D., Steeves, T., Weaver, M. and Russell, M. (1977). The localization within plant cells of enzymes involved in arginine biosynthesis *Can. J. Biochem.* **56**, 273–9.

Simcox, P. D., Reid, E. E., Canvin, D. T. and Dennis, D. T. (1977). Enzymes of the glycolytic and pentose phosphate pathways in plastids from the developing endosperm of *Ricinus communis* L. *Plant Physiol.* **59**, 1128–32.

Stumpf, P. K. (1984). Fatty acid biosynthesis in higher plants. In *Fatty Acid Metabolism and Its Regulation*, ed. S. Numa, Elsevier, Amsterdam, New York, Oxford, pp. 155–79.

Suzuki, A., Oaks, A., Jacquot, J.-P., Vidal, J. and Gadal, P. (1985). An electron transport system in maize roots for reactions of glutamate synthesis and nitrite reductase *Plant Physiol.* **78**, 374–78.

Tyson, R. H. and ap Rees, T. (1988). Starch synthesis by isolated amyloplasts from wheat endosperm. *Planta* **175**, 33–8.

Vezinas, L.-P., Hope, H. J. and Joy, K. W. (1987). Isoenzymes of glutamine synthetase in roots of pea (*Pisum sativum* L. cv. Little Marvel) and alfalfa (*Medicago media* Pers. cv. Saranac). *Plant Physiol.* **83**, 58–62.

Wallsgrove, R. M., Keys, A. J., Lea, P. J. and Miflin, B. J. (1983). Photosynthesis, photorespiration and nitrogen metabolism *Plant, Cell Environ.* **6**, 301–9.

Washitani, I. and Sato, S. (1978). Studies on the function of proplastids in the metabolism of *in vitro* cultured tobacco cells V. Primary transamination. *Plant Cell Physiol.* **19**, 43–50.

Zilkey, B. F. and Canvin, D. T. (1971). Localization of oleic acid biosynthesis enzymes in proplastids of developing castor bean endosperm *Can. J. Bot.* **50**, 323–6.

Cytosolic Carbon Metabolism

II

Cytosolic Carbon Metabolism

5
Carbohydrate synthesis and degradation
Nicholas J. Kruger

Introduction

The aim of this chapter is to discuss the pathways of carbohydrate synthesis and breakdown in plants. It is limited to a consideration of metabolism to and from hexose phosphates, since the latter form a pool of intermediates in which the reactions of carbohydrate synthesis and degradation converge and interact with other pathways. This chapter will concentrate primarily, though not exclusively, on the metabolism of sucrose and starch. Together these two compounds provide the major substrates for respiration in many cells and so may be considered to dominate the carbohydrate metabolism of higher plants. Structural polysaccharides are included in this discussion since they, too, are synthesized directly from hexose phosphates.

During the oxidation of carbohydrates to provide energy, substantial proportions of many intermediates may be withdrawn for biosynthetic reactions. In this way, carbohydrates can contribute to the synthesis of proteins, lipids and organic acids. Although such compounds are often quantitatively important products of carbohydrate metabolism, they are not considered here. This is because, in general, such compounds are derived from carbohydrates indirectly through intermediates in the pathways of hexose phosphate oxidation. These pathways are discussed in subsequent chapters.

The central role of hexose phosphates

Hexose phosphates are not only intermediates common to the pathways of synthesis and degradation of most carbohydrates, they are the principal site at which these pathways converge. Hexose phosphates are derived either from the breakdown of sugars and polysaccharides or from triose phosphates formed during photosynthesis and gluconeogenesis. They may be used for the synthesis of carbohydrates or for metabolism through the glycolytic and pentose phosphate pathways. The intermediates in this pool are glucose 1-phosphate, glucose 6-phosphate and fructose 6-phosphate, which are interconverted by phosphoglucomutase (EC 5.4.2.2; reaction [5.1]) and glucose 6-phosphate isomerase (EC 5.3.1.9; reaction [5.2]).

glucose 1-phosphate $\rightleftharpoons$ glucose 6-phosphate   [5.1]
$\Delta G^{\circ\prime} = -7.11$ kJ mol^{-1}

glucose 6-phosphate $\rightleftharpoons$ fructose 6-phosphate   [5.2]
$\Delta G^{\circ\prime} = 1.67$ kJ mol^{-1}

The hexose phosphate pool is generally regarded to be close to equilibrium in plant cells. Evidence for this is four-fold. Firstly, the low standard free energy change of the two reactions indicates that both are readily reversible *in vitro*. Secondly, in many tissues the maximum catalytic activities of the two enzymes are high relative to the anticipated flux through these steps. Thirdly, comparison of the apparent equilibrium constants for the reactions with mass–action ratios calculated from measurements of the amount of each intermediate *in vivo* indicate that both enzymes probably operate close to equilibrium. Finally, arguments based on comparative biochemistry and the far more extensive investigations of carbohydrate metabolism in other organisms suggest that the hexose phosphates in plants are near equilibrium. Data to support these statements are presented in previous reviews (ap Rees, 1977; Turner and Turner, 1980).

Much of the above evidence is weakened by the recent appreciation that carbohydrate metabolism in plants is uniquely organized (ap Rees, 1985). In particular, hexose phosphates are not confined to a single compartment but occur in both the cytoplasm and plastids. This subcellular compartmentation may considerably distort our interpretation of measurements of enzymes and metabolites obtained from whole tissues. Moreover, analogy with other systems in which the organization of hexose phosphate metabolism is considerably simpler may be misleading.

The crucial feature of the above criticism is the extent to which hexose phosphates may exchange between the pools in the cytoplasm and plastids. There is strong evidence that in photosynthetic tissues these pools are independent. One indication of this is that the envelope of chloroplasts isolated from a range of plants is largely impermeable to hexose phosphates (see Chapter 21). Further evidence is provided by measurements of the subcellular distribution of metabolites following rapid fractionation of plant protoplasts (Stitt et al., 1980). Such studies reveal that the level of hexose phosphates inside and outside the chloroplasts varies independently. On the other hand, discrete pools of hexose phosphates do not necessarily occur in all tissues. Very recently it has been demonstrated that intact plastids isolated from wheat endosperm are capable of incorporating radioactivity into starch when incubated in [^{14}C]-glucose l-phosphate (Tyson and ap Rees, 1988). This implies that glucose 1-phosphate is taken up by these plastids and that, at least in this tissue, there is a direct link between the hexose phosphate pools in the cytoplasm and amyloplasts. Until more data are available from a greater range of tissues, we cannot decide whether this is a general feature of non-photosynthetic cells.

Our uncertainty about the extent of compartmentation of hexose phosphates in most tissues contributes to the problem of interpreting the evidence that these metabolites are close to equilibrium. This problem can be overcome by considering separately the cytoplasm and plastids in tissues in which the two pools of hexose phosphates are known to be independent. Data from such systems are extremely limited. In the few photosynthetic tissues that have been studied in sufficient detail, both the cytoplasm and chloroplasts contain significant activity of phosphoglucomutase and glucose 6-phosphate isomerase (ap Rees, 1985). This suggests that both compartments have the capacity to interconvert hexose phosphates. Stronger evidence is provided by measurements of metabolites in chloroplasts rapidly isolated from spinach protoplasts (Stitt et al., 1980). The mass-action ratios calculated from these measurements are sufficiently similar to the apparent equilibrium constants to suggest that the hexose phosphates are close to equilibrium in these organelles.

On balance, it is likely that generally glucose 1-phosphate, glucose 6-phosphate and fructose 6-phosphate are maintained near equilibrium. However, there is very little strong evidence to support this view. This is because in some tissues hexose phosphates exist in two independent pools. Consequences of this compartmentation on the organization of carbohydrate metabolism are considered in later sections.

Sucrose metabolism

The importance of sucrose

The significance of sucrose in plant metabolism has been discussed in detail in previous reviews (ap Rees, 1984; Hawker, 1985) which provide the reference for many of the statements made in this section. Sucrose has three interrelated roles in plants. Firstly, it is a principal product of photosynthesis and can account for a large proportion of the total CO_2 absorbed by a plant during photosynthesis. The kinetics of labeling of sucrose in illuminated leaves exposed to $^{14}CO_2$ indicate that this compound is an important product of photosynthesis rather than an intermediate.

Secondly, sucrose is a major form in which carbon is translocated in plants. This is demonstrated by analysis of the content of sieve tubes. The importance of sucrose in translocation is not simply a reflection of its dominant role as a product of photosynthesis since it is also the principal form in which carbon reserves are exported from non-photosynthetic tissues. During seed germination the mobilization of starch, lipid and, to a lesser extent even protein, involves their conversion to sucrose prior to translocation from the storage tissues to the developing embryo.

Thirdly, sucrose is the main storage sugar in plants. Chemical analysis reveals that in many plants sucrose is a major component of storage organs, such as the swollen tap root of carrot and sugar beet. However, sucrose is not restricted to such specialized storage tissues and high levels can accumulate in other organs. For example, this sugar may constitute up to 25% of the dry weight of ivy leaves.

The importance of sucrose in plant metabolism is emphasized by the fact that not only does it contribute to the three functions described above, but that generally it dominates each of these roles. In certain species sugar alcohols or oligosaccharides are formed during photosynthesis and are transported and stored. However, even where such compounds are of major importance they never entirely replace sucrose.

Sucrose synthesis

Although most apparent in photosynthetic and gluconeogenic tissues, the ability to synthesize sucrose is a widespread, possibly universal, characteristic of higher plant cells. Sucrose is derived from hexose phosphates through the combined activities of UDP-glucose pyrophosphorylase (EC 2.7.7.9; reaction [5.3]), sucrose phosphate synthase (EC 2.4.1.14; reaction [5.4]) and sucrose phosphatase (EC 3.1.3.24; reaction [5.5]). Figure 5.1 provides an overview of the pathways of sucrose metabolism. Our knowledge of the properties of sucrose phosphate synthase and sucrose phosphatase have been reviewed in detail by Avigad (1982) and Hawker (1985)

glucose 1-phosphate + UTP $\rightleftharpoons$ UDP-glucose + PP$_i$ [5.3]
$\Delta G^{\circ\prime} = -2.88$ kJ mol^{-1}

UDP-glucose + fructose 6-phosphate $\rightleftharpoons$ sucrose 6^F-phosphate + UDP [5.4]
$\Delta G^{\circ\prime} = -1.46$ kJ mol^{-1}

sucrose 6^F-phosphate + H$_2$O $\rightarrow$ sucrose + P$_i$ [5.5]
$\Delta G^{\circ\prime} = -18.4$ kJ mol^{-1}

The evidence that sucrose is synthesized via this pathway, and not through an alternative route involving sucrose synthase, comes from three sources. First, the large overall free energy change of the reaction catalyzed by sucrose phosphatase ensures that the production of sucrose is irreversible and allows synthesis to continue in tissues that already contain large amounts of sucrose. Secondly, the kinetics of labeling of pathway intermediates in tissues fed ^{14}CO$_2$ or ^{14}C-glucose are consistent with the proposed pathway. During sucrose synthesis hexose phosphates, UDP-glucose and sucrose phosphate are labeled, and both fructose 6-phosphate and the fructosyl moiety of sucrose are labeled before free fructose. The labeling in these latter compounds argues strongly against the involvement of sucrose synthase in sucrose

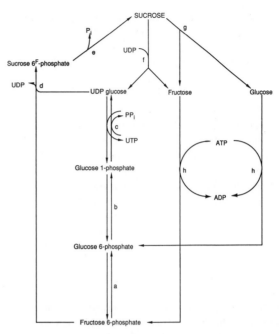

Fig. 5.1 Pathways of sucrose metabolism. The letters denote the following enzymes: a, glucose 6-phosphate isomerase [5.2]; b, phosphoglucomutase [5.1]; c, UDP-glucose pyrophosphorylase [5.3]; d, sucrose phosphate synthase [5.4]; e, sucrose phosphatase [5.5]; f, sucrose synthase [5.6]; g, invertase [5.7]; h, hexose kinase [5.15]. The numbers in brackets refer to reactions described in the text.

production. Thirdly, in a range of tissues that make sucrose, the activity of sucrose phosphate synthase greatly exceeds that of sucrose synthase. The significance of this correlation is strengthened by the observation that in some tissues the activity of sucrose phosphate synthase, but not that of sucrose synthase, is sufficient to account for the rate of sucrose production.

There is little doubt that sucrose synthesis from hexose phosphates occurs in the cytoplasm. The inability of purified isolated chloroplasts to form labeled sucrose from ^{14}CO$_2$, and the relative impermeability of the chloroplast inner envelope to sucrose, indicates that sucrose is synthesized outside the chloroplast. This view is confirmed by studies on photosynthesis from ^{14}CO$_2$ by isolated protoplasts in which labeled sucrose appears initially in the extrachloroplastic fraction. Careful fractionation of both photosynthetic and gluconeogenic cells has revealed that all of the enzymes necessary for this pathway are present in the cytoplasm. Furthermore, both sucrose phosphate synthase and UDP-glucose, a metabolic intermediate in the pathway, are restricted to the cytoplasm.

While it is clear from the above results that sucrose is made in the cytoplasm, what is less certain is whether sucrose synthesis is confined exclusively to this compartment. The major weakness of the evidence is that it is restricted largely to photosynthetic and gluconeogenic tissues where sucrose is predominantly exported from the cells. However, sucrose is an important reserve carbohydrate and in many tissues it accumulates in the vacuole rather than being exported. Despite this, the ability of vacuoles isolated from such tissues to absorb sucrose is generally poor. The apparent absence of a clearly defined active sucrose uptake mechanism in the tonoplast membrane has led to proposals that sucrose accumulation may involve uptake of a precursor which is converted to sucrose either during or after transport through the tonoplast (Hawker, 1985). Two lines of evidence are consistent with this view. One is the recent demonstration that in several storage tissues most of the sucrose phosphatase is associated with vacuoles (Hawker et al., 1987). The other is the apparent ability of isolated vacuoles and tonoplast vesicles from sugar cane and beet to catalyze the production of labeled sucrose from exogenous UDP-[^{14}C]-glucose in the absence of added fructose 6-phosphate (Maretzki and Thom, 1986). The latter observation has led to the proposal that sucrose is synthesized from UDP-glucose by an enzyme complex in the tonoplast and then released into the vacuole. As yet this system is only poorly characterized and we need to know much more about the enzymes and intermediates involved and their precise intracellular location. In addition, from the limited available data it seems that this system does not contribute significantly to the accumulation of sucrose in vacuoles. When asymmetrically labeled sucrose is fed to several tissues, including beet, there is only limited redistribution of label between the glucosyl and fructosyl moieties of sucrose (Thorne and Giaquinta, 1984). This suggests that the vast majority of the sucrose is taken up and stored in the vacuole without prior metabolism to hexose phosphates or UDP-glucose. Although the exact mechanism of sucrose uptake by vacuoles in such tissues remains unclear, evidence for the involvement of a sucrose/H^+ antiport system is accumulating.

Our understanding of the regulation of the only established pathway of sucrose production is limited almost exclusively to photosynthetic tissues. In spinach leaves, changes in the levels of metabolites when the rate of sucrose synthesis is varied indicate that this pathway is regulated, at least partly, by sucrose phosphate synthase or sucrose phosphatase or both. For example, inhibition of sucrose synthesis by a reduction in the light intensity or CO_2 concentration may be accompanied by an increase in the level of UDP-glucose or hexose phosphates (Stitt et al., 1988). This demonstrates that the activity of at least one of these two enzymes is dependent on some factor other than substrate concentration. However, the relative contribution of these two enzymes to the regulation of sucrose synthesis is still only poorly understood.

The large standard free energy change of the reaction catalyzed by sucrose phosphatase suggests that it is far from equilibrium *in vivo* and consequently susceptible to regulation by allosteric effectors. This enzyme is distinguished from other phosphatases by its requirement for Mg^{2+} and its inhibition by several sugars, including sucrose. The inhibition by sucrose is often invoked as the explanation for feedback regulation of sucrose synthesis by its product (Hawker, 1985). Although this inhibition may contribute to such control, it is unlikely to account completely for the regulation of sucrose production by sucrose. Studies of partially purified sucrose phosphatase from both sugar cane and carrot indicate that sucrose is a partial competitive inhibitor. With this type of effect the enzyme is never completely inactive, even at extremely high concentrations of sucrose. Thus other factors must be involved if sucrose synthesis is ever to be completely inhibited.

More recently, attention has focused on the potential regulatory role of sucrose phosphate synthase. Three lines of evidence suggest that this enzyme may contribute to the control of sucrose production in leaves. Firstly, in a range of tissues the activity of sucrose phosphatase is far greater than that of sucrose phosphate synthase (Hawker, 1985). This suggests that any sucrose phosphate formed will be rapidly hydrolyzed, maintaining the reaction catalyzed by sucrose phosphate synthase far from equilibrium *in vivo* even though the equilibrium constant of this reaction indicates that it may be readily reversible *in vitro*. Secondly, there is a close correlation between the rate of sucrose synthesis and the extractable activity of sucrose phosphate synthase (Huber et al., 1985). This correlation is observed between species, between genotypes within a species, and in plants in which the rate of sucrose synthesis is experimentally manipulated by varying factors such as photoperiod, temperature and nutritional status. Particularly significant in this context are the diurnal changes in sucrose phosphate synthase activity which follow closely the rate of sucrose production during a normal photoperiod. Thirdly, the known regulatory properties of sucrose phosphate synthase are entirely consistent with this enzyme having an important role in the regulation of sucrose synthesis (Stitt et al., 1988). Recent work on spinach and barley leaves suggests that sucrose phosphate synthase occurs in two kinetically distinct forms, the proportions of which can be modified by factors related to the availability of phosphate and the accumulation of sucrose. These two forms are distinguished by differences in their substrate affinities, sensitivity to inhibition by phosphate, and probably also the extent of their activation by glucose 6-phosphate. A consideration of changes in the proportion of the enzyme in a kinetically active form and variations in the levels of allosteric effectors suggests that both factors contribute to the regulation of sucrose phosphate synthase, and that this is sufficient to account for the observed variation in the rate of sucrose synthesis throughout a 24-hour of these enzymes have been thoroughly reviewed by Avigad (1982) and Hawker (1985).

period. The way in which such factors may interact to coordinate sucrose production with the supply of photosynthate and the demand for sucrose is described in detail in Chapter 21.

Sucrose breakdown

Plants contain two types of enzyme capable of cleaving sucrose. One is sucrose synthase (EC 2.4.1.13) which catalyzes the reaction shown in reaction [5.6]. Evidence from a wide range of tissues suggests that this enzyme is confined to the cytoplasm. The other type of enzyme is invertase (EC 3.2.1.26) which catalyzes the essentially irreversible hydrolysis of sucrose to glucose and fructose (reaction [5.7]). Both acid and alkaline invertases are found in plants and are distinguished by having pH optima of about 5 and 7.5, respectively. Acid invertase exists in vacuoles and associated with plant cell walls, whereas alkaline invertase is probably restricted to the cytoplasm. The properties

$$\text{sucrose} + \text{UDP} \rightleftharpoons \text{UDP-glucose} + \text{fructose} \qquad [5.6]$$
$$\Delta G^{\circ\prime} = -3.99 \text{ kJ mol}^{-1}$$

$$\text{sucrose} + H_2O \rightarrow \text{glucose} + \text{fructose} \qquad [5.7]$$
$$\Delta G^{\circ\prime} = -29.3 \text{ kJ mol}^{-1}$$

The contribution of these enzymes to the pathway of sucrose breakdown may, in part, be determined by the source of sucrose. Sucrose obtained through translocation can enter a cell via the symplast or the apoplast. Several studies using asymmetrically labeled sucrose have demonstrated that sugar obtained in this way moves primarily through the symplast and is not cleaved into glucose and fructose during transport. It seems likely that most cells receive their sucrose via this route. However, there is strong evidence that in certain tissues sucrose is supplied through the apoplast. This must be the route used by developing seeds in which there are no protoplasmic connections between the maternal and embryonic tissues. Studies on the pathway of sucrose uptake from the apoplast have failed to reveal a consistent pattern. Hydrolysis of sucrose precedes uptake by developing seeds of maize, sorghum and pearl millet; whereas in wheat, rye and barley sucrose is apparently transferred without prior hydrolysis (Thorne and Giaquinta, 1984). Even for tissues in which extensive hydrolysis occurs in the apoplast, such extracellular cleavage is not necessarily an essential prerequisite for sugar uptake. Recent studies on maize kernels have demonstrated that 1-fluorosucrose, a sucrose analog resistant to hydrolysis by invertase, is absorbed by the developing seeds at rates similar to those of sucrose (Schmalstig and Hitz, 1987b). Thus, although hydrolysis of sucrose in the apoplast clearly occurs, its role in sugar uptake is uncertain.

The second major source of sucrose for metabolism is that stored in the vacuole. At present there is not enough information to exclude the possibility that sucrose itself moves out of the vacuole. However, the more likely route for the mobilization of stored sucrose is through hydrolysis by acid invertase in the vacuole producing hexoses that are released into the cytoplasm. The main evidence for this proposal is the well established inverse relationship between intracellular acid invertase activity and sucrose content in many plants (ap Rees, 1984). Detailed studies indicate that in beetroot both acid invertase and sucrose are largely confined to the vacuole and that the decline in sucrose during aging of tissue slices is associated with a corresponding increase in vacuolar invertase activity. These correlations suggest that the level of invertase can determine the amount of sucrose in vacuoles. Increasing invertase activity in the vacuole may prevent sucrose accumulation and make stored sugar available for metabolism (ap Rees, 1984).

Sucrose entering the cytoplasm can be metabolized by either alkaline invertase or sucrose synthase. Although the reaction catalyzed by sucrose synthase is readily reversible there is good evidence that this enzyme is involved primarily in the breakdown of sucrose. Firstly, the distribution of sucrose synthase in different tissues is that expected for an enzyme concerned in sucrose breakdown rather than synthesis. The activity is generally low in photosynthetic and gluconeogenic cells, and is often high in actively growing tissues that rely on sucrose as their respiratory substrate (ap Rees, 1984). Secondly, in some tissues the activity of the invertases are far lower than that of sucrose synthase, and are insufficient to catalyze the observed rates of sucrose metabolism. A good example of this is potato tubers in which the activities of both acid and alkaline invertase are so low that sucrose synthase must be the dominant route of sucrose breakdown (ap Rees, 1988). Thirdly, analysis of the *shrunken* mutant of maize reveals that reduction in the level of sucrose synthase in the developing endosperm restricts the ability of this tissue to metabolize sucrose (Boyer, 1985).

At present, the relative roles of alkaline invertase and sucrose synthase in sucrose degradation are only poorly understood. With the exception of potato tubers and a few other tissues, studies based on comparing enzyme activities with the rate of sucrose breakdown have been unable to establish the route of degradation. This is because generally the capacities of sucrose synthase and alkaline invertase are sufficient for both to contribute significantly to sucrose metabolism. An alternative method to establish the relative roles of these two enzymes *in vivo* is to compare the rates of metabolism of sucrose and 1-fluorosucrose. The latter compound is an extremely poor substrate for invertase, but can be cleaved by sucrose synthase. Such experiments are not simple. Adequate evidence must be provided that the sugars are taken up by the tissue at the same rate. In addition, appropriate allowances must be made for differences in the rates of metabolism of the two substrates and the extent to which the enzymes discriminate between these two compounds. All of these points have been addressed in a recent study on developing soybean leaves (Schmalstig and Hitz, 1987a). The results described in this report show that the contribution of the two enzymes to metabolism varies. In the youngest leaves sucrose was cleaved almost exclusively by sucrose synthase, whereas invertase accounted for about half of the sucrose metabolism in older leaves. This technique probably offers the best approach to defining the precise roles of sucrose synthase and alkaline invertase. However, we need to establish that neither fluorosucrose nor its products modify metabolism before we can be sure that such results are an accurate estimate of the two pathways *in vivo*.

The preceding discussion suggests that in most conditions both invertase and sucrose synthase contribute to sucrose breakdown. The free glucose and fructose produced in these reactions are probably phosphorylated by the range of hexose kinases found in plants. These enzymes are discussed in a subsequent section. The fate of UDP-glucose produced by sucrose synthase is less certain. A previous proposal that it may provide the sugar nucleotide donor for starch biosynthesis now seems unlikely. The reasons for this are two-fold. Firstly, despite the broad substrate specificity of sucrose synthase *in vitro*, there is good evidence that UDP-glucose is the predominant, if not exclusive, sugar nucleotide produced by this enzyme *in vivo* (ap Rees, 1984). Secondly, data summarized in a later section indicate that in both photosynthetic and non-photosynthetic cells ADP-glucose, not UDP-glucose, is the precursor of starch. An alternative suggestion is that UDP-glucose is used for the synthesis of cellulose and other structural polysaccharides that are made outside the plastids. Some UDP-glucose may be metabolized in this way, but the demands of such pathways are likely to be small relative to the amount of UDP-glucose produced by sucrose synthase. Instead, the vast majority of UDP-glucose is probably converted to hexose phosphates. Detailed consideration of the fate of sucrose metabolized by potatoes, peas and maize seeds reveals that the activity of invertase is inadequate to support the observed rates of respiration and starch synthesis. This deficit is so great that much of the UDP-glucose produced during sucrose degradation must enter the hexose phosphate pool (ap Rees, 1988).

The most likely route for the conversion of UDP-glucose to hexose phosphate is through the reaction catalyzed by UDP-glucose pyrophosphorylase. Of the enzymes capable of fulfilling the role, UDP-glucose pyrophosphorylase is the only one known to be present in the cytoplasm of a wide range of tissues at activities sufficient to support the estimated rates of metabolism (ap Rees, 1988). The main problem with this proposal is that, although the reaction itself is readily reversible, the enzyme is often considered to operate only in the direction of UDP-glucose synthesis *in vivo*. This opinion is based on the assumption that any pyrophosphate produced is rapidly hydrolyzed by the extremely high activities of pyrophosphatase that are present in plants. However, two lines of evidence suggest that this assumption is wrong. Firstly, the available data indicate that acid and alkaline pyrophosphatase are largely restricted to the vacuole and plastids, respectively. Secondly, recent measurements suggest that plant tissues contain appreciable amounts of pyrophosphate in the cytoplasm. Comparison of the mass–action ratio with the equilibrium constant for UDP-glucose pyrophosphorylase suggests that, at least in spinach leaf and pea embryos, the reaction is close to equilibrium. This means that the enzyme is equally likely to catalyze the reaction in either direction and that the net flux through this step *in vivo* will depend on the relative levels of the reactants (Weiner *et al.*, 1987; ap Rees, 1988).

The source of pyrophosphate required for the conversion of UDP-glucose to glucose 1-phosphate is unknown. Although pyrophosphate is produced in the biosynthesis of protein, isoprenoid lipids and most polysaccharides it is doubtful that such pathways could provide sufficient pyrophosphate to meet the demands of sucrose metabolism. This opinion is strengthened by the likelihood that much of the pyrophosphate produced in plastids is hydrolyzed by the alkaline pyrophosphatase activity found in these organelles.

One possibility is that pyrophosphate is produced by pyrophosphate:fructose 6-phosphate phosphotransferase operating in the direction of fructose 6-phosphate synthesis (EC 2.7.1.90; reaction [5.8]).

fructose 1,6-bisphosphate + P_i ⇌ fructose 6-phosphate + PP_i [5.8]

$\Delta G^{\circ\prime} = 2.93$ kJ mol^{-1}

Fructose 1,6-bisphosphate could be resynthesized from the fructose 6-phosphate formed in this reaction by phosphofructokinase. The outcome of this substrate cycle would be the provision of pyrophosphate at the expense of ATP. Although such a scheme is still speculative, much of our knowledge of pyrophosphate:fructose 6-phosphate phosphotransferase is consistent with this idea. The wide distribution of the enzyme in plants suggests that it is involved in a common metabolic process. In addition it is confined to the cytoplasm which is the site of sucrose cleavage by sucrose synthase. Finally, its activity is generally in excess of that required to meet the demands for pyrophosphate by sucrose metabolism (ap Rees, 1985, 1988).

There are two criticisms of the above proposal. One is that the scheme cannot account for the high activities of pyrophosphate:fructose 6-phosphate phosphotransferase found in gluconeogenic tissues that are unlikely to break down sucrose. This observation does not necessarily disprove the proposal, because the enzyme may have functions besides the provision of pyrophosphate. The other objection is that this scheme conflicts with the general view that pyrophosphate:fructose 6-phosphate phosphotransferase is analogous to phosphofructokinase and operates in the glycolytic direction. This criticism is weak because at present there is no compelling evidence to support the latter view. Most data are equally compatible with the enzyme operating in either direction (ap Rees, 1985). Indeed, recent estimates of metabolite levels suggest that the reaction catalyzed by the enzyme is close to equilibrium *in vivo* (Weiner *et al.*, 1987; ap Rees, 1988). If this is correct, then the net flux through this reaction will depend on the relative amounts of the reactants. Thus, it is entirely conceivable that pyrophosphate:fructose 6-phosphate phosphotransferase is capable of providing pyrophosphate for the metabolism of UDP-glucose in the cytoplasm.

Starch metabolism

The importance of starch

The pre-eminence of sucrose in plant carbohydrate metabolism is challenged only by the claims of starch. This is by far the dominant storage polysaccharide in plants and is a major metabolic substrate. The importance of starch as a storage compound is revealed by its wide distribution. It is present in all major organs of most higher plants and, in certain tissues, can accumulate to high levels. For example, starch typically accounts for 65–75% of the dry weight of cereal grains and about 80% of that of mature potato tubers. In such tissues starch provides the principal respiratory substrate, and is capable of supporting high rates of metabolism. A particularly dramatic example of this is the extremely high rate of respiration in the inflorescence of *Arum maculatum* during thermogenesis. This occurs principally, if not exclusively, at the expense of starch stored in the spadix (ap Rees, 1977).

Like sucrose, starch is also a major immediate product of photosynthesis. In the light up to 30% of the $^{14}CO_2$ fixed by leaves is incorporated into starch. However, in general there is an appreciable delay after the onset of photosynthesis before starch begins to accumulate. Moreover, this starch is usually mobilized in the ensuing dark period and can be converted to sucrose for export. Such observations have led to the concept that starch accumulation in chloroplasts is primarily a mechanism for storing reduced carbon when the rate of photosynthesis exceeds the capacity of the leaf to export sucrose (Stitt, 1984). Although this view is helpful in explaining some of the aspects of sucrose and starch metabolism in leaves, we should not consider starch solely as a buffer for sucrose. There is ample evidence that the extent of starch accumulation in leaves is controlled. In tissues exhibiting CAM metabolism, the ability to fix CO_2 at night is critically dependent on a continuous supply of phosphoenolpyruvate. The latter is often derived from starch accumulated in the preceding light period. Thus, in CAM plants, starch synthesis during the day must be strictly controlled to provide appropriate amounts of phosphoenolpyruvate for CO_2 fixation in the ensuing dark period. Even in plants exhibiting C3 photosynthesis there is evidence that the level of starch is regulated. In many species the percentage of photosynthate retained in starch actually rises in plants grown for a few days at lower irradiance or in shorter daylength. Such observations are not consistent with starch being simply an overflow product, since under these conditions less, not more, starch should be made. These and other points are considered in more detail by Stitt (1984), and argue strongly that starch, in its own right, is an important product of photosynthesis.

Starch synthesis

Starch is found predominantly, if not exclusively, as water-insoluble granules that are confined to the plastids. These granules contain a mixture of amylose and amylopectin. Amylose is primarily a linear molecule containing between 600 and 3000 1,4-α-glucosyl residues, although 1,6-α-glucosyl branch points occur about every 1000 residues. Amylopectin is generally much larger and more highly branched, containing between 6000 and 60 000 glucosyl residues with an average of one 1,6-α-glucosyl linkage every 20 to 26 units. The structure and properties of these components of starch have been discussed in detail recently (Manners, 1985).

The pathway of starch synthesis from hexose phosphates involves the following steps. First, ADP-glucose is produced from glucose 1-phosphate by ADP-glucose pyrophosphorylase (EC 2.7.7.27; reaction [5.9]). The glucosyl unit is then transferred from ADP-glucose to the non-reducing end of an α-glucan primer by starch synthase, forming an additional 1,4-α-glucosidic bond (EC 2.4.1.21; reaction [5.10]). Finally, the 1,6-α-glucosyl branch points of amylopectin are introduced by branching enzyme which hydrolyzes a 1,4-α-glucosyl bond and transfers the resulting short oligosaccharide to a primary hydroxyl group in a similar glucan chain (EC 2.4.1.24; reaction [5.11]). Figure 5.2 provides an outline of the pathways involved.

glucose 1-phosphate + ATP ⇌ ADP-glucose + PP$_i$ [5.9]
$\Delta G°' = -2.88$ kJ mol^{-1}

ADP-glucose + α-glucan$_{(n)}$ → α-glucan$_{(n+1)}$ + ADP   [5.10]
$\Delta G°' = -13.8$ kJ mol^{-1}

linear 1,4-α-glucan → branched 1,6-α-1,4-α-glucan   [5.11]
$\Delta G°' = -9.6$ kJ mol^{-1}

The principal evidence for the above pathway comes from studies of the activity and location of the relevant enzymes. The activities of ADP-glucose pyrophosphorylase and starch synthase are sufficient to account for the estimated rates of starch production in tissues that differ widely in their metabolism (Table 5.1). Detailed studies have shown that in maize kernels changes in the activities of these two enzymes mirror the increase in the rate of starch accumulation during development (Ozbun *et al.*, 1973). Similar though less complete data are available for other tissues. Moreover, for soybean suspension culture and spinach leaf, evidence from careful fractionation of protoplasts indicates that the

Carbohydrate synthesis and degradation

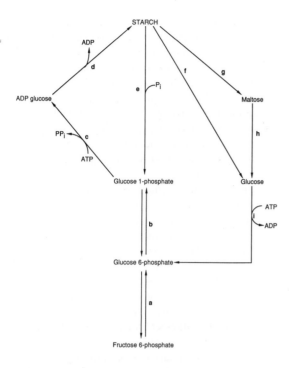

Fig. 5.2 Pathways of starch metabolism. The letters correspond to the following enzymes: a, glucose 6-phosphate isomerase [5.2]; b, phosphoglucomutase [5.1]; c, ADP-glucose pyrophosphorylase [5.9]; d, starch synthase [5.10]; e, starch phosphorylase [5.12]; f, α-amylase; g, β-amylase; h, glucosidase; i, hexose kinase [5.15]. The numbers in brackets denote reactions referred to in the text.

vast majority of the activity of these two enzymes is confined to plastids, the site of starch accumulation (Okita *et al.*, 1979; ap Rees *et al.*, 1984).

Further evidence is provided by analysis of mutant lines. In maize, two endosperm mutants, *shrunken-2* and *brittle-2*, are characterized by having only about 25% of the normal amount of starch. Correlated with this is a reduction in ADP-glucose pyrophosphorylase activity to less than 10% of that normally found in maize endosperm (Boyer, 1985). More recently, a starchless *Arabidopsis* mutant has been isolated which is deficient in ADP-glucose pyrophosphorylase but which contains levels of UDP-glucose pyrophosphorylase and starch phosphorylase that are the same as those in wild-type plants (Lin *et al.*, 1988). These mutants indicate that starch is synthesized primarily through a route involving ADP-glucose.

The possibility that UDP-glucose rather than ADP-glucose may provide the glucosyl donor for starch synthesis in some tissues seems unlikely. In addition to the above evidence implicating a dominant role for ADP-glucose the following considerations argue strongly against the immediate involvement of UDP-glucose. In most tissues starch synthase is specific for ADP-glucose (Preiss, 1982). Only in some reserve tissues is the enzyme capable of using UDP-glucose in addition to ADP-glucose. However, generally this is only a minor activity associated with starch granules, its affinity for UDP-glucose is extremely low, and its solubilization from

Table 5.1 Rate of starch accumulation and activity of enzymes of starch synthesis in plant tissues

Tissue	Rate of starch accumulation (nmol hexose equivalent g^{-1} fresh wt min^{-1})	Enzyme activity		Reference
		ADP-glucose pyrophosphorylase (nmol hexose equivalent g^{-1} fresh wt min^{-1})	Starch synthase	
Arum spadix	168	1626	588	ap Rees *et al.* (1984)
Soybean suspension culture	3.3	46	22	ap Rees *et al.* (1984)
Maize endosperm	92.3	735	160	Ozbun *et al.* (1973)
Pea embryo	73.3	1957	83.3	Edwards *et al.* (1988)
Spinach leaf	44[a]	700[a]	160[a]	Heldt *et al.* (1977) Okita *et al.* (1979)

[a]Values are nmol mg chlorophyll^{-1} min^{-1}.

the starch granules results in a decrease in activity with UDP-glucose (Macdonald and Preiss, 1985). The significance of this activity is unknown, but it is unlikely to contribute to starch synthesis *in vivo*. Futhermore, the available evidence suggests that UDP-glucose together with the two main enzymes responsible for its synthesis, UDP-glucose pyrophosphorylase and sucrose synthase, are confined to the cytoplasm (ap Rees, 1984, 1988). Our current understanding of the properties of the plastid envelope suggests that such UDP-glucose is not directly available for starch biosynthesis via starch synthase inside the plastids (see Chapter 21).

In photosynthetic tissues the hexose phosphates required for starch synthesis are provided primarily by CO_2 fixation during photosynthesis. A clear demonstration of this is the ability of isolated intact chloroplasts to incorporate label from $^{14}CO_2$ into starch at rates approaching those of starch accumulation *in vivo* (Heldt *et al.*, 1977). For non-photosynthetic cells the ultimate source of carbon is almost certainly translocated sucrose. The principal difficulty with these tissues is in establishing the pathway by which hexose phosphates are formed in the plastids from those that are produced in the cytoplasm during sucrose breakdown. By analogy with the transport properties of the chloroplast envelope, it is often assumed that carbon is imported by amyloplasts mainly via the phosphate translocator (Boyer, 1985). In such a scheme, cytoplasmic hexose phosphates must be converted to triose phosphates prior to uptake. Once inside the amyloplast, triose phosphates could be reconverted to hexose phosphates by fructose 1,6-bisphosphatase. The justification for this analogy is weak since the metabolism of amyloplasts and chloroplasts differs dramatically and there is no compelling reason why the transport properties of these organelles should be identical. Indeed, three recent findings provide strong evidence that, at least in developing wheat endosperm, glucose l-phosphate is imported directly from the cytoplasm for starch synthesis. Firstly, intact isolated amyloplasts incorporate labeled glucose 1-phosphate into starch (Tyson and ap Rees, 1988). Secondly, these amyloplasts apparently lack fructose 1,6-bisphosphatase activity which is essential for the resynthesis of hexose phosphates for starch production if carbon is imported into the plastid as triose phosphates (Entwistle and ap Rees, 1988). Thirdly, labeled starch formed in developing wheat grains fed with either [1-^{13}C]-glucose or [6-^{13}C]-glucose shows little randomization (Keeling *et al.*, 1988). This contrasts with the almost complete equilibration between these two positions we would anticipate from a pathway involving triose phosphates, since the latter are rapidly interconverted by triose phosphate isomerase. At present it is uncertain whether the ability to import hexose phosphates from the cytoplasm is a general feature of plastids from all non-photosynthetic cells or whether it is restricted to specific tissues that accumulate massive amounts of starch.

Our understanding of the control of the pathway of starch synthesis from hexose phosphates is confined primarily to the contribution made by ADP-glucose pyrophosphorylase. In the many tissues that have been studied, this enzyme has consistently been reported to be inhibited by phosphate, which induces sigmoidal kinetics, and to be activated by 3-phosphoglycerate, which relieves the inhibition by phosphate. These properties form the basis for a hypothesis of the regulation of starch synthesis (Preiss, 1982). The extremely high activity of alkaline pyrophosphatase in plastids (Weiner *et al.*, 1987) implies that the pyrophosphate produced during the synthesis of ADP-glucose is rapidly hydrolyzed. Consequently, the reaction catalyzed by ADP-glucose pyrophosphorylase is almost certainly far from equilibrium *in vivo* and therefore susceptible to regulation by allosteric effectors. However, this point cannot be firmly established until we have authenticated measurements of the level of ADP-glucose in the plastids.

In chloroplasts the response of ADP-glucose pyrophosphorylase to 3-phosphoglycerate and phosphate may be the main factor regulating starch synthesis. This view is supported by the observation that in isolated chloroplasts starch synthesis is enhanced when the stromal 3-phosphoglycerate: phosphate ratio is increased (Heldt *et al.*, 1977). Similar experiments designed to manipulate the relative levels of these metabolites in intact leaves have the predicted effect on the rate of starch synthesis (Preiss, 1982). This ratio probably reflects the availability of fixed carbon for starch synthesis in the light. Thus, the regulation of ADP-glucose pyrophosphorylase by these metabolites allows the rate of starch production to match the supply of substrate provided by photosynthesis. The importance of the stromal 3-phosphoglycerate: phosphate ratio in coordinating carbon metabolism in the cytoplasm and chloroplasts is considered in more detail in Chapter 21.

Less is known about the regulation of starch synthesis in non-photosynthetic cells. ADP-glucose pyrophosphorylase from such tissues responds to 3-phosphoglycerate and phosphate in the same way as the enzyme from chloroplasts (Preiss, 1982). This observation implies that control of the pathway in non-photosynthetic tissues is similar to that in leaves. However, since carbon for starch synthesis does not enter the amyloplast as triose phosphates, the significance of regulation by the 3-phosphoglycerate : phosphate ratio is uncertain. Moreover, evidence is accumulating that other reactions in the pathway may contribute to its regulation. The reduced level of starch found in wrinkled peas containing a mutation at the *Ra* locus can be attributed to a reduction in the activity of branching enzyme (Edwards *et al.*, 1988). This suggests that under certain conditions branching enzyme can regulate starch synthesis. There is also similar evidence indicating that mutations affecting starch synthase in several species can influence the accumulation of starch (Boyer, 1985). Thus, particularly in non-photosynthetic tissues, the control of starch synthesis may be shared between the different steps of the pathway.

Consideration of the relative contribution of each step to the regulation of starch synthesis is complicated by the observation that plants contain multiple forms of starch synthase and that in most tissues a substantial proportion of the starch synthase is tightly associated with starch granules (Preiss, 1982). The starch synthases that have been studied in greatest detail are those from developing maize endosperm. This tissue contains two forms of soluble activity and a further two forms can be resolved in the granule-bound activity following solubilization. An additional minor granule-bound form has been identified in the endosperm of *waxy* maize. Each of these forms is distinguished by its physical and kinetic properties. Immunochemical studies indicate that the two soluble forms and at least the major granule-bound activity are unrelated (Macdonald and Preiss, 1985). Therefore, the granule-bound activity is due to specific forms of the enzyme and is not merely a fraction of the soluble activity that has become associated with the enlarging starch granule.

Branching enzyme also exists in multiple forms (Preiss, 1982). Three types of this enzyme have been identified in maize endosperm. The two major forms, IIa and IIb, differ from a minor form, I, in the ratio of activities obtained from two different assay methods. This presumably reflects differences in the substrate specificity of the various forms. Immunochemical studies, as well as analysis of the amino acid composition and peptide fragments produced by proteolytic digestion of the purified polypeptides, have been used to compare the three activities. The results indicate that branching enzyme I is quite different from forms IIa and IIb which are very similar and may differ only in the amount of carbohydrate non-covalently associated with the enzyme (Singh and Preiss, 1985).

Our understanding of the role of the multiple forms of starch synthase and branching enzyme in starch synthesis is limited. One possibility worth consideration is that the separate forms of each enzyme act on different substrates. The components of a starch granule are heterogeneous. Neither amylose nor amylopectin contain a unique structure, and at least within amylopectin there is a marked dichotomy in the length of the component 1,4-α-glucan chains (Manners, 1985). Since the separate forms of starch synthase and branching enzyme differ in their substrate specificity, it is conceivable that such forms may contribute to the synthesis of different components of the starch. Studies of mutants, particularly those of maize, support this view (Boyer, 1985). For example, in *waxy* maize endosperm the almost complete absence of amylose is correlated with a massive reduction in the activity of the two forms of starch synthase normally associated with the starch granule. By contrast, another maize mutant, *amylose extender*, is deficient in branching enzyme IIb. This line is characterized by an increase in the proportion of amylose in the granule from less than 30% in the non-mutant to about 60%. Furthermore, the amylopectin from this mutant is far less branched than that from normal maize. A similar reduction in the activity of branching enzyme in pea due to a mutation at the *Ra* locus produces seeds with comparable high levels of amylose (Edwards *et al.*, 1988).

The above evidence is consistent with the idea that different forms of these enzymes may help to determine both the fine structure and relative proportions of amylose and amylopectin within starch. However, the temptation to assign specific roles to individual forms of starch synthase and branching enzyme should be resisted until we have more detailed information on how such enzymes interact in the synthesis of starch granules.

Starch breakdown

There is a range of enzymes in plants that can contribute to starch breakdown. 1,4-α-Glucosyl bonds can be cleaved hydrolytically by α-amylase (EC 3.2.1.1) or other endoamylases resulting in the production of a mixture of linear and branched oligosaccharides and, ultimately, glucose, maltose, maltotriose and a range of branched α-limit dextrins. Starch can also be hydrolyzed by β-amylase (EC 3.2.1.2) which catalyzes the removal of successive maltose units from the non-reducing end of α-glucan chains. The maltose and other short maltosaccharides produced by these enzymes may be further hydrolyzed to glucose by α-glucosidase (EC 3.2.1.20) and subsequently phosphorylated by hexokinase, as described in the next section. Alternatively, 1,4-α-glucosyl bonds can be cleaved phosphorolytically by starch phosphorylase (EC 2.4.1.1; reaction [5.12]). This enzyme produces glucose 1-phosphate from successive glucosyl residues at the non-reducing end of an α-glucan chain. Although phosphorylase can only attack oligosaccharides larger than maltotetraose, further phosphorolytic cleavage may occur due to the action of a glucosyltransferase such as D-enzyme (EC 2.4.1.25; reaction [5.13]). By increasing the degree of polymerization of short oligosaccharides this enzyme can make a greater proportion of the glucan accessible to phosphorylase, and hence contribute to starch degradation.

$$\alpha\text{-glucan}_{(n)} + P_i \rightleftharpoons \alpha\text{-glucan}_{(n-1)} + \text{glucose 1-phosphate} \quad [5.12]$$
$$\Delta G^{o\prime} = 2.98 \text{ kJ mol}^{-1}$$

$$\alpha\text{-glucan}_{(m)} + \alpha\text{-glucan}_{(n)} \rightleftharpoons \alpha\text{-glucan}_{(m+n-1)} + \text{glucose} \quad [5.13]$$

Branch points in starch are removed exclusively by de-branching enzyme (EC 3.2.1.41) which cleaves 1,6-α-glucosyl bonds hydrolytically releasing linear oligosaccharides for further metabolism. The properties and roles of these enzymes have been reviewed in detail by Preiss (1982) and Manners (1985).

Recent attention has focused on two aspects of starch breakdown. One is the initial degradation of the insoluble starch granule. It is generally believed that this step is catalyzed only by an endoamylase and that the soluble oligosaccharides released in the reaction provide the substrate for the other enzymes of starch breakdown (Stitt and Steup, 1985). This view is based on several reports in which α-amylase was the only enzyme capable of degrading isolated starch grains, and on the fact that starch phosphorylase has a much higher affinity for linear low molecular weight oligosaccharides than for branched polyglucans. This evidence is limited and is considerably weakened by the following considerations. First, in general the rates of starch breakdown in these studies were low and obtained using α-amylase from bacterial and mammalian sources rather than plants. The characteristics of starch breakdown in such systems may have little relevance to the mechanism of starch breakdown *in vivo*. Second, we know little about the hydrolysis of starch granules by the Ca^{2+}-dependent heat-labile endoamylase that replaces conventional α-amylase in several tissues. Third, the belief that starch phosphorylase is unable to attack starch granules is questionable. Phosphorylase from pea chloroplasts can release labeled glucose 1-phosphate from ^{14}C-labelled starch granules (Kruger and ap Rees, 1983a). Similarly, appreciable amounts of glucose 1-phosphate are produced when spinach leaf starch grains are incubated with extra chloroplastic phosphorylase from the same source (Steup *et al.*, 1983). Finally, starch phosphorylase has been studied from far too few tissues for us to generalize about its kinetic properties, nor do such properties necessarily indicate the substrate for the enzyme *in vivo*. Together these criticisms emphasize that at present there is no compelling evidence to support the idea that the initial degradation of starch is due solely to hydrolysis by endoamylase although it remains an attractive hypothesis.

The second aspect of starch breakdown to have received attention is the relative importance of hydrolytic and phosphorylytic routes of degradation. This point is difficult to resolve because there is no clear distinction between the two pathways, and oligosaccharides released by hydrolysis may themselves be subsequently metabolized by either amylases or phosphorylase. In addition, the pathway of mobilization may to some extent depend on the specific metabolic requirements of the tissue. Since starch accumulates in many tissues that have quite different physiological roles, it is unlikely that in each the route of starch degradation will be the same. For example, during germination of cereal grains, starch breakdown coincides with the destruction of the endosperm. There is strong evidence that in such instances starch degradation is extracellular and occurs via a hydrolytic route (Boyer, 1985). Glucose produced by the action of α-amylase, de-branching enzyme and α-glucosidase is absorbed by the scutellum, converted to sucrose and then transported to the embryo. In this type of seed the mobilization of starch is controlled primarily by alterations in the levels of the relevant enzymes. During germination the activity of some enzymes increases by *de novo* synthesis, whereas for others the increase is due to activation of previously latent enzymes associated with protein bodies. These processes are generally under environmental and hormonal control and often occur in response to changes in the level of gibberellic acid (Halmer and Bewley, 1982).

Other seeds have been studied in which the major storage tissue remains intact throughout germination. In such examples starch breakdown occurs intracellularly, although in some the amyloplast membrane is destroyed making the starch grains accessible to cytoplasmic enzymes. The pathway of starch metabolism in these tissues is variable. The available data suggest that in pea cotyledons starch breakdown is largely phosphorylytic, whereas in soybean and lentil α-amylase is the predominant activity. Although many seeds contain high activity of β-amylase, its role in starch degradation during germination is uncertain. Studies on both rye and soybean have shown that varieties which lack β-amylase germinate normally and mobilize starch at rates similar to those of varieties containing this enzyme. Thus, although β-amylase may contribute to the degradation of oligosaccharides released by α-amylase, it is not essential for starch breakdown (ap Rees, 1988).

In the examples described above, starch degradation is essentially irreversible and continues until all of the starch is metabolized. However, this is not so in all tissues. In leaves, starch forms a temporary reserve which accumulates or is mobilized depending on the carbohydrate status of the cell. In such instances cellular integrity must be maintained during starch breakdown and therefore at least the initial stages of starch degradation must occur in the plastids. Studies of the subcellular distribution of the enzymes of starch metabolism suggest that in leaves starch is mobilized by the combined activities of endoamylase and starch phosphorylase (Stitt, 1984; Stitt and Steup, 1985). Characterization of the products of starch degradation in isolated chloroplasts reveal that glucose 1-phosphate produced by phosphorylase is metabolized within the chloroplast to triose phosphates and 3-phosphoglycerate, whereas the other products, glucose and maltose, are probably exported to the cytoplasm. Although maltose is generally assumed to result from hydrolytic cleavage of starch, the following arguments suggest that this is not so. Firstly, β-amylase in leaves is apparently confined to the vacuole (Ziegler and Beck, 1986) and consequently unlikely to contribute to starch breakdown. Secondly, the specificities of endoamylase (Okita and Preiss, 1980) and D-enzyme (Lin and Preiss, 1988) are such that maltose is unlikely to be a major product of these enzymes. Thirdly, the proportions of glucose and maltose produced during starch breakdown vary between species. In spinach glucose is the dominant product, whereas in pea maltose is the major sugar released (Stitt and Steup, 1985). This difference is difficult to explain if both are formed by the same hydrolytic pathway. Finally, there is good evidence that at least some of the maltose produced during starch mobilization in isolated pea chloroplasts is formed via maltose phosphorylase (EC 2.4.1.8; reaction [5.14]) (Kruger and ap Rees, 1983b).

$$\text{glucose + glucose 1-phosphate} \rightleftharpoons \text{maltose} + P_i \quad [5.14]$$
$$\Delta G^{\circ\prime} = -1.6 \text{ kJ mol}^{-1}$$

These considerations are important, because until the source of maltose is established the pathway of starch breakdown in leaves is uncertain.

Without a clear understanding of the pathways of starch degradation our knowledge of its control is limited. Diurnal variation in the level of starch in leaves suggests that the mobilization of starch is regulated. In principle this control could be achieved solely by variation in the rate of starch synthesis, with the rate of starch breakdown being constant. However, the absence of detectable starch turnover in leaves from both pea and sugar beet in the light argues strongly against this possibility and implies that the pathway of starch degradation is regulated directly (Stitt and Steup, 1985). Although starch breakdown is normally restricted during the day, such regulation is not achieved by light *per se*. This view is supported by the observation that net starch degradation can occur in leaves during continuous illumination, and by the demonstration that, under certain conditions illuminated spinach chloroplasts catalyze the concomitant synthesis and degradation of starch (Stitt and Steup, 1985).

Several studies have demonstrated that the breakdown of starch at the beginning of the night is often delayed until the level of sucrose in the leaf has declined, suggesting that starch mobilization may be controlled by the requirement of the cell for a respiratory substrate. Experiments with isolated chloroplasts from spinach and pea demonstrate that the production of phosphorylated intermediates from starch is stimulated by phosphate. These results raise the possibility that the rate of starch phosphorolysis is determined by the availability of phosphate. Although this may control the production of hexose phosphates in the stroma it cannot, by itself, explain the regulation of starch breakdown. At least in spinach chloroplasts, the reduction in the rate of phosphorylytic breakdown of starch in the absence of phosphate is offset by a corresponding increase in the accumulation of glucose and maltose so that the overall rate of starch degradation is unchanged. This interaction indicates that the rate of starch breakdown is largely independent of the pathway of degradation and implies that both hydrolysis and phosphorylysis are controlled. To date, studies of the kinetic properties of the enzymes involved in starch degradation have failed to reveal any obvious regulatory features. Thus, although there is good evidence that starch mobilization in leaves is regulated, we have little idea of how such control is achieved.

The metabolism of free hexose

The cleavage of sucrose and the hydrolysis of starch are the major sources of free hexose in plants. The glucose and fructose produced by these pathways are metabolized principally through conversion to the corresponding hexose 6-phosphate by a hexose kinase (EC 2.7.1.1; reaction [5.15]).

hexose + ATP → hexose 6-P + ADP   [5.15]
$\Delta G^{o\prime}$ glucose = -16.7 kJ mol^{-1}
$\Delta G^{o\prime}$ fructose = -14.6 kJ mol^{-1}

This reaction is essentially irreversible, but the precise standard free energy change depends on the sugar that is phosphorylated.

The few tissues that have been studied in any detail each contain a variety of hexose kinases (Turner and Turner, 1980). These are distinguished by differences in their affinity and maximum activity with different sugars. Some are general hexokinases that can phosphorylate a wide range of sugars, such as glucose, fructose, mannose and galactose. In contrast, the kinetic characteristics of other forms are such that *in vivo* they are likely to be specific for individual sugars. Differences in the properties of hexokinases are typified by those presented in Table 5.2 for the four activities identified in pea seeds.

Table 5.2 Kinetic constants of pea seed hexose kinases

Hexokinase form	Substrate	K_m (mM)	V_{max} (nmol mg protein^{-1} min^{-1})	V_{max}/K_m
I	Glucose	0.07	16	229
	Fructose	30	10	0.3
	Mannose	0.5	11	22
II	Glucose	0.048	15.6	325
	Fructose	10	24	2.4
	Mannose	0.076	11.9	157
III	Glucose	0.14	8.6	61
	Fructose	0.06	98	1 633
IV	Glucose	0.4	1.6	4
	Fructose	0.057	60	1 053
	Mannose	1.0	1.4	1.4

Data from Turner and Copeland (1981) and references therein.

The multiplicity of enzymes capable of phosphorylating different hexoses may be due, at least partly, to three features of carbohydrate metabolism in plants. One is the degree to which such metabolism is compartmented. The second is the variety of hexoses that can be metabolized and the range of sources from which these sugars are derived. The third is the consideration that the two dominant sugars in metabolism, glucose and fructose, are produced in variable amounts. These features make it extremely unlikely that a single enzyme would be suitable for the phosphorylation of the range of sugars produced under different conditions.

At present we do not know the precise physiological role of each hexose kinase. However, their functions are likely to depend partly on their subcellular location. In the plants that have been investigated the majority of the fructokinase activity is in the cytoplasm (ap Rees, 1985). This enzyme is probably responsible for the metabolism of fructose derived from the cleavage of sucrose by sucrose synthase. The observation that the activity of fructokinase greatly exceeds that of glucokinase in many tissues is consistent with the view that sucrose synthase rather than invertase is the major route of sucrose degradation and that consequently the amount of fructose produced is larger than that of glucose.

In some tissues a large proportion of the activity that preferentially phosphorylates glucose is associated with mitochondria (ap Rees, 1985). Subfractionation studies of mitochondrial preparations and protection experiments using exogenous proteases provide strong evidence that this hexokinase is located on the outer mitochondrial membrane. The significance of this association is unclear, although conceivably this arrangement may allow newly synthesized ATP to be used preferentially for the phosphorylation of hexose in the cytoplasm. In mammals the extent to which a similar hexokinase binds to mitochondria is dependent on the levels of glucose 6-phosphate and ATP, and there is evidence that variation in the amount of binding may contribute to the regulation of glucose phosphorylation. Whether a similar system operates in plants is uncertain. In castor bean endosperm, mitochondrial hexokinase can be released by specific hexose phosphates and nucleotides (Miernyk and Dennis, 1983). However, these metabolites have no effect on the much larger proportion of hexokinase that is associated with mitochondria in pea leaves and young stems (Tanner *et al.*, 1983).

The route by which any glucose produced during starch hydrolysis is phosphorylated is unknown. This glucose may be exported from the plastid and phosphorylated in the cytoplasm. However, isolated chloroplasts from both pea and spinach have the potential to metabolize glucose rapidly (Stitt, 1984). The enzyme responsible for this phosphorylation has not been identified. Although chloroplast preparations contain appreciable hexokinase activity, there is good evidence that, at least in pea leaves, this activity is attached to the outer membrane and probably cannot contribute to the phosphorylation of stromal sugars (Stitt, 1984). Similarly, although one of the forms of hexokinase in castor bean endosperm is apparently confined to the plastid stroma, kinetic studies suggest that this form is a specific fructokinase and is unlikely to phosphorylate glucose efficiently *in vivo* (Miernyk and Dennis, 1983). More detailed studies are needed to see whether plastids contain additional hexokinases.

Synthesis of structural polysaccharides

The biosynthesis of cell walls is one of the most complex areas of plant biochemistry and is a quantitatively important aspect of carbohydrate metabolism, especially during cell growth. Our understanding of the synthesis of the structural polysaccharides has been described in detail elsewhere (Fincher and Stone, 1982; Delmer, 1983) and this discussion concentrates on identifying the precursors for the synthesis of these polysaccharides.

Although the precise details of cellulose biosynthesis in higher plants are not established, UDP-glucose is almost certainly the immediate glucosyl donor for this polymer (Delmer, 1983). Detailed analysis of the levels and labeling of intermediates in developing cotton fibers supplied with radioactive glucose are entirely consistent with UDP-glucose, rather than some other sugar nucleotide such as GDP-glucose, being the precursor of cellulose. Furthermore, UDP-glucose has recently been shown to support 1,4-β-glucan synthesis in a cell-free system derived from *Phaseolus aureus* seedlings at rates approaching those of cellulose synthesis *in vivo* (Benziman and Callaghan, 1985).

Most precursors for the other major cell wall polysaccharides are probably derived from UDP-glucose through a series of sugar nucleotide transformations. The principal hexose and pentose residues required for the synthesis of pectins and hemicelluloses can be produced by the following series of reactions: firstly, oxidation of UDP-glucose to UDP-glucuronic acid; secondly, decarboxylation of UDP-glucuronic acid to UDP-xylose; and thirdly, epimerization of these three compounds to form UDP-galactose, UDP-galacturonic acid and UDP-arabinose. Evidence that these reactions contribute to polysaccharide synthesis is provided by the demonstration that plants contain the necessary enzymes and that each of these sugar nucleotides is readily incorporated into appropriate polymers in cell-free extracts (Fincher and Stone, 1982).

On the evidence presented above, UDP-glucose may be considered the principal precursor for structural polysaccharide synthesis. This compound is derived from two sources. In cells in which sucrose is metabolized by sucrose synthase, UDP-glucose is produced during the initial cleavage of sucrose and may be used directly for polysaccharide synthesis. In other tissues, UDP-glucose can be formed from the hexose phosphate pool by UDP-glucose pyrophosphorylase. As described earlier, this reaction is probably close to equilibrium *in vivo*. Therefore, the conversion of glucose 1-phosphate to UDP-glucose will be determined largely by the extent to which the latter intermediate is metabolized to form the structural polysaccharides.

Entry of hexose phosphates into glycolysis

In many plant tissues glycolysis is the dominant pathway of carbohydrate oxidation and, under most conditions, it is a major drain on the hexose phosphate pool. The first committed step in this sequence is the conversion of fructose 6-phosphate to fructose 1,6-bisphosphate. Plants contain two enzymes capable of catalyzing this reaction: phosphofructokinase (EC 2.7.1.11; reaction [5.16]) and pyrophosphate:fructose 6-phosphate phosphotransferase (EC 2.7.1.90; reaction [5.17]).

fructose 6-phosphate + ATP → fructose 1,6-bisphosphate + ADP [5.16]
$\Delta G^{o\prime} = -14.2$ kJ mol^{-1}

fructose 6-phosphate + PP$_i$ → fructose 1,6-bisphosphate + P$_i$ [5.17]
$\Delta G^{o\prime} = -2.93$ kJ mol^{-1}

There is strong evidence that phosphofructokinase contributes to the entry of hexose phosphates into glycolysis. Firstly, the reaction catalyzed by this enzyme is essentially irreversible *in vitro*, and a comparison of the equilibrium constant for this reaction with mass-action ratios calculated from the amounts of the relevant metabolites in several tissues indicate that reaction is probably far from equilibrium *in vivo*. Secondly, there is a good correlation between phosphofructokinase activity and glycolysis in a wide range of plants. For example, during the development of pea roots, variation in the maximum catalytic activity of phosphofructokinase reflects the changes that occur in the rate of respiration. More detailed investigations on the club of the developing spadix of *Arum maculatum* have shown a significant, coordinated increase in the maximum catalytic activity of many glycolytic enzymes, including phosphofructokinase, prior to the massive increase in glycolysis that occurs during thermogenesis (ap Rees, 1988). Thirdly, in all tissues in which careful measurements have been made, the activity of phosphofructokinase is sufficient to catalyze the estimated glycolytic flux. Fourth, studies of carbohydrate metabolism in isolated pea chloroplasts and root plastids provide evidence for the direct involvement of phosphofructokinase in glycolysis in plastids. Since these organelles lack pyrophosphate:fructose 6-phosphate phosphotransferase, the observed glycolytic flux must be catalyzed by phosphofructokinase. Finally, the known kinetic properties of phosphofructokinase are sufficient to construct a coherent hypothesis for the regulation of glycolysis (Turner and Turner, 1980; Dennis and Greyson, 1987).

In contrast, evidence for the direct involvement of pyrophosphate:fructose 6-phosphate phosphotransferase in glycolysis is limited. Recently, Hatzfeld *et al.* (1989) have shown that a large increase in the rate of glycolysis in a suspension culture of *Chenopodium rubrum* is accompanied by an increase in the amount of fructose 2,6-bisphosphate and a decrease in pyrophosphate. These results show that activation of pyrophosphate:fructose 6-phosphate phosphotransferase by an increase in fructose 2,6-bisphosphate can accompany an increase in the rate of glycolysis. This supports the view that the enzyme can operate in the glycolytic direction when a large, rapid increase in the rate of glycolysis occurs. However, this observation does not imply that pyrophosphate:fructose 6-phosphate phosphotransferase always operates in this direction *in vivo*. In the experiments described above glycolysis was stimulated by simultaneously adding the metabolic uncoupler CCCP (carbonylcyanide *m*-chlorophenylhydrazone) and increasing the pH of the medium. Such a profound disturbance of metabolism may not reflect normal plant processes. Furthermore, studies on the spadix of *Arum maculatum* indicate that pyrophosphate:fructose 6-phosphate phosphotransferase is not an essential prerequisite for a large, rapid increase in glycolysis. In this tissue the enormous increase in respiration that occurs during thermogenesis is not accompanied by any detectable change in either fructose 2, 6-bisphosphate or pyrophosphate, and the maximum catalytic activity of the enzyme is well below that required to accommodate the observed rate of glycolysis (ap Rees, 1988). Thus, at least in this one specialized tissue, pyrophosphate:fructose 6-phosphate phosphotransferase is not essential for glycolysis. This means that the precise function of the enzyme remains an enigma. Until this is resolved the pathway by which hexose phosphates enter glycolysis in plants will remain uncertain.

Acknowledgment

I am grateful to Drs T. ap Rees and P. L. Keeling for discussing their work prior to publication.

References

ap Rees, T. (1977). Conservation of carbohydrate by the non-photosynthetic cells of higher plants. *Symp. Soc. Exp. Biol.* **31**, 7–32.

ap Rees, T. (1984). Sucrose metabolism. In *Storage Carbohydrates in Vascular Plants*, ed. D. H. Lewis, Cambridge University Press, Cambridge, pp. 53–73.

ap Rees, T. (1985). The organization of glycolysis and the pentose phosphate pathway in plants. In *Encyclopedia of Plant Physiology*, Vol. 18, eds R. Douce and D. Day, Springer-Verlag, Berlin, pp. 391–417.

ap Rees, T. (1988). Hexose phosphate metabolism by non-photosynthetic tissues of higher plants. In *The Biochemistry of Plants*, Vol. 14, ed. J. Preiss, Academic Press, New York, pp. 1–33.

ap Rees, T., Leja M., Macdonald, F. D. and Green, J. H. (1984). Nucleotide sugars and starch synthesis in spadex of *Arum maculatum* and suspension cultures of *Glycine max*. *Phytochemistry* **23**, 2463–8.

Avigad, G. (1982). Sucrose and other disaccharides. In *Encyclopedia of Plant Physiology*, Vol. 13A, eds F. A. Loewus and W. Tanner, Springer-Verlag, Berlin, pp. 217–347.

Benziman, M. and Callaghan, T. (1985). Biosynthesis of 1,4-β-D-glucan in cell-free preparations from mung beans. In *Biochemistry of Plant Cell Walls*, eds C. T. Brett and J. R. Hillman, Cambridge University Press, Cambridge, pp. 243–57.

Boyer, C. D. (1985). Synthesis and breakdown of starch. In *Biochemical Basis of Plant Breeding*, Vol. 1, ed. C. A. Neyra, CRC Press, Boca Raton, pp. 133–53.

Delmer, D. P. (1983). Biosynthesis of cellulose. *Adv. Carbohydr. Chem. Biochem.* **41**, 105–53.

Dennis, D. T. and Greyson, M. F. (1987). Fructose 6-phosphate metabolism in plants. *Physiol. Plant.* **69**, 395–404.

Edwards, J., Green, J. H. and ap Rees, T. (1988). Activity of branching enzyme as a cardinal feature of the Ra locus in *Pisum sativum*. *Phytochemistry* **27**, 1615–20.

Entwistle, G. and ap Rees, T. (1988). Enzymic capacities of amyloplasts from wheat endosperm. *Biochem. J.* **255**, 391–6.

Fincher, G. B. and Stone, B. A. (1982). Metabolism of noncellulosic polysaccharides. In *Encyclopedia of Plant Physiology*, Vol. 13B, eds W. Tanner and F. A. Loewus, Springer-Verlag, Berlin, pp. 68–132.

Halmer, P. and Bewley, J. D. (1982). Control of external and internal factors over the mobilization of reserve carbohydrates in higher plants. In *Encyclopedia of Plant Physiology*, Vol. 13A, eds F. A. Loewus and W. Tanner, Springer-Verlag, Berlin, pp. 784–93.

Hatzfeld, W.-D., Dancer, J. and Stitt, M. (1989). Direct evidence that pyrophosphate:fructose 6-phosphate phosphotransferase can act as a glycolytic enzyme in plants. *FEBS Lett.* **254**, 215–8.

Hawker, J. S. (1985). Sucrose. In *Biochemistry of Storage Carbohydrates in Green Plants*, eds P. M. Dey and R. A. Dixon, Academic Press, New York, pp. 1–51.

Hawker, J. S., Smith, G. M., Phillips, H. and Wiskich, J. T. (1987). Sucrose phosphatase associated with vacuole preparations from red beet, sugar beet and immature sugarcane stem. *Plant Physiol.* **84**, 1281–5.

Heldt, H. W., Chong, J. C., Maronde, D., Herold, A., Stankovic, Z. S., Walker, D. A., Kraminer, A., Kirk, M. R. and Heber, U. (1977). Role of orthophosphate and other factors in the regulation of starch formation in leaves and isolated chloroplasts. *Plant Physiol.* **59**, 1146–55.

Huber, S. C., Kerr, P. S. and Kalt-Torres, W. (1985). Regulation of sucrose formation and movement. In *Regulation of Carbon Partitioning in Photosynthetic Tissue*, eds R. L. Heath and J. Preiss, Waverley Press, Baltimore, pp. 199–214.

Keeling, P. L., Wood, J. R., Tyson, R. H. and Bridges, I. G. (1988) Starch biosynthesis in the developing wheat grain. *Plant Physiol.* **87**, 311–19.

Kruger, N. J. and ap Rees, T. (1983a). Properties of α-glucan phosphorylase from pea chloroplasts. *Phytochemistry* **23**, 1981–8.

Kruger, N. J. and ap Rees, T. (1983b). Maltose metabolism by pea chloroplasts. *Planta* **158**, 179–84.

Lin, T.-P. and Preiss, J. (1988). Characterization of D-enzyme (4-α-glucanotransferase) in *Arabidopsis* leaf. *Plant Physiol.* **86**, 260–5.

Lin, T.-P., Caspar, T., Somerville, C. and Preiss, J. (1988). Isolation and characterization of a starchless mutant of *Arabidopsis thaliana* (L.) Heynh lacking ADPglucose pyrophosphorylase activity. *Plant Physiol.* **86**, 1131–5.

Macdonald, F. D. and Preiss, J. (1985). Partial purification and characterization of granule-bound starch synthases from normal and waxy maize. *Plant Physiol.* **78**, 849–52.

Manners, D. J. (1985). Starch. In *Biochemistry of Storage Carbohydrates in Green Plants*, eds P. M. Dey and R. A. Dixon, Academic Press, New York, pp. 149–203.

Maretzki, A. and Thom, M. (1986). A group translocator for sucrose assimilation in tonoplast vesicles of sugarcane cells. *Plant Physiol.* **80**, 34–7.

Miernyk, J. A. and Dennis, D. T. (1983). Mitochondrial, plastid and cytosolic isozymes of hexokinase from developing endosperm of *Ricinus communis*. *Arch. Biochem. Biophys.* **226**, 458–68.

Okita, T. W. and Preiss, J. (1980). Starch degradation in spinach leaves. Isolation and characterization of amylase and R-enzyme of spinach leaves. *Plant Physiol.* **66**, 870–6.

Okita, T. W., Greenberg, E., Kuhn, D. N. and Preiss, J. (1979). Subcellular localization of the starch degradative and biosynthetic enzymes of spinach leaves. *Plant Physiol.* **64**, 187–92.

Ozbun, J. L., Hawker, J. S., Greenberg, E., Lammel, C. and Preiss, J. (1973). Starch synthase, phosphorylase ADPglucose pyrophosphorylase, and UDPglucose pyrophosphorylase in developing maize kernels. *Plant Physiol.* **51**, 1–5

Preiss, J. (1982). Biosynthesis of starch and its regulation. In *Encyclopedia of Plant Physiology*, Vol. 13A, eds F. A. Loewus and W. Tanner, Springer-Verlag, Berlin, pp. 397–417.

Schmalstig, J. D. and Hitz, W. D. (1987a). Contribution of sucrose synthase and invertase to the metabolism of sucrose in developing leaves. *Plant Physiol.* **85**, 407–12.

Schmalstig, J. D. and Hitz, W. D. (1987b). Transport and metabolism of a sucrose analog (1'-fluorosucrose) into *Zea mays* L. endosperm without invertase hydrolysis. *Plant Physiol.* **85**, 902–5.

Singh, B. K. and Preiss, J. (1985). Starch branching enzymes from maize. *Plant Physiol.* **79**, 34–40.

Steup, M., Robenek, H. and Melkonian, M. (1983). *In vitro* degradation of starch granules isolated from spinach chloroplasts. *Planta* **158**, 428–36

Stitt, M. (1984). Degradation of starch in chloroplast: a buffer to sucrose metabolism. In *Storage Carbohydrates in Vascular Plants*, ed. D. H. Lewis, Cambridge University Press, Cambridge, pp. 205–29.

Stitt, M. and Steup, M. (1985). Starch and sucrose degradation. In *Encyclopedia of Plant Physiology*, Vol. 18, eds R. Douce and D. Day, Springer-Verlag, Berlin, pp. 347–90.

Stitt, M., Wirtz, W. and Heldt, H. W. (1980). Metabolite levels during induction in the chloroplast and extrachloroplast compartments of spinach protoplasts. *Biochim. Biophys. Acta* **593**, 85–102.

Stitt, M., Wilke, I., Feil, R. and Heldt, H. W. (1988). Coarse control of sucrose-phosphate synthase in leaves: alterations of the kinetic properties in response to the rate of photosynthesis and the accumulation of sucrose. *Planta* **174**, 217–30.

Tanner, G. J., Copeland, L. and Turner, J. F. (1983). Subcellular localization of hexose kinases in pea stems: mitochondrial hexokinase. *Plant Physiol.* **72**, 659–63.

Thorne, J. H. and Giaquinta, R. T. (1984). Pathways and mechanisms associated with carbohydrate translocation in plants. In *Storage Carbohydrates in Vascular Plants*, ed. D. H. Lewis, Cambridge University Press, Cambridge, pp. 75–96.

Turner, J. F. and Copeland, L. (1981). Hexokinase II of pea seeds. *Plant Physiol.* **68**, 1123–7.

Turner, J. F. and Turner, D. H. (1980). The regulation of glycolysis and the pentose phosphate pathway. In *The Biochemistry of Plants*, Vol. 2, ed. D. D. Davies, Academic Press, New York, pp. 279–316.

Tyson, R. H. and ap Rees, T. (1988). Starch synthesis by isolated amyloplasts from wheat endosperm. *Planta* **175**, 33–8.

Weiner, H., Stitt, M. and Heldt, H. W. (1987). Subcellular compartmentation of pyrophosphate and alkaline pyrophosphatase in leaves. *Biochim. Biophys. Acta* **893**, 13–21.

Ziegler, P. and Beck, E. (1986). Exoamylase activity in vacuoles isolated from pea and wheat protoplasts. *Plant Physiol.* **82**, 1119–21.

6 Glycolysis, the oxidative pentose phosphate pathway and anaerobic respiration
Jan A. Miernyk

Glycolysis

Introduction

The classical definition of glycolysis is the conversion of one mole of glucose into two moles of pyruvate (Fig. 6.1). Under anaerobic conditions pyruvate may be reduced to either lactate or ethanol. Chemical bond energy and reducing power inherent in the carbohydrate molecule is conserved in the form of ATP and NADH. Monosaccharides other than glucose are catabolized through glycolysis after conversion to a glycolytic intermediate. In this section I will describe a truncated glycolytic sequence, from fructose 1,6-bisphosphate to pyruvate, as it occurs in plant cells. The reactions from storage polymers to fructose 1,6-bisphosphate are considered in the preceding chapter. Under normal aerobic conditions pyruvate is the endpoint of the glycolytic sequence in most plant cells. Pyruvate is then oxidatively converted to acetyl-CoA for use in biosynthetic reactions. The fermentative metabolism of pyruvate is described later in this chapter.

The enzymes of glycolysis

Aldolase

Aldolase (fructose 1,6-bisphosphate D-glyceraldehyde 3-phosphate lyase, EC 4.1.2.13) catalyzes the aldol cleavage of fructose 1,6-bisphosphate yielding two triose phosphates, dihydroxyacetone phosphate and glyceraldehyde 3-phosphate. There are two classes of fructose bisphosphate aldolases. Class I aldolases are relatively large homotetramers (M_r approximately 160 000), have an essential sulfhydryl group and a Schiff base reaction mechanism. Class II aldolases are smaller (M_r approximately 65 000), monomeric proteins that require potassium and divalent cations, and do not have a Schiff base mechanism. Animal aldolases are class I, while class II aldolases occur in bacteria and fungi. Higher plants through ferns and mosses have class I aldolases, while those from eukaryotic algae can be either class I or II.

Aldolases have been purified to homogeneity from wheat, pea, spinach, and maize leaves, rice bran, and the green alga *Chara foetida*. The physicochemical properties (molecular weight, isoelectric point, amino acid composition) of plant aldolases are similar to those of mammalian class I aldolases, as are the catalytic properties (pH optima around 7, Michaelis constants of approx. 20 μM for fructose 1,6-bisphosphate). There have even been active hybrid enzymes formed from wheat and rabbit subunits (Heil and Lebherz, 1978). The amino acid sequence of anaerobically induced maize root cytoplasmic aldolase has been determined from cDNA clones (Kelley and Tolan, 1986). These clones encode a 355 amino acid protein of 38 611 molecular weight. There is considerable sequence homology between the maize protein and those from animal sources, being as high as 55% when compared with rabbit muscle aldolase A. Sequence homology is even higher in specific regions, such as that which comprises the catalytic site.

78 Cytosolic Carbon Metabolism

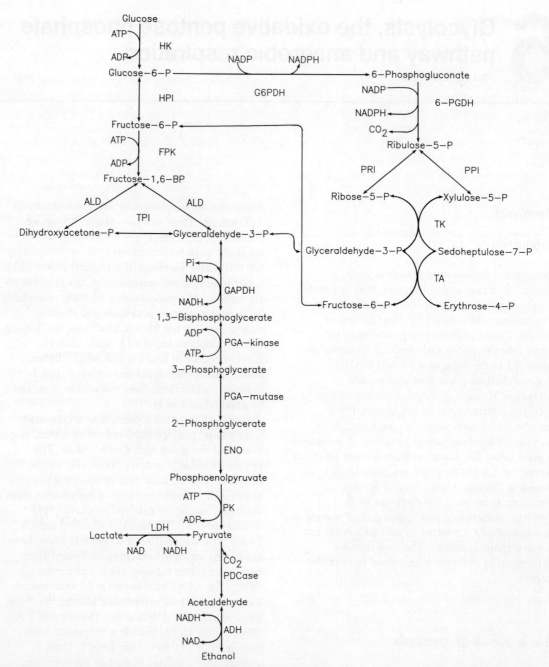

Fig. 6.1 Interrelationships between glycolysis and the pentose phosphate pathway. Abbreviations used: HK, hexokinase; HPI, hexose phosphate isomerase; PFK, phosphofructokinase; ALD, fructose 1,6-bisphosphate aldosale; GAPDH, NAD glyceraldehyde 3-phosphate dehydrogenase; PGA, phosphoglycerate; PK, pyruvate kinase; PDCase, pyruvate decarboxylase; ADH, alcohol dehydrogenase; LDH, lactate dehydrogenase; G6PDH, glucose 6-phosphate dehydrogenase; 6PGDH, 6-phosphogluconate dehydrogenase; PRI, phosphoriboisomerase; PPE, phosphopentose epimerase; TK, transketolase; TA, transaldolase.

Triose phosphate isomerase

Triose phosphate isomerase (D-glyceraldehyde 3-phosphate ketol isomerase, EC 5.3.1.1) catalyzes the reversible isomerization of glyceraldehyde 3-phosphate and dihydroxyacetone phosphate. In addition to glycolysis, the enzyme has a role in gluconeogenesis, fatty acid metabolism, the pentose phosphate pathway, and photosynthetic carbon metabolism. First characterized in 1935, triose phosphate isomerases from bacterial, fungal, and animal systems have been extensively studied. Having an extremely high turnover rate, limited only by diffusion, triose phosphate isomerase is often described as having evolved 'perfect' enzymatic activity. In general, triose phosphate isomerases are homodimeric proteins with each subunit having an M_r 27 000. Higher eukaryotes generally have multiple isozymes of triose phosphate isomerase, and there have been detailed genetic studies of the isozymes. While there are data consistent with multiple structural genes for some triose phosphate isomerase isozymes, it is also clear that in some instances post-translational alterations are responsible for multiple forms. Complete primary sequences, from either direct amino acid sequencing or nucleotide sequencing of cDNAs, have been published for triose phosphate isomerase from bacteria, fungi, invertebrates, and higher animals. Analysis of the sequences reveals extensive homology and a low rate of evolutionary divergence. Functional homology has also been demonstrated (e.g. the cloned chicken muscle triose phosphate isomerase gene can complement an *E. coli* mutant lacking enzyme activity; Straus and Gilbert, 1985).

Among higher plants, triose phosphate isomerase has been partially purified from pea seeds, and purified to homogeneity from spinach, lettuce, celery, and rye leaves. Overall, the physicochemical and catalytic properties of plant triose phosphate isomerases are very similar to those of the enzyme from other sources. Plants also contain multiple isozymes of triose phosphate isomerase, and the genetics of these isozymes have been addressed by Gottlieb and his associates (e.g. Pichersky and Gottlieb, 1983). While there is as yet relatively little primary sequence information for plant triose phosphate isomerases, substantial immunochemical homology has been demonstrated, and it is possible to form active hybrid enzymes made up of subunits from different sources (Pichersky *et al.*, 1984). Marchionni and Gilbert (1986) reported the nucleotide sequences of several members of the small multigene family of maize triose phosphate isomerases, and showed that there is a high degree of homology with the previously published animal and microbial sequences. The maize root triose phosphate isomerase gene has eight introns. Gilbert and his associates have proposed an evolutionary tree based upon the conservation of structure and location of introns in triose phosphate isomerase genes. This proposal shows the divergence of prokaryotes and eukaryotes approximately 1.5 billion years ago, with plants and animals diverging approximately 0.5 billion years later.

Glyceraldehyde 3-phosphate dehydrogenase

Glyceraldehyde 3-phosphate dehydrogenase (D-glyceraldehyde 3-phosphate:NAD^+ oxidoreductase [phosphorylating], EC 1.2.1.12) catalyzes the only oxidative reaction occurring in the glycolytic pathway. NAD^+-specific glyceraldehyde 3-phosphate dehydrogenase has been purified from a variety of sources, both prokaryotic and eukaryotic, and in most instances the catalytic and physicochemical properties are very similar. The enzyme is a homotetramer made up of subunits with an M_r 34 000–37 000. Each subunit can bind one molecule of NAD^+. Maximal *in vitro* catalytic activity occurs at pH values of 8.5 to 9.0. Michaelis constants for glyceraldehyde 3-phosphate, NAD^+, and P_i are 0.2 mM, 0.15 mM, and 0.4 mM, respectively. There is a readily acylated reactive cysteine located within the active site. The first step in the reaction is the formation of a hemithioacetal between this cysteine and glyceraldehyde 3-phosphate. The carbonyl group is thus converted directly to an alcohol which is readily dehydrogenated. There have been many reports of tissue or developmental stage-specific isozymes of glyceraldehyde 3-phosphate dehydrogenase. Holland and associates have isolated and sequenced three distinct, non-tandemly repeated structural genes from bakers yeast (Holland et al., 1983). Expression of the isozymes is apparently related to external carbon nutrition. Primary sequence data are available for glyceraldehyde 3-phosphate dehydrogenases from several bacterial, fungal, and animal sources. The size of the subunits remains essentially constant at 333 ± 4 amino acids. Overall there is more than 40% homology in the sequences from *E. coli* to man, with nearly 80% homology in certain catalytic domains, and blocks of amino acids which are conserved in all sequences.

Plants can contain two additional forms of glyceraldehyde 3-phosphate dehydrogenase, one specific for $NADP^+$ (EC 1.2.1.13) which functions in the Calvin cycle, and a non-phosphorylating species. Neither of these forms will be considered here. The glycolytic NAD^+-glyceraldehyde 3-phosphate dehydrogenase has been purified and characterized from several plant seeds (pea, green gram, castor oil), and leaves (spinach, white mustard, tobacco). In all cases the properties of the plant enzymes are similar to those of other organisms. Steady-state kinetic studies of glyceraldehyde 3-phosphate dehydrogenases have led to the proposal of several different mechanisms. The generally accepted kinetic mechanism originally derived from the pea seed enzyme is a bi uni uni uni ping-pong with NAD^+ binding first followed by P_i; 1,3-bisphosphoglycerate is then released, and subsequently there is a Theorell–Chance displacement between glyceraldehyde 3-phosphate and NADH (Duggleby and Dennis, 1982). Shih et al. (1986) derived the amino acid sequence for tobacco leaf cytoplasmic NAD^+-glyceraldehyde 3-phosphate dehydrogenase from the nucleotide sequence of a cloned cDNA. This sequence has a higher homology with the *E. coli* or animal enzymes than it does with the chloroplast $NADP^+$ form. Their results indicate that $NADP^+$ and NAD^+-specific glyceraldehyde 3-phosphate dehydrogenases diverged before the plant–animal divergence, and provide support for the symbiotic origin of the chloroplast.

Phosphoglycerate kinase

Phosphoglycerate kinase (ATP:3-phospho-D-glycerate 1-phosphotransferase, EC 2.7.2.3) is a key enzyme for the generation of ATP during glycolysis. The PGA kinases from all sources thus far examined are monomeric proteins with M_r values around 45 000. Like most glycolytic enzymes the structure of phosphoglycerate kinase has been widely conserved. Early X-ray diffraction studies and later experiments employing immunochemical methods demonstrated the conservation between the yeast and mammalian enzymes. The homology, as high as 64%, has now been verified by both direct protein sequencing and nucleotide sequencing of cDNAs. Phosphoglycerate kinases have relatively broad pH optima from 6 to 9, and K_m values of 0.5 mM, 1.3 mM, 0.2 mM and 0.002 mM for MgATP, 3-PGA, MgADP, and 1,3-bisPGA, respectively. The enzyme appears to have an ordered sequential kinetic mechanism, and it has been suggested that one or more histidine residues participates in catalysis.

Phosphoglycerate kinase has been partially purified from pea seeds and developing castor oil seed endosperm, and purified to homogeneity from silver beet (Cavell and Scopes, 1976) and spinach leaves. Kuntz and Krietsch (1982) also reported purification of the enzyme from two cyanobacteria, *Spirulina platensis* and *Spirulina geitleri*. In general, the catalytic and physicochemical properties of phosphoglycerate kinases from the plant and cyanobacterial sources are very much like those from yeast or mammals. The pH optima for the higher plant enzymes were from 6.5 to 9.0. K_m values were 0.2 mM, 1 mM, 0.3 mM, and 0.0002 mM for MgATP, 3-PGA, MgADP, and 1,3-bisPGA, respectively. In some early reports an assay method which involved trapping of 1,3-bisPGA with NH_2OH was employed. When K_m values for MgATP were determined using these assays, the values were approximately 10-fold higher than those determined in enzymatically coupled assays. The divalent cation specificity of the beet leaf enzyme was $Mg^{2+}>Co^{2+}>Mn^{2+}>Ca^{2+}>Ni^{2+}>Zn^{2+}$. Levels of cations in excess of those required for adenylate binding were inhibitory. The nucleoside triphosphate specificity of the spinach leaf enzyme was ATP>>ITP=UTP>dATP (Kuntz and Krietsch, 1982). Other nucleoside triphosphates were not substrates. The M_r values for purified plant PGA kinases are from 46 000 to 48 000. There are as yet no primary sequence data for phosphoglycerate kinase from any higher plant source.

Phosphoglycerate mutase

There are two classes of phosphoglycerate mutases, those which catalyze an intermolecular phosphoryl group transfer, utilizing 2,3-bisphosphoglycerate as a cofactor (EC 2.7.5.3), and those which catalyze an intramolecular phosphoryl group transfer and have a cofactor-independent mechanism (EC 5.4.2.1) (Carreras et al., 1982). In both instances, however, a phosphoryl–enzyme intermediate is involved in the catalytic mechanism. All plant PGA mutases thus far examined have the 2,3-bisphosphoglycerate independent mechanism. There is as yet no sequence data for any plant PGA mutase, but when such information becomes available it will make an interesting comparison with the sequence of the cofactor-dependent enzyme.

Phosphoglycerate mutase has been purified to homogeneity from only three plant sources: wheat and rice germ, and developing *Ricinus communis* endosperm (Botha and Dennis, 1986), but within this small group there is considerable consistency in catalytic and physicochemical properties. In each instance the enzyme is monomeric with subunit M_r values from 62 000 to 64 000. It seems likely that plant phosphoglycerate mutases are metalloenzymes containing 1 mol of tightly bound divalent cation per mol of protein. Michaelis constants are 0.3–0.4 mM for 3-phosphoglycerate and 0.06–0.1 mM for 2-phosphoglycerate. In each instance optimum *in vitro* catalytic activity occurred at pH values from 7.0 to 7.5.

Enolase

Enolase (2-phospho-D-glycerate hydro-lyase, EC 4.2.1.11) is a metalloenzyme which catalyzes the freely reversible interconversion between 2-PGA and phosphoenolpyruvate (Brewer, 1981). Divalent cations are required both for maintenance of a conformation necessary for enzymatic activity, and for catalysis. Enolases from all sources thus far studied are dimeric proteins composed of identical subunits of approximately M_r 50 000. Enolases from most sources have a pH optimum in the forward direction around 7.5, and have Michaelis constants for 2-PGA of approximately 0.03 mM. Isozymes of enolase are common, being tissue or developmental stage-specific in mammalian systems, and differentially expressed in response to external carbon nutrition in microbial systems. Bakers yeast enolase, one of the most extensively studied glycolytic enzymes, is encoded by one of two non-tandemly repeated structural genes (Holland et al., 1981). The structural genes encode two proteins of 436 amino acids which are more than 95% homologous. A third, heterodimeric isozyme is formed during yeast cell disruption. There are at least three distinct isozymes of enolase in mammalian systems: a 'neuron-specific' form, and alpha–alpha and gamma–gamma homodimers. As with yeast, alpha–gamma heterodimers can be formed during cell disruption.

While enolase activity has been detected in homogenates from many plant tissues, the enzyme has been purified and characterized from only potato tubers, spinach leaves, and developing *Ricinus communis* endosperm (Miernyk and Dennis, 1984). The physicochemical properties of *Ricinus* cytoplasmic protein are similar to those of the much more thoroughly studied yeast enzyme. The pH optima are near 7.6 both in the direction of phosphoenolpyruvate formation and the back reaction. Michaelis constants are 0.025 to 0.06 mM and 2.5 to 7 mM for 2-phosphoglycerate and phosphoenolpyruvate, respectively. The *Ricinus* enzyme also requires magnesium for maximum activity and is inhibited by fluoride in the presence of inorganic phosphate.

Pyruvate kinase

Pyruvate kinase (ATP:pyruvate 2-*O*-phosphotransferase, EC 2.7.1.40) catalyzes the transfer of phosphate from phosphoenolpyruvate to ADP, yielding ATP and pyruvate. The reaction has a free energy change of approximately −7.5 Kcal which, coupled with very high K_m values for ATP and pyruvate, determines that the reaction is irreversible under biological conditions. Pyruvate kinase has been purified and extensively characterized from several mammalian sources. In general, pyruvate kinases are homotetramers (made up of subunits of M_r 56 000–60 000), have pH optima around 7, and have K_m values of 100–600 μM and 80–300 μM for ADP and PEP, respectively. Pyruvate kinases typically require both magnesium and potassium ions for maximum activity. In most mammalian systems there are tissue-specific isozymes of pyruvate kinase which have distinct kinetic and regulatory properties. Most data point towards a common genetic origin for the isozymes. Liver pyruvate kinase occupies a key position for the control of carbohydrate metabolism, and is subject to hormonal and metabolic regulation (Blair, 1980). Much of the hormonal regulation is by means of reversible phosphorylation catalyzed by a cAMP-dependent protein kinase. The phosphorylated enzyme is inactive. Much of the metabolic regulation is the function of effectors binding to allosteric sites and affecting the cooperativity of substrate interactions. The steady-state expression of liver pyruvate kinase mRNA and protein is also under long-term dietary and hormonal control. Except that there is no evidence for phosphorylation, the catalytic and physicochemical properties of fungal pyruvate kinases are in general similar to those of the mammalian liver isozyme (e.g. Hohn and Paznokas, 1987).

In contrast to the extensive research done on mammalian pyruvate kinase, relatively little has been reported on this important enzyme from plant sources. Some kinetic and regulatory properties have been reported for pyruvate kinase from cotton and castor oil seeds. There is no evidence for positive cooperativity in substrate, cofactor, or effector binding to plant pyruvate kinases, and Michaelis constants of 50 μM and 30–300 μM, for PEP and ADP, respectively, have been reported. All animal, fungal, and higher plant pyruvate kinases have an ordered sequential kinetic mechanism. Interestingly, Ireland *et al*. (1980) reported that while both isozymes of pyruvate kinase from developing *Ricinus* endosperm have an ordered sequential mechanism, the order of product release is different. For the cytosolic isozyme pyruvate is released before ATP, while ATP is released first by the plastid isozyme. The cytosolic isozyme of pyruvate kinase from developing castor oil seed endosperm has recently been purified to homogeneity (Plaxton, 1988). The native relative molecular mass of this isozyme is 240 000, in the same range as values reported for non-plant pyruvate kinases. When analyzed under denaturing conditions there appear to be two subunits, one of M_r 57 000 and the other of M_r 56 000. Subsequent work suggests that both these subunits exist *in vivo* implying a heterotetrameric structure (Plaxton, 1988).

Summary

As a pathway of such crucial importance, it is perhaps not surprising that the structures and catalytic mechanisms of most of the component enzymes of glycolysis are highly conserved across all phylogenetic boundaries. In most instances the glycolytic enzymes from plant cells are similar to the more thoroughly studied mammalian and yeast enzymes. It is interesting to note, however, that plant aldolases are more like the mammalian enzyme than the enzyme from fungi and bacteria, while plant phosphoglycerate mutases are dissimilar to both the bacterial and mammalian enzymes. There appear to be some significant differences in the regulation of glycolysis in plants. Some of these are addressed in the preceding chapter, and others later in this chapter.

The oxidative pentose phosphate pathway

Introduction

The pentose phosphate pathway is an alternative to glycolysis for the metabolism of glucose (Fig. 6.1). While this pathway is capable of the complete conversion of glucose to CO_2, it is probably rare that complete conversion occurs. In plants and some bacteria there are two pentose phosphate cycles, the so-called reductive and oxidative pathways. The reductive pentose phosphate pathway participates in the photosynthetic fixation of CO_2, and will not be further considered in this chapter (see Chapter 17). It is widely regarded that the primary role of the oxidative pentose phosphate pathway is the production of NADPH to be used in biosynthetic reactions. Wood (1986) has proposed another 'primary' function, that of providing pentose phosphates for the synthesis of nucleotides, nucleic acids, and cell wall polymers. Additionally, in plants and some microorganisms erythrose 4-phosphate can serve as a precursor of the aromatic amino acids. Thus, while metabolism of glucose through the pentose phosphate cycle is conceptually similar to metabolism through glycolysis, the primary roles, production of biosynthetic intermediates and reducing power, are substantially different.

Enzymes of the oxidative pentose phosphate pathway

Glucose 6-phosphate dehydrogenase

Glucose 6-phosphate dehydrogenase (D-glucose 6-phosphate:NADP oxidoreductase, EC 1.1.1.49) catalyzes the oxidation of glucose 6-phosphate to δ-glucono-1,5-lactone-6-phosphate with the concomitant reduction of $NADP^+$. While glucose 6-phosphate dehydrogenase is a key enzyme in the pentose phosphate pathway, it is also associated with the Entner-Doudoroff pathway in some bacteria. The subunit M_r values of microbial glucose 6-phosphate dehydrogenases are in the range of 50 000–60 000, while those of the mammalian enzyme are somewhat larger at 58 000–67 000. The minimum enzymatically active structure for glucose 6-phosphate dehydrogenase is a dimer, although tetrameric and higher order oligomeric forms have been reported (Levy, 1979). Catalytic activity is generally maximal around pH 7.4, with Michaelis constants of 0.86 mM and 0.042 mM for glucose 6-phosphate and $NADP^+$, respectively. The natural substrate, D-glucose 6-phosphate, gives the highest V_{max} and the lowest K_m values for all glucose 6-phosphate dehydrogenases thus far examined. Compounds with minor changes in structure can be oxidized (e.g. mannose 6-phosphate, glucose 6-sulfate), but always less well than the natural substrate. The stereospecificity of hydrogen transfer is from glucose 6-phosphate to the *si* face at carbon-4 of the nicotinamide ring. There have been many reports of multiple forms of glucose 6-phosphate dehydrogenase. In microbial systems these isozymes often appear in response to changes in external carbohydrate nutrition. Some of the 'isozymes' reported in mammalian systems may be conformational isomers.

There are, however, two true isozymes of glucose 6-phosphate dehydrogenase in plant cells, one located in the cytoplasm and the other in the plastids (e.g. Schnarrenberger *et al.*, 1973; Simcox and Dennis, 1978a). While there have been several reports of partial purification of the isozymes, they have been purified to homogeneity only from light-grown *Pisum sativum* seedlings. The pea cytoplasmic isozyme is a M_r 244 000 homotetramer with subunits of M_r 60 000 (Fickenscher and Scheibe, 1986). The K_m values for $NADP^+$ and glucose 6-phosphate are 14 and 120 μM, respectively, and the enzyme is absolutely specific for these substrates. Among a wide range of potential effectors tested, only NADPH had any significant effect upon catalytic activity with a K_i of 11 μM. Antibodies prepared against the cytoplasmic isozyme of glucose 6-phosphate dehydrogenase do not cross-react with the plastid form (Fickenscher and Scheibe, 1986). Srivastava and Anderson (1983) reported the purification and characterization of the pea chloroplast isozyme. Like the cytoplasmic form, the plastid isozyme is a homotetramer; however, both the native M_r (224 000) and the subunit M_r (56 000) are slightly smaller. The pH optimum of the chloroplast isozyme is around 8.2 and the kinetic constants for $NADP^+$, glucose 6-phosphate and NADPH are 2 μM, 370 μM, and 19 μM, respectively. The pea chloroplast enzyme is also absolutely specific for NADP, but there was a low level of activity with glucose 6-sulfate. Chloroplast glucose 6-phosphate dehydrogenase can be converted from an active disulfide form to an inactive, reduced sulfhydryl form. This mechanism is thought to form the basis for regulation of enzyme activity during the light to dark transition (Buchanan, 1980). There has not yet been any primary sequence data published for either isozyme of glucose 6-phosphate dehydrogenase from any plant source.

6-Phosphogluconate dehydrogenase

6-Phosphogluconate dehydrogenase (6-phospho-D-gluconate:NADP oxidoreductase [decarboxylating], EC 1.1.1.44) catalyzes the irreversible oxidative decarboxylation of 6-phosphogluconate yielding ribulose 5-phosphate plus CO_2, with the concomitant reduction of NADP. The enzyme from most sources is specific for $NADP^+$, and requires divalent cations for maximal activity. It is thought that 3-keto-6-phosphogluconate is a reaction intermediate, and that the mechanism of oxidative decarboxylation is similar to that of malic enzyme and isocitrate dehydrogenase. The enzyme from mammalian and fungal sources is a homodimer with native M_r values from 100 000 to 110 000 and subunit M_r values around 51 000. The K_m values range from 15 to 26 μM for NADP and 30 to 160 μM for 6-phosphogluconate, while K_i values are around 20 μM for NADPH. Through the use of group-specific directed inhibitors, it has been shown that 6-phosphogluconate dehydrogenase from *Candida* has both a cysteine and a lysine residue within the active site; however, modification of the lysine with pyridoxal phosphate does not affect catalytic activity.

Purification to homogeneity has been achieved for 6-phosphogluconate dehydrogenases from several plant tissues (e.g. Simcox and Dennis, 1978a; Al-Quadan *et al.*, 1981), and the physicochemical and catalytic characteristics of the enzymes from plant sources are similar to those of the mammalian and fungal enzymes.

There are multiple isozymes of 6-phosphogluconate dehydrogenase in plant cells (Schnarrenberger *et al.*, 1973; Simcox and Dennis, 1978b), with at least one isozyme present in the plastid and cytoplasmic compartments. Al-Quadan *et al.* (1981) compared the plastid and cytosolic isozymes from tobacco by peptide mapping, and while there were similarities each protein also had some unique peptides. There has been considerable genetic analysis of the isozymes, including the chromosomal localization of the genes encoding both plastid and cytoplasmic isozymes, and of allozymes and intergenic hybrid enzymes within both compartments (e.g. Hsam *et al.*, 1982; Tanksley and Kuehn, 1985). There has not yet, however, been any published report on the molecular genetics of plant 6-phosphogluconate dehydrogenase.

Transaldolase

Transaldolase (EC 2.2.1.2) catalyzes the freely reversible interconversion between sedoheptulose 7-phosphate plus glyceraldehyde 3-phosphate, and erythrose 4-phosphate plus fructose 6-phosphate. The mechanism of transaldolase, a base catalyzed aldol cleavage reaction in which a Schiff base intermediate is formed, is similar to that of class I aldolases (Tsolas and Horecker, 1972). In addition to the pentose phosphate pathway, the enzyme participates in the bacterial ribulose monophosphate cycle. Transaldolase from mammalian and some fungal species (e.g. *Candida utilis*) is a dimeric enzyme with native M_r values between 65 000 and 75 000, and subunit M_r values near 33 000. In other fungi (e.g. *Saccharomyces cerevisiae*) and bacteria, transaldolase is a monomeric protein with native and subunit M_r values from 50 000 to 65 000 (Levering and Dijkhuizen, 1986). In most cases the pH optimum for the transaldolase catalyzed reaction is around 8.0, with K_m values of 50 to 200 μM for sedoheptulose 7-phosphate, 200 μM for glyceraldehyde 3-phosphate, 300 to 800 μM for fructose 6-phosphate, and 20 to 60 μM for erythrose 4-phosphate. Inorganic phosphate inhibits enzyme activity, but only at higher than physiological concentrations (e.g. K_i of 50 mM for the *C. utilis* enzyme). There have been numerous reports of isozymes from both mammalian and microbial systems, the expression being related to external carbon nutrition in the cases of the microbes.

While transaldolase activity has been assayed in homogenates from several different tissues (Tsolas and Horecker, 1972), the plant enzyme has not yet been purified or characterized. The results from cell fractionation studies indicate that plants contain two isozymes of transaldolase, one cytoplasmic and the other within the plastids (Simcox *et al.*, 1977; Emes and Fowler, 1978).

Transketolase

Transketolase (EC 2.2.1.1) is a glycoaldehyde transferase, reversibly transferring the two carbon ketol moiety from sedoheptulose 7-phosphate and glyceraldehyde 3-phosphate to give ribose 5-phosphate plus xylulose 5-phosphate. Thiamine pyrophosphate and divalent cations are tightly bound cofactors required for catalytic activity. Both the mammalian and yeast enzymes are homodimers with a native M_r of 140 000 and a subunit M_r of 69 000 (Paoletti and Aldinucci, 1986). The mammalian enzyme has a pH optimum near 8.0 and kinetic constants of 25 μM and 66 μM for xylulose 5-phosphate and ribose 5-phosphate, respectively. Multiple isozymes of both mammalian and yeast transketolases have been reported, possibly having different substrate specificities (Paoletti and Aldinucci, 1986).

In photosynthetic tissues transketolase also participates in the reductive pentose phosphate pathway, and can additionally catalyze the interconversion between fructose 6-phosphate plus glyceraldehyde 3-phosphate to xylulose 5-phosphate plus erythrose 4-phosphate. Transketolase was purified to homogeneity from spinach leaves by Villafranca and Axelrod (1971), and more recently from spinach and wheat leaves by Murphy and Walker (1982). There are both cytoplasmic and plastid isozymes of transketolase in plant cells, although, in contrast to the distribution for other glycolytic and pentose phosphate pathway enzymes, it appears that the majority of activity is located within the plastids (e.g. Feierabend and Gringel, 1983). The chloroplast isozyme has a less alkaline pH optimum (7.5 to 8.5) than the cytoplasmic form (8.5–9.0), and the pH/activity relationship of the chloroplast isozyme is altered by magnesium while activity of the cytosolic isozyme is largely unaffected. Kinetic constants are around 100 μM for xylulose 5-phosphate, erythrose 4-phosphate, and ribose 5-phosphate. It appears that TPP binds less tightly to wheat leaf transketolase than to the mammalian or yeast enzymes. The native M_r is 150 000, while the subunit M_r, as determined by SDS-PAGE, is 37 600. Apparently the apo-monomer binds TPP and forms a catalytically active dimeric form which can then further associate to give the typical holo-homotetrameric structure (Murphy and Walker, 1982). There have not yet been any published reports of primary sequence or immunochemical data for transketolase from any plant source.

Pentose phosphate isomerase

Pentose phosphate isomerase (D-ribose ketol isomerase, EC 5.3.1.6) catalyzes the interconversion between ribose 5-phosphate and ribulose 5-phosphate. The interconversion is generally thought to proceed via an intramolecular proton transfer involving the formation of a cis-enediol anion intermediate, although detailed spectral analyses of the reaction are not entirely consistent with this mechanism (Knowles et al., 1980). The enzyme has been variously reported to exist in dimeric, trimeric, and tetrameric structures, with native M_r values ranging from 40 000 to 228 000 and subunit M_r values from 20 000 to 58 000 (Noltmann, 1972). The pH optima for the isomerase reaction range from 7.0 to 8.5, with K_m values of 0.5 to 5.0 mM for D-ribose 5-phosphate. Ribose 5-phosphate isomerase is inhibited by sulfhydryl reagents and non-physiological concentrations of glycolytic and pentose phosphate cycle intermediates and inorganic phosphate. The Rhodospirillum rubrum enzyme is inhibited by D-ribulose 1,5-bisphosphate with a K_m of 36 μM. Isozymes of ribose 5-phosphate isomerase have been reported from mammalian cells and microorganisms, generally with distinct physicochemical and catalytic properties.

Ribose 5-phosphate isomerase has been partially purified from several plant sources and in general the physicochemical and catalytic properties are similar to those from non-plant sources. The enzyme has been purified to homogeneity only from alfalfa shoots (Axelrod and Jang, 1954) and spinach leaf chloroplasts (Knowles et al., 1980). In both instances the enzymes are homodimers of subunit M_r 26 000 and native M_r 53 000, and have pH optima around 7.0 and K_m values for ribose 5-phosphate around 2 mM. Plant cells typically contain two isozymes of ribose 5-phosphate isomerase, one in the cytoplasm and the other within the plastids (e.g. Anderson, 1971). There is not yet any published primary sequence data or immunochemical analysis of ribose 5-phosphate isomerase from any plant source.

Pentose phosphate epimerase

Pentose phosphate epimerase (D-ribulose 5-phosphate 3-epimerase, EC 5.1.3.1) catalyzes the interconversion between ribulose 5-phosphate and xylulose 5-phosphate. The reaction mechanism is thought to proceed through an enediol intermediate, and the results from studies using stable isotopes indicate that the enzyme must have two hydrogen donor/acceptor sites which interact with the hydrogen atom at C3 of ribose 5-phosphate and xylulose 5-phosphate. The epimerase has been purified from several animal and microbial sources (Wood, 1985). It was reported that both the yeast and mammalian enzymes have a native M_r of 45 000, and that the mammalian protein is a homodimer made up of M_r 23 000 subunits. Maximal activity of the epimerase occurs in the range 7.0 to 8.0, with Michaelis constants ranging from 0.2 to 2.0 mM for ribulose 5-phosphate and xylulose 5-phosphate. The mammalian enzyme is irreversibly inactivated by reaction with sulfhydryl-directed inhibitors.

There have been a few published measurements of ribulose 5-phosphate epimerase activity in homogenates of plant tissues, but the enzyme has not yet been purified or characterized from any plant source.

The 'F-type' or classical pentose phosphate pathway versus the 'L-type' pathway

J. F. Williams and his associates were unable to reconcile the results of their experiments on the pentose phosphate pathway in rabbit liver with predictions based upon the classical F-type (*Fat* cell) pathway, and introduced an alternative, the L-type (*Liver*) pathway (reviewed by Williams *et al.*, 1987). While the L-type pathway can explain a number of apparent inconsistencies in results of metabolic studies, it has not yet been widely accepted and remains controversial (e.g. Scofield *et al.*, 1985).

Research on the pentose phosphate pathway in plant tissues has contributed relatively little to clarification of the F-type versus L-type controversy. In studying short-term metabolism in *Chlorella sorokiniana*, Heath (1984) reported a pattern of release of $^{14}CO_2$ from differentially labeled glucose which was inconsistent with the classical F-type pathway, but more in agreement with the proposed L-type formulation. The L-type results were found only in short-term experiments, being completely obscured by extension of incubation times to two minutes or more. Williams *et al.* (1987) have reinterpreted the labeling data of other workers as supporting the existence of the L-type pathway in plants; however, in no instance have the original authors addressed this interpretation of their data. Williams has additionally made reference to the occurrence in plant preparations of the two enzymes unique to the L-type pentose phosphate pathway, arabinose 5-phosphate ketol isomerase, which interconverts ribulose 5-phosphate and arabinose 5-phosphate, and the phosphotransferase, which participates in the synthesis of octulose 1,8-bisphosphate, in various reviews. There have not yet been any primary research publications on these enzymes from plant sources. In a detailed study of the compartmentation of the pentose phosphate pathway in the developing endosperm of *Ricinus communis*, a tissue with a very active pathway, attempts to measure activities of the ketol isomerase and phosphotransferase in clarified homogenates or isolated organelle fractions were unsuccessful (Miernyk *et al.*, in preparation). Thus, support for the L-type pentose phosphate pathway in plant tissues is meager. There have, however, been very few studies, and much additional research is necessary in order to fully resolve this issue.

Summary

While the pentose phosphate pathway operates within plant cells, it has received relatively little detailed attention. There are clearly two discreet pathways, one in the cytoplasm and the other within the plastids. Distinct isozymes are responsible for the activities in the two compartments. Of the pathway enzymes, only the two dehydrogenases have been extensively studied. It has been proposed that the L-type pathway may occur in plants but there is no conclusive evidence for this.

Anaerobic metabolism

Introduction

As obligate aerobes, higher plants are able to tolerate only short-term anaerobiosis. While plant mitochondria can survive anaerobic conditions for several days, in the absence of molecular oxygen as a terminal electron acceptor there can be no mitochondrial respiration. Thus, anaerobiosis induces a shift from normal respiration to the fermentative production of ethanol and lactic acid.

In most plant cells ethanol and CO_2 are the main products of anaerobic metabolism, with smaller amounts of lactic acid formed. If NADH, produced during the glyceraldehyde 3-phosphate dehydrogenase reaction, is not reoxidized, then glycolysis will be completely inhibited. Under anaerobic conditions pyruvate can be converted either to lactate, by lactate dehydrogenase, or ethanol, by pyruvate decarboxylase plus alcohol dehydrogenase, with a concomitant oxidation of NADH. Maintenance of a portion of the total nicotinamide nucleotides in the oxidized state allows some degree of substrate-level phosphorylation and ATP production.

Unlike higher animals, plant glycolysis during anaerobiosis results primarily in ethanol production. It then seems likely that lactate dehydrogenase plays a different role in plants than it does in animals. It has been proposed that in the early stages of oxygen depletion constitutive lactate dehydrogenase activity results in the accumulation of lactate. The accumulated lactate lowers the cytoplasmic pH and activates pyruvate decarboxylase, thus triggering ethanolic glycolysis. The proposed pH fluctuations have been verified by ^{31}P-NMR studies (Roberts, 1985). In addition to the constitutive enzyme activity, however, many plants produce hypoxically-induced lactate dehydrogenase. This suggests some additional role for lactate dehydrogenase in the survival of long-term anaerobiosis.

Enzymes of anaerobic metabolism

Lactate dehydrogenase

Lactate dehydrogenase (L-(+)-lactate:NAD^+ oxidoreductase, EC 1.1.1.27) converts pyruvate to lactate using NADH as the reductant. The enzyme has been purified and characterized from numerous mammalian sources. Typically, mammalian lactate dehydrogenases are tetrameric enzymes composed of M_r 35 000 subunits. The *in vitro* catalytic activity in the direction of pyruvate reduction is maximal at pH values from 6.0 to 7.0 while the pH optimum is 8.0 to 9.5 in the direction of lactate oxidation. Michaelis constants for NADH and pyruvate are 0.01 mM and 0.10 mM, respectively. Lactate dehydrogenase is not specific for pyruvate and will reduce glyoxylate and other α-keto acids. There have been numerous reports of the tissue-specific expression of lactate dehydrogenase isozymes in mammalian systems.

There appear to be two types of lactate dehydrogenases in plant tissues, one constitutive and the other induced by oxygen deficit. Not all plants have the constitutive activity, but if it is present there does not seem to be any tissue specificity in expression. The inducible lactate dehydrogenases are most commonly reported in preparations from roots, developing or imbibed seeds, and large, bulky storage tissues. The catalytic and physicochemical properties of plant lactate dehydrogenases are similar to those from mammalian tissues. The pH optima for *in vitro* activity are 7.0 in the direction of pyruvate reduction and from 8.4 to 9.5 in the direction of lactate oxidation. Michaelis constants for pyruvate, NADH, lactate, and NAD are 0.18–0.72 mM, 0.01–0.08 mM, 11–39 mM, and 0.2–0.76 mM, respectively. All plant lactate dehydrogenases appear to be tetrameric proteins with native M_r values of 150 000–157 000 and subunit M_r values of 37 500–40 000. Isozymes of lactate dehydrogenase are common in plant tissues, the number of electrophoretic forms ranging from one to six (e.g. Boyle and Yeung, 1983). In most cases it appears that the products of two structural genes yield five heterotetrameric isozymes. While plant and animal lactate dehydrogenases are similar on the basis of subunit size and catalytic properties, further comparisons will require determination of the complete primary sequence of the plant enzyme. A highly conserved region among lactate dehydrogenases is the 'Arg$_6$' peptide, derived from the substrate binding site of the enzyme. Mayr et al. (1982) have isolated and sequenced the Arg$_6$ peptide from potato tuber lactate dehydrogenase and reported a high degree of homology with the animal and bacterial sequences, supporting the proposal of a common ancestral enzyme.

Pyruvate decarboxylase

A key enzyme in alcoholic fermentation is pyruvate decarboxylase (2-oxoacid carboxylase, EC 4.1.1.17) which catalyzes the irreversible decarboxylation of pyruvate yielding acetaldehyde and CO_2. Magnesium and thiamine pyrophosphate are required cofactors. This enzyme, which is not present in animal tissues, was first discovered in yeast. The yeast enzyme is a homotetramer with a native M_r of 230 000 and a subunit M_r of 60 000. Maximum in vitro catalytic activity occurs at pH values around 6.5. At alkaline pH values the decarboxylase dissociates into catalytically inactive dimers. Thiamine pyrophosphate is required for reassociation into the active tetrameric form. Initial rate studies at varying pyruvate concentrations resulted in sigmoidal kinetics. It has been reported that inorganic phosphate has a negative homotropic cooperative effect upon pyruvate binding, increasing the $S_{0.5}$ value for pyruvate from 1 mM to 10 mM, and increasing the Hill coefficient from 1.4 to 2.1. Both the dissociation/reassociation and the phosphate inhibition could be important in the regulation of ethanolic glycolysis.

Pyruvate decarboxylase has been purified from wheat germ, sweet potato roots, and maize kernels and roots (Lee and Langston-Unkefer, 1985). Both the native M_r of 240 000, and the subunit M_r of 60 000, are similar to the yeast enzyme. The in vitro pH optimum is 6.1 to 6.6. There is also cooperativity in pyruvate binding, with Hill coefficients of 1.3 to 3.2. It was initially reported that inorganic phosphate was a negative effector of pyruvate binding, but was later reported that potassium ions, rather than phosphate, were inhibitory. When the yeast protein is assayed in vitro there is a pronounced lag in pyruvate decarboxylase activity, and the hysteretic activation of the enzyme could correspond to the reassociation/activation of the enzyme. There is not yet any primary sequence data for pyruvate decarboxylase from any plant source.

Alcohol dehydrogenase

In the final reaction of alcoholic fermentation, alcohol dehydrogenase (alcohol:NAD oxidoreductase, EC 1.1.1.1) catalyzes the further reduction of the acetaldehyde produced by the pyruvate decarboxylase reaction, NADH providing the reducing equivalents. The equilibrium position of this reaction is very pH dependent. The reduction of acetaldehyde is favored at pH 7 while the oxidation of ethanol is favored at pH 9.5. Alcohol dehydrogenases from animals, fungi, and insects have been purified and extensively studied (Branden et al., 1975). In all cases the enzyme is a homodimer with subunits of M_r 32 000–38 000. Michaelis constants are 0.075 mM, 0.15 mM, 13.5 mM, and 30 mM for NAD, NADH, ethanol, and acetaldehyde, respectively. Alcohol dehydrogenase has a relatively sharp pH optimun of 8.5 in the direction of acetaldehyde reduction, and a broad optimum of 8.0 to 10.5 in the direction of ethanol oxidation. Most, if not all, alcohol dehydrogenases are metalloenzymes, often containing 2 or more moles of zinc per mole of enzyme. There are several isozymes which are expressed in a tissue or developmental stage-specific fashion.

Alcohol dehydrogenase from maize is the most thoroughly studied of the plant enzymes. The maize enzyme is a homodimer with subunits of M_r 40 000. The pH optimum is 8.5 in the direction of acetaldehyde reduction and 8.5 to 10 in the direction of ethanol oxidation. The Michaelis constants are 0.025–0.075 mM, 0.085 mM, 8–14 mM and 30–50 mM for NAD, NADH, ethanol, and acetaldehyde, respectively. There are two alcohol dehydrogenase structural genes in the maize genome, *Adh1* and *Adh2*. In terms of structure, expression, and control, maize *Adh1* is one of the best understood genes in higher organisms (Freeling and Bennett, 1985). The product of the *Adh1* gene is the major form in seeds, seedlings, and pollen, but is absent from mature leaves. The product of the *Adh2* gene is expressed at a lower level in germinating seedlings. When exposed to anaerobiosis, the synthesis of three isozymes of alcohol dehydrogenase is induced. The three isozymes are due to the random dimerization of ADH1 and ADH2 subunits. Zn^{2+} is required for the dimerization. While ADH2 gene is highly homologous to ADH1, the catalytic efficiency of ADH2 is much lower. The amino acid sequences of ADH1 and ADH2 have been derived from the nucleotide sequences of the cloned cDNAs, and there is an 87% homology in the amino acid sequences. The sequence of ADH1 is 50% homologous to horse liver alcohol dehydrogenase, but only 20% homologous with that of the yeast enzyme. These data suggest a common evolutionary origin for animal and plant *Adh* genes, and a more distant relationship with yeast.

The anaerobic proteins

Plants may be classified as either anoxia-tolerant or anoxia-intolerant, based upon responses to continuous anaerobic conditions. Generally, tolerant plants are those which survive continuous anoxia in the dark for extended periods (up to a week) and then are able to resume growth when re-exposed to air. Specialized metabolism is necessary in order to tolerate the lack of molecular oxygen. Changes in carbohydrate metabolism may be crucial in surviving anaerobiosis. Hypoxic 'training' of maize roots confers an improved energy status and tolerance of anaerobic conditions and is linked to induction of an effective ethanolic fermentation pathway. In nature, O_2 depletion is gradual, over periods of hours or days, allowing acclimation. Induction of synthesis of the anaerobic proteins (ANPs) is part of acclimation. When plant cells are exposed to anaerobic conditions there is a rapid dissociation of polysomes, a reduction in total protein synthesis, and a major shift in the pattern of polypeptides synthesized (Fig. 6.2). Maize roots are the plant system in which the effects of anoxia upon protein synthesis have been most thoroughly studied. During the first hour after the initiation of anaerobiosis there is a shift from the translation of aerobic messages to the synthesis of a few transition proteins. This is followed during the second hour by a further shift to the exclusive synthesis of approximately 20 polypeptides, the ANPs (Sachs *et al.*, 1980). This behavior is in many ways analogous to the heat-shock response (Sachs and Ho, 1986).

A few of the ANPs have now been identified through a combination of protein chemistry and molecular genetics. Those identified include sucrose synthase (EC 2.4.1.13), hexose phosphate isomerase (Kelley and Freeling, 1984a), aldolase (Kelley and Freeling, 1984b), pyruvate decarboxylase, and alcohol dehydrogenase isozymes 1 and 2. It seems likely that other ANPs will be enzymes involved in carbohydrate metabolism. The activities of phosphoglucomutase (EC 5.4.2.6), NAD^+-linked glyceraldehyde 3-phosphate dehydrogenase, phosphoglycerate mutase, enolase, and lactate dehydrogenase increase under anaerobic conditions (J. Bailey-Serres, B. Kloeckener-Gruissem and M. Freeling, unpublished data, 1988). Given the homology between heat shock and the anaerobic response, it would seem that at least some heat-shock proteins would be glycolytic enzymes. However, in only one case has this been demonstrated. Enolase has been unequivocally identified as a heat-shock protein in yeast (Iida and Yahara, 1985).

Interestingly, activities of the key regulatory glycolytic enzymes hexokinase, phosphofructokinase, and pyruvate kinase do not increase during anaerobiosis. The glycolytic ANPs are located in the cytoplasm, and may be gene products distinct from the normal respiratory glycolytic enzymes. A similar response to anaerobiosis has been found in other plant tissues and in bacteria. While providing an excellent system for studying the coordinated expression of a family of genes, it remains as yet unclear how synthesis of the ANPs might directly increase the probability of surviving anoxia.

Cytosolic Carbon Metabolism

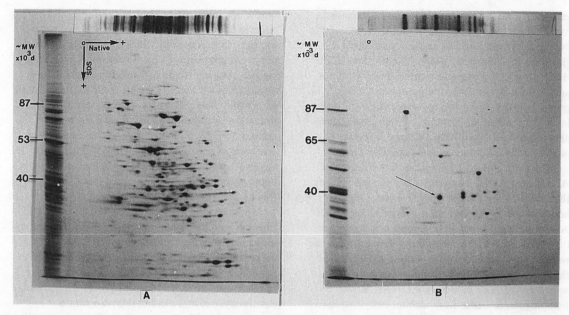

Fig. 6.2 Fluorographs of native/SDS two-dimensional polyacrylamide gels of maize seedling primary root proteins labelled during incubation with [^{35}S]-methionine. (A), Incubation with methionine for one hour under aerobic conditions; (B), incubation with methionine during the 12th to 17th hours of anaerobiosis (argon atmosphere). The arrow indicates the position of the ADH1 subunit (unpublished data from M. M. Sachs, 1988).

Summary

A well-defined biochemical sequence allows higher plants to survive short-term anaerobiosis. The decrease in pH resulting from the activity of constitutive lactate dehydrogenase serves to activate pyruvate decarboxylase, allowing ethanolic fermentation. This supports the reoxidation of NADH. Anaerobiosis causes an inhibition of normal protein synthesis, and stimulates the synthesis of a small family of anaerobic proteins. All of the anaerobic proteins thus far identified are enzymes of carbohydrate metabolism.

Interrelationships between pathways

Introduction

Glycolysis and the pentose phosphate pathway cannot be considered as separate, independent entities. The enzymes of the two pathways are co-localized within the same cellular compartments, and there are several intermediates, glucose 6-phosphate, fructose 6-phosphate, and glyceraldehyde 3-phosphate, common to both pathways (Fig. 6.1). There is the additional confounding point that there is often substantial gluconeogenesis at the same time, and in the same subcellular compartment, as glycolysis and the pentose phosphate pathway. Carbohydrate metabolism should then be viewed as a continuously variable, interactive process.

It has long been considered that the ratio of flux through glycolysis relative to flux through the pentose phosphate pathway is an indicator of the physiological state of a tissue. A relatively high flux through the pentose phosphate pathway is considered to support a high rate of biosynthesis in an actively growing or differentiating tissue. Quantitation of the flux through the two pathways is most commonly evaluated by the release of $^{14}CO_2$ from specifically labeled glucose. If only glycolysis is functioning in carbohydrate breakdown then there should be an equal release of $^{14}CO_2$ from the C1 and C6 positions of glucose. Any release of $^{14}CO_2$ from C1 in excess of that from C6 is considered to result from flux through the pentose phosphate pathway. For example, using this strategy Thomas and ap Rees (1972) were able to quantitate the contribution of the pentose phosphate pathway to overall carbohydrate metabolism during germination of *Cucurbita* seeds. In this instance it was observed that while the pentose phosphate pathway was operative at all times, glycolysis was the predominant pathway during early post-germinative growth.

Control of the partitioning of carbon through the two pathways comes primarily through regulation of the early reactions. If the activity of phosphofructokinase is inhibited *in vitro*, then more carbon would be diverted through the pentose phosphate pathway. Conversely, if the activities of glucose 6-phosphate and 6-phosphogluconate dehydrogenase are inhibited, then there will be an increase in flux through glycolysis. Possible mechanisms for regulation of these enzyme activities are discussed in another section.

Summary

Glycolysis and the pentose phosphate pathway, as well as gluconeogenesis (in part the reversal of glycolysis), operate at the same time and within the same subcellular compartment, resulting in a complex interactive relationship. The relative contribution of each to the whole is controlled through the regulation of a few key enzymes.

Organization and compartmentation

Introduction

It has long been considered that carbohydrate metabolism proceeds via non-associated enzymes localized in the cytoplasm. It is becoming increasingly clear that this is not true in mammalian cells, and it is certainly not the case in plants. Two converging lines of evidence support the compartmentation of glycolysis in animal and microbial cells. Firstly, there have been several reports of particulate glycolytic activity, and of the specific association of glycolytic enzymes with supramolecular structures (e.g. Plaxton and Storey, 1986). Secondly, there are numerous examples where the co-immobilization of sequential glycolytic enzymes has led to coupled kinetic behavior which much more closely mimics tissue glycolysis than do the reactions occurring in dilute aqueous solution (e.g. De Luca and Kricka, 1983). Advantages of enzyme association into what has been termed 'metabolons' include greater efficiency through the channeling of substrates between sequential enzymes, and the potential for regulation of overall activity through alteration of the degree of association. It remains to be shown that the association of glycolytic enzymes with one another, and with other cell structures, takes place *in situ*, but it is obvious that this possibility must be considered.

Localization

There are examples of the specific association of glycolytic enzymes with organellar fractions in mammalian cells (e.g. the ambiquitous association of an isozyme of hexokinase with the mitochondrial outer membrane). There are also rare examples of organelles specialized for glycolysis. In trypanosomes there is a peroxisome-like organelle, the glycosome, which contains the glycolytic enzymes. Plant cells are unique, however, in that they contain two discrete, spatially separated glycolytic and pentose phosphate pathways, one in the cytoplasm and the other in the plastids (reviewed by Dennis and Miernyk, 1982). The role of plastid glycolysis is different from that of the classical cytoplasmic sequence, and is directed toward the production of pyruvate. In plant cells *de novo* fatty acid synthesis takes place exclusively in the plastids (Ohlrogge *et al.*, 1979). The pyruvate produced by plastid glycolysis is converted to acetyl-CoA by an 'isozyme' of the pyruvate dehydrogenase complex (Miernyk *et al.*, 1985), and this acetyl-CoA then serves as the substrate for fatty acid synthetase. In addition to fatty acid synthesis, the plastid pentose phosphate pathway provides NADPH for other biosynthetic activities such as nitrogen assimilation (Emes and Fowler, 1983).

Plastid glycolysis has been most thoroughly characterized in non-green tissues, where the demand for pyruvate is higher than in leaves. Complete glycolytic pathways have been detailed in plastids isolated from developing castor oil (Simcox et al., 1977), mustard (Liedvogel and Bäuerle, 1986), wheat (Entwistle and ap Rees, 1988) and maize (Miernyk et al., in preparation) seeds, and cauliflower florets (Journet and Douce, 1985). In a few reported instances, isolated plastids lack the activities of one or more glycolytic enzymes. One such example is the lack of phosphoglycerate mutase activity in isolated pea seedling chloroplasts (Stitt and ap Rees, 1979). It was subsequently shown that the amyloplasts from developing pea seeds do contain phosphoglycerate mutase (Denyer and Smith, 1988), so it is not yet clear whether the lack of enzyme activity in pea leaves is an artefact, or whether it represents some tissue or developmental stage-specific difference in metabolism.

It has been recently reported that a significant portion of the total activities of the pentose phosphate pathway enzymes in mammalian tissues is located within the endoplasmic reticulum (ER) (Bublitz and Steavenson, 1988). While animal cells lack plastids, it could be that the ER serves in some instances as the metabolic equivalent. There has not yet been any detailed examination of the ER from plant cells for the possible occurrence of glycolytic or pentose phosphate pathway activities.

Isozymes

Two or more separable proteins which catalyze the same reaction are by definition isozymes (reviewed by Markert, 1975). This original description of the term isozyme was intended as operational only, and not as a definition of the basis of molecular heterogeneity. After a system has been substantially characterized, 'isozymes' can then be appropriately modified through the use of terms such as allelic, epigenetic, conformational, etc.

Plant cells contain two separate glycolytic and

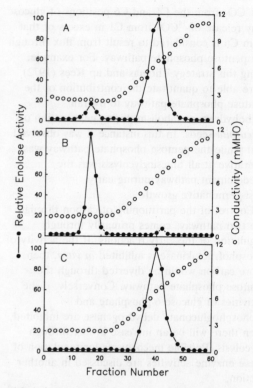

Fig. 6.3 Anion-exchange chromatography of a clarified homogenate (A); a purified chloroplast fraction (B); and a cytosol fraction (C) prepared from young, rapidly expanding *Ricinus communis* leaves and assayed for enolase activity.

pentose phosphate pathways, one in the cytoplasm and the other within the plastids. In each instance the catalytic activities in the two compartments can be attributed to distinct isozymes (see Fig. 6.3) (Dennis and Miernyk, 1982). The differences in molecular and catalytic properties between the pairs of isozymes range from the relatively minor (e.g. phosphoglycerate mutase, Botha and Dennis, 1986; 6-phosphogluconate dehydrogenase, Simcox and Dennis, 1978b), through substantial (e.g. enolase, Miernyk and Dennis, 1984), to major (e.g. ATP-phosphofructokinase [EC 2.7.1.11], Garland and Dennis, 1980). In many instances the differences between the isozymes are an evolutionary response to the differences between metabolism within the cytoplasm and that of the plastids.

There are many potential mechanisms for the synthesis of multiple isozymes. These range from the relatively simple scenario where two distinct structural genes give rise to two distinct enzymes, through the differential processing of primary transcripts, to the complex post-translational modification of a single transcript. There have been examples of each of these mechanisms reported for the glycolytic isozymes from various non-plant organisms (e.g. Noguchi et al., 1986). To date there is only a single instance where there is sufficient data to distinguish among the various possibilities for the existence of multiple forms of a plant carbohydrate metabolizing enzyme. Gottlieb and associates recently reported distinct structural genes for the plastid and cytosolic isozymes of hexose phosphate isomerase (EC 5.3.1.9) from *Clarkia* (Tait et al., 1988). It is obviously premature, however, to assume that all plant isozymes can be explained by the expression of distinct structural genes.

Synthesis and assembly

Although the compartmentation of glycolysis and the pentose phosphate pathway within plastids has been firmly established, virtually nothing is known with regard to the synthesis and assembly of the organellar components. While plastids contain their own unique DNA, the plastome is too small to encode all of the proteins which are found in this class of organelles. In a few instances plastid-localized proteins are synthesized through cooperation between the nuclear and plastid genomes. The best understood example of this is ribulose 1,5-bisphosphate carboxylase/oxygenase. The large, catalytic subunit of this enzyme is encoded by the plastome and synthesized within the organelle. The small subunit is encoded by the nuclear genome, synthesized in the cytoplasm as a high molecular weight precursor, recognized by the plastid, taken up post-translationally, proteolytically processed to the mature size, and finally assembled into the holoenzyme (see Chapters 16 and 20). For the majority of proteins, synthesis and import into the plastid is probably analogous to the pathway followed by the small subunit of ribulose bisphosphate carboxylase.

Preliminary evidence, from genetic studies and the use of mutants, is consistent with the encoding of all of the glycolytic and pentose phosphate pathway enzymes by the nuclear genome. Unequivocal proof of a nuclear location awaits the isolation of specific structural genes. If the plastid carbohydrate metabolizing enzymes are nuclear in origin, it will be interesting to see how the interactions between the transit peptides/receptors/processing peptidases for these proteins compare with those of ribulose bisphosphate carboxylase small subunit and the photosynthetic membrane proteins. It should be noted that the activities of carbohydrate metabolizing enzymes in the cytoplasm are generally much different from the corresponding plastid activities. A sophisticated regulatory mechanism is obviously necessary so that not only are the proper enzymes directed to the correct final location, but also in appropriate quantities.

Summary

With regard to plant metabolism, there is no validity to the outmoded textbook generalization that carbohydrate metabolism results from the activities of non-associated, soluble, cytoplasmic enzymes. There is an increasing body of evidence for organization of cytoplasmic glycolytic enzymes into a metabolon. The activities of carbohydrate metabolizing enzymes are located both within the cytoplasm and plastids of plant cells. In all cases the differentially compartmentalized activities are due to distinct isozymes. The results of initial experiments are most consistent with a nuclear location of multiple structural genes encoding the isozymes.

Regulation

Introduction

In order to respond to external stimuli, there must be mechanisms for altering the flux through a metabolic pathway. Enzymology can define the kinetic constants for the individual steps, and this can provide some insight into the overall control. This reductionist approach does not, however, take into account major variables such as substrate availability and changes in the cellular milieu. More appropriate approaches such as measurements of flux control coefficients are covered in Chapter 3. Additionally, the extremely important concept of metabolic regulation via compartmentation is addressed in detail in Chapter 4.

Acute regulation

A number of techniques have been used in studying the potential regulation of glycolysis in plant cells, ranging from the measurement of mass–action ratios for substrates, cofactors, and effectors, through rapid shifts from aerobic to anaerobic conditions, to theoretical considerations based upon thermodynamic constants. The consensus, based upon results from each of these types of experiments, is that short-term, acute regulation of glycolysis occurs at the reactions catalyzed by hexokinase (EC 2.7.1.1), phosphofructokinase, and pyruvate kinase. The kinetic and regulatory properties of hexokinase and phosphofructokinase are addressed in Chapter 5. While it seems obvious that there must be regulation of the activity of pyruvate kinase in plant cells, the mechanism of this regulation remains, for the most part, elusive. There is no convincing general evidence in plants for reversible phosphorylation, or cooperative substrate and effector binding, strategies which dominate the regulation of mammalian pyruvate kinase (Blair, 1980). There have been suggestions of metabolite regulation by citrate, 2-oxoglutarate, and ATP, but these proposals appear inadequate for true *in vivo* control. The first really interesting data on the regulation of pyruvate kinase activity in plants comes from recent studies of the green alga *Selenastrum minutum*.

In higher plants and algae, ammonium enrichment shifts the flow of newly fixed carbon into amino acids and intermediates of mitochondrial metabolism, with a concurrent reduction in the flow to sucrose and starch. An increase in pyruvate kinase activity has been implicated in this shift (Elrifi and Turpin, 1987). A plastid and cytosolic isozyme of pyruvate kinase has recently been separated and purified from *S. minutum* (Lin et al., 1989a). In general, the physicochemical characteristics of the isozymes are typical of those of eukaryotes. The *in vitro* activity of the cytoplasmic isozyme is inhibited by inorganic phosphate and glutamate and stimulated by dihydroxyacetone phosphate (Lin et al., 1989b). The effects of inorganic phosphate and dihydroxyacetone phosphate seem to reciprocally antagonize one another. Inhibition by inorganic phosphate is linear competitive with respect to phosphoenolpyruvate, and appears to induce negative cooperativity in substrate binding, with the Hill coefficient increased to near two (Fig. 6.4). Addition of 0.5 mM dihydroxyacetone phosphate to the reaction prevents the phosphate inhibition and shifts the Hill coefficient back to one (Lin et al., 1989b). These results are the first clear demonstration of allosteric behavior of pyruvate kinase from a plant source. The results from the combined *in vivo* and *in vitro* studies of

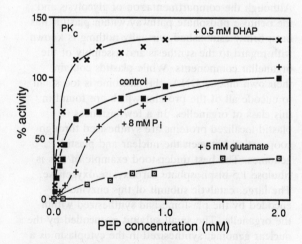

Fig. 6.4 Activity of the cytoplasmic isozyme of pyruvate kinase from *S. minutum* as a function of phosphoenolpyruvate concentration at fixed concentrations of inorganic phosphate (P_i), dihydroxyacetone phosphate (DHAP) and glutamate. Maximum activity is 17.8 μmol min^{-1}. At saturating concentrations of ADP, incubation with 8 mM P_i shifts the Hill coefficient from 1.0 to 1.8 (from Lin et al., 1989b).

pyruvate kinase from *S. minutum* provide some insight into the regulation of plant pyruvate kinases, but much additional research will be necessary before there is any clear understanding of this important control point.

While studies on regulation of the oxidative pentose phosphate pathway have not been nearly as extensive as those on glycolysis, there is a similar consensus in results. Short-term acute regulation of the pentose phosphate pathway occurs at the reactions catalyzed by the first two enzymes, glucose 6-phosphate dehydrogenase and 6-phosphogluconate dehydrogenase. In general, regulation of the activities of the two dehydrogenases is via product inhibition by NADPH. Inhibition is simple linear competitive with respect to NADP, with both K_m values for NADP and K_i values for NADPH approximately 15 μM. In photosynthetic tissues there is light-mediated regulation of glucose 6-phosphate dehydrogenase activity via an oxidation/reduction cycle (Buchanan, 1980). Transfer of reducing equivalents is probably via the ferredoxin/thioredoxin couple, although alternatives, such as the Light Effect Mediator, have been proposed. There are few data supporting metabolite regulation of the pentose phosphate pathway enzymes, although it has been demonstrated that hexose bisphosphates can inhibit the *in vitro* activity of 6-phosphogluconate dehydrogenase (e.g. Miernyk *et al.*, 1984).

Long-term regulation

While there is some evidence for long-term developmental regulation of glycolytic and oxidative pentose phosphate pathway activities in plants (Ireland and Dennis, 1981; Botha and Dennis, 1986), the mechanism(s) of this regulation are unknown. Presumably control of both protein synthesis and degradation is involved. Developmentally expressed promoter sequences have been described as part of the genes for seed storage proteins, and it will be of considerable interest to see if similar sequences are part of the flanking regions of genes for the plastid enzymes involved in carbohydrate metabolism.

Summary

The consensus points of short-term, acute regulation of plant carbohydrate metabolism, hexokinase, phosphofructokinase, pyruvate kinase, glucose 6-phosphate dehydrogenase, and 6-phosphogluconate dehydrogenase, are the same as those of mammalian and microbial systems. In some instances, such as regulation of the pentose phosphate pathway by the product inhibitor NADPH, the mechanism of regulation is the same. In others, such as the light-driven oxidative activation of chloroplast glucose 6-phosphate dehydrogenase and the regulation by compartmentation in the plastids as well as cytoplasm, the regulatory mechanisms are unique to plants. While there have been some indications of long-term developmental regulation of the enzymes of plant carbohydrate metabolism, there is as yet no information on the mechanism(s) of this regulation.

Primary sequence modifications

Introduction

The folding of a linear polymer polypeptide into a highly efficient catalyst occurs in response to the physical properties of the component amino acids. The most efficient conformation can then be stabilized by salt bridges, disulfide bonds, etc. There is, however, relatively little flexibility in the information contained within the primary sequence. In complex organisms there must be, among other things, mechanisms by which proteins can be targeted to specific locations, by which catalytic activities can be regulated, and by which protein lifetimes can be controlled. Much of the responsiveness necessary for efficient protein function is the result of post-synthetic primary sequence modifications.

N-terminal modifications

The initiation of protein synthesis in all living cells is dependent upon the assembly of the translational apparatus on mRNA at a specific codon for methionine. In plastids, mitochondria, and bacteria the initiator methionine is formylated. The N-formyl group is subsequently removed by a formate lyase. In the cytoplasm of eukaryotic cells protein synthesis is initiated at an unmodified methionine residue. The initiator methionine can be removed from the nascent polypeptide by a methionyl aminopeptidase, but this removal is not necessary.

The enzymes of the plastid glycolytic and pentose phosphate pathways are most likely encoded by the nuclear genome, synthesized within the cytoplasm, recognized, and imported post-translationally by the organelle. Synthesis of precursor forms of proteins with an N-terminal transit sequence is typical for proteins compartmentalized in this fashion. The transit sequence is then removed after import by a specific endopeptidase, yielding the mature N-terminal sequence. There are some examples of N-terminal modification of proteins encoded by the plastome (e.g. the 33K herbicide-binding protein), but there have not been any reports of further N-terminal modifications of nuclear encoded plastid proteins.

N-acylation is a common post-translational modification of cytoplasmic proteins, with acetate being the most common acyl group added. Serine, alanine and methionine residues are the N-terminal amino acids most commonly acetylated. N-terminal glycine is seldom acetylated, but can be acylated with myristic acid. Structural features other than only the N-terminal residue are required for recognition by the acetyltransferases. Other N-terminal modifications (i.e. cyclization yielding pyroglutamate residues) are rare relative to acylation.

Reversible covalent modifications

One of the most common post-translational protein modifications is phosphorylation, and the most common function of this modification is the modulation of catalytic activity. There is a considerable body of literature concerning phosphorylation of mammalian glycolytic enzymes. While it has been firmly established that mammalian phosphofructokinase can be phosphorylated *in vivo*, this modification does not appreciably alter the *in vitro* catalytic characteristics and the role of this phosphorylation is not well understood. Both *in vivo* and *in vitro* phosphorylation of other glycolytic enzymes (e.g. glyceraldehyde 3-phosphate dehydrogenase, enolase, phosphoglycerate mutase) from various mammalian tissues have been reported (Reiss *et al.*, 1986; Nettelblad and Engstrom, 1987; Ashmarina *et al.*, 1988). There are also recent reports of the phosphorylation of glycolytic enzymes (e.g. aldolase, enolase) in bacteria (Babul and Fraenkel, 1988; Dannelly and Reeves, 1988). In many instances there is a decrease in maximal catalytic activity upon phosphorylation. There are also reports of altered rates of protein turnover in response to phosphorylation. It is not yet apparent, however, how the phosphorylation of these proteins can be integrated into our overall understanding of carbohydrate metabolism. In contrast, regulation of mammalian liver pyruvate kinase by reversible phosphorylation has been studied intensively and is well understood (Blair, 1980).

Relatively little research has been done with regard to phosphorylation of plant glycolytic enzymes. It has been suggested that *Ricinus* endosperm cytoplasmic enolase might be phosphorylated, and this phosphorylation could be a signal for proteolytic degradation (Miernyk and Dennis, unpublished observation). In recent experiments, homogeneous cytosolic pyruvate kinase from germinating castor bean endosperm was phosphorylated, *in vitro*, by an endogenous protein kinase (Plaxton, unpublished results). This finding raises the possibility that plant pyruvate kinases may be at least partially regulated, *in vivo*, by the reversible incorporation of covalently bound phosphate.

At this time there have not been any published reports of phosphorylation of enzymes of the pentose phosphate pathway from any organism.

There are many covalent protein modifications other than phosphorylation, for example, glycosylation, methylation, sulfation, etc. In most instances these modifications serve either to directly modulate catalytic activity, or to indirectly affect activity by altering protein half-life. In some instances it has been proposed that these modifications play some role in the targeting of proteins from the site of synthesis to the final location within the cell. There are not any reliable reports of these types of modification to the carbohydrate metabolizing enzymes.

Other primary sequence modifications

There have been reports of physiologically important post-synthetic proteolytic modifications of glycolytic enzymes (e.g. Rigiani *et al.*, 1987) in mammalian systems. These modifications apparently result from interaction with a carboxypeptidase-type activity in serum. Modification of human enolase in this fashion results in significant changes in the catalytic and physicochemical properties of the enzyme.

Summary

In recent years there has been extensive study of the modification of primary translation products. A myriad of modifications can occur both co- and post-translationally, including N- and C-terminal processing and covalent modification. To date there have been no reports of any primary sequence modifications to any plant carbohydrate metabolizing enzyme, other than anecdotal references to 'N-terminal blockage'. It is likely that modification of plant enzymes does occur, but has not been described because of the lack of research in this area.

Prospectus

Given the importance of carbohydrate metabolism, and the length of time that the basic pathways have been known, it is somewhat surprising that our understanding of glycolysis and the pentose phosphate pathway in plant cells is so meager. There is clearly a need for much additional study, and it seems obvious that a multi-disciplinary approach would be the most productive. I hope that some of the areas deserving of future research effort are apparent from this review. There is much basic physiological research which must be done in order to understand interactions between carbohydrate metabolism and other metabolic pathways, and the role(s) of carbohydrate metabolism in plant growth and development. Relatively few of the enzymes of carbohydrate metabolism have been purified and characterized from plant sources, and future efforts by protein chemists and enzymologists will add to our overall understanding. While there have been preliminary descriptions of the subcellular and tissue-specific localization of glycolytic and pentose phosphate pathway isozymes, understanding the underlying mechanisms awaits the efforts of future plant cell biologists. Finally, the rapid development of techniques in molecular biology provides us all with powerful new tools. Before there can be any realistic efforts towards the genetic engineering of higher plants there must be a more thorough understanding of what should be changed. This understanding will come only from additional basic research into plant metabolism.

References

Al-Quadan, F., Wender, S. H., and Smith, E. C. (1981). Comparison of tryptic peptide maps of two isoenzymes of 6-phosphogluconate dehydrogenase from tobacco tissue cultures. *Phytochemistry* **20**, 1201–3.

Anderson, L. E. (1971). Chloroplast and cytoplasmic enzymes III. Pea leaf ribose 5-phosphate isomerases. *Biochim. Biophys. Acta* **235**, 237–44.

Ashmarina, L. I., Louzenko, S. E., Severin Jr, S. E., Muronetz, V. I. and Nagradova, N. K. (1988). Phosphorylation of D-glyceraldehyde 3-phosphate dehydrogenase by Ca^{2+}/calmodulin-dependent protein kinase II. *FEBS Lett.* **231**, 413–16.

Axelrod, B. and Jang, R. (1954). Purification and properties of phosphoriboisomerase from alfalfa. *J. Biol. Chem.* **209**, 847–55.

Babul, J. and Fraenkel, D. G. (1988). Phosphate modification of fructose-1,6-bisphosphate aldolase in *Escherichia coli*. *Biochem. Biophys. Res. Commun.* **151**, 1033–8.

Blair, J. B. (1980). Regulatory properties of hepatic pyruvate kinase. In *The Regulation of Carbohydrate Metabolism in Mammals*, ed. C. Veneziale, University Park Press, Baltimore, pp. 121–51.

Botha, F. C. and Dennis, D. T. (1986). Isozymes of phosphoglyceromutase from the developing endosperm of *Ricinus communis*: Isolation and kinetic properties. *Arch. Biochem. Biophys.* **245**, 96–103.

Boyle, S. A. and Yeung, E. C. (1983). Embryogeny of *Phaseolus*: developmental pattern of lactate and alcohol dehydrogenases. *Phytochemistry* **22**, 2413–16.

Branden, C-I., Jornvall, H., Eklund, H. and Furugren, B. (1975). Alcohol dehydrogenases. In *The Enzymes*, 3rd edn, Vol 11A, ed. P. D. Boyer, Academic Press, New York, pp. 103–90.

Brewer, J. M. (1981). Yeast enolase: mechanism of activation by metal ions. *CRC Crit. Rev. Biochem.* **11**, 209–54.

Bublitz, C. and Steavenson, S. (1988). The pentose phosphate pathway in the endoplasmic reticulum. *J. Biol. Chem.* **263**, 12849–53.

Buchanan, B. B. (1980). Role of light in the regulation of chloroplast enzymes. *Ann. Rev. Plant Physiol.* **31**, 341–74.

Carreras, J., Mezquita, J., Bosch, J., Bartrons, R. and Pons, G. (1982). Phylogeny and ontogeny of the phosphoglycerate mutases – IV. Distribution of glycerate-2,3-P_2 dependent and independent phosphoglycerate mutases in algae, fungi, plants and animals. *Comp. Biochem. Physiol.* **71B**, 591–7.

Cavell, S. and Scopes, R. K. (1976). Isolation and characterization of the 'photosynthetic' phosphoglycerate kinase from *Beta vulgaris*. *Eur. J. Biochem.* **63**, 483–90.

Dannelly, H. K. and Reeves, H. C. (1988). In vivo phosphorylation of enolase from *Escherichia coli*. *Abstr. Annu. Meeting Amer. Soc. Microbiol.* K-43.

De Luca, M. and Kricka, L. J. (1983). Coimmobilized multienzymes: an *in vitro* model for cellular processes. *Arch. Biochem. Biophys.* **226**, 285–91.

Dennis, D. T. and Miernyk, J. A. (1982). Compartmentation of nonphotosynthetic carbohydrate metabolism. *Ann. Rev. Plant Physiol.* **33**, 27–50.

Denyer, K. and Smith, A. M. (1988). The capacity of plastids from developing pea cotyledons to synthesize acetyl CoA. *Planta* **173**, 172–82.

Duggleby, R. G. and Dennis, D. T. (1982). Glyceraldehyde-3-phosphate dehydrogenase from pea seeds. *Methods Enzymol.* **89**, 319–25.

Elrifi, I. R. and Turpin, D. H. (1987). The path of carbon flow during NO_3-induced photosynthetic suppression in N-limited *Selenastrum minutum*. *Plant Physiol.* **83**, 97–104.

Emes, M. J. and Fowler, M. W. (1978). Location of transketolase and transaldolase in apical cells of pea roots. *Biochem. Soc. Trans.* **6**, 203–5.

Emes, M. J. and Fowler, M. W. (1983). The supply of reducing power for nitrite reduction in plastids of seedling pea roots (*Pisum sativum* L.). *Planta* **158**, 97–102.

Entwistle, G. and ap Rees, T. (1988). Enzymatic capacities of amyloplasts from wheat (*Triticum aestivum*) endosperm. *Biochem. J.* **255**, 391–6.

Feierabend, J. and Gringel, G. (1983). Plant transketolases: subcellular distribution, search for multiple forms, site of synthesis. *Z. Pflanzenphysiol.* **110**, 247–58.

Fickenscher, K. and Scheibe, R. (1986). Purification and properties of the cytoplasmic glucose-6-phosphate dehydrogenase from pea leaves. *Arch. Biochem. Biophys.* **247**, 393–402.

Freeling, M. and Bennett, D. C. (1985). Maize *Adh1*. *Ann. Rev. Genet.* **19**, 297–323.

Garland, W. J. and Dennis, D. T. (1980). Plastid and cytosolic phosphofructokinases from the developing endosperm of *Ricinus communis*. II. Comparison of the kinetic and regulatory properties of the isoenzymes. *Arch. Biochem. Biophys.* **204**, 310–17.

Heath, R. L. (1984). A new type of hexose monophosphate shunt in *Chlorella sorokiniana*. $^{14}CO_2$ release from differentially labeled glucose. *Plant Physiol.* **75**, 964–7.

Heil, J. A. and Lebherz, H. G. (1978). 'Hybridization' between aldolase subunits derived from mammalian and plant origin. *J. Biol. Chem.* **253**, 6599–605.

Hohn, T. M. and Paznokas, J. L. (1987). Purification and properties of two isozymes of pyruvate kinase from *Mucor racemosus*. *J. Bacteriol.* **169**, 3525–30.

Holland, M. J., Holland, J. P., Thill, G. P. and Jackson, K. A. (1981). The primary structures of two yeast enolase genes. *J. Biol. Chem.* **256**, 1385–95.

Holland, J. P., Labieniec, L., Swimmer, C. and Holland, M. J. (1983). Homologus nucleotide sequences at the 5' termini of messenger RNAs synthesized from the yeast enolase and glyceraldehyde-3-phosphate dehydrogenase gene families. *J. Biol. Chem.* **258**, 5291–9.

Hsam, S. L. K., Zeller, F. J. and Huber, W. (1982). Genetic control of 6-phosphogluconate dehydrogenase (6-PGD) isoenzymes in cultivated wheat and rye. *Theor. Appl. Genet.* **62**, 317–20.

Iida, H. and Yahara, Y. (1985). Yeast heat-shock protein of M_r 48 000 is an isoprotein of enolase. *Nature* **315**, 688–90.

Ireland, R. J. and Dennis, D. T. (1981). Isoenzymes of the glycolytic and pentose-phosphate pathway during the development of the castor oil seed. *Can. J. Bot.* **59**, 1423–5.

Ireland, R. J., De Luca, V. and Dennis, D. T. (1980). Characterization and kinetics of isozymes of pyruvate kinase from developing castor bean endosperm. *Plant Physiol.* **65**, 1188–93.

Journet, E. and Douce, R. (1985). Enzyme capacities of isolated cauliflower bud plastids for lipid synthesis and carbohydrate metabolism. *Plant Physiol.* **79**, 458–67.

Kelley, P. M. and Freeling, M. (1984a). Anerobic expression of maize glucose phosphate isomerase I. *J. Biol. Chem.* **259**, 673–7.

Kelley, P. M. and Freeling, M. (1984b). Anaerobic expression of maize fructose-1,6-bisphosphate aldolase. *J. Biol. Chem.* **259**, 14180–3.

Kelley, P. M. and Tolan, D. R. (1986). The complete amino acid sequence for the anaerobically induced aldolase from maize derived from cDNA clones. *Plant Physiol.* **82**, 1076–80.

Knowles, F. C., Chanley, J. D. and Pon, N. G. (1980). Spectral changes arising from the action of spinach chloroplast ribosephosphate isomerase on ribose 5-phosphate. *Arch. Biochem. Biophys.* **202**, 106–15.

Kuntz, G. W. K. and Krietsch, W. K. G. (1982). Phosphoglycerate kinase from spinach, blue–green algae, and yeast. *Methods Enzymol.* **90**, 110–14.

Lee, T. C. and Langston-Unkefer, P. J. (1985) Pyruvate decarboxylase from *Zea mays* L. I. Purification and partial characterization from mature kernels and anaerobically treated roots. *Plant Physiol.* **79**, 242–7.

Levering, P. R. and Dijkhuizen, L. (1986). Regulation and function of transaldolase isozymes involved in sugar and one-carbon metabolism in the ribulose monophosphate cycle methylotroph *Arthrobacter* P1. *Arch. Microbiol.* **144**, 116–23.

Levy, H. R. (1979). Glucose-6-phosphate dehydrogenases. *Adv. Enzymol.* **48**, 97–192.

Liedvogel, B. and Bäuerle, R. (1986). Fatty acid synthesis in chloroplasts from mustard (*Sinapis alba* L.) cotyledons: formation of acetyl coenzyme A by intraplastid glycolytic enzymes and a pyruvate dehydrogenase complex. *Planta* **169**, 481–9.

Lin, M., Turpin, D. H. and Plaxton, W. C. (1989a). Pyruvate kinase isozymes from the green alga *Selenastrum minutum*. I. Purification and physical and immunological characterization. *Arch. Biochem. Biophys.* **269**, 219–27.

Lin, M., Turpin, D. H. and Plaxton, W. C. (1989b). Pyruvate kinase isozymes from the green alga *Selenastrum minutum*. II. Kinetic and regulatory properties. *Arch. Biochem. Biophys.* **269**, 228–38.

Marchionni, M. and Gilbert, W. (1986). The triosephosphate isomerase gene from maize: introns antedate the plant–animal divergence. *Cell* **46**, 133–41.

Markert, C. L. (1975). Biology of isozymes. In *Isozymes*, Vol. 1, Molecular Structure, ed. C. L. Markert, Academic Press, New York, pp. 1–9.

Mayr, U., Hensel, R. and Kandler, O. (1982). Subunit composition and substrate binding region of potato L-lactate dehydrogenase. *Phytochemistry* **21**, 627–31.

Miernyk, J. A. and Dennis, D. T. (1984) Enolase isozymes from *Ricinus communis*: partial purification and characterization of the isozymes. *Arch. Biochem. Biophys.* **233**, 643–51.

Miernyk, J. A., Camp, P. J. and Randall, D. D. (1985). Regulation of plant pyruvate dehydrogenase complexes. *Curr. Topics Plant Biochem. Physiol.* **4**, 175–90.

Miernyk, J. A., MacDougal, P. S. and Dennis, D. T. (1984). *In vitro* inhibition of the plastid and cytosolic isozymes of 6-phosphogluconate dehydrogenase from developing endosperm of *Ricinus communis* by fructose 2,6-bisphosphate. *Plant Physiol.* **76**, 1093–4.

Murphy, D. J. and Walker, D. A. (1982). The properties of transketolase from photosynthetic tissue. *Planta* **155**, 316–20.

Nettelblad, F. A. and Engstrom, L. (1987). The kinetic effects of *in vitro* phosphorylation of rabbit muscle enolase by protein kinase C. *FEBS Lett.* **214**, 249–52.

Noguchi, T., Inoue, H. and Tanaka, T. (1986). The M1- and M2-type isozymes of rat pyruvate kinase are produced from the same gene by alternative RNA splicing. *J. Biol. Chem.* **261**, 13807–12.

Noltmann, E. A. (1972). Aldose-ketose isomerases. In *The Enzymes*, 3rd edn, Vol. 6, ed. P. D. Boyer, Academic Press, New York, pp. 271–354.

Ohlrogge, J. B., Kuhn, D. N. and Stumpf, P. K. (1979). Subcellular localization of acyl carrier protein in leaf protoplasts of *Spinacia oleracea*. *Proc. Natl Acad. Sci. USA* **76**, 1194–8.

Paoletti, F. and Aldinucci, D. (1986). Immunoaffinity purification of rat liver transketolase: evidence for multiple forms of the enzyme. *Arch. Biochem. Biophys.* **245**, 212–19.

Pichersky, E. and Gottlieb, L. D. (1983). Evidence for duplication of the structural genes coding plastid and cytosolic isozymes of triose phosphate isomerase in diploid species of *Clarkia*. *Genetics* **105**, 421–36.

Pichersky, E., Gottlieb, L. D. and Higgins, R. C. (1984). Hybridization between subunits of triose phosphate isomerase isozymes from different subcellular compartments of higher plants. *Mol. Gen. Genet.* **193**, 158–61.

Plaxton, W. C. (1988). Purification of pyruvate kinase from germinating castor bean endosperm. *Plant Physiol.* **86**, 1064–9.

Plaxton, W. C. and Storey, K. B. (1986). Glycolytic enzyme binding and metabolic control in anaerobiosis. *J. Comp. Physiol.* **156**, 635–40.

Reiss, N., Kanety, H. and Schlessinger, J. (1986). Five enzymes of the glycolytic pathway serve as substrates for purified epidermal-growth-factor-receptor kinase. *Biochem. J.* **239**, 691–7.

Rigiani, N. R., Wevers, R. A., Rijk, E. and Soons, J. B. (1987). Postsynthetic modification of human enolase isozymes. *Clin. Chem.* **33**, 757–60.

Roberts, J. K. M. (1985). Use of high resolution ^{31}P nuclear magnetic resonance spectroscopy to study metabolism in living plant tissues. *Curr. Topics Plant Biochem. Physiol.* **4**, 207–17.

Sachs, M. M. and Ho, T-H. D. (1986). Alteration of gene expression during environmental stress in plants. *Ann. Rev. Plant Physiol.* **37**, 363–76.

Sachs, M. M., Freeling, M. and Okimoto, R. (1980).

The anaerobic proteins of maize. *Cell* **20**, 761–7.

Schnarrenberger, C., Oser, A. and Tolbert, N. E. (1973). Two isozymes each of glucose-6-phosphate dehydrogenase and 6-phosphogluconate dehydrogenase in spinach leaves. *Arch. Biochem. Biophys.* **154**, 438–48.

Scofield, R. F., Kosugi, K., Chandramouli, V., Kumaran, K., Schumann, W. C. and Landau, B. R. (1985). The nature of the pentose pathway in liver. *J. Biol. Chem.* **260**, 15439–44.

Shih, M-C., Lazar, G. and Goodman, H. M. (1986). Evidence in favour of the symbiotic origin of chloroplasts: primary structure and evolution of tobacco glyceraldehyde-3-phosphate dehydrogenases. *Cell* **47**, 73–80.

Simcox, P. D. and Dennis, D. T. (1978a). Isoenzymes of the glycolytic and pentose phosphate pathways in proplastids from the developing endosperm of *Ricinus communis* L. *Plant Physiol.* **61**, 871–7.

Simcox, P. D. and Dennis, D. T. (1978b). 6-Phosphogluconate dehydrogenase isoenzymes from the developing endosperm of *Ricinus communis* L. *Plant Physiol.* **62**, 287–90.

Simcox, P. D., Reid, E. E., Canvin, D. T. and Dennis, D. T. (1977). Enzymes of the glycolytic and pentose phosphate pathways in proplastids from the developing endosperm of *Ricinus communis* L. *Plant Physiol.* **59**, 1128–32.

Srivastava, D. K. and Anderson, L. E. (1983). Isolation and characterization of light- and dithiothreitol-modulatable glucose-6-phosphate dehydrogenase from pea chloroplasts. *Biochim. Biophys. Acta* **724**, 359–69.

Stitt, M. and ap Rees, T. (1979). Capacities of pea chloroplasts to catalyze the oxidation pentose phosphate pathway and glycolysis. *Phytochemistry* **18**, 1905–11.

Straus, D. and Gilbert W. (1985). Chicken triose-phosphate isomerase complements an *Escherichia coli* deficiency. *Proc. Natl Acad. Sci. USA* **82**, 2014–18.

Tait, R. C., Froman, B. E., Laudencia-Chingcuanco, D. L. and Gottlieb, L. D. (1988). Plant phosphoglucose isomerase genes lack introns and are expressed in *Escherichia coli*. *Plant Mol. Biol.* **11**, 381–8.

Tanksley, S. D. and Kuehn, G. D. (1985). Genetics, subcellular localization, and molecular characterization of 6-phosphogluconate dehydrogenase isozymes in tomato. *Biochem. Genet.* **23**, 441–54.

Thomas, S. M. and ap Rees, T. (1972). Glycolysis during gluconeogenesis in cotyledons of *Cucurbita* pepo. *Phytochemistry* **11**, 2187–94.

Tsolas, O. and Horecker, B. L. (1972). Transaldolase. In *The Enzymes*, 3rd edn, Vol. 7, ed. P. D. Boyer, Academic Press, New York, pp. 259–80.

Villafranca, J. J. and Axelrod, B. (1971). Heptulose synthesis from nonphosphorylated aldoses and ketoses by spinach transketolase. *J. Biol. Chem* **246**, 3126–31.

Williams, J. F., Arora, K. K. and Longenecker, J. P. (1987). The pentose pathway: a random harvest. Impediments which oppose acceptance of the classical (F-type) pentose cycle for liver, some neoplasms and photosynthetic tissue. The case for the L-type pentose pathway. *Int. J. Biochem.* **19**, 749–817.

Wood, T. (1985). *The Pentose Phosphate Pathway*, Academic Press, New York.

Wood, T. (1986). Physiological functions of the pentose phosphate pathway. *Cell Biochem. Funct.* **4**, 241–7.

Mitochondrial Metabolism

III

Mitochondrial Metabolism

7 Mitochondrial structure
William Newcomb

Introduction

Mitochondria are organelles with a double membrane that occur in all eukaryotic cells. Mitochondria are the sites of oxidative phosphorylation, the process in which energy released by the oxidation of pyruvate, fatty acids and other compounds is coupled to the phosphorylation of ADP to form ATP. Mitochondria can usually be easily distinguished from small plastids because the inner mitochondrial membrane is invaginated, forming long narrow lobes called cristae. In addition, the outer membrane of mitochondria usually stains very faintly in contrast to that of plastids. Two distinct compartments are present in mitochondria (Fig. 7.1). These are the intermembrane spaces, located between the inner and outer mitochondrial membranes, and the matrix, which is located within the confines of the inner membrane. Some authors also distinguish the intracristal space which lies within the invaginations of the cristae and thus is an extension of the intermembrane space. In contrast, three distinct compartments are present within chloroplasts.

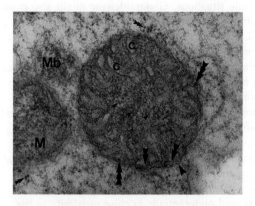

Fig. 7.1 Transmission electron micrograph of a mitochondrion in an infected cell within a root nodule of *Myrica gale*. Shown are the outer mitochondrial membrane (single large arrows), inner mitochondrial membrane (double large arrows), cristae (c), mitochondrial ribosomes (single small arrows), cytoplasmic ribosomes (double small arrows) and electron-dense particles (presumably calcium phosphate) (triple large arrows). Also shown are a portion of another mitochondrion (M), a microbody (Mb) and a small vacuole (V). × 43 200.

Shape, distribution and numbers

Mitochondria typically appear as rod-shaped organelles, 0.5 μm × 1–2 μm, in most transmission electron micrographs of eukaryotic cells. Certainly, in higher plant cells most mitochondria appear to be rod-shaped with hemispherically-shaped ends. However, this may not always be the case. For example, some plant mitochondria appear to be cup- or filament-shaped. Phase-contrast microscopic studies show that mitochondria change shape, varying from globular to threadlike or branched. Observations of certain animal cells with high voltage transmission electron microscopy have shown that the mitochondria can have a long filamentous morphology. In certain algae (*Chlorella* and *Chlamydomonas*) the mitochondria form a single branched organelle, the mitochondrial reticulum. Ultrathin sections of these cells usually show small profiles of circular or rod-shaped mitochondria. However, three-dimensional reconstructions using micrographs of serial ultrathin sections of these cells revealed the mitochondrial reticulum. It is generally believed that mitochondrial reticula are infrequent in higher plant cells because profiles of branched mitochondria are rarely observed in ultrathin sections of botanical specimens. It is nevertheless important to realize that glutaraldehyde may cause mitochondria to swell and this should be taken into consideration in certain situations. Glutaraldehyde might also cause mitochondria to fragment.

Unlike the situation in some animal cells, there does not appear to be a particular spatial arrangement of mitochondria within the cytoplasm of higher plant cells. Transfer cells which contain numerous wall ingrowths believed to facilitate short distance transport of solutes have numerous mitochondria, presumably for the energy requirements of solute movement across the plasma membrane. While the mitochondria appear to be more numerous in the portion of the cytoplasm near the wall ingrowths, they are not arranged preferentially in close proximity to the plasma membrane bounding the cell wall ingrowths. In some plant cells mitochondria lie near chloroplasts, and in living cells mitochondria can be seen moving toward and away from larger plastids, suggesting that this movement might be associated with an exchange of metabolites. In some plant cells strands of ER surround mitochondria and plastids; again this association may indicate an exchange of solutes between these organelles.

The number of mitochondria per unit volume of cytoplasm appears to remain similar in different developmental stages of the same cell type even though the total number of mitochondria in the cell increases. For example, in young cells of the central zone of the *Zea mays* root cap, about 200 mitochondria are present in each cell; in the mature cells that are greatly enlarged about 2000 to 3000 mitochondria are present. In very active cells such as transfer cells, companion cells of the phloem and secretory cells in nectaries, a large fraction of the cytoplasm (up to *c*. 20%) may be occupied by mitochondria. Some tiny flagellated algae (e.g. *Micromonas*) may have only one mitochondrion per cell. *Chlorella* has one mitochondrial reticulum per cell, as does the yeast *Saccharomyces*. The alga *Chlamydomonas* has 10–15 branched mitochondria per cell.

Outer membrane

The inner and outer membranes are distinct in their appearance. The outer membrane is smooth and occasionally appears to be connected to profiles of smooth ER. In addition, the outer membrane contains less protein and more lipid that the inner membrane. The outer membrane is permeable to all molecules of 10 000 daltons or less because of the presence of a large channel-forming protein.

Inner membrane

The inner membrane, which consists of an unusually high amount of protein (70% protein and 30% lipid by weight), is invaginated forming cristae and contains particles on its inner surface which bounds the mitochondrial matrix. The cristae increase the surface area of the inner membrane. Negative staining has revealed spherically-shaped particles of ~9 nm diameter. These particles are not apparent in freeze-fracture electron microscopic studies, as the fracture plane most likely passes through the middle of the inner membrane rather than through the surface where the particles are located. The factors or particles have been isolated and shown to have the ability to break down ATP. They were originally referred to as the coupling factor because *in vivo* they are thought to couple the energy release during electron transport to the synthesis of ATP. These particles are now known as ATP synthetase (also called F_0F_1ATPase) and constitute about 15% of the total inner membrane protein. This protein complex contains a transmembrane proton channel. When protons flow down the channel according to their electrochemical potential, ATP is synthesized. Other integral components of the inner membrane include: succinic dehydrogenase, the only tricarboxylic acid cycle enzyme that is tightly bound into the membrane; the NADH dehydrogenase complex which accepts electrons carried by NADH from tricarboxylic acid cycle enzymes located in the matrix; ubiquinone, which receives hydrogen atoms from succinic dehydrogenase and the NADH dehydrogenase complex; the bc_1 complex that accepts electrons from ubiquinone and transfers them to cytochrome; and the cytochrome oxidase complex which transfers electrons from cytochrome c to oxygen (see Chapter 9).

Mitochondrial matrix

The mitochondrial matrix is enclosed by the inner mitochondrial membrane and consists of a ground substance of fine particles: nucleoids are electron transparent regions containing fine fibrils (mitochondrial DNA) which are digestible by deoxyribonuclease; mitochondrial ribosomes which are smaller than cytoplasmic ribosomes; and electron-dense particles that consist largely of calcium phosphate. Usually more than one nucleoid is present per mitochondrion but there is at least one exception, *Beta vulgaris* (beet) leaf mitochondria, which contain only one nucleoid per mitochondrion. Many of the fine particles within the matrix are likely enzymes associated with the oxidation of pyruvate and fatty acids, the tricarboxylic acid cycle (see Chapter 8), and the expression of mitochondrial DNA.

Mitochondrial development

Mitochondria are self-replicating organelles and only rise by division from pre-existing mitochondria. The numbers of mitochondrion per cell along with mitochondrial fine structure may be indicative of the rates of cellular metabolic activity. For example, root cortical cells have only a few mitochondria, while transfer cells contain many. In addition, the number of cristae per mitochondrion may change during development. In the *Arum* flower spadix the number of cristae per mitochondria increases as the spadix develops and the respiration rate rises.

Further reading

Alberts, B., Bray, D., Lewis, J., Raff, M., Roberts K. and Watson, J. D. (1989). *Molecular Biology of the Cell*, 2nd edn, Garland Publishing Inc., New York.

Gunning, B. E. S. and Steer, M. W. (1975). *Ultrastructure and the Biology of Plant Cells*, Edward Arnold, London.

8 Carbon metabolism in mitochondria
T. ap Rees

Introduction

The citric acid cycle (also termed the tricarboxylic acid (TCA) cycle and the Krebs cycle) is the dominant component of carbon metabolism in the mitochondria of higher plants. This section will concentrate entirely on this cycle and its relationship to the rest of metabolism. Other important aspects of mitochondrial metabolism, such as photorespiration, are dealt with in other chapters. At the outset it must be appreciated that plant metabolism is so closely integrated that we divide it at our peril, because any subdivision of it will lack essential elements of the whole. Division according to intracellular location is traditional but dangerous, as so many pathways involve complex interactions between two or more components of the plant cell. The case for treating mitochondrial metabolism separately rests on the fact that the whole of the citric acid cycle is confined to the mitochondria and on the crucial and universal role that the cycle plays in plant metabolism.

Any account of the citric acid cycle begins with pyruvate, although it is not a component of the cycle. This is done because pyruvate is the major link with the preceding pathway, glycolysis, and because early versions of the cycle envisaged pyruvate reacting with oxaloacetate. In 1937 Krebs inserted 'pyruvic acid (?)' into the cycle, and three years later wrote that 'we can now omit the query'. Krebs has subsequently argued that his demonstration that suspensions of minced flight muscle of pigeons produced citrate from oxaloacetate was his main contribution to the elucidation of the cycle as this 'one crucial fact provided the links between all the other observations' (Krebs, 1981). This is a very modest assessment compared to the brilliance and importance of Krebs' recognition that the sequence operates as a catalytic cycle, and the subsequent insight that this brought to our understanding and investigation of biochemistry as a whole.

Reactions of the citric acid cycle

The reactions of the cycle are shown in Fig. 8.1.

Enzymes

A detailed knowledge of the mechanism of each reaction and the properties of each enzyme should form the basis of our understanding of the cycle. Reaction mechanisms have barely been studied with plant enzymes. As these mechanisms are unlikely to vary between organisms, then we may rely on studies of the mechanisms of enzymes from other organisms. Investigations of the properties of plant enzymes lag behind those made of the cycle enzymes from animals and microbes. As regulation of the cycle is likely to vary with the organism, we may not assume that the plant enzymes have the same regulatory properties as the mammalian or microbial enzymes. There is an urgent need to extend the excellent example of the work of Randall and his colleagues on plant pyruvate dehydrogenase to the plant enzymes of the cycle itself.

Carbon metabolism in mitochondria

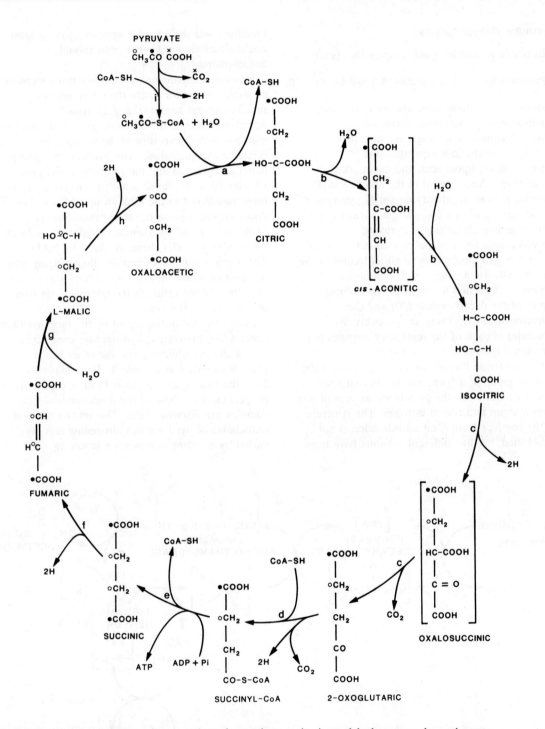

Fig. 8.1 The citric acid cycle: a, citrate synthase; b, aconitase; c, isocitrate dehydrogenase; d, oxoglutarate dehydrogenase; e, succinate-CoA ligase; f, succinate dehydrogenase; g, fumarate hydratase; h, malate dehydrogenase. The role of pyruvate dehydrogenase in providing acetyl-CoA is illustrated in reaction i.

Pyruvate dehydrogenase

This is a large multienzyme complex that catalyzes

Pyruvate + NAD^+ + CoA → acetyl-CoA + NADH + H^+ + CO_2

The complex contains pyruvate decarboxylase, dihydrolipoamide acetyltransferase and dihydrolipoamide dehydrogenase with their respective prosthetic groups, thiamine pyrophosphate, lipoic acid, and flavin adenine dinucleotide. Also present in the mitochondrial complex are two regulatory enzymes, pyruvate dehydrogenase kinase and phosphopyruvate dehydrogenase phosphatase. Pyruvate dehydrogenase from animal mitochondria has an M_r of $7-9 \times 10^6$ and a core of 60 molecules of the acetyltransferase arranged with icosahedral symmetry. Attached to this core are multiple copies of the decarboxylase (20) and the dehydrogenase (5). There are probably five molecules of each of the regulatory enzymes per complex. Lipoic acid is bound to the acetyltransferase through an amide bond with the ε-amino group of a lysine residue so that lipoic acid is connected to the protein via an arm of nine carbon atoms and one of nitrogen. The structure of the complex from plant mitochondria is not established, but five different subunits have been identified and the complex appears to be at least qualitatively similar to that from animal mitochondria.

Pyruvate dehydrogenase catalyzes three steps in a coupled sequence where the intermediates remain enzyme bound (Fig. 8.2). Initial decarboxylation of the pyruvate gives CO_2 and the α-hydroxyethyl derivative of the thiazole ring of thiamine pyrophosphate. The hydroxyethyl group is dehydrogenated and the resulting acetyl group transferred to the lipoic acid. This acetyl group is then transferred to the thiol group of CoA. The dihydrolipoic acid on the acetyltransferase is oxidized by dihydrolipoamide dehydrogenase, with the resulting $FADH_2$ being oxidized by NAD^+. The whole sequence exemplifies the swinging arm mechanism where the long lipoyl-lysyl arm of the acetyltransferase channels the intermediates from one enzyme to the next.

Plant cells are distinguished by the fact that they contain two pyruvate dehydrogenase complexes, one in the mitochondria and the other in the plastids (Miernyk et al., 1985). The complexes from the two organelles differ in their regulatory properties; only those of the mitochondrial complex are discussed here. The activity of plant mitochondrial pyruvate dehydrogenase can be varied by covalent modification involving

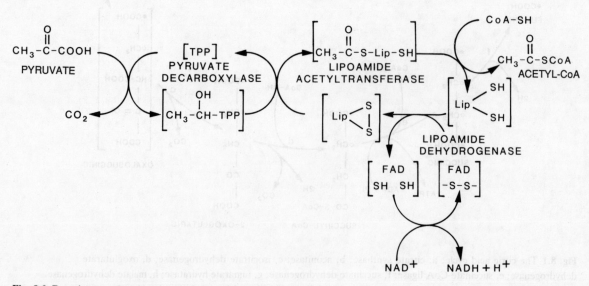

Fig. 8.2 Reactions catalyzed by the pyruvate dehydrogenase complex.

phosphorylation and dephosphorylation via pyruvate dehydrogenase kinase and phosphopyruvate dehydrogenase phosphatase respectively. The kinase catalyzes multisite phosphorylation by ATP onto threonine residues of the decarboxylase subunit. This results in complete inactivation: reactivation is achieved by the phosphatase. In animal mitochondria the kinase is inhibited by pyruvate and stimulated by NADH and CoA, whilst the phosphatase is stimulated by Ca^{2+}. The detailed properties of the plant enzymes are still not known but it is likely that pyruvate dehydrogenase in plant mitochondria is at least partially regulated by a cascade similar in principle to that found in animal mitochondria.

The importance of enzyme cascades in regulation has been discussed by Stadtman et al. (1981). At a given steady state, the proportion of the pyruvate dehydrogenase complex that is phosphorylated will determine the activity of pyruvate dehydrogenase. This proportion will be a function of the relative activities of the kinase and the phosphatase, which will, in turn, be determined by the concentrations of substrates and effectors of the kinase and phosphatase. This system of control provides a means whereby pyruvate dehydrogenase can be modulated by a wide range of effectors, and also permits amplification of a small change in effector to a large change in dehydrogenase activity.

Plant mitochondrial pyruvate dehydrogenase is also subject to direct inhibition by NADH and acetyl-CoA. Both are linear competitive inhibitors with respect to NAD^+ and CoA, respectively. The K_i for NADH, 18 μM, is appreciably lower than the K_m for NAD^+, 126 μM. Thus the enzyme is particularly sensitive to an increase in the ratio $NADH:NAD^+$. The relationship between these direct inhibitions and that caused by phosphorylation is not known, but the data available suggest that in vivo high ATP, NADH and acetyl-CoA will reduce the activity of pyruvate dehydrogenase, whereas activity will be increased if the level of these effectors falls and if the concentration of pyruvate rises.

Citrate synthase

This enzyme catalyzes an aldol condensation between the methyl group of acetyl-CoA and the carbonyl group of oxaloacetate. The condensation gives citroyl-CoA as the intermediate and the thioester bond is immediately hydrolyzed to yield citrate and CoA. The mammalian enzyme is made of two identical subunits (M_r 49 $\times$ 10^3), each divided into a large and a small domain. The two independent active sites are located at the interface of the large and small domains.

In plants citrate synthase is confined to the mitochondria, except in gluconeogenic tissues where it is also found in the glyoxysomes. The enzymes from the two compartments are immunologically distinct. The sensitivity of the mitochondrial enzyme to inhibition by ATP is probably not an important means of regulation. A high concentration, 3.5 to 5 mM, is needed to produce 50% inhibition, and similar inhibition is shown by the enzyme from cyanobacteria that do not use the citric acid cycle as a source of energy. The inhibition is probably due to the structural similarity between acetyl-CoA and ATP leading to isosteric binding at the active site.

Aconitase

This enzyme catalyzes the reversible interconversion of citrate and isocitrate via the enzyme-bound intermediate cis-aconitate. Dehydration is followed by hydration and the net effect is an interchange of H and OH. Aconitase contains non-heme iron as a 4Fe–4S cluster. Studies with sycamore cells showed that 60% of the aconitase was in the mitochondria and 40% in the cytosol. Cytosolic aconitase is also found in mammalian cells.

Isocitrate dehydrogenase

Plants contain both NAD- and NADP-linked isocitrate dehydrogenases. The former is confined to the mitochondria. The latter is found in the cytosol and the chloroplast but there is not enough data to decide whether it is also present in the mitochondria. The likelihood is that any mitochondrial activity is low and that it is the NAD-linked enzyme that is involved in the citric acid cycle. This enzyme catalyzes two reactions, the oxidation of isocitrate to enzyme-bound oxalosuccinate which is then decarboxylated to give CO_2 and 2-oxoglutarate. The enzyme from pea mitochondria shows Michaelian kinetics at neutral and slightly acid pH, but sigmoid kinetics at more alkaline pH. The few plant enzymes that have been examined differ from the mammalian enzyme in that they were not inhibited by ATP. The plant enzymes are inhibited by NADH. As with pyruvate dehydrogenase, the affinity for NADH is greater than that for NAD^+. Thus the enzyme is very sensitive to the ratio NADH:NAD.

2-Oxoglutarate dehydrogenase

The oxidation of 2-oxoglutarate to succinyl-CoA is analogous to conversion of pyruvate to acetyl-CoA and is achieved by an enzyme complex completely comparable to pyruvate dehydrogenase (Perham *et al.*, 1987). 2-Oxoglutarate dehydrogenase is not modulated by phosphorylation. The enzyme has been isolated from cauliflower and shown to have an M_r of 2.1×10^6. Further details are awaited.

Succinate-CoA ligase

This enzyme catalyzes the only substrate-level phosphorylation in the cycle: the free energy of hydrolysis of the thioester bond of succinyl-CoA permits the synthesis of ATP. The likely mechanism is given below.

Succinyl-CoA + P_i + Enz $\rightleftharpoons$ Enz-succinyl phosphate + CoA
Enz-succinyl phosphate $\rightleftharpoons$ Enz-phosphate + succinate
Enz-phosphate + ADP $\rightleftharpoons$ ATP + Enz

Little is known of the plant enzyme. All succinate-CoA ligases examined consist of two types of subunit: α, M_r $29-34 \times 10^3$ and β, M_r $41-45 \times 10^3$. The eukaryotic enzyme is an $\alpha\beta$ dimer. There is now evidence that mammalian cells contain two ligases, one specific for ADP and one for GDP, and that the latter catalyzes the synthesis of succinyl-CoA during ketone body metabolism. Thus we should expect plants to contain only the ADP-specific enzyme; this expectation holds for the few plant enzymes studied.

Succinate dehydrogenase

This enzyme is a component of both the citric acid cycle and, as complex II, the respiratory electron transport chain (see Chapter 9). Detailed studies of the mammalian enzyme show it to be formed of a large (Fp) subunit of M_r 70 000 that contains covalently bound FAD and two binuclear 2Fe–2S iron–sulfur clusters; a second large (Ip) subunit of M_r 27 000 that contains one tetranuclear 4Fe–4S iron–sulfur cluster; and a number of smaller subunits. The Ip subunit spans the inner membrane of the mitochondria and is attached to the Fp subunit that is in the matrix space. The smaller subunits are largely buried in the membrane. The higher plant enzyme appears to be similar to that described for mammals.

Fumarase

This enzyme catalyzes the *trans* addition of H and OH to the double bond of fumarate. The addition is stereospecific in that the OH is added to only one side of the double bond so that only L-malate is formed. We know little of the enzyme from plants except that it is confined to the mitochondrial matrix. The enzyme from pig heart is a tetramer of identical subunits, each of M_r 48 500.

Malate dehydrogenase

The final step in the cycle is catalyzed by NAD-linked malate dehydrogenase. The equilibrium of the step favors malate formation but *in vivo* the level of oxaloacetate and oxidation of NADH permit malate oxidation. Isoenzymes of NAD-linked malate dehydrogenase are found in the mitochondria, cytosol and glyoxysomes in plants. The mitochondrial enzyme consists of two identical subunits, M_r $37-38 \times 10^3$ and is distinguishable from those elsewhere in the cell immunologically and in respect of isoelectric point and heat stability.

NAD-malic enzyme

L-malate + NAD^+ $\rightarrow$ pyruvate + CO_2 + H^+ + NADH

This enzyme is very closely associated with the citric acid cycle in plants, is confined to the mitochondrial matrix, and catalyzes the oxidative decarboxylation of malate *in vivo*. The plant enzyme consists of two different subunits, has a configuration of $(\alpha\beta)_n$, and can exist as a dimer, tetramer or octamer. Malate moves the equilibrium towards the tetramer, which has the highest affinity for malate and the highest intrinsic activity. The enzyme requires a divalent cation and is activated by CoA, fumarate and sulfate.

Organization of the citric acid cycle

The complete operation of the cycle involves the coordination and sequential interaction of eight enzymes, seven of which are in the mitochondrial matrix and the eighth has access to it. There is increasing evidence that confinement of the cycle enzymes to the matrix is, on its own, not enough to allow the cycle to function effectively, and that there is some form of supramolecular organization of the enzymes (Srere et al., 1987).

The concentration of protein in the matrix is probably close to 15 mM, which approaches that in protein crystals. The implication from this is that enzymes and substrates are unlikely to be able to diffuse freely in such a viscous milieu. Thus it seems likely that substrates will be metabolized mainly, if not entirely, by neighboring enzymes, and sequential enzymes will need to be in close contact. This arrangement would permit channeling of metabolites with passage of substrates from one enzyme to another without equilibration with the bulk-phase of the mitochondrial matrix.

Direct evidence of association of cycle enzymes is available from studies of animal mitochondria. Specific interactions between purified enzymes of the cycle have been demonstrated *in vitro*. This has been done for six of the eight steps in the cycle. In addition, there is also evidence of an association of the matrix enzymes with the inner membrane. A number of the matrix enzymes have been shown to bind preferentially to the inner membrane. The binding of citrate synthase has been shown to be a saturable process and to be specific to the inner surface of the membrane. Finally, light sonication of rat liver mitochondria has been shown to produce a sedimentable preparation, called a metabolon, in which the membranes had been disrupted but the enzymes remained bound to the inner membrane.

Evidence for the operation of the citric acid cycle in plants

The central role of the citric acid cycle in plant metabolism is well established (Beevers, 1961). Only the crucial evidence is considered.

Enzyme activities

There is adequate evidence that plant cells can catalyze the reactions of the citric acid cycle at rates comparable to that of tissue respiration. This evidence comprises isolation of physiologically competent mitochondria from a wide range of tissues and measurements of the activities of individual enzymes of the cycle from an equal range of tissues. Nonetheless, there is a dearth of quantitative evidence of this type, as for no tissue do we know the maximum catalytic activities of each of the enzymes involved in the cycle. Such information is essential for understanding control of the cycle. The major difficulty here is that, as we have seen, many of the cycle enzymes have counterparts outside the mitochondria. Thus measurement of the total activity of a tissue does not measure the activity associated with the cycle. Further, activity in a mitochondrial preparation does not measure the capacity of a tissue as the yield of mitochondria is not 100%. The estimates in Table 8.1 were obtained by measuring the total activity in the tissues and then correcting these values after making independent estimates of the proportion of each enzyme that was confined to the mitochondria. Our results show that for these tissues, at least, these enzymes of the cycle have maximum catalytic activities well in excess of the rates of respiration. The activity of malate dehydrogenase is at least 12 times that of the next highest value, but the activities of the other enzymes are fairly comparable.

Substrate contents

There is perfectly adequate evidence that plant tissues as a whole contain each of the substrates of the citric acid cycle. As with the enzymes, quantitative data are very rare. I know of no

Table 8.1 Estimates of the maximum catalytic activities of enzymes catalyzing the citric acid cycle in cauliflower florets and the developing club of *Arum maculatum* (unpublished data of A.J. MacDougall and T. ap Rees)

Enzyme	Activity ($\mu mol\ g^{-1}$ fresh wt min^{-1})			
	Cauliflower	α stage Arum	Pre-thermogenic Arum	Thermogenic Arum
Citrate synthase	1.05	18.6	60.8	92.4
Aconitase	0.73	15.6	42.4	45.8
Isocitrate dehydrogenase (NAD)	0.44	4.3	24.8	18.9
Fumarase	2.41	10.0	31.4	34.3
Malate dehydrogenase (NAD)	55	247	889	1 584

Pyruvate production from carbohydrate breakdown in thermogenic clubs of *Arum* reaches rates of up to 20 μmol pyruvate g^{-1} fresh wt min^{-1}.

tissue for which authenticated measurements for each of the cycle intermediates have been made. For the measurements that are available, exemplified by Table 8.2, it is clear that a very striking feature of plant tissues is that their total content of cycle intermediates is very high indeed. On a protein basis, the malate content of carrot storage tissue is 1500 times that of cow liver. The precise acids that accumulate vary with the tissue. Malate is almost always high, and considerable levels of citrate and aconitate are also found. The probable significance of this accumulation of acid is that it provides a metabolically cheap solute for the adjustment of osmotic pressure and that it serves to balance any cation excess, caused, for example, by extensive metabolism of anions such as NO_3^-.

We are now faced with the problem of how a catalytic cycle manages to operate in the presence of such large quantities of some of its intermediates. The answer is that the accumulated acids are stored largely in the vacuole and are compartmented from the very small pools that are present in the mitochondria. The clearest demonstration of this are the classic experiments of MacLennan *et al.* (1963), who fed [1-^{14}C]-acetate to a range of plant tissues and showed that the specific activity of the respired $^{14}CO_2$ exceeded those of the acids extracted from the tissue. The differences were greatest for the

Table 8.2 Amounts of pyruvate and citric acid cycle intermediates in plants

Acid	Content ($\mu mol\ g^{-1}$ fresh wt)[a]						
	Club of Arum maculatum	Apical cm of maize root	Maize coleoptile	Wheat leaf	Carrot root	Bryophyllum leaf	Rat liver
Pyruvate	0.07	–	–	–	–	–	0.13–0.25
Citrate	16.6	1.5	0.8	0.6	1.2	8.0	0.17–0.26
Aconitate	–	3.4	4.2	1.0	–	–	–
Isocitrate	0.11	–	–	–	–	60.0	0.01–0.02
2-Oxoglutarate	0.35	–	–	–	–	–	0.14
Succinate	–	0.2	0.2	0.2	–	–	0.74
Fumarate	0.90	–	–	–	–	–	0.08
Malate	21.5	7.5	2.7	1.7	15.9	19.0	0.28–0.50
Oxaloacetate	0.06	–	–	–	–	–	0.008–0.004

[a]Data from MacLennan *et al.* (1963) except that for *Arum* and rat liver which is from ap Rees *et al.* (1981) and Williamson and Brosnan (1974), respectively.

acids that accumulated in greatest quantities. For example, carrots accumulate malate (Table 8.2); the specific activity of the respired $^{14}CO_2$ was 31 times greater than that of the [^{14}C]-malate extracted from the tissue. Thus the pool of [^{14}C]-malate that gave rise to the $^{14}CO_2$ must have been much smaller than the total tissue content of malate. It is most unlikely that the actual concentrations of the acids taking part in the cycle in plants are appreciably different from those in other mitochondria.

Metabolism of labeled substrates

The presence of adequate amounts of enzyme and substrates is a necessary condition for proposing a pathway but does not prove that such a pathway exists and operates *in vivo*. This is best done by determining whether intact cells or tissues metabolize the proposed substrates in the manner required by the hypothesis. For the citric acid cycle the expected distribution of label has been obtained with a sufficiently wide range of tissues for us to accept that it is general. The most definitive of these experiments are those in which Harley and Beevers (1963) fed [1-^{14}C] and [2-^{14}C]-acetate to maize roots. They demonstrated the expected sequential labeling of the successive acids of the cycle (Table 8.3) and also showed that label appeared as $^{14}CO_2$ after a lag that corresponded with the time taken for the label to move round the cycle. In addition, they showed that the intramolecular labeling of both citrate and malate was precisely that predicted from the

Table 8.3 Specific activities of citric acid cycle acids after supplying [2-^{14}C]-acetate to the apical cm of maize roots (from Harley and Beevers, 1963). The specific activity of the [^{14}C]-acetate supplied was in the range 2.5 to 14.3 × 10^4 cpm μmol^{-1}.

Acid	Specific activity (cpm μmol^{-1}) after incubation for				
	1 min	2 min	5 min	10 min	15 min
Citrate	59	539	921	1 724	2 525
Succinate	13	29	170	509	533
Malate	0	2	10	63	96

reactions of the citric acid cycle (Fig. 8.1).

The major remaining question about the universality of the operation of the citric acid cycle in plants is the extent to which the cycle functions in photosynthetic cells in the light. For the leaves of C3 plants it has been argued that during photosynthesis cytosolic ATP is provided by the chloroplasts and that this increases the ratio ATP:ADP in the cytosol and so suppresses mitochondrial metabolism. This now seems unlikely. Recent experiments in which [^{14}C]-succinate and [^{14}C]-acetate were supplied to wheat leaves in the light establish that the cycle can operate in photosynthesizing cells and suggest that in wheat the flux through the cycle in the light is 75–85% of that in the dark (McCashin *et al.*, 1988).

The situation in the leaves of C4 and CAM plants is more complex and not yet resolved (see Chapter 19). C4 leaves that use either of the malic enzymes as their decarboxylase generate large amounts of pyruvate. The level of this pyruvate in the C4 cycle must be maintained to allow CO_2 fixation to continue. Thus we should expect mechanisms to restrict the consumption of this pyruvate by the citric acid cycle, particularly in plants that use NAD-malic enzyme and thus produce the pyruvate in the mitochondria. Complex effects of light on the cycle may also occur in C4 plants that use phosphoenolypyruvate carboxykinase as their decarboxylase; the carboxykinase requires ATP and in *Urochloa panicoides*, at least, this is provided by the mitochondria.

In CAM plants that depend upon malic enzyme to release CO_2 from malate in the light the question arises as to whether the pyruvate so formed is converted directly to sugar or after oxidation to CO_2 via the citric acid cycle. The available evidence suggests that most of the pyruvate produced is metabolized directly to carbohydrate via pyruvate orthophosphate dikinase, but there is at least some citric acid cycle activity in the light in both malic enzyme and phosphoenolpyruvate carboxykinase types of CAM leaf. For both C4 and CAM plants we need more detailed studies of the extent to which the cycle operates in the light. The need for carbon from the cycle for biosynthesis makes it likely that in both CAM and C4 leaves the cycle will operate to some extent in the light and that this operation will have to be carefully regulated to allow it to be compatible with photosynthesis.

The catabolic role of the citric acid cycle

Classically, the cycle is portrayed as the penultimate stage in the respiration of all of the significant substrates oxidized by cells. The large number of cycle intermediates is shown as affording a range of entry points and thus allowing all the major oxidative pathways to converge on the cycle. In heterotrophic organisms the cycle obtains its substrate from carbohydrates, lipids and proteins. The former feed in acetyl-CoA; breakdown of amino acids produces both acetyl-CoA and cycle intermediates. Essentially the cycle catalyzes:

Acetyl-CoA + oxaloacetate → oxaloacetate + $2CO_2$ + CoASH

The intermediates cannot be directly oxidized by the cycle as it is only capable of oxidizing acetyl-CoA. Thus the oxidation of cycle intermediates requires their conversion to acetyl-CoA. This can be done by converting the intermediate to malate, which is then metabolized to pyruvate by NAD-malic enzyme. The pyruvate then enters the cycle via acetyl-CoA in the usual way. This view of the cycle in catabolism needs some modification for plants. The latter are autotrophs and so produce their own respiratory substrate, mainly sucrose or starch. This is reflected by an almost complete dominance of carbohydrate as the source of substrate for the cycle in plants, with lipid and protein making rather small contributions.

Respiration of lipid

Lipids are widely stored but infrequently respired. Utilization of storage lipid involves its conversion to sucrose to permit transport from the site of storage to that of use. This complex conversion (Fig. 8.3) involves β-oxidation, the glyoxylate cycle and gluconeogenesis. During these processes the emphasis is on keeping the acetyl-CoA, formed by breakdown of fat, out of the citric acid cycle. The location of β-oxidation and much of the glyoxylate cycle in the glyoxysome ensures that the products of fat breakdown are kept away from the steps of the citric acid cycle that result in CO_2 production. The effectiveness of this arrangement is demonstrated by the fact that, as predicted from Fig. 8.3, 75% of the lipid carbon is converted to sucrose in the endosperm of germinating castor bean. In seeds where fat is stored, not in the endosperm that senesces, but in cotyledons that become photosynthetic, some of the acetyl-CoA from fat enters the citric acid cycle. Even here the fraction that is respired to CO_2 is small, about 6% in marrow.

Inspection of Fig. 8.3 shows that, regardless of whether the products of β-oxidation are channeled entirely to sucrose, the citric acid cycle does make an important contribution to the mobilization of storage lipid. The cycle mediates the conversion of succinate, formed from isocitrate by isocitrate lyase, to the oxaloacetate required to keep the glyoxylate cycle going. The essential feature of this role of the citric acid cycle is that it is anabolic rather than catabolic. With, I think, one known exception, higher plants have not been shown to fuel the citric acid cycle by net breakdown of storage lipid through β-oxidation. The exception is the thermogenic tissue of the small number of Araceae that store fat rather than starch, e.g. *Philodendron selloum*, a minority in what is already an exceptional group of plants.

Even where there is no net breakdown of lipid, turnover of the latter might provide acetyl-CoA for the citric acid cycle. The available evidence suggests that any such contribution is minor. First, the RQ (CO_2 released in respiration/O_2 consumed in respiration) of plant tissues is generally close to unity (see Chapter 9 for detailed discussion). Second, although the enzymes of β-oxidation are found in non-fatty, non-gluconeogenic tissues of plants such as pea leaves and maize roots the reported activities are low compared to those of enzymes of the citric acid cycle. It has also been argued that in such tissues β-oxidation is confined to the peroxisomes (Gerhardt, 1986). This is not so. Thomas, Wood and Burgess have provided

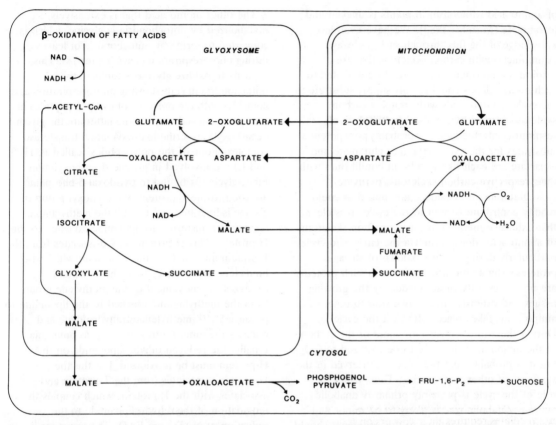

Fig. 8.3 Relationship between mitochondrion, glyoxysome and cytosol during the conversion of fat to sucrose in germinating seedlings.

evidence that mitochondria in the cotyledons of germinating peas contain the enzymes of β-oxidation, and that in this tissue the capacity for β-oxidation is divided roughly equally between the peroxisomes and the mitochondria. This evidence (Thomas et al., 1988) is good enough to withstand earlier charges that the mitochondrial preparations were not pure. It is clear that plant mitochondria are not necessarily devoid of the enzymes of β-oxidation but in general their activity is likely to be too low to make a major contribution to the citric acid cycle.

Respiration of protein and amino acids

There is still no convincing evidence that net breakdown of protein under natural conditions produces significant substrate for oxidation by the citric acid cycle in plants. Where net breakdown of protein occurs, such as in senescent leaves or during xylogenesis, the amino acids appear to be translocated elsewhere and stored or used at once for protein synthesis rather than respired. As with lipid, turnover of protein could provide a supply of amino acids for respiration. Again, this is unlikely to be a significant source of substrate for the cycle. This is because the protein content of plant cells and its rate of turnover is small compared to the rate of respiration (ap Rees, 1980).

Plants contain appreciable amounts of free amino acids, including many that are never found in proteins. These are primarily stores of nitrogen but are also a potential source of substrate for the citric acid cycle. Our knowledge of the pathways of amino acid catabolism in plants is uneven and incomplete (Mazelis, 1980). We have even less knowledge of the flux through such pathways. Amino acids with carbon skeletons that are respiratory intermediates or are closely related to such intermediates could readily supply substrate to the cycle. Amino acids with complex carbon skeletons formed by extensive metabolism of respiratory intermediates are probably poor potential substrates for the cycle. Alanine, glutamate and aspartate are each likely to be in equilibrium with their respective carbon skeletons, pyruvate, 2-oxoglutarate and oxaloacetate, and thus could readily contribute carbon to the cycle. It is clear that extensive metabolism of the carbon skeletons of amino acids does occur via the citric acid cycle, particularly during the breakdown of storage proteins. The amino acids present in such proteins are not necessarily those needed by the growing tissues and extensive interconversion to other amino acids takes place, often via the cycle. During these interconversions some of the carbon of the amino acids is likely to be oxidized to CO_2 but it is probable that the bulk is conserved in the synthesis of other amino acids. Thus, again the role of the cycle is probably primarily anabolic and any catabolic role is likely to be minor and incidental. Two exceptions to this argument should be considered.

Proline accumulates in many plant tissues subjected to the stresses of drought or cold. This proline disappears rapidly when the stress is removed. During this removal proline could make a significant but transitory contribution to the citric acid cycle. In barley leaves recovering from drought, consumption of the accumulated proline takes about eight hours (Stewart and Voetberg, 1985) and could contribute carbon to the cycle at about 20% of the rate of respiration. This metabolism of proline occurs in the mitochondria and involves conversion to Δ'-pyrroline 5-carboxylic acid by proline oxidase, conversion of the latter to glutamate via Δ'-pyrroline 5-carboxylic acid dehydrogenase and metabolism of the glutamate to 2-oxoglutarate by NAD-linked glutamate dehydrogenase. The first two enzymes are located on the matrix surface of the inner membrane of the mitochondria and the glutamate dehydrogenase is in the matrix.

The other amino acid that is extensively metabolized by mitochondria is glycine, which is converted to serine by mitochondria of leaves during photorespiration (see Chapter 18). This capacity to oxidize glycine is confined to mitochondria in cells showing photorespiration and coincides with the presence of five polypeptides in the mitochondrial matrix. The latter are the glycine decarboxylase–serine hydroxymethyl transferase complex. Four of the polypeptides (called P, H, T and L) are involved in glycine decarboxylation. First, glycine is bound to pyridoxal 5-phosphate attached to the P-protein: the carboxyl group of glycine is liberated as CO_2 and the methylamine residue is transferred to lipoamide attached to the H-protein. This H-protein now dissociates from the P-protein and binds to the T-protein, which also binds tetrahydrofolate. Next the T-protein catalyzes the movement of the methylene group from the methylamine attached to the lipoamide to produce 5′,10′-methylenetetrahydrofolate and releases the amino nitrogen group as ammonia. Finally, the reduced dihydrolipoamide on the H-protein must be reoxidized. For this the H-protein dissociates from the T-protein and associates with the L-protein, which couples the reoxidation of the dihydrolipoamide to the reduction of NAD^+ via FAD. The sequence is shown below; the proteins that catalyze each reaction are shown above the arrows.

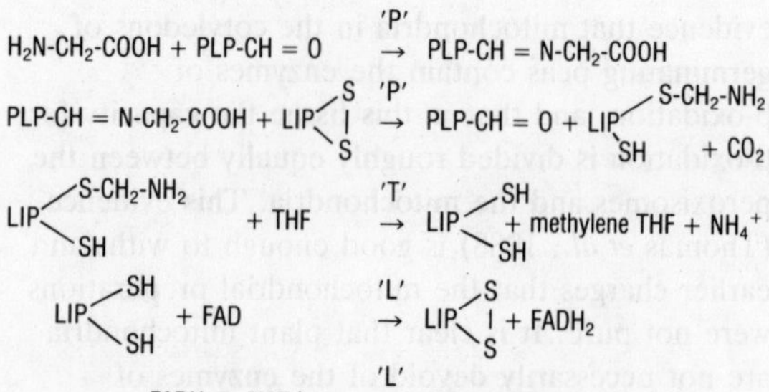

$$H_2N\text{-}CH_2\text{-}COOH + PLP\text{-}CH=O \xrightarrow{\text{'P'}} PLP\text{-}CH=N\text{-}CH_2\text{-}COOH$$

$$PLP\text{-}CH=N\text{-}CH_2\text{-}COOH + LIP\genfrac{}{}{0pt}{}{S}{S} \xrightarrow{\text{'P'}} PLP\text{-}CH=O + LIP\genfrac{}{}{0pt}{}{S\text{-}CH_2\text{-}NH_2}{SH} + CO_2$$

$$LIP\genfrac{}{}{0pt}{}{S\text{-}CH_2\text{-}NH_2}{SH} + THF \xrightarrow{\text{'T'}} LIP\genfrac{}{}{0pt}{}{SH}{SH} + \text{methylene THF} + NH_4^+$$

$$LIP\genfrac{}{}{0pt}{}{SH}{SH} + FAD \xrightarrow{\text{'L'}} LIP\genfrac{}{}{0pt}{}{S}{S} + FADH_2$$

$$FADH_2 + NAD^+ \xrightarrow{\text{'L'}} FAD + NADH + H^+$$

Serine hydroxymethyl transferase catalyzes the synthesis of serine from the methylene THF, formed in the above reactions, and another molecule of glycine.

The ability of leaf mitochondria to oxidatively decarboxylate glycine does not, in fact, involve the citric acid cycle directly. The reaction does not involve a cycle acid and the product, serine, moves to the peroxisome for further metabolism. However, the production of large quantities of NADH during glycine metabolism in the light is likely to affect the cycle both through competition between glycine decarboxylase and the cycle dehydrogenases for NAD^+ and through competition for NADH-oxidizing capacity. It is probable that the precise effect of glycine oxidation on cycle activity will vary with the metabolic state of the cell, particularly the rates of photosynthesis and the demands made on the citric acid cycle. Present evidence favors the view that at least some of the NADH produced from glycine metabolism is oxidized via the respiratory chain, although a malate:oxaloacetate shuttle may also contribute (Dry et al., 1987).

Respiration of carbohydrate

Proof that the products of glycolysis and the oxidative pentose phosphate pathway are metabolized via the citric acid cycle has been available for years and was summarized by Beevers (1961). Pyruvate is the product of carbohydrate oxidation, is readily taken up by plant mitochondria (Brailsford et al., 1986), and, as shown in Fig. 8.4, is readily metabolized by plant tissues in a manner that suggests almost exclusive entry into the citric acid cycle. The high yields of $^{14}CO_2$ from [1-^{14}C]-pyruvate strongly suggest conversion of all of the added [^{14}C]-pyruvate to acetyl-CoA. The lags in the release of pyruvate carbons 2 and 3 are consistent with the operation of the cycle. Interference with

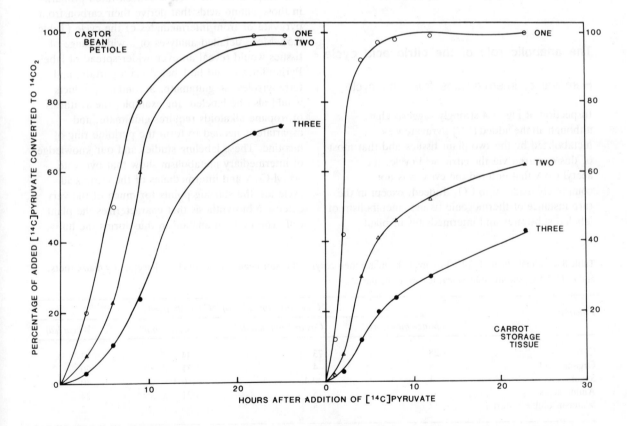

Fig. 8.4 Patterns of $^{14}CO_2$ production from specifically labeled [^{14}C]-pyruvate supplied to replicate samples of mature petiole of castor bean (from Neal and Beevers, 1960) and aged disks of carrot storage tissue (from ap Rees and Beevers, 1960). One, two and three denote position of label in [^{14}C]-pyruvate.

the operation of the cycle, by inhibitors of pyruvate and 2-oxoglutarate dehydrogenases or by anoxia, results in the expected diversion of pyruvate to products of fermentation. Finally, there is a very close relationship between the distribution of ^{14}C from specifically labeled pyruvate and that from specifically labeled glucose.

The catabolic role of the citric acid cycle in plants is primarily the oxidation of the products of glycolysis and the oxidative pentose phosphate pathway. Together the three pathways constitute the path of carbon in plant respiration, with little contribution from any other pathway. It appears that the autotrophy of plants has led to the citric acid cycle becoming the third component of a linear sequence in plant respiration rather than the point at which a wide range of catabolic pathways converge as occurs in heterotrophic cells.

The anabolic role of the citric acid cycle

Removal of intermediates from the cycle

Inspection of Fig. 8.4 strongly suggests that although all the added [^{14}C]-pyruvate was metabolized by the two plant tissues and that most of this occurred via the citric acid cycle, the acetyl-CoA that entered the cycle was not completely oxidized to CO_2. Indeed, except in the rare instance of thermogenic tissues, metabolism of labeled substrates and intermediates of plant respiration always reveals that a significant proportion of the carbon that enters the respiratory pathways is not converted to CO_2. The proportion that is retained in the tissue varies and may amount to 50% of the carbon that initially entered the respiratory pathways but is highest in growing tissues.

Analysis of tissues supplied with labeled pyruvate or citric acid cycle intermediates reveals the fate of carbon that enters the cycle but does not appear as CO_2 (Table 8.4). The bulk of the retained label is recovered as organic acids, amino acids (free and protein-bound), and, to a lesser extent, lipid. Analysis of the detailed distribution of label shows that the label in the organic acid fraction is concentrated in those acids that are accumulated in large amounts (Table 8.2). Within the amino acids the label is concentrated primarily in those amino acids that derive their carbon from pyruvate or from intermediates of the cycle (Table 8.5). More detailed analyses of a wider range of tissues would reveal an even wider spread of label. Pyrimidines would be labeled via aspartate, and tetrapyrroles via glutamate. Secondary products would also be labeled, for example, the synthesis of tropane alkaloids requires glutamate, and aspartate is needed to form the pyridine ring of nicotine. These labeling studies and our knowledge of intermediary metabolism show that pyruvate, acetyl-CoA and intermediates of the citric acid cycle are the starting points for much of the very extensive biosynthesis that characterizes the plant cell. The cycle in anabolism thus forms the hub

Table 8.4 Metabolism of [3-^{14}C]-pyruvate by maize mesocotyl and castor bean petiole, of [1-^{14}C]-acetate by maize roots, and of [1,4-^{14}C]-succinate by wheat leaves in the dark.

Fraction	^{14}C per fraction as % of ^{14}C metabolized			
	Maize mesocotyl[a]	Castor bean petiole[b]	Maize root[b]	Wheat leaf[c]
CO_2	28	73	44	53
Organic acids	9	4	31	20
Sugars	1	0.4	0.4	0.3
Amino acids	41	5	21	24
Water-insoluble material	17	5	–	2

[a]Incubated for 28 h (Neal and Beevers, 1960).
[b]Incubated for 2 h, data expresssed as % ^{14}C in total from fractions listed (MacLennan et al., 1963).
[c]Incubated for 2 h (McCashin et al., 1988).

Table 8.5 Metabolism of [U-^{14}C]-acetate to amino acids by suspension cultures of rose cells after a pulse of 10 min and a subsequent chase of 2 h. For each amino acid ^{14}C in the soluble and protein-bound form has been summed. From Fletcher and Beevers (1971).

Amino acid	^{14}C ($cpm \times 10^{-3}$) per amino acid	
	Pulse	Chase
Glu + Gln	143.3	30.3
Pro	4.73	8.57
Arg	0.93	5.16
Asp + Asn	21.7	15.8
Met	1.66	1.72
Thr	1.14	4.65
Ile	0.67	3.51
Lys	0.39	4.12
Ala	3.23	4.67
Ser	0.63	1.11
Val	0.47	1.84
Leu	20.89	23.05
Gly	0.25	0.02
Phe	0.06	0
Tyr	0.03	0.15
His	0	0

from which many of the pathways of biosynthesis radiate.

Anaplerotic fixation of CO_2

The accumulation by plants of citrate, aconitate, isocitrate and malate, and the use of 2-oxoglutarate and oxaloacetate to supply, respectively, glutamate and aspartate for synthesis means that net synthesis of at least six intermediates of the citric acid cycle must occur in plants. Given a supply of acetyl-CoA, any one intermediate in the cycle can be converted into another. Thus the need for the synthesis of cycle intermediates could be met by the net formation of just one intermediate. However, this need cannot be met by the cycle as it is incapable of catalyzing a net synthesis of any of its intermediates. Thus, if the cycle operates in anabolism, there must be a means of replenishing it through what is know as an anaplerotic ('filling-up') pathway.

The replenishment of the cycle is achieved through the carboxylation of glycolytically produced phosphoenolypyruvate to give oxaloacetate. This step is catalyzed by phosphoenolypyruvate carboxylase. This enzyme is universally distributed in plants in appreciable activities, confined to the cytosol and uses bicarbonate as substrate in a most effective irreversible carboxylation. Oxaloacetate formed in the cytosol by the carboxylase is rapidly metabolized either to malate via cytosolic malate dehydrogenase or to aspartate via transaminase. The malate could enter the mitochondrion via the dicarboxylate translocator; in addition there is a specific translocator for the movement of oxaloacetate into plant mitochondria (Heldt and Flugge, 1987).

Evidence for the anaplerotic role of phosphoenolypyruvate carboxylase in plants is provided by the universal occurrence of CO_2 fixation in the dark and by the labeling patterns produced by such fixation from $^{14}CO_2$. Initially, there is labeling of oxaloacetate and confinement of most of the label to malate and aspartate. With time, the label spreads to other acids of the citric acid cycle and to amino acids, particularly those that obtain their carbon skeletons from the cycle (Table 8.6). Further, the extent of this dark fixation of CO_2 and the precise distribution of label from $^{14}CO_2$ vary according to the biosynthetic demands being made on the citric acid cycle. In nodules of lupin roots, CO_2 fixation is greater than in primary or secondary root tissue and increases three-fold at the onset of N_2 fixation when the demand for carbon skeleton increases. In barley roots, accumulation of cations in excess of anions is accompanied by an increase in CO_2 fixation and organic acid content.

Relationship between products of anaplerotic CO_2 fixation and the citric acid cycle

The precise relationship between the C4 acids produced by dark fixation of CO_2 in the cytosol and the intermediates of the cycle in the mitochondria is not established. The presence of phosphoenolpyruvate carboxylase and malate dehydrogenase means that the cytosol possesses

Table 8.6 Labeling of suspension cultures of rose cells after exposure to [^{14}C]-bicarbonate for 90 min in the dark. From Nesius and Fletcher (1975).

Component	^{14}C (cpm × 10^{-3}) per component	
Lipid	220	
Organic acids	1 623	
Malate	1 552	
Citrate	73	
Succinate	2	
Amino acids	Free	Protein-bound
Total	411	630
Asp	85	141
Thr	7	35
Met	5.5	15
Leu + Ile	2	10
Glu	98	80
Pro	4	16
Ser	6.5	4
Gly	3	9
Ala	5.5	9
Tyr	1.5	0
Val	0.5	4
Phe	0.5	3
Asn	4	–
Gln	6	–

the capacity to synthesize both malate and aspartate independently of the cycle. Thus malate accumulation and the provision of aspartate for biosynthesis need not involve the cycle directly. As far as we know, the cytosol cannot convert the products of CO_2 fixation to citrate, isocitrate or 2-oxoglutarate. Thus the net synthesis of citrate from malate formed in the cytosol must involve the citric acid cycle. However, the presence of aconitase and isocitrate dehydrogenase in the cytosol could provide a cytosolic route for the subsequent conversion of citrate to aconitate and 2-oxoglutarate.

There is experimental support for a dominant role for the cytosol in providing the carbon for malate accumulation and for the synthesis of aspartate. Lips and Beevers (1966) gave a sample of maize roots a pulse of [^{3}H]-acetate and [^{14}C]-bicarbonate and followed the labeling of malate in a cold chase. The [^{3}H]-malate lost label rapidly as it was metabolized via the cycle. No change over a 3-hour period was detected in the amount of [^{14}C]-malate present. Thus the product of CO_2 fixation was kept quite separate from the malate involved in the cycle. Similarly, the labeling of aspartate and its derivatives shown in Table 8.6 is unlikely to have involved the citric acid cycle. Phosphoenolpyruvate carboxylase will not label carbons 2 and 3 of oxaloacetate. These are the only two carbons of oxaloacetate that survive passage through the cycle (Fig. 8.1). Thus the labeled oxaloacetate that gave rise to the aspartate did not come from the cycle.

Rigid separation of the products of dark fixation from the intermediates of the citric acid cycle is not the general rule. The labeling of citrate and of glutamate and its derivatives by $^{14}CO_2$ in the dark (Table 8.6) provides very good evidence that products of CO_2 fixation enter the citric acid cycle. Further, in a range of tissues ^{14}C fixed into C4 acids from a pulse of $^{14}CO_2$ in the dark is rapidly lost as $^{14}CO_2$ in a subsequent cold chase (Bryce and ap Rees, 1985). The distribution of label in these tissues suggests that glycolytically produced phosphoenolpyruvate is carboxylated to yield malate which enters the mitochondria where it is converted to pyruvate by NAD-malic enzyme.

The precise relationship between cytosolic C4 acids and the citric acid cycle probably varies with the tissue and the conditions. It is clear that a considerable fraction of the phosphoenolpyruvate formed in cytosolic glycolysis is converted to C4 acids in the cytosol. Thus phosphoenolpyruvate is an important branch point in glycolysis and pyruvate is not the unique product of aerobic glycolysis in plants. The fate of the C4 acids varies and seems likely to include use in the cytosol to provide carbon for biosynthesis, storage in the vacuole, and entry into the mitochondria where they can either be used to replenish the level of oxaloacetate or be converted to pyruvate via NAD-malic enzyme. We urgently need to discover the role of the latter enzyme and the mechanisms responsible for the partitioning of C4 acids amongst the above pathways. The relationship between the cycle and anabolism is complicated even further in a way unique to plants, as many of the biosyntheses occur in the plastids. The latter play a dominant role in the synthesis of amino acids, pyrimidines, purines, fatty acids, porphyrins and some secondary products (ap Rees, 1987). Thus mitochondrial metabolism must be closely integrated with that of the plastids as well as the cytosol.

Provision of acetyl-CoA for biosyntheses

Acetyl-CoA is the starting point for the synthesis of fatty acids, isoprenoids and a number of secondary products, e.g. the A-ring of flavonoids. Fatty acids are made in the plastid; the initial steps in the synthesis of isoprenoids probably occur in the cytosol (Gray, 1987). Thus it is probable that there is a need for acetyl-CoA in the plastids and the cytosol as well as the mitochondria. We do not know how these needs are met.

At least some of the acetyl-CoA in plastids is almost certainly formed by plastidic pyruvate dehydrogenase using pyruvate either imported from the cytosol or produced locally by plastidic glycolysis. A further possibility is that acetyl-CoA in the mitochondria is hydrolyzed by a mitochondrial short-chain acyl-CoA hydrolase to yield free acetate which moves to the plastid for conversion to acetyl-CoA via a plastidic acetyl-CoA synthetase. We still need proof of the existence of acetate in plant cells as none of the available measurements is accompanied by adequate evidence that the acetate measured was not produced artefactually during killing and extraction of the tissue.

As pyruvate dehydrogenase and acetyl-CoA synthetase appear to be absent from the cytosol, neither of the routes of acetyl-CoA synthesis described for plastids will provide cytosolic acetyl-CoA. One possibility (Kaethner and ap Rees, 1985) is that citrate from the citric acid cycle moves to the cytosol and is metabolized there by ATP citrate lyase to produce acetyl-CoA, and oxaloacetate which returns to the cycle.

ATP + citrate + CoA $\rightarrow$ acetyl-CoA + ADP + P_i + oxaloacetate

Until we know whether these hypotheses are right we cannot assess the role of the mitochondria in providing the acetyl-CoA required for anabolism. Any such assessment must also take into account the presence in at least some plant mitochondria of a carnitine:acylcarnitine translocator (Thomas et al., 1988).

Control of the citric acid cycle

The major and diverse roles of the cycle in plants suggest that its control is stringent, to maximize the use of carbon and energy; flexible and coordinated, to meet changes in these demands; and complex because of the range of these demands. Whatever our expectations, the complex role of the cycle in plant metabolism and the duplication of many of the enzymes and substrates in compartments of the cell outside the mitochondrion have led to an almost total ignorance of how the cycle is controlled. Regulation of mitochondrial behavior has been reviewed in depth recently (see list for further reading). Most of these reviews are in compendia that amount to 1324 pages: the indices, collectively, make no more than three single page references to control of the citric acid cycle.

To understand how the cycle is controlled, we need to characterize the enzymes fully, propose hypotheses of control, test these hypotheses *in vivo*, and determine the quantitative contribution of each step in the cycle to the control of the cycle as a whole. As we have seen, there has been very modest progress towards characterizing the enzymes. We do not know whether future research will reveal additional regulatory properties or whether the present information is the total for the cycle enzymes. If the latter is true, then it would appear that the control of the cycle will be dominated by pyruvate dehydrogenase controlling the availability of acetyl-CoA, and to a lesser extent by the effect of the $NADH:NAD^+$ ratio on isocitrate dehydrogenase. Such a mechanism would place a premium on the $NADH:NAD^+$ ratio in regulation of the cycle and it is not immediately clear how this would allow the cycle to respond to demands of carbon for anabolism.

Hypotheses of control have been put forward but none has yet been tested rigorously by comparing changes in flux through the cycle with changes in the concentrations of the proposed effectors. This requires a method for measuring the concentration of metabolites in the mitochondria and cytosol. The development of non-aqueous fractionation techniques and also methods of rapid cell fractionation offer hope in this area.

No progress at all has been made towards determining the relative contributions of the different reactions of the cycle to control of flux through the cycle as a whole. It cannot be emphasized too strongly that the demonstration that an enzyme responds to an effector *in vitro* does not in any way prove that this enzyme exerts significant control *in vivo*. Progress in this question is not the forlorn hope it seems. Kacser (1987) has developed a theory of investigation of control in which the contribution of a step to control *in vivo*, the control coefficient, is given by

$$\frac{\delta J}{J} \bigg/ \frac{\delta V}{V}$$

where J is the pathway flux and V is the maximum catalytic activity of the enzyme in question. The key questions in practice are the development of methods to measure flux and of ways to change the maximum catalytic activity. Some progress in flux measurement can be made with careful use of isotopes (McCashin *et al.*, 1988). Genetic engineering can be used to alter enzyme activities. We ought to be able to create transgenic plants that differ in the expression of a single gene for a cycle enzyme and then observe the effect of this alteration in maximum catalytic activity on flux through the cycle. The changes in activity could be achieved by using *Agrobacterium*-mediated transformation to introduce full-length cDNA or genomic clones in appropriate constructs into plants so as to produce sense or antisense RNA that would lead to increased or decreased amounts of enzymes. The power of combining molecular biology and biochemistry in the study of the control of the cycle is borne out by the work of Walsh *et al.* (1987), who used recombinant DNA techniques to cause under- and over-expression of citrate synthase in *E. coli*.

Acknowledgment

I thank Dr John Lunn for his constructive criticism.

Further reading

ap Rees, T. (1987). Compartmentation of plant metabolism. In *The Biochemistry of Plants*, Vol. 12, ed. D. D. Davies, Academic Press, New York, pp. 87–115.

Dry, I. B., Bryce, J. H. and Wiskich, J. T. (1987). Regulation of mitochondrial metabolism. In *The Biochemistry of Plants*, Vol. 11, ed. D. D. Davies, Academic Press, New York, pp. 213–52.

Krebs, H. (1981). *Reminiscences and Reflections*, Clarendon Press, Oxford.

Moore, A. L. and Beechey, R. B. (1986). *Plant Mitochondria*, Plenum Press, New York.

Wiskich, J. T. and Dry, I. B. (1985). The tricarboxylic acid cycle in plant mitochondria. In *Encyclopedia of Plant Physiology*, New Series, Vol. 18, eds R. Douce and D. A. Day, Springer-Verlag, Berlin, pp. 281–313.

References

ap Rees, T. (1980). Assessment of the contributions of metabolic pathways to plant respiration. In *The Biochemistry of Plants*, Vol. 2, ed. D. D. Davies, Academic Press, New York, pp. 1–29.

ap Rees, T. and Beevers, H. (1960). Pentose phosphate pathway as a major component of induced respiration of carrot and potato slices. *Plant Physiol.* **35**, 839–47.

ap Rees, T., Fuller, W. A. and Green, J. H. (1981). Extremely high activities of phosphoenol-pyruvate carboxylase in thermogenic tissues of Araceae. *Planta* **152**, 79–86.

ap Rees, T. (1987). Compartmentation of plant metabolism. In *The Biochemistry of Plants*, Vol. 12, ed. D. D. Davies, Academic Press, New York, pp. 87–115.

Beevers, H. (1961). *Respiratory Metabolism in Plants*, Row, Peterson, Evanston.

Brailsford, M. A., Thompson, A. G., Kaderbhai, N. and Beechey, R. B. (1986). Pyruvate metabolism in castor-bean mitochondria. *Biochem. J.* **239**, 355–61.

Bryce, J. H. and ap Rees, T. (1985). Rapid decarboxylation of the products of dark fixation of CO_2 in roots of *Pisum and Plantago*. *Phytochemistry* **24**, 1635–8.

Dry, I. B., Bryce, J. H. and Wiskich, J. T. (1987). Regulation of mitochondrial metabolism. In *The Biochemistry of Plants*, Vol. 11, ed. D. D. Davies, Academic Press, New York, pp. 213–52.

Fletcher, J. S. and Beevers, H. (1971). Influence of cycloheximide on the synthesis and utilization of amino acids in suspension cultures. *Plant Physiol.* **48**, 261–4.

Gerhardt, B. (1986). Basic metabolic function of the higher plant peroxisome. *Physiol. Veg.* **24**, 397–410.

Gray, J. C. (1987). Control of isoprenoid biosynthesis in higher plants. *Adv. Bot. Res.* **14**, 25–91.

Harley, J. L. and Beevers, H. (1963). Acetate utilization by maize roots. *Plant Physiol.* **38**, 117–23.

Heldt, H. W. and Flugge, U. I. (1987). Subcellular transport of metabolites in plant cells. In *The Biochemistry of Plants*, Vol. 12, ed. D. D. Davies, Academic Press, New York, pp. 49–85.

Kacser, H. (1987). Control of metabolism. In *The Biochemistry of Plants*, Vol. 11, ed. D. D. Davies, Academic Press, New York, pp. 39–67.

Kaethner, T. M. and ap Rees, T. (1985). Intercellular location of ATP citrate lyase in leaves of *Pisum sativum* L. *Planta* **163**, 290–4.

Lips, S. H. and Beevers, H. (1966). Compartmentation of organic acids in corn roots. *Plant Physiol.* **41**, 709–12.

MacLennan, D. H., Beevers, H. and Harley, J. L. (1963). 'Compartmentation' of acids in plant tissues. *Biochem J.* **89**, 316–27.

Mazelis, M. (1980). Amino acid catabolism. In *The Biochemistry of Plants*, Vol. 5, ed. B. J. Miflin, Academic Press, New York, pp. 542–67.

McCashin, B. G., Cossins, E. A. and Canvin, D. T. (1988). Dark respiration during photosynthesis in wheat leaf slices. *Plant Physiol.* **87**, 155–61.

Miernyk, J. A., Camp, P. J. and Randall, D. D. (1985). Regulation of plant pyruvate dehydrogenase complexes. *Curr. Top. Plant Biochem. Physiol.* **4**, 175–90.

Neal, G. E. and Beevers, H. (1960). Pyruvate utilization in castor-bean endosperm and other tissues. *Biochem. J.* **74**, 409–16.

Nesius, K. K. and Fletcher, J. S. (1975). Contribution of nonautotrophic CO_2 fixation to protein synthesis in suspension cultures of Paul's scarlet rose. *Plant Physiol.* **55**, 643–5.

Perham, R. N., Packman, L. C. and Radford, S. E. (1987). 2-Oxo acid dehydrogenase multi-enzyme complexes: in the beginning and halfway there. *Biochem. Soc. Symp.* **54**, 67–81.

Srere, P. A., Sumegi, B. and Sherry, A. D. (1987). Organizational aspects of the citric acid cycle. *Biochem. Soc. Symp.* **54**, 173–82.

Stadtman, E. R., Chock, P. B. and Rhee, S. G. (1981). Interconvertible enzyme cycles in cellular regulation. *Curr. Top. Cell. Regul.* **18**, 79–94.

Stewart, C. R. and Voetberg, G. (1985). Relationship between stress-induced ABA and proline accumulations and ABA-induced proline accumulation in excised barley leaves. *Plant Physiol.* **79**, 24–7.

Thomas, D. R., Wood, C. and Masterson, C. (1988). Long-chain acyl CoA synthetase, carnitine and β-oxidation in the pea-seed mitochondrion. *Planta* **173**, 263–6.

Walsh, K., Schena, M., Flint, A. J. and Koshland, D. E. (1987). Compensatory regulation in metabolic pathways – responses to increases and decreases in citrate synthase levels. *Biochem. Soc. Symp.* **54**, 183–95.

Williamson, D. H. and Brosnan, J. T. (1974). Concentrations of metabolites in animal tissues. In *Methods of Enzymatic Analysis*, 2nd edn, Vol. 4, ed. H. U. Bergmeyer, Verlag-Chemie, Weinheim, pp. 2266–305.

9 Oxidation of mitochondrial NADH and the synthesis of ATP
Hans Lambers

Introduction

Depending on species and the age of a plant, 30–50% of all the carbohydrates fixed in photosynthesis are respired in the same day. A significant part of this respiration proceeds via a non-phosphorylating pathway, so that energy conservation is less than maximal. Furthermore, improvements in crop yield have been obtained by selecting genotypes with low rates of dark respiration (Lambers, 1985). It is therefore imperative to obtain a good understanding of this important aspect of plant metabolism.

The recent appearance of an excellent monograph (Douce, 1985), Volume 18 in the *Encyclopedia of Plant Physiology* (Douce and Day, 1985) and the proceedings of the international meeting on plant mitochondria, held in Aberystwyth in 1986 (Moore and Beechey, 1987), have been of great help in writing some sections of this chapter. For further information on most of the topics treated here, the reader is referred to these books.

This chapter will first provide the basic information on the structure and functioning of respiratory components, including the biochemistry of substrate oxidation and ATP formation. It will then discuss the stoichiometry of substrate consumption, carbon dioxide release and ATP formation *in vivo*. Finally, the problem of the regulation of respiration *in vivo*, both in the light and in the dark will be addressed. Notwithstanding the rapid progress made in this field in the last decade, the reader will be confronted with questions which cannot yet be answered.

The location and organization of the mitochondrial electron transport pathways

Respiration involves the transfer of electrons from organic molecules to oxygen. This transfer occurs via an electron transport chain in the inner mitochondrial membrane. The mitochondrial electron transport chain is a complicated arrangement of some 40 redox centers, 50 polypeptides and significant amounts of phospholipids. One single chain has a 'molecular weight' of approximately 1.52×10^6. The respiratory components of the electron transport pathways are arranged into discrete multiprotein units.

A general description of the components of the respiratory chains in plant mitochondria

In mammalian mitochondria four major complexes (I-IV) are associated with electron transfer and one with the production of ATP (complex V) (Fig. 9.1). The four electron transport complexes are: complex I (catalyzes the transfer of electrons from internal NADH to ubiquinone), complex II (responsible for the transfer of electrons from succinate to ubiquinone), complex III (generally called the bc_1 complex and transfers electrons from ubiquinol to cytochrome c), and complex IV (also called cytochrome c oxidase, catalyzes the transfer of electrons from cytochrome c to oxygen). All of these complexes are located in the inner

Oxidation of mitochondrial NADH and the synthesis of ATP

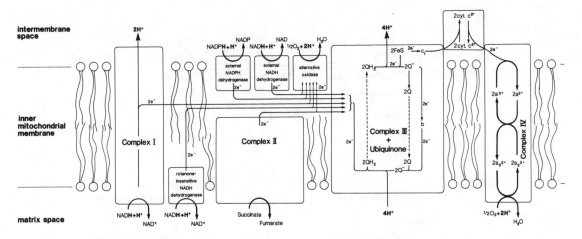

Fig. 9.1 The organization of the four electron transporting complexes (I–IV) of the respiratory chain in higher plant mitochondria. All components are located in the inner mitochondrial membrane. Some are transmembranous; others face the matrix or the space between the inner and the outer mitochondrial membrane (intermembrane space). The exact location of the alternative oxidase in the inner membrane is not certain, but it may face the matrix side. Q (ubiquinone) is a mobile pool of quinone and quinol molecules. This figure is based on information in Douce (1985), Lance *et al.* (1985), Moore and Rich (1985). For further explanation see text.

mitochondrial membrane (Fig. 9.1). Complexes I, III and IV are thought to be embedded in the membrane in such a way that they are in contact with both the intermembrane space and the matrix whereas complex II is only associated with the matrix side. The binding site for NADH of complex I faces the matrix.

Two small redox molecules must be added to the list of electron transferring multiprotein complexes to define the full electron transfer sequence. First, ubiquinone (Q) connects both complex I and complex II to complex III. The quinone and quinol molecules (ubiquinol is the fully reduced form of ubiquinone; the half-reduced form is called ubisemiquinone; Fig. 9.2) operate in the hydrophobic phase of the membrane and are closely associated with complex III. Second, cytochrome *c* facilitates electron transfer from complex III to complex IV.

In higher plant mitochondria some additional components occur. One of these is a second NADH dehydrogenase, also located in the inner membrane, but with its binding site for NADH facing the intermembrane space. A second is a NADPH dehydrogenase, accepting electrons from external NADPH; this NADPH is mainly produced in the oxidative pentose phosphate pathway. Then there is the cyanide-resistant alternative pathway, present in mitochondria from a wide range of plant species. This pathway is also located in the inner membrane, possibly facing the intermembrane space. Finally, there are good arguments to add yet another component: the rotenone-insensitive NADH dehydrogenase, distinct from complex I. It is not blocked by inhibitors of complex I (e.g. rotenone, amytal and piericidin A). Since the dehydrogenase is reduced by internal NADH only, it is presumably facing the matrix. Figure 9.1 summarizes the organization of the components and their location in the inner mitochondrial membrane. A further description is given below.

Complex I and the rotenone-insensitive dehydrogenase

Complex I is the main entry point of electrons from 'internal' NADH, that is, NADH produced in the matrix space during the oxidation of malate, pyruvate, oxoglutarate, isocitrate, and glycine (Fig. 9.1). Both a flavin mononucleotide (FMN) and

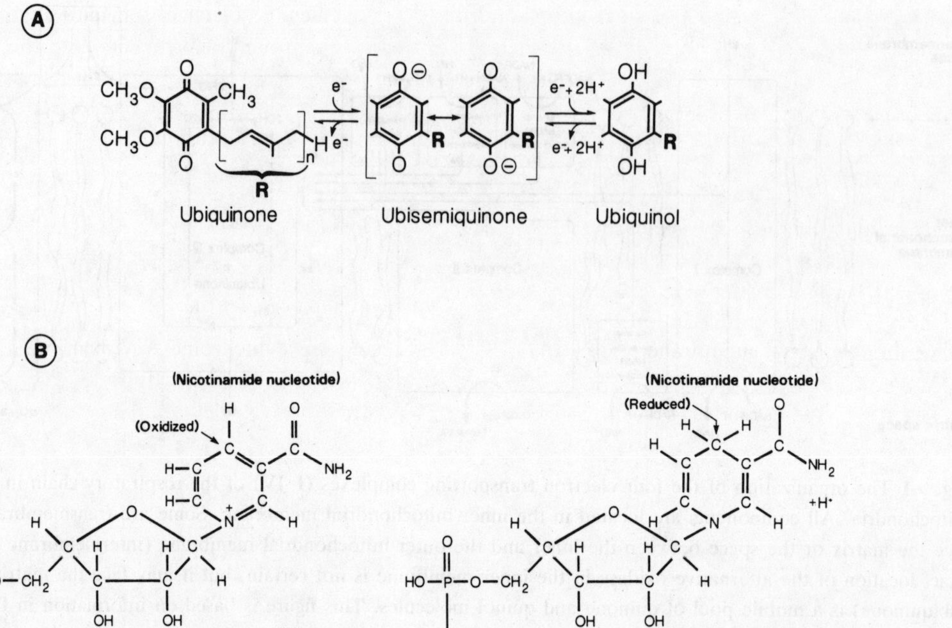

Fig. 9.2 The structure and redox states of ubiquinone (A) and nicotinamide adenine dinucleotide (NAD) (B). Ubiquinone is a substituted 1,4 benzoquinone; the side-chain (R) is composed of 10 isoprenyl units. Reduction of ubiquinone to ubiquinol requires two electrons and two protons. Reduction with only one electron gives ubisemiquinone. Reduction of NAD (NAD$^+$) to NADH + H$^+$ occurs at the 4-position of the ring in the nicotinamide portion of the molecule, as indicated by arrows.

several iron–sulfur proteins are involved in the transfer of electrons from NADH to ubiquinone. This transfer is coupled to the translocation of H$^+$ from the matrix to the intermembrane space, as will be further explained below. Complex I is therefore called the 'first coupling site' or 'site 1' of proton extrusion. The rotenone-insensitive dehydrogenase, unlike complex I, is not linked with proton extrusion.

Complex II

Unlike all other intermediates of the TCA cycle, which are oxidized by matrix enzymes, succinate is oxidized by the membrane-bound enzyme succinate dehydrogenase. Electrons from succinate enter the respiratory chain via complex II. Flavin adenine dinucleotide (FAD), several non-heme iron proteins, and iron–sulfur proteins are involved in the electron transfer to ubiquinone. Unlike complex I, complex II is not connected with the translocation of H^+ across the inner mitochondrial membrane.

The external NAD(P)H dehydrogenases

External, or cytosolic NAD(P)H, like succinate, feeds its electrons into the chain at the level of ubiquinone. Unlike complex I, and similar to complex II, the external dehydrogenases are not connected with the translocation of H^+ across the inner mitochondrial membrane (Fig. 9.1).

Complex III

Complex III (cytochrome c reductase) is responsible for the transfer of electrons from ubiquinol to cytochrome c. Ubiquinones are closely associated with complex III, which contains cytochromes b and c_1 and an iron–sulfur center, named the Rieske iron–sulfur protein after the person who first described it 25 years ago. Both the Rieske protein and cytochrome c_1 are exposed to the intermembrane side. The mechanism of electron transfer, which is coupled to the translocation of four protons per electron pair from the matrix to the intermembrane space, is still poorly understood. It is assumed that a 'Q cycle' allows the observed proton translocation in the following manner. Electrons from the dehydrogenases or complex II are donated to ubisemiquinone (Q^-), which is thought to be fairly immobile in the membrane. Ubisemiquinone reduction to ubiquinol (QH_2) requires two protons per electron, which are taken up from the matrix. The ubiquinol then diffuses to the intermembrane side of the inner membrane, where one electron is donated to the Rieske protein, producing ubisemiquinone. Two protons are then released into the intermembrane space. The chemistry of this redox reduction is outlined in Fig. 9.2. Ubisemiquinone donates an electron to a transmembranous cytochrome b, forming ubiquinone. Q diffuses to the matrix side of the inner membrane where it accepts an electron from cytochrome b so that Q^- is regenerated and the cyle completed (Fig. 9.1). Hence, one electron traverses the electron transport chain and one cycles through cytochrome b. In the process two protons are transported for every electron passing through complex III. Complex III constitutes 'site 2' of proton extrusion.

Complex IV

Complex IV (cytochrome c oxidase) is the terminal oxidase of the cytochrome pathway. It contains cytochromes (cytochrome a and cytochrome a_3) and copper. Cytochrome a, which faces the intermembrane space, accepts electrons from cytochrome c. The electrons are then donated to cytochrome a_3, which reacts with oxygen. Cytochrome a_3 faces the matrix side, so that the reaction with oxygen removes four protons per O_2 from the matrix. This makes complex III the third coupling site, although the protons are not extruded from the matrix as at sites 1 and 2, but are removed via covalently binding to oxygen.

The cyanide-resistant path

Respiratory O_2 consumption of many higher plant tissues is not fully inhibited by inhibitors of the cytochrome path such as KCN (Table 9.1). It is now widely accepted (Lance *et al.*, 1985) that this component of respiration is due to a cyanide-resistant, alternative electron transport pathway firmly embedded in the inner mitochondrial membrane. Triton X-100, an anionic detergent, affects the alternative path at much lower concentrations than the cytochrome path. This suggests that the alternative path, or some component(s) thereof, faces the intermembrane space, where it is easily accessible to detergent. Since the effect of inhibitors of complex IV on mitochondrial oxygen uptake and that of antimycin, which inhibits complex III (Table 9.2), is the same, the branching point of the alternative path from the cytochrome path must be before complex III. The observation that succinate oxidation, and in some tissues also oxidation of external NAD(P)H, is also cyanide-sensitive leads

Table 9.1 A comparison of the cyanide resistance of respiration of roots and leaves of a number of species and of oxygen uptake by mitochondria isolated from these tissues. The percentage cyanide resistance of tissue respiration was calculated from the rate measured in the presence of 0.2 mM cyanide and the rate measured in the presence of 0.1 μM FCCP, which uncouples oxidative phosphorylation and electron transport; this was done to obtain a rate of electron transfer through the cytochrome path similar to the state 3 rate. Cyanide resistance of isolated mitochondria was calculated from the rate in the presence and absence of 0.2 mM KCN. Mitochondrial substrates were 10 mM malate plus 10 mM succinate and a saturating amount of ADP. Cyanide-resistant oxygen uptake by isolated mitochondria was fully inhibited by SHAM and disulfiram; in the presence of both KCN and SHAM approximately 10% of the control respiration proceeded in some of the tissues ('residual respiration'). Intact tissue respiration and oxygen uptake by mitochondria was measured on the same batch of plants. (Data from Lambers et al., 1983.)

Species	Tissue	Cyanide-resistance (%)	
		Whole tissue	Mitochondria
Gossypium hirsutum	Roots	36	22
Phaseolus vulgaris	Roots	61	41
Spinacea oleracea	Roots	40	34
Triticum aestivum	Roots	38	35
Zea mays	Roots	47	32
Pisum sativum	Leaves	39	30
Spinacea oleracea	Leaves	40	27

Table 9.2 A compilation of effectors of respiratory metabolism. Those used most frequently in current research or to be preferred because of their specificity are italicized. (Based on information included in Douce, 1985; Douce and Day, 1985; Moore and Beechey, 1987.) Of those compounds which are naturally found in organisms, (one of) the source(s) is included in the last column (mainly based on information in Harborne, 1982; Douce, 1985).

Inhibitor of	Compound	Source
Complex I	*Rotenone*	*Derris* roots
	Amytal	
	Piericidin A	*Streptomyces*
	Benzyladenine	*mobaraensis*
	Glyceollin	*Glycine max*
Complex II	Malonate	
Complex III	*Antimycin*	
	Myxothiazol	
	2-n-heptyl-1-hydroxy-quinoline-N-oxide (HQNO)	
Complex IV	*Cyanide*	
	Azide	
	Carbon monoxide	
Alternative path	Disulfiram	
	Benzhydroxamic acid (BHAM)	
	m-Chlorohydroxamic acid (m-CLAM)	
	Salicylhydroxamic acid (SHAM)	
	Propyl gallate	
	8-Hydroxyquinoline (8-OHQ)	
Oxidative phosphorylation (uncouplers)	Carbonyl cyanide-m-chlorophenyl hydrazone (CCCP)	
	Carbonyl cyanide-p-trifluoromethoxy-phenyl-hydrazone (FCCP)	
	Dinitrophenol (DNP)	
	Pinosylvin monomethylether	*Pinus* sp.
Membrane potential	Valinomycin + K$^+$	*Streptomyces* sp.
H$^+$/P$_i$-symporter	N-ethylmaleimide (NEM)	
Complex V (ATP synthetase)	Oligomycin	*Streptomyces diastatochrogenes*
Adenine nucleotide translocator	Carboxy-atractyloside	*Atractylis gummifera*

to the conclusion that the branching point is after complex I. This has implicated ubiquinone as a component common to both pathways (Fig. 9.1). It is likely that the alternative path contains one or more proteins (Elthon and McIntosh, 1986), one of which was found to have a molecular weight of approximately 38 000 (Berthold et al., 1987). Complete identification of the protein(s) has yet to occur, despite the wide acceptance of the existence of this path for over a decade.

Transport of electrons from ubiquinone to oxygen via the alternative path is not coupled to the extrusion of protons from the matrix to the intermembrane space. This is a major difference in comparison with the cytochrome path. In searching for a physiological function of the alternative path this difference has to be taken into consideration.

Substrate oxidation, proton extrusion and oxidative phosphorylation

The major substrates

In mammalian cells, pyruvate is the major endproduct of glycolysis and serves as the major substrate for mitochondrial respiration. In addition to pyruvate, plant mitochondria use malate as a major TCA cycle substrate (see Chapter 8; Bryce and ap Rees, 1985). In this case malate can be viewed as an alternative glycolytic endproduct. It is formed from the carboxylation of phosphoenolpyruvate (PEP) by PEP carboxylase which yields oxaloacetate (OAA). The OAA is then reduced to malate by malate dehydrogenase. The malate formed in the cytosol is transported across the inner membrane of the mitochondria where some may be oxidized in a conventional manner via the TCA cycle and some via malic enzyme, producing pyruvate and CO_2. The pyruvate is then available for subsequent oxidation via pyruvate dehydrogenase and the TCA cycle (see Chapter 8). In addition to the oxidation of TCA cycle intermediates, glycine produced via the photorespiratory pathway is also oxidized in the mitochondria. A discussion of this process is presented in Chapter 18. Glycine has rapid access to complex I; its oxidation tends to be favored over that of TCA cycle intermediates (see Chapter 18).

Figure 9.3 outlines the major pathways of mitochondrial carbon oxidation and the associated production of NADH and $FADH_2$.

The mitochondrial 'states'

Freshly isolated intact mitochondria do not

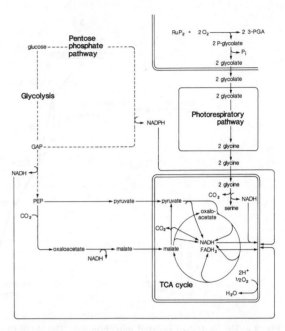

Fig. 9.3 The major substrates for the electron transport pathways are malate, pyruvate, intermediates of the TCA cycle, NADH from glycolysis (oxidized via the external NADH dehydrogenase only), NADPH from the oxidative pentose phosphate pathway (also oxidized via an external dehydrogenase), and glycine (in tissues with an operative photorespiratory pathway).

consume an appreciable amount of oxygen. Upon addition of one of the substrates included in Fig. 9.3, there is a small increase in oxygen uptake. As soon as ADP is added, a rapid consumption of oxygen can be measured. This 'state' of the mitochondria is called 'state 3'; the two earlier ones are referred to as 'state 1' (mitochondria, but no substrate) and 'state 2' (substrate present), respectively. Upon conversion of all ADP into ATP, the respiration rate of the mitochondria declines again to the rate found before addition of ADP. The state is referred to as 'state 4'. Upon addition of more ADP, the mitochondria go into state 3 again, followed by state 4 upon depletion of ADP. This can be repeated until all oxygen in the cuvette has been consumed (Fig. 9.4). Thus the respiratory activity is effectively controlled by the availability of ADP. This phenomenon is called 'respiratory control' and is quantified by the

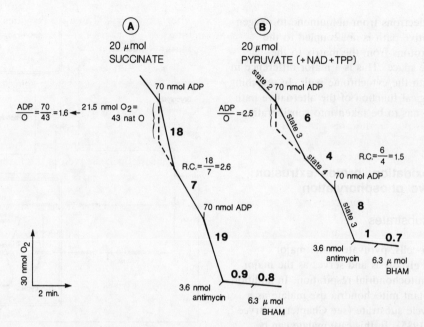

Fig. 9.4 The 'states' of isolated mitochondria. The ADP:O ratio is calculated from the oxygen consumption during the phosphorylation of a known amount of added ADP (state 3 rate) after correction for the rate of oxygen in state 4. The respiratory control ratio (RC) is the ratio of the state 3 and the state 4 rate of oxygen uptake. State 1 is the respiration rate in the absence of substrate and ADP; it was not determined in this experiment. Mitochondria were isolated from *Iris hollandica* bulbs as described by Hemrika-Wagner *et al.* (1982); rates are expressed as nmol O_2 min^{-1} (mitochondria isolated from one g fresh weight)$^{-1}$. Final volume of the assay mixture was 0.9 ml. Unpublished data from A. Marissen, Free University of Amsterdam.

'respiratory control ratio': the ratio of the rate of O_2 consumption at substrate saturation in the presence of ADP to that after ADP has been depleted (Fig. 9.4).

The oxygen consumed during the phosphorylation of ADP, after subtraction of the state 4 rate, can be related to the total amount of ADP added. Thus the ADP:O ratio can be calculated. This ratio varies for different substrates. In the absence of inhibitors, it is approximately 3 for NAD-linked substrates and 2 for succinate and external NAD(P)H (Fig. 9.4). The rationale for the observed ADP:O values and effects of inhibitors are discussed below.

Inhibitors, uncouplers and other effectors of respiratory metabolism

Respiratory inhibitors (Table 9.2) have been an important tool in elucidating the organization of the respiratory pathways, as summarized in Fig. 9.1. First, there are the inhibitors of complex I, of which rotenone is the most specific. There are others, such as amytal, piericidin A, and benzyladenine (a synthetic cytokinin). A range of inhibitors can also be isolated from plant tissues, e.g. glyceollin (from *Glycine max*). However, even when complex I is completely inhibited, internal NADH may still be oxidized. This results from the presence of a bypass, generally called the 'rotenone-insensitive dehydrogenase'. Succinate oxidation is not inhibited by inhibitors of complex I, and this is one line of evidence supporting the position of succinate dehydrogenase in Fig. 9.1.

Succinate oxidation is competitively inhibited by malonate, a dicarboxylic acid containing three carbon atoms as compared with succinate which contains four. Malonate has no effect on the oxidation of NAD-linked substrates or external NADH.

Antimycin binds to cytochrome b and inhibits complex III. Cyanide, azide and carbon monoxide bind to cytochrome a_3 and inhibit complex IV. These compounds have been termed inhibitors of the cytochrome path. When the cytochrome path is blocked, the oxidation of NAD-linked substrates, succinate and external NAD(P)H may not be fully inhibited. This is due to the presence of the alternative path. Both complex III and the alternative path accept electrons from ubiquinone. In the presence of an inhibitor of the cytochrome path and an inhibitor of the alternative path, oxygen uptake by isolated mitochondria is virtually completely inhibited (Fig. 9.4). Oxidation of external NADH and succinate are not always affected to the same extent by inhibitors of the cytochrome path. This is somewhat surprising since it is thought that the oxidation of both substrates is connected with ubiquinone.

Substituted hydroxamic acids, e.g. salicylhydroxamic acid (SHAM), chlorohydroxamic acid (m-CLAM) and benzhydroxamic acid (BHAM), are frequently used with isolated mitochondria as specific inhibitors of the alternative path. They can also be used *in vivo*, but care has to be taken to choose the correct concentration (Møller *et al.*, 1988). Disulfiram, though active in lower concentrations, penetrates isolated mitochondria very slowly and is not widely used. It is not an appropriate inhibitor for experiments with intact tissues. Propyl gallate has also been used as an inhibitor of the alternative path.

Uncouplers are useful tools in the study of plant respiration. They make membranes, including the inner mitochondrial membrane, permeable to protons, therefore inhibiting oxidative phosphorylation, as will be explained below. Many compounds belonging to this category are found as 'secondary compounds' in plant tissues.

Valinomycin, in the presence of K^+, also increases the conductance for protons across the inner mitochondrial membrane, because valinomycin allows rapid entry of K^+ ions into the mitochondria. K^+ ions are then exchanged for protons. Thus, in the presence of K^+, valinomycin dissipates the membrane potential. In the absence of K^+, valinomycin has no effect on the membrane potential, indicating that its effect differs from that of other uncouplers. Table 9.2 provides a summary of effectors of respiratory metabolism.

Proton extrusion linked with electron transport

During the transfer of electrons via the cytochrome electron transport chain, some of the energy released is conserved in the form of ATP (Fig. 9.4). However, ATP formation is not directly linked to electron transport, but rather to the generation of the proton gradient across the inner mitochondrial membrane as described by the chemiosmotic theory (Mitchell, 1966). The transport of protons out of the mitochondrial matrix can be demonstrated by adding a small amount of oxygen to a suspension of intact mitochondria in a lightly buffered medium that contains a suitable substrate (Mitchell and Moyle, 1967). Acidification of the medium does not occur when the inner membrane is damaged by detergents. These observations are in agreement with the chemiosmotic theory (Mitchell, 1966).

The basic features of this now widely accepted chemiosmotic model are that (1) protons are transported outwards, coupled to the transfer of electrons down the electron transport chain. This gives rise to both a proton gradient (ΔpH) and a membrane potential ($\Delta \psi$); (2) the inner membrane is impermeable to protons, hydroxyl ions and other ions, except by special transport systems; and (3) there is an ATP synthetase, which transforms the energy of the electrochemical gradient, generated by the proton extruding system, into ATP.

The pH gradient, ΔpH, and the membrane potential, $\Delta \psi$, are interconvertible, and it is the combination of the two which forms the protonmotive force (Δp), the driving force for ATP synthesis:

$$\Delta p = \Delta \psi - 2.3\, RT/F\, \Delta pH$$

where F is Faraday's number, so that all components in the above equation are expressed in mV.

Δp is estimated by separate determination of $\Delta \psi$ and ΔpH. This is commonly done by measuring the steady-state distribution of a permeant ion, i.e. a compound which diffuses freely across the inner membrane. Once the concentration of the permeant ion inside and outside the mitochondria is known, $\Delta\psi$ can be calculated using the Nernst equation, and ΔpH can be calculated using the Henderson–Hasselbalch equation. In the presence of valinomycin, potassium ions are freely transported across the inner membrane and thus this system can be used to estimate $\Delta\psi$ (the system is called a $\Delta\psi$-probe). Alternatively, synthetic lipophylic cations (e.g. methyltriphenylphosphonium (TPMP$^+$) or tetraphenylphosphonium (TPP$^+$)), can be used as $\Delta\psi$-probes. Permeative weak acids (e.g. acetate) are used to measure ΔpH (ΔpH-probes). For a description of other techniques used to determine Δp, the reader is referred to Moore and Rich (1985). Higher plant mitochondria in state 4 have Δp ranges from 153 to 262 mV. $\Delta\psi$ contributes 126–250 mV, whereas the ΔpH contributes 12–36 mV (the latter is the equivalent of 0.5 pH units) (Douce, 1985; Moore and Rich, 1985).

The ratio of H$^+$ extrusion to oxygen uptake has been determined for various substrates (Table 9.3). Since some of the protons flow back in exchange for inorganic phosphate, the H$^+$/P$_i$ symporter has to be inhibited to allow a proper estimation of the stoichiometry. Unfortunately, the inhibition of the H$^+$/P$_i$ symporter by N-ethylmaleimide (NEM) also reduces the H$^+$/O stoichiometry when malate is the substrate (Table 9.3).

The determination of the 'sites' of proton extrusion, as indicated in Fig. 9.1, involved the use of different substrates in combination with inhibitors and artificial electron acceptors. Table 9.4 shows that the H$^+$/O ratio for malate oxidation is 6.96 in the absence of rotenone and 4.88 in the presence of this inhibitor of complex I. Subtracting the value in the presence of rotenone from that in its absence leads to the conclusion that the H$^+$/O ratio of site 1 is approximately 2. The H$^+$/O ratio of succinate oxidation is 6.27 in the absence and 0.0 in the presence of antimycin, an inhibitor of complex III. This indicates that the H$^+$/O ratio of sites 2 + 3 is 6.27 and that the alternative path is not coupled to proton extrusion. Oxidation of ascorbate, which donates electrons to complex IV

Table 9.3 The stoichiometry of proton extrusion and electron transfer associated with oxidation of various substrates in *Phaseolus aureus* (mung bean) and rat liver mitochondria. Values were determined in the absence and presence of N-ethylmaleimide (NEM), which inhibits proton uptake associated with inorganic phosphate (H$^+$/P$_i$ symport). Note that NEM was found to reduce the stoichiometry for NAD-linked substrates; the reason for this is unknown, but it complicates the calculation of the stoichiometry per site (see text and Table 9.4). In the presence of N, N, N', N'-tetramethyl-p-phenylene diamine (TMPD), ascorbate donates electrons to cytochrome c. (Based on data in Mitchell and Moore, 1984.)

Source	Substrate	Treatment	H$^+$/O ratio
Mung bean	Malate	−NEM	6.96
	Malate	+NEM	5.87
	Succinate	−NEM	4.58
	Succinate	+NEM	6.27
	NADH	−NEM	4.64
	NADH	+NEM	5.43
Rat liver	Ascorbate + TMPD	−NEM	2.58
	Succinate	−NEM	4.84
	Succinate	+NEM	6.30

Table 9.4 The stoichiometry of H$^+$/O in mung bean and *Sauromatum guttatum* (last line only) mitochondria, calculated per site, based on information given in Table 9.3. Note that the stoichiometry for sites 2 + 3 cannot be derived from measurements with malate as substrate in the absence of rotenone, since NEM cannot be used in this case so that some protons were allowed to move back to the matrix side with phosphate via the H/P$_i$-symporter. For the same reason the stoichiometry of site 1 may have been somewhat underestimated. (Based on data from Mitchell and Moore, 1984.)

Substrate	Treatment	Sites	H$^+$/O ratio
Malate	Control	1 + 2 + 3	6.96
	+ rotenone	2 + 3	4.88
	(by difference)	1	2.08
Succinate	Control	2 + 3	6.27
	+ antimycin	alt. ox.	0
Ascorbate + TMPD	+ antimycin	3	2.58
	(by difference)	2	3.69
Malate	+ antimycin	1	2.57

in the presence of tetramethyl-p-phenylene diamine (TMPD), gives a H^+/O ratio of 2.58, the stoichiometry for site 3. Combined with the value for the H^+/O ratio of sites 2 + 3, a H^+/O ratio for site 2 of 3.69 is found. The H^+/O ratio for site 1 can also be determined using malate in the presence of an inhibitor of the cytochrome path, e.g. antimycin. Following this approach, a value of 2.57 was found for *Sauromatum guttatum* mitochondria (Table 9.4).

Thus, the H^+/O ratio for site 1 is around 2 (2.08–2.57); the H^+/O ratio for site 2 is close to 4 (3.69); and that of site 3 is close to 2 (2.58) (Table 9.4). For complex I this stoichiometry can be explained by the redox reactions involved. The complex is reduced by NADH (or $NADH + H^+$, producing NAD^+). The electron transfer from NADH to complex I is thus coupled to the removal of two protons from the matrix, which are released in the intermembrane space upon donation of electrons to ubiquinone. The stoichiometry of site 2 can be explained by the operation of a 'Q cycle', described above (Figs 9.1 and 9.2). Reduction of oxygen (concomitantly with the uptake of two protons to form water) at the matrix side of the inner membrane explains the stoichiometry of site 3. The absence of proton translocation when external NADH is oxidized is due to the fact that the protons donated to ubiquinone are taken up from the intermembrane space (to which they are released again thereafter), rather than from the matrix, as is the case with internal NADH. Similarly, succinate oxidation in complex II is not linked with proton extrusion. Any protons taken up from the matrix during the oxidation of succinate are released to the matrix again upon donation of electrons to ubiquinone.

ATP formation linked with proton re-entry

According to the chemiosmotic theory (Mitchell, 1966) proton extrusion, coupled to electron transport in the respiratory chain, leads to a protonmotive force (Δp). Δp is the driving force for ATP synthesis. Thus ATP synthesis in mitochondria is associated with the re-entry of protons to the matrix, as mediated by complex V

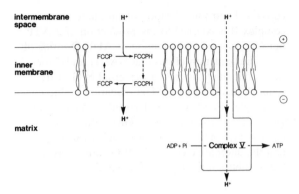

Fig. 9.5 A model of complex V (coupling factor or ATP synthetase). This complex is embedded in the inner mitochondrial membrane. It allows proton re-entry, coupled to the phosphorylation of ADP at the inside surface of the inner membrane. Approximately three protons re-enter per ATP formed. Uncouplers such as FCCP (Table 9.2) inhibit oxidative phosphorylation by allowing proton re-entry without the ATP synthetase.

(coupling factor or ATP synthetase; Fig. 9.5). This is a reversible reaction, as ATP hydrolysis leads to the extrusion of protons. Although the exact stoichiometry of proton re-entry to ATP formation is still under debate, values around three are widely accepted (Moore and Rich, 1985).

The stoichiometry of H^+/O per site, in combination with the stoichiometry of ATP formation per proton, explains the difference in ADP:O ratios found for different substrates (Fig. 9.4). For pyruvate oxidation, involving all sites of proton extrusion, an ADP:O ratio of (2.57 + 3.69 + 2.58)/3 = 2.94 is expected. For succinate oxidation, involving site 2 and site 3 only, an ADP:O ratio of (3.69 + 2.58)/3 = 2.90 is expected. For a number of reasons the ADP:O values found in isolated mitochondria are often lower than calculated, but the relative values tend to be 3.0:2.0 for internal NADH and succinate, respectively (Fig. 9.4). Recent *in vivo* studies employing ^{31}P-NMR have allowed the calculation of ADP:O ratios in intact tissue (Roberts, 1984; Roberts *et al.*, 1984). This work supports the contention that NADH oxidation by complex I is coupled to the production of 3 ATP per O consumed and the oxidation of succinate via complex II is coupled to the production of 2 ATP per O consumed. Observation of ADP:O ratios of 3 for the oxidation of internal NADH and 2 for the oxidation of succinate or external NAD(P)H requires that internal NADH oxidation be linked to complex I and that all electron transport occurs via the cytochrome pathway. Given the presence of both the rotenone-insensitive dehydrogenase and the alternative chain, ADP:O ratios may in fact exhibit a high degree of variability.

ATP formation and CO_2 production *in vivo*

From information discussed above the relationship between ATP production and CO_2 release can be calculated as can the consumption of O_2 during the complete oxidation of hexose and other substrates in the plant cell. This theoretical value of ATP production can then be compared with some experimental data that have recently become available. These calculations assume an ADP:O ratio of 3.0 for internal NADH and 2.0 for succinate or external NAD(P)H.

ATP production coupled to hexose oxidation

Since ADP:O ratios for the oxidation of TCA cycle intermediates and external NADH have been established, it should be possible to calculate the ATP production per hexose molecule when it is fully oxidized to CO_2 and H_2O in glycolysis and the TCA cycle. We can carry out these calculations assuming glycolysis ends with either the production of 2 molecules of pyruvate or 2 molecules of malate per hexose.

If pyruvate is the endproduct the following analysis applies (see Chapter 6):

Hexose + $2NAD^+$ + 2ADP + $2P_i \rightarrow$
2 pyruvate + 2NADH + $2H^+$ + 2ATP [9.1]

If the two molecules of NADH are oxidized via the externally facing dehydrogenase of the electron transport chain, 4ATP will be produced. Glycolysis will therefore yield 6ATP. The subsequent oxidation of pyruvate via the TCA cycle would occur as follows (see Chapter 8):

2 pyruvate + 2FAD + $8NAD^+$ + 2ADP + $2P_i \rightarrow$
$6CO_2$ + $2FADH_2$ + 8NADH + $8H^+$ + 2ATP [9.2]

Oxidation of 2 molecules of $FADH_2$ via complex II will yield 4ATP while oxidation of 8 molecules of NADH via complex I will yield 24ATP. The result is that TCA cycle oxidation of 2 molecules of pyruvate generates 30ATP. Hence, the ATP yield for complete oxidation of a hexose via glycolysis and the TCA cycle is 36ATP/hexose. The net reaction becomes:

Hexose + $6O_2$ + 36ADP + $36P_i \rightarrow 6CO_2$ + $6H_2O$ + 36ATP [9.3]

If malate is considered the endproduct of glycolysis the analysis differs slightly but the final ATP yield remains unchanged. In this case PEP, rather than being converted to pyruvate via pyruvate kinase, is converted to OAA via PEP carboxylase. The OAA is then reduced to malate. The overall reaction for glycolysis becomes:

Hexose + $2CO_2 \rightarrow$ 2 malate [9.4]

The reason there is no net ATP production is that pyruvate kinase has been bypassed. The lack of net NADH production results from the consumption of the NADH produced by glyceraldehyde 3-phosphate dehydrogenase in the reduction of OAA to malate. The malate thus produced is imported into the mitochondrion where it may be decarboxylated by malic enzyme to form pyruvate.

2 malate + $2NAD^+ \rightarrow$ 2 pyruvate + 2NADH + $2H^+$ + $2CO_2$ [9.5]

The net effect is that NADH generated in the cytosol during glycolysis is imported to the mitochondrion via the sequential action of three enzymes (PEP carboxylase, malate dehydrogenase and malic enzyme). The oxidation of these 2NADH via complex I yields 6ATP, unlike the 4ATP generated from the oxidation of 2 molecules of cytosolic NADH. This gain of 2ATP offsets exactly the 2ATP lost in bypassing pyruvate kinase. As the pyruvate produced in this pathway is available for TCA cycle oxidation according to eqn [9.2], the final ATP yield for complete oxidation of hexose via 'malate glycolysis' is also 36ATP/hexose (eqn [9.3]).

If the oxidative pentose phosphate pathway, rather than glycolysis, is employed, the ATP yield per hexose is lower. The first part of this oxidation can be described as the cytosolic production of one molecule of glyceraldehyde 3-phosphate (GAP) from hexose:

1 hexose + 1 ATP → 1 hexoseP + 1ADP [9.6]

3 hexoseP + 6NADP$^+$ → 2 hexoseP + 1GAP + 6NADH + 3CO$_2$ [9.7]

Glyceraldehyde 3-phosphate can be further transformed, starting with the reactions of glycolysis and ending with pyruvate. The sum of these reactions is:

1GAP + NAD$^+$ + ADP + P$_i$ → 1 pyruvate + ATP + NADH [9.8]

Combining eqns [9.6], [9.7] and [9.8] yields:

1 hexose + 1NAD$^+$ + 6NADP$^+$ → 1 pyruvate + 1NADH + 6NADPH + 3CO$_2$ [9.9]

The oxidation of this NADH and NADPH by the externally facing dehydrogenase (ADP:O ratio of 2) yields 14ATP. The subsequent oxidation of the pyruvate according to the stoichiometry of eqn [9.2] would yield 15ATP. As a result, complete oxidation of the hexose via the pentose phosphate pathway and the TCA cycle would yield 29ATP. The balanced equation can be written as:

Hexose + 29ADP + 29P$_i$ + 6O$_2$ → 6CO$_2$ + 29ATP + 6H$_2$O [9.10]

O$_2$ consumption and CO$_2$ release coupled to the oxidation of hexose and other substrates: the respiratory quotient

The molar ratio of CO$_2$ released to O$_2$ consumed is called the respiratory quotient (RQ). From reaction eqns [9.3] and [9.10] it is calculated that the RQ for the complete oxidation of hexose to CO$_2$ + H$_2$O is 1.0, irrespective of the pathways involved in the oxidation. Complete oxidation of substrates which are more oxidized than hexose yields an RQ greater than 1. From reaction eqns [9.2] and [9.5] it can be calculated that the complete oxidation of malate yields an RQ of 1.3. The oxidation of more reduced substrates, e.g. of fatty acids in some seeds, gives an RQ smaller than 1. Oxidation of fatty acids produces 1 NADH per acetyl-CoA, the oxidation of which produces 2CO$_2$, 3NADH and 1FADH$_2$. Fatty acid oxidation thus gives an RQ of 0.8. Conversely, when a net synthesis of malate occurs in the cell or when fatty acids are synthesized, the RQ is smaller than 1 and greater than 1, respectively. Transfer of electrons to acceptors other than oxygen, e.g. nitrate, also increases the RQ. In most tissues which do not rapidly metabolize fatty acids or organic acids, the RQ is slightly above 1.0.

Regulation of the partitioning of electrons between the cytochrome and the alternative pathways and between the two internal NADH dehydrogenases

The existence of two respiratory pathways, both transporting electrons from ubiquinol to oxygen, and two internal NADH dehydrogenases raises the question how electron flow is regulated between the two paths. This is particularly relevant, since the cytochrome path and the internal rotenone-sensitive dehydrogenase are coupled to proton extrusion, whilst both the rotenone-resistant internal dehydrogenase and transport of electrons from ubiquinol to oxygen via the alternative path are not coupled to the generation of a protonmotive force.

The cytochrome path and the alternative path

Bahr and Bonner (1973) were the first to conclude that simple competition for electrons between the alternative pathway and the cytochrome pathway cannot explain the experimental data. Further evidence for this has come from 'titration experiments' carried out both with isolated mitochondria and intact tissues (Theologis and Laties, 1978; Fig. 9.6). In these experiments oxygen uptake was measured at a range of concentrations of an inhibitor of the cytochrome path in both the absence and presence of an inhibitor, which fully blocks the alternative path. If inhibition of the alternative path increased the flow of electrons through the cytochrome path, inhibitors of the cytochrome path would be less

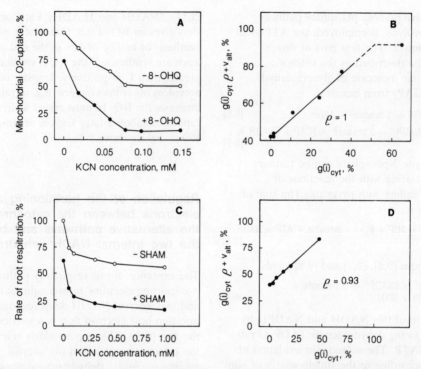

Fig. 9.6 Oxygen uptake by mitochondria isolated from callus-forming discs of *Solanum tuberosum* (A and B) and by intact roots from *Carex acutiformis* (C and D) at a range of concentrations of a specific inhibitor of the cytochrome path, in the absence and presence of a concentration of an inhibitor which fully blocks the alternative path. In B and D, the rates obtained in the absence of an inhibitor of the cytochrome path are plotted against those obtained in the presence of an inhibitor blocking the cytochrome path. A straight line with a slope of 1 indicates that the alternative path does not become engaged until the cytochrome path reaches saturation. In A and B, the applied inhibitor of the alternative path was 8-hydroxyquinoline (8-OHQ); in C and D it was salicylhydroxamic acid (SHAM). For further explanation, see text and Theologis and Laties (1978). (Data on isolated mitochondria: unpublished information provided by A.M. Wagner; data on intact roots: redrawn from information in Van der Werf *et al.*, 1988.)

effective under these conditions than in the absence of inhibitors of the alternative path. This has not been found to be the case indicating that inhibition of the alternative path has no effect on the activity of the cytochrome path. Thus, the alternative path does not compete for electrons with the cytochrome path, and only becomes active when the cytochrome path is almost fully saturated with electrons. In the absence of definitive information on the nature of components of the alternative path, it is impossible to describe the biochemical nature of such a regulatory mechanism.

The two internal dehydrogenases

Though the two internal dehydrogenases have not been physically separated, some of their kinetic parameters have been studied. Møller and Palmer (1982) measured the K_m of the internal rotenone-sensitive dehydrogenase to be 8 μM, which is an order of magnitude lower than that of the resistant one. The lower affinity of the rotenone-resistant dehydrogenase, and the lack of any proton extrusion, suggests that it is presumably operative when the NADH/NAD ratio is high, or when the availability of ADP is low,

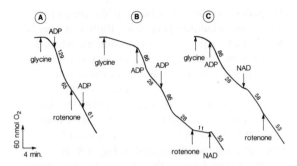

Fig. 9.7 Oxygen uptake by mitochondria isolated from *Pisum sativum* leaves. Trace A was obtained with untreated mitochondria, traces B and C with mitochondria depleted of NAD. Added NAD is rapidly taken up by mitochondria and converted into NADH in the matrix, in the presence of NAD-linked substrates, e.g. glycine. Assay conditions were: 10 mM glycine, 0.4 μmol ADP, 0.2 mM NAD, 15 μM rotenone, 0.33 (A) and 0.27 (B and C) mg of mitochondrial protein. (Data from Day *et al.*, 1987.)

such as under state 4 conditions. Figure 9.7 shows that rotenone had very little effect on oxygen uptake of freshly isolated pea leaf mitochondria in state 4 (Day *et al.*, 1987). However, when these mitochondria were deprived of their endogenous NAD by incubating them for 4 h at 4°C, rotenone had a marked effect on the state 4 rate of oxygen uptake. Addition of NAD, which is readily taken up and converted into NADH in the presence of an NAD-linked substrate such as glycine, increased oxygen uptake in the presence of rotenone. When NAD was added prior to rotenone, this inhibitor had very little effect. The data in Fig. 9.7 lead to the following conclusions: (1) inhibition by rotenone demonstrates that the rotenone-insensitive path has insufficient capacity to take over the role of the sensitive path; and (2) rotenone only has an appreciable effect when the NADH level in the mitochondria is low. Thus it appears that the rotenone-insensitive path can only become active in the presence of a high level of NADH in the mitochondria. Note, however, that the results in Fig. 9.7 do not provide information about the operation of the resistant dehydrogenase in the absence of inhibitors. Furthermore, evidence for the operation of the rotenone-insensitive path in intact tissues is lacking. In the absence of any suitable inhibitor of the rotenone-insensitive bypass such evidence will be hard to obtain.

Regulation of electron transport through the cytochrome path

Respiratory control, as referred to in Fig. 9.4, can be explained by the chemiosmotic theory. In the absence of ADP, the protonmotive force increases and restricts the flow of electrons to O_2. Information such as that included in Fig. 9.4 gives the impression that there is an abrupt transition from state 3 to state 4. For a long time it has been generally believed that mitochondria do not operate in a state somewhere between state 3 and state 4. More subtle experimentation with isolated mitochondria has revealed that mitochondria can operate in such an intermediate state. More importantly, it has become apparent that, *in vivo*, mitochondria often operate in this intermediate state.

Isolated mitochondria

As shown in Fig. 9.4, addition of ADP to isolated mitochondria causes a rapid increase in respiration until all ADP has been depleted and transition to state 4 occurs. When small doses of ADP are continually added, this does not occur. This can be done with the 'hexokinase–glucose system'. ATP is added to the mitochondrial suspension, in conjunction with hexose and the enzyme hexokinase. This enzyme catalyzes the formation of hexose phosphate, thus releasing ADP. The rate at which ADP is released can be carefully controlled by varying the added activity of hexokinase.

When ADP is added with the 'hexokinase–glucose system', a range of oxidation rates can be obtained (Fig. 9.8). Simultaneously with the decrease in succinate oxidation the $\Delta\psi$ increases. This is to be expected, since only in the presence of ADP is the re-entry of protons via complex V possible. When the rate of ADP phosphorylation is controlled by varying degrees of

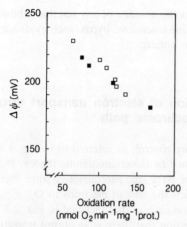

Fig. 9.8 The relationship between $\Delta\psi$ and the rate of succinate oxidation in mitochondria isolated from potato tubers. The phosphorylation was limited by the rate of supply of ADP (closed symbols), using the 'hexokinase–glucose system'; the enzyme activity was varied from 0–17 nkat. The phosphorylation was also controlled by varying the concentration of carboxyatractyloside from 0 to 0.2 μM (open symbols). This compound inhibits the import of ADP into the mitochondria. $\Delta\psi$ was measured with the TPP$^+$ method. (Data from Diolez and Moreau, 1987.)

inhibition of ADP import (using carboxyatractyloside), the same relationship between $\Delta\psi$ and the rate of succinate oxidation is found (Fig. 9.8). Clearly, intermediate states of mitochondrial respiration are possible.

The control of the activity of the cytochrome path in intact tissues

Having established that mitochondrial states, intermediate between state 3 and state 4, can be experimentally induced *in vitro*, it is of interest to find out in which state the mitochondria operate *in vivo*. This can be done by using combinations of specific inhibitors, uncouplers and substrates providing electrons for the respiratory chain.

If the addition of an uncoupler, e.g. DNP or FCCP (Table 9.2), stimulates the rate of respiration in the absence of an inhibitor of the alternative path, respiration must have been limited by adenylates. However, this does not necessarily indicate limitation of oxidative phosphorylation by adenylates, as illustrated for isolated mitochondria (Fig. 9.8). Rather, it may be the substrate supply to the mitochondria which is restricted by adenylates, e.g. via the control of glycolysis. However, if the respiration is first shown to be reduced by an inhibitor of the alternative path (e.g. SHAM), and still stimulated by uncoupler in the presence of SHAM, it must be electron transport *per se* that is limited by ADP, in a manner illustrated in Fig. 9.8. If uncoupler stimulates in the absence but not in the presence of SHAM, adenylates control respiration via the substrate supply to the mitochondria. Comparison of the effects of an exogenous substrate which escapes the control of glycolysis (e.g. glycine oxidation during photorespiration) with one that is oxidized via glycolysis (e.g. sucrose) allows us to discriminate between the two possibilities of control by adenylates.

Table 9.5 shows that indeed adenylates control the rate of oxygen uptake in a number of intact tissues and that this control is sometimes exerted via an effect on the substrate supply (Table 9.5a), and sometimes through limitation of the flux of electrons through the respiratory chain (Table 9.5b). The relief of adenylate control of the cytochrome path can actually decrease the flux through the alternative path (Azcón-Bieto *et al.*, 1983). However, respiration is not invariably controlled by adenylates. Saglio and Pradet (1980) found that respiration of *Zea mays* root tips, depleted of sugars, could be stimulated by exogenous sugars, but not by DNP. The uncoupler was only effective in the presence of a high endogenous level of sugars or when glucose was exogenously supplied at 0.2 M. A similar observation was made on *Triticum aestivum* leaves, harvested at the end of the night (low carbohydrate levels) and after several hours of photosynthesis (high carbohydrate levels (Azcón-Bieto *et al.*, 1983; Table 9.5c).

From current information it appears that adenylate control of electron transport and glycolysis often coincides with a regulation by the availability of substrate. Clearly, substrate not only controls respiration in 'starved' cells, but also in tissues going through a normal diel cycle (Table 9.5c). Increased substrate levels tend to increase the activity of the alternative path, rather than that of the cytochrome path (Lambers, 1985).

Table 9.5 Control of respiration of intact tissues: (a) by adenylates, via their effect on glycolysis; (b) by adenylates, via their effect on the cytochrome path; and (c) by the substrate supply to the mitochondria. The reasoning behind the conclusion of which factors exert the major control is further explained in the text.
(a) Mature leaves of *Lolium perenne*. Rates are expressed as percentages of the basal rate (7.3 μmol g^{-1} fresh wt h^{-1}). Applied concentrations: CCCP, 2 μM; sucrose and glycine, 40 mM; SHAM, 8 mM; KCN, 0.2 mM. (Data from Day et al., 1985.)

Addition	Oxygen consumption (%)
A None	100
+ CCCP	122
+ CCCP + sucrose	136
+ CCCP + sucrose + SHAM	135
+ CCCP + sucrose + SHAM + KCN	26
B None	100
+ sucrose	102
+ sucrose + CCCP	123
C None	100
+ glycine	119
+ glycine + CCCP	132

(b) Roots of *Zea mays* (A) and *Triticum aestivum* (B). Rates are expressed as percentages of the basal rate (230 and 200 μmol g^{-1} dry wt h^{-1}) for *Z. mays* and *T. aestivum*, respectively. Applied concentrations: CCCP, 2 μM; SHAM, 10 mM; KCN, 0.5 mM. (Data from Day and Lambers, 1983.)

Addition	Oxygen consumption (%)
A None	100
+ KCN	47
+ SHAM	70
+ CCCP	141
+ SHAM + CCCP	91
B None	100
+ KCN	36
+ SHAM	69
+ CCCP	117
+ SHAM + CCCP	86

(c) Mature leaves of *Triticum aestivum*, harvested at the end of the night (A and B) and after 6 hours of photosynthesis (C and D). Rates are expressed as percentages of the basal rate (0.7 and 1.0 μmol m^{-2} s^{-1}) for leaves at the end of the night and after 6 hours of photosynthesis, respectively. SHAM had no effect on the respiration of control leaves harvested at the end of the night and inhibited 29% when leaves were harvested after 6 hours of photosynthesis. KCN slightly stimulated and inhibited 13%, respectively. Applied concentrations: FCCP, 1 μM; sucrose, 60 mM; glycine, 100 mM; SHAM, 5 mM; KCN, 0.3 mM. (Data from Azcón-Bieto et al., 1983.)

Sequential addition	Oxygen consumption (%)
A None	100
+ sucrose	143
+ SHAM	118
+ KCN	0
B None	100
+ FCCP	110
+ sucrose	183
C None	100
+ sucrose	100
+ FCCP	120
+ SHAM	98
+ KCN	0
D None	100
+ glycine	103
+ FCCP	130

The physiological function of the alternative path

In trying to understand the physiological role of the cyanide-resistant path, it is important to take into account its non-phosphorylating nature and its operation during high rates of substrate oxidation. A second point is that there may be plant cultivars lacking the alternative path (Musgrave et al., 1987) suggesting it is not essential for metabolism. However, no matter how exciting these recently discovered cultivars are, it should be kept in mind that their respiration may not invariably show a lack of alternative path capacity (R. Welschen and H. Lambers, unpublished).

Heat production

In most of the recent textbooks on plant physiology, the role of the alternative path in heat production in the spadix of *Arum* lilies is mentioned. During its 'respiratory crisis' the respiration rate of this reproductive organ approaches the rate found in the flying muscles of hummingbirds, and the temperature rises to approximately 10 °C above ambient (Meeuse, 1975). Consequently, odoriferous amines are volatilized and pollinators are attracted. During the respiratory crisis, respiration appears to be largely cyanide resistant. However, not only the respiration of these reproductive organs but also that of roots and leaves is often at least partly cyanide resistant. The application of specific inhibitors has demonstrated that the alternative path is often active, particularly in roots where it may be responsible for up to 40% of all respiration. The role of the alternative path in roots and leaves appears not to be one of heat production (McNulty and Cummins, 1987) and a number of hypotheses have been put forward to explain its role in these organs.

Energy overflow

Titration experiments, as described in Fig. 9.6, demonstrate that the alternative path is only engaged when the cytochrome path becomes saturated. Furthermore, increases in carbohydrate supply to cells appear to increase alternative path engagement. These observations have led to the 'energy overflow hypothesis' (Lambers, 1980), which states that respiration via the alternative path only proceeds in the presence of high concentrations of respiratory substrate. It considers the alternative path as a sort of 'coarse control' of carbohydrate metabolism, but not as an alternative to the finer control by adenylates discussed above (Day and Lambers, 1983). It is quite possible that the continuous oxidation of part of the substrate via a non-phosphorylating electron transport path allows the plant to increase the availability of carbon for sinks which suddenly arise (Lambers, 1985), but this interpretation remains speculative.

Energy overcharge

As already alluded to, electron flow via the cytochrome path may increase at the expense of the alternative path activity if ADP availability increases. On the other hand, an increased availability of ADP in cells in which the cytochrome path operates at maximum capacity may lead to increased engagement of the alternative path. This model was termed the 'energy overcharge model' (De Visser *et al.*, 1986) and is likely to be valid under non-steady-state conditions and/or when the activity of the cytochrome path is no longer controlled by adenylates.

NADH oxidation in the presence of a high energy charge

If cells require a large amount of carbon skeletons (e.g. oxoglutarate or succinate) yet have no large need for ATP, the operation of the alternative path could prove useful (Palmer, 1976). Though this suggestion is theoretically sound, it is hard to envisage such a situation *in vivo*. Whenever the rate of carbon skeleton production is high, there tends to be a great need for ATP to further metabolize and incorporate these skeletons. Also, when the carbon skeletons are used for the synthesis of amino acids, significantly more NADH is required for the reduction of nitrate (if this is the source of N) than is generated in the production of carbon skeletons. Lance *et al.* (1985) have suggested the need for a non-phosphorylating path to allow rapid oxidation of malate in the absence of a large need for ATP. Although this suggestion has a sound biochemical rationale, it remains hard to envisage a physiological situation where a rapid rate of NADH oxidation is required in the absence of ATP consumption.

Dark respiration in green cells during the light

In the light there are some fundamental differences between a photosynthetically active and a heterotrophic cell. A photosynthetic cell has two different systems to produce ATP: photophosphorylation and oxidative phosphorylation. For a long time it was believed that photophosphorylation would drain all the ADP away from the mitochondria and so restrict oxidative phosphorylation. A number of recent developments have changed this view. First, the discovery of the non-phosphorylating path in leaves has made this contention more difficult to accept; the path would allow electron transport to continue even at low levels of available ADP. Second, it seems likely that the level of ADP in the mitochondria is not as low in the light as originally believed. And third, the observation that glycine has preferred access to the respiratory chain also places this hypothesis in doubt. It is now no longer a question whether dark respiration continues during the light, but rather, to what extent (Gardeström and Edwards, 1985).

Adenylate levels in cell compartments during light and dark

Using an ingenious technique for rapid (within 0.1 s) preparation of subcellular fractions, Stitt et al. (1982) were able to measure adenine nucleotide levels in the cytosol, chloroplasts and mitochondria from wheat leaf protoplasts. Levels were measured in protoplasts exposed to light or kept in the dark.

Light did not have a significant effect on the total level of adenylates in the protoplasts. However, it decreased the level of ADP and increased that of ATP in the mitochondria, whereas the levels in the cytosol were virtually unaffected. These results cannot be reconciled with the previous belief that the lack of ADP in the cytosol, due to rapid import into the chloroplasts, restricts the flow of electrons through the mitochondrial electron path. Moreover, the ATP/ADP ratio at which electron flow through the cytochrome path becomes restricted exceeds 20 (Wiskich and Dry, 1985). This is considerably higher than the ratio of 2.3 which Stitt et al. (1982) found in mitochondria during photosynthesis. Clearly, there is no reason to believe that mitochondrial electron flow in the light is restricted by the lack of ADP in the cytosol and/or by the high ATP/ADP ratio in the mitochondria.

Effects of inhibitors of oxidative phosphorylation on photosynthesis

Oligomycin, a specific inhibitor of oxidative phosphorylation (Table 9.2), inhibits the rate of photosynthetic O_2 evolution in barley leaf protoplasts (Ebbighausen et al., 1987). The same inhibitor has no effect on the rate of photosynthesis in disrupted protoplasts. The lack of inhibition in disrupted protoplasts is because mitochondria and chloroplasts no longer interact in the broken cells. The inhibition of photosynthesis by oligomycin in the intact protoplasts shows that interruption of oxidative phosphorylation inhibits the rate of photosynthesis. This is taken as evidence for the occurrence of oxidative phosphorylation, and thus for operation of the cytochrome path, during the light. It also suggests a role for dark respiration in the supply of ATP for CO_2 assimilation during photosynthesis.

Concluding remarks

In the past several decades, considerable progress has been made in improving our understanding of both the biochemistry and physiology of mitochondrial energy metabolism. In spite of this progress there are still large gaps in our knowledge. Given the role of respiratory losses in determining plant productivity, further insights are required into the problems associated with plant respiration. Only with the combined efforts of plant physiologists, biochemists and molecular biologists will major progress be realized. This will lead to interesting applications of fundamental knowledge to the rational improvement of crop productivity.

Acknowledgments

I wish to thank a number of colleagues and students who have greatly contributed to this manuscript by their constructive criticism. In alphabetical order they are: Ad Borstlap, Adrie van der Werf, Ronald Hetem, Hendrik Poorter, Linus van der Plas and Anneke Wagner.

Further reading

Douce, R. (1985). *Mitochondria in Higher Plants. Structure, Function, and Biogenesis*, Academic Press, Orlando.

Douce, R. and Day, D. A. (eds) (1985). *Encyclopedia of Plant Physiology*, New Series, Vol. 18, Springer-Verlag, Berlin.

Moore, A. L. and Beechey, R. B. (eds) (1987). *Plant Mitochondria. Structural, Functional and Physiological Aspects*, Plenum Press, New York.

Nicholson, D. G. (1982). *Bioenergetics. An Introduction to the Chemiosmotic Theory*, Academic Press, London.

References

Azcón-Bieto, J., Day, D. A. and Lambers, H. (1983). The regulation of respiration in the dark in wheat leaf slices. *Plant Sci. Lett.* **32**, 313–20.

Bahr, J. T. and Bonner, W. D. (1973). Cyanide-insensitive respiration. 2. Control of the alternate pathway. *J. Biol. Chem.* **248**, 3446–50.

Berthold, D. A., Fluke, D. J. and Siedow, J. N. (1987). A determination of the molecular weight of the aroid alternative oxidase by radiation inactivation analysis. In *Plant Mitochondria. Structural, Functional and Physiological Aspects*, eds A. L. Moore and R. B. Beechey, Plenum Press, New York, pp. 351–9.

Bryce, J. H. and ap Rees. T. (1985). Rapid decarboxylation of the products of dark fixation of CO_2 in roots of *Pisum* and *Plantago*. *Phytochemistry* **24**, 1635–8.

Day, D. A. and Lambers, H. (1983). The regulation of glycolysis and electron transport in roots. *Physiol. Plant.* **58**, 155–60.

Day, D. A., De Vos, O. C., Wilson, D. and Lambers, H. (1985). The regulation of respiration in the leaves and roots of two *Lolium perenne* populations with contrasting mature leaf respiration rates and yield. *Plant Physiol.* **78**, 678–83.

Day, D. A., Wiskich, J. T. and Dry, I. B. (1987). Regulation of ADP-limited respiration in isolated plant mitochondria. In *Plant Mitochondria. Structural, Functional and Physiological Aspects*, eds A. L. Moore and R. B. Beechey, Plenum Press, New York, pp. 59–66.

De Visser, R., Spreen Brouwer, K. and Posthumus, F. (1986). Alternative path mediated ATP synthesis in roots of *Pisum sativum* upon nitrogen supply. *Plant Physiol.* **80**, 295–300.

Diolez, P. and Moreau, F. (1987). Relationship between membrane potential and oxidation rate in potato mitochondria. In *Plant Mitochondria. Structural, Functional and Physiological Aspects*, eds A. L. Moore and R. B. Beechey, Plenum Press, New York, pp. 17–25.

Ebbighausen, H., Hatch, M. D., Lilley, R. McC., Kromer, S., Stitt, M. and Heldt, H. W. (1987). On the function of malate–oxaloacetate shuttles in a plant cell. In *Plant Mitochondria. Structural, Functional and Physiological Aspects*, eds A. L. Moore and R. B. Beechey, Plenum Press, New York, pp. 171–80.

Elthon, T. E. and McIntosh, L. (1986). Characterization and solubilization of the alternative oxidase of *Sauromatum guttatum* mitochondria. *Plant Physiol.* **82**, 1–6.

Gardeström, P. and Edwards, G. E. (1985). Leaf mitochondria (C_3 + C_4 + CAM). In *Encyclopedia of Plant Physiology*, New Series, eds R. Douce and D. A. Day, Springer-Verlag, Berlin, pp. 315–46.

Harborne, J. B. (1982). *Introduction to Ecological Biochemistry*, Academic Press, London.

Hemrika-Wagner, A. M., Kreuk, K. C. M. and van der Plas, L.H.W. (1982). Influence of growth temperature on respiratory characteristics of mitochondria from callus-forming potato tuber discs. *Plant Physiol.* **70**, 602–5.

Lambers, H. (1980). The physiological significance of cyanide-resistant respiration in higher plants. *Plant Cell Environ.* **3**, 293–302.

Lambers, H. (1985). Respiration in intact plants and tissues: Its regulation and dependence on environmental factors, metabolism and invaded organisms. In *Encyclopedia of Plant Physiology*, New Series, eds R. Douce and D. A. Day, Springer-Verlag, Berlin, pp. 418–73.

Lambers, H., Day, D. A. and Azcón-Bieto, J. (1983). Cyanide-resistant respiration in roots and leaves. Measurements with intact tissues and isolated mitochondria, *Physiol. Plant.* **58**, 148–54.

Lance, C., Chauveau, M. and Dizengremel, P. (1985). The cyanide-resistant pathway of plant mitochondria. In *Encyclopedia of Plant Physiology*, New Series, eds R. Douce and D. A. Day, Springer-Verlag, Berlin, pp. 202–47.

McNulty, A. K. and Cummins, W. R. (1987). The relationship between respiration and temperature in leaves of the arctic plant *Saxifraga cernua*. *Plant Cell Environ.* **10**, 319–25.

Meeuse, B. J. D. (1975). Thermogenic respiration in Araoids. *Ann. Rev. Plant Physiol.* **26**, 117–26.

Mitchell, J. A. and Moore, A. L. (1984). Proton stoichiometry of plant mitochondria. *Biochem. Soc. Trans.* **12**, 849–50.

Mitchell, P. (1966). Chemiosmotic coupling in oxidative and photosynthetic phosphorylation. *Biol. Rev.* **41**, 445–502.

Mitchell, P. and Moyle, J. (1967). Respiration-driven proton translocation coupled to ATP hydrolysis in rat liver mitochondria. *Biochem. J.* **105**, 1147–62.

Møller, I. M. and Palmer, J. M. (1982). Direct evidence for the presence of a rotenone-resistant NADH dehydrogenase on the inner surface of the inner membrane of plant mitochondria. *Physiol. Plant.* **54**, 267–74.

Møller, I. M., Berczi, A., van der Plas, L. H. W. and Lambers, H. (1988). Measurement of the activity and capacity of the alternative pathway in intact plant tissues: Identification of problems and possible solutions. *Physiol. Plant.* **72**, 642–9.

Moore, A. L. and Rich, P. R. (1985). Organization of the respiratory chain and oxidative phosphorylation. In *Encyclopedia of Plant Physiology*, New Series, eds R. Douce and D. A. Day, Springer-Verlag, Berlin, pp. 134–72.

Musgrave, M. E., Strain, B. R. and Siedow, J. N. (1987). Response of two pea hybrids to CO_2 enrichment: A test of the energy overflow hypothesis for alternative respiration. *Proc. Natl Acad. Sci. USA* **83**, 8157–61.

Palmer, J. M. (1976). The organization and regulation of electron transport in plant mitochondria. *Ann. Rev. Plant Physiol.* **27**, 133–57.

Roberts, J. K. M. (1984). Study of plant metabolism *in vivo* using NMR spectroscopy. *Ann. Rev. Plant Physiol.* **35**, 375–86.

Roberts, J. K. M., Wemmer, D. and Jardetzky, O. (1984). Measurements of mitochondrial ATP-ase activity in maize root tips by saturation transfer ^{31}P nuclear magnetic resonance. *Plant Physiol.* **74**, 632–9.

Saglio, P. H. and Pradet, A. (1980). Soluble sugars, respiration, and energy charge during aging of excised maize root tips. *Plant Physiol.* **66**, 516–19.

Stitt, M., Lilley, R. McC. and Heldt, H. W. (1982). Adenine nucleotide levels in the cytosol, chloroplasts, and mitochondria of wheat leaf protoplasts. *Plant Physiol.* **70**, 971–7.

Theologis A. and Laties, G. G. (1978). Relative contribution of cytochrome-mediated and cyanide-resistant electron transport in fresh and aged potato slices. *Plant Physiol.* **62**, 232–7.

van der Werf, A., Kooijman, A., Welschen, R. and Lambers, H. (1988). Respiratory costs for the maintenance of biomass, for growth and for ion uptake in roots of *Carex diandra* and *Carex acutiformis*. *Physiol. Plant.* **72**, 483–91.

Wiskich, J. T. and Dry, I. B. (1985). The tricarboxylic acid cycle in plant mitochondria: Its operation and regulation. In *Encyclopedia of Plant Physiology*, New Series, eds R. Douce and D. A. Day, Springer-Verlag, Berlin, pp. 281–313.

IV

Mitochondrion–Cytosol Interaction

VI

Mitochondrion–Cytosol Interaction

10 The mitochondrial genome and its expression
Michael W. Gray

The genetic function of mitochondrial DNA

In 1948, Boris Ephrussi described the *petite mutation* of yeast (*Saccharomyces cerevisiae*), providing the first genetic evidence for the existence of mitochondrial genes. The petite mutation is notable because (1) it is inherited in a non-Mendelian fashion, and (2) petite mutants are unable to grow on non-fermentable substrates such as ethanol or glycerol, whose utilization depends on a functional respiratory chain. These respiratory-deficient mutants grow on glucose, but form small ('petite') colonies compared to wild-type cells. Ephrussi postulated that the petite mutation was localized in an extrachromosomal or cytoplasmic element, termed the rho (ρ) factor; this was later identified as mitochondrial DNA (mtDNA).

The role of mtDNA in the biogenesis of mitochondria, that of supplying a small number of essential polypeptide components of the mitochondrial respiratory chain, is now well established (Tzagoloff and Myers, 1986). Mitochondrial DNA has this same function in all eukaryotes that possess mitochondria. A dozen or so mtDNA-encoded mRNAs, which specify parts of several of the respiratory complexes (Table 10.1), are transcribed from mtDNA and translated by a distinctive mitochondrial protein synthesizing system, the rRNA and tRNA components of which are also encoded by mtDNA. Mitochondrial DNA carries <10% of the genetic information needed to assemble a mitochondrion, with most mitochondrial proteins being encoded in nuclear DNA, synthesized in the cytosol, and imported into the organelle (Chapter 11). Nevertheless, the correct expression of genes encoded in mtDNA, and the coordination of mitochondrial and nuclear gene function, are critical for the assembly of an active electron transport chain, which allows the mitochondrion to carry out coupled oxidative phosphorylation (Chapter 9).

Although the basic function of mtDNA is constant, the precise set of genes encoded in mtDNA is not (Table 10.1). The ND genes, which encode subunits of the NADH dehydrogenase of complex I, are completely absent in the mtDNA of *S. cerevisiae* and some other yeasts (Tzagoloff and Myers, 1986). Genes encoding subunits of ATP synthase are also variably present. It is assumed that 'missing' mitochondrial genes are encoded in the nucleus, presumably having been moved there in the course of evolution (see below).

Three mammalian mtDNAs (human, mouse and ox) have been sequenced in their entirety, and their genetic function has been defined completely (Table 10.1). Where sequence information is still scarce (as in the case of plant mtDNA), additional genes may yet be discovered. Isolated plant mitochondria synthesize ~20 polypeptides (Forde *et al.*, 1978), suggesting that plant mtDNA encodes a greater number of proteins than animal mtDNA. Indeed, several genes unique to plant mtDNA have already been identified by sequence analysis (Table 10.1), as have novel open reading frames (ORFs) that appear to have no counterparts in the mtDNA of other eukaryotes. It seems certain, however, that the genetic information content of plant mtDNA is *not* an order of magnitude greater than that of animal

Table 10.1 Identified genes in mitochondrial DNA

	Human	Yeast[a]	Plants
Ribosomal RNAs			
large subunit	16S	21S	26S
small subunit	12S	15S	18S
5S	–	–	+
Transfer RNAs	22	25	[b]
Protein components of the respiratory chain			
Complex I: NADH-ubiquinol oxidoreductase (NADH dehydrogenase)			
ND1	+	–	(+)[c]
ND2	+	–	(+)[c]
ND3	+	–	+
ND4	+	–	
ND4L	+	–	
ND5	+	–	+
ND6	+	–	
Complex III: Ubiquinol-cytochrome c oxidoreductase			
COB	+	+	+
Complex IV: Cytochrome c oxidase			
COXI	+	+	+
COXII	+	+	+
COXIII	+	+	+
Complex V: H⁺-ATPase (ATP synthase)			
ATP6	+	+	+
ATP8	+	+	–
ATP9	–	+	+
ATPA	–	+	+
Small subunit ribosomal protein		VAR1	S12, S13, S14

[a] Other genetic loci identified in yeast (*S. cerevisiae*) mtDNA include a tRNA synthesis locus (*tsl*), a gene for an intron transposition factor (*fit1*), and maturase ORFs found in the introns of the COXI (aI1–6) and COB (bI2–4) genes (Tzagoloff and Myers 1986).
[b] An unknown number of tRNA genes is encoded by plant mtDNA.
[c] Identification based on sequence analysis and/or heterologous hybridization experiments, but whether loci encode functional genes remains to be established.

mtDNA, in spite of the 15 to 150-fold greater size of plant mtDNA. Although the smallest plant mtDNA is larger than the chloroplast genome of higher plants (Section V), the latter appears to encode ten times as many genes as the former.

Mitochondrial genome diversity

Variation in size and physical form

One of the most remarkable aspects of mtDNA is the tremendous diversity in its size and physical form in different eukaryotes (Sederoff, 1984; Gray, 1989). This structural diversity (Table 10.2) contrasts sharply with the basic genetic conservatism of mtDNA.

The smallest mtDNAs (14–39 kbp) are found in vertebrate and invertebrate animals, although small (<20 kbp) mtDNAs are also present in some fungi (e.g. *Schizosaccharomyces pombe* (17–19 kbp) and *Torulopsis glabrata* (19–20 kbp)) and protists (e.g. the alga *Chlamydomonas reinhardtii* (16 kbp)). Animal mitochondrial DNA varies within a relatively narrow size range, but size differences are more pronounced within fungi and protists. Plants have relatively enormous mtDNAs, which range from a minimum of about

Table 10.2 Size range of mitochondrial DNA[a]

		kbp
Animals		
Vertebrates		14–24
Invertebrates		14–39
Fungi		17–176
Protists		16–75
Plants		
Brassica campestris	turnip	218[b]
Brassica hirta	white mustard	208[b]
Brassica nigra	black mustard	231[b]
Citrullus vulgaris	watermelon	330
Cucumis melo	muskmelon	2 400
Cucumis sativus	cucumber	1 500
Cucurbita pepo	zucchini	840
Pisum sativum	pea	360
Raphanus sativa	radish	242[b]
Spinacia oleracea	spinach	327[b]
Triticum aestivum	wheat	430[b]
Vicia faba	broad bean	285
Zea mays	maize	570[b]

[a] Data are summarized from Sederoff (1984) and supplemented with more recent data from the current literature (see Gray, 1989).
[b] Size of 'master chromosome', estimated from complete physical mapping of the genome.

200 kbp to at least 2400 kbp in size; the latter approximates the size of a bacterial genome (e.g. 4700 kbp in *E. coli*). Even within a single family of plants (the Cucurbitaceae), mtDNA is seen to vary over a 7-fold size range (compare watermelon and muskmelon, Table 10.2).

Supercoiled, covalently closed circular DNA can be isolated in high yield from animal mitochondria. Most fungal mtDNAs are also circular, at least as judged from their physical and genetic maps; only rare intact circular molecules, approximating the estimated genome size, have ever actually been visualized by electron microscopy of *S. cerevisiae* mtDNA. A few mtDNAs are linear, the best characterized being those of the ciliated protozoa *Paramecium aurelia* and *Tetrahymena pyriformis*, as well as that of the unicellular green alga, *C. reinhardtii*. Circular maps have been proposed for several plant mtDNAs; however, as discussed below, most plant mtDNAs are physically heterogeneous, and display an unusual structural complexity.

Variation in genome organization

Given the variation in mtDNA size, it is not surprising that there is also marked diversity in the way in which genes in mtDNA are arranged and expressed. In order to put the plant mitochondrial genome into perspective, we will first consider some of the important features of mitochondrial molecular biology in other eukaryotes. The human and yeast (*S. cerevisiae*) mitochondrial genomes represent two very different organizational patterns, which are briefly summarized here and described more fully elsewhere (Grivell, 1983; Tzagoloff and Myers, 1986).

Human mtDNA, a 'parsimonious' genome, is a model of genetic economy. A circular molecule of 16 569 bp, it encodes 2 rRNAs, 22 tRNAs and 13 proteins (Table 10.1). Over most of its length, there is little or no non-coding sequence; genes are essentially joined end-to-end, with few or no intervening nucleotides. Coding sequences, which do not contain introns, are reduced to a minimum: protein genes are not flanked by 5'- and 3'-non-coding regions, and genes for rRNAs (16S and 12S) are considerably smaller than their bacterial homologs (which specify 23S and 16S rRNAs, respectively). Only the so-called 'D-loop' (displacement loop) region, which varies in length (up to ~1000 bp) in different animal species, does not have a coding function; instead, this region contains important regulatory signals controlling replication and transcription.

Mitochondrial gene order is conserved among closely related animal species (e.g. it is the same in human, mouse, bovine and frog mtDNA), but diverges in more distant comparisons (e.g. between vertebrate and invertebrate mtDNAs). A striking feature of mammalian mtDNA is the interspersion of tRNA genes with rRNA and protein coding genes. The presence of tRNA genes at the beginning and end of most non-tRNA genes provides the basis of a processing model, discussed below, in which tRNA sequences are thought to provide the signals for precise endonucleolytic cleavage of long co-transcripts.

In the 'profligate' yeast (*S. cerevisiae*) mitochondrial genome, genetic information is arrayed more spaciously. Here, coding sequences are separated by AT-rich non-coding spacer sequences, which constitute half of the genome. Introns interrupt a number of genes, some of which are highly split: in one strain of *S. cerevisiae*, for example, the COXI gene contains 9 introns, which together comprise ~85% of the length of the gene. Most tRNA genes are clustered together in *S. cerevisiae* mtDNA; however, a few are scattered around the genome, and these may play a role in processing (Tzagoloff and Myers, 1986). Gene order is quite variable among those fungal mtDNAs studied to date.

Variation in mode of expression

Transcription in mammalian mitochondria reflects the close physical linkage of genes in mammalian mtDNA. Transcription in this system is symmetric and complete, with the two complementary strands ('H' and 'L') being copied in their entirety (most of the genes in mammalian mtDNA are encoded by the H-strand; L-strand transcripts are largely unstable). RNA synthesis from each template strand is initiated at a promoter located within the D-loop region, with single polycistronic transcripts being produced. These transcripts are processed by precise endonucleolytic cleavages, many of which appear to be positioned just before and after the interspersed tRNA sequences. The mRNAs liberated in this way start at or very close to the AUG initiation codon (i.e. they contain few or no 5'-non-coding nucleotides). These nascent mRNAs also lack any 3'-untranslated region, and many do not even have a complete translation termination codon; however, post-transcriptional polyadenylation creates a termination codon in those mRNA transcripts lacking one.

In yeast mtDNA, all genes but one (a tRNA^{Thr} gene) are on the same strand; however, in contrast to mammalian mtDNA, yeast mitochondrial genes are organized into a number of separate transcriptional units (perhaps 10–20), each with its own promoter (and presumably terminator). Transcription in yeast mitochondria begins within a conserved nonanucleotide motif that represents the essential promoter element; known primary transcripts contain two or more coding sequences, and rather more complex processing than occurs in mammalian mitochondria is necessary to generate the mature 5'- and 3'-termini. Unlike mammalian mitochondrial mRNAs, those of yeast are not polyadenylated and have long 5'-untranslated leaders.

Because there are introns in some yeast mitochondrial genes, there must be a mechanism for the post-transcriptional removal of intron sequences from the primary transcript, with splicing together of exons. In yeast mitochondria, an elaborate processing pathway exists in which intron sequences themselves play an essential role in their own removal (Grivell, 1983; Tzagoloff and Myers, 1986). Many of the yeast introns contain ORFs that are contiguous (in frame) with the preceding exon. Translation of such 'fused' reading frames gives rise to hybrid proteins carrying an N-terminal portion of the exon-encoded protein (e.g. COB or COXI) and a C-terminal portion contributed by the intronic ORF. Such proteins, termed *maturases*, act in an autocatalytic fashion to facilitate the removal of the intron sequence that encodes them, thereby destroying the template from which they are made and promoting maturation of the primary transcript.

The plant mitochondrial genome

Physical complexity and heterogeneity

Plant mtDNA is not only exceptionally large (Table 10.2), it displays an unusual physical complexity (Leaver and Gray, 1982; Lonsdale, 1984; Pring and Lonsdale, 1985). Restriction endonuclease patterns of plant mtDNA are characterized by a large number of fragments, some of which are present in submolar or multimolar amounts. These observations initially suggested that plant mtDNA may consist of a physically heterogeneous collection of molecules (Leaver and Gray, 1982).

As isolated from normal tissue, plant mtDNA consists largely of linear molecules with no fixed size distribution. Circular molecules have been found, but their proportion is usually vanishingly small; such molecules are also heterogeneous, although discrete circular size classes have been observed in the mtDNA of some plants. The proportion of circular molecules is observed to be substantially higher in mtDNA prepared from plant cells grown in tissue culture. It is not clear whether this reflects a real difference in the physical form of plant mtDNA in normally grown versus cultured tissue, or is due to a stabilization of the mtDNA during its preparation from tissue culture cells.

Plant mitochondria may also contain small circular and linear DNA species that may or may not share sequence homology with the main mitochondrial genome (Pring and Lonsdale, 1985).Because these plasmid-like DNAs can be lost without any apparent phenotypic effect, they seem to have no essential genetic function. Their origins are obscure, but they do contribute to the physical heterogeneity of plant mtDNA.

Because of the complexity of restriction patterns, and the difficulty in deciding which fragment(s) represent single copy sequence, accurate estimates of the size of individual plant mitochondrial genomes have been difficult to obtain. However, both restriction analysis and second order renaturation kinetics yield values in excess of 200 kbp (Table 10.2), well beyond the size of the largest mtDNAs in other eukaryotes. In eukaryotes, moderately and highly repetitive DNA can constitute a major part of the nuclear genome; however, kinetic complexity measurements provide little evidence of highly repetitious sequences that could account for the large size of plant mtDNA. Mapping and sequencing studies *have* revealed the presence of repeated sequences, some of which are >10 kbp long, that occur only a few times in the plant mitochondrial genome. As discussed below, these repeated sequences appear to play an important role in generating much of the heterogeneity that characterizes plant mtDNA.

Repeated sequences, reciprocal recombination, and a structurally fluid plant mitochondrial genome

A number of models have been proposed to explain the physical complexity and heterogeneity of plant mtDNA (Lonsdale, 1984). However, the greatest insights into the structure and organization of the plant mitochondrial genome have come from physical maps of several plant mtDNAs (Lonsdale *et al.*, 1984; Palmer and Shields, 1984; Stern and Palmer, 1986; Makaroff and Palmer, 1987; Palmer and Herbon, 1987). In both *Brassica campestris* (Chinese cabbage, turnip) and *Zea mays* (maize), the mitochondrial genome can be represented as a single, circular molecule, termed the *master chromosome*, which contains the entire genetic complexity of the organelle DNA. A feature of the master chromosome is the presence of repeated sequences that appear to mediate reciprocal recombination. A result of such recombination is the dispersion of genetic information among a number of subgenomic, circular molecules.

The consequences of recombination between inverted or direct repeats in a circular DNA molecule are illustrated in Fig. 10.1. Recombination between two inverted repeats (Fig. 10.1A) inverts the single copy region between them; this 'flip-flop' phenomenon occurs in chloroplast DNA, which typically has a single pair of inverted repeats containing the rRNA genes. Recombination between two direct repeats in a circular molecule leads to its resolution into two smaller circular molecules, each containing

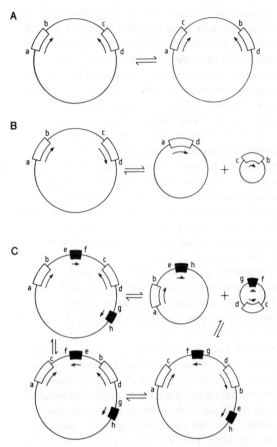

Fig. 10.1 Possible arrangements for repeated sequences in the mitochondrial genome. (A) Single pair of inverted repeats, recombination leads to sequence inversion ('flip-flop'); (B) single pair of direct repeats, recombination leads to two circular products ('loop-out'); (C) a single pair of inverted and direct repeats which are nested. This gives four possible genome configurations. The larger the number of direct and inverted repeats, the more complex the genome organization becomes. Reprinted with permission from Lonsdale (1984).

one copy of the repeat (Fig. 10.1B); reversal of this process results in co-integration of the subgenomic circles to re-form the original circle. Such a tripartite structure, in which three circular chromosomes are related by a co-integration–resolution pathway mediated by reciprocal recombination, has been proposed for *Brassica campestris* mtDNA (Palmer and Shields, 1984). In this model, a master chromosome, 218 kbp long, contains two copies of an ~2 kbp element present as direct repeats separated by 135 and 83 kbp. A similar model describes the spinach mitochondrial genome: in this case, a 327 kbp master circle contains two ~6 kbp direct repeats that participate in an intragenomic recombination event, reversibly generating two subgenomic circles of 93 and 234 kbp (Stern and Palmer, 1986).

As the number of direct and/or inverted repeats increases in a circular molecule, so do the possibilities for recombination (Fig. 10.1C). Maize mtDNA contains six major sequence reiterations, each present twice, ranging in size from 1 to 14 kbp. Five of these repeats are direct, while the 14 kbp repeat is inverted. The master chromosome, which contains one set of each repeat, is estimated to be 570 kbp in size (Lonsdale et al., 1984). All of the repeats except a 10 kbp pair appear to be recombinationally active. As a result, numerous subgenomic circular molecules are postulated to exist in maize mitochondria, in addition to the 570 kbp master circle. A number of the proposed recombination products do, in fact, correspond closely to discrete size classes that had previously been visualized in the population of rare circular molecules isolated from maize mtDNA. Because the postulated master circle and subgenomic circular molecules are very large, they may not easily survive isolation, which could account for the high proportion of heterogeneous linear molecules and low proportion of circular molecules in isolated plant mtDNA.

The *multipartite model* of master chromosome and subgenomic products is attractive because it explains the existence and structural interrelationships of many of the repeated sequences identified in plant mtDNA, and because it can account for much, if not most, of the observed complexity and physical heterogeneity of this genome. It does, however, raise a number of questions about how plant mtDNA is replicated and segregated in the course of cell division. Moreover, in a few instances, pairs of recombining sequences do not seem to be in the predicted equimolar amounts in isolated mtDNA (Falconet et al., 1984); in these cases, the actual structure of the mitochondrial genome may be more complex than the master chromosome model suggests.

Although most plant mtDNAs so far characterized show evidence of high frequency recombination and a multipartite organization, it has recently been reported (Palmer and Herbon, 1987) that the mitochondrial genome of *Brassica hirta* (white mustard) has only one copy of the *B. campestris* 2 kbp repeat, and so exists in the form of a single, circular, 208 kbp chromosome.

Promiscuous DNA in the plant mitochondrial genome

Among many unique characteristics of plant mtDNA, one of the most intriguing is its ability to incorporate and maintain sequences derived from other genomes: so-called *promiscuous DNA*. Stern and Lonsdale (1982) first described a 12 kbp DNA sequence common to both the mitochondrial and chloroplast genomes of maize. The sequence in maize mtDNA obviously originated from maize chloroplast DNA, because it contains the 16S rRNA gene which is found in the inverted repeat of chloroplast DNA. Such promiscuous chloroplast sequences are widely distributed in plant mtDNA, seemingly in random fashion, both with respect to the plant species and the portion of the chloroplast genome that is incorporated into the mitochondrial genome (Pring and Lonsdale, 1985). The finding of promiscuous chloroplast DNA in plant mtDNA was quickly followed by the discovery of mitochondrial and chloroplast sequences in nuclear DNA (Pring and Lonsdale, 1985), firmly establishing the concept of inter-organellar transfer of genetic information.

Ribosomal RNA, transfer RNA and protein coding genes are all represented in the promiscuous chloroplast sequences that have found their way into plant mtDNA. Except in the case of some tRNA genes (see below), there is no evidence to suggest that these chloroplast sequences actually function in plant mitochondria. Promiscuous DNA accounts for part of the 'extra' DNA that the plant mitochondrial genome contains; however, it does not appear that the larger plant mtDNAs contain appreciably more promiscuous DNA than the smaller ones. Perhaps the most significant thing about promiscuous DNA is that it illustrates, in graphic fashion, the structural plasticity and genetic adaptibility of the plant mitochondrial genome.

Genes in plant mitochondrial DNA

Plant mitochondrial genes that have been characterized to date are listed in Table 10.1; some of these (e.g. genes for 5S rRNA, the α-subunit of the F_1-ATPase, and homologs of *E. coli* ribosomal proteins S12, S13 and S14) have not been found in the mtDNA of other eukaryotes. Whether plant mtDNA contains the full set of protein coding genes found in animal mtDNA remains to be determined. Also, it is too early to tell whether different plant mtDNAs encode all of the same genes.

Ribosomal RNAs and their genes

Plant mitochondrial ribosomes sediment at about 78S and contain 18S (small subunit) and 26S (large subunit) rRNA species, which are closer in size to the corresponding eukaryotic cytoplasmic rRNAs than to their homologs in bacteria (16S and 23S), and much larger than animal mitochondrial small subunit (12S) and large subunit (16S) rRNAs (Leaver and Gray, 1982). Plant mitochondrial ribosomes also contain a distinctive 5S rRNA (absent in the mitochondrial ribosomes of other eukaryotes), but lack the separate 5.8S rRNA species of cytoplasmic ribosomes (Leaver and Gray, 1982).

Plant mitochondrial 26S, 18S and 5S rRNAs are all encoded by the plant mitochondrial genome. The rRNA genes were the first genes to be identified in plant mtDNA, and proved to have a novel arrangement (Bonen and Gray, 1980). Genes for 18S and 5S rRNAs are very close together in plant mtDNA, whereas 18S and 26S rRNA genes are far apart. A close physical linkage of 18S and 5S rRNA genes, in the same transcriptional orientation, has been found in all angiosperm mtDNAs examined to date; in wheat, the 5S rRNA coding sequence begins 114 bp downstream of the end of the 18S rRNA gene (Gray and Spencer, 1983). In maize mtDNA, the 26S gene is in the opposite transcriptional orientation from the 18S–5S couple (see Fig. 10.2), with the 18S and 26S genes about 16 kbp apart. The organization of rRNA genes in plant mtDNA contrasts sharply with that in eubacterial and chloroplast genomes, where rRNA genes are arrayed and co-transcribed in the order 16S–23S–5S.

In wheat mtDNA, rRNA genes are located in two of the ten sets of repeat sequences that this genome

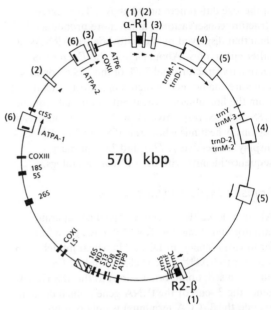

Fig. 10.2 Location of genes on the 570 kbp master circle of the mitochondrial genome of maize. The approximate size and orientation (inner arrows) of repeated DNA sequences are shown (numbered open boxes). The black boxes represent sequences whose origin and function have been identified; these include known mitochondrial genes (rRNA and protein coding), designated as in Table 10.1, as well as individual tRNA (trn) genes (C, cysteine; D, aspartic acid; F, phenylalanine; fM, methionine (initiator); M, methionine (elongator); S, serine; Y, tyrosine). Numeric suffixes (-1, -2, -3) denote different members of the same gene family. Chloroplast DNA-homologous sequences are also shown: LS, gene for large subunit of ribulose 1,5-bisphosphate carboxylase; 16S, small subunit rRNA gene, part of a 12 kbp chloroplast-homologous sequence (hatched box); ct5S, chloroplast 5S rRNA gene. In addition, sequences related to the R1 and R2 'episomal DNAs' are indicated. Figure updated from Dawson *et al.* (1986) and kindly supplied by Dr D. M. Lonsdale.

contains. In maize and a number of other plant mtDNAs, the rRNA genes are single copy. It is not yet known how expression of the same genes, differing in gene dosage in different plant mtDNAs, is regulated.

Plant mitochondrial rRNAs show a striking structural resemblance (both in primary sequence and potential secondary structure) to their eubacterial homologs (Spencer *et al.*, 1984), in spite of the size differences noted above. The degree of structure conservation is much more pronounced than that displayed by the mitochondrial rRNAs of other eukaryotes. About 84% of the *E. coli* 16S secondary structure and 78% of the 23S secondary structure can be reconstructed in almost exact detail from the homologous wheat mitochondrial sequences (18S and 26S, respectively); within these conserved regions, wheat mitochondrial 18S and 26S rRNAs display, respectively, 77% and 73% primary sequence identity with their *E. coli* homologs.

Transfer RNAs and their genes

About a dozen tRNA genes, including separate initiator and elongator tRNA^{Met}, have been found so far in various plant mtDNAs. In wheat mtDNA, an initiator tRNAs^{Met} gene is closely linked (in the same transcriptional orientation) to the 18S rRNA gene: the 3'-end of the tRNA gene (which does not encode the 3'-CCA terminus) is only one bp removed from the 5'-end of the 18S gene (Gray and Spencer, 1983). This is a somewhat surprising finding, given the spaciousness of the plant mitochondrial genome. In maize mtDNA, however, the same tRNA gene is located elsewhere in the genome. For the most part, plant mitochondrial tRNA genes have been found to be solitary, not closely linked to other tRNA or to protein coding genes; these lone tRNA genes may therefore constitute individual transcription units.

Like the rRNAs, tRNAs in plant mitochondria are more similar in sequence to their eubacterial and chloroplast counterparts than to the same tRNAs in other mitochondria. Nor do plant mitochondrial tRNAs display any of the primary sequence and secondary structure aberrations (Grivell, 1983) that characterize some fungal and especially animal mitochondrial tRNAs. Some plant mitochondrial tRNA genes bear an especially striking similarity to their chloroplast homologs, i.e. they appear to represent promiscuous chloroplast DNA. There is increasing evidence, however, that at least some of these potentially functional genes do in fact give rise to stable tRNAs that presumably participate in mitochondrial protein synthesis (Marechal *et al.*, 1987); such genes may therefore be viewed as having been recruited from chloroplast DNA.

Mammalian mtDNA encodes only 22 tRNAs (and only a single tRNA^{Met}), while yeast (*S. cerevisiae*) mtDNA specifies 25 tRNAs (Tzagoloff and Myers, 1986). This is less than the number of tRNAs required to read the genetic code using the classical 'wobble' pairing rules proposed by Crick. In mammalian and fungal mitochondria, there is an expanded codon recognition pattern in which a single tRNA species is able to decode all four codons for the same amino acid in a four-codon family. What the situation is in plant mitochondria remains to be determined, and efforts are underway in several laboratories to catalog the complete set of tRNA genes encoded by a given plant mtDNA.

Protein coding genes

Sequence analysis has provided considerable insight into the structure and organization of protein coding genes in plant mtDNA, both in monocotyledons and dicotyledons. The maize COXII gene, the first to be sequenced, proved to have a 794 bp intron (Fox and Leaver, 1981), an unexpected result because the same gene is not split in animal and fungal mtDNA. This finding suggested that plant mitochondrial protein genes might be even more highly split than yeast mitochondrial protein genes, and that numerous long introns in plant mtDNA might at least partially account for its large size. However, no introns have been found in those plant mitochondrial COXI and COB genes that have been sequenced to date, even though the homologous yeast mitochondrial genes are highly split. Other sequenced protein genes from plant mitochondria also lack introns, which therefore must not be abundant in plant mtDNA.

Some interesting points emerge from comparison of the structure of the same gene from different plant mtDNAs. The wheat COXII gene, for instance, has the same intron as the maize COXII gene; however, the wheat intron contains a 422 bp intron-insert. The rice COXII gene has the same intron *and* intron-insert as the wheat COXII gene, but the rice intron-insert has an additional insert of 52 bp. Maize, wheat and rice are all monocots; two dicots, *Oenothera* (evening primrose) and pea, have a COXII gene that lacks an intron. On the other hand, a restriction map of the COXII gene in carrot (another dicot) is coincident with that of the rice COXII gene, indicating that the carrot COXII gene has the rice intron and intron-insert arrangement. This suggests that the plant mitochondrial COXII gene possessed an intron before the divergence of monocots and dicots, and that the intron was subsequently lost in some dicot species. In the wheat and rice COXII genes, the intron-insert sequences are flanked by almost perfect inverted repeats, which in turn are flanked by short direct repeats; these features are characteristic of transposable elements.

Sequence comparisons have established that plant mitochondrial protein coding genes are highly conserved in primary structure, but that similarities disappear, abruptly, outside of the coding region. Comparing wheat and maize COXII genes, for example, exon and intron sequences are equally well conserved (98.9% and 99.3% nucleotide identity, respectively). Immediately upstream of the AUG initiation codon, there is no significant similarity between the maize and wheat sequences, whereas very high similarity continues downstream of the COXII coding sequence for at least 100 nucleotides. This pattern of extremely high conservation of coding sequence, with abrupt disappearance of similarity in flanking regions, is a consistent feature of plant mitochondrial genes, and may be indicative of rearrangements that have moved coding regions into new sequence contexts.

In plant mitochondrial protein genes, about 40% of the codons have T in the third position; this value is rather constant from gene to gene, and may well be diagnostic for bona fide protein coding genes. The only apparent divergence from the universal genetic code in plant mitochondria is the use of CGG to encode tryptophan rather than arginine (Fox and Leaver, 1981); TGA is used as a normal termination codon in plant mitochondria, rather than coding for tryptophan as it does in animal, fungal and protozoan mitochondria.

Gene maps

A number of genes have now been located on physical maps of several plant mtDNAs. An updated genetic map of the maize mitochondrial genome is shown in Fig. 10.2. Genes are distributed around the circular map, on both strands, with little evidence of tight clustering. Gene maps have also been published for spinach (Stern and Palmer, 1986) and *Brassica campestris* (Makaroff and Palmer, 1987) mtDNAs; these maps are completely different from each other and from the maize map in Fig. 10.2. Genes that are single copy in one mitochondrial genome may be located in recombination repeats in another; examples of such duplicated genes are ATPA in maize mtDNA, COXII in *B. campestris* mtDNA, and the rRNA genes in wheat mtDNA. These observations prompted Makaroff and Palmer (1987) to declare that 'the order and arrangement of genes in plant mitochondrial genomes is irrelevant to their proper expression and function'.

Expression of the plant mitochondrial genome

As the molecular architecture of the plant mitochondrial genome has unfolded, attention has begun to shift to questions of expression. In a situation in which the same gene may be flanked by totally different sequences in different mtDNAs, and where gene order as a whole is not conserved, we would obviously like to know what motifs are used to signal the beginning and end of transcription, and how these promoter and terminator sequences are arrayed. As well, we would like to know what proportion of the large plant mitochondrial genome is, in fact, transcribed.

Transcription

Because of its relative simplicity, the *Brassica* mitochondrial genome offers some distinct advantages in trying to answer such questions. In a complete transcriptional mapping of the 218 kbp *B. campestris* mitochondrial genome, Makaroff and Palmer (1987) found 24 abundant and positionally distinct transcripts of size >500 nucleotides. This number corresponds closely to the estimated number (~20) of polypeptides synthesized by isolated plant mitochondria (Forde *et al.*, 1978). The abundant transcripts accounted for ~30% of the *B. campestris* mitochondrial genome; in addition a number of less abundant transcripts, many of which overlapped each other and the major transcripts, were also detected. Although *B. campestris* mtDNA contains a number of promiscuous chloroplast sequences, none of these appeared to be transcribed within the mitochondrion.

The results of gene and transcript mapping are consistent with coding regions being randomly distributed throughout the plant mitochondrial genome, unlinked and independently transcribed. Both simple and complex transcription patterns have been observed for the same gene in different plants. In wheat, for example, a *COXII* exon-specific probe detects one abundant transcript of size 1.5 kbp; in maize, a number of discrete transcripts are revealed by a *COXII* exon probe. These differences undoubtedly relate to the fact (noted previously) that 5'-flanking sequences, which presumably carry transcriptional control signals, are very different in the two cases. However, whether complex transcript patterns reflect multiple promoter sites, a multiplicity of processing sites, or recombinations between transcribed regions of different mitochondrial genes, remains to be determined.

In view of the variability in gene arrangement in different plant mtDNAs, there is considerable interest in the biochemical mechanism of transcription in plant mitochondria. Plant mitochondrial mRNAs contain 5'- and 3'-non-coding regions that may range up to several hundred nucleotides in length; these messages do not, however, appear to be polyadenylated. In *Oenothera*, some mitochondrial mRNAs have 5'-ends that map to a consensus motif, AAGTGAGG . . ., that could represent the promoter (Hiesel *et al.*, 1987). However, capping experiments with guanylyl transferase need to be done to distinguish nascent termini (carrying a 5'-tri- or diphosphate) from processed termini (carrying 5'-monophosphate); i.e. to distinguish putative transcription initiation sites from processing sites. Potential transcription terminator sites, sharing nucleotide sequence identity in the vicinity of 3'-termini, have also been identified in plant mitochondrial mRNAs (Schuster *et al.*, 1986); these regions can be folded into potential secondary structures similar to bacterial terminators. Once efficient and accurate *in vitro* transcription and processing systems are developed, it will be possible to test putative promoter, terminator and processing signals in a functional assay.

The complexity of the plant mitochondrial genome raises other fundamental questions about RNA synthesis in plant mitochondria. Are identical promoters used for rRNA, tRNA and protein coding genes? Is a single RNA polymerase responsible for all transcription in plant mitochondria? Is transcriptional activity modified by *trans*-acting factors? If such factors exist, are there different ones for different plant mitochondrial genes or classes of genes? As of the present, these questions remain to be answered.

Translation

Although isolated plant mitochondria synthesize a discrete array of polypeptides in an energy-dependent fashion (Forde *et al.*, 1978), an *in vitro* plant mitochondrial translation system has not yet been developed. Consequently, little is known about the details of translation in plant mitochondria. In view of the strongly eubacterial character of plant mitochondrial rRNAs (Spencer *et al.*, 1984), it is anticipated that the translation system as a whole will exhibit eubacteria-like properties. One such property is antibiotic sensitivity, with protein synthesis in plant mitochondria being strongly inhibited by chloramphenicol but resistant to cycloheximide (Forde *et al.*, 1978). The existence of a eubacteria-like 'Shine–Dalgarno' interaction between plant mitochondrial mRNAs and 18S rRNA has been proposed, but is still controversial (Boer *et al.*, 1985) and has yet to be demonstrated directly.

Cytoplasmic male sterility

Cytoplasmic male sterility (CMS) is a widely distributed, non-Mendelian trait that prevents the production of viable pollen. Because CMS is valuable for the production of hybrid seed, the molecular biology of this phenomenon is being studied intensively. Although CMS has been reported in over 140 plant species, its cause has been investigated in only a few; such studies suggest that CMS is a complex phenomenon, with perhaps a number of different etiologies. However, substantial evidence does implicate the mitochondrial genome as the determinant of the CMS trait. Here we will consider a specific case, involving the Texas (T−) cytoplasm of maize, as an example of the alterations in mtDNA and mitochondrial function that are seen to accompany CMS and reversion from it; more comprehensive and detailed discussions of the molecular basis of CMS can be found elsewhere (Leaver and Gray, 1982; Hanson and Conde, 1985).

The observation that CMS can result from interspecific crosses first suggested that the trait may be a consequence of some kind of incompatibility between the nuclear genome of one species and the cytoplasm of another. In maize, three types of CMS are recognized (*cms-T, cms-C* and *cms-S*), which are uniquely distinguished by nuclear restorer genes that suppress the cytoplasmically-determined CMS phenotype. These three maize types are also distinguished by characteristic differences in mtDNA restriction profiles, and by a characteristic spectrum of variant polypeptides synthesized by isolated mitochondria. In particular, an additional, 13 kD polypeptide is made in *cms-T* mitochondria. Nuclear suppression of *cms-T*, resulting in fertility restoration, leads to suppression of the synthesis of the 13 kD polypeptide, which is also missing in spontaneous fertile revertants. Maize plants carrying the *cms-T* cytoplasm are preferentially susceptible to the fungal pathogen *Helminthosporium maydis* race T, which produces a toxin that selectively disrupts the functioning of *cms-T* mitochondria. Sensitivity to the toxin is strictly correlated with the presence or absence of the variant 13 kD mitochondrial polypeptide: nuclear suppression of *cms-T* or spontaneous reversion to fertility both result in resistance to the *H. maydis* toxin, at the level of isolated mitochondria as well as whole plants. This linkage between toxin sensitivity and male sterility suggests that in *cms-T* mitochondria, a single defective gene specifies an altered mitochondrial membrane polypeptide that determines the disease susceptibility–male sterility syndrome.

The techniques of restriction mapping, cloning and sequencing have been used to advantage in investigating the molecular basis of *cms-T* in maize. Dewey *et al.* (1986) sequenced a 3547 bp DNA region (TURF 2H3) of *cms-T* mtDNA, a region selected because it gives rise to abundant transcripts unique to *cms-T*. This region contains two ORFs, ORF13 and ORF25, encoding 13 kD and 25 kD polypeptides, respectively. An ORF25 probe hybridizes to transcripts from all maize cytoplasms (normal and male sterile), and therefore appears to be a normal, active mitochondrial gene; on the other hand, an ORF13 probe hybridizes only to transcripts from *cms-T* mitochondria. ORF13 is a remarkable, mosaic gene that apparently originated from multiple rearrangement events that brought together portions of the flanking and/or coding regions of the maize mitochondrial 26S rRNA gene, the *ATP6* gene, and a promiscuous chloroplast tRNA^{Arg} gene. That the 13 kD, *cms-T*-specific polypeptide is indeed the product of the ORF13 gene is indicated by immunoprecipitation experiments using antibodies raised to a synthetic peptide derived from the ORF13 sequence (see Wise *et al.*, 1987).

Other evidence linking the TURF 2H3 region to the *cms-T* trait is the fact that TURF 2H3 transcripts appear to be uniquely altered in *cms-T* plants restored to fertility by nuclear restorer genes. Moreover, the TURF 2H3 region contains a 6.7 kbp *Xho*I fragment, specifically associated with *cms-T* mtDNA, that was found to be absent in 33 of 34 male-fertile, toxin-resistant *cms-T* revertants. In the remaining revertant, the 6.7 kbp *Xho*I fragment remains, but contains a G→A transition adjacent to a 5 bp insertion not found in TURF 2H3. The insertion is internal to the ORF13 reading frame and generates a frameshift, resulting in a premature stop codon that truncates the ORF13 gene product from 13 kD to 8.3 kD (Wise *et al.*, 1987). While it still must be proven that the 13 kD polypeptide is the direct cause of the *cms-T* phenotype, the evidence so far strongly supports this idea.

Why should a defect in mitochondrial function, apparently only manifested in the presence of a fungal toxin, specifically affect male reproductive development in maize? This fundamental issue must still be settled. It has been speculated that an anther-specific substance, mimicking the *H. maydis* toxin in its effects on *cms-T* mitochondria, may be produced during normal pollen development (Leaver and Gray, 1982). In any event, an emerging theme in CMS is rearrangement of plant mtDNA, with the creation of novel functional regions. The products of nuclear restorer genes may then serve to modify transcripts produced from these novel functional regions, so as to minimize or eliminate their deleterious effects. Further detailed investigation of transcription in these mosaic gene regions seems certain to throw further light on the control of normal gene function in the structurally fluid plant mitochondrial genome.

Evolution of plant mitochondrial DNA and the origin of mitochondria

As well as presenting two radically different patterns of genome organization and expression, animal and plant mtDNAs show essentially opposite modes of evolution. Gene arrangement is strongly conserved in animal (at least mammalian) mtDNA, whereas gene sequence is not: in fact, at the level of primary sequence, animal mitochondrial genes are among the most rapidly evolving genes known. In contrast, as we have seen, plant mitochondrial genes diverge extremely slowly in primary sequence, and in this respect are among the most conservative genes known. On the other hand, gene arrangement is extremely variable in plant mtDNA, even among closely related plant species. A high rate of rearrangement, a hallmark of plant mtDNA, is apparently the dominant factor in its evolution.

In spite of the marked variability that characterizes mitochondrial genome organization and expression (Gray, 1989), it has been possible, using rRNA sequence comparisons, to adduce evidence for a eubacterial, endosymbiotic origin of all known mitochondrial genomes (Gray et al., 1984). Because plant mitochondrial rRNAs so strikingly resemble their eubacterial homologs (Spencer et al., 1984), they have provided some of the most compelling support for the endosymbiont hypothesis (Gray, 1983). Using the wheat mitochondrial 18S rRNA sequence, Yang et al. (1985) were able to trace the origin of mitochondria to a specific group of eubacteria, the α-subdivision of the so-called purple bacteria. This is precisely the group of eubacteria that had earlier been proposed, on biochemical grounds, to contain the nearest contemporary relatives of mitochondria. What has happened to the mitochondrion and its genome since the endosymbiotic event(s) that gave rise to this organelle, and why the evolutionary pathways taken have been so different in different eukaryotes, are matters of continuing keen interest and not a little speculation (Gray, 1989).

Summary and future prospects

Our current picture of plant mtDNA is one of a structurally fluid genome of modest coding capacity, containing much non-coding DNA, with genes largely unlinked and able to exist in many different sequence contexts without major effects on their expression. The techniques of cloning, mapping and sequencing, coupled with more biochemical approaches such as *in vitro* transcription, will continue to expand our understanding of this large and complex genome. Further detailed comparisons of closely and distantly related plant mtDNAs should contribute additional insights into how genomes evolve, and the functional constraints that must temper the seemingly rampant rearrangements that characterize plant mtDNA. Finally, higher plants offer a unique opportunity to study developmental aspects of mitochondrial biogenesis and gene expression; this area is sure to attract attention in the near future.

References

Boer, P. H., McIntosh, J. E., Gray, M. W. and Bonen, L. (1985). The wheat mitochondrial gene for apocytochrome b: absence of a prokaryotic ribosome binding site. *Nucl. Acids Res.* **13**, 2281–92.

Bonen, L. and Gray, M. W. (1980). Organization and expression of the mitochondrial genome of plants I. The genes for wheat mitochondrial ribosomal and transfer RNA: evidence for an unusual arrangement. *Nucl. Acids Res.* **8**, 319–35.

Dawson, A. J., Hodge, T. P., Isaac, P. G., Leaver, C. J. and Lonsdale, D. M. (1986). Location of the genes for cytochrome oxidase subunits I and II, apocytochrome b, α-subunit of the F_1 ATPase and the ribosomal RNA genes on the mitochondrial genome of maize (*Zea mays* L.). *Curr. Genet.* **10**, 561–4.

Dewey, R. E., Levings, C. S. III and Timothy, D. H. (1986). Novel recombinations in the maize mitochondrial genome produce a unique transcriptional unit in the Texas male-sterile cytoplasm. *Cell* **44**, 439–49.

Falconet, D., Lejeune, B., Quetier, F. and Gray, M. W. (1984). Evidence for homologous recombination between repeated sequences containing 18S and 5S ribosomal RNA genes in wheat mitochondrial DNA. *EMBO J.* **3**, 297–302.

Forde, B. G., Oliver, R. J. C. and Leaver, C. J. (1978).

Variation in mitochondrial translation products associated with male-sterile cytoplasms in maize. *Proc. Natl Acad. Sci. USA* **75**, 3841–5.

Fox, T. D. and Leaver, C. J. (1981). The *Zea mays* mitochondrial gene coding cytochrome oxidase subunit II has an intervening sequence and does not contain TGA codons. *Cell* **26**, 315–23.

Gray, M. W. (1983). The bacterial ancestry of plastids and mitochondria. *BioScience* **33**, 693–9.

Gray, M. W. (1989). Origin and evolution of mitochondrial DNA. *Ann. Rev. Cell. Biol.* **5**, 25–50.

Gray, M. W. and Spencer, D. F. (1983). Wheat mitochondrial DNA encodes a eubacteria-like initiator methionine transfer RNA. *FEBS Lett.* **161**, 323–7.

Gray, M. W., Sankoff, D. and Cedergren, R. J. (1984). On the evolutionary descent of organisms and organelles: a global phylogeny based on a highly conserved structural core in small subunit ribosomal RNA. *Nucl. Acids Res.* **12**, 5837–52.

Grivell, L. A. (1983). Mitochondrial DNA. *Sci. Amer.* **248**, 78–89.

Hanson, M. R. and Conde, M. F. (1985). Functioning and variation of cytoplasmic genomes: lessons from cytoplasmic–nuclear interactions affecting male fertility in plants. *Int. Rev. Cytol.* **94**, 213–67.

Hiesel, R., Schobel, W., Schuster, W. and Brennicke, A. (1987). The cytochrome oxidase subunit I and subunit III genes in *Oenothera* mitochondria are transcribed from identical promoter sequences. *EMBO J.* **6**, 29–34.

Leaver, C. J. and Gray, M. W. (1982). Mitochondrial genome organization and expression in higher plants. *Ann. Rev. Plant Physiol.* **33**, 373–402.

Lonsdale, D. M. (1984). A review of the structure and organization of the mitochondrial genome of higher plants. *Plant Mol. Biol.* **3**, 201–6.

Lonsdale, D. M., Hodge, T. P. and Fauron, C. M.-R. (1984). The physical map and organization of the mitochondrial genome from the fertile cytoplasm of maize. *Nucl. Acids Res.* **12**, 9249–61.

Makaroff, C. A. and Palmer, J. D. (1987). Extensive mitochondrial specific transcription of the *Brassica campestris* mitochondrial genome. *Nucl. Acids Res.* **15**, 5141–56.

Marechal, L., Runeberg-Roos, P., Grienenberger, J. M., Colin, J., Weil, J. H., Lejeune, B., Quetier, F. and Lonsdale, D. M. (1987). Homology in the region containing a tRNA^{Trp} gene and a (complete or partial) tRNA^{Pro} gene in wheat mitochondrial and chloroplast genomes. *Curr. Genet.* **12**, 91–8.

Palmer, J. D. and Herbon, L. A. (1987). Unicircular structure of the *Brassica hirta* mitochondrial genome. *Curr. Genet.* **11**, 565–70.

Palmer, J. D. and Shields, C. R. (1984). Tripartite structure of the *Brassica campestris* mitochondrial genome. *Nature* **307**, 437–40.

Pring, D. R. and Lonsdale, D. M. (1985). Molecular biology of higher plant mitochondrial DNA. *Int. Rev. Cytol.* **97**, 1–46.

Schuster, W., Hiesel, R., Isaac, P. G., Leaver, C. J. and Brennicke, A. (1986). Transcript termini of messenger RNAs in higher plant mitochondria. *Nucl. Acids Res.* **14**, 5943–54.

Sederoff, R. R. (1984). Structural variation in mitochondrial DNA. *Adv. Genet.* **22**, 1–108.

Spencer, D. F., Schnare, M. N. and Gray, M. W. (1984). Pronounced structural similarities between the small subunit ribosomal RNA genes of wheat mitochondria and *Escherichia coli*. *Proc. Natl Acad. Sci. USA* **81**, 493–7.

Stern, D. B. and Lonsdale, D. M. (1982). Mitochondrial and chloroplast genomes of maize have a 12-kilobase DNA sequence in common. *Nature* **299**, 698–702.

Stern, D. B. and Palmer, J. D. (1986). Tripartite mitochondrial genome of spinach: physical structure, mitochondrial gene mapping, and locations of transposed chloroplast DNA sequences. *Nucl. Acids Res.* **14**, 5651–66.

Tzagoloff, A. and Myers, A. M. (1986). Genetics of mitochondrial biogenesis. *Ann. Rev. Biochem.* **55**, 249–85.

Wise, R. P., Pring, D. R. and Gengenbach, B. G. (1987). Mutation to male fertility and toxin insensitivity in Texas (T)-cytoplasm maize is associated with a frameshift in a mitochondrial open reading frame. *Proc. Natl Acad. Sci USA* **84**, 2858–62.

Yang, D., Oyaizu, Y., Oyaizu, H., Olsen, G. J. and Woese, C. R. (1985). Mitochondrial origins. *Proc. Natl Acad Sci., USA* **82**, 4443–7.

11 Protein import into the mitochondrion
Karl B. Freeman, Susanna Reichling and Alexander Balogh

Introduction

The formation of new mitochondria occurs by growth and division of pre-existing mitochondria, but of the hundreds of mitochondrial proteins, at most about 20 are coded by plant mitochondrial DNA (mtDNA) and synthesized on mitochondrial ribosomes (see Chapter 10). The rest are coded by nuclear DNA, synthesized on cytosolic ribosomes and imported into mitochondria. To ensure efficient formation of functional mitochondria there must be coordination between the two genetic systems, especially since the respiratory and H^+-ATPase protein complexes of the mitochondrial inner membrane are composed of subunits, some of which are coded by mtDNA and some by nuclear DNA. Although little is known about how this coordination is achieved, there is considerable understanding of the genetic role of mtDNA and on the import of proteins by mitochondria, especially in mammals and fungi. A start has been made on studies of protein import by plant mitochondria. These studies will be considered here, but most of the emphasis will be on results from mammals, yeast (*Saccharomyces cerevisiae*) and *Neurospora crassa*. Further, far more is known about import of proteins by chloroplasts (see Chapter 20) than by plant mitochondria.

Proteins destined for all cellular and extracellular locations are synthesized on cytosolic polysomes. This means that each precursor protein must contain information that targets the protein to a specific site or allows non-specific insertion into various cellular membranes. Specific cellular targeting usually, but not always, reflects the presence of an amino-terminal presequence in the precursor. Proteins that lack targeting information would remain, by default, as soluble proteins in the cytosol. Acting in concert with the targeting signals, the cytosol and/or each type of organelle must have machinery (for example, membrane receptors) that recognizes this information to ensure specific targeting. There must also be a system for integrating the protein into the organelle, and the protein must contain information that ensures its direction to a specific sub-organellar site. Finally, as many of these processes involve translocation of proteins across membranes, it would be expected that energy should be required. In this chapter, all of these points will be considered as they apply to the import of proteins into mitochondria.

There are many excellent comprehensive reviews on uptake of proteins by mitochondria (Chua and Schmidt, 1979; Hay *et al.*, 1984; Harmey and Neupert, 1985; Reid, 1985; Doonan *et al.*, 1984; Douglas *et al.*, 1986; Pfanner and Neupert, 1987; Nicholson and Neupert, 1988). Although some reviews, and especially the recent one of Nicholson and Neupert (1988), are referred to routinely, all contain much the same information. Two shorter reviews (Hurt and van Loon, 1986; Hurt, 1987) are useful for models of import. Recently, many studies have indicated a commonality of how targeted proteins are translocated across membranes of different organelles, and several reviews address this question (see Wickner and Lodish (1985) and Zimmermann and Meyer (1986) and references therein for details).

Methodology

Most studies on import of proteins into mitochondria

use *in vitro* approaches which now follow a standard pattern. These approaches have proved exceptionally flexible, and when coupled with modern molecular biological techniques and some studies *in vivo*, have provided a detailed but still incomplete picture on import. An excellent early review on general approaches is available (Gasser and Hay, 1983).

Nature and site of synthesis of imported proteins

Studies on an imported protein begin with its purification and production of antibodies against it. The antibodies are used to recover the protein precursor after translation of polysomes or mRNAs in the presence of [^{35}S]-methionine, usually using a reticulocyte lysate protein synthesizing system. A few yeast precursor proteins have also been recovered with antibodies after pulse-labeling of cells or protoplasts. The precursor's size is compared with that of the mature protein by sodium dodecyl sulfate-polyacrylamide gel electrophoresis. By using free or membrane-bound polysomes, or mRNAs from these or from the mitochondrial fraction, the cellular site of synthesis can be determined. Definitive characterization of the size of the precursor requires determination of its primary sequence. This is done by cDNA cloning and sequencing; in yeast, genetic approaches have allowed the selection of genomic clones with great facility. Comparison of the derived amino acid sequence of the precursor to the sequence of the mature protein immediately identifies any extra sequences.

Import by isolated mitochondria

Detailed studies on the mechanism of import have been possible because isolated mitochondria take up mitochondrial protein precursors synthesized *in vitro*. The approach has been facilitated by *in vitro* expression of cloned cDNAs to produce single protein precursors (Douglas *et al.*, 1986). Usually the precursor and mitochondria used are from the same species. For reasons that are becoming understood only now, import requires the simultaneous presence of both mitochondria and the reticulocyte lysate translation system in the incubation medium. The articles by Chien and Freeman (1984) and Shore *et al.* (1983) on mammalian systems, and White and Scandalios (1987) on a plant system, illustrate the methods used to examine the size and import *in vitro* of precursors.

What types of questions have been and are being asked on the mechanism of import using the isolated system? These are considered in detail below and include the nature of the targeting sequence, the presence of mitochondrial receptors for the precursors, the energy requirement for import, the path followed by the imported protein, the mechanism of translocation through or integration into membranes, and the nature of precursor processing such as proteolytic cleavage.

Isolation of components involved in import

An understanding of the detailed mechanism of import can be achieved only by isolating the components involved and characterizing their nature and mode of action. Studies to date have given us a general description of import but details on the components of the import machinery, including putative cytosolic factors, outer membrane receptors, and some of the proteases involved in processing, are lacking.

General model of import

The current model of import of matrix proteins into mitochondria of mammals, yeast and *Neurospora* is shown in Fig. 11.1. This can serve as a focus on the areas of knowledge and ignorance on import.

1. Imported mitochondrial proteins are synthesized on cytosolic free polysomes, which means that they are taken up by a post-translational mechanism.
2. Most imported matrix proteins are synthesized with an amino-terminal presequence (see insert in Fig. 11.1). Presequences have been found to contain most, if not all, of the information for targeting proteins to mitochondria. These presequences have been called targeting, addressing, signal, transport or translocation sequences or, more generally, prepieces or

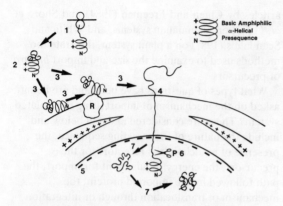

Fig. 11.1 Model of import of matrix proteins into mitochondria. Details are given in the text. An insert diagram shows the amphiphilic α-helical presequence. Presequences have a high content of basic (+) and hydroxyl (·) amino acids, hydrophobic residues which are not present in a continuous stretch, and lack acidic residues. The arrangement of the residues is such that the hydrophilic amino acids are on one side of the α-helix and hydrophobic amino acids on the other side; that is, the helix is amphiphilic (cf. Fig. 11.2). P, protease; R, receptor.

presequences. Some other details on presequences are given in the figure.
3. After passage of the precursor through the cytosol the targeting sequence either reacts with the outer membrane phospholipid bilayer and then with a putative receptor (R), or reacts with a receptor directly. The precursor or precursor/receptor complex is then positioned to permit passage of the precursor through the membranes. The conformation of the precursor that binds to mitochondria and is taken up is not known, but it is suggested that uptake requires an unfolded form of the protein. Cytosolic factors, including ATP or GTP, appear necessary for import. The energy might be used to keep the precursor in an unfolded state.
4. Translocation across mitochondrial membranes is thought to occur at contact points at which the precursor probably crosses the membranes via a proteinaceous pore rather than via the phospholipid bilayer. For precursors with targeting presequences, the amino terminus enters the matrix ahead of the rest of the protein which can extend beyond the outer membrane during import.
5. Translocation requires the energy of the mitochondrial electrochemical gradient and in particular the membrane potential ($\Delta\psi$).
6. In the matrix, the presequence is proteolytically removed by a metal-requiring protease (P).
7. After import, the protein achieves its final conformation and interacts with other proteins to form homo- or hetero-oligomers.

The major variant on these themes is that additional signals and mechanisms are necessary to direct proteins to the mitochondrial outer or inner membrane or to the intermembrane space.

Nature of the precursor

Site of synthesis

Protein synthesis in the eukaryotic cell cytosol occurs on two classes of polysomes, those that are free and those that are bound to the endoplasmic reticulum. The latter class of polysomes synthesizes proteins destined for export or for the plasma membrane. Since synthesis is mechanistically linked to translocation across or into the membrane of the rough endoplasmic reticulum, this process is referred to as co-translational translocation. Post-translational translocation across the endoplasmic reticulum has recently been described and may require an unfolded state of the protein (Zimmermann and Meyer, 1986). Given these two mechanisms of translocation, how does import into the mitochondrion occur? The demonstration of cytosolic pools of mitochondrial proteins in yeast and *Neurospora* that could be imported when protein synthesis was blocked, pointed to post-translational import *in vivo* (Reid, 1985; Douglas *et al.*, 1986). The ability of isolated mitochondria to import proteins in the absence of protein synthesis clearly shows this also. Nevertheless, in yeast, ribosomes synthesizing mitochondrial proteins are found bound to the outer membrane of mitochondria (Reid, 1985). This probably does not indicate a mechanistically linked co-translational process. Rather, it seems to reflect the relative kinetics of synthesis and binding of an amino-terminal targeting sequence to mitochondria.

Targeting and sorting sequences

Precursors of larger size

The most striking difference between mature and precursor forms of imported mitochondrial proteins is the larger size of the precursor. Detailed lists of the sizes of mature and precursor forms of proteins are available (Hay et al., 1984; Harmey and Neupert 1985). Based on some 60 examples, three general conclusions can be made: (1) proteins destined for the matrix, inner membrane or intermembrane space, but not the outer membrane, are synthesized as larger precursors containing an amino-terminal presequence; (2) the length of the presequence varies from about 20 to 80 amino acids; and (3) a small number of proteins of the matrix, inner membrane and intermembrane space are not synthesized as precursors of larger size.

Characteristics of presequences

Presequences of imported mitochondrial proteins of yeast, *Neurospora* and mammals do not have extensive identity of sequence but do share a number of characteristics (Nicholson and Neupert, 1988). These include the presence of a high content of randomly distributed basic (positively-charged) amino acids, particularly arginine, and hydroxylated amino acid residues, particularly serine, the presence of hydrophobic residues which are not found in a continuous stretch, and a lack of acidic (negatively-charged) residues. It has been hypothesized that presequences can form amphiphilic (amphipathic) α-helical structures (Fig. 11.1) and that these may be important in import. An example is shown in Fig. 11.2 for the amino terminus of sweet potato cytochrome *c* oxidase subunit Vc (Nakagawa et al., 1987). (This is not a presequence, but illustrates the nature of an amphiphilic α-helix in a convincing manner.) Examples of presequences and sites of cleavage are given in Table 11.1. A detailed list is given by Nicholson and Neupert (1988).

As discussed in the next section, these presequences act as the targeting signal. They presumably react with receptors (Gillespie, 1987) and enter mitochondria ahead of the rest of the precursor (Schleyer and Neupert, 1985). This suggests that the presequence is important in enabling the translocation of the precursor across

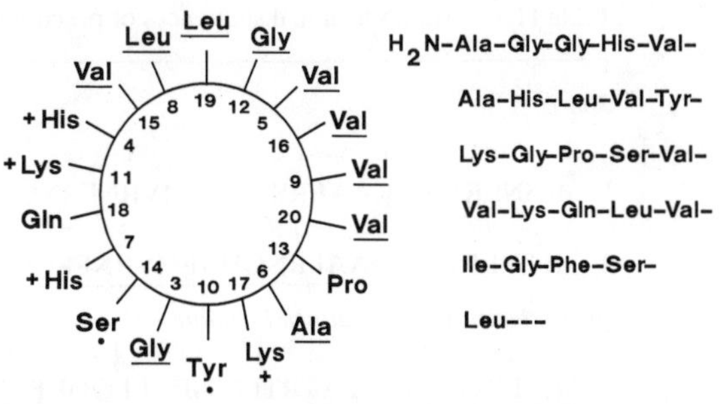

Fig. 11.2 The amino terminus of sweet potato cytochrome *c* oxidase subunit Vc and its arrangement on a helical wheel (modified from Nakagawa et al., 1987). The sequence was determined from the protein. Although the protein and its precursor have the same mobility on sodium dodecyl sulfate-polyacrylamide gel electrophoresis, the lack of an amino-terminal methionine in the mature sequence leaves open the possibility that the protein is actually made with a presequence. However, the sequence does yield an amphiphilic α-helix when plotted on a helical wheel as shown. The region with the maximum hydrophobic moment was plotted by Nakagawa et al. (1987). The basic (+), hydroxyl (·) and hydrophobic (−) residues are shown.

membranes. Positive charge appears important for this (Allison and Schatz, 1986; Horwich et al., 1987). Since a membrane potential (negative inside) is necessary for import, the presequence might be pulled 'electrophoretically' into the mitochondrial matrix. The presumptive amphiphilic α-helix could aid in translocation, since such helices can bind to and penetrate phospholipid bilayers (Nicholson and Neupert, 1988). However, artificial or actual presequences, constructed to maintain a positive charge but lacking an obvious potential for forming amphiphilic α-helices, still target proteins to mitochondria (Allison and Schatz, 1986; Horwich et al., 1987). Amphilicity itself, however, does appear to be necessary. One possibility is that the presequence may aid in import by binding to the outer membrane phospholipid bilayer, which would permit rapid diffusion to presumptive receptors (Skerjanc et al., 1987). Electrophoretic penetration of the membrane dependent on the membrane potential might then be at proteinaceous sites.

Table 11.1 Amino-terminal sequences of precursors of some imported mitochondrial proteins

Matrix

```
    +    +    +   ++   ..    +  +   +     ↓.
1. MLSNLRILLNKAALRKAHTSMVRNFRYGKPVQSQVQ
    +    +    +         +  +      ↓ +
2. MLFNLRILLNNAAFRNGHNFMVRNFRCGQPLQNKVQ
```

Inner Membrane – unknown location

```
    +  . +. + +       .+ ↓  . +    ↓    +
3. MLSLRQSIRFFKPATRTLCSSRYLLQQKPVVKTAQ
```

Inner Membrane – cytosolic face

```
   + .  . ++       +. . +.  . . .   .+. +.   +                     −. . −↓
4. MFSNLSKRWAQRTLSKSFYSTATGAASKSGKLTQKLVTAGVAAAGITASTLLYADSLTAEAM
```

```
    +  .  . +   +     + .+ ↓ . . . .    ↓ . . . . . −.
5. MAPVSIVSRAAMRAAAAPARAVRALTTSTALQGSSSSTFESP
```

Outer Membrane

```
   ++ .  .+ +.           . .                         + ++ .  +−−++
6. MKSFITRNKTAILATVAATGTAIGAYYYYNQLQQQQQRGKKNTINKDEKK
```

The sequences given in the single letter code are taken from Horwich *et al.* (1987), Nicholson and Neupert (1988) and Hurt and van Loon (1986) and are as follows: 1 and 2, rat and human ornithine carbamoyltransferase, respectively; 3, yeast cytochrome *c* oxidase subunit IV; 4, yeast cytochrome c_1; 5, *Neurospora* Fe/S protein of the bc_1 complex; 6, yeast 70 kD protein. Basic (+), acidic (−) and hydroxylated (·) amino acyl residues are indicated above the sequence as are points of proteolytic cleavage (↓). The first underlined portion is the intracellular- and matrix-targeting domain and the second is the proposed stop-transport (stop-transfer) domain. The dashed portions are areas of less certainty. There have been no studies for those sequences in which a targeting domain has not been indicated. The line over a portion of the cytochrome c_1 precursor is the position within which a proteolytic cleavage occurs in the matrix.

Targeting role: fusion proteins

The presence of a presequence on imported mitochondrial proteins, and the fact that the signal for targeting proteins to the endoplasmic reticulum was a presequence, focused attention on these sequences as putative mitochondrial targeting signals. The role of the presequence can be examined directly using recombinant DNA technology to join it, or parts of it, to a non-mitochondrial 'passenger' protein resulting in a fusion protein whose coding sequence is contained in an expressible plasmid (Douglas *et al.*, 1986; Nicholson and Neupert, 1988). Usually, soluble cytosolic proteins such as dihydrofolate reductase or β-galactosidase have been used as the so-called 'passenger' protein. *In vitro* transcription of plasmid vectors containing such fusion proteins followed by translation and import into isolated mitochondria or import *in vivo* in yeast demonstrates that the presequences contain the intracellular targeting information (Douglas *et al.*, 1986; Hurt and van Loon, 1986; Nicholson and Neupert, 1988). Further, the targeting sequence of matrix protein precursors and of amino-terminal portions of precursors for other compartments is sufficient to direct proteins to the mitochondrial matrix as described in the section on intramitochondrial sorting.

The whole of the presequence is not needed for targeting. In yeast, the most amino-terminal portion only is essential (Hurt and van Loon, 1986; Nicholson and Neupert, 1988). For example, the first

12 amino acids of the 25 amino acid presequence of cytochrome c oxidase subunit IV (Table 11.1) will direct import of fused dihydrofolate reductase into the matrix of yeast mitochondria. Removal of the first four amino acids prevents targeting. In contrast, the midportion (amino acids 8 to 25) of the 32 amino acid presequence of human ornithine carbamoyltransferase (Table 11.1) appears sufficient for targeting (Horwich et al., 1987). It should be noted that mature forms of mitochondrial proteins are usually not imported by isolated mitochondria; but there is evidence in yeast (Bedwell et al., 1987), and perhaps mammals (see Doonan et al., 1984), that some imported proteins contain redundant targeting information in their mature sequences which permits import of proteins from which the main targeting sequence has been removed.

The precise targeting sequence of proteins of inner mitochondrial compartments that are not synthesized as larger precursors is still not known. Their amino-terminal sequences usually differ from presequences. The targeting information of the ADP/ATP carrier of the yeast mitochondrial inner membrane is within the amino-terminal 115 amino acids. The targeting information for apocytochrome c may be in the carboxyl-terminal portion of the protein. In contrast, the outer membrane 70 kD protein of yeast (Table 11.1) has an amino-terminal sequence similar to the basic portion of presequences and acts as a targeting sequence but is not cleaved (Hurt and van Loon, 1986). This leads to the question of intra-mitochondrial sorting, discussed in a later section.

Conformation of precursors

It has not been possible to characterize the conformation of precursors *in vivo* but many imported proteins, synthesized in a reticulocyte lysate, have been shown to exist in aggregates up to eight times the monomeric size (Hay et al., 1984; Nicholson and Neupert, 1988). In some cases, there is evidence that these are homo-oligomers but usually it has not been determined whether the aggregates are homo- or hetero-oligomers. Hetero-oligomers could arise if the targeted protein interacted with cytosolic factors necessary for import. A related problem is that many membrane proteins are hydrophobic and would be relatively insoluble in the cytosol. In the case of the hydrophobic, 81 amino acid H^+-ATPase subunit IX (proteolipid) of *Neurospora*, aqueous solubility may be promoted by a long polar presequence of 66 amino acids and the formation of aggregates by interaction of the hydrophobic portion of the precursor.

The import process

Processing

Proteolytic processing

Proteolytic processing of precursors of larger size is clearly necessary to achieve the mature length and possibly the mature conformation of the protein. Processing *in vivo* occurs rapidly as the half-lives of cytosolic precursors are very short, being on the order of less than 5 minutes (Hay et al., 1984; Nicholson and Neupert, 1988). Precursors cannot normally be detected in mitochondria. Early studies showed that matrices of fungal and mammalian mitochondria contain a protease which was dependent on zinc or cobalt ions for activity. Since the membrane permeable chelator *o*-phenanthroline inhibited processing by isolated mitochondria, it was concluded that the protease activity isolated from the matrix is responsible for the cleavage. The protease cleaves a variety of *in vitro* synthesized mitochondrial precursors to mature size, but will not cleave *in vitro* synthesized non-mitochondrial proteins (Hay et al., 1984). Cleavage does not appear to depend on the species of origin of the matrix protease or precursor. This indicates a commonality of mechanism but, as seen in Table 11.1, there is no identity of sequences at the cleavage site. Therefore, some conformation of the precursor must be necessary for protease recognition.

Although the matrix protease accounts for the total proteolytic processing of mitochondrial matrix protein precursors, a number of proteins of the inner membrane and intermembrane space are proteolytically processed in two steps. In one case, subunit IX of the H^+-ATPase of *Neurospora*, the matrix protease, is responsible. However, there is evidence for a total of three mitochondrial proteases being involved in processing. The sites of sequential processing in the mitochondrion might relate to the mechanism by which proteins are sorted into submitochondrial compartments. For example, the second site of processing of some proteins in which the presequence is removed in two stages is in a location exposed to the intermembrane space (Hartl et al., 1987; van Loon and Schatz, 1987; Nicholson and Neupert, 1988). This will be considered in more detail in a later section.

Other types of processing

A number of precursors are modified by steps that occur in addition to, or without, processing (Nicholson and Neupert, 1988). The addition of heme to mitochondrial cytochromes is the most obvious example. In the case of cytochromes c and c_1, this requires covalent attachment to cysteine residues. The second proteolytic processing step of the precursor of cytochrome c_1 occurs only after the attachment of heme. The import of the cytochrome c precursor has been studied extensively, and in yeast the enzyme adding heme has been isolated. Iron–sulfur centers are formed in subunits of three of the respiratory complexes by attaching iron to cysteine residues. This process appears to require energy from NADH. Finally, a number of coenzymes and cofactors such as NAD^+, FAD and Cu^{2+} have tight non-covalent binding to mitochondrial enzymes.

Energy requirement

Three types of energy could be utilized for import: the mitochondrial inner membrane electrochemical potential, high energy compounds, and the difference in energy of the precursor and mature form of the protein in the cytosolic and mitochondrial compartments due to energy released on folding and interaction of subunits. Nothing is known of the last class but the first two are required in quite different ways.

Electrochemical potential

Early studies showed that import of proteins destined for the matrix or inner membrane could be blocked by uncouplers of oxidative phosphorylation which collapse the electrochemical potential. In contrast, import of outer membrane proteins and at least one protein, apocytochrome c, of the intermembrane space was unaffected (Hay et al., 1984; Nicholson and Neupert, 1988).

The electrochemical potential is composed of two components, the proton gradient (ΔpH) and membrane potential of all charged species ($\Delta \psi$). These are generated by proton pumping during electron transport down the respiratory chain or by the action of the H^+-ATPase. The problem was to distinguish which of the two components, ΔpH or $\Delta \psi$, was necessary. A series of experiments using ion transporters along with inhibitors of the respiratory chain, the H^+-ATPase and the ADP/ATP carrier (to block ATP uptake into mitochondria) effectively showed that it was the membrane potential ($\Delta \psi$) that was the key component (Hay et al., 1984; Nicholson and Neupert, 1988).

It should be noted that energy is not required for proteolytic processing nor does inhibition of the matrix protease prevent import. This is to be expected since import of proteins of the inner membrane and matrix that are not processed still requires the energy in the form of the membrane potential (Nicholson and Neupert, 1988).

What is the role of the energy of the membrane potential? It was found that import of the β-subunit of the F_1-ATPase of *Neurospora* could be stopped short of complete transfer (Nicholson and Neupert, 1988). At 7 °C, the amino-terminal presequence was removed by the matrix protease, but the bulk of the protein was accessible to externally added protease. By raising the temperature to 25 °C, import was completed in the absence of a membrane potential. Thus the membrane potential is needed for an initial phase of import; that is, the interaction of the precursor's presequence with the inner membrane, perhaps with electrophoretically driven transport of the presequence across it.

High energy compounds

The result of two types of experiments have both indicated that a high energy compound (ATP, GTP, etc.) is necessary for import. Energy might be required to maintain the protein to be imported in an unfolded state in the cytosol and during the import process until the protein is within the outer membrane (Zimmermann and Meyer, 1986; Hurt, 1987; Nicholson and Neupert, 1988). The first type of experiment showed that, even in the presence of a membrane potential, import would not take place unless ATP or GTP was present. Non-hydrolyzable ATP analogs would not support the uptake. Since it had been shown that proteins 'frozen' in mature conformation could not be imported, attempts were made to determine whether the energy was used to maintain an import-competent conformation. It was shown that, in the presence of ATP or GTP, precursor proteins were in a form which was more sensitive to proteases. This indicates that the import-competent form is likely to be a relatively unfolded conformation.

Components of the import process

Cytosolic factors

Import of proteins into isolated mitochondria requires the presence of the complete reticulocyte lysate translation mixture suggesting that there may be cytosolic factors in this mixture that are necessary for import. Several studies have attempted to determine whether such cytosolic proteins exist. At present, even though one such factor has been isolated from yeast, no clear picture has emerged (Nicholson and Neupert, 1988). It may be that cytosolic components act non-specifically, or specifically to ensure that the precursor is kept in a conformation competent for import as described above.

Receptors

Import of proteins into the mitochondrion requires not only that imported proteins have a targeting sequence but that this information is recognized by cytosolic factors or mitochondria resulting in import specifically into the mitochondria. Attention has been focused on putative mitochondrial-specific protein receptors for recognizing the precursor targeting sequence. Since it is unlikely that mitochondria have a specific receptor for each imported protein (how would each of the receptors be specifically imported?), there must be only a small number of types of protein receptors, and possibly only one. It is conceivable that mtDNA codes for these receptors, thus forming a unity between the nuclear/cytosolic and mitochondrial compartments. However, there is no evidence in favor of this (cf. Chapter 10). Unless receptors have a unique import mechanism, it would appear that receptors are targeted to mitochondria by the same mechanism as are other imported proteins. In this case, receptors must always be present, reflecting perhaps the fact that mitochondria arise only from pre-existing mitochondria.

Most models place receptors in the mitochondrial outer membrane. Pretreatment of mitochondria with low concentrations of proteases which are excluded from inner mitochondrial compartments blocks binding and import of precursors (Nicholson and Neupert, 1988). This indicates that import is dependent on proteinaceous components in the outer membrane. Further, when the uptake of precursors is blocked by incubation at 0 °C or with uncouplers so that the precursor remains bound to the mitochondrial membrane, the uptake commences as soon as the block is removed. Binding is specific for the outer membrane and occurs with the precursor but not with the mature form of the protein. The presence of more than one type of receptor has been indicated by differential effects of proteases on the binding of precursors. Competition studies on binding by different precursors have supported this conclusion. In particular, apocytochrome c appears to have a distinct receptor that might be located in the intermembrane space. Scratchard plot analyses indicate from 1 to 100 pmol of binding sites per mg mitochondrial protein for a variety of precursors. So far, a purified receptor has not been isolated and shown to be functional in reconstituted liposomes. Antibodies against a 45 kD outer membrane protein from yeast mitochondria blocked import. A synthetic presequence of the rat ornithine carbamoyl-transferase precursor (Table 11.1) could be specifically cross-linked to a 30 kD rat mitochondrial outer membrane protein (Gillespie, 1987; Nicholson and Neupert, 1988). These studies indicate that isolation of receptors is feasible and may be achieved in the near future.

Intramitochondrial sorting: translocation across or integration into mitochondrial membranes

Nature of the problem

The hydrophobic portion of the phospholipid bilayer of membranes presents a thermodynamic barrier for the translocation of hydrophilic proteins. The difficulty this presents for the transport of such proteins would be overcome if they were to move through a proteinaceous pore, as proposed in many models of translocation (e.g. Singer et al., 1987). Any model for the import of mitochondrial proteins must allow not only for translocation across but also their integration into membranes. For precursors that cross or integrate into the inner membrane, it is not known whether the outer and inner membranes act as a unit or whether interaction with the inner membrane requires its own signals.

The above problems form a background to the fundamental question of how imported mitochondrial proteins are sorted into the various submitochondrial compartments. Studies with fusion proteins in which a presequence has been fused to a non-mitochondrial protein, clearly implicate the presequence as the targeting signal for import as far as the matrix. However, how do imported proteins get sorted into different compartments? This and the above problems will now be examined. Note that most data have been obtained from the laboratory of W. Neupert for *Neurospora* and of G. Schatz for yeast.

Outer and inner membrane translocation contact sites

Electron microscopy of mitochondria from rat liver and *Neurospora* show the presence of sites of apposition of the inner and outer membranes (Nicholson and Neupert, 1988). In rat liver mitochondria, they appear to be stable structures as they are present in different conformational states of the mitochondrion. Given that such contact sites exist, does import occur at these points? The evidence in favor of this, obtained by Neupert's group (Schwaiger et al., 1987), is as follows.

1. As discussed above, import intermediates of the precursor of the $F_1\beta$ subunit of H^+-ATPase or cytochrome c_1 that span the membrane can be trapped at low temperatures. In this situation, the precursor has entered the mitochondrial matrix by its amino-terminal presequence but the remainder of the molecule extends outside the outer membrane. This shows that the inner and outer membranes are sufficiently close at the point of import for the precursor to span both membranes.
2. These points of import have been shown to be in the same location as morphological contact sites identified using the antibody-bound $F_1\beta$ precursor tagged with protein A-gold, which makes the antibody visible in the electron microscope. All the membrane spanning intermediates were at contact sites.
3. The contact sites appear to be stable structures that pre-exist in mitochondria. Thus, mitochondria in which most of the outer membrane had been removed by digitonin could still import precursors, but only at contact sites as visualized by electron microscopy.
4. These contact sites have been isolated.

These studies indicate that proteins are imported into the inner membrane or matrix at translocation contact sites and that an intermediate in the intermembrane space does not exist. Whether the contact sites represent an actual physical contact or semi-fusion of the two membranes or whether proteins are involved is not known. However, there is evidence that proteins are imported through a hydrophilic membrane environment (Pfanner et al., 1987) which could indicate the presence of a proteinaceous 'pore' at the translocation contact sites. It is not known if the presumed receptors are part of the contact region. The protease-sensitive component on the outer membrane that is necessary for import of the $F_1\beta$ precursor is near the contact site, but that for the ADP/ATP carrier appears more widely spread on the outer membrane (Schwaiger et al., 1987).

Stop-transport domain model

Sequencing of imported mitochondrial proteins, in particular from yeast, and the use of fusion proteins to study the role of amino-terminal portions of precursors have led to a unified model of intramitochondrial sorting (Hurt and van Loon, 1986). The model is based on the idea that hydrophobic sequences stop translocation of proteins through a membrane by interacting with the hydrophobic portion of the phospholipid bilayer. These sequences are called stop-transfer or stop-transport domains. In this model the imported mitochondrial protein precursors have one of three types of amino-terminal presequences, as shown in Table 11.1. Matrix and perhaps some integral inner membrane proteins have a mitochondrial targeting presequence that is basic, i.e. positively charged followed by a cleavage site at the amino terminus of the mature protein. These proteins are targeted to the matrix unless there are sequences that result in them being inserted into the inner membrane during the translocation process. However, this has so far not been demonstrated definitively for integral inner membrane proteins (van Loon and Schatz, 1987; cf. Nicholson and Neupert, 1988). In contrast, the ADP/ATP carrier precursor of *Neurospora*, which does not have a presequence, is integrated into the inner membrane as it moves into the mitochondrion (Nicholson and Neupert, 1988). Proteins residing in the intermembrane space or on the cytosolic side of the inner membrane have longer presequences, as shown in Table 11.1. In these cases, there is an intracellular and matrix-targeting domain and an initial cleavage site acted upon by the matrix protease. This is followed by a stretch of about 20 hydrophobic amino acids (stop-transport domain) which stops translocation of the protein to the matrix. A protease in the intermembrane space releases the mature protein into the intermembrane space where it stays as a soluble protein (e.g. cytochrome b_2) or binds to the inner membrane (e.g. cytochrome c_1). Proteins of the outer membrane should have a non-cleaved basic targeting domain followed by a stop-transport domain. Subtle differences in spacing of the matrix-targeting and stop-transport domains or of the amino acid composition of the stop-transport sequences would dictate whether translocation stops in the inner or outer membrane. For example, as shown in Table 11.1, the stop-transport domain of the yeast 70 kD outer membrane protein is closer to the targeting domain than in the precursor of cytochrome c_1 of the intermembrane/inner membrane compartment.

These ideas have some experimental support (Hurt and van Loon, 1986; van Loon and Schatz, 1987) including: (1) the 61 amino acid presequence of yeast cytochrome c_1, which is on the cytosolic face of the inner membrane, will when fused to cytosolic dihydrofolate reductase direct it to the intermembrane space; (2) the amino-terminal portion of the yeast 70 kD outer membrane protein, which has the basic and hydrophobic sequences adjacent to each other, will direct soluble proteins to the cytosolic face of the outer membrane; (3) the isolated basic portions of the targeting sequences of these two proteins will direct passenger proteins to the matrix; (4) two step processing, with the second protease being in the intermembrane space, has been demonstrated; and (5) in yeast, the import of a passenger protein attached to the cytochrome c_1 presequence did not involve passage of the whole protein through the matrix before it was localized in the intermembrane space/inner membrane compartment.

Despite the simplicity and unifying nature of the model, there are some difficulties with it. Although the nature of the amino-terminal intracellular- and matrix-targeting domain is now well known and fits the model, there are several proteins of the matrix, inner membrane and intermembrane space that are not synthesized as larger precursors and their amino termini do not have a standard basic targeting sequence. More importantly, the import of some proteins of the intermembrane space/inner membrane compartment involves initial passage of the proteins through the matrix. This has lead to the following model to account for these phenomena.

Evolutionary model

Neupert and his coworkers have obtained convincing evidence that the import of the Fe/S protein of the bc_1 complex of *Neurospora* involves passage of the protein through the matrix before it is targeted to the outer face of the inner membrane (Nicholson and Neupert, 1988). A peptide of 24 amino acids is cleaved from the presequence of this protein by the matrix protease (Table 11.1). The rest of the protein is then directed back across the inner membrane. A further peptide of 8 amino acids is removed but the mitochondrial location of this protease is not known. Neupert and his coworkers have suggested that the latter part of this transport sequence follows a similar path to that seen in bacteria where the homologous protein is exported and resides on the extracytoplasmic face of the cytoplasmic membrane (Nicholson and Neupert, 1988). Indeed, assuming that this mitochondrial protein was once encoded by the original endosymbiotic bacterial precursor of the mitochondrion, transfer of the gene to the nucleus would have required the acquisition of DNA sequence which would code for a presequence targeting the protein to the matrix. This would have to be added to the amino terminus of the original signal that directed the protein across the bacterial cytoplasmic membrane (cf. mitochondrial inner membrane) to the external surface of the bacterium. In support of this idea, it has been observed that random sequences from *Escherichia coli* or mouse dihydrofolate reductase DNA can code for sequences that target to mitochondria (Baker and Schatz, 1987). This idea could clearly apply to many imported proteins.

The route of import and sorting of the *Neurospora* Fe/S protein described above is not unique. It has also been observed for *Neurospora* cytochrome c_1 and yeast cytochrome b_2 (Hartl et al., 1987) and the Fe/S protein and cytochrome c oxidase subunit IV of yeast (van Loon and Schatz, 1987). At this stage, there is a contradiction in the results for the pathway of import of cytochrome c_1 in yeast (van Loon and Schatz, 1987) and in *Neurospora* (Hartl et al., 1987), and only further work will resolve the applicability of the stop-transport and evolutionary models to intramitochondrial protein sorting.

Final steps and coordination

The final steps in import include folding of the protein into its mature conformation and integration into soluble or membrane hetero- or homo-oligomers. Little is known of these processes, which are not unique to the problem of import into the mitochondrion but are a general biochemical question. One aspect that is unique to organelles is that they contain their own genetic system. This requires that the expression of nuclear and mitochondrial genes be coordinated. In addition, protein complexes formed from proteins from the two genetic systems must also be coordinated. A review of this area is beyond the scope of this chapter, and the review by Nicholson and Neupert (1988) can be consulted for a short overview and further references.

Import into the plant mitochondrion

Studies on import into plant mitochondria have only just begun. Since the mechanism of import into fungal and mammalian mitochondria appear to be essentially the same, it would be expected that the mechanism in plants would also be similar. The three studies published so far indicate strongly that this is the case. Briefly, it has been found that the targeting sequence is at the amino-terminal end of the precursor and that it is rich in basic and hydroxylated amino acids and lacks acidic residues. It is in the form of a presequence in two of the three cases examined. Protein import requires a mitochondrial electrochemical gradient and the presequence is removed on import. The three studies are: (1) maize Mn-superoxide dismutase import in which the precursor which is more basic than the mitochondrial protein was taken up by isolated maize mitochondria, a process which was dependent on an electrochemical potential. The presequence was cleaved to release the mature sized protein (White and Scandalios, 1987); (2) the smallest subunit (Vc) of sweet potato cytochrome c oxidase is synthesized at its mature size but contains a basic amino terminus (see Fig. 11.2) which has the potential to form an amphiphilic α-helix (Nakagawa et al., 1987); (3) the derived sequence (from the *atp 2-1* gene) for the β-subunit of mitochondrial ATP synthase of *Nicotiana plumbaginifolia* shows that the protein is synthesized with a presequence of about 9 kD and this presequence will direct bacterial chloramphenicol acetyltransferase to mitochondria in transgenic plants (Boutry et al., 1987).

The experiments of Boutry et al. have also addressed the problem of how proteins are targeted specifically to the mitochondrion or to the chloroplast. This is particularly interesting because the presequences that target proteins to each organelle are generally both rich in basic and hydroxylated amino acids and lack acidic residues (see Hurt et al., 1986 for references). Indeed 31 residues out of the 45 residue transit presequene of the small subunit of ribulose 1,5-bisphosphate carboxylase/oxygenase (Rubisco) (a chloroplast protein) of *N. plumbaginifolia* could target proteins to yeast mitochondria although less efficiently than specific mitochondrial presequences (Baker and Schatz, 1987). However, Boutry *et al.* (1987) found that the presequences of the β-subunit of mitochondrial ATP synthase and of the small subunit of Rubisco, both from *N. plumbaginifolia*, although they had very similar sequences would specifically direct bacterial chloramphenicol acetyltransferase to either the mitochrondrion or chloroplast respectively. This is despite the fact that both had the same overall amino acid characteristics that are found in presequences of proteins that are directed to the mitochondria in fungi and mammals. Clearly there are unknown subtle differences between the two types of targeting presequences. One difference between mitochondrial and chloroplast presequences is that the mitochondrial targeting sequence has more basic residues (5 compared with 2 in the first 30 amino acids). Fortunately, the techniques of modern molecular biology will help provide the solution to this problem. Further, there is still a virtually untapped area for research on the detailed mechanism of import of proteins into the plant mitochondrion.

References

Allison, D. S. and Schatz, G. (1986). Artificial mitochondrial presequences. *Proc. Natl Acad. Sci. USA* **83**, 9011–5.

Baker, A. and Schatz, G. (1987). Sequences from a prokaryotic genome or the mouse dihydrofolate reductase gene can restore the import of a truncated precursor protein into yeast mitochondria. *Proc. Natl Acad. Sci. USA* **84**, 3117–21.

Bedwell, D. M., Klionsky, D. J. and Emr, S. D. (1987). The yeast F_1-ATPase β subunit precursor contains functionally redundant mitochondrial protein import information. *Mol. Cell. Biol.* **7**, 4038–47.

Boutry, M., Nagy, F., Poulsen, C., Aoyagi, K. and Chua, N.-H. (1987). Targeting of bacterial chloramphenicol acetyltransferase to mitochondria in transgenic plants. *Nature* **328**, 340–2.

Chien, S.-M. and Freeman, K. B. (1984). Import of rat liver mitochondrial malate dehydrogenase: synthesis, transport and processing *in vitro* of its precursor. *J. Biol. Chem.* **259**, 3337–42.

Chua, N.-H. and Schmidt, G. W. (1979). Transport of proteins into mitochondria and chloroplasts. *J. Cell. Biol.* **81**, 461–83.

Doonan, S., Marra, E., Passarella, S., Saccone, C. and Quagliariello, E. (1984). Transport of proteins into mitochondria. *Int. Rev. Cytol.* **91**, 141–86.

Douglas, M. G., McCammon, M. T. and Vassarotti, A. (1986). Targeting proteins into mitochondria. *Microbiol. Rev.* **50**, 166–78.

Gasser, S. M. and Hay, R. (1983). Overview of mitochondrial protein import. Assessing import of proteins into mitochondria: an overview. *Methods Enzymol.* **97**, 245–54.

Gillespie, L. L. (1987). Identification of an outer mitochondrial membrane protein that interacts with a synthetic signal peptide. *J. Biol. Chem.* **262**, 7939–42.

Harmey, M. A. and Neupert, W. (1985). Synthesis and intracellular transport of mitochondrial proteins. In *The Enzymes of Biological Membranes*, Vol. 4, ed. A. Martinosi, Plenum, New York, pp. 431–64.

Hartl, F.-U., Ostermann, J., Guiard, B. and Neupert, W. (1987). Successive translocation into and out of the mitochondrial matrix: targeting of proteins to the intermembrane space by a bipartite signal peptide. *Cell* **51**, 1027–37.

Hay, R., Böhni, P. and Gasser, S. (1984). How mitochondria import proteins. *Biochim. Biophys. Acta* **779**, 65–87.

Horwich, A. L., Kalousek, F., Fenton, W. A., Furtak, K., Pollock, R. A. and Rosenberg, L. E. (1987). The ornithine transcarbamylase leader peptide directs mitochondrial import through both its midportion structure and net positive charge. *J. Cell Biol.* **105**, 669–77.

Hurt, E. C. (1987). Unravelling the role of ATP in post-translational protein translocation. *Trends Biochem. Sci.* **12**, 369–70.

Hurt, E. C. and van Loon, A. P. G. M. (1986). How proteins find mitochondria and intramitochondrial compartments. *Trends Biochem. Sci.* **11**, 204–7.

Hurt, E. C., Soltanifar, N., Goldschmidt-Clermont, M., Rochaix, J.-D. and Schatz, G. (1986). The cleavable pre-sequence of an imported chloroplast protein directs attached polypeptides into yeast mitochondria. *EMBO J.* **5**, 1343–50.

Nakagawa, T., Maeshima, M., Muto, H., Kajiura, H., Hattori, H. and Asahi, T. (1987). Separation, amino-terminal sequence and cell-free synthesis of the smallest subunit of sweet potato cytochrome *c* oxidase. *Eur. J. Biochem.* **165**, 303–7.

Nicholson, D. W. and Neupert, W. (1988). Synthesis and assembly of mitochondrial proteins. In *Protein Transfer and Organelle Biogenesis*, eds R. C. Das and P. W. Robbins, Academic Press, New York, pp. 677–746.

Pfanner, N. and Neupert, W. (1987). Biogenesis of mitochondrial energy transducing complexes. *Curr. Topics Bioenerg.* **15**, 177–219.

Pfanner, N., Hartl F.-U., Guiard, B. and Neupert, W. (1987). Mitochondrial precursor proteins are imported through a hydrophilic membrane environment. *Eur. J. Biochem.* **169**, 289–93.

Reid, G. A. (1985). Transport of proteins into mitochondria. In *Current Topics in Membranes and Transport*, Vol. 24, Membrane protein biosynthesis and turnover, eds P. Kanuf and J. Cook, Academic Press, New York, pp. 295–336.

Schleyer, M. and Neupert, W. (1985). Transport of proteins into mitochondria: translocation intermediates spanning contact sites between outer and inner membranes. *Cell* **43**, 339–50.

Schwaiger, M., Herzog, V. and Neupert, W. (1987). Characterization of translocation contact sites involved in the import of mitochondrial proteins. *J. Cell Biol.* **105**, 235–46.

Shore, G. C., Rachubinski, R., Argan, C., Rozen, R., Pouchelet, M., Lusty, C. J. and Raymond, Y. (1983). Synthesis and intracellular transfer of mitochondrial matrix proteins in rat liver: studies *in vivo* and *in vitro*. *Methods Enzymol.* **97**, 396–408.

Singer, S. J., Maher, P. A. and Yaffe, M. P. (1987). On the translocation of proteins across membranes. *Proc. Natl Acad. Sci. USA* **84**, 1015–9.

Skerjanc, I. S., Shore, G. C. and Silvius, J. R. (1987). The interaction of a synthetic mitochondrial signal peptide with lipid membranes is independent of transbilayer potential. *EMBO J.* **6**, 3117–23.

van Loon, A. P. G. M. and Schatz, G. (1987). Transport of proteins to the mitochondrial intermembrane space: the 'sorting' domain of the cytochrome c_1 presequence is a stop-transfer sequence specific for the mitochondrial inner membrane. *EMBO J.* **6**, 2441–8.

White, J. A. and Scandalios, J. G. (1987). *In vitro* synthesis, importation and processing of Mn-superoxide dismutase (SOD-3) into maize mitochondria. *Biochim. Biophys. Acta* **926**, 16–25.

Wickner, W. T. and Lodish, H. F. (1985). Multiple mechanisms of protein insertion into and across membranes. *Science* **230**, 400–7.

Zimmermann, R. and Meyer, D. I. (1986). 1986: A year of new insights into how proteins cross membranes. *Trends Biochem. Sci.* **11**, 512–5.

12 Metabolite exchange between the mitochondrion and the cytosol
Roland Douce and Michel Neuburger

Introduction

The tricarboxylic acid (TCA) cycle provides reducing equivalents to the electron chain for ATP synthesis and also, via ancillary reactions, provides numerous substrates for biosynthetic reactions in the cytoplasm. The relative importance of these roles in the overall plant cell metabolism will depend on (1) the particular tissue (leaf, root, etc.); (2) stage of development, the orderly coordinated sequence of events that attend the change from a single-celled zygote to a multicelled adult; and (3) environmental factors (light, temperature, water deficits, etc.). These multiple and changing demands necessitate a rather complex detailed regulation of the individual enzymes of the TCA cycle in order to allow a non-uniform flux through the various segments of the cycle. The velocity with which this cycle turns is mainly determined by three quantities: availability of substrates, redox state (NADH/NAD$^+$), and energy state (ATP/ADP, energy charge, etc.). These three controlling factors interact in a highly coordinated and cooperative manner. In addition, these controls are complemented by input and output through the various anion translocators, according to demand and supply of metabolites.

Within a plant cell the rate of respiration is controlled primarily by the rates of reactions feeding electrons to the respiratory chain and by the rates of reactions consuming or producing ATP. In addition, there are several points on the TCA cycle where important intermediates are removed to provide the building blocks for the biosynthesis of proteins, nucleic acids and specialized molecules such as chlorophylls. For mitochondria to function as integral components of the plant cell there must be movement of numerous anions including phosphate ($H_2PO_4^-$, HPO_4^{2-}), ADP^{3-}, ATP^{4-}, various TCA cycle intermediates and amino acids between the matrix of the mitochondria and the cytoplasm. In this chapter we will concentrate on the transport systems of plant mitochondria and their importance in plant cell carbon metabolism.

Transport systems of plant mitochondria

The outer membrane of the mitochondrion isolated by rather gentle procedure has been found to be permeable to sucrose, nucleotides, and NAD$^+$ but not to cytochrome c although the possibility that this was a result of membrane damage during isolation could not be completely ruled out (Douce, 1985). However, it is now accepted that the outer membrane of mitochondria acts as a molecular sieve which permits the passage of small hydrophilic molecules. This sieving property, which allows uptake of small substrate molecules into the intermembrane space but excludes enzymes, is due to a 31 kD channel-forming protein called mitochondrial porine or voltage-dependent anion-selective channel, because all mitochondrial porines investigated so far show a decreasing single channel conductance as a result of increasing voltage (Mannella, 1985). Electron microscope image reconstruction indicates that the mitochondrial pore is a cylinder normal to the membrane plane with a diameter of about 2.5 nm. How many subunits are required to form a pore is not known. Interestingly, it has been shown that the mitochondrial pore exhibits several closed states of different magnitudes suggesting that porine can exert a control on the exact exclusion limit for the penetration of the outer membrane (Mannella, 1985).

In contrast, the inner mitochondrial membrane is impermeable to glucose and larger uncharged molecules, H^+, mono-, di-, and trivalent cations, anions, amino acids, nucleotides and phosphate (P_i). Indeed, the hydrophobic bilayers devoid of porine creates an effective barrier to the passage of charged species. The high activation energy required to insert an ion into a hydrophobic region is the reason for the extremely low ion permeability of bilayer regions. This overall unspecific impermeability of the inner membrane towards hydrophilic solutes is overcome by specific translocators (Chappell and Croft, 1966). The kinetic properties of the translocators may vary from species to species and even from tissue to tissue in the same species. These carriers are essential for the maintenance of a normal mitochondrial function. If anions could freely cross the membrane they would be ejected by the high electric field generated by respiration, and essential substrates would be lost. On the other hand, the inner membrane is highly permeable to small uncharged molecules such as H_2O, NH_3, O_2 and lipophilic molecules. Available evidence indicates that CO_2 also readily passes through the mitochondrial membrane whereas the bicarbonate anion (HCO_3^-) does not. The inner membrane is permeable to several monocarboxylic acids of low molecular weight with relatively high pK values (such as acetate) which pass, after protonation, through the membrane as undissociated acids (Chappell and Croft, 1966).

Phosphate and adenine nucleotide translocation in plant mitochondria

The mitochondrial F_1-ATPase, responsible for ATP synthesis in oxidative phosphorylation, is located on the inner surface of the inner mitochondrial membrane (Douce, 1985). Consequently, the transport of P_i (influx), ADP (influx) and ATP (efflux) between the cytosol and the matrix are essential for the continued synthesis of ATP.

Adenine nucleotide transport

The adenine nucleotide exchange is highly specific for ADP and ATP and is basically electrogenic; free ATP^{4-} and ADP^{3-} are translocated but not their Mg^{2+} complexes. This is highly significant because it is probable that most ATP molecules in the cytosol and in the mitochondria are chelated with Mg^{2+}. A specific system for the removal of Mg^{2+} ions in the vicinity of the carrier could be involved. Since ADP has three and ATP has four negative charges at physiological pH, the ADP^{3-} – ATP^{4-} exchange causes a net movement of charges across the mitochondrial membrane. However, since this carrier operates as an antiport it does not change the intramitochondrial concentration of adenine nucleotides (Vignais, 1976).

Three specific and powerful inhibitors of the adenine nucleotide transport system are known: atractyloside, carboxyatractyloside (gummiferin, the toxic principles of an Algerian thistle) and bongkrekic acid, an antibiotic formed by a mold in decaying coconut meals. One of the peculiarities of the mitochondrial nucleotide carrier is its ability to bind atractyloside and bongkrekic acid, in an asymmetric manner. The atractylosides bind to the carrier on the cytosolic surface of the inner membrane and bongkrekic acid on the matrix facing surface (Vignais, 1976). The nucleotide carrier has been isolated as a carboxyatractylate protein complex that is probably a dimer, each subunit having a M_r value of 30 000. Interestingly, the nucleotide carrier is the most abundant integral protein in the inner membrane of mitochondria.

The transfer of electrons from substrate to O_2 via the cytochrome electron transport pathway is coupled to an electrogenic translocation of protons across the inner mitochondrial membrane (Mitchell, 1969). Protons have a positive charge and, consequently, proton translocation from the matrix to the cytosol, without cotransport of anions, generates both a proton gradient (ΔpH) and a membrane potential ($\Delta \psi$) in which the matrix side is negatively charged and the cytoplasmic side positively charged. The two factors are additive, both contributing to a 'protonmotive force' differential across the membrane according to the following equation:

$$\Delta p = \Delta\psi - 2.303\, RT/F\Delta pH$$

(at 25 °C the equation can be written as $\Delta p = \Delta\psi - 59\Delta pH$).

Measurements of $\Delta\psi$, determined by the distribution of a non-physiological lipophilic cation across the inner membrane, have indicated a value of approximately 250 mV for state 4 respiration in plant mitochondria (Diolez and Moreau, 1985). This high value results from the presence in the inner mitochondrial membrane of a powerful K^+/H^+ antiporter which partially collapses the pH gradient replacing it with a potential gradient thereby increasing $\Delta\psi$ (Hanson and Koeppe, 1975). This transmembrane potential is a potent driving force for an electrophoretic asymmetric exchange of ATP^{4-} and ADP^{3-} which drives ADP^{3-} inside the mitochondria and ATP^{4-} outside (Fig. 12.1). Accordingly, at equilibrium, the cytosol should have a high ATP/ADP ratio, and this has been found by ^{31}P-NMR spectra obtained *in vivo* from plant cells or tissues (Roby *et al.*, 1987). Also, *in situ* plant leaf mitochondria resemble mitochondria from non-green cells in that they are highly energized and maintain a high cytosolic ATP/ADP that is independent from chloroplasts because the inner membrane of the envelope surrounding mature chloroplasts does not possess a nucleotide translocator.

The results of Chance and Williams (1956) for mammalian mitochondria strongly suggest that the most plausible explanation of respiratory control *in vivo* is the availability of ADP and the kinetics of its transport by the nucleotide carrier. Since the cytosolic ADP concentration as measured by ^{31}P-NMR is extremely low in a plant cell, it is very likely that the rate of O_2 consumption is dependent upon the rate of ADP delivery to the mitochondria during the course of metabolism. Mitochondria *in vivo* are often described as being intermediate between states 4 and 3 since respiration can be increased (to state 3) with uncouplers or decreased (to state 4) by inhibiting ATP production with inhibitors such as

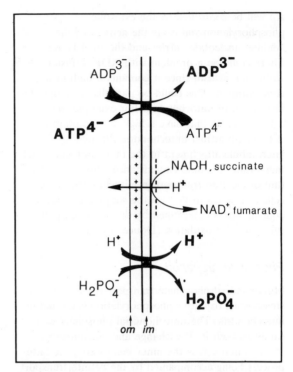

Fig. 12.1 Effect of the electrochemical gradient of protons across the membrane generated by the electron chain on the electrophoretic asymmetric nucleotide exchange and on the net movement of phosphate ($H_2PO_4^-$) across the inner membrane.

oligomycin. In this intermediate state, an increase in the activity of the adenine nucleotide carrier will strongly affect respiration. The exchange of ATP for ADP in the matrix will lower Δp which will affect the redox state of electron chain components. This lowering of Δp will stimulate respiration by increasing the rate of the proton pumps and hence increase the rate of electron transport[*]. Decreases in Δp could also occur following increased flow of protons through the F_1-ATPase during ATP synthesis. In other words,

---

[*] In the resting state (state 4) control of O_2 consumption in isolated mitochondria is exerted by the leak of protons through the inner membrane (a high proton conductance pathway occurs at high Δp), whereas in more active phosphorylating states (up to state 3) control is distributed between a number of steps including the proton leak, the adenine nucleotide carrier and cytochrome oxidase.

Δp will be controlled by the cytosolic phosphorylation status via the activity of the adenine nucleotide carrier and the F_1-ATPase. There could be a problem with ADP diffusion within the cytosol since it contains a high protein concentration. This could be partially overcome by movement of mitochondria. For example, the cytosolic areas that have a high consumption of ATP might attract mitochondria through microtubule attachment to the outer mitochondrial membrane. In some circumstances, however, the rate of electron transport in mitochondria within cells can be controlled via the supply of reducing equivalents as determined by the availability of substrates for oxidation (Douce, 1985).

Phosphate transport

Mitochondrial phosphate transport is essential for steady-state oxidative phosphorylation catalyzed by mitochondria. The mitochondrial phosphate carrier catalyzes both P_i–P_i exchange and net movement of the anion across the inner membrane, the latter process being accompanied by the countertransport of OH^-. In fact, it is usually impossible to distinguish between a symport (i.e. a transport process involving the obligatory coupling of two ions in parallel) of a species with H^+ and an antiport of the species with OH^-. A ΔpH is maintained across the mitochondrial inner membrane by the electrogenic efflux of protons which is associated with the function of the electron chain. Consequently, proton cotransport provides a means by which the phosphate carrier can utilize the large electrochemical potential gradient of protons as an energy source and to give directionality to the transport process; that is, P_i transport is driven by a pre-established proton gradient (Fig. 12.1). This transport, which is electroneutral, is sensitive to thiol reagents such as N-ethylmaleimide or mersalyl. This carrier protein present in all the mitochondria isolated so far has been partially purified and its kinetic properties have been investigated in reconstituted proteoliposomes (Wohlrab, 1986).

Plant mitochondria swell spontaneously when suspended in a solution of ammonium phosphate. N-Ethylmaleimide inhibits this swelling by more than 80%. Furthermore, when the ΔpH across the inner mitochondrial membrane is collapsed (e.g. by nigericin which permits entry of H^+ in exchange for K^+ (K^+, H^+ antiport) across the mitochondrial inner membrane and collapses only ΔpH), rapid passive efflux of P_i occurs which is inhibited by mersalyl. These results indicate that a phosphate carrier similar to that of animal mitochondria is present in plant mitochondria, and that high matrix P_i content is maintained only so long as respiration maintains a ΔpH (Day and Wiskich, 1984). In our laboratory, P_i transport into potato mitochondria has been measured using a new rapid filtration technique, which offers fast time resolution (starting from 10 ms), thus allowing fast kinetic measurements at room temperature (Kathryn Wright, Michel Neuburger and Roland Douce, unpublished data). The initial rates of P_i accumulation measured within the first 150 ms and at 10 °C give a K_m (total P_i) of 0.6 mM

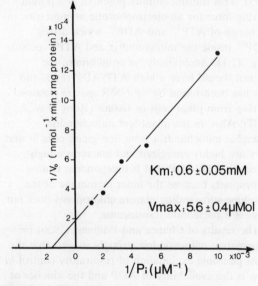

Fig. 12.2 A Lineweaver–Burk plot showing the dependence of rates of the accumulation of internal P_i on external P_i concentration in plant mitochondria. Time courses were obtained at 10 °C using a rapid-filtration technique (Biologic Rapid Filtration System, Grenoble, France) using P_i at concentrations of 100 μM, 200 μM, 300 μM, 500 μM and 800 μM. The initial velocity for each concentration was measured from points obtained in the first 150 ms. The kinetic parameters were calculated to be $K_m(P_i)$ = 600–700 μm and V = 5000–60000 nmol mg protein^{-1} min^{-1}.

and a maximum velocity (V) equal to 5000–6000 nmol P_i mg protein^{-1} min^{-1} (Fig. 12.2). Potato tuber and sycamore cell mitochondria oxidize the substrate succinate at a rate between 400–500 nmol O_2 consumed mg protein^{-1} min^{-1} at 25 °C. Assuming a P/O ratio of 2, this would require that P_i enters the mitochondria at a rate of 1600–2000 nmol P_i mg protein^{-1} min^{-1} if it is not to limit the synthesis of ATP. Furthermore, as the K_m value of P_i for the translocator is approximately 0.6 mM and the cytosolic P_i concentration determined by ^{31}P-NMR is between 1 and 2 mM (Roby et al., 1987), the translocator could operate at full capacity, and thus in physiological situations the activity of the phosphate carrier is far higher than is necessary to supply phosphate to the H$^+$-ATPase even at maximum rates of oxidative phosphorylation (Douce, 1985). Thus, in the case of the phosphate translocator, neither the capacity of the carrier nor the rate of P_i supply to the electron transport chain limits the in vivo rate of respiration.

The mitochondrial phosphate carrier accounts for the bulk of mitochondrial P_i transport. A much smaller fraction of P_i may be transported by the dicarboxylate carrier because this carrier can operate as a dicarboxylate $^{2-}$–HPO_4^{2-} antiport system, as will be discussed below.

Tricarboxylic acid cycle anion translocation in plant mitochondria

The endproduct of glycolysis under aerobic conditions in plant is the pyruvate ion and very likely malate and oxaloacetate ions. Furthermore, the tricarboxylic acid cycle serves as an important source of carbon skeletons for synthetic processes occurring in the cytoplasm and mechanisms for the movement of TCA cycle anions into and out of mitochondria must occur and have been demonstrated for mitochondria isolated from a wide range of tissues. Mitochondria from different plants and even from different tissues of the same plant vary in their complement of exchange–diffusion carriers (or antiports). The precise carrier composition presumably is a reflection of the function of the tissue from which the mitochondria are isolated. For example, the metabolic pathway for the conversion of fat to carbohydrate in castor bean endosperm or for the conversion of glycolate to glycerate in leaves involves the mitochondria in a number of necessary oxidations (Huang et al., 1983).

The kinetic properties of a TCA cycle substrate carrier can be examined, free of the constraints of limiting and undefined internal substrate levels, by preloaded mitochondria with a saturating concentration of a transportable substrate (e.g. citrate, malate, succinate, etc.). The exchange reaction is then carried out under defined conditions.

Pyruvate carrier

The availability or access of pyruvate to the pyruvate dehydrogenase complex in the mitochondrial matrix is of great metabolic importance for respiration. Pyruvate, although it is a monocarboxylate, requires a carrier because of its relatively low pK. Pyruvate uptake in mammalian mitochondria occurs in exchange for OH$^-$ (or in conjunction with H$^+$), and is inhibited by SH-group poisons, and strongly and specifically by α-cyano-4-hydroxycinnamate (Day and Wiskich, 1984). Consequently, pyruvate transport is electroneutral, and at equilibrium the ratio of pyruvate across the inner membrane is a direct function of ΔpH. At high non-physiological concentrations pyruvate is, however, no longer only transported via the translocator, but 'passive diffusion' through the inner membrane also becomes important.

The pyruvate transport system in plant mitochondria is very similar to that found in animal mitochondria. Pyruvate uptake is inhibited by α-cyano-4-hydroxycinnamic acid with 50% inhibition occurring at 2μM (Day and Wiskich, 1984). In addition, pyruvate uptake can also occur by 'passive diffusion' at high pyruvate concentrations. Day and Hanson (1977) have suggested that in corn mitochondria the rate of pyruvate transport (estimated to be 20 nmol mg protein^{-1} min^{-1}) is too low to support rapid respiration under conditions of high energy demand. Since pyruvate can be generated intramitochondrially from malate via malic enzyme, Day and Hanson (1977) postulated that rapid malate transport may make up the difference, supplying pyruvate via malic enzyme. This would be particularly important for the replenishment of losses of TCA cycle intermediates. However, higher rates of pyruvate transport have been found in mung bean hypocotyl and potato tuber mitochondria (200 nmol mg protein^{-1} min^{-1} at 10 °C), indicating that in these cases of uptake it is not rate-limiting. In fact, the continued uptake of pyruvate by plant mitochondria appears to be linked to the metabolism of pyruvate (Brailsford et al., 1986).

Dicarboxylate carrier

Transporters for the movement of dicarboxylic acids across the mitochondrial membrane in mitochondria that have been isolated from a wide range of tissues have been demonstrated (Wiskich, 1977). In mammalian mitochondria, dicarboxylate transport is mediated by a HPO_4^{2-}–dicarboxylate^{2-} exchange and by a dicarboxylate^{2-}–dicarboxylate^{2-} exchange, both of which are inhibited by pentylmalonate and by 2-N-butylmalonate (Chappell and Croft, 1966). Work with plants indicates that a similar exchange occurs in plant mitochondria. Many workers using indirect techniques such as back exchange experiments or osmotic swelling in ammonium salts have established a requirement of P_i (HPO_4^{2-} species) for uptake of dicarboxylates in a variety of plant mitochondria (Wiskich, 1977). Osmotic swelling of plant mitochondria in ammonium malate and back exchange are inhibited by pentylmalonate and by 2-N-butylmalonate. Inhibition of malate or succinate oxidation by these inhibitors has been taken as evidence for the operation of the dicarboxylate carrier (Wiskich, 1977). In contrast to mammalian mitochondria, freshly isolated plant mitochondria appear to contain low levels of endogenous exchangeable metabolites suggesting an inherent leakiness of the inner membrane. Depleted mitochondria lose the ability to take up dicarboxylates, an ability which can be restored if a dicarboxylate and an energy source are provided (Douce, 1985). Under these conditions a potent dicarboxylate–dicarboxylate exchange can be triggered. It is clear, therefore, that respiration which provides an energy source profoundly affects dicarboxylate uptake and this dependence of uptake on mitochondrial respiration is probably the major factor involved in conditioning. This is a phenomenon that is observed with many mitochondrial preparations in which there is an increase in state 3 respiration rates during several state 3/state 4 cycles (Douce, 1985).

The sensitivity of dicarboxylate accumulation to uncouplers and electron transport chain inhibitors is thought to occur by directly affecting P_i uptake, by discharging the large electrochemical potential gradient of protons and thus lowering the amount of internal P_i available for exchange with external dicarboxylate. Although mitochondria normally exhibit very low electrophoretic permeabilities to physiologically important anions such as dicarboxylates, Zoglowek et al. (1988) have made the provocative suggestion that the major route of net malate uptake in plant mitochondria is not by malate–P_i antiport, as in mammalian mitochondria, but by a powerful electrogenic uniport. The physiological role of this dicarboxylate anion channel is unknown. The presence of such an anion uniport in the inner membrane can easily explain the well-known fact that isolated plant mitochondria contain low concentrations of endogenous dicarboxylates. Depletion of anionic substrates can occur via this unusual carrier during the course of mitochondrial isolation.

The metabolic pathway for the conversion of fat to carbohydrate in castor bean endosperm involves the mitochondria in a number of oxidations (Huang et al., 1983). Succinate generated in the glyoxysomes is thought to be further metabolized in the mitochondria, providing malate for gluconeogenesis in the cytosol. For dicarboxylates to be metabolized within the mitochondria, they must be able to cross the inner mitochondrial membrane. Chappell and Beevers (1983) have used a centrifugation method to investigate features of dicarboxylate transport in mitochondria purified from castor bean endosperm. They indicated that the maximum rates of dicarboxylate–dicarboxylate exchange ($V = 250$ nmol mg protein^{-1} min^{-1} at 4 °C) are much greater than those measured for dicarboxylate accumulation ($V = 15$ nmol mg protein^{-1} min^{-1} at 25 °C) and are sufficient to account for the observed rates of succinate oxidation. This oxidation proceeds to malate, which accumulates in the bathing medium because of the fast succinate–malate exchange. Hence, initially succinate is taken into the mitochondrion in exchange for P_i. It is then oxidized to malate which in turn is exchanged for more external succinate. In general, when plant mitochondria are supplied with a single substrate, its oxidation is frequently limited to a few steps. The resulting intermediates do not remain confined to the mitochondria but are lost and often subsequent re-entry into the mitochondria is observed (Douce, 1985).

α-Ketoglutarate carrier

Intact plant mitochondria that are oxidizing α-ketoglutarate in the presence of ADP excrete malate and there is transient accumulation of succinate in the bathing medium. However, in the presence of malonate, a potent inhibitor of succinate dehydrogenase, mitochondria excrete almost exclusively succinate. During the course of α-ketoglutarate oxidation only that part of the TCA cycle from α-ketoglutarate to malate is operative and the dicarboxylates for the most part pass into the external medium (Douce, 1985). These experiments suggest that α-ketoglutarate transport can occur by a dicarboxylate–α-ketoglutarate exchange.

The presence of an α-ketoglutarate transporter has been demonstrated in isolated plant mitochondria by osmotic swelling in ammonium salts and by back exchange experiments (Wiskich, 1977). The exchangeability of α-ketoglutarate for dicarboxylate is strongly inhibited by phenylsuccinate and is mersalyl-insensitive (Wiskich, 1977). Unlike mammalian mitochondria, however, α-ketoglutarate transport at concentrations above 5 mM is not inhibited by phthalonate in plant mitochondria. This suggests that the animals and plant α-ketoglutarate–dicarboxylate anion exchangers are fundamentally different (Day and Wiskich, 1984). Finally, in plant mitochondria, monocarboxylates, glutamate and P_i are not exchanged on this carrier.

Oxaloacetate carrier

Oxaloacetate has been found to rapidly traverse the inner membrane of all the plant mitochondria

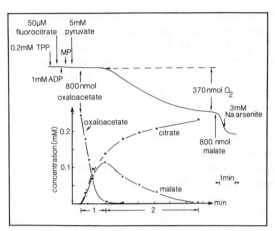

Fig. 12.3 Pyruvate oxidation triggered by oxaloacetate in intact potato tuber mitochondria. Upper trace: O_2 consumption during pyruvate oxidation. Lower trace: disappearance of added oxaloacetate, production of malate and production of citrate during pyruvate oxidation. Added oxaloacetate readily penetrates the matrix space through the inner membrane. Part of the oxaloacetate, at least one-half, reacts with acetyl-CoA under the control of citrate synthase, as indicated by citrate accumulation. Simultaneously, the other part of the oxaloacetate is converted into malate by malate dehydrogenase, a conversion that depends on NADH generated by the thiamine pyrophosphate (TPP)-linked pyruvate dehydrogenase complex. Under these conditions, O_2 consumption by the respiratory chain is inhibited (upper trace). When all the oxaloacetate is consumed, previously-accumulated malate re-enters the mitochondria and is rapidly oxidized to form oxaloacetate. Under these conditions, NADH from malate dehydrogenase and pyruvate dehydrogenase is available for oxidation by the respiratory chain (upper trace). Note that during the course of pyruvate oxidation triggered by oxaloacetate, extensive loss of malate and subsequent re-entry into the mitochondria is observed. Fluorocitrate, a potent inhibitor of aconitase, has been added to prevent citrate oxidation. MP: purified mitochondria.

studied so far (Fig. 12.3). For example, at an extramitochondrial oxaloacetate concentration of 50 μM, the influx of oxaloacetate is so rapid that NAD^+-linked O_2 consumption that is dependent on TCA cycle substrates stops because of the competition for NADH that is used in the reduction of oxaloacetate by malate dehydrogenase (the equilibrium of the malate dehydrogenase reaction lies far towards malate formation; K_{eq} 3 × 10^{-5}) (Fig. 12.3). This respiratory inhibition is subsequently relieved as the oxaloacetate becomes reduced. While this 'swamping' effect by excess oxaloacetate at best affords only a coarse type of control over electron transport, it amply demonstrates how a metabolic imbalance ensues when those enzymes normally limited by the supply of oxaloacetate are presented with a relative surfeit (Douce, 1985). The uptake of oxaloacetate is half-saturated at micromolar concentrations. Consequently, malate dehydrogenase and aspartate aminotransaminase external to the mitochondrial inner membrane can provide oxaloacetate to the matrix space.

In pea leaf mitochondria, malate transport is sensitive to 2-N-butylmalonate while that of oxaloacctatc appears not to be. Conversely, phthalonate had little effect on malate transport but severely inhibited oxaloacetate transport. These results demonstrate that malate efflux and oxaloacetate influx occur on separate carriers (Day and Wiskich, 1984). In support of this suggestion, the uptake of oxaloacetate is not inhibited by a 1000-fold excess of malate. Cauliflower bud mitochondria swell spontaneously when suspended in high (non-physiological) concentrations of ammonium oxaloacetate, implying exchange for hydroxyl ions (or cotransport of a proton along with the anion). In this case oxaloacetate transport should be increased at lower pH. Oliver (1987) has shown, however, that the highest rates of oxaloacetate uptake occurred at alkaline pH values. On the other hand, pea leaf mitochondria did not swell when suspended in isotonic solutions of ammonium or K^+ oxaloacetate until valinomycin was added, suggesting that oxaloacetate is taking up by electrogenic uniport which is facilitated by the valinomycin-mediated uptake of the compensating cation (Zoglowek et al., 1988). For the present, the details of oxaloacetate transport in plant mitochondria remain unknown, but it is thought that the carrier is specific for oxaloacetate. The question whether both malate and oxaloacetate are transported by a single transport protein or by two different ones cannot be answered at present, although Zoglowek et al. (1988) have provided evidence that a malate–oxaloacetate shuttle across the inner mitochondrial membrane is catalyzed by an electrogenic uniport of malate and of oxaloacetate linked to a counter exchange.

The extraordinarily low half-saturation of oxaloacetate transport makes it possible for a very active malate–oxaloacetate shuttle to occur between the mitochondria and the cytosolic compartment of a plant cell under physiological conditions. For example, it has been proposed that the NADH generated during glycine oxidation by mitochondria isolated from green leaves (C3 plants) may be reoxidized via a malate–oxaloacetate shuttle which may be linked *in vivo* to hydroxypyruvate reduction in the peroxisomes (Ebbighausen et al., 1985) (Fig. 12.4). Alternatively, it has also been proposed that the NADH generated during glycine oxidation may be reoxidized via a malate–aspartate shuttle (Douce, 1985). However, in view of the low levels of peroxisomal aspartate aminotransferase, this shuttle is unlikely to play a major role in the reoxidation and export of NADH from leaf mitochondria. According to Ebbighausen et al. (1985) malate–oxaloacetate shuttle activity could account for the reoxidation of all the NADH produced during photorespiratory glycine oxidation *in vivo* (see, however, Dry et al., 1987). In our laboratory, oxaloacetate transport into pea leaf mitochondria has been measured using a rapid filtration technique. The initial rates of oxaloacetate accumulation measured within the first 150 ms and at 10 °C gave a K_m of 5 μM and V equal to 2 μmol mg protein^{-1} min^{-1} both values being adequate for the maintenance of observed *in vivo* photorespiratory shuttle rates. These data support the mechanism proposed by Ebbighausen et al. (1985) that suggests that a malate–oxaloacetate shuttle has the capacity to account for the total reoxidation of NADH produced during glycine oxidation *in vivo*. Furthermore, it is very likely that green cells *in vivo* will generally be respiring between states 4 and 3, especially in the light, because the rate at which a green cell respires is limited by the availability of ADP for oxidative phosphorylation. This limitation will lead to an increase in the matrix NADH/NAD$^+$ ratio which in turn will impede the cleavage of glycine*. It is highly

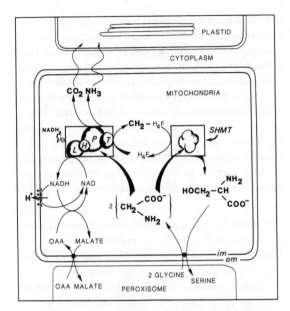

Fig. 12.4 Schematic representation of glycine oxidation in green leaf mitochondria. During photorespiration, glycine is cleaved in the matrix by the glycine cleavage enzyme system (containing four protein components which have been tentatively named P-protein, H-protein, T-protein and L-protein) to CO_2, NH_3 and 5,10-methylene-tetrahydropteroyl-L-glutamic acid (5,10-CH_2-H_4F). The latter compound reacts with a second mole of glycine to form serine and tetrahydropteroyl-L-glutamic acid (H_4F) in a reaction catalyzed by serine hydroxymethyltransferase (SHMT). NADH produced during the course of glycine oxidation is oxidized either by the respiratory chain or in the conversion of oxaloacetate to malate catalyzed by malate dehydrogenase that is located in the matrix. A rapid malate–oxaloacetate transport shuttle appears to play an important role in the photorespiratory cycle by facilitating the transfer of reducing equivalents generated in the mitochondria during glycine oxidation to the peroxisomal compartment for the reduction of β-hydroxypyruvate. Note the stoichiometry of two glycine molecules entering the mitochondrial matrix in exchange for one serine leaving.

---

* The steady-state activity of the glycine cleavage complex is sensitive to the NADH/NAD^+ molar ratios because NAD^+ (in stimulating) and NADH (in inhibiting) act directly on L-protein (lipoamide dehydrogenase) via lipoyl moiety covalently bound to the H-protein, a constituent element of the glycine cleavage system.

probable then, that, *in vivo*, NADH produced during the course of glycine oxidation is reoxidized very rapidly by malate dehydrogenase located in the matrix space (Fig. 12.4). It is possible that the transfer of oxaloacetate and malate between the mitochondrion and peroxisome could be greatly facilitated by close proximity of their boundary membranes. It is not known whether the rate of oxaloacetate delivery to the peroxisome during the course of malate oxidation limits the overall rate of this malate–oxaloacetate shuttle.

Since all the plant mitochondria isolated so far possess a high capacity oxaloacetate carrier (Douce, 1985), carbon input from the cytosol to the TCA cycle could also occur as oxaloacetate, thus bypassing conventional operation of the TCA cycle (Fig. 12.5). This anion arises either from malate oxidation catalyzed by cytosolic malate dehydrogenase and/or β-carboxylation of phosphoenolpyruvate. The latter reaction is catalyzed by a cytosolic phosphoenolpyruvate carboxylase present in all plant cells examined so far. This enzyme appears to be a non-biotin containing carboxylase which uses HCO_3^- rather than CO_2 as substrate. Unfortunately information as to the exact role of phosphoenolpyruvate carboxylase in the replenishment of TCA cycle intermediates (in the form of oxaloacetate or malate) remains sparse. Further experimental studies on the role of phosphoenolpyruvate carboxylase in the anaplerotic function of the TCA cycle are needed. An alternative source of oxaloacetate is from the degradation of proteins.

In some circumstances, intact mitochondria can export oxaloacetate. For example, when malate is being oxidized under state 3 conditions, mitochondria excrete oxaloacetate via the oxaloacetate carrier when the malic enzyme does not operate. As malate oxidation proceeds, the concentration of oxaloacetate in the medium slowly increases up to an equilibrium concentration. When this is achieved, the efflux of oxaloacetate stops and the malate dehydrogenase reaction in the matrix is reversed. The concentration of oxaloacetate attained at equilibrium is dependent on the malate concentration and on the metabolic state. The excretion of oxaloacetate, which is severely inhibited by phthalonate, occurs under state 3

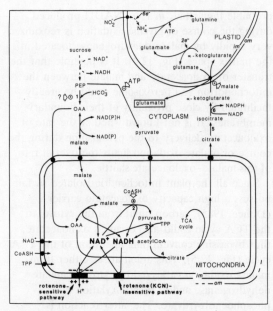

Fig. 12.5 Schematic representation of malate oxidation in plant mitochondria and citrate delivery to the cytoplasm for the fixation of NH_4^+. This scheme emphasizes the flexibility of plant mitochondria and indicates that there exists a concerted action of malate dehydrogenase, NAD^+-linked malic enzyme (2) and pyruvate dehydrogenase (3) to provide citrate in the anaplerotic function for the TCA cycle. The scheme also indicates that matrix NADH produced by the dehydrogenases, including NAD^+-linked malic enzyme, can be oxidized equally well by either the respiratory chain or by malate dehydrogenase for the reduction of oxaloacetate. Transport of electrons to O_2 can be either phosphorylating (via the rotenone-sensitive pathway) or non-phosphorylating (via the rotenone (KCN)-insensitive pathway). The latter mechanism may be important when TCA cycle intermediates such as citrate are exported from the mitochondrion for use elsewhere in the cell (see text). Note that carbon input to the TCA cycle can occur in the form of cytosolic oxaloacetate and malate (produced by the combined operation of phosphoenolpyruvate carboxylase (1) and malate dehydrogenase in the cytosol). Pyruvate can be provided either by the action of pyruvate kinase in the cytosol or by operation of malic enzyme in the matrix utilizing malate generated either in the matrix or the cytosol. Finally, this scheme illustrates that isolated intact plant mitochondria actively accumulate NAD^+, thiamine pyrophosphate (TPP) and coenzyme A (CoA) from the external medium. This leads to a substantial increase in the matrix concentration of these cofactors which would stimulate TCA cycle dehydrogenases, the NAD^+-linked malic enzyme and rotenone (KCN)-insensitive pathway. Other enzymes shown in this diagram are: (5), cytosolic aconitase; (6), cytosolic α-ketoglutarate dehydrogenase; (4), citrate synthase (the first step in the TCA cycle catalyzed by citrate synthase is an aldol condensation of acetyl-CoA and oxaloacetate to give an enzyme-bound thioester intermediate. Hydrolysis of this thioester gives the citrate ion and coenzyme A. The equilibrium position of this reaction is very much in favor of the products because the free energy of hydrolysis of the thioester drives the reaction to completion).

conditions (i.e. at high $NAD^+/NADH$ ratios) and in the presence of high concentrations of malate. The high concentrations of NAD^+ and malate shift the reaction slightly towards oxaloacetate production which facilitates oxaloacetate excretion (Douce, 1985). Likewise, during the course of succinate oxidation under state 3 conditions, plant mitochondria also excrete large amounts of oxaloacetate (Douce et al., 1986). Consequently, it is the concentration of oxaloacetate on both sides of the inner mitochondrial membrane that appears to govern the efflux or influx of oxaloacetate. Irrespective of its role, the low K_m of the oxaloacetate carrier for its substrate allows it to compete successfully with cytosolic or matrix malate dehydrogenase.

Citrate transport

It is accepted that the transport of citrate by plant mitochondria is by a tricarboxylate carrier similar to the one found in animal mitochondria. This carrier or citrate carrier catalyzes an electroneutral exchange of a tricarboxylate anion (e.g. citrate) with a dicarboxylate anion (Chappell and Croft, 1966). At physiological pH the citrate anion carries three negative charges so that transport of a proton is required to achieve electrical balance. The principal evidence for this carrier has been obtained from exchange studies and in passive swelling studies with isosmotic solutions of NH_4^+ salts (Wiskich, 1977). The primary conclusion from this work is that three transporters operate in sequence namely: (1) the phosphate transporter which requires a pH gradient across the inner mitochondrial membrane and imports $H_2PO_4^-$; (2)

the dicarboxylate carrier which exchanges the phosphate for dicarboxylate; and finally (3) the citrate carrier which exchanges dicarboxylate for citrate. Inhibition of any one of the carriers should inhibit citrate influx. Unfortunately, in many plant mitochondria exchange fluxes are not inhibited by mersalyl, a potent inhibitor of the phosphate carrier, and 2-N-butylmalonate, an inhibitor of the dicarboxylate carrier. Furthermore, 1,2,3-benzenetricarboxylate, a specific inhibitor of citrate transport in mammalian mitochondria, does not inhibit citrate uptake in mitochondria from potato tubers (Day and Wiskich, 1984). A cotransport of citrate and H^+ has been described for the uptake of citrate into isolated corn shoot mitochondria, although a citrate-malate exchange was also found in the same organelles (Day and Wiskich, 1984). It therefore appears that several different mechanisms might exist in plant mitochondria for the uptake of citrate. These all have one thing in common: rates of uptake are very low. According to Day and Wiskich (1984), these low uptake rates may account for the general low rate of citrate oxidation by isolated plant mitochondria. More work is required to understand citrate transport in plant mitochondria since potential utilization of citrate as a mitochondrial respiratory substrate from organic acid pool has been suggested. Since malate is usually present in the cytoplasm at higher concentrations than citrate, one would expect a malate-citrate exchange to function in exchanging mitochondrial citrate for cytoplasmic malate *in vivo*.

A fundamental metabolic requirement for the fixation of NH_4^+ involving glutamine synthetase and glutamate synthase is the maintenance of a sufficiently large (and continuously supplied) pool of α-ketoglutarate in the cytoplasm (Miflin and Lea, 1980). The presence of powerful $NADP^+$-linked isocitrate dehydrogenase and aconitase activities in the cytosol (Brouquisse *et al.*, 1987) may be the means by which citrate is converted to α-ketoglutarate for net ammonia assimilation. The majority (or all) of the extramitochondrial citrate is originally synthesized in the mitochondrion. Malate that is synthesized in the cytosol by phosphoenolpyruvate carboxylase and malate dehydrogenase can be imported into the mitochondrion and then converted into citrate by the NAD^+-linked malic enzyme and citrate synthase (Fig. 12.5). There might be endogenous controls polarizing the malate-citrate exchange facilitating citrate efflux from the mitochondrial matrix. If this exchange is 100% electrogenic (since citrate and malate carry three and two negative charges respectively at physiological pH, the hypothetical citrate^{3-}-malate^{2-} exchange would cause a net movement of charges across the mitochondrial membrane), the driving force would be provided by the electrochemical potential difference of the proton gradient. These conclusions demonstrate the central importance of the citrate carrier to plant cell metabolism, since the extent to which plant mitochondria are capable of citrate export might dramatically affect the supply of cytosolic α-ketoglutarate available to fast growing cells for amino acid metabolism.

Amino acid transport

The experiments of Leaver and coworkers (for review see Leaver and Gray, 1982) first demonstrated that amino acids were taken up and incorporated into proteins by plant mitochondria. This incorporation is dependent upon the mitochondria being intact and possessing coupled oxidative phosphorylation. It is sensitive to inhibitors of respiration and organellar protein synthesis. This requires that during the biogenesis of mitochondria net influx of amino acids into the matrix space must occur, since almost all amino acids are synthesized inside plastids. Unfortunately, the mechanism by which amino acids are transported into plant mitochondria is not clear.

Studies on the swelling of mitochondria from a number of different plant tissues strongly suggest that the movement of many amino acids into the mitochondrial matrix is not carrier-mediated and occurs by passive diffusion (Proudlove and Moore, 1982). Whilst several amino acids are reported to cross the inner membrane of plant mitochondria by diffusion alone, there also appear to be a number of carrier proteins involved in the transport of specific amino acids. These include carriers for glutamate, glycine and perhaps proline.

Glutamate and aspartate transport

In plant mitochondria that have been purified on a Percoll gradient, glutamate metabolism can be triggered by the addition of oxaloacetate in the presence of ADP and thiamine pyrophosphate (Fig. 12.6). Under these conditions, part of the oxaloacetate is metabolized by transamination whereas the other part is rapidly converted to malate. This conversion requires reduced pyridine nucleotides that are generated by the thiamine pyrophosphate-linked α-ketoglutarate dehydrogenase localized in the matrix space. The ratio of the flux rates of the two pathways is strongly dependent on the glutamate concentration in the medium. At low glutamate concentrations (up to 1 mM), the flux rates in the pathways are identical and aspartate, malate and succinate are formed equally. On the other hand, at high glutamate concentration (above 1 mM), a greater percentage of the oxaloacetate is metabolized through the transaminase pathway, and the excess of α-ketoglutarate thus formed is excreted from the matrix space by the α-ketoglutarate carrier (Douce, 1985). These results therefore suggest that glutamate probably enters the matrix space in exchange for aspartate. In plant mitochondria, however, there are data to indicate (see above) that exchange in the opposite direction (entry of aspartate and efflux of glutamate) also occurs when aspartate is metabolized by the transamination pathway (Douce, 1985).

Unfortunately, information on the transport of glutamate and aspartate in plant mitochondria is limited. Exchange of glutamate with a dicarboxylate has been reported in cauliflower and beet root mitochondria, because the swelling of mitochondria in ammonium glutamate requires the presence of both P_i and a permeant dicarboxylate anion (Day and Wiskich, 1984). On the other hand, back exchange experiments with castor bean mitochondria carried out by Millhouse et al. (1983) have failed to show influx of glutamate in exchange for dicarboxylate anion. They found instead a glutamate–aspartate exchange. The latter exchange was freely reversible and appeared not to be electrogenic, in contrast with mammalian mitochondria. In this latter case, a proton is cotransported on the carrier with glutamate and the exchange becomes therefore virtually unidirectional, glutamate entering the mitochondrial matrix space and aspartate leaving. Finally, Proudlove and Moore (1982) have used the classical silicone oil separation technique to directly determine uptake kinetics of glutamate and aspartate into mitochondria isolated from pea leaves. Influx of each amino acid appeared to be by diffusion because substrate saturation was not observed and compensatory or counter ions were not required for uptake.

A more detailed knowledge of glutamate and aspartate transport in plant mitochondria would be advantageous. However, it is clear that, in contrast with the situation observed in mammals, there is no evidence for a glutamate–OH^- or an electrogenic glutamate–aspartate transporter. That

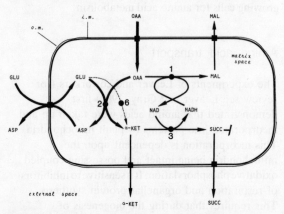

Fig. 12.6 Scheme of the metabolic processes involved in glutamate oxidation triggered by oxaloacetate in plant mitochondria. Added oxaloacetate (OAA) readily penetrates to the matrix. At least one-half of the OAA reacts with glutamate (GLU) under the control of glutamate–oxaloacetate transaminase (2); the α-ketoglutarate produced is either oxidized by the α-ketoglutarate dehydrogenase (3) or excreted into the external space. Since succinate dehydrogenase is strongly inhibited by OAA, succinate (SUCC) accumulates in the external space. Simultaneously, the other part of the OAA is converted into malate (MAL) by malate dehydrogenase (1), a conversion that depends on NADH generated by α-ketoglutarate dehydrogenase. In these conditions, O_2 consumption by the respiratory chain is inhibited. The activity of glutamate dehydrogenase (6) is negligible.

plant mitochondria do not exhibit an electrogenic glutamate–aspartate antiporter driving glutamate inside the mitochondria and aspartate outside can be explained by the existence of an NADH dehydrogenase located on the outside of the inner membrane (Douce, 1985). Consequently, a complex exchange system for the transfer of reducing equivalents from cytosol to matrix involving an electrogenic antiport that is driven by the membrane potential across the inner mitochondrial membrane is not required as it is in the case of mammalian mitochondria. Instead, fluxes of aspartate and glutamate will respond to removal of each compound and will be determined by movement of their corresponding ketoacids and also by the local concentration of each throughout the cells.

Serine and glycine transport

The photorespiratory carbon cycle requires the participation of three organelles in the plant cell: the chloroplast, the peroxisome and the mitochondrion. In the course of the cycle, one molecule of serine leaves the mitochondrion, and two molecules of glycine are taken up (Fig. 12.4). Despite the evidence that glycine and serine transporters are common in a number of biological systems, there is no consensus that the much larger flux of glycine across the inner membrane of plant leaf mitochondria is carrier-mediated (Douce, 1985). According to several groups, influx of glycine is not related to the energy status of the mitochondrion and appears diffusional because transport rates increase linearly up to very high glycine concentrations (Proudlove and Moore, 1982). This is surprising in view of the very rapid rates of decarboxylation of glycine in isolated leaf mitochondria (Douce, 1985). In contrast, according to Oliver (1987), glycine oxidation is sensitive to the non-penetrating sulfhydryl reagents, mersalyl and p-chloromercuribenzoate, suggesting that a glycine transporter does exist. Furthermore, the data of Oliver (1987) strongly suggest that glycine can exchange directly with serine and hydroxyl ions. This would allow a flexible stoichiometry for the exchange reaction that would accommodate the 2:1 ratio needed for the reaction by which glycine is metabolized and results in a stoichiometry of two glycine molecules entering the mitochondrial matrix in exchange for one serine leaving.

Cofactor uptake by plant mitochondria

Intact and well coupled plant mitochondria appear capable of the uptake of several important enzyme cofactors. Metabolic pathways that have factor-dependent enzymes localized in the matrix might be subject to modulation in this manner.

Net import of adenine nucleotides

The adenine nucleotide content (ATP + ADP + AMP) of freshly isolated animal mitochondria is fairly constant and in the range of 11–13 nmol mg protein^{-1}. Rapid leakage of nucleotides across the inner mitochondrial membrane does not usually occur. Consequently, addition of ADP to the medium triggers the maximum rate of nucleotide exchange. In contrast, the total amount of adenine nucleotides present in intact plant mitochondria is very low (1–2 nmol mg protein^{-1}). Consequently, the initial rates of substrate oxidation catalyzed by plant mitochondria in the presence of ADP are limited by the internal concentration of adenine nucleotides that can be exchanged by the nucleotide carrier. This is also the case for ATP hydrolysis by intact mitochondria facilitated by an uncoupler (Douce, 1985) (Fig. 12.7). However, Abou-Khalil and Hanson (1977) have demonstrated the existence in plant mitochondria of a mechanism for net uptake of ADP or ATP which is insensitive to carboxyatractyloside. This net uptake is strongly accelerated if an electrochemical gradient of protons across the membrane is established. Conversely, this net uptake is strongly inhibited by uncouplers. We can speculate, therefore, that *in vivo* the matrix adenine nucleotide pool is maintained against a concentration gradient by virtue of a one-way movement of adenine nucleotides. This net uptake would also explain how the total adenine nucleotide content is established and maintained during mitochondrial proliferation.

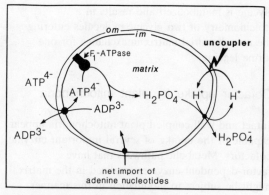

Fig. 12.7 Mechanism of ATP hydrolysis by intact mitochondria triggered by an uncoupler. Mitochondria are first loaded with adenine nucleotides in an energy-dependent manner (net import) in order to trigger a rapid ATP–ADP exchange. Note that under these conditions the ADP^{3-}–ATP^{4-} exchange associated with a net movement of $H_2PO_4^-$ does not cause a movement of charges across the mitochondrial membrane. ΔpH across the inner membrane generated by the symport of $H_2PO_4^-$ with H^+ is continuously collapsed by the rapid leak of protons through the inner membrane induced by an uncoupler. om: outer membrane; im: inner membrane.

Net import of NAD^+

Complex I, which is the segment of the respiratory chain responsible for electron transfer from NADH to ubiquinone, operates in close relationship with all of the NAD^+-linked TCA cycle dehydrogenases and utilizes a common pool of NAD^+ that is present in the matrix. The NADH formed by the TCA cycle dehydrogenases diffuses to the inner membrane where it is oxidized. The diffusion rate of pyridine nucleotides within the matrix space, which contains a high concentration of proteins, can limit the overall rate of mitochondrial respiration. In order to overcome this problem, the mitochondrial matrix space usually contains a high concentration of pyridine nucleotides (i.e. above 6–7 nmol mg protein^{-1}).

Stimulation of respiration in isolated plant mitochondria by exogenous NAD^+ is well known (Palmer, 1976) and although it occurs during the oxidation of all NAD^+-linked substrates, it is especially the case with malate. However, not all plant mitochondria respond to added NAD^+, and those that do not respond generally have high endogenous NAD^+ contents and rapid rates of respiration (Douce, 1985). Very often, the amount of NAD^+ present in plant mitochondria is sufficient to allow state 3 rates that are approximately 90% of the maximal rate that can be obtained; that is, adding NAD^+ slightly stimulates state 3 rates. However, the rate of O_2 uptake in the presence of rotenone, an inhibitor of electron flow through complex I, is almost completely dependent on NAD^+ (Douce, 1985). Plant mitochondria have the ability to bypass complex I during oxidation of endogenous NADH. This bypass is coupled to only two sites of energy transduction and may be catalyzed by a second internal NADH dehydrogenase (Chapter 9; Palmer, 1976). This second dehydrogenase has been shown to have a much lower affinity for NADH than its rotenone-sensitive counterpart. Hence, plant mitochondria need to maintain a sufficiently high internal pool of NAD^+ to satisfy the requirements of the internal rotenone-insensitive pathway.

The mitochondrial inner membrane is generally considered to be impermeable to nicotinamide nucleotides (Douce, 1985). The mechanism by which exogenous NAD^+ can stimulate internally located enzymes such as NAD^+-linked malic enzyme and NADH dehydrogenases is therefore not obvious. However, Neuburger and Douce (1983) have shown that isolated intact plant mitochondria actively accumulate NAD^+ from the external medium. This leads to a substantial increase in the matrix concentration of the cofactor which stimulates matrix dehydrogenases and electron transport activities, especially the rotenone-resistant respiration. Hence, plant mitochondria appear to possess a specific NAD^+ carrier, since NAD^+ uptake is concentration-dependent and exhibits Michaelis–Menten kinetics. The maximum velocity of the carrier is strongly affected by the initial concentration of NAD^+ present in the mitochondrion. Furthermore, the rate of NAD^+ transport is temperature-dependent and the analog N-4-azido-2-nitrophenyl-4-aminobutyryl-3′-NAD^+ (NAP_4-NAD^+) inhibits (almost completely) net NAD^+ import.

Neuburger and Douce (1983) have also found NAD^+ efflux from intact isolated mitochondria. Under these conditions, the concentration of NAD^+ in the mitochondrion progressively decreases, and oxidation of NAD^+-linked substrates becomes increasingly dependent on added NAD^{+*}. As the level of NAD(H) falls progressively in the mitochondria it becomes too low to act as a substrate for rotenone-resistant respiration, but is still sufficient for the rotenone-sensitive pathway. The rate of NAD^+ efflux from the matrix space is dependent upon the intramitochondrial NAD^+ concentration and is inhibited by the above analog that inhibits NAD^+ uptake, indicating that a protein is required for flux in both directions. It seems likely, therefore, that in plant mitochondria the matrix NAD^+ concentration is tightly controlled (Fig. 12.8).

The physiological role of this NAD^+ carrier still remains uncertain. Since the intramitochondrial concentration of NAD^+ has such a profound influence on the NAD^+-linked malic enzyme and O_2 uptake via the rotenone-insensitive pathway by isolated mitochondria, it is potentially a very powerful regulator of malate oxidation *in vivo* and could play an important role in the coarse control of metabolism, particularly during the transition from dormancy to active growth and vice versa. In this context, the mitochondria from young growing tissues, including the shoots of sprouting potato tubers, have higher matrix NAD^+ contents than those from storage tissues (Neuburger *et al.*, 1985). It is also interesting to note that, over the time period of potato tuber storage, the endogenous NAD^+ content of the mitochondria declines progressively and thereafter must rise again during sprouting. It is possible, therefore, that the rates of respiration in this tissues would be affected by the concentration of NAD^+ in the cytoplasm and this concentration might differ significantly from one tissue to the other, or even

---

* The fact that intact purified mitochondria progressively lose their NAD^+ content and that this leads to a dramatic decrease in the O_2 uptake rates strongly suggests that most of the NAD^+ is not bound to the inner membrane or to the various dehydrogenases. This is another proof that NAD^+ molecules diffuse freely in the matrix space.

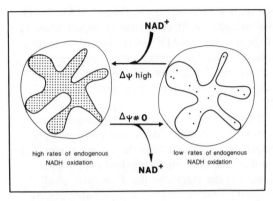

Fig. 12.8 Transport of NAD^+ in plant mitochondria. When suspended in a medium that avoids rupture of the outer membrane purified mitochondria progressively lose their NAD^+ content by passive diffusion as the electrochemical potential gradient of protons across the inner membrane is low. This leads to a slow decrease of endogenous NADH oxidation. Conversely, addition of NAD^+ to the medium restores the initial rate of endogenous NADH oxidation, insofar as the electrochemical potential gradient of protons across the inner membrane is high.

between different physiological states of the same tissue. Finally, if NAD^+ rises in the matrix space without a simultaneous requirement for ATP, then the matrix NAD^+ pool will become more reduced. This might increase respiration via the rotenone-insensitive pathway without the need for gross changes in the cytoplasmic phosphorylation status.

Net import of coenzyme A

Although a large number of synthetic and degradative reactions in all tissues depend on coenzyme A (CoA), little is known about the regulation of CoA levels in the cell and even less about the intracellular distribution of CoA between the cytosol and mitochondrial matrix.

Plant mitochondria isolated from a number of tissues have a relatively low endogenous CoA content. These mitochondria are capable of actively accumulating CoA content in a manner sensitive to uncouplers and low temperature. This net uptake is catalyzed by a specific transport system and leads to large increases in the CoA content of the matrix. NAD^+ and CoA transport does not share a common carrier, because NAP_4-NAD^+ has virtually no effect on CoA transport whereas it strongly inhibits NAD^+ influx. This CoA uptake follows saturation kinetics with an apparent K_m of 0.2 mM and a maximum velocity of 4–6.5 nmol mg protein^{-1} min^{-1} (Neuburger et al., 1984).

The physiological function of a mitochondrial transport system for CoA would be to move the intact CoA molecule from the cytosol where it is synthesized, to the mitochondrial matrix where it is used in the entry into the TCA cycle of all major fuel substrates (pyruvate, α-ketoglutarate and malate). Plant mitochondria, in contrast with mammalian mitochondria, readily oxidize malate without the necessity of removing oxaloacetate because they possess a specific NAD^+-linked malic enzyme. This enzyme has an absolute requirement for Mn^{2+} and is characterized by a low affinity for substrates. It is inhibited by bicarbonate, which accumulates in the matrix space at alkaline pH and this inhibition is fully relieved by CoA. Consequently, since the cytosolic pH in vivo is around 7.5, NAD^+-linked malic enzyme will have a low activity unless a rather large pool of CoA is present in the matrix. The CoA transport system might therefore function in vivo to control the activity of the NAD^+-linked malic enzyme (Day et al., 1984). This may be very important in vivo since malic enzyme, which is not very sensitive to high matrix NADH concentrations, can readily pass electrons to the non-phosphorylating rotenone-insensitive NADH dehydrogenase, which would lead to the TCA cycle functioning anaplerotically. In other words, the reaction catalyzed by NAD^+-linked malic enzyme compensates for the drain on the TCA cycle that occurs when intermediates that lead to citrate, oxaloacetate and α-ketoglutarate are removed from the cycle for biosynthesis.

Net import of thiamine pyrophosphate

There is also evidence that thiamine pyrophosphate (TPP) can enter isolated plant mitochondria. It is essential to add this cofactor for the oxidation of α-ketoglutarate and pyruvate (Fig. 12.9) (Lance et al., 1967): the isolated organelles appear to be depleted of endogenous

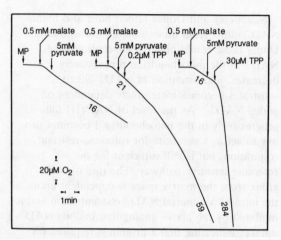

Fig. 12.9 Effect of thiamine pyrophosphate (TPP) on pyruvate oxidation by potato tuber mitochondria. Note that TPP added to the medium even at very low concentration (0.2 μM) triggers, after a lag phase, a rapid rate of O_2 consumption. This experiment demonstrates that TPP molecules penetrate the inner membrane of plant mitochondria. MP: purified mitochondria.

TPP (presumably during their isolation, although this has yet to be demonstrated) but rapidly accumulate it when it is provided externally. To determine the kinetics of TPP uptake, mitochondria have been incubated with labeled TPP at concentrations in the range 1–100 μM, and the initial rates of uptake determined by rapid filtration experiments (in the millisecond time range). The K_m value for TPP is 50 μM and the maximum velocity is 0.8 nmol mg protein^{-1} min^{-1}. Furthermore, the extent of TPP accumulation is unaffected by the addition of 50 μM NAP_4-NAD^+ to the incubation medium, indicating that NAD^+ and TPP do not share a common carrier.

Conclusions

The control of plant respiration in vivo is likely exerted through the finely regulated supply of substrates to the mitochondrial electron transport chain and phosphorylating system. As shown in this chapter, the mechanisms involved in the regulation of anion transport across the inner mitochondrial membrane, other than adenine nucleotides and P_i, are still poorly defined. Likewise, nothing is known about the chemical nature of respiratory substrates (pyruvate, malate, oxaloacetate, NADH) utilized by the mitochondrion inside plant cells. The carriers, particularly the tri- and dicarboxylate ones are not specific and a number of factors including pH and charges on the molecules affect the kinetics and equilibrium of the exchange reactions. The recent discovery by Zoglowek et al. (1988) showing the presence in plant mitochondria of a very powerful electrophoretic uniport system especially for oxaloacetate and malate (anion channel ?) does little to clarify the mechanisms of anion transport regulation in plant mitochondria. For example, electrophoretic ejection of an anion such as malate, followed by electroneutral re-uptake, represents a futile cycle and a drain on respiratory energy.

It is now apparent that plant mitochondria possess several distinct carriers for the net uptake of cofactors. These specific carriers would allow regulation of the matrix cofactors pool size *in vivo*, either maintaining them at a constant level or allowing changes in response to specific signals. In this way, metabolic pathways with cofactor-dependent enzymes localized in the matrix might be subject to modulation leading to considerable modification of carbon traffic between the cytosol and matrix. For example, in the presence of a large excess of free CoA in the matrix, NAD^+-linked malic enzyme can operate, allowing the conversion of C4 acids into acetyl-CoA, the normal respiratory substrate, without the necessity of supplying pyruvate from glycolysis. Likewise, an excess of NAD^+ in the matrix space is required to engage the non-phosphorylating rotenone-resistant electron pathway, releasing partially the constraint exerted by the protonmotive force across the inner membrane on internal NADH oxidation. Interestingly, it is only at high matrix NAD^+ concentrations that a functional link between malic enzyme (exhibiting a low affinity for NAD^+) and the non-phosphorylating rotenone-insensitive NADH dehydrogenase (exhibiting a low affinity for NADH) is observed. Under these conditions malic enzyme and the TCA cycle could play a significant role in the metabolism of fast growing cells by providing numerous substrates for biosynthetic reactions in the cytoplasm. There may be an advantage, therefore, in having a non-phosphorylating electron transport chain to enable this role to be fulfilled in the presence of a high protonmotive force across the inner mitochondrial membrane. There is circumstantial evidence that some of these changes in matrix cofactor pool size occur *in vivo*. However, until clearer evidence is forthcoming, their role in the control of respiration in intact cells and tissues cannot be properly evaluated.

Finally, this article indicates that the only way to either replenish pools of depleted TCA intermediates or metabolize intermediates present in excess is through anaplerotic (literally: filling up) reactions which provide short-cuts across the cycle.

References

Abou-Khalil, S. and Hanson, J. B. (1977). Net adenosine diphosphate accumulation in mitochondria. *Arch. Biochem. Biophys.* **183**, 581–7.

Brailsford, M. A., Thomson, A. G., Kaderbhai, N. and Beechey, R. B. (1986). Pyruvate metabolism in castor-bean mitochondria. *Biochem. J.* **239**, 355–61.

Brouquisse, R., Nishimura, M., Gaillard, J. and Douce, R. (1987). Characterization of a cytosolic aconitase in higher plant cells. *Plant Physiol.* **84**, 1402–7

Chance, B. and Williams, G. R. (1956). The respiratory chain and oxidative phosphorylation. *Adv. Enzymol.* **17**, 65–134.

Chappell, J. and Beevers, H. (1983). Transport of dicarboxylic acids in castor bean mitochondria. *Plant Physiol.* **72**, 434–40.

Chappell, J. B. and Croft, A. R. (1966). Ion transport and reversible volume changes of isolated mitochondria. In *Regulation of Metabolic Processes in Mitochondria*, Vol. 17, ed. J. M. Tager, Elsevier, New York, pp. 293–316.

Day, D. A. and Hanson, J. B. (1977). Pyruvate and malate transport and oxidation in corn mitochondria. *Plant Physiol.* **59**, 630–5.

Day, D. A. and Wiskich, J. T. (1984). Transport processes of isolated mitochondria. *Physiol. Veg.* **22**, 241–61.

Day, D. A., Neuburger, M. and Douce, R. (1984). Activation of NAD-linked malic enzyme in intact plant mitochondria by exogenous coenzyme A. *Arch. Biochem. Biophys.* **231**, 233–42.

Diolez, P. and Moreau, F. (1985). Correlation between ATP synthesis, membrane potential and oxidation rate in plant mitochondria. *Biochim. Biophys. Acta* **806**, 56–64.

Douce, R. (1985). *Mitochondria in Higher Plants. Structure, Function, and Biogenesis,* Academic Press, Orlando.

Douce, R., Neuburger, M. and Givan, C. V. (1986). Regulation of succinate oxidation by NAD^+ in mitochondria purified from potato tubers. *Biochim. Biophys. Acta* **850**, 64–71.

Dry, I. B., Dimitriadis, E., Ward, A. D. and Wiskich, J. T. (1987). The photorespiratory shuttle. Synthesis of phthalonic acid and its use in the characterization of the malate/aspartate shuttle in pea (*Pisum sativum*) leaf mitochondria. *Biochem. J.* **245**, 669–75.

Ebbighausen, H., Jia, C. and Heldt, H. W. (1985). Oxaloacetate translocator in plant mitochondria. *Biochim. Biophys. Acta* **810**, 184–99.

Hanson, J. B. and Koeppe, D. E. (1975). Mitochondria. In *Ion Transport in Plant Cell and Tissues*, eds D. A. Baker and J. L. Hall, North-Holland Publishing, Amsterdam, pp. 79–99.

Huang, A. H. C., Trelease, R. N. and Moore, T. S., Jr. (1983). *Plant Peroxisomes*, Academic Press, New York.

Lance, C., Hobson, G. E., Young, R. E. and Biale, J. B. (1967). Metabolic processes in the cytoplasmic particles from the avocado fruit. IX. The oxidation of pyruvate and malate during the climacteric cycle. *Plant Physiol.* **42**, 471–8.

Leaver, C. J. and Gray, M. W. (1982). Mitochondrial genome organization and expression in higher plants. *Ann. Rev. Plant Physiol.* **33**, 373–402.

Mannella, C. A. (1985). The outer membrane of plant mitochondria. In *Encyclopedia of Plant Physiology*, Vol. 18, eds R. Douce and D. Day, Springer-Verlag, Berlin, pp. 106–33.

Miflin, B. J. and Lea, P. J. (1980). Ammonia assimilation. In *The Biochemistry of Plants*, Vol. 5, ed. B. J. Miflin, Academic Press, New York, pp. 169–99.

Millhouse, J., Wiskich, J. T. and Beevers, H. (1983). Metabolite oxidation and transport in mitochondria of germinating castor bean endosperm. *Aust. J. Plant Physiol.* **10**, 167–77.

Mitchell, P. (1969). Chemiosmotic coupling and energy transduction. In *Theoretical and Experimental Biophysics*, ed. A. Cole, Dekker, New York, pp. 160–216.

Neuburger, M. and Douce, R. (1983). Slow passive diffusion of NAD^+ between intact isolated plant mitochondria and suspending medium. *Biochem. J.* **216**, 443–50.

Neuburger, M., Day, D. and Douce, R. (1984). Transport of coenzyme A in plant mitochondria. *Arch. Biochem. Biophys.* **299**, 253–8.

Neuburger, M., Day, D. A. and Douce, R. (1985). Transport of NAD^+ in percoll purified tuber mitochondria: inhibition of NAD influx and efflux by N-4-azido-2-nitrophenyl-4-aminobutyryl-3'-NAD^+. *Plant Physiol.* **78**, 405–10.

Oliver, D. J. (1987). Glycine uptake by pea leaf mitochondria: a proposed model for the mechanism of glycine–serine exchange. In *Plant Mitochondria. Structural, Functional and Physiological Aspects,* eds A. L. Moore and R. B. Beechey, Plenum Press, New York and London, pp. 219–26.

Palmer, J. M. (1976). The organization and regulation of electron transport in plant mitochondria. *Ann. Rev. Plant Physiol.* **27**, 133–57.

Proudlove, M. O. and Moore, A. L. (1982). Movement of amino acids into isolated plant mitochondria. *FEBS Lett.* **147**, 26–30.

Roby, C., Martin, J. B., Bligny, R. and Douce, R. (1987). Biochemical changes during sucrose deprivation in higher plant cells. Phosphorus-31 nuclear magnetic resonance studies. *J. Biol. Chem.* **262**, 5000–7.

Vignais, P. V. (1976). Molecular and physiological aspects of adenine nucleotide transport in mitochondria, *Biochim. Biophys. Acta* **456**, 1–38.

Wiskich, J. T. (1977). Mitochondrial metabolite transport. *Ann. Rev. Plant Physiol.* **28**, 45–69.

Wohlrab, H. (1986). Molecular aspects of inorganic phosphate transport in mitochondria. *Biochim. Biophys. Acta* **853**, 115–34.

Zoglowek, C., Krömer, C. and Heldt, H. W. (1988). Oxaloacetate and malate transport of plant mitochondria. *Plant Physiol.* **87**, 109–15.

Photosynthesis

V

Photosynthesis

13 Plastid structure and development
William Newcomb

Introduction

Plastids are self-replicating organelles, surrounded by an envelope comprised of two membranes. They are only found in cells of photosynthetic eukaryotes. With the exception of the generative and sperm cells of certain angiosperms, all plant cells possess plastids which occur in a variety of types, sizes, shapes and colors. It is easiest to categorize plastids on the basis of their color: green, other colors (red, orange and yellow), and colorless. Green-colored plastids are chloroplasts, which contain the chlorophyll pigments and are the site of photosynthesis. Red-, orange- and yellow-colored plastids are chromoplasts which contain carotenoid pigments and are commonly found in flowers, fruits, senescing leaves and certain roots. Chromoplasts function mainly to attract pollinators and animals, which aid in the distribution of pollen and seeds respectively. Colorless plastids lack chlorophyll and carotenoid pigments and may be referred to by an old term, leucoplast. However, it is more common and precise to name these various plastids on the basis of their principal storage products; amyloplasts store starch, while proteins and lipids are stored in proteinoplasts and elaioplasts respectively. Etioplasts are either colorless or pale yellow in color; these plastids usually form in dark-grown seedlings and are arrested in their development; upon exposure to light they synthesize chlorophyll and become chloroplasts. Proplastids, which have been regarded as the progenitor for all other types of plastids, are also colorless. J. Whatley has distinguished three main types of proplastids involved in chloroplast development, namely eoplasts, amyloplasts and amoeboid plastids.

Common features

All plastids have a number of features in common. They are surrounded by an envelope comprised of inner and outer membranes which are separated by a 10–20 nm gap. Plastids contain DNA which appears as fibrils in a matrix-free region (the so-called nucleoid) of the stroma, which is the background matrix material of the plastid. The stroma contains numerous ribosomes that are smaller than the cytoplasmic ribosomes. The plastid ribosomes may be arranged singularly or in polyribosomes. Plastoglobuli, small lipid droplets which are not membrane bound, are also present in the stroma of many plastids.

Plastid inheritance

In most angiosperms plastid inheritance is maternal because the sperm cells lack plastids. This situation arises during the first pollen mitosis when all of the plastids go into the vegetative cell resulting in an aplastidic generative cell. In some species (e.g. *Antirrhinum majus*, *Beta vulgaris*, *Hyoscyamus niger* and *Solanum tuberosum*), although most of the plastids pass to the vegetative cell, during the first pollen mitosis a few plastids pass into the generative cell.

However, during the development of the sperm cells in these plants the plastids are lost and as a result plastid inheritance remains maternal. In a few angiosperm genera, namely *Oenothera* and *Pelargonium*, plastid inheritance is biparental because plastid distribution is equal during the first pollen mitosis and subsequent development does not eliminate plastids resulting in plastids being present in the sperm cells.

Plastid ultrastructure

Proplastids

Proplastids, the progenitors of other plastids, are found in root and shoot meristems, embryos, endosperm and in young developing leaves. In view of the fact that most plastids may differentiate into another type of plastid, the progenitor function of proplastids should be regarded as secondary to their simple structure which is a primary distinguishing feature. Proplastids are not structurally complex. They consist of an envelope, a nucleoid with DNA fibrils, a small number of ribosomes, some flattened membrane sacs called thylakoids, which may appear to be continuous with the inner membrane of the envelope, and sometimes small starch granules (Fig. 13.1). Proplastids vary in shape and can be ellipsoidal, spherical, cylindrical and branched, with or without bulged or invaginated regions. They are usually only slightly larger than mitochondria. Within meristematic cells, the division of proplastids keeps pace with mitosis and cytokinesis. Proplastids may also divide after cytokinesis has ceased in certain root cells where they usually develop into amyloplasts (Fig. 13.2). In light-grown plants, the proplastids in stems and leaves usually differentiate into chloroplasts. Proplastids may also have important biochemical functions such as the synthesis and storage of starch and other metabolites.

Chloroplasts

Chloroplasts are green photosynthetic plastids that have an elaborate arrangement of interconnected membranous sacs in the plastid stroma. Chloroplasts typically measure about $5 \times 2 \times 1-2$ μm. The internal membrane system of a typical higher plant chloroplast includes grana (also known as grana thylakoids) which are stacks of flattened disk-shaped membranous sacs. About 50 grana are present in a typical angiosperm chloroplast. Membranous channels, that have been called thylakoids, frets, stroma thylakoids or stroma lamellae by various authors, traverse the stroma and interconnect the grana. The concentration of photosystems I and II differ in the thylakoids from the stroma and grana. The degree of stacking of the thylakoids into grana varies with the physiological requirements of the cell (see Chapters 15 and 19). The chlorophyll pigments and light reactions of photosynthesis are associated with the thylakoid membrane system. The dark reactions of photosynthesis occur in the

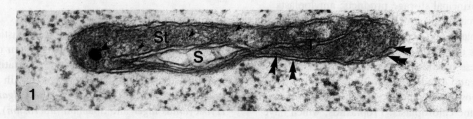

Fig. 13.1 Transmission electron micrograph of a proplastid in an epidermal cell of an embryo of chickpea (*Cicer arietinum* L.). Shown are the two membranes comprising the bounding membrane envelope (double large arrowheads), an osmiophilic droplet (single large arrowhead), a starch (S) granule, a thylakoid (T) and ribosomes (small arrowheads) in the stroma (St). × 74 800.

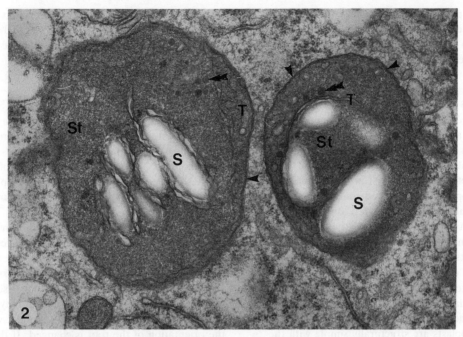

Fig. 13.2 Transmission electron micrograph of two amyloplasts in a pericycle cell adjacent to a root nodule of the garden pea (*Pisum sativum*). Shown within the plastids are the bounding membrane envelope (single arrowheads), osmiophilic droplets (double arrowheads), starch (S) granules and stroma (St). × 33 300.

stroma which also contains numerous ribosomes and large proteinaceous particles. These particles may be ribulose 1,5-bisphosphate carboxylase/oxygenase which makes up more than 50% of the protein in the chloroplast stroma.

Plastids have three compartments, the intramembrane space between the outer and inner membranes, the stroma and the thylakoid space. In contrast, mitochondria possess only the intramembrane space and a matrix which corresponds to the plastid stroma. In chloroplasts, light reactions cause a proton gradient to develop across the thylakoid membrane. This gradient is discharged during the synthesis of ATP by the ATP synthetase which is located in the membrane. Concomitantly with ATP synthesis, $NADP^+$ is reduced to NADPH and water is oxidized with liberation of oxygen (see Chapter 15). In the dark reactions of photosynthesis, the ATP and NADPH synthesized during the light-dependent reactions are used for carbon fixation, the chemical reduction of carbon dioxide to carbohydrate (see Chapter 17).

Chloroplast development

Leaves develop from primordia initiated along the flanks of the shoot apical meristem. These meristematic cells possess proplastids which contain a few thylakoids in their stroma. Some authors have referred to such a plastid as a spherically-shaped eoplast. The eoplast accumulates starch, becoming an amyloplast which subsequently loses the starch deposits and becomes amoeboid. Most authors have not distinguished among these types and have referred to all three of these forms as being proplastids. The amoeboid plastids grow in size and develop stroma lamellae before grana appear. When these plastids mature, the stroma lamellae align and the disk-shaped thylakoids of the grana become stacked. The mature chloroplasts eventually senesce and the stroma lamellae and grana become disorganized. Some of these senescent chloroplasts (termed gerontoplasts by Whatley) may synthesize carotenoid pigments and thus can be termed chromoplasts. In some tissue cultures and evergreen plants, chloroplasts may dedifferentiate and then redifferentiate as chromoplasts. This in part led to the development of Whatley's ideas on the occurrence of a plastid cycle.

Etioplasts

Etioplasts are found in plants which are grown in continuous darkness, which results in these tall etiolated plants that are yellow due to the presence of protochlorophyll in the etioplasts. While etioplasts are regarded commonly as putative chloroplasts whose development has been arrested, it is important to note that etioplasts are not a normal stage in the development of chloroplasts. In some cases, etioplasts represent an interesting developmental alternative. Most seedlings receive a regular diurnal sequence of light and dark periods and etioplasts do not develop under these conditions.

The etioplasts are structurally simple with the most distinctive feature being a paracrystalline arrangement of membranes, termed the prolamellar body. Upon exposure to light the etioplasts differentiate into chloroplasts during which the protochlorophyll becomes converted into chlorophyll and the prolamellar body reorganizes into grana and stroma lamellae. There is considerable variation in plastid development in various species. For example, prolamellar bodies do not form in the etioplasts of dark-grown *Avena* coleoptiles.

Chromoplasts

Chromoplasts are red-, orange- and yellow-colored due to the presence of carotenoid pigments and commonly occur in flower petals, fruits and certain roots such as carrots. Often the chromoplasts develop from chloroplasts but chromoplasts may also be formed from proplastids and amyloplasts. During the transition of a chloroplast into a chromoplast, the chlorophyll is destroyed and the grana and stroma lamella become reorganized. Chromoplasts may also revert back to chloroplasts; examples of this include carrot roots and orange fruit skins. Osmiophilic droplets or plastoglobuli, filamentous pigmented bodies, and crystals are normally found in many chromoplasts and are the sites of carotenoid synthesis and/or storage.

Leucoplasts

Leucoplasts are colorless plastids which have lost their progenitor function, i.e. are distinct from proplastids. Amyloplasts, elaioplasts and proteinoplasts are colorless and are sites of the synthesis of starch, lipids and proteins respectively. Some leucoplasts have deposits of starch and proteins.

Amyloplasts are characterized by the presence of one or more starch granules which distend and alter the shape of the organelle. Only a few thylakoids are normally found within the stroma. Amyloplasts are most common in root tissues and may be involved in the detection of gravity. Amyloplasts of the root cap will sediment when the orientation of the root is changed. If the root cap is removed, the root's ability to detect gravity is lost. When the root cap reforms, the ability to detect gravity is restored. Amyloplasts of roots may develop into chloroplasts if the roots are exposed to light.

Leucoplasts with specialized functions occur in certain cells and tissues. Some but not all of these plastids have unique ultrastructural features. For example, the cells of some suspensors of angiosperm embryos contain large plastids which contain extensive profiles of small diameter membranous tubules (Fig. 13.3). Specialized leucoplasts also occur in sieve elements and the uninfected cells of certain leguminous root nodules which export ureides.

Further reading

Albert, B., Bray, D., Lewis, J., Raff, M., Roberts, K. and Watson, J. D. (1989). *Molecular Biology of the Cell*, 2nd edn, Garland Publishing Inc., New York.

Gunning, B. E. S. and Steer, M. W. (1975). *Ultrastructure and the Biology of Plant Cells*, Edward Arnold, London.

Newcomb, E. H. (1967). Fine structure of protein-storing plastids in bean root tips. *J. Cell Biol.* **33**, 143–63.

Possingham, J. V. (1980). Plastid replication and

Plastid structure and development 197

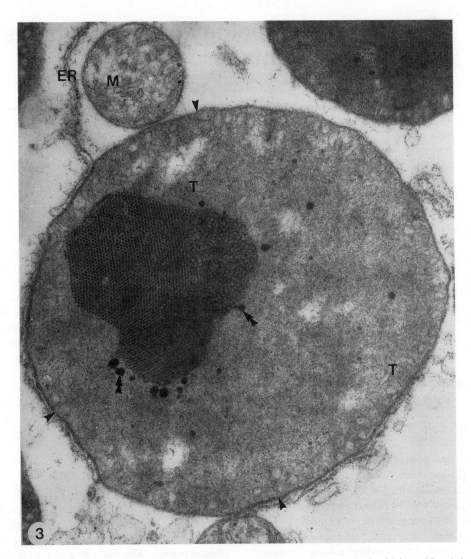

Fig. 13.3 Transmission electron micrograph of a specialized leucoplast in a suspensor cell of a chickpea (*C. arietinum*) embryo. Shown in the plastid are the bounding membrane envelope (single arrowheads), several electron-dense osmiophilic droplets (double arrowheads), several thylakoids (T) and a large region of electron-dense, regularly-arranged, membranous tubules. Shown outside the plastid are a mitochondrion (M) and profiles of endoplasmic reticulum (ER). × 36 400.

development in the life cycle of higher plants. *Ann. Rev. Plant Physiol.* **31**, 113–29.

Thomson, W. W. and Whatley, J. M. (1980). Development of nongreen plastids. *Ann. Rev. Plant Physiol.* **31**, 375–94.

Whatley, J. B. (1974). Chloroplast development in primary leaves of *Phaseolus vulgaris*. *New Phytol.* **73**, 1097–110.

Whatley, J. B. (1977). Variations in the basic pathway of chloroplast development. *New Phytol.* **78**, 407–20.

Whatley, J. B. (1978). A suggested cycle of plastid development interrelationships. *New Phytol.* **80**, 489–502.

Whatley, J. B. (1979). Plastid development in the primary leaf of *Phaseolus vulgaris*: variations between different types of cells. *New Phytol.* **82**, 1–10.

14 Molecular biology of photosynthesis in higher plants
John E. Mullet

Introduction

This chapter will summarize the organization and regulation of genes involved in photosynthesis in higher plants. Not all aspects of plastid function/gene expression will be covered and additional background information can be obtained in several books (see Further reading). Citations on specific topics covered in this chapter can be found in recent reviews which are listed with the relevant topic.

Plastid function

In higher plants, photosynthetic electron transport and CO_2 fixation occur within chloroplasts. Chloroplasts, as described in the previous chapter, are a specialized type of plastid and are so-named when chlorophyll accumulates within this organelle. Chloroplasts consist of a double outer envelope, a stromal phase where soluble proteins such as Rubisco are located and an inner membrane, termed the thylakoid. Four major photosynthetic complexes are integrated into the thylakoid: Photosystem I (PSI), Photosystem II (PSII), a cytochrome complex, and ATP synthetase. These complexes mediate light-dependent vectorial electron flow and the resultant generation of ATP and reducing power. The end products of photosynthetic electron transport are used in CO_2 fixation and other plastid functions. In addition to photosynthesis *per se*, the plastid carries out numerous other functions. For example, steps in amino acid, fatty acid, pyrimidine, terpene and tetrapyrrole biosynthesis occur in plastids. This organelle is also a site of nitrite and sulfur reduction. Some steps in plant growth regulator synthesis are also localized in plastids. The interdependence of cell and plastid function in photosynthetic as well as non-photosynthetic tissue makes the regulation of genes encoding plastid polypeptides involved in photosynthesis inseparable from the more general problem of cell biogenesis and function. On the other hand, the genes encoding photosynthetic enzymes have special features and these will be dealt with after a broader consideration of plastid origin and gene expression.

Origin of plastids

Plastids in higher plants contain a genome which has up to 123 genes. However, this is only a fraction of the total number of genes required for photosynthesis. The remaining genes are encoded in nuclear DNA. It is believed that the presence of DNA in plastids reflects the origin of this organelle from a free-living oxygen-evolving photosynthetic prokaryote (Gray and Doolittle, 1982). This concept, which states that plastids are derived from prokaryotic endosymbionts, is based on a large body of information which shows that the genes and mechanisms for decoding the genetic information within plastids are strikingly similar to those found in prokaryotes. For example, plastids contain 70S ribosomes and uncapped RNAs which are not polyadenylated. Many plastid transcripts also have sequences homologous to prokaryotic ribosome binding sites which precede protein coding regions. In contrast, the cytoplasm in eukaryotes has 80S ribosomes which translate capped and polyadenylated mRNAs lacking ribosome binding sites. Plastid transcription units are often polycistronic as in bacteria. In addition, transcription initiation in plastids is specified by DNA sequence elements which are similar to bacterial promoter elements. Furthermore, the presence of gene clusters in plastid genomes which resemble operons found in *E. coli* and homology between plastid and *E. coli* genes, provide strong support for the endosymbiont hypothesis. In summary, gene organization and decoding mechanisms found in plastids are consistent with the hypothesis that these organelles arose from free-living photosynthetic prokaryotes.

Integration of plastid and cell function

Plastids isolated from higher plants do not replicate and survive only a short period of time *in vitro*. This indicates that these organelles are not autonomous and are dependent on cytoplasmic functions and nuclear gene products. The interdependence of plastids and the cells in which they reside is relatively specific and transfer of plastids into foreign plant cells results in impaired function and loss of the transferred organelles. It is clear that extensive changes have occurred in the photosynthetic endosymbiont during the integration process. For example, plastids of higher plants do not have peptidylglycan walls, which are found in bacteria. Presumably, the genes required for wall biosynthesis have been lost or inactivated. Interestingly, the cyanelles of *Cyanophora paradoxa*, which resemble plastids and perform a photosynthetic function in this organism, retain a partial peptidylglycan wall (Herdman and Denman, 1977). Some plastid functions are not found in prokaryotes and must have been acquired during the process of host–endosymbiont coevolution. The involvement of plastids in plant growth regulator biosynthesis is an example of this type of change. Biosynthetic pathways which were present in the host and endosymbiont have also become integrated. For example, some steps in lipid biosynthesis in higher plants occur on ER membranes whereas other steps are localized on the plastid envelope.

The integration of cell and plastid function must have involved the transfer of many genes from the plastid to the nucleus. The fact that all known enzymes in chlorophyll and carotenoid biosynthesis are encoded by nuclear genes underscores this point. These functions were presumably at one time exclusively localized in the plastid genome. In parallel with changes in gene localization, there occurred a restructuring of regulatory circuits. The expression of genes involved in photosynthesis in higher plants is regulated developmentally and shows organ, tissue and cell specificity. Furthermore, plastid populations in higher plants may acquire specialized functions such as starch storage (amyloplasts), carotenoid (chromoplasts) or lipid (leucoplasts) biosynthesis and storage. Plastids with specialized functions are usually localized in specific organs or tissues such as tubers (amyloplasts) or epidermal layers of fruit and petals (chromoplasts). This implies that the host now regulates the expression of specific biosynthetic pathways, a control that was once localized in plastids.

The discussion above concerning the origin of plastids provides a useful perspective when trying to understand the organization and regulation of genes encoding plastid functions in higher plants. First, we should anticipate that plastid gene organization and decoding mechanisms will have prokaryotic features. There is already ample justification noted above for this view. Second, the transfer of genetic information from the plastid to the nucleus raises several questions. Why were some genes transferred but not others? What signals are needed to direct nuclear gene products to the plastid compartment? How are genes in the two genomes coactivated during chloroplast development and what additional mechanisms are used to coordinate gene expression and protein complex assembly? These questions and others will be addressed in the remainder of this chapter. The following three sections will deal with the mechanisms of plastid and nuclear gene expression and the final section will focus on coordination of gene expression during chloroplast and leaf biogenesis.

The plastid genome

Coding capacity

The complete sequence of the tobacco plastid DNA and the identification of genes by numerous investigators makes it possible to estimate that the higher plant plastid genome has the coding capacity for 123 genes (Shinozaki et al., 1986).

Table 14.1 Genes coded by plastid DNA

	Gene product
I. Genes required to decode plastid DNA	
Transcription	
rpoA, B, C	α, β, β' subunits of RNA polymerase
Translation	
rDNA	16S, 23S, 4.5S, 5S ribosomal RNA
trn (30 genes)	tRNAs
rps (12 genes)	30S ribosomal subunits
rpl (7 genes)	50S ribosomal subunits
infA	initiation factor 1, putative secretory factor
II. Genes involved in photosynthesis	
rbcL	large subunit of Rubisco
psaA, psaB	80–85 kD PSI chlorophyll apoproteins
psaC	9 kD PSI Fe–S protein
psbA, psbD	32–34 kD PSII reaction centre chlorophyll apoproteins
psbB, psbC	47 kD, 43kD PSII chlorophyll apoproteins
psbE–psbF	cytochrome 559 (PSII)
psbH	10 kD PSII phosphoprotein
atpA, B, E, F, H, I	ATP synthetase subunits
18 > 100 codons	unidentified
18 < 100 codons	unidentified

These genes fall into three categories (Table 14.1). The first category consists of genes that are involved in decoding the plastid genetic information. These genes encode three putative subunits of the chloroplast RNA polymerase, ribosomal RNAs, tRNAs, ribosomal proteins and proteins which may function in translation initiation (IF1) and protein targeting (secX gene product). The second category of genes encode proteins involved in photosynthesis. The large subunit of Rubisco and several subunits of each protein complex involved in photosynthetic electron transport (PSI, PSII, cytochrome complex, ATP synthetase) are encoded by plastid DNA. The third group of genes include sequences for a putative NADH oxidoreductase complex and 18 open reading frames (>100 codons) of unknown function.

There have been many speculations about why the small group of genes described above is still present in plastids whereas most of the endosymbiont's genetic capacity has been lost or transferred to the nucleus. One clue to this puzzle may be found in the functions encoded by plastid genes. The plastid genes identified thus far either participate in decoding the plastid genome or encode proteins involved in photosynthesis. Clearly, these two groups of genes are coupled, because to express genes involved in photosynthesis the plastids' transcription and translation apparatus must be synthesized. This latter process is no small undertaking which is shown by the fact that mesophyll leaf cells accumulate between 10^6 and 10^7 plastid ribosomes. Plastid ribosomes and mRNAs are present at much reduced levels in non-photosynthetic plastids such as proplastids, chromoplasts or amyloplasts. Furthermore, functions carried out by these non-photosynthetic plastids (plastid replication, carotenoid biosynthesis, starch accumulation) are not known to require plastid gene expression and most likely involve primarily nuclear genes. The localization of these genes in the nucleus means that differentiation of specialized non-photosynthetic plastids does not require concomitant build-up of plastid transcription and translation capacity. The plastid compartment in non-photosynthetic cells can therefore function as other non-DNA containing organelles such as glyoxysomes or microbodies. Perhaps the advantages of this situation contributed to the present distribution of genes in the nucleus and plastid.

The rationale described above poses the question as to why all of the plastid genes have not been relocated to the nucleus. It has been suggested that some genes have been retained in the plastid because the proteins they encode would be difficult to transport through the chloroplast envelope. This idea, although ingenious, does not explain why some ribosomal proteins remain encoded by plastid DNA but not others of similar size and solubility. Further, it has recently been shown that when *psbA*, a plastid gene, is moved to the nucleus of a transgenic plant and provided with appropriate expression signals, it will direct a functional *psbA* gene product into plastids (Cheung *et al.*, 1988). These data suggest that it is possible to move extant plastid genes to the nucleus without loss of function. Further support for this idea comes from the observation that although EF-Tu is encoded by a plastid gene in *Euglena*, in higher plants it is a nuclear gene. Similarly, *rpoA* is a nuclear gene in geranium but a plastid gene in other higher plants. This suggests that gene transfer is still occurring although slowly, possibly because gene transfer disrupts complex regulatory circuits which now coordinate plastid and nuclear gene expression.

DNA organization

The plastid genome in higher plants is circular and varies in size from 120 to 217 kbp. Most of this size variation can be accounted for by the presence or absence of a portion of the plastid genome which has been duplicated and is present in an inverted orientation in the plastid DNA molecule. The location of this inverted repeat is relatively fixed with respect to other genes and it separates a small single copy region from a large single copy DNA region. In most higher plants, the inverted repeat is 22 to 26 kbp within which the rRNA transcription unit is located. In geranium the repeated DNA is larger and genes such as *psbB*, *petB*, *petD*, *petA* and *rbcL* are included in the inverted repeat. Finally, some plastid genomes, such as those in pea and mung bean, lack the inverted repeat.

Plastid gene content in higher plants is very constant and many polycistronic transcription units are conserved. Several gene pairs such as *psaA–psaB*, *psbD–psbC*, *atpB–atpE*, are cotranscribed in all higher plant plastid genomes examined to date (see Fig. 14.1 for *psbD–psbC* gene organization in barley). The cotranscription of genes may ensure that the synthesis of subunits is stoichiometric and/or could promote protein–protein interactions required for assembly of functional complexes. For example, *psaA* and *psaB* encode polypeptides which are tightly associated in the reaction center of PSI. Other genes, such as *rbcL* and some genes encoding tRNAs, are not part of polycistronic transcription units.

While the plastid gene content of higher plants is very constant, there is variation in gene order which results primarily from DNA inversions. The DNA inversions have reshuffled plastid genomes such that distances between genes, and relative orientation of transcription units, varies considerably in genomes of higher plants. For example, *rps16* is proximal to *trnK* in barley (Fig. 14.1), whereas in pea, *rbcL* occupies this position. The greatest variation in gene order is found in peas (at least 12 rearrangements) perhaps due to the lack of an inverted repeat in this plastid genome which may stabilize the genome (Palmer, 1985).

DNA copy number and localization

Multiple copies of the plastid genome are found in each plastid. The amount of DNA per plastid varies with the stage of leaf and chloroplast development. Proplastids with as few as 22 copies of DNA have been reported whereas chloroplasts in general contain 200 to 300 DNA molecules. The polyploid nature of the plastid genome is even more striking when one considers leaf cells. In pea, barley and spinach, each leaf cell contains 9000 to 13 000 copies of plastid DNA which are dispersed in 40 to 120 plastids. This means that *rbcL*, a single copy plastid gene which encodes the large subunit of Rubisco, is present in about

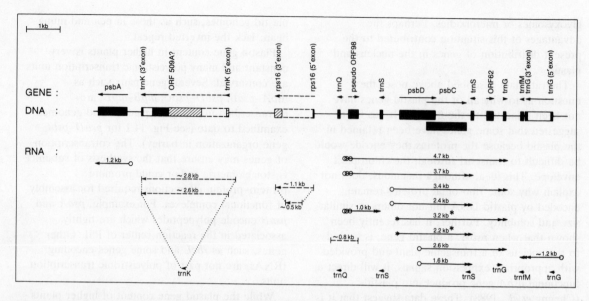

Fig. 14.1 Map of genes and transcripts from a 14 kbp region of the barley chloroplast genome which encodes 7 tRNAs and 6 proteins. Gene names are located above solid boxes which correspond to protein or tRNA coding regions. Introns are shown as open boxes. Arrows indicate the direction of transcription and transcripts which hybridize to each DNA region are designated by arrows in the lower part of the figure. Open circles correspond to RNA 5′-termini; the 5′-end of light-induced *psbD–psbC* transcripts contains an asterisk. Dashed or hatched lines indicate that sequence analysis and transcript mapping is not complete for these regions.

10 000 copies in a mesophyll cell. The small subunit of Rubisco, which accumulates in stoichiometric amounts relative to the large subunit, is encoded by a small gene family (5 to 10 members) within the nucleus. Therefore, the number of *rbcL* genes outnumber *rbcS* genes by 100 to 1, although the gene products accumulate in equal amounts. *In vivo* translation studies show that the large and small subunits are synthesized at approximately equal rates and although the *rbcL* and *rbcS* transcript levels are not known precisely, both mRNAs are abundant in leaves. This suggests that either the *rbcL* genes are transcribed less frequently than *rbcS* genes or that *rbcL* mRNA is less stable than *rbcS* mRNA. Further quantitative data needs to be gathered to resolve this question definitively.

The large copy number difference between some plastid genes and their counterparts located in the nucleus raises the question why there are so many copies of plastid DNA per cell. Bendich (1987) has argued that synthesis of rRNA could limit ribosome accumulation during rapid chloroplast development. He proposes that the plastid genome is amplified to provide more copies of rDNA thereby stimulating rRNA and ribosome synthesis.

Plastid DNA is associated with the chloroplast envelope or thylakoid membranes in aggregates of 5 to 20 DNA molecules. In proplastids of wheat, a single aggregate of DNA is observed. As chloroplasts develop, additional DNA accumulates and it becomes arranged first in a ring at the periphery of the plastid and later in a more dispersed arrangement within the organelle (Miyamura *et al.*, 1986). The compaction of plastid DNA can also change during plastid differentiation. For example, chromoplast DNA is more condensed than chloroplast DNA and this difference is correlated with reduced transcription activity in chromoplasts. The mechanism of plastid DNA compaction is unknown but may be mediated by a 'histone-like' protein found in plastids which is homologous to a similar protein HU in *E. coli*.

Decoding the plastid genome

Transcription of plastid genes

Transcription of plastid genes resembles transcription in prokaryotes in many respects (Briat et al., 1986). In fact, when E. coli is transformed with the plastid genes rbcL and psbA, transcription is initiated in E. coli at a site similar or identical to one that used by the chloroplast RNA polymerase. This result is consistent with studies showing that the DNA sequence elements which direct transcription initiation of plastid genes (TTGaca and TAtaaT; which are located 35 and 10 nucleotides upstream of the site of transcription initiation, respectively) are similar to prokaryotic promoter elements. Furthermore, as in bacteria, these promoter elements precede rRNA, tRNA and protein coding genes. The only exceptions reported thus far are two tRNA genes ($trnS_1$, $trnR_1$) which lack these sequences and may use internal promoter elements similar to eukaryotic tRNA genes. Transcription of plastid genes by E. coli RNA polymerase, however, is not identical to that found with the chloroplast RNA polymerase. The activity and initiation accuracy of the chloroplast enzyme are greatly enhanced by supercoiled templates (Lam and Chua, 1987). E. coli RNA polymerase on the other hand, will transcribe linear templates readily. E. coli RNA polymerase also initiates transcription at sites not used by chloroplast RNA polymerase. For example, E. coli RNA polymerase initiates transcription within the rbcL leader region at a site not recognized by the chloroplast transcription apparatus. Furthermore, E. coli RNA polymerase transcribes atpB more efficiently than rbcL whereas the reverse is true for the chloroplast enzyme. Finally, transcription in E. coli is sensitive to rifampicin whereas chloroplast transcription is not. These results show that transcription initiation in E. coli and chloroplasts have some overall similarities but differ significantly in specific ways.

It is possible that differences in E. coli and chloroplast transcription result from differences in RNA polymerase composition. The E. coli transcription apparatus consists of an RNA polymerase containing four subunits (β, β', α, σ) and accessory factors (nusA, nusB, rho, tau). Chloroplast RNA polymerase preparations contain from 5 to 14 subunits. At least four of these proteins are homologous to E. coli RNA polymerase subunits (β, β', α, σ). Sequence analysis indicates that plastid DNA encodes proteins homologous to the β, β' and α subunits of E. coli RNA polymerase (on the rpoA, B and C genes) (Shinozaki et al., 1986). The analysis of chloroplast rpoC revealed an open reading frame which could encode an additional chloroplast RNA polymerase subunit. This extra subunit (γ; 66 kD) has been reported in the Anabaena RNA polymerase (Schneider et al., 1987). In summary, it is likely that the chloroplast RNA polymerase contains at least five subunits (β, β', γ, α, σ); the presence of proteins homologous to nusA, nusB, rho and tau have not yet been reported.

Two lines of evidence indicate that the chloroplast RNA polymerase population may be heterogeneous. First, it was reported that all subunits of an isolated RNA polymerase preparation were synthesized on cytoplasmic ribosomes. The fact that genes encoding α, β and β' subunits are localized in plastids raises the possibility that some RNA polymerase subunits may be encoded by both nuclear and plastid genes. Second, transcriptionally active complexes of DNA and RNA polymerase have been isolated which show preferential transcription of rRNA. Based on the properties of these preparations it has been proposed that different RNA polymerases may transcribe rRNA and protein genes.

Transcription termination of plastid genes has not been studied extensively. Analysis of RNA 3'-ends reveal sequences with dyad symmetry proximal to RNA termini. In E. coli, factor-independent transcription terminators typically contain a GC-rich region of dyad symmetry followed by an AT-rich stretch that often contains a run of thymidines. In vitro experiments indicate that DNA sequences with dyad symmetry located at the 3'-end of rbcL do not cause transcription termination. Transcription was terminated with low efficiency by 3'-regions of spinach psbA, rrnB and petD. In contrast, trnS and trnH caused termination with high efficiency (80%) although these genes are not followed by regions of dyad symmetry. These results suggest that the stem–loop structures found at RNA 3'-ends may not play a significant role in factor-independent transcription termination. Presently available data, on the other hand, do not rule out a role for these sequences in factor-mediated transcription termination or in transcript stability.

Transcription from plastid gene promoters is stimulated *in vitro* when supercoiled DNA templates are used instead of relaxed circular or linear templates. Furthermore, DNA conformation affected the relative ratio of transcription of two adjacent promoters (*atpB, rbcL*) *in vitro*. Chloroplasts of higher plants contain gyrase and topoisomerase I activity which could alter the superhelicity of plastid DNA *in vivo*. At present, however, no direct test of this possibility has been reported in higher plants. In *Chlamydomonas*, inhibitors of gyrase such as novobiocin change the relative transcription rates of several plastid genes.

RNA processing

tRNAs

Transcripts containing unprocessed tRNA sequences are produced from the individual tRNA transcription units, from the ribosomal DNA transcription unit [tRNA-Ala(UGC), tRNA-Ile(GAU)] and from transcription units which encode proteins (see Fig. 14.1). The production of mature tRNAs involves 5' and 3'-endonucleolytic cleavage of primary transcripts, addition of CCA to the 3'-end of the processed RNA and base modification. Interestingly, plastid RNAse P, the enzyme responsible for 5'-end cleavage of tRNA precursor RNAs, does not contain an associated RNA (Wang *et al.*, 1988). This is in contrast with the situation in other eubacterial RNAse P enzymes which consist of a 377 to 400 nucleotide RNA plus a 14 kD protein. The requirement for RNA in eukaryotic RNAse P activity is not established.

Base modification of chloroplast tRNAs is similar to that found in bacteria. For example, tRNA-Glu contains several modified bases including four pseudouridines, a 5-methylcytosine and a 5-methylaminomethyl-2-thiouridine. This latter modification has been only reported in plastids and prokaryotes providing additional support for the endosymbiont theory of prokaryotic origin of plastids.

Introns

Introns are DNA sequences within genes which disrupt protein coding regions termed exons. Introns and exons are cotranscribed and the resultant precursor RNAs are spliced to remove introns. Four classes of introns have been recognized. One intron type is found in nuclear genes which encode proteins. Introns of this type are characterized by invariant GU and AG dinucleotides at the intron boundaries. A second intron type is found in nuclear tRNA genes. Two additional intron classes, termed Group I and Group II introns, have been distinguished on the basis of conserved sequence elements and potential folding patterns. Group I introns have been found in chloroplasts, mitochondria and ribosomal genes of *Tetrahymena* and Group II introns are present in mitochondria of yeast and chloroplasts.

Euglena plastid protein genes contain numerous introns whereas tRNAs contain none. In contrast, 6 tRNA genes and 9 genes which encode proteins contain introns in higher plants. One plastid gene intron has been identified as a Group I intron (*trnL*(UUA)). In contrast, the gene encoding tRNA-Lys(UUU) contains a 2.5 kbp Group II intron (Fig. 14.1). An open reading frame within this gene's intron encodes a protein homologous to mitochondrial RNA maturases (Neuhaus and Link, 1987). This raises the interesting possibility that the putative maturase located in *trnK* could facilitate intron processing in plastids. An even more remarkable observation was made recently with regard to *rps12* (Koller *et al.*, 1987). This gene is encoded by three exons. The 5'-exon is located 28 kbp from the two 3'-exons and *trans*-splicing is needed to produce functional RNA. *Trans*-splicing has also been reported for *psaA* in *Chlamydomonas*.

Other RNA processing activities

Although plastid RNAs are not capped or

polyadenylated, RNA maturation pathways can be very complex (see Fig. 14.1; Westhoff and Herrmann, 1988). In addition to intron removal, primary transcripts are often cleaved to remove a portion of the untranslated RNA proximal to open reading frames. The function of the 5'-end RNA processing is unknown but it may alter RNA stability or transcript translatability. Polycistronic transcripts are also processed at internal sites. Here again the role RNA processing plays in gene expression is unclear but one result is differential accumulation of RNA from some parts of long transcription units.

Protein translation, secretion and assembly

Transcription, RNA processing and translation all occur within the plastid compartment suggesting that expression of plastid genes could involve coupled transcription and translation. This situation also raises the question how ribosomes distinguish spliced from unspliced transcripts (if they do). Perhaps, as has been found in mitochondria, tRNA synthetases or other components of the translation apparatus facilitate the processing of plastid transcripts.

Protein synthesis in plastids is sensitive to chloramphenicol, occurs on 70S ribosomes and, in general, is similar to bacterial translation. Plastids contain initiation factors with activities similar to bacterial initiation factors (IF1, IF2, IF3). These factors, plus presumed interaction between the 30S ribosomal subunit and a putative ribosome binding site which precedes the open reading frame in a plastid transcript, direct the initial events in plastid protein translation. Following incorporation of formyl-methionine, translation elongation continues until a stop codon is reached. Translation elongation is mediated by EF-Tu, EF-G and a pool of 30 different amino-acylated tRNAs. The universal code is used to specify amino acids although codon usage shows a clear preference for A or U in the third position. In addition, codon use frequency is correlated with the concentration of isoaccepting tRNAs.

Many of the proteins encoded by plastid genes are localized in the thylakoid membrane. The secretion pathway which allows correct localization of these proteins has not been studied in detail but it is known that the proteins can be inserted cotranslationally into the thylakoids. Some membrane proteins (i.e. cytochrome f) contain an N-terminal cleavable signal sequence which facilitates their movement through a membrane (Rothstein et al., 1985). Other membrane proteins must contain non-cleavable signal domains (i.e. proteins encoded by *psaA–psaB*). In one case, C-terminal cleavage and palmitylation are involved in gene product translation and/or assembly (*psbA* gene product) (Mattoo and Edelman, 1987).

Polysomes encoding soluble proteins such as the large subunit of Rubisco are sometimes found associated with thylakoids. The functional significance of the membrane localization is unknown although it has been reported that large subunits will precipitate readily when expressed in *E. coli*. In plastids, newly synthesized large subunits are found associated with a 60 kD large subunit binding protein (Hemmingsen and Ellis, 1986). This interaction maintains the large subunit in an assembly competent form during the interim between translation and association with the small subunit of Rubisco which is imported from the cytoplasm.

Nuclear genes encoding plastid proteins

The majority of plastid-localized proteins are encoded by nuclear genes. These genes are transcribed by RNA polymerase II and the resulting transcripts are spliced, capped and polyadenylated in the nucleus. The mRNAs then are translated by 80S ribosomes in the cytoplasm to produce proteins which can be transported into plastids post-translationally. Following uptake into the chloroplast, the proteins are assembled with cofactors and other proteins to form functional complexes. Below, the organization and expression of two miltigene families (*rbcS* and *cab*) which encode plastid proteins are discussed. The *rbcS* genes encode the small subunit of Rubisco and *cab* genes encode chlorophyll *a/b* binding proteins which form a light-harvesting complex (LHCII) associated with PSII.

Gene organization

Three *cab* genes which encode identical proteins are present in *Arabidopsis*. These genes are located within a 6.5 kbp DNA region. In most other higher plants, the *cab* gene family is much larger. For example, there are 16 *cab* genes in *Petunia* and these have been divided into 5 subfamilies based on sequence homology. In tomato, two clusters of *cab* genes have been reported; one cluster contains three genes and the second, four genes. An even larger gene family is found in *N. plumbaginifolia* where two *cab* subfamilies can be distinguished; one subfamily consists of 19 members. Multiple *rbcS* genes are also found in higher plant genomes. Tomato contains five *rbcS* genes and at least five *rbcS* genes are present in potato.

Most *cab* genes do not contain introns. However, in *Lemna gibba*, an intron was reported in one *cab* gene and it was suggested that this intervening sequence may have arisen from a transposable element (Karlin-Neumann et al., 1985). In *Petunia*, only one of the 16 *cab* genes contains an intron. In contrast, most higher plant *rbcS* genes have introns. Three introns are found in some Solanaceae *rbcS* genes but only two in other dicots. Monocot *rbcS* genes contain a single intron. It has been shown that prokaryotic *rbcS* genes differ from higher plant genes because the latter has an N-terminal amino acid sequence which directs the small subunit to plastids. There is also an additional stretch of internal amino acids. These additional sequences could have been added to a prokaryotic progenitor sequence by exon shuffling. The retention of various numbers of introns after exon addition may explain the variable distribution of introns observed in higher plant *rbcS* genes (Wolter et al., 1988). Inefficient splicing of monocot introns in transgenic dicots also indicates divergence of splicing mechanisms (Keith and Chua, 1986).

Sequence homologous to 'CCAAT' and 'TATA' boxes which are important transcription elements in animal nuclear genes, have been found at the 5'-end of *rbcS* and *cab* genes. Furthermore, sequences which resemble animal gene polyadenylation signals are sometimes found at the 3'-end of the *rbcS* and *cab* transcription units. However, the presence of homologous sequence elements alone does not constitute proof of function. In fact, sequences known to promote polyadenylation of animal transcripts function only with low efficiency in higher plants (Hunt et al., 1987).

Regulation of transcription

The *rbcS* and *cab* genes are highly expressed in photosynthetic tissue such as leaves but only at low levels in roots. Within leaves, expression of these genes is high in mesophyll cells but reduced in epidermal and vascular cells. Further differential expression is found in C4 plants such as maize where *rbcS* is highly expressed in bundle sheath cells but at much reduced levels in mesophyll cells. The *cab* gene expression overall shows the reverse pattern in these plants. In addition, members of multigene families often exhibit different expression patterns and each gene may contribute a large or small fraction of the total output from the gene family (Fluhr et al., 1986).

Expression of *rbcS* and *cab* is attenuated in callus growing on sucrose, in meristematic cells and in photobleached tissue. These tissue lack chloroplasts and contain only undifferentiated plastid precursors, proplastids. This could imply that a factor provided by developing chloroplasts is in some way required for *rbcS* and *cab* transcription (Mayfield and Taylor, 1987). Finally, transcription of *cab* and *rbcS* is regulated by light through the action of phytochrome, a red-light absorbing photoreceptor (Tobin and Silverthorne, 1985).

Regulation of *cab* and *rbcS* transcription is complex showing organ and cell specificity and regulation by developmental state and light. In the cases examined so far, proximal DNA sequences within one kbp of the *cab* and *rbcS* open reading frames are sufficient for gene regulation. In pea, a 250 bp DNA region functions as a phytochrome responsive enhancer for *rbcS* (Green et al., 1987). This element can function in either orientation and its position can vary up to 500 bp without altering its activity. A wheat *cab* gene contains a similar phytochrome enhancer element which is located at position −124 to −357 (Simpson et al., 1986; Nagy et al., 1987). Within the rbcS enhancer element, three DNA-binding domains have been localized (Green et al., 1987). Two of these domains function as negative regulatory elements in dark-grown plants. Interestingly, the protein(s) which bind to this region are present in both dark-grown and illuminated plants which suggests that light-activation of gene transcription may be due to protein modification rather than light-induced synthesis of a new protein factor.

Protein transport and assembly

The small subunit and chlorophyll a/b apoproteins are synthesized in the cytoplasm as precursors containing 33–36 N-terminal amino acids not present in the mature proteins. These additional amino acids form a positively charged 'transit' sequence which directs the secretion of these proteins into plastids. Transit sequences are important for binding precursor proteins to protein receptors on the chloroplast envelope. The receptors can be inactivated by protease treatment or in the case of small subunit by antibodies formed against the transit sequence. Binding does not require ATP but subsequent steps in protein transport are ATP dependent. The actual process of protein translocation through the chloroplast envelope is unclear at present. However, transported proteins are processed in one or two steps by soluble proteases and become localized in the stroma, thylakoid membrane or the thylakoid loculus. Recent evidence suggests that the association of thylakoid proteins such as the chlorophyll a/b proteins and thylakoids is facilitated by an as yet uncharacterized protein(s). Furthermore, membrane localization and assembly of these proteins into LHCII can occur without removal of transit sequences (see Chapter 20 for additional information).

The accumulation of rbcS and cab gene products is dependent not only transcription, translation and protein transport but also on the presence of cofactors and other protein subunits with which they associate in plastids. For example, mutants which do not synthesize the large subunit of Rubisco, also do not accumulate the small subunit because this protein is unstable when not assembled into the Rubisco holoenzyme. In a similar way, the chlorophyll a/b apoproteins do not accumulate in chlorophyll-deficient plants because the apoproteins are proteolyzed in the absence of cofactor binding (Cuming and Bennett, 1981).

Regulation of gene expression during chloroplast development

Introduction

Proplastids, the progenitors of chloroplasts, are inherited maternally by the plant zygote. These organelles are small (1–2 μm diameter), non-photosynthetic and present in low numbers in meristematic cells (Mullet, 1988). Leaf and chloroplast development lead to the formation of mesophyll cells which contain 40 to 150 chloroplasts which are 6 to 8 μm in diameter. How chloroplasts are formed from proplastids and what genes and signal transduction pathways are involved in this process are as yet unknown.

The biogenesis of chloroplasts is similar in all higher plants. Basically, two overall changes occur during chloroplast/leaf development: (1) the number of plastids in mesophyll cells increases; and (2) plastid and nuclear genes are activated which leads to the accumulation of the photosynthetic apparatus in the organelles. Light plays a central role in regulating chloroplast development. In dicots, cell division, cell enlargement and early phases of chloroplast development are light dependent. In contrast, primary foliage leaf development in monocots such as barley, is not light dependent and only later stages of chloroplast development require illumination.

Phases of chloroplast development

Chloroplast development in leaves begins by formation of leaf primordia from cells of the shoot apex. In monocots, meristematic activity soon becomes localized in a basal intercalary meristem and most of the cells of the leaf are derived from this region. Consequently, in developing monocot leaves, older cells are located at the leaf apex and undifferentiated cells at the leaf base. The spatial separation of sequential phases of cell and plastid development within the monocot leaf provides a useful frame of reference for analysis of chloroplast development and gene expression.

Activation of cell division in the barley leaf basal meristem is paralleled by increased plastid replication and DNA synthesis. When cells stop dividing, cell enlargement begins and cells are displaced apically due to growth of new cells produced by the basal meristem. Plastid division and DNA synthesis are not synchronized with cell division; these processes continue in enlarging mesophyll cells until approximately 60 plastids each with 150 to 200 copies of plastid DNA have accumulated. This phase of chloroplast development is not known to require plastid gene expression but does exhibit cell and organ specificity.

The second phase of chloroplast development, the build-up of plastid transcription/translation capacity, requires both plastid and nuclear gene expression. For example, 19 of 52 ribosomal proteins are plastid encoded, the remainder being encoded by nuclear genes. It is not clear how the sets of genes in these two compartments which encode the plastids transcription and translation apparatus are co-activated. On the other hand, once the build-up process is initiated, stoichiometric production of subunits could be accomplished through the action of autogenous translational control loops as has been observed for ribosomal protein synthesis in bacteria.

The accumulation of the chloroplast photosynthetic apparatus begins during build-up of the plastids transcription and translation capacity. Interestingly, transcription of some nuclear genes such as rbcS appear regulated by a 'factor' produced when plastid transcription and translation is activated. In barley, plastid transcription increases five-fold per plastid (20-fold per cell) soon after cell division stops and cell enlargement begins. Thirty-six to forty-eight hours later a full set of mature chloroplasts is present in mesophyll cells of illuminated plants. Once mature chloroplasts have been produced, the build-up process stops and plastid transcription and translation activity decline.

Role of light

Higher plants contain at least three photoreceptors: a blue light receptor, and two red light photoreceptors, phytochrome and protochlorophyllide holochrome (Schneider et al., 1987). These three photoreceptors have been shown to influence nearly all phases of chloroplast development although the extent of each photoreceptors influence varies with plant species, and gene examined. The role of the red light photoreceptors will be discussed below with reference to the biogenesis of PSII in barley.

Photosystem II is a large protein complex which carries out light-dependent charge separation, O_2 evolution and reduction of plastoquinone. This unit consists of at least 15 different proteins, seven of which are encoded by plastid DNA and the remainder by nuclear genes (Fig. 14.2). The seven plastid PSII genes are encoded on four transcription units. These genes encode cytochrome 559 (subunits E–F), a phosphoprotein of unknown function (subunit H), two proteins which bind the reaction center chlorophylls (subunits A and D) and two chlorophyll apoproteins which serve a light-harvesting function (subunits B and C). The expression of these genes is regulated in an incompletely defined way by regulatory nuclear genes (reg 1-n, Fig. 14.2). In addition to this 'core' of plastid-encoded subunits there are nuclear-encoded polypeptides which facilitate oxygen evolution (OEE 1–3) and form a chlorophyll a/b light-harvesting antennae (LHCII, encoded by cab genes). As shown schematically in Fig. 14.2, proteins derived from the two gene pools are brought to the thylakoid membrane by their respective secretion pathways where assembly with cofactors (i.e. chlorophyll, heme, carotenoid) and other subunits takes place.

In monocots such as barley, leaf and chloroplast development proceeds to a large extent in darkness. In dark-grown barley, plastid number per cell reaches that of illuminated seedlings and many plastid proteins accumulate. At least three PSII subunits (cyt. 559, OEE proteins 1 and 3) can be detected in plastids of dark-grown plants. In contrast, subunits A, B, C, D and LHCII proteins (cab gene products) do not accumulate under these conditions. When dark-grown plants

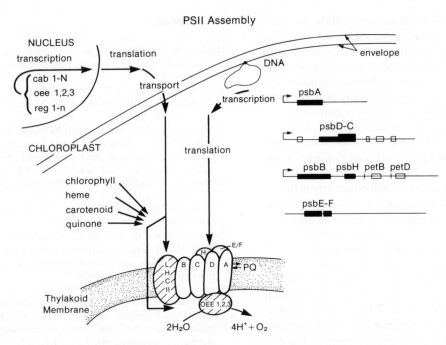

Fig. 14.2 Diagram showing genes which encode PSII proteins and the pathway by which these subunits are synthesized, transported to the chloroplast thylakoid membrane and assembled with cofactors and other PSII subunits. *Cab* genes encode the chlorophyll *a/b* apoproteins of LHCII; *oee* 1–3 encode proteins involved in oxygen evolution; *psbA* and *psbD* encode chlorophyll apoproteins of the reaction center; *psbC* and *psbB* encode chlorophyll apoproteins; *psbH* corresponds to a phosphoprotein and *psbE–F* encode subunits of cyt 559.

are illuminated, these proteins accumulate and functional PSII units are produced.

The absence of subunits A, B, C, D in dark-grown plants is not due to lack of transcription or transcript availability. In fact, transcripts for these genes are abundant and are localized on plastid polysomes. Furthermore, pulse-labelling studies show radiolabel accumulation in subunit D but not subunits A, B or C. These results suggest that the absence of subunits A, B and C in dark-grown seedling is due to a block in polypeptide translation or very rapid turnover of these proteins.

Light-induced accumulation of subunits A, B and C is mediated by the protochlorophyllide holochrome and requires synthesis of chlorophyll *a*. Protochlorophyllide, a precursor to chlorophyll, accumulates in dark-grown higher plants. Illumination allows reduction of protochlorophyllide and subsequent formation of chlorophyll. Since subunits A, B and C bind chlorophyll, this cofactor may bind to nascent chains to stimulate translation or stabilize the apoproteins.

The apoproteins encoded by *cab* genes also bind chlorophyll but in this case *cab* transcripts are not present at high levels in dark-grown plants. Instead, light-induced conversion of a red light absorbing form of phytochrome to a far-red absorbing form leads to increased transcription of the *cab* genes and eventually to the import of chlorophyll *a/b* apoproteins into plastids. Here a second level of control is exerted by chlorophyll availability. In the absence of chlorophyll, the LHCII apoproteins are degraded.

Maintenance of PSII in mature chloroplasts

The transition from the rapid build-up of the photosynthetic apparatus to maintenance of the accumulated structures presents a different regulatory problem. In mature plastids, proteins which are unstable or damaged need to be turned-over and resynthesized. The proteins which comprise the PSII reaction center (subunits A, D) appear especially labile in illuminated plants. As a consequence, the rate of synthesis of these proteins in mature chloroplasts remains high in contrast to most other plastid proteins. The mechanism of differential synthesis of these two proteins involves the maintenance of mRNA for these proteins in mature plastids even though overall chloroplast transcription activity is low relative to developing chloroplasts. Expression of *psbA* (encodes subunit A) is maintained by a strong promoter and high RNA stability. Expression of *psbD*, on the other hand, is maintained in mature barley leaves in a more complex way (Gamble *et al.*, 1988). *PsbD* is transcribed along with several additional genes in barley chloroplasts (Fig. 14.1). As chloroplasts mature, the *psbD* transcript population changes and two specific transcripts accumulate. These transcripts are found only in illuminated plants and have a common 5'-end. The accumulation of these RNAs could be due to preferential transcription of *psbD* in mature chloroplasts or to RNA processing which produces stable transcripts.

In summary, genes encoding proteins involved in photosynthesis exhibit cell-specific expression. The activation of genes occurs sequentially in a process which is regulated developmentally and by light. Although the sequence of events leading to chloroplast formation is now known in some detail, we lack important insight into the signal transduction pathways and mechanisms of gene regulation involved.

Further reading

Baker, N. R. and Barber, J. (1984). *Chloroplast Biogenesis*, Vol. 5, Elsevier/North Holland Biomedical Press, 380 pp.

Hatch, M. D. and Boardman, N. K. (eds) (1987). Photosynthesis. In *The Biochemistry of Plants*, Vol. 10, Academic Press, 409 pp.

Kirk, J. T. O. and Tilney-Bassett, R. A. E. (1978). *The Plastids: Their Chemistry, Structure, Growth and Inheritance*, 2nd edn, Elsevier/North Holland Biomedical Press, Amsterdam/New York, 650 pp.

References

Bendich, A. J. (1987). Why do chloroplasts and mitochondria contain so many copies of their genome? *BioEssays* **6**, 279–82.

Briat, F. J., Lescure, A. M. and Mache, R. (1986). Transcription of the chloroplast DNA: a review. *Biochimie* **68**, 981–90.

Cheung, A. Y., Bogorad, L., Van Montagu, M. and Schell, J. (1988). Relocating a gene for herbicide tolerance: a chloroplast gene is converted into a nuclear gene. *Proc. Natl Acad. Sci. USA* **85**, 391–5.

Cuming, A. C. and Bennett, J. (1981). Biosynthesis of the light-harvesting chlorophyll *a/b* protein. Control of messenger RNA activity by light. *Eur. J. Biochem.* **118**, 71–80.

Fuhr, R., Moses, P., Morelli, G., Coruzzi, G. and Chua, N.-H. (1986). Expression dynamics of the pea *rbcS* multigene family and organ distribution of the transcripts. *EMBO J.* **5**, 2063–71.

Gamble, P. E., Sexton, T. B. and Mullet, J. E. (1988). Light-dependent changes in *psbD* and *psbC* transcripts of barley chloroplasts: accumulation of two transcripts maintains *psbD* and *psbC* translation capability in mature chloroplasts. *EMBO J.* **7**, 1289–97.

Gray, M. W. and Doolittle, W. F. (1982). Has the endosymbiont hypothesis been proven? *Microbiol. Rev.* **46**, 1–42.

Green, P. J., Kay, S. A. and Chua, N.-H. (1987). Sequence-specific interactions of a pea nuclear factor with light-responsive elements upstream of the *rbcS*-3A gene. *EMBO J.* **6**, 2543–9.

Hemmingsen, S. E. and Ellis. R. J. (1986). Purification and properties of ribulosebisphosphate carboxylase large subunit binding protein. *Plant Physiol.* **80**, 269–76.

Herdman, M. and Denman, S. (1977). The cyanelle: chloroplast or endosymbiotic prokaryote? *FEMS Lett*, **1**, 7–12.

Hunt, A. G., Chu, N. M., Odell, J. T., Nagy, F. and Chua, N.-H. (1987). Plant cells do not properly recognize animal gene polyadenylation signals. *Plant Mol. Biol.* **8**, 23–35.

Karlin-Neumann, G. A., Kohorn, B. D., Thornber, J. P. and Tobin, E. M. (1985). A chlorophyll *a/b*-protein encoded by a gene containing an intron with characteristics of a transposable element. *J. Mol. Appl. Genet.* **3**, 45–61.

Keith, B. and Chua, N.-H. (1986). Monocot and dicot pre-mRNAs are processed with different efficiencies in transgenic tobacco. *EMBO J.* **5**, 2419–25.

Koller, B., Fromm, H, Galun, E. and Edelman, M. (1987). Evidence for *in vivo trans* splicing of pre-mRNAs in tobacco chloroplasts. *Cell* **48**, 111–19.

Lam, E, and Chua, N.-H. (1987). Chloroplast DNA gyrase and *in vitro* regulation of transcription by template topology and novobiocin. *Plant Mol. Biol.* **8**, 415–24.

Mattoo, A.K. and Edelman, M. (1987). Intramembrane translocation and posttranslation palmitoylation of the chloroplast 32-kDa herbicide-binding protein. *Proc. Natl Acad. Sci. USA* **84**, 1497–1501.

Mayfield, S. P. and Taylor, W. C. (1987). Chloroplast photooxidation inhibits the expression of a set of nuclear genes. *Mol. Gen. Genet.* **208**, 308–14.

Miyamura, S., Nagata, T. and Kuroiwa, T. (1986). Quantitative fluorescence microscopy on dynamic changes of plastid nucleoids during wheat development. *Protoplasma* **133**, 66–72.

Mullet, J. E. (1988). Chloroplast development and gene expression. *Ann. Rev. Plant Physiol.* **39**, 475–502.

Nagy, F., Boutry, M., Hsu, M.-Y., Wong, M. and Chua, N.-H. (1987). The 5'-proximal region of the wheat cab-1 gene contains a 268-bp enhancer-like sequence for phytochrome response. *EMBO J.* **6**, 2537–42.

Neuhaus, H. and Link, G. (1987). The chloroplast tRNA^{Lys}(UUU) gene from mustard (*Sinapis alba*) contains a class II intron potentially coding for a maturase-related polypeptide. *Curr. Genet.* **11**, 251–7.

Palmer, J. D. (1985). Comparative organization of chloroplast genomes. *Ann. Rev. Genet.* **19**, 325–54.

Rothstein, S. J., Gatenby, A. A., Willey, D. L. and Gray, J. C. (1985). Binding of pea cytochrome *f* to the inner membrane of *Escherichia coli* requires the bacterial *secA* gene product. *Proc. Natl Acad. Sci. USA* **82**, 7955–9.

Schneider, G. J., Tumer. N. E., Richaud, C., Borbely, G. and Haselkorn, R. (1987). Purification and characterization of RNA polymerase from the cyanobacterium *Anabaena* 7120. *J. Biol. Chem.* **262**, 14633–9.

Shinozaki, K., Ohme, M., Tanaka, M., Wakasugi, T., Hayshida, N., Matsubayashi, T., Zaita, N., Chunwongse, J., Obokata, J., Yamaguchi-Shinozaki, H., Ohto, C., Torazawa, K., Meng, B. Y., Sugita, M., Deno, H., Kamogashira, T., Yamada, K., Kusuda, J., Takaiwa, F., Kato, A., Tohdoh, N., Shimada, H. and Sugiura, M. (1986). The complete nucleotide sequence of the tobacco chloroplast genome: its gene organization and expression. *EMBO J.* **5**, 2043–9.

Simpson, J., Schell, J., Van Montagu, M. and Herrera-Estrella, L. (1986). Light-inducible and tissue-specific pea *lhcp* gene expression involves an upstream element combining enhancer- and silencer-like properties. *Nature* **323**, 551–4.

Tobin, E. M. and Silverthorne, J. (1985). Light regulation of gene expression in higher plants. *Ann. Rev. Plant Physiol.* **36**, 569–93.

Wang, M. J., Davis, N. W. and Gegenheimer, P. (1988). Novel mechanisms for maturation of chloroplast transfer RNA precursors. *EMBO J.* **7**, 1567–74.

Westhoff, P. and Herrmann, R. G. (1988). Complex RNA maturation in chloroplasts. The *psbB* operon from spinach. *Eur. J. Biochem.* **171**, 551–64.

Wolter, F. P., Fritz, C. C., Willmitzer, L., Schell, J. and Schrier, P. H. (1988). *rbcS* Genes in *Solanum tuberosum*: conservation of transit peptide and exon shuffling during evolution. *Proc. Natl Acad. Sci USA* **85**, 846–50.

15 The formation of ATP and reducing power in the light
Barbara B. Prezelin and Norman B. Nelson

Introduction

Photosynthesis begins with the absorption of light by pigment molecules within plant cells. Absorbed light energy drives the series of photosynthetic reactions that ultimately lead to the formation of new organic carbon. In this chapter we will introduce the mechanisms by which light is absorbed and its excitation energy funneled into photochemical reactions that result in the formation of chemical energy (ATP) and reducing power (NADPH). The light-dependent formation of ATP and NADPH is generally referred to as the light reactions of photosynthesis, described by the partial reaction:

$$2H_2O + 2NADP^+ + 3ADP^{3-} + 3HPO_4^{2-} + H^+$$
$$\xrightarrow{h\nu} O_2 + 2NADPH + 3ATP^{4-} + 3H_2O$$

The present chapter describes the structural and functional organization of the photosynthetic components which comprise the light reactions and which are found entirely within photosynthetic membranes (Fig. 15.1). Also discussed are the regulatory mechanisms which ensure the efficient use of light energy to photochemically generate the ATP and NADPH required for carbon fixation. Following chapters will address the enzyme systems which catalyze the reduction of inorganic carbon with photochemically produced ATP and NADPH.

Light absorption and energy transfer mechanisms

In oxygen-evolving (oxygenic) photosynthesis carried out by higher plants, algae and cyanobacteria, only light in the violet to red part of the visible spectrum (from 400 to 700 nm) can be absorbed by photosynthetic pigments. Light is absorbed in packets of energy known as quanta. The energy of a quantum is directly proportional to its frequency (ν) and inversely proportional to its wavelength (λ), the latter being expressed in nanometers (nm). Quanta at shorter wavelengths, near the blue-violet region of the visible spectrum, have higher excitation energy for photosynthesis than quanta at the longer wavelengths found near the red end of the visible spectrum.

Not all visible light reaching a plant will be used for photosynthesis. When light strikes a plant, variable fractions of the photosynthetically available radiation (PAR) are scattered off the plant surface, transmitted through the plant or absorbed by molecular components within the plant. Not all the molecules that absorb light within a plant cell function in photosynthesis. Photosynthetic pigments are localized within the photosynthetic membranes; these can absorb light throughout the visible spectrum and can transfer that absorbed energy with speed and efficiency to the photochemical reaction centers of photosynthesis.

There are three general classes of photosynthetic pigments, including chlorophylls (Chl), phycobilins and carotenoids (Table 15.1). Of the three types of green-colored chlorophylls, only Chl *a* is found in all plants and cyanobacteria, and is an essential component of the light reactions of photosynthesis. The other two types of chlorophyll, Chl *b* and *c*, function as light-harvesting pigments whose absorbed light energy is passed on to Chl *a* to help drive the photochemical events of photosynthesis. These *accessory* chlorophylls are

The formation of ATP and reducing power in the light

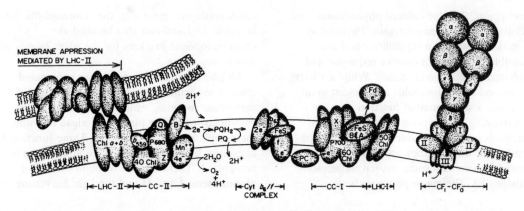

Fig. 15.1 A stylized model of the four structural units of the thylakoid membrane, which catalyze the light-harvesting, electron transport, and energy-coupling of photosynthesis. PSII consists of two parts: a LH pigment–protein complex, (LHCII) and a core complex (CCII) which catalyzes electron transfer from water to plastoquinone (PQ). A cyt. b_6/f complex (also containing a Rieske iron–sulfur (Fe–S) center) oxidizes the plastohydroquinone (PQH_2) and donates electrons to plastocyanin (PC). PSI, consisting of a LH pigment–protein complex (LHCI) and a core complex (CCI) oxidizes plastocyanin and transfers electrons to $NADP^+$ (via ferredoxin). The ATPase, which uses a proton gradient to drive ATP synthesis, consists of a hydrophobic CF_0 and a surface-bound CF_1 (by permission, Kaplan and Arntzen, 1982).

differentially distributed among plant groups and can be used as chemotaxonomic markers to identify the presence of certain plant groups in natural communities (Table 15.1). There are also three subcategories of light-harvesting phycobilin pigments, which include the red-colored

Table 15.1 The major plant groups, and their principal photosynthetically active chlorophylls, biliproteins, and carotenoids. A (+) indicates the presence of the pigment in the plant group

	Chl a	Chl b	Chl c_1	Chl c_2	phycoerythrin	phycocyanin	allophycocyanin	major xanthophylls
Higher plants	+	+						lutein
Green algae	+	+						lutein/siphona-xanthin
Diatoms	+		+	+				fucoxanthin
Brown algae	+		+	+				fucoxanthin
Chrysophytes	+		+	+				butanoyl-fucoxanthin
Prymnesiophytes	+		+	+				hexanoyl-fucoxanthin
Dinoflagellates	+			+				peridinin
Cryptophytes	+			+	+	+		alloxanthin
Red algae	+				+	+	+	zeaxanthin
Cyanobacteria (= blue–green algae)	+				+	+	+	zeaxanthin, myxoxanthin

phycoerythrin, the blue-colored phycocyanin, and the violet-colored allophycocyanin. Phycobilins vary in their absorption capabilities and are responsible for the coloration of red algae and cyanobacteria (blue–green algae). While a variety of orange-colored carotenoids are present in all plants, only a small number function as light-harvesting pigments for photosynthesis. Carotenoids can be subdivided into two groups depending on whether they are hydrocarbons (carotenes) or contain some oxygen molecules (xanthophylls). The major photosynthetic carotenoids are generally the xanthophylls listed in Table 15.1 and can also be used as chemotaxonomic markers for the plant groups in which they occur.

All photosynthetic pigments are conjugated colored molecules and are bound to thylakoid membrane proteins, which are localized in the chloroplast of higher plants and algae or within the cytoplasm of cyanobacteria. Three functional categories of pigment–protein complexes can be recognized. These are the reaction centers of photosystems I and II and the light harvesting

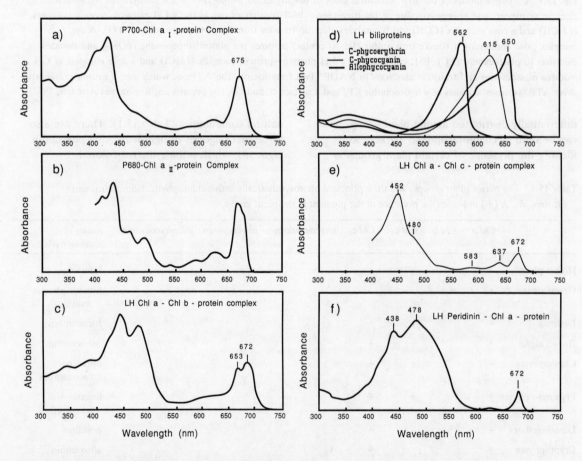

Fig. 15.2 Representative absorption spectra for major pigment–protein complexes in plants, including: (a) P700–Chl a–protein complexes, representing PSI (by permission, Prézelin and Alberte, 1978); (b) P680–Chl a–protein complexes, representing PSII (by permission, Satoh and Butler, 1978); (c) LH Chl a/b–protein complexes (by permission, Thornber and Alberte, 1977); (d) LH phycobiliproteins (by permission, O'Carra and O'hEocha, 1976); (e) LH Chl a/c–protein complexes (by permission, Boczar, 1985); and (f) LH peridinin–chlorophyll a–protein complexes (by permission, Prézelin and Haxo, 1976).

complexes. The reaction center of photosystem I (PSI) is the Chl *a*–protein complex in plants that absorbs at the longest wavelength (700 nm) (Fig. 15.2a), receiving electrons from a cytochrome carrier via a plastocyanin–protein complex and donating electrons to the enzymatic ferredoxin complex which reduces NADP. The reaction center of photosystem II (PSII) is a shorter wavelength absorbing (680 nm) Chl *a*–protein complex (Fig. 15.2b) which catalyzes the oxidation of water to oxygen and donates electrons via a plastoquinone pool to PSI. The composition of the reaction center core complexes within PSI and PSII are highly conserved in all plant groups, but only account for a small fraction of total plant chlorophyll *a*. Most light-harvesting (= accessory) chlorophylls, carotenoids and phycobilins are localized in light-harvesting (LH) pigment–protein complexes associated with PSI and PSII. These LH complexes absorb light at wavelengths where PSI and PSII absorption is weakest, and include the Chl *a*–Chl *b*–protein complexes of green algae and higher plants (Fig. 15.2c), the phycobiliprotein complexes of cyanobacteria and red and cryptophyte algae (Fig. 15.2d), and the carotenoid–Chl *a*–Chl *c*–protein complexes of chromophytic (Chl *c*-containing) algae (Fig. 15.2e,f). Their added presence around PSI and PSII core complexes broadens the range of visible light used to drive photosynthesis.

Chlorophylls will be used to illustrate the basic features of how light energy is captured by photosynthetic pigments and efficiently transferred from LH pigment–protein complexes to PSI and II. Chlorophylls are all characterized by a porphyrin ring structure where magnesium is chelated in the center and liganded at four sites to pyrrole nitrogen atoms (Fig. 15.3). It is with this polar head group that most bonding of apoproteins occurs. Only Chl *a* and Chl *b* have phytol, a long-chain lipophilic terpenoid side-group, attached to the porphyrin ring. Phytol may provide additional hydrophobic binding sites for apoproteins (Fenna and Mathews, 1979) or lie to the outside of apoproteins, thereby providing hydrophobic bonding with the integral lipophilic components of the thylakoid membrane (Anderson, 1975) (Fig. 15.1). The structure and conformational constraints of the porphyrin ring

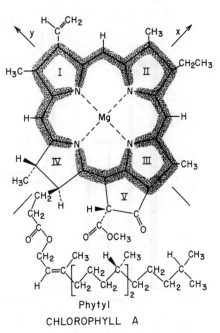

Fig. 15.3 The molecular structure of chlorophyll *a*. The porphyrin pi-electrons are indicated by the shading. Absolute configurations are shown for the asymmetrically substituted porphyrin carbon atoms and for phytol. Chlorophyll *b* differs from chlorophyll *a* in the replacement of the methyl group on ring II by a formyl group. Neither chlorophyll c_1 or c_2 has phytol, and the propionic group of ring IV is replaced by $HC= CH_2-COOH$. Chlorophyll a_2 has a vinyl group ($-CH=CH_2$) instead of the ethyl group on ring II of chlorophyll *a* (by permission, Sauer 1975).

bound to proteins give different chlorophyll–protein complexes their specific absorption properties (Fig. 15.2).

Porphyrin rings are essentially planar complexes surrounded by dense clouds of pi-electrons which become redistributed when a quanta of light is absorbed, thereby promoting the chlorophyll molecule to a higher excited electronic state (Fig. 15.4). The energized pigment molecule represents an excited *singlet* state whose energy is greater than that of the initial or ground state molecule. The absorbed quanta must be of such a vibrational frequency that the light energy matches possible energy transition states of the pigment molecule. Thus the absorption spectrum of any pigment is a signature of the light energy levels

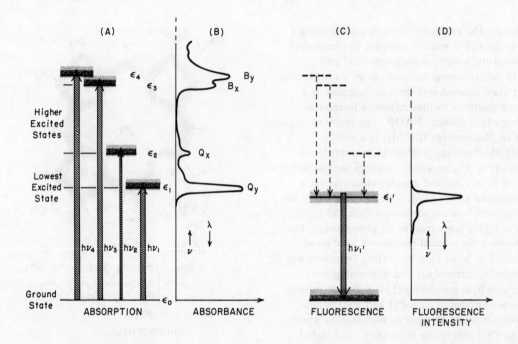

Fig. 15.4 Absorption and fluorescence of chlorophylls. (A) Energy level diagram, showing spectral transitions (vertical arrows). The energy levels are broadened (shading) by vibrational sublevels that are not usually resolved in solution spectra. (B) Absorption spectrum corresponding to energy levels of part (A). This spectrum is turned 90° from the usual orientation in order to show the relationship to the energy levels. Conventions for designating the spectral transitions (Q_y, Q_x, B_x, B_y, where x and y refer to the axes shown in Fig. 15.2) are shown. (C) Diagram showing radiationless relaxation (dashed arrows) and fluorescence (shaded arrow). (D) Fluorescence emission spectrum corresponding to part (C). Note the red shift of fluorescence compared with the corresponding Q_y absorption illustrated in parts (A) and (B). This Stokes shift owes to vibrational relaxation in the excited electronic state prior to fluorescence emission and in the ground electronic state after emission (by permission, Sauer 1975).

that can be most effectively absorbed (Fig. 15.4A,B) (Sauer, 1975).

Individual pigment molecules can absorb a range of light energies, but packets of excitation energy (excitons) are transferred to other pigments only from the lowest excited singlet state of the pigment molecule (Fig. 15.4C). For instance, when Chl *a* absorbs the high energy of blue light (440 nm), the molecule is promoted to a higher excited state than it would when excited by absorbed red light (680 nm) (Fig. 15.4). But a blue light excited Chl *a* molecule will quickly relax to its lowest excited state, losing the extra excitation energy through non-radiative dissipative processes (i.e. thermal decay). Thus the energy transferred to another pigment molecule by a blue light or red light excited Chl *a* molecule will be the same.

Resonance transfer of vibrational energies occurs between neighboring pigments when their lowest energy absorption bands overlap. Exciton transduction from peripheral LH pigment–protein complexes to integrally-bound PSI and PSII is dependent on the unidirectional coupling of lowest excitation states between neighboring pigment molecules. The speed and efficiency of energy transduction depends on the orientation and distance between pigment molecules, which needs to be less than 10 nm to effectively compete with dissipative processes. Due to some energy loss during the time of energy transfer, the excitation energy decreases as excitons are transduced from

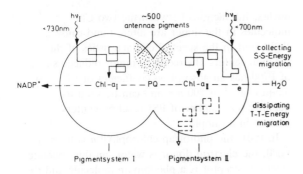

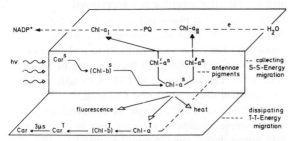

Fig. 15.5 Collecting and dissipating energy migrations in the antenna pigments of photosynthesis (by permission, Witt 1979).

LH pigments to the reaction centers of PSI and PSII (Fig. 15.5), which represent the lowest energy traps in photosynthesis.

The absorption of quanta and subsequent transduction of excitons to PSI or PSII must be completed within a nanosecond if photochemistry is to compete successfully against other dissipative processes, i.e. further radiationless (thermal) de-excitation and the luminescing (light-emitting) processes of fluorescence, phosphorescence and delayed light emission. The fraction of excitons funneled into photochemical events or lost through dissipative processes is highly variable and is largely regulated by thylakoid membrane state changes which alter the coupling between LH complexes and the photosystems. Excitons which never reach reaction centers, or reach reaction centers but are not permanently trapped, can migrate back to LH complexes and be dissipated as luminescence. Fluorescence (Fig. 15.4d) is a light-emitting process by which pigments in the excited singlet state return to ground state if their excess energy is not funneled into photochemistry within the fluorescence lifetime of the molecule.

Fluorescence only occurs from the lowest excited singlet state, so the wavelength of the fluorescence maximum often is a few nanometers longer than the longest absorption maximum of the pigment. For example, chlorophylls absorb light energy throughout the visible spectrum (Fig. 15.4b), but only emit fluoresced light energy at the red end (lowest energy) of the visible spectrum (Fig. 15.4d).

The majority (>90%) of *in vivo* fluorescence at room temperature arises from back reactions of the primary photochemical events occurring in the reaction centers and LH chlorophyll of PSII (Fig. 15.4). In some instances, excitons re-emitted from PSII can be directed to PSI before excitation energy of the absorbed quanta is lost. It is this ability to redirect unused excitation energy that enables PSII to spill over excess excitons to PSI. Spillover is a unidirectional transfer of excitons from PSII to PSI. Excitons are not transferred out of PSI and into PSII because PSI is of a lower energy state than PSII and because it dissipates its excess excitons via radiationless de-excitation. Thus, all excitons reaching PSI have a very low probability of gaining the additional energy required to overcome the energy barrier which exists between the reaction center and surrounding LH chlorophyll. However, as temperatures are lowered, so are the energy barriers, and fluorescence intensity from PSI in higher plants and some algae increases dramatically at liquid nitrogen temperatures (77 K).

Photochemistry

An exciton is considered trapped when photochemical charge separation of electrons and protons occurs across the thylakoid, allowing transmembrane electron flow to begin. The charge separation reaction can be visualized as the following:

$$DPA \xrightarrow{h\nu} D(P^*A_1) \xrightarrow{5\ ps} D(P^+A_1^-) \xrightarrow{200\ ps} D^+(PA_1)A_2^-$$

where P is the phototrap chlorophyll *a*, and D and A_1 A_2 represent electron donors and acceptors, respectively. In the excited singlet state (represented by *), phototrap chlorophyll *a* molecules eject an electron which is rapidly transferred (5 ps) to an associated primary electron acceptor (A_1). After 200 ps the primary electron acceptor has passed the electron to a secondary electron acceptor (A_2) and the phototrap chlorophyll has been reduced by the oxidation of an electron donor (D). In both photosystems, the primary electron donors lie near the inner surface of the thylakoid, the reaction centers are buried in the thylakoid membrane, and the electron acceptors are found near the outer stroma surface of the thylakoids (Fig. 15.1). Thus, the photochemical events transfer one electron from each phototrap across the thylakoid membrane. In this manner, light energy is stabilized as stored chemical potential to be used in the formation of oxidizing and reducing compounds, transmembrane electric fields, ion gradients, and membrane conformational changes.

In PSI, the phototrap chlorophyll *a* is termed P700, the electron donor is usually the copper-containing protein plastocyanin (PC) and the electron acceptors are the two iron–sulfur (FeS) proteins ferredoxin A and B (Fig. 15.1). The structure of P700 is not known, but is generally believed to represent a Chl *a* dimer. Other possibilities include an enolized form of a single Chl *a* molecule that is protein-stabilized or a chlorophyll molecule that is structurally distinct from Chl *a*. Each PSI core complex has a molecular weight of 67 kD and contains about 40 accessory Chl a_I molecules which pass their absorbed light energy exclusively to one P700 reaction center. It is the accessory chlorophylls within the P700–Chl a_I–protein complex which give it the longest wavelength (675–677 nm) red absorption maximum of any Chl–protein complex in plant cells (Fig. 15.2a). The P700–Chl *a*–protein complexes of higher plants are characterized by a photobleaching signal which has a maximum around 700 nm (Fig. 15.6a). This figure represents the difference between the absorbance spectrum of P700 before and after oxidation induced by light or chemical means. The negative signal at around 700 nm and 438 nm arises in less than 20 ns as a result of the charge separation reaction due to photo-oxidation. It has been suggested that when one Chl *a* in the dimer ejects an electron, the nuclear interactions between the two Chl molecules in the dimer are altered and cause a slight shift in the spectral characteristics of the reaction center. The oxidized-minus-reduced signal of P700 has an absorption coefficient of 64 to 70 mequiv cm^{-1} at 700 nm and can be used to quantify the number of PSI reaction centers present in a sample.

In PSII, the phototrap chlorophyll *a* is termed P680, the electron donor is water, an intermediate electron acceptor is a pheophytin molecule and the primary electron acceptor is a specialized protein-embedded plastoquinone molecule (Q) (Fig. 15.1). The PSII reaction center complex (≥600 kD) has been more difficult to isolate than PSI because of its lability in the presence of harsh detergents. The phototrap of PSII is thought to be a special Chl dimer or ligated Chl monomer. It has been termed P680 because of characteristic spectroscopic absorption changes seen in light-minus-dark (oxidized-minus-reduced) difference spectra which are inhibited by specific

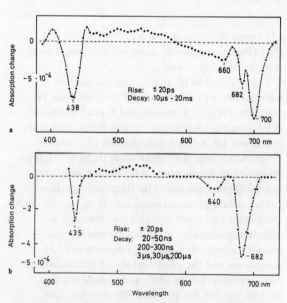

Fig. 15.6 Difference spectra of absorbance changes due to the turnover of Chl a_I (P700) (top) and Chl a_{II} (bottom). The fast rise noted in the figure panel is due to photo-oxidation. The decay is multiphasic and is interpreted as the reduction of reaction centers by electron donors (by permission, Renger 1983).

system II inhibitors (Fig. 15.6b). Each P680-containing core complex of PSII contain one phototrap and 50–60 light-harvesting chlorophyll a_{II} molecules. Unlike P700, the P680 signal is not used to quantify PSII reaction centers as it is not resolved simply, occurring in a part of the visible spectrum where secondary optical signals from P700 complexes and LH chlorophyll a molecules interfere. Rather, a diagnostic indicator of PSII is an absorbance change at 550 nm (C-550) which presumably occurs when a secondary electron acceptor for PSII, i.e. a plastoquinone, is reduced and thereby modifies the absorption properties of nearby beta-carotene molecules which are sensitive to the redox changes of the electron acceptor. Two distinct classes of PSII complexes have been identified. The alpha type of PSII are found in appressed regions of thylakoid membranes and have more LH Chl a interconnecting them than the beta type of PSII located in non-appressed regions of thylakoid membranes.

Electron transport

Electrons flow from the electron donor side of PSII to the electron acceptor side of PSI, and can be represented by an interpretation of the gross organization of photosynthetic components seen in Fig. 15.1. The two photosystems are linked in series by a transport chain of electron and hydrogen carriers and the flow of electrons from water to $NADP^+$ is termed 'non-cyclic' electron transport. This pattern of light-driven electron transport is described by the Z-scheme shown in Fig. 15.7, where the ordinate shows the redox potential of the electron carriers. In oxygenic photosynthesis, electrons are transferred from a redox potential sufficiently positive to oxidize water to a redox potential sufficiently negative to reduce $NADP^+$. The electron transport carriers are arranged in an organized manner across the thylakoids, which have a distinct sidedness. While the water-splitting reactions occur on the inner side of the thylakoid which is in contact with an internal aqueous phase (the lumen or intrathylakoid space), $NADP^+$ reduction occurs on the exterior of the thylakoid which is in contact

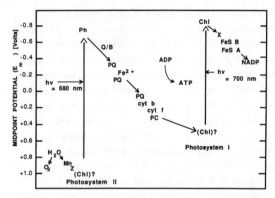

Fig. 15.7 The 'Z' scheme. Electron transfer components and their redox potentials in the photosynthetic systems of green plants. The primary donor species in both PSI and PSII are Chl species. The question marks indicate that their exact nature is not known (by permission, modified from Okamura et al., 1982).

with the surrounding stroma. A 1:1 ratio of electron carriers per reaction center is generally assumed, with the exception of the plastoquinone (PQ) pool linking PSI and PSII. However, the relative proportion of reaction center components and electron carriers can vary with changes in growth conditions.

The donor side of PSII is represented by a manganese-containing water-splitting enzyme complex (34 kD) which generally donates electrons to P680 within 30 ns. Being slower than the few picoseconds required for the initial charge separation leading to $P680^+$ and reduced pheophytin, the donor side of P680 is the rate-limiting step determining the turnover time of PSII. Facilitated by the enzyme complex, one electron is extracted from two water molecules during each of four successive photochemical acts and eventually results in the release of four protons and one oxygen molecule to the intrathylakoid space.

$$2H_2O \xrightarrow{4h\nu} O_2 + 4H^+ + 4e^-$$

When PSI and PSII are present in equal ratios, the minimum quantum requirement for photosynthesis is eight. For each molecule of oxygen evolved, four photoacts must occur within each paired photosystem to transfer four electrons from water to the carbon dioxide reduction systems.

Between PSI and II are a series of chemical electron carriers. There is some controversy over the order and composition of electron acceptors in the chain, but the most favored simplified pathway is shown in Fig. 15.6. With each photoact, one electron is passed from P680 to pheophytin to a bound quinone, Q, which passes its reducing equivalents to plastoquinone (PQ) via the B protein (Figs. 15.1 and 15.6). The secondary acceptor B is a rapidly turned over plastoquinone–protein complex which accumulates two electrons before passing the electron pair on to PQ. It is between Q and B that the photosynthetic inhibitor 3-(3,4-dichlorophenyl)-1,1-dimethylurea (DCMU or diuron) acts to block electron flow. When B accepts two electrons, it also takes up two protons from the stroma, and both electrons and protons are passed along the electron transport chain when B is subsequently oxidized by the neighboring plastoquinone pool (PQ). The transmembrane PQ pool, with five to ten PQ molecules per electron chain, interconnects more than one electron transport chain and thereby regulates electron flow between several photosystems.

When oxidized, PQ donates two electrons to PSI electron acceptors and releases two protons to the inside of the thylakoids into the lumen space. It is the light-driven proton build-up in the lumen that drives ATP formation, thereby linking rates of photophosphorylation to rates of electron transport. Mediating electron flow between the PQ pool and the plastocyanin (PC) molecule, which serves as the primary electron donor to PSI, is a cytochrome b_6/cytochrome f (cyt b/f) complex (Fig. 15.1). The oxidation site for reduced PQH_2 is at an iron–sulfur site (Rieske Fe–S), where protons are released and electrons are passed rapidly to cyt f within the complex. The cyt f reduces PC, which then reduces P700.

On the acceptor side of PSI, electrons from P700 reduce an iron-containing acceptor, X, which passes electrons onto two iron–sulfur centers, A and B (Fig. 15.1). From here electrons are transferred one at a time to ferredoxin (Fd) which in turn reduces $NADP^+$ with the aid of ferrodoxin-NADP oxidoreductase, a FAD-containing flavoprotein. The ferredoxin-NADP reductase is located on the stromal surface of the thylakoid, in protected sites near membrane-bound ATPases, and is light-dependent in its ability to catalyze the two-electron reduction of $NADP^+$ to NADPH.

As $NADP^+$ becomes reduced, the ratio of $NADP^+$ to NADPH is altered in the chloroplast. The $NADP^+$:NADPH pool is one of several important feedback mechanisms that regulate the rate of photosynthetic electron transport. When $NADP^+$ concentrations are lowered, then electrons reaching the top of PSI can be recycled back to the PQ pool via ferredoxin. These electrons once more flow to P700 by transfer through the cyt b/f complex and PC, all the while contributing to greater proton transport into the lumen to drive additional ATP formation. This is known as *cyclic electron flow* or *cyclic photophosphorylation*, and is especially important when ATP demand for phosphorylating sugar intermediates is high or when environmental CO_2 limitation slows the rate of carbon fixation and thus slows the rate of NADPH turnover.

Photophosphorylation

The light-driven proton build-up, brought on by both PQ oxidation and water-splitting reactions during electron transport, lowers the intrathylakoid pH significantly. It appears that the internal proton concentration can reach magnitudes 10^4 times as great as the external proton concentration, with the pH of the intrathylakoid space being as low as 4.0. The transmembrane gradient represents a protonmotive force (PMF) which drives transmembrane photophosphorylation reactions. Thus, the rate of cyclic and non-cyclic electron transport controls the rate of proton build-up, which in turn determines the rate of ATP formation. The phosphorylation of ADP is stimulated by an enzyme complex which couples ATP synthesis to electron transport and is termed 'coupling factor' (CF, or ATPase). CF is a large (325 kD) protein complex consisting of a hydrophobic protein complex (CF_0) which spans the thylakoid membrane and is the binding site for

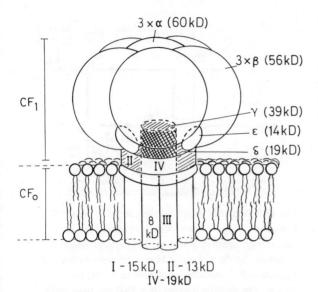

Fig. 15.8 The coupling factor (CF). A schematic model of ATP synthetase (CF_0–CF_1 complex). The shaded components are coded for in the nuclear genome while unshaded polypeptides are chloroplast encoded (by permission, Barber and Baker, 1985).

a large hydrophilic protein complex (CF_1) containing the active site for ATP synthesis. CF_1 exists as a particle about 10 nm in diameter, and is attached to the outer, non-appressed stromal surfaces of thylakoid membranes (Fig. 15.8). The entire CF complex is a reversible ATPase. The chemiosmotic coupling theory first proposed by Mitchell (1974) supports the view that when transmembrane gradients in pH are generated during electron transport, thylakoid membrane state changes occur which stimulate proton passage through the proteolipid channel of CF_0. The magnitude of proton movement depends on the light intensity (the driving force), the pH gradient in steady state, and the internal buffer capacity of the membrane. It is estimated that between 2.5 and 3.0 protons are pumped for each ADP phosphorylated.

The actual mechanism by which ATP is formed from the dehydration of ADP and the binding of P_i is not known. One hypothesis is that hydroxyl ions and protons are released in ADP dehydration reactions, with the hydroxyl ions drawn toward the proton-rich interior to combine with the protons to form water. As a result, the protons from the dehydration reaction are released in a 1:1 stoichiometric ratio with each proton hydrated on the interior, and thus a net proton release across the thylakoid membrane results. The reverse process is hypothesized to occur when ATP reserves are mobilized in the stroma, to allow the hydrolysis of ATP to ADP and P_i.

Under ideal conditions of non-cyclic electron transport, when eight quanta are absorbed (four per PSI and four per PSII), eight protons ($4H^+$ from H_2O splitting reactions in PSII and $4H^+$ from electron transport by PQ pool between PSI and PSII) will be deposited in the intrathylakoid space and two NADPH are generated in the stroma. The eight protons in the intrathylakoid space can be used as a protonmotive force to phosphorylate three ADP to three ATP. The 3ATP:2NADPH yielded in non-cyclic electron transport is what is required to fix one molecule of carbon dioxide into a hexose sugar via the reductive pentose phosphate pathway (Chapter 17). Generally, however, the quantum requirement for CO_2 reduction is usually greater than eight, especially when synthetic endproducts of photosynthesis are organic molecules other than simple sugars. Metabolic demands can often require that light-dependent ATP production be increased relative to NADPH reduction.

State 1/State 2 transitions

A balance in exciton distribution between PSI and PSII is required if these interdependent reaction centers are to operate efficiently. However, PSI and PSII do not absorb all wavelengths of light equally. Difference in spectral absorption of PSI and PSII occurs throughout the visible spectrum but is most pronounced at wavelengths greater than 680 nm (far red) (State 1 light), where light is absorbed predominantly by PSI. Under far-red light conditions, there is a decrease in the activity of PSII, termed the red drop. If light of shorter wavelengths (State 2 light) is superimposed on far-red illumination, then there will be a marked *enhancement* of PSII activity.

A number of mechanisms exist to keep a

balanced exciton distribution between the two photosystems. For long-term (hours to days) exposure to light environments which favor light absorption by one photosystem over another, plants can adapt by changing the amount or composition of LH pigments servicing each phototrap, by varying the PSI/PSII ratio and/or by altering the density or arrangement of thylakoids within the chloroplast. To accommodate short-term (seconds to hours) biases in the light field, plants can vary the distribution of excitation energy between photosystems in two ways. Plants can redirect some absorbed light out of PSII into PSI (i.e. spillover) and/or rearrange the existing LH pigment–protein complexes among PSI and PSII complexes so as to alter the absorbance cross-section of photosystems. In either instance, exposure to State 2 light will favor absorption by and over-excitation of PSII and induce changes that lead to the redistribution of excitation energy away from PSII and toward PSI. Exposure to State 1 light will favor light absorption and over-excitation of PSI, quickly leading to an inhibition of the transfer of excitation energy from PSII to PSI and/or the inhibition of realignment of PSII light-harvesting complexes with PSI. Molecular mechanisms are now being proposed to explain how changes in environmental light conditions induce an immediate and effective redistribution of excitation energy between PSI and PSII, referred to as State 1/State 2 transitions (Fig. 15.9).

In higher plants and green algae, the light-harvesting capabilities of PSII are determined by the presence of LH Chl a/b protein complexes (LHCII) (Fig. 15.2c). These LHCII complexes can be phosphorylated by a kinase enzyme, whose activity is regulated by the redox state of the PQ pool situated between PSI and PSII (Fig. 15.9). Under State 2 light conditions, the PQ pool becomes largely reduced and activates the kinase system, which in turn phosphorylates LHCII complexes. Phosphorylation is thought to increase the net negative charge on surfaces of LHCII complexes which are associated with PSII and are located mainly in the appressed regions of grana. As a result of the change in net charge, phosphorylated LHCII complexes are thought to dissociate from PSII within the grana stacks and

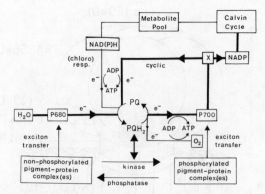

Fig. 15.9 A general model for the control of State 1/State 2 transitions in green plants. A highly reduced PQ pool catalyzes phosphorylation of the LH pigment–protein complexes by the action of a kinase, allowing spillover of excitons to PSI. Reduction potential in the PQ pool is increased by electron transfer from PSII and from cyclic photophosphorylation and photorespiration, and is reduced by reduction of P700 in PSI (by permission, Williams and Allen, 1987).

diffuse to non-appressed regions of stroma lamellae, where they associate with the LH Chl a_I–protein complexes of PSI. Overall, the absorption cross-section of PSII is reduced, while that of PSI is increased, thereby leading to a better balance of excitation energy between the two photosystems.

Under State 1 light conditions, the PQ pool becomes oxidized, kinase activity decreases, and phosphorylation of LHCII ceases. Phosphorylated LHCII complexes are probably dephosphorylated by a phosphatase system. The resultant loss of a net negative charge would make it possible for LHCII complexes to return to the stacked regions of the thylakoid and reassociate with PSII. Thus, one model of State 1/State 2 transitions is based on the light-regulated phosphorylation of light-harvesting pigment–protein complexes.

As models develop to describe the control mechanisms underlying state transitions in photosynthesis, additional factors regulating kinase activation via the redox state of the PQ pool are being incorporated (Fig. 15.9). Aside from the initial distribution of excitation energy between PSI and PSII, excess cyclic electron flow around PSI would also lead to a net reduction of the PQ pool and the phosphorylation of LH pigment–protein complexes. As previously discussed, cyclic electron flow can vary in response to changes in the availability of NADP and ATP, which in turn are influenced by carbon limitation and metabolic demands within the chloroplast. Alternate views regarding mechanisms underlying state transitions do exist, and the pigment – protein phosphorylation model has not been fully accepted for all plant groups in which thylakoid arrangements, composition of LH complexes, and metabolic regulation within chloroplasts can vary broadly.

References

Anderson, J. M. (1975). The molecular organization of chloroplast thylokoids. *Biochim. Biophys. Acta* **416**, 191–235.

Barber, J. and Baker, N. R. (1985). *Photosynthetic Mechanisms and the Environment*. Topics in Photosynthesis, Vol. 6, Elsevier, Amsterdam.

Boczar, B. A. (1985). Functional organization of chlorophyll in chlorophyll *c*-containing marine phytoplankton, PhD thesis, University of California, Santa Barbara.

Fenna, R. E. and Mathews, B. W. (1979). Bacterial chlorophyll-proteins from green photosynthetic bacteria. In *The Porphyrins*, Vol. 7, ed. D. Dolphin, Academic Press, New York, pp. 473–94.

Foyer, C. H. (1984). *Photosynthesis*, Wiley-Interscience, New York.

Geacintov, N. E. and Breton, J. (1987). Energy transfer and fluorescence mechanisms in photosynthetic membranes. *CRC Crit. Rev. Plant Sci.* **5**, 1–44.

Govindjee (ed.) (1982). *Photosynthesis: Energy Conversion by Plants and Bacteria*, Vol. 1, Academic Press, New York.

Govindjee, Amesz, J. and Fork, D. C. (1986). *Light Emission by Plants and Bacteria*, Academic Press, New York.

Kaplan, S. and Arntzen, C. J. (1982). Photosynthetic membrane structure and function. In *Photosynthesis: Energy Conversion by Plants and Bacteria*, Vol. 1, ed. Govindjee, Academic Press, New York, pp. 65–140.

Larkum, A. W. D. and Barrett, J. (1983). Light-harvesting processes in algae. In *Advances in Botanical Research*, ed. H. W. Woolhouse, Academic Press, New York, pp. 3–222.

Mitchell, P. (1974). A chemiosmotic molecular mechanism for proton translocating adenosine triphosphatases. *FEBS Lett.* **43**, 189–94.

O'Carra, P. and O'heocha, C. (1976). Algal biliproteins and phycobilins. In *Chemistry and Biochemistry of Plant Pigments*, ed. T. W. Goodwin, Academic Press, London, pp. 328–76.

Okamura, M. Y., Feher, G. and Nelson, N. (1982). Reaction centers. In *Photosynthesis: Energy Conversion by Plants and Bacteria*, Vol. 1, ed. Govindjee, Academic Press, New York, pp. 195–272.

Ort, D. R. and Melandri, B. A. (1982). Mechanism of ATP synthesis. In *Photosynthesis: Energy Conversion by Plants and Bacteria*, Vol. 1, ed. Govindjee, Academic Press, New York, pp. 537–87.

Prézelin, B. B. (1981). Light reactions in photosynthesis. In *Physiological Bases of Phytoplankton Ecology*, ed. T. Platt, Can. Bull. Fish. Aquat. Sci. Vol. 210, pp. 1–43.

Prézelin, B. B. and Alberte, R. S. (1978). Photosynthetic characteristics and organization of chlorophyll in marine dinoflagellates. *Proc. Natl Acad. Sci. USA* **75**, 1801–04.

Prézelin, B. B. and Boczar, B. A. (1986). Molecular bases of cell absorption and fluorescence in phytoplankton: potential applications to studies in optical oceanography. In *Progress in Phycological Research*, Vol. 4, eds F. E. Round and D. J. Chapman, Biopress, Bristol, pp. 349–464.

Prézelin, B. B. and Haxo, F. T. (1976). Purification and characterization of peridinin-chlorophyll *a*-proteins from the marine dinoflagellate *Glenodinium* sp. and *Gonyaulax polyedra*. *Planta* **128**, 133–41.

Raven, J. A. (1984). *Energetics and Transport in Aquatic Plants*, MBL Lectures in Biology, Vol. 4, Alan R. Liss, New York.

Renger, G. (1983). Photosynthesis. In *Biophysics*, eds. W. Hoppe, W. Lohmann, H. Markl, and H. Ziegler, Springer-Verlag, Berlin, pp. 515–42.

Satoh, K. and Butler, W. L. (1978). Low-temperature spectral properties of subchloroplast fractions purified from spinach. *Plant Physiol.* **61**, 373–9

Sauer, K. (1975). Primary events and the trapping of energy. In *Bioenergetics of Photosynthesis*, ed. Govindjee, Academic Press, New York, pp. 116–75.

Thornber, J. P. and Alberte, R. S. (1977). The organization of chlorophyll *in vivo*. In *Photosynthesis I: Photosynthetic Electron Transport and Photophosphorylation*, eds A. Trebst and M. Avron, Springer-Verlag, Berlin, pp. 574–82.

Williams, W. P. and Allen, J. F. (1987). State 1/State 2 changes in higher plants and algae. *Photosynth. Res.* **13**, 19–45.

Witt, H. T. (1979). Energy conversion in the functional membrane of photosynthesis. Analysis by light pulse and electric pulse methods: the central role of the electric field. *Biochim. Biophys. Acta* **505**, 355–427.

16 Ribulose 1, 5-bisphosphate carboxylase/oxygenase: mechanism, activation, and regulation
Richard G. Jensen

Introduction

During plant photosynthesis, CO_2 is fixed in the leaf by the enzyme, ribulose 1,5-bisphosphate carboxylase/oxygenase, which will be referred to as Rubisco. The physiological measurement of photosynthesis, determined as the rate of CO_2 gas exchange with intact leaves, can be modeled closely by understanding the activity of Rubisco. Initially, the overall properties of this enzyme will be discussed before considering in depth its structure and assembly, its kinetic characteristics and the regulation of its activity in the leaf. This should demonstrate the magnitude of complexity and the capability for fine regulation of Rubisco as well as its influence on the entire process of photosynthesis.

Indicative of Rubisco's major role in photosynthesis is the observation that if CO_2 is increased in the atmosphere surrounding a plant held at constant moderate to high illumination, there is also an increase in CO_2 uptake. This indicates that CO_2 availability and its incorporation by Rubisco, and not the light reactions, are the limiting step in carbon fixation.

Properties of Rubisco

Melvin Calvin and his associates at the University of California, Berkeley were among the first to record the reaction catalyzed by Rubisco. Using $^{14}CO_2$, they reported that the first product of photosynthetic CO_2 fixation with algal suspensions was 3-phosphoglycerate (PGA) (Bassham and Calvin, 1957). The enzyme catalyzing the fixation of CO_2 was named 'carboxydismutase' and it proved to be the same as 'fraction I protein' found in ultracentrifugation analysis of the soluble proteins of plant leaf extracts (Wildman and Bonner, 1947). Rubisco catalyzes the covalent attachment of CO_2 to a five carbon sugar, ribulose 1,5-bisphosphate (RuBP), and the simultaneous hydrolysis of the six carbon intermediate to form two molecules of PGA, of which one bears the carbon introduced from CO_2 (Fig. 16.1).

When RuBP is bound to an active site of Rubisco it can also be attacked by O_2 with the products being 2-phosphoglycolate (P-glycolate) and PGA. P-glycolate is the first intermediate of the photorespiratory pathway (Bowes et al., 1971). The two processes of carboxylation and oxygenation are competitive. At air levels at CO_2 (0.03%) and O_2 (21%) two to three moles of CO_2 are fixed per mole of O_2, indicating that Rubisco has a higher affinity for CO_2 than O_2. The presence of the oxygenation may in part reflect that this enzyme evolved at a time when the atmosphere contained much less O_2.

This large enzyme has a most interesting structure with a molecular weight of 550 000. It is located in the stroma of the chloroplast and can be up to 50% of the total chloroplast soluble protein. Because this enzyme is found in all green tissue, it is the most abundant single protein in the biosphere (Ellis, 1979). Of several enzymes involved in CO_2 fixation this is the only enzyme whose product leads to net fixation of CO_2 into carbohydrate; the other carboxylases, such as phosphoenolpyruvate carboxylase, form organic acids which if serving as

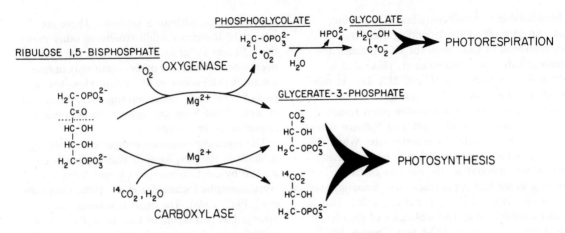

Fig. 16.1 The reactions catalyzed by ribulose 1,5-bisphosphate carboxylase/oxygenase (Rubisco).

a substrate for carbohydrate synthesis must be decarboxylated.

Rubisco is made up of two types of subunit or protomer, designated large (L) and small (S). There are eight L subunits each with a molecular weight of 51 000–58 000 and eight S subunits of 12 000–18 000 forming a structure of L_8S_8 (Jensen and Bahr, 1977). The hexadecamer is dissociated into subunits under a variety of denaturing conditions such as 8 M urea, sodium dodecyl sulfate and extreme pH. The L subunit contains the active site. The exact function of the S subunit is not understood but its presence in the holoenzyme is essential for activity.

The L subunit is encoded in the chloroplast genome: its mRNA is present as part of the chloroplast RNA and it is synthesized on chloroplast ribosomes. The S subunit is encoded in the nuclear genome: its mRNA is associated with cytoplasmic ribosomes. This subunit is synthesized as a precursor protein having a polypeptide leader sequence which directs the S subunit to be transported across the chloroplast envelope. Inside the chloroplasts the two subunits are assembled with the assistance of an assembly or L subunit binding protein (LSBP).

The large amount of Rubisco in the chloroplast stroma results in a high concentration of active sites, estimated to be 4–10 mM (Walker *et al.*, 1986). Thus the concentration of enzyme catalytic sites is greater than many of the intermediates involved in carbon assimilation. Yet the chloroplast RuBP in light is generally 2 to 4 times the catalytic site concentration and not limiting for CO_2 fixation.

The activity exhibited by Rubisco is modulated in the plant leaf by a number of factors, most notably its response to light intensity. The eight substrate binding sites on Rubisco can be activated to catalyze carboxylation or oxygenation by binding at each site of an effector or 'activator' CO_2 ($^ACO_2$). This $^ACO_2$ is bound as a carbamate to the epsilon amino group of lysine (Lys-201) of an L subunit and requires Mg^{2+}. The Mg^{2+} concentration and pH in the stroma are altered by changes in light intensity and this affects the amount of $^ACO_2$ which is bound to the active sites. However, the major regulator of activation is a second protein, known as Rubisco activase. A description of this regulatory protein and its action will be given later. Once the enzyme is activated by formation of the carbamate, the substrate, RuBP, binds and can be attacked by either CO_2 or O_2. Changes in Rubisco activation involve changes in the proportion of binding sites which are carbamylated. The number of such active sites is a major factor modulating the rate of CO_2 fixation in the leaf.

As stated above, CO_2 assimilation in the intact leaf follows closely the kinetic properties of Rubisco, including its activation. Support for this observation comes from several sources, including a well-defined physiological property, the CO_2 compensation point. When a C3 plant is illuminated in a closed container, it will either take up or evolve CO_2 depending upon the concentration of CO_2, until the CO_2 reaches a stable steady-state concentration. This value is about 45 ppm CO_2 at 25 °C and 21% O_2. At this point the rate of net CO_2 assimilation is zero. The phenomenon of the compensation point results from the catalytic bifunctionality of Rubisco where O_2 and CO_2 compete at the active site. When the velocity of the oxygenase reaction becomes twice that of the carboxylase, the net rate of CO_2 fixation in the leaf approaches zero, assuming dark respiration is small (Farquhar et al., 1980). This is understandable when two molecules of glycolate yield one CO_2 and one PGA (see Chapter 18).

Along with changes in activation of Rubisco by carbamate formation, there are other mechanisms which control the activity of the eight binding sites. These mechanisms involve inhibition of the enzyme; namely the tight binding of RuBP to deactivated binding sites and the tight binding of an inhibitor, 2-carboxyarabinitol 1-phosphate (CA1P) to carbamated sites. The modulation of activity by these and their control by Rubisco activase will be discussed later.

Regulation of carbon assimilation during photosynthesis occurs either by the availability of the substrates RuBP and CO_2; or by regulation of the activity of Rubisco. For example, in the dark no CO_2 fixation occurs because RuBP is unavailable. In the light the major control is either by regulation of Rubisco activity or the limited amount of CO_2 in air. The substrate, RuBP, is generally at a greater concentration than the binding sites of Rubisco, except at very low light intensities and/or high CO_2.

Structure and assembly of the enzyme

Structure

In nature there are two general types of subunit structure for Rubisco. The most common, which occurs in all eukaryotes and many of the prokaryotes, is a hexadecameric structure composed of eight copies each of the L subunit and the S subunit, noted as L_8S_8. The second structure is restricted to the purple, non-sulfur, photosynthetic bacteria and consists as a dimer of L subunits, L_2, without S subunits. There are reports of Rubisco subunit structures other than the L_8S_8 and L_2 structures, but some have not been substantiated and others occur only in rare instances (McFadden et al., 1986). However, a pair of L subunits associated together, either with or without the S subunit appears to be a minimum requirement for activity.

The three-dimensional arrangement of the subunits in the L_8S_8 form of Rubisco has been studied by electron microscopy and X-ray crystallography (Schneider et al., 1986; Chapman et al, 1987, 1988). The eight L subunits are arranged as two layers of four subunits each in square array. It had originally been proposed that the two layers might slide against each other; however, recently this has been shown to be incorrect. The shape of the molecule resembles a keg which is 105 angstroms along the 4-fold axis and 132 angstroms in diameter at the widest point. All eight L subunits are elongated and interdigitated in pairs (Fig. 16.2).

The 477 residues (the number depends on the organism) of the L subunit can be thought of being made up of two domains. The first, approximately 168 residues in the higher plant L

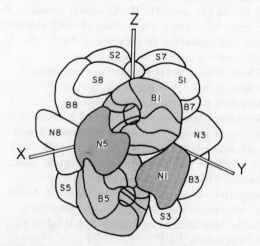

Fig. 16.2 The three-dimensional relationship between two L subunits and associated S subunits of Rubisco, from Chapman et al. (1987, 1988). The active site is formed by the N domain of one L subunit overlapping a pocket at the mouth of the barrel of the B domain of the second L subunit.

subunit, form the N domain with most of these (residues 5 to 134) forming a four-stranded antiparallel beta sheet of an 'open-face sandwich' type of topology with one surface exposed to solvent. Residues 169 to 477 fold into an alpha/beta barrel or B domain. The barrel is formed from eight hydrogen-bonded, twisted, but roughly parallel beta strands. The carboxy-terminal connector (residues 135 to 168) forms inside the sheet of the N domain, passing completely underneath the barrel forming a short helix before linking to the barrel. The mouth of the barrel at the carboxy-termini of the beta strands opens into a pocket that is covered partially by the N domain of the second protein of the L subunit pair. This pocket at the mouth is the active site. Three residues shown to be involved in catalysis are within 10 angstroms at the mouth (Lys-175, Lys-201 and Glu-60). Glu-60 is located in the N domain of the second L subunit. Indeed, three segments of the N domain pass within 13 angstroms of the epsilon amino group of Lys-201, where the enzyme is activated by the formation of a covalent bond with $^ACO_2$ (Chapman et al., 1987). The presence of Mg^{2+} tends to retain the bound $^ACO_2$ and to prepare the active site so that effective binding of RuBP can occur. Similar to a few other enzyme examples, the active site in Rubisco is formed at the interface of two subunits. Previous studies involving site-directed mutagenesis of the L_2 of *Rhodospirillum rubrum* reached a very similar conclusion (Schneider et al., 1986), confirming that the overall conformation of the N and B domains of the L_2 Rubisco of *R. rubrum* is similar to the L_8S_8 Rubisco's.

The S subunits cluster as two tetramers, one near the top and one near the bottom of the four interdigitating L pairs. The S subunits have approximately 123 residues which fold into brain-shaped domains with their carboxy-termini peaking near the four-fold axis (Chapman et al., 1988). The contribution of the S subunits to the conformation of the active site remains obscure.

It may be that variations in the relative position of the N domain are the cause of the changes in the relative affinity for CO_2 and O_2 (specificity factor) which distinguish various plant species. The CO_2 molecule is bulkier than the O_2 molecule and its diffusion to the bound enediol-RuBP might be hindered under conditions where the N domain is closer to the pocket (Andrews and Lorimer, 1987).

The amino acid sequences of the L subunits appears closely conserved between various L_8S_8 Rubisco molecules with more than 80% homology, but only 25% homology to the L_2 of *R. rubrum*. However, the residues implicated in the active sites of the L_8S_8 and L_2 Rubisco proteins are almost completely conserved, suggesting that they probably have a common origin.

Comparison of sequences for the S subunit from L_8S_8 molecules reveals considerably less homology than that observed with the L subunit. There are two strongly conserved regions in all S subunits. The first, residues 10 to 21, occurs both in cyanobacterial and in higher plant S subunits. The second region, residues 61 to 76, is absolutely conserved in the plant enzyme but has an interesting difference in cyanobacteria. The cyanobacterial subunit is 12 residues shorter than that from higher plants, and this deletion occurs in such a position that the first five residues of the second conserved region, residues 61 to 65, are missing. Residues 66 to 76 are identical for all chloroplast S subunits and are strong homologs to corresponding regions of the cyanobacterial S subunit.

At present, some aspects of the model of the L_8S_8 structure are still preliminary and are based upon the decarbamated (non-activated) enzyme. Currently, structural analysis is continuing using the $^ACO_2$-carbamated enzyme to which is bound Mg^{2+} and 2-carboxyarabinitol 1,5-bisphosphate (CABP). CABP is a close structural mimic of the hydrated gem diol form of the reaction intermediate, 3-keto 2-carboxyarabinitol 1,5-bisphosphate. The structure of the quaternary complex that will emerge from these studies will be of particular relevance to the mechanisms of activation and catalysis. These studies should pinpoint the nature of the interaction of the S subunits with the L subunits and perhaps suggest why the presence of the S subunit is essential for activity of the L_8S_8 structure.

Synthesis and assembly of Rubisco

In higher plants, the complex pathways of synthesis and assembly of the two subunit L_8S_8

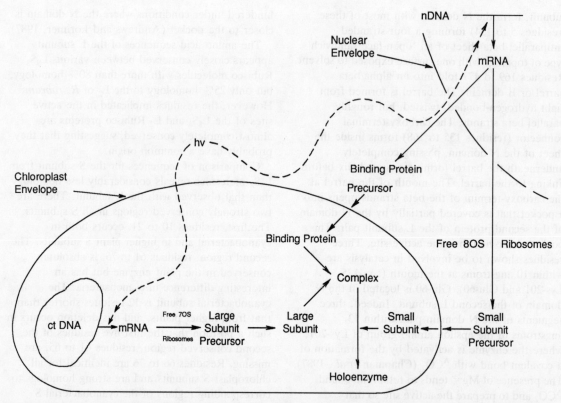

Fig. 16.3 The routes of synthesis, processing and assembly of Rubisco in higher plants. Redrawn from Ellis and van der Vies, (1988).

Rubisco has been studied in some detail. The elements of this process are diagrammed in Fig. 16.3 (taken from Ellis and van der Vies (1988)). Assembly of the Rubisco holoenzyme occurs in the chloroplast stroma from L subunits synthesized in the chloroplast plus S subunits made as a precursor protein in the cytoplasm, transported into the plastid, and processed to the mature S subunit. There must be a coordinated participation of the chloroplast and cytoplasmic genomes. In each chloroplast, the circular DNA molecule bears a single gene for the L subunit, which is generally without introns in higher plants. Because of the polyploid nature of the chloroplast DNA with 50 to 1000 genome copies per chloroplast and with 40 to 200 plastids per higher plant cell, there may be up to several hundred thousand presumably identical copies for the L subunit in each photosynthetic cell.

In the nucleus there are multiple S subunit genes, comprising a multigene family of 6 to 12 members per nuclear genome. Although the gene products differ little in the amino acid sequence of the mature peptide, there is considerable divergence at the nucleotide level. Introns are found in S subunit genes. These vary in DNA length and number both between genera and species. Large differences in both the 5' and 3' non-coding regions have been found which presumably control the differences in expression of different S genes in different tissues and during different stages of development. Particular members of the gene family are expressed to different extents in various tissues. It has also been found that some of the members of the gene family are not expressed in any tissue and are therefore considered pseudogenes.

The dual coding sites for the two subunits L and S, along with a transmembrane import of the S subunit precursor into the chloroplast, indicate that this complex assembly mechanism is regulated in plants. The manner in which this regulation occurs is only partially understood. However, several observations have indicated possible mechanisms. When isolated intact chloroplasts are incubated in a medium containing radioactive amino acids, the synthesis of chloroplast encoded proteins can be observed. These *in vitro* systems use light energy to drive protein synthesis. Under these conditions, one of the major products synthesized is the Rubisco L subunit. The incorporation of radiolabel into holoenzyme is increased in the presence of precursor S subunit. The pool sizes of unassembled S subunit are usually quite small and the half-life appears to be less than ten minutes, suggesting that the unassembled S subunit is very labile.

Both of the subunits have an amino-terminal fragment removed at some stage in the assembly process. This processing of the S subunit is an integral part of the uptake process into the chloroplast. The S subunit is synthesized as a 20 kD precursor protein and has a 'transit peptide' of 46 to 57 amino acids. This transit peptide acts like a key, identifying an attachment site on the chloroplast envelope and facilitates uptake of the S subunit into the chloroplast by a process that requires ATP. During or after transport the transit peptide is excised. The transit peptide has been fused to unrelated proteins and these proteins are then imported into the chloroplast. Indeed, even the L subunit has been transported into isolated chloroplasts and assembled into holoenzyme using the S subunit transit peptide. Thus, it seems that only the transit peptide is required to target cytoplasmic proteins into the chloroplast.

Once inside the chloroplast the S subunit forms a complex with the L subunit, most likely aided by a third protein, a 'binding protein', to eventually form the L_8S_8 holoenzyme. This binding or assembly protein is also nuclear encoded as a precursor protein and is transported into the chloroplast by a mechanism similar to the transport of the S subunit (Musgrove and Ellis, 1986). Whether this binding protein actually participates in the assembly of the holoenzyme is not apparent. It certainly does sequester L subunits and may act as a buffer against mismatches in the rate of synthesis of the two subunits. Unassembled L subunits are insoluble and there is a possibility that a disruptive accumulation of insoluble L subunits might occur if they are not stored by the binding protein until assembly can occur.

The assembly of the Rubisco subunits is known to be regulated by light. It is clear that, at least in part, this control acts at the level of transcription and is very likely mediated by phytochrome and a photoreceptor for blue light. Transcription of the S subunit gene is induced by light and causes an accumulation of the S subunit mRNA. The translation of the mRNA is also induced by light, as the synthesis of the S subunit is initiated very rapidly when dark grown cotyledons of amaranth are transferred to light and precedes the increase in the levels of mRNA. On transfer back to darkness, the synthesis is rapidly depressed in the absence of any reduction in the levels of mRNA. These data suggest that regulation of the synthesis of the S subunit can occur both at transcription and translation. As expected, expression of the S subunit gene is tissue specific and largely restricted to the green parts of plants, although mRNA for the S subunit may be found in other tissues, e.g. in roots, but at much reduced amounts. At the DNA level, a 33 bp sequence immediately preceding the transcriptional initiation site of the S subunit gene along with an enhancer-like element upstream have been implicated in light regulation of transcription.

Transcriptional control cannot account for the level of the L subunit in the chloroplast. While the mRNA level for the S subunit closely parallels the level of the mature peptide during greening, accumulation of the mRNA for the L subunit precedes that of the L peptide. This suggests that there are other, as yet, unknown post-transcriptional mechanisms for coordinating the rate of synthesis of the L subunit to that of the appearance of the S subunit in the chloroplast.

Because the L and S subunits are coded and transcribed in different compartments, transcription and translation of the genes require different types of RNA polymerase and ribosomes. Expression of the S subunit gene involves the use of eukaryotic RNA polymerase and ribosomes whereas expression of the L subunit gene in the chloroplast is more prokaryotic in nature. All of the S subunit genes examined today contain introns, although the number differs between monocots (1) and dicots (2–3) (Gutteridge and Gatenby, 1987).

The chloroplast L subunit gene is transcribed and translated using promoter sequences, RNA polymerase and ribosomes that are prokaryote-like. Many chloroplast genes also possess prokaryotic 'Shine–Dalgarno' sequences which may function like their bacterial counterparts as ribosome binding sites. In etiolated pea seedlings light may induce the accumulation of L subunit mRNA and the amount of Rubisco protein present is proportional to the amount of L subunit mRNA in the tissue. However, this has not been found in other plants such as mustard or amaranth. It is possible that the mechanism of regulation of the L subunit gene varies between different plant species. In maize and other C4 plants leaves, the mRNA for the L subunit has been detected in bundle sheath chloroplasts but is absent or at very low levels in the mesophyll cells. This agrees with the observation that Rubisco is found in the bundle sheath chloroplast but not in the mesophyll chloroplast.

Mechanism of activation and catalysis

With all forms of Rubisco, irrespective of their origin, the active site is catalytically competent when it is complexed with CO_2 and Mg^{2+} ($E-{}^ACO_2-Mg^{2+}$) (Fig. 16.4). This active ternary complex is formed by reversible attachment of an activator CO_2 (${}^ACO_2$) to the epsilon amino group of lysine-201 to form a carbamate. This ${}^ACO_2$ is distinct from the substrate CO_2. Carbamylation is slow and takes several minutes for completion at room temperature. It is followed by rapid binding of Mg^{2+}. ^{13}C-Nuclear magnetic resonance studies of this ternary complex in the presence of 2-carboxyarabinitol 1,5-bisphosphate (CABP) a substrate analog has demonstrated that the ${}^ACO_2$, Mg^{2+} and the carboxyl group of CABP are contiguous (Pierce and Reddy, 1986). The presence of Mg^{2+} stabilizes the carbamate and is required for binding of the substrate, RuBP. Other metal cations, such as Ca^{2+} or Mn^{2+}, can be substituted to stabilize the carbamate, although Mg^{2+} is probably the cation active in nature and is more effective than the others.

There is physical and chemical evidence to suggest that the formation of the ternary complex is accompanied by a change in the conformation of the enzyme (Andrews and Lorimer, 1987). The enzyme reacts with affinity labels that attack amino acids at the active site much faster when the enzyme is carbamylated. The extent of subunit cross-linking of the carbamylated enzyme is also enhanced compared with enzyme that is not carbamylated. The binding of CABP is tight when the enzyme is carbamylated and weak when it is not.

The interaction of either the carbamylated or non-carbamylated L_8S_8 Rubisco with RuBP and other phosphorylated effectors is complex. The L_8S_8 Rubisco contains eight catalytic binding sites with one site per protomer. The enzyme is highly specific for RuBP, although a wide variety of compounds can bind to the catalytic site. These include most of the intermediates of the Calvin cycle, with differing affinities, as well as di- and tricarboxylic acid compounds. Indeed, even P_i and PGA bind rather tightly to the active site.

RuBP and various phosphorylated effectors interact competitively at the catalytic site. The degree of ${}^ACO_2$ carbamylation with purified Rubisco is markedly influenced by the presence of these effectors. Thus, if high affinity effectors such as 6-phosphogluconate or NADPH are bound at a catalytic site, subsequent binding of RuBP at this site would be inhibited.

Contrary to earlier assumptions, the eight sites do not behave independently. Binding at one catalytic site influences the properties of the other seven. In the presence of less than binding site amounts of 6-phosphogluconate, the affinity of the other seven sites for ${}^ACO_2$ is increased. The binding of the inhibitor CABP in the enzyme shows negative cooperativity supporting the

$$E_{(inactive)} + {}^ACO_2 \underset{}{\overset{slow}{\rightleftharpoons}} E \cdot {}^ACO_{2\,(inactive)}$$

$$E \cdot {}^ACO_2 + Mg^{++} \underset{}{\overset{fast}{\rightleftharpoons}} E \cdot {}^ACO_2 \cdot Mg^{++}_{(active)}$$

Fig. 16.4 Mechanism of activation of Rubisco at each substrate binding site, E.

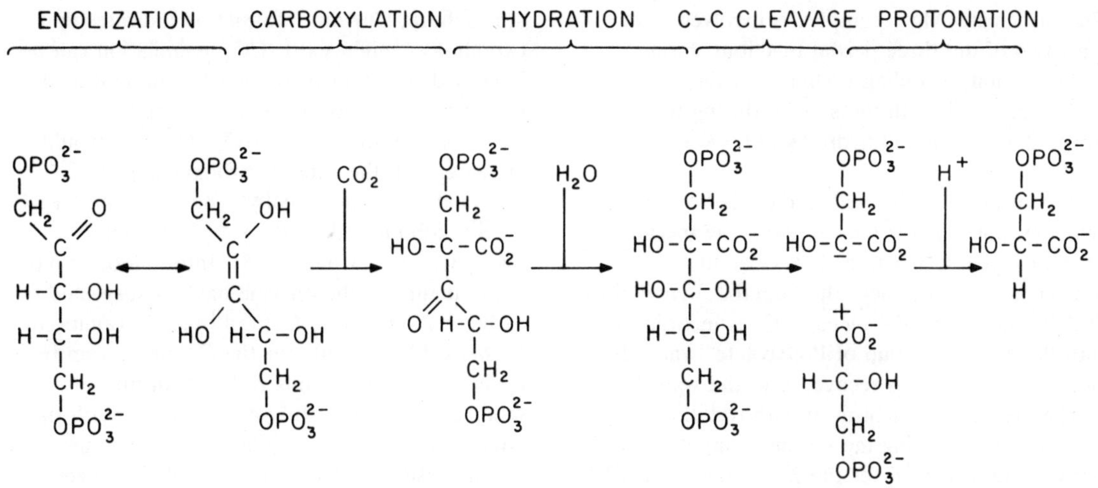

Fig. 16.5 Mechanism of carboxylation of RuBP bound to Rubisco. The mechanism, originally proposed by Calvin (1956), has been extensively studied by Lorimer and others since then (Andrews and Lorimer, 1987).

conclusion that there is interaction between the catalytic sites (Johal et al., 1985). Many of the Calvin cycle intermediates in the chloroplast are present at concentrations considerably less than the concentration of the active sites of Rubisco. When cooperativity effects are taken into consideration, these compounds could affect the amount of carbamylated sites on the enzyme at air levels of CO_2. The loss of one site might be the price paid to insure that the other seven sites are carbamylated at air levels of CO_2 and leaf amounts of Mg^{2+}. Without effectors, the CO_2 and Mg^{2+} present would be inadequate to support the observed level of carbamylation.

The mechanism involved in catalysis is ordered as shown in Fig. 16.5. The reaction involves several intermediate steps at the $E-^ACO_2-Mg^{2+}$ active site. Bound RuBP undergoes deprotonation and enolization to generate the 2,3-enediol. Carboxylation of the 2,3-enediol creates a six carbon beta-keto intermediate, 3-keto 2-carboxyarabinitol 1,5-bisphosphate. The reaction intermediate then undergoes hydration to the gem diol form. Protonation of the 3-oxygen of the gem diol initiates carbon–carbon cleavage, yielding a molecule of PGA and the C-2 carbanion form of PGA. Specific protonation of this carbanion results in the second PGA which completes the reaction.

This detailed scheme is the essence of the mechanism originally proposed by Calvin in 1956. It has now been substantiated by studies using radiolabeling, NMR and EPR (Andrews and Lorimer, 1987).

After binding of RuBP and formation of the 2,3-enediol, attack by CO_2 or O_2 forms 3-keto-carboxyarabinitol 1,5-bisphosphate (3-keto-CABP). There do not appear to be any sites on the enzyme which bind either of the two gaseous substrates. The rate of reaction depends upon their concentration and the temperature which influences the rate of diffusion.

The evidence for the 3-keto-CABP as the bound intermediate is compelling. The fact that CABP itself acts as a potent inhibitor of carboxylation suggests strongly that the 3-keto-CABP is the six carbon intermediate. The 3-keto-CABP has been released from the enzyme and identified. It is more stable in solution than had previously been anticipated (Lorimer et al., 1986). Whether the enzyme catalyzes hydrolysis or decarboxylation of this intermediate is greatly dependent on the metal ion. No decarboxylation occurs with Mg^{2+}, demonstrating that with this metal ion the enzyme is fully committed to the formation of PGA. When Mn^{2+} or Co^{2+} are used as the activator metal, the reaction equilibrium shifts toward decarboxylation.

The carbamate–Mg^{2+} complex is involved in at least two of the three partial reactions during carboxylation, including exchange of the C-3 proton of RuBP with the solvent during formation of the 2,3-enediol and hydrolysis of 3-keto-CABP to PGA.

In contrast with the well-studied carboxylase reaction, not much is known about the mechanism of the oxygenase reaction. It is clear that molecular oxygen attacks the bound 2,3-enediol of RuBP. One atom of molecular O_2 is incorporated into the carboxyl group of P-glycolate, while the other oxygen atom is recovered in the water.

The ordered mechanism of carboxylation and oxygenation involving the mutual competition between O_2 and CO_2 for the 2,3-enediol of RuBP requires that the relative rates of carboxylation and oxygenation (v_c/v_o) must vary in proportion to the ratio of their concentrations $[CO_2]/[O_2]$. The partitioning coefficient or relative specificity factor, S_{rel}, expresses this relationship as the ratio of the maximum velocities, V_c and V_o, to the K_m values, K_c and K_o, for enzyme utilizing CO_2 and O_2 respectively (Jordan and Ogren, 1983).

$$S_{rel} = (V_c/K_c)/(V_o/K_o)$$

In order for the specificity factor to increase, that is, carboxylation is increased relative to oxygenation, the rate of reaction with CO_2 must increase compared with that in O_2. Since Rubisco from different species has different specificity factors, attempts at altering the relative reactivity of CO_2 and O_2 towards the 2,3-enediol of RuBP is justified. The three-dimensional structure of the enzyme will be of great assistance in designing experiments to accomplish this goal, although such experiments could well lead to the opposite result. For example, if at the active site the distance between the N domain of a neighboring L subunit and that active pouch at the barrel was reduced, the diffusion of the smaller substrate (O_2) would be favored. This could increase the ratio of oxygenation to carboxylation.

The pressing question involving Rubisco is whether oxygenation can be reduced and/or carboxylation increased by any mechanism. This could possibly be achieved by increasing the relative diffusion of CO_2 to the active site, without changing the concentration of CO_2 or O_2 in the atmosphere. Current information indicates that two choices might exist. CO_2 assimilation can be increased by either increasing the amount of the reactive 2,3-enediol, or by increasing the concentration of CO_2 over O_2. The first would increase both the rate of carboxylation and oxygenation proportionally. The latter option has been accomplished by the C4 plants and microalgae (see Chapter 18). Information about the structure of the enzyme has not suggested how carboxylation can be favored over oxygenation. Perhaps this accounts for the C4 mechanism being evolved in higher plants rather than further modification of the Rubisco structure. With the advent of modern molecular biology, site-directed mutagenesis will determine the roles of specific residues in the active site of Rubisco. This will provide more detailed information about the nature of the active site and may provide insights into ways to enhance the specificity factor. In the end, this may not be possible. Nature has solved this problem not by altering the active site, but rather by development of CO_2 concentrating mechanisms (Chapter 18).

Regulation of Rubisco activity in the plant

The properties, exhibited by the leaf during photosynthetic CO_2 gas exchange, are similar to those of purified Rubisco. This is not unexpected, as the enzyme is the 'gate-keeper' which stands between the internal metabolic systems of the mesophyll cells and CO_2 in the environment. Indeed, comparison of the properties of the purified enzyme with those from leaf CO_2 gas exchange experiments has brought some valuable insights into the operation and metabolic activity of Rubisco in the intact plant.

The CO_2 compensation point (Γ) is an excellent example where the physiological responses of intact leaves have been corroborated by the biochemical studies of Rubisco. When a C3 plant is illuminated in a closed space, the leaves take up CO_2 until the CO_2 concentration in the enclosed atmosphere reaches a stable steady-state value at which the rate of net CO_2 assimilation (A) becomes zero. Γ is about 45 μbar CO_2 at 25 °C and

210 mbar O_2. Net CO_2 exchange can be expressed by:

$$A = v_c(1 - 0.5\, v_o/v_c) - R_d$$

where v_c and v_o are the rates of carboxylation and oxygenation, and R_d is the rate of dark or mitochondrial respiration in the light. From this equation it can be deduced that V_o/V_c becomes 2 at concentrations of CO_2 (Γ) and O_2 where A approaches $- R_d$. As the respiration term (R_d) is relatively small, τ approaches one-half of the O_2 concentration divided by S_{rel}, the specificity factor.

The kinetic properties of Rubisco give an indication of the manner in which this enzyme operates in the plant. In Table 16.1, a list of kinetic constants are presented by Woodrow and Berry (1988) (compare with Keys, 1986). Most of these constants were determined with enzyme preparations that exhibit somewhat more activity than was found with the enzyme that had been purified, since the activity of Rubisco is lowered as a result of damage during preparation or aging. It can be assumed that the inhibition constants of O_2 on carboxylation is the same as the substrate constants, i.e. that $K_i(O_2) = K_m(O_2)$ and likewise

Table 16.1 Kinetic constants of Rubisco as exhibited in C3 plants.

Constant	Symbol	Y_{25} (values at 25 °C)[a]	Q_{10}
CO_2/O_2 specificity	S_{rel}	88 MM^{-1}	0.74
		2360 bar bar^{-1}	0.67
$K_m(CO_2)$	K_c	9 μM	1.8
		270 μbar	2.1
$K_m(O_2)$	K_o	535 μM	1.0
		400 mbar	1.2
K_m (RuBP)		28 μM	1.9
Activity	V_c	3.6 μmol mg^{-1} min^{-1}	2.4
	k_{cat}	3.3 s^{-1} site^{-1}	

[a] These values are based upon measurements conducted with crude leaf extracts or purified Rubisco and were selected by Woodrow and Berry (1988), assuming that the activity is greater than or equal to that observed *in vitro*. The Q_{10} values approximating the temperature dependence of these constants may be used to calculate the value (Y_T) at any temperature between 5 and 35 °C according to $Y_T = Y_{25} \times Q_{10}^{(T-25)/10}$

$K_i(CO_2) = K_m(CO_2)$. To calculate V_c and V_o all that is required are the $K_m(CO_2)$, the $K_m(O_2)$, the K_m(RuBP) and the turnover for carboxylase as the catalytic constant (k_{cat}). The gaseous substrates have been included in Table 16.1 in terms of their dissolved concentrations and on an equivalent partial pressure basis. In interpreting gas exchange measurements, one should note that the apparent k_{cat} values decline rapidly after extraction of Rubisco from the leaf. Indeed, the values of purified Rubisco seldom approach this activity. Further experiments will increase the accuracy of these constants so that they will account for the *in vivo* response.

In gas exchange measurements the concentrations of CO_2 and O_2 in the leaf can present a problem since they can only be estimated after determining the diffusion resistances of these gases (von Caemmerer and Farquhar, 1981). The partial pressures of CO_2 and O_2 are estimated for air surrounding the leaf. The mole fraction of the CO_2 must, of course, be lower and that of O_2 must be higher in the chloroplast stroma that in the surrounding air. The CO_2 diffusion gradient can be calculated from the net flux, knowing the conductance of the diffusion pathway. Simultaneous measurements of assimilation rate and transpiration permit measurement of conductance of both the stroma and boundary layer. These may be used to obtain reliable estimates of the concentration of CO_2 in the intercellular air spaces. It has been estimated that the concentration of CO_2 in the stroma may be as much as 60 μbar less than that in the intercellular air spaces during active CO_2 uptake from which it can be deduced that the aqueous-phase conductance of wheat leaves (g_{aq}) is 0.5–1.0 mol m^{-2} s^{-1} (Evans *et al.*, 1986). This conductance should be a constant for a given leaf. There is usually some uncertainty concerning the value to use for g_{aq}, but if determined on one leaf on a given plant it often can be very similar to other leaves of that plant if they are of similar age and shape.

Because Rubisco is present in large amounts in the chloroplast, the concentration of active sites is consistently determined to be between 4 to 10 mM. The Rubisco activity is adequate to account for the measurements of CO_2 response kinetics from an intact leaf. However, an estimation of the concentration of Rubisco catalytic sites is needed to determine the significance of the RuBP pool size and to calculate the apparent k_{cat} of Rubisco. The concentration of Rubisco catalytic sites in a leaf extract can be obtained by incubating the enzyme with ^{14}C-carboxyarabinitol 1,5-bisphosphate which binds very tightly and specifically to the carbamylated sites of Rubisco. The term k_{cat} gives the amount of activity per active site of Rubisco. The activation level of Rubisco under any given environmental condition can be estimated by rapidly cooling or quick freezing the leaf. Processing of the extract has to be carried out on ice in buffers containing low CO_2. The apparent Rubisco activity in the extract is measured immediately and gives an estimate of the amount of activity present in the leaf. A second aliquot of the extract is incubated with high concentration of Mg^{2+} and CO_2 at room temperature to activate all sites and the potential or total activity can then be measured (Perchorowicz et al., 1981).

In some plants, Rubisco activity is regulated by the production of 2-carboxyarabinitol 1-phosphate (CA1P), which is a strong inhibitor of Rubisco activity (Berry et al., 1987). When a plant is capable of synthesizing CA1P, diurnal variations in the total activity of Rubisco can be found. The concentration of this compound is determined by titrating the activity of a known quantity of Rubisco with the solution containing CA1P, providing the CA1P concentration is less than that of Rubisco. Analysis of the quantity of Rubisco and the availability of catalytic sites in a leaf extract can be used to estimate the CA1P concentration. Further information of this mode of regulating Rubisco is given later.

The observed rate of CO_2 gas exchange is a function of the number of available active sites multiplied by the k_{cat}. This activity is influenced by four other factors: (1) the concentration of the gaseous substrates, CO_2 and O_2; (2) the total concentration of RuBP and the presence of inhibitors that compete with RuBP binding; (3) the stroma concentration of the tight binding inhibitor, CA1P; and (4) other factors such as Rubisco activase that influence the activation of Rubisco. These four factors will be discussed in terms of how they influence the observed rate of CO_2 fixation.

Influence of CO_2 and O_2 concentration

The influence of the gases substrates on the observed CO_2 exchange rate has been assumed to follow the classical Michaelis–Menten expression of an enzyme with a single substrate (CO_2), in the presence of a competitive inhibitor (O_2). This expression assumes that the concentration of RuBP is saturating and is appropriate only under those conditions. Rubisco does not need to be fully activated. One can assume that activated sites are fully active, and those that are not activated do not participate in the reaction. This approximation works well. RuBP must bind first to the enzyme, so that it is highly improbable that the apparent $K_m(CO_2)$ is dependent on the level of RuBP. K_m values are not dependent on the amount of enzyme. The kinetic constants for Rubisco in the complete or intact photosynthetic system agree fairly well with those measured with the purified enzyme. This is interesting because the kinetic constants for CO_2 fixation in the intact leaf were estimated well before the activation process of the purified Rubisco had been understood. Similar kinetic constants were observed with isolated spinach chloroplasts when the CO_2 response was compared with that obtained from the purified enzyme at saturating RuBP levels (Bahr and Jensen, 1977).

Concentration of RuBP

The second factor is the RuBP concentration. This becomes important where RuBP is less than the number of enzyme binding sites. Even though some sites may not be active, they usually have bound RuBP. Those that are active need at least an equal concentration of RuBP to be saturated. Studies of the level of RuBP in intact leaves have shown that under most light intensities the concentration of RuBP is, at least, several times that of the binding site. There is little response of the RuBP levels as the rate of CO_2 fixation declines from a light-saturated level to a rate that is limited by light intensity. This decrease is due to regulation of Rubisco activation rather than changes in RuBP (Perchorowicz et al., 1981).

The change in the activation state of Rubisco in the leaf during the transition from low to high light occurs with a half-life of one to two minutes. Conversely, the change in the activation state when going from high to low light is much slower with a half-life of 5 to 15 minutes. Changes in light intensity have been used in experiments on the response of Rubisco in the leaf and its effect on the rate of CO_2 assimilation. Measurements of the leaf RuBP levels taken during a transition of high to low light showed that the immediate change in fixation was due to a rapid drop in the concentration of RuBP to limiting levels rather than being a result of a change in the activation state of the enzyme (Mott et al., 1984). Enzyme activation declined over a longer time period and became the limiting factor when RuBP concentration became higher than binding site concentrations.

Low amounts of RuBP limit CO_2 assimilation in the dark, in low light and with high CO_2 levels. Some RuBP is bound to inactive sites on Rubisco and some is also bound to Mg^{2+}, as the dissociation constant for the $[Mg-RuBP]^{2+}$ is 1.6 mM (von Caemmerer and Edmondson, 1986). Thus the RuBP concentration has to be about one and one-half to two times that of the concentration of the active sites to be certain that it is at a saturating level for CO_2 fixation. The presence of the inhibitor, CA1P, or large amounts of inactive binding sites must be considered when estimating the amount of RuBP necessary for saturation.

Inhibition by carboxyarabinitol 1-phosphate

In many plants, including bean and tobacco, an inhibitor of Rubisco, carboxyarabinitol 1-phosphate (CA1P), has been shown to affect the availability of Rubisco active sites. CA1P binds tightly to the activated form of Rubisco with a dissociation constant (K_D) of 32 nM (Berry et al., 1987). In leaves held overnight in darkness, the concentration of this compound may approach or exceed that of the binding sites. The activity of Rubisco in extracts from these leaves may only be 10% or less of the normal value found for the enzyme extracted from illuminated leaves. Upon illumination of the leaf, CA1P is metabolized and declines to a small concentration, the value of which depends on the light intensity. Under saturating light, CA1P usually cannot be detected, although its presence has been measured during the late afternoon and twilight hours. Little is known about the mechanism of its synthesis but it appears that the rate of its breakdown depends upon the turnover of photosystem II. In plants such as bean, a major part of the regulation of Rubisco appears to be achieved via CA1P, whereas in plants like spinach, which do not produce CA1P, the control is generally by changes in activation (Woodrow and Berry, 1988). Under all these conditions, where the activity of Rubisco is controlled either by activation state or CA1P, the concentration of RuBP is usually high and saturating. CA1P is usually made in the absence of RuBP.

Rubisco activase

Comparing mutants of *Arabidopsis* in which activation of Rubisco does not occur, Ogren and coworkers (Ogren et al., 1986) identified a protein in the chloroplast which maintains Rubisco in the activated form. They have suggested that the protein catalyzes an ATP-dependent activation of the inactive E–RuBP complex to the active E–$^ACO_2$–Mg^{2+}–RuBP complex. Those sites not activated in light will have bound RuBP, as the binding of RuBP to inactive sites is tighter than the binding to active sites (Jordan and Chollet, 1983). It has not been clear how increased light intensity could free these sites of RuBP so that the carbamate or $^ACO_2$ might be formed.

The details of the mechanism of the activase reaction is not yet understood. The activase reacts with Rubisco, which has bound RuBP at inactive sites and, in the presence of ATP, causes the sites to become activated, producing the enzyme–CO_2–Mg^{2+}–RuBP complex plus ADP and inorganic phosphate. Activase functions to link activation of Rubisco to changes in light intensity. The [ATP]/[ADP] ratio *in vivo* is affected by the activity of Rubisco and the light intensity. As this ratio increases so does the activity of activase, causing more Rubisco sites to be carbamated, increasing

CO_2 fixation and producing PGA which utilizes the ATP. Understanding this role of the activase explains the observation that increasing CO_2 surrounding a leaf can result in decreasing the activation state of Rubisco.

Rubisco activity is regulated so that a number of factors are balanced. These are as follows: (1) at low light intensities Rubisco activity is modulated to balance the input of quantum energy. This modulation is mostly by way of changes in activation or the proportion of carbamylated to decarbamylated binding sites; (2) under high light, energy absorbed exceeds the ability of Rubisco to fix and assimilate CO_2. The mechanisms whereby the excess energy is dissipated involves fluorescence emission and loss as heat; and (3) if the rate of energy input and its utilization by Rubisco activity are both high and approach the maximum activity of reactions involved in product synthesis, then the rate of energy input and the Rubisco activity must be modulated so that they do not exceed the ability to process the products, PGA and P-glycolate.

A complete discussion of the regulation of the photosynthetic carbon reduction cycle is beyond the scope of this chapter, and the reader is referred to the following chapter. However, there are some properties of that pathway which directly affect the activity of Rubisco. Due to compartmentation of the photosynthetic process in the chloroplast, the chloroplast maintains the total phosphate level constant, which includes the esterified phosphate plus P_i. This is maintained by the phosphate translocator located in the outer envelope. In chloroplasts of intact leaves 80% of the total stromal phosphate is in the form of RuBP, PGA, and P_i. Carbon exits the Calvin cycle to the cytoplasm, either as triose phosphates or PGA, and eventually contributes the carbons for sucrose synthesis. In exchange, P_i or PGA are transported into the chloroplast maintaining total phosphate constant. During the formation of starch from hexose phosphates, P_i is left behind. Because of the already high levels of RuBP and PGA in the chloroplast stroma, it is important that sufficient P_i be available to produce ATP by photophosphorylation. Current information suggests that normal changes in P_i levels probably do not effect ATP synthase activities, but in some manner changes in stromal P_i do affect Rubisco activity, either by feedback inhibition of the activase or by direct inhibitory effects of P_i on Rubisco activity. Evidence from CO_2 exchange measurements with intact leaves supports the hypothesis that Rubisco activity is modulated by the availability of P_i.

Since Rubisco activity serves as a pacemaker between utilization of absorbed light energy and the assimilation of carbon, its activity can act as a feedback control on energy input. This is especially important where the intensity of illumination changes between light-limited to light-saturated rates or where CO_2 varies from saturation to limitation. At this transition point, fluorescence and light scattering at 540 nm increase in intact leaves, indicating that a pH gradient has been developed across the thylakoids. This suggests that, as the absorbed flux of quantum energy approaches the maximum capacity of Rubisco to sustain the flux, there is a marked rise in the swelling and pH gradient across the chloroplast thylakoids as the ATP synthase is regulated.

Light intensity, especially at limiting levels, exerts feedforward regulation on enzyme activity of the photosynthetic carbon reduction cycle. The rate is not regulated by RuBP concentration but by a decrease in activation or by synthesis and binding of CA1P.

Because of the complexity of the regulatory systems which affect Rubisco, it is difficult to develop kinetic models to draw quantitative conclusions about the amount this enzyme limits photosynthesis. The principal experimental approach towards this has been to vary by known amounts the internal CO_2 and O_2 concentrations and irradiation levels. In this manner it is possible to simulate the effect on the photosynthetic system by changing the observed activity of Rubisco. The advantage of this approach depends upon an understanding of the kinetics of Rubisco. The $K_m(CO_2)$ and the $K_m(O_2)$ values do not vary much with changes in the level of RuBP or the degree of activation. Thus, it is possible to predict relatively accurately the fractional effects of CO_2 and O_2 changes on both the carboxylase and oxygenase activities over a wide range of conditions.

Concerning the regulation of Rubisco activity by other metabolites, it is clear that all of the enzymes of the photosynthetic carbon reduction cycle, the starch and sucrose synthetic pathways, and likely the photosystems, have the potential to affect Rubisco activity and therefore modulate the rate of CO_2 assimilation. This will keep the overall process of photosynthesis balanced. At low light intensities Rubisco activity is clearly highly regulated by feedback processes. In view of the almost linear relationship between irradiance and CO_2 uptake under these conditions, the rate of photosynthesis is, most probably, highly limited by quantum efficiencies of the photosystems. Under intermediate light intensities both Rubisco and the photosystems may be subject to feedback inhibition. The degree to which the other enzymes limit photosynthesis under such light intensities is not clear.

Although many of the details are missing concerning the complexity of the feedback processes during photosynthesis and metabolism and the conditions under which they operate, one can still appreciate the abundance of information which has accumulated on Rubisco over the last 40 years. Rubisco is a major pacemaker of metabolism during photosynthesis and is capable of regulating the rate of CO_2 assimilation. It is regulated by the light reactions through the activation–deactivation of the reaction sites by means of the activase. The synthesis of CA1P firmly inhibits binding of other metabolites, until light is abundant and RuBP is in high concentrations. The following questions are still in need of an answer. What are the advantages of producing CA1P? Why is the enzyme not more efficient? It has one of the lower turnover rates of any enzyme. If under moderate to high light intensities, the capacity of RuBP synthesis is in excess over its utilization, why does Rubisco not fully operate as fast as RuBP is produced, especially to eliminate RuBP excesses which can sequester much of the phosphate and cause the tight binding of RuBP to inactivate reaction sites? Although these questions have partial answers, there remains much that is unclear about the function of Rubisco in photosynthesis.

Further reading and references

Andrews, T. J. and Lorimer, G. H. (1987). Rubisco: structure, mechanisms, and prospects for improvement. In *The Biochemistry of Plants*, Vol. 10, eds M. D. Hatch and N. K. Boardman, Academic Press, San Diego, pp. 131–218.

Bahr, J. T. and Jensen, R. G. (1974). Ribulose diphosphate carboxylase from freshly ruptured spinach chloroplasts having an *in vivo* $K_m(CO_2)$. *Plant Physiol.* **53**, 39–44.

Bassham, J. A. and Calvin, M. (1957). *The Path of Carbon in Photosynthesis*, Prentice-Hall, Inc., New Jersey.

Berry, J. A., Lorimer, G. H., Pierce, J., Seemann, J. R., Meek, J. and Freas, S. (1987). Isolation, identification, and synthesis of 2-carboxyarabinitol 1-phosphate, a diurnal regulator of ribulose bisphosphate carboxylase activity. *Proc. Natl Acad. Sci. USA* **84**, 734–8.

Bowes, G., Ogren, W. L. and Hageman, R. (1971). Phosphoglycolate production catalyzed by ribulose diphosphate carboxylase. *Biochem. Biophys. Res. Comm.* **45**, 716–22.

Braenden, R. (1987). The active site of ribulose-1,5-bisphosphate carboxylase/oxygenase. *Life Chem. Rep.* **5**, 235–47.

Calvin, M. (1956). The photosynthetic carbon cycle. *J. Chem. Soc.* pp. 1895–1915.

Chapman, M. S., Suh, S. W., Cascio, D., Smith, W. W. and Eisenberg, D. (1987). Sliding-layer conformational change limited by the quaternary structure of plant RuBisCO. *Nature* **329**, 354–6.

Chapman, M. S., Suh, S. W., Curmi, P. M. G., Cascio, D., Smith, W. W. and Eisenberg, D. S. (1988). Tertiary structure of plant RuBisCO: Domains and their contacts. *Science* **241**, 71–4.

Ellis, R. J. (1979). The most abundant protein in the world. *Trends Biochem. Sci.* **4**, 241–4.

Ellis, R. J. and van der Vies, S. M. (1988). The Rubisco subunit binding protein. *Photosyn. Res.* **16**, 101–15.

Evans, J. R., Sharkey, T. D., Berry, J. A. and Farquhar, G. D. (1986). Carbon isotope discrimination measured concurrently with gas exchange to investigate CO_2 diffusion in leaves of higher plants. *Aust. J. Plant Physiol.* **13**, 281–92.

Farquhar, G. D., von Caemmerer, S. and Berry, J. A. (1980). A biochemical model of photosynthetic CO_2 assimilation in leaves of C3 species. *Planta* **149**, 78–90.

Gutteridge, S. and Gatenby, A. A. (1987). The molecular analysis of the assembly, structure and function of Rubisco. *Oxford Surv. Plant Mol. Cell Biol.* **4**, 95–135.

Jensen, R. G. and Bahr, J. T. (1977). Ribulose 1,5-bisphosphate carboxylase-oxygenase. *Ann. Rev.*

Plant Physiol. **28**, 379–400.

Johal, S., Partridge, B. E. and Chollet, R. (1985). Structural characterization and the determination of negative cooperativity in the tight binding of 2-carboxyarabinitol bisphosphate to higher plant ribulose bisphosphate carboxylase. *J. Biol. Chem.* **260**, 9894–904.

Jordan, D. B. and Chollet, R. (1983). Inhibition of ribulose bisphosphate carboxylase by substrate ribulose 1,5-bisphosphate. *J. Biol. Chem.* **258**, 13752–8.

Jordan, D. B. and Ogren, W. J. (1983). Species variation in the kinetic properties of ribulose-1,5-bisphosphate carboxylase/oxygenase. *Arch. Biochem. Biophys.* **227**, 425–33.

Keys, A. J. (1986). Rubisco: its role in photorespiration. *Phil. Trans. R. Soc. London, B* **313**, 325–36.

Lorimer, G. H., Andrews, T. J., Pierce, J. and Schloss, J. V. (1986). 2′-Carboxy-3-keto-D-arabinitol 1,5-bisphosphate, the six-carbon intermediate of the ribulose bisphosphate carboxylase reaction. *Phil. Trans. R. Soc. London, B* **313**, 397–407.

McFadden, B. A., Torres-Ruiz, J., Daniell, H. and Sarojini, G. (1986). Interaction, functional relations and evolution of large and small subunits in Rubisco from Prokaryota and Eukaryota. *Phil. Trans. R. Soc. London, B* **313**, 347–58.

Mott, K. A., Jensen, R. G., O'Leary, J. W. and Berry, J. A. (1984). Photosynthesis and ribulose bisphosphate concentrations in intact leaves of *Xanthium strumarium* L. *Plant Physiol.* **76**, 968–71.

Musgrove, J. E. and Ellis, R. J. (1986). The Rubisco large subunit binding protein. *Phil. Trans. R. Soc. London, B* **313**, 419–28.

Ogren, W. L., Salvucci, M. E. and Portis, A. R. J. (1986). The regulation of Rubisco activity. *Phil. Trans. R. Soc. London, B* **313**, 337–46.

Perchorowicz, J. T., Raynes, D. A. and Jensen, R. G. (1981) Light limitation of photosynthesis and activation of ribulose bisphosphate carboxylase in wheat seedlings. *Proc. Natl Acad. Sci. USA* **78**, 2985–9.

Pierce, J. and Reddy, G. S. (1986). The sites for catalysis and activation of ribulosebisphosphate carboxylase share a common domain. *Arch. Biochem. Biophys.* **245**, 483–93.

Schneider, G., Lindquist, Y., Braenden, C. I. and Lorimer, G. H. (1986). The three dimensional structure of ribulose-1,5-bisphosphate carboxylase/oxygenase from *Rhodospirillum rubrum* at 2.9 Ang. resolution. *EMBO J.* **5**, 3409–15.

von Caemmerer, S. and Edmondson, D. L. (1986). Relationship between steady-state gas exchange, *in vivo* ribulose bisphosphate carboxylase activity and some carbon reduction cycle intermediates in *Raphanus sativus. Aust. J. Plant Physiol.* **13**, 669–88.

von Caemmerer, S. and Farquhar, G. D. (1981). Some relationships between the biochemistry of photosynthesis and the gas exchange of leaves. *Planta* **153**, 376–87.

Walker, D. A., Leegood, R. C. and Sivak, M. N. (1986). Ribulose bisphosphate carboxylase-oxygenase: its role in photosynthesis. *Phil. Trans. R. Soc. London B* **313**, 305–24.

Wildman, S. G. and Bonner, J. (1947). The proteins of green leaves. I. Isolation, enzymic properties, and auxin content of spinach cytoplasmic proteins. *Arch. Biochem.* **14**, 381–413.

Woodrow, I. E. and Berry, J. A. (1988). Enzymatic regulation of photosynthetic CO_2 fixation in C3 plants. *Ann. Rev. Plant Physiol.* **39**, 533–94.

17 The reductive pentose phosphate pathway and its regulation

Fraser D. Macdonald and Bob B. Buchanan

Introduction

Life on our planet obtains its substance and energy through the process of photosynthesis, a grand device by which photosynthetic organisms use the electromagnetic energy of sunlight to synthesize carbohydrates (CH_2O) and other cellular constituents from carbon dioxide and water.

$$CO_2 + 2H_2O^* \xrightarrow{\text{light}} (CH_2O) + O_2^* + H_2O$$

Photosynthesis may be broadly divided into two phases: a light phase, in which the electromagnetic energy of sunlight is trapped and converted into ATP and NADPH (see chapter 15), and a synthetic phase, in which the ATP and NADPH generated by the light phase are used, in part, for biosynthetic carbon reduction. As described below, light also functions in the regulation of the synthetic or carbon reduction phase of photosynthesis and in related biochemical processes of chloroplasts.

In most plants, the major products of photosynthesis are starch (formed in chloroplasts), and sucrose (formed in the cytosol). Both of these products (collectively called photosynthate) are formed from photosynthetically generated dihydroxyacetone phosphate (DHAP) via pathways that in some respects are similar to the gluconeogenic pathway of animal cells. In the first case, DHAP is converted to hexose phosphates, which, in turn, are converted to starch within the chloroplast. In sucrose synthesis, DHAP (or a derivative) is transported to the cytosol and there is converted to sucrose.

All oxygenic (oxygen evolving) organisms from the simplest prokaryotic cyanobacteria to the most complicated land plants have a common pathway for the reduction of CO_2 to sugar phosphates. This pathway is known as the reductive pentose phosphate (RPP), Calvin–Benson or C3 cycle (Calvin and Bassham, 1962; Bassham and Buchanan, 1982).

Although the RPP cycle is the fundamental carboxylating mechanism, a number of plants have evolved adaptations in which CO_2 is first fixed by a supplementary pathway and then released in the cells in which the RPP cycle operates. One of these supplementary pathways, the C4 pathway, involves special leaf anatomy and a division of biochemical labor between cell types. Plants endowed with this pathway, through greater efficiency, are able to flourish under conditions of high light intensity and elevated temperatures. A second supplementary pathway was found in species of the Crassulaceae and is called Crassulacean acid metabolism (CAM). These plants are often found in dry areas and fix CO_2 at night into C4 acids. During the day, the leaves can close their stomata to conserve water while CO_2 released from the C4 acids is converted to sugar phosphates by the RPP cycle using absorbed light energy (see Chapter 19).

CO_2 fixation is also found in many bacteria, both photosynthetic and non-photosynthetic. The purple sulfur and purple non-sulfur bacteria employ the RPP cycle, as do plants. The photosynthetic green sulfur bacteria, however, use a group of ferredoxin-linked carboxylases in a pathway known as the reductive carboxylic acid cycle (Bassham and Buchanan, 1982). The ferredoxin-linked carboxylases also function in CO_2 assimilation in diverse types of fermentative and methanogenic bacteria.

In the following sections, we will first describe the RPP cycle and then discuss its regulation and the way in which its activity is coordinated with the utilization of photosynthate.

The reductive pentose phosphate cycle

The RPP cycle is the primary carboxylating mechanism in plants. In C3 plants, the entire process of photosynthesis (the light reactions and the RPP cycle) occurs within chloroplasts. The enzymes which catalyze steps in the RPP cycle are water soluble and are located in the soluble portion of the chloroplast (the stroma).

Elucidation of the pathway was chiefly the work of Calvin, Benson, Bassham and coworkers, although there were important contributions by others. In their experiments, the group of Calvin used green algae, *Chlorella* and *Scenedesmus,* but their results have since been confirmed in a wide variety of higher plants.

The crux of the pathway (Fig. 17.1) is the carboxylation of ribulose 1,5-bisphosphate (Rbu-1,5-P_2) to produce two molecules of 3-phosphoglycerate (3-PGA). The next reactions (the reductive phase of the cycle) are those in which ATP and NADPH, produced by the light reactions of photosynthesis, are consumed during the reduction of 3-PGA to glyceraldehyde 3-phosphate (G3P). To complete the cycle (the regeneration phase), intermediates formed from G3P are utilized via a series of isomerizations, condensations and rearrangements resulting in the conversion of five molecules of triose phosphate to three of pentose phosphate, eventually ribulose 5-phosphate (Rbu-5-P). Phosphorylation of Rbu-5-P by ATP regenerates the original carbon acceptor Rbu-1,5-P_2, thus completing the cycle.

The three phases of the RPP cycle will now be described in greater detail. The individual reactions and the characteristics of the enzymes which catalyze each step have been reviewed in more detail elsewhere (Robinson and Walker, 1981).

Carboxylation phase (3CO_2 + 3Rbu-1,5-P_2 → 6PGA)

$$
\begin{array}{c}
CH_2O-P \\
| \\
C=O \\
| \\
3\ HCOH \\
| \\
HCOH \\
| \\
CH_2OH-P \\
\text{Ribulose 1,5-bisphosphate} \\
\text{(Rbu-1,5-}P_2\text{)}
\end{array}
+ 3*CO_2 + 3H_2O \xrightarrow{\text{Ribulose 1,5-bisphosphate carboxylase/oxygenase}}
\begin{array}{c}
*COOH \\
| \\
6\ HCOH \\
| \\
CH_2O-P \\
\text{3-Phosphoglycerate (3-PGA)}
\end{array}
\quad [17.1]
$$

Carboxylation is catalyzed by ribulose 1,5-bisphosphate carboxylase/oxygenase, Rubisco, probably the most abundant protein on earth. Carboxylation of Rbu-1,5-P_2 at the C-2 carbon gives rise to a short-lived, six carbon intermediate which breaks down to give two molecules of 3-PGA. A stable analog of this intermediate (carboxyarabinitol 1,5-bisphosphate) is a very tight binding inhibitor of Rubisco that has been of great importance in analyzing the Rubisco reaction. A similar inhibitory analog is also now known to occur naturally in some plants (see later). Note that in following carbon through the cycle, we begin with three molecules of CO_2 to give eventually a net synthesis of one molecule of triose phosphate.

Reduction phase (6PGA + 6ATP + 6NADPH → 6G3P + 6ADP + 6NADP$^+$ + 6P_i)

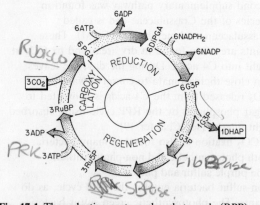

Fig. 17.1 The reductive pentose phosphate cycle (RPP). Abbreviations: RuBP, ribulose 1,5-bisphosphate; PGA, 3-phosphoglycerate; DiPGA, 1,3-diphosphoglycerate; Ru5P, ribulose 5-phosphate; G3P, glyceraldehyde 3-phosphate.

$$\begin{array}{c} \text{COOH} \\ | \\ 6 \text{ HCOH} + 6\text{ATP} \\ | \\ \text{CH}_2\text{O} - \text{P} \\ \text{3-PGA} \end{array} \xrightarrow{\text{Phosphoglycerate kinase}} \begin{array}{c} \overset{\text{O}}{\overset{\|}{\text{C}}} - \text{O} - \text{P} \\ | \\ 6 \text{ HCOH} + 6\text{ADP} \\ | \\ \text{CH}_2\text{O} - \text{P} \\ \text{1,3-Diphosphoglycerate} \end{array} \qquad [17.2]$$

$$\begin{array}{c} \overset{\text{O}}{\overset{\|}{\text{C}}} - \text{O} - \text{P} \\ | \\ 6 \text{ HCOH} + 6\text{NADPH} \\ | \\ \text{CH}_2\text{O} - \text{P} \\ \text{1,3-Diphosphoglycerate} \end{array} \xrightleftharpoons{\begin{array}{c}\text{Glyceraldehyde}\\ \text{3-phosphate}\\ \text{dehydrogenase}\end{array}} \begin{array}{c} \text{CHO} \\ | \\ 6 \text{ HCOH} + 6\text{NADP}^+ + 6\text{P}_i \\ | \\ \text{CH}_2\text{O} - \text{P} \\ \text{Glyceraldehyde} \\ \text{3-phosphate (G3P)} \end{array} \qquad [17.3]$$

The two reactions of the reduction phase (Equations [17.2] and [17.3]) utilize ATP and NADPH produced by the light reactions of photosynthesis to synthesize G3P from 3-PGA. These steps resemble part of the gluconeogenic pathway in the cytoplasm, except that glyceraldehyde 3-phosphate dehydrogenase (G3PDH) in chloroplasts is specific for NADPH rather than NADH.

Regeneration phase (5G3P + 3ATP → 3Rbu-1,5-P_2 + 3ADP + 2P_i)

Of the six PGA molecules produced above, one represents a net synthesis of fixed carbon and the other five must be used to regenerate Rbu-1,5-P_2.

$$\begin{array}{c} \text{CHO} \\ | \\ 1 \text{ HCOH} \\ | \\ \text{CH}_2\text{O} - \text{P} \\ \text{G3P} \end{array} \xrightleftharpoons{\text{Triose-P isomerase}} \begin{array}{c} \text{CH}_2\text{OH} \\ | \\ 1 \text{ C} = \text{O} \quad (\text{1G3P used here}) \\ | \\ \text{CH}_2\text{O} - \text{P} \\ \text{Dihydroxyacetone-P} \\ \text{(DHAP)} \end{array} \qquad [17.4]$$

$$\begin{array}{c} \text{CHO} \\ | \\ 1 \text{ HCOH} \\ | \\ \text{CH}_2\text{O} - \text{P} \\ \text{G3P} \\ \text{(from reaction [17.4]} \end{array} + \begin{array}{c} \text{CH}_2\text{OH} \\ | \\ 1 \text{ C} = \text{O} \\ | \\ \text{CH}_2\text{O} - \text{P} \\ \text{DHAP} \end{array} \xrightleftharpoons{\text{Aldolase}} \begin{array}{c} \text{CH}_2\text{O} - \text{P} \\ | \\ \text{C} = \text{O} \\ | \\ \text{HO} - \text{C} - \text{H} \quad (\text{2nd G3P used here}) \\ | \\ \text{HCOH} \\ | \\ \text{HCOH} \\ | \\ \text{CH}_2\text{O} - \text{P} \\ \text{Fructose 1,6-bisphosphate} \\ \text{(Fru-1,6-}P_2) \end{array} \qquad [17.5]$$

$$\begin{array}{c} \text{CH}_2\text{O} - \text{P} \\ | \\ \text{C} = \text{O} \\ | \\ \text{HO} - \text{C} - \text{H} + \text{H}_2\text{O} \\ | \\ \text{H} - \text{C} - \text{OH} \\ | \\ \text{H} - \text{C} - \text{OH} \\ | \\ \text{CH}_2\text{O} - \text{P} \\ \text{Fru-1,6-}P_2 \end{array} \xrightleftharpoons{\text{fructose 1,6-bisphosphatase}} \begin{array}{c} \text{CH}_2\text{OH} \\ | \\ \text{C} = \text{O} \\ | \\ \text{HO} - \text{C} - \text{H} + \text{P}_i \\ | \\ \text{H} - \text{C} - \text{OH} \\ | \\ \text{H} - \text{C} - \text{OH} \\ | \\ \text{CH}_2\text{O} - \text{P} \\ \text{Fructose-6-P} \\ \text{(Fru-6-P)} \end{array} \qquad [17.6]$$

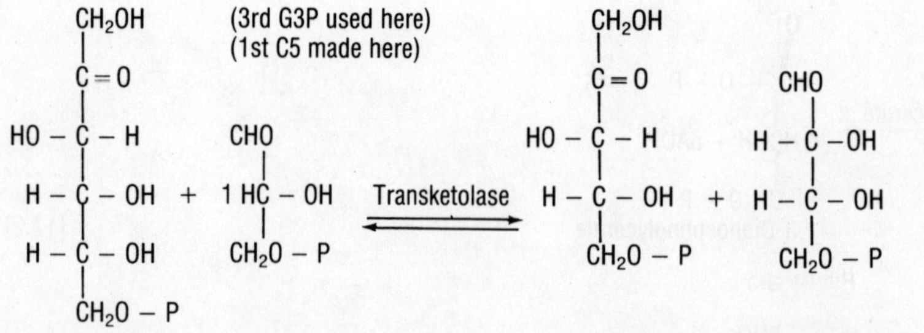

Fru-6-P + G3P ⇌ Xylulose-5-P + Erythrose-4-P (Transketolase)

Fru-6-P: CH₂OH–C(=O)–HO–C–H–H–C–OH–H–C–OH–CH₂O–P

G3P (3rd G3P used here) (1st C5 made here): CHO–HC–OH–CH₂O–P

Xylulose-5-P: CH₂OH–C(=O)–HO–C–H–H–C–OH–CH₂O–P

Erythrose-4-P: CHO–H–C–OH–H–C–OH–CH₂O–P

[17.7]

G3P ⇌ DHAP (Triose-P isomerase)

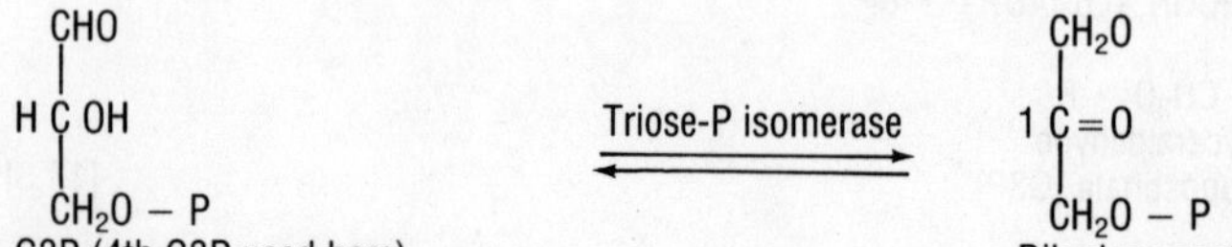

G3P (4th G3P used here): CHO–HCOH–CH₂O–P

Dihydroxyacetone-P (DHAP): CH₂O–C(=O)–CH₂O–P

[17.8a]

DHAP + Erythrose-4-P ⇌ Sedoheptulose 1,7-bisphosphate (Aldolase)

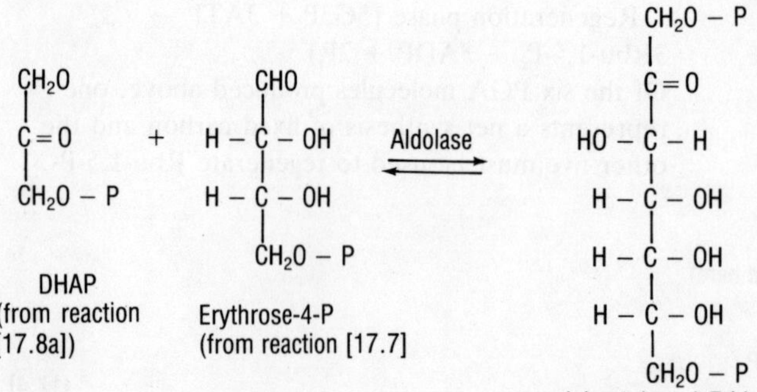

DHAP (from reaction [17.8a]): CH₂O–C(=O)–CH₂O–P

Erythrose-4-P (from reaction [17.7]): CHO–H–C–OH–H–C–OH–CH₂O–P

sedoheptulose 1,7-bisphosphate: CH₂O–P–C(=O)–HO–C–H–H–C–OH–H–C–OH–H–C–OH–CH₂O–P

[17.8b]

Sedoheptulose 1,7-bisphosphate + H₂O —Sedoheptulose 1,7-bisphosphatase→ Sedoheptulose 7-phosphate + P$_i$ [17.9]

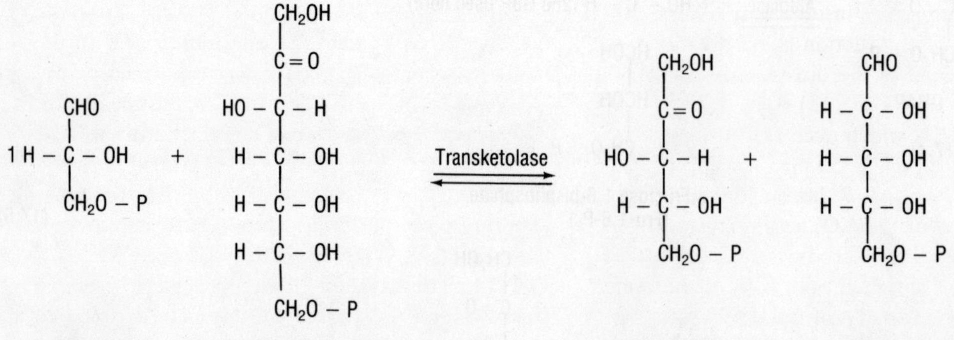

G3P + Sedoheptulose-7-P ⇌ Xylulose-5-P + Ribose-5-P (Transketolase)

G3P (5th G3P used here): CHO–H–C–OH–CH₂O–P

Sedoheptulose-7-P: CH₂OH–C(=O)–HO–C–H–H–C–OH–H–C–OH–H–C–OH–CH₂O–P

Xylulose-5-P: CH₂OH–C(=O)–HO–C–H–H–C–OH–CH₂O–P

Ribose-5-P: CHO–H–C–OH–H–C–OH–H–C–OH–CH₂O–P

[17.10]

$$\begin{array}{c}CH_2OH\\|\\C=O\\|\\2\ HO-C-H\\|\\H-C-OH\\|\\CH_O-P\end{array} \xrightarrow[\text{Phosphopentoepimerase}]{} \begin{array}{c}CH_2OH\\|\\C=O\\|\\2\ H-C-OH\\|\\H-C-OH\\|\\CH_2O-P\end{array}$$

Xyluose-5-P → Ribulose-5-P

(One xylulose-5-P is from reaction [17.7]; the other is from reaction [17.10]). [17.11]

$$\begin{array}{c}CHO\\|\\H-C-OH\\|\\H-C-OH\\|\\H-C-OH\\|\\CH_2O-P\end{array} \xrightarrow[\text{Phosphoriboisomerase}]{} \begin{array}{c}CH_2OH\\|\\C=O\\|\\H-C-OH\\|\\H-C-OH\\|\\CH_2O-P\end{array}$$

Ribose-5-P → Ribulose-5-P   [17.12]

Ribose-5-P is from reaction [17.10]. The one ribulose-5-P formed in reaction [17.12] and the two formed in reaction [17.11] comprise the three used in reaction [17.13].

$$3\text{Ribulose-5-P} + 3\text{ATP} \xrightarrow{\text{Phosphoribulokinase}} 3\text{Ribulose 1,5-bisphosphate} + 3\text{ADP} \qquad [17.13]$$

$$3CO_2 + 9ATP + 6NADPH_2 + 5H_2O \rightarrow \begin{array}{c}CHO\\|\\H-C-OH\\|\\CH_2O-P\\G3P\end{array} + 8P_i + 9ADP + 6NADP \qquad [17.14]$$

The RPP cycle displays four features which are necessary for its role as a fundamental carboxylating system (Robinson and Walker, 1981): (1) the Rubisco reaction has a highly favorable equilibrium in the direction of PGA synthesis ($\Delta G' = -35.1$ kJ); (2) Rubisco has a high affinity for CO_2 which occurs at a relatively low concentration in air ~350 ppm or ~10μM in an air saturated aqueous medium); (3) there is a cyclic regeneration of the CO_2 acceptor Rbu-1,5-P_2 from the products of the carboxylation reaction, thus enabling the continued operation of the cycle; and (4) the cycle is capable of the net production of fixed carbon in the form of triose phosphate. For every three turns of the cycle during which six molecules of triose phosphate are formed, five molecules must be utilized to reform three molecules of Rbu-1,5-P_2 while the sixth triose phosphate molecule is available as an endproduct (photosynthate) for biosynthetic reactions, predominantly starch synthesis in the chloroplast and sucrose synthesis in the cytosol.

In addition to its carboxylase activity Rubisco can act as an oxygenase (see Chapter 18). In this reaction, molecular O_2 is bound and reacts with Rbu-1,5-P_2 to give 3-PGA and 2-phosphoglycolate. 2-Phosphoglycolate cannot be utilized in the RPP cycle and thus represents a loss of fixed carbon. This loss is partly compensated by the process of photorespiration during which three-quarters of the lost carbon is returned to the chloroplast as 3-PGA. The oxygenase reaction is greatly reduced by lowered O_2 or raised CO_2 pressure (compared to air levels), and hence photorespiration is greatly reduced in C4 plants, CAM plants, algae and cyanobacteria which possess CO_2-concentrating mechanisms. The oxygenase activity of Rubisco may be necessary to protect the chloroplast against photo-oxidation damage when CO_2 is limiting. Alternatively, it has been suggested that

Rbu-1,5-P_2 oxygenation is an inevitable consequence of the carboxylation mechanism of Rubisco (see Chapters 16 and 18).

Regulation of the reductive pentose phosphate cycle.

The principal and ultimate regulator of chloroplast carbohydrate metabolism is light. In fulfilling its regulatory role, light is absorbed by chlorophyll and then is converted to regulatory signals that modulate selected enzymes. Such regulation is essential because enzymes for degrading carbohydrates coexist in chloroplasts with enzymes of carbohydrate synthesis. Selected biosynthetic enzymes are light-activated, whereas degradative enzymes are light-deactivated. In this way chloroplasts minimize the concurrent functioning of enzymes or pathways that operate in opposing directions ('futile cycling') and thereby maximize the efficiency of temporally disparate metabolic processes. In C3 plants, the regulatory function of light thus maintains 'enzyme order' by assuring that carbon dioxide assimilation takes place during the day, and carbohydrate degradation occurs primarily at night (Buchanan, 1980; Cséke and Buchanan, 1986). Through the provision of DHAP formed either from newly fixed carbon or the breakdown of stored starch, chloroplasts are able to supply carbon for the cytosolic synthesis of sucrose – the transport carbohydrate in most plants (see Chapters 21 and 29) – and thereby meet the energy needs of non-photosynthetic (heterotrophic) tissues at all times.

Identification of the sites of regulation

The sensitivity of a metabolic pathway to regulation resides principally in only a small number of the total steps in the pathway (ap Rees, 1980). Such regulatory steps characteristically have large, negative free energy changes (ΔG) and thus are essentially irreversible. The physiological free energy changes (ΔG^S) for the reactions of the RPP cycle were calculated by Bassham and Krause (1969) from measurements of the steady-state levels of radioactive compounds in photosynthesizing *Chlorella*. The reactions shown to be substantially displaced from equilibrium and therefore potential sites of metabolic regulation were those catalyzed by Rubisco, fructose 1,6-bisphosphatase (FBPase), sedoheptulose 1,7-bisphosphatase (SBPase) and phosphoribulokinase (PRK) (equations [17.1], [17.6], [17.9] and [17.13], respectively).

Further evidence as to the importance of these sites in the regulation of the pathway comes from the analysis of light–dark and dark–light transient changes in levels of metabolites. It would be expected that increasing flux through a regulated step would lead to depletion of the substrate for that step of the pathway, and decreasing flux would lead to a rise in the concentration of that substrate. The kinetic analysis of such experiments is complicated by the cyclic nature of the pathway, since the production of substrate for one reaction may be affected by the regulation of a subsequent step. Nevertheless, analysis confirms that the reactions catalyzed by Rubisco, FBPase, SBPase, and PRK are of greatest significance in controlling the flux through the RPP pathway (see Bassham and Buchanan, 1982 for review). As recalled from Fig. 17.1, Rubisco is a member of the carboxylation phase and FBPase, SBPase and PRK are members of the regeneration phase of the cycle.

In contrast, the enzymes involved in the reduction of 3-PGA to triose phosphate (the reduction phase) together catalyze a freely reversible oxidation/reduction the direction of which, *in vivo*, is largely determined by the levels of ATP and ADP, NADPH and $NADP^+$. In the light, with high levels of ATP and of NADPH, the reactions proceed in the direction of triose phosphate because of sustained production of 3-PGA and consumption of triose phosphate. In steady-state photosynthesis this provides for a coordination of the activity of other parts of the cycle. Any component tending to increase the activity of PRK, for example, will cause the consumption of ATP and production of ADP. The resulting deficiency of ATP in turn will slow the rate of 3-PGA reduction, leading to decreased synthesis of Rbu-5-P and bringing the cycle back into balance. It should be noted, however, that one enzyme of the reduction phase, NADP-G3PDH, is also regulated directly by light (see below).

Mechanisms of regulation

Regulation of ribulose 1,5-bisphosphate carboxylase/oxygenase

The capacity of Rubisco to carboxylate Rbu-1,5-P_2 is determined by the concentration of substrates (Rbu-1,5-P_2, CO_2 and O_2), and the amount and activity of enzyme. Under conditions of low CO_2 and high light, it is possible to show a direct correlation between Rubisco content and CO_2 fixation of spinach leaves (Seeman, 1986). During short-term changes in the rate of photosynthesis, however, modulation of the activity of the enzyme occurs (Perchorowicz et al., 1981). Activation of the enzyme involves the formation of a complex with CO_2 and the subsequent addition of a divalent metal ion (Mg^{2+}, in vivo) to form the activated ternary complex (Miziorko and Lorimer, 1983). The equilibrium of this reaction is sensitive to pH, and low pH in the stroma would be expected to lead to deactivation. Upon illumination, protons move rapidly from the stromal compartment into the thylakoids causing an increase in stromal pH from about 7.0 to 8.0. The efflux of H^+ is countered by an influx of other cations, possibly including Mg^{2+} and Ca^{2+} (Barber, 1976), and thus both the pH and cation concentration in the stroma have been proposed as being favorable for Rubisco activation in the presence of CO_2. Two other mechanisms for the activation/deactivation of Rubisco have recently been reported. Work with a mutant of *Arabidopsis* that requires a high CO_2 concentration for growth has led to the proposal of a mechanism whereby Rubisco is activated by light (Ogren et al., 1986). The mechanism, which presently appears to be unrelated to other systems of enzyme regulation, involves a newly identified protein, Rubisco activase, that links light to enzyme activity. While details of the activation mechanism are yet to be established, it presently appears that ATP and light-induced changes in the electrochemical potential of thylakoid membranes are involved (Streusand and Portis, 1987). Such a mechanism for the regulation of Rubisco by light could explain results obtained over the years by a number of different laboratories.

A second novel mechanism of Rubisco regulation involves a phosphorylated inhibitor of catalysis which can occupy the catalytic site of the enzyme. The discovery of this inhibitor in some but not other plant species followed the observation that Rubisco extracted from *Phaseolus* leaves in the light was significantly more active than from darkened leaves, despite optimal *in vitro* activation of the enzyme with CO_2 and Mg^{2+}. Several studies have shown that phosphorylated compounds can be effective inhibitors of Rubisco *in vitro*. The results with *Phaseolus*, however, are the first to document the importance *in vivo* of a compound which is light modulated and present in sufficient amounts to reduce the dark activity of Rubisco to close to zero (Seeman, 1986). The inhibitor has been identified as carboxyarabinitol 1-phosphate, an analog of the six carbon intermediate formed during catalysis (Berry et al., 1986; Gutteridge et al., 1986).

The ferredoxin/thioredoxin system

Among the more general mechanisms of light-mediated enzyme regulation the ferredoxin/thioredoxin system, involving ferredoxin, ferredoxin–thioredoxin reductase (FTR), and a thioredoxin (Buchanan, 1980; Cséke and Buchanan, 1986), is important. Thioredoxins are proteins, typically with a molecular weight of 12 kD, that are widely distributed in the animal, plant, and bacterial kingdoms. Thioredoxins undergo reversible reduction and oxidation through changes in a disulfide group (S–S → 2SH). In the ferredoxin/thioredoxin system, a thioredoxin (Td) is reduced via the iron–sulfur protein FTR, by ferredoxin (Fd), which itself is reduced by the chlorophyll system of illuminated chloroplast thylakoid membranes (equations [17.15] and [17.16]).

$$4Fd_{ox} + 2H_2O \xrightarrow{\text{Light}} 4Fd_{red} + O_2 + 4H^+ \quad [17.15]$$

$$2Fd_{red} + 2H^+ + Td_{ox} \xrightarrow{\text{FTR}} 2Fd_{ox} + Td_{red} \quad [17.16]$$

Two different thioredoxins, designated thioredoxin f and thioredoxin m, are a part of the ferredoxin/thioredoxin system in oxygenic photosynthetic organisms. In the reduced state, the two thioredoxins selectively activate enzymes of carbohydrate biosynthesis, including FBPase, SBPase and PRK. In addition, thioredoxins have been shown to activate NADP-G3PDH' and deactivate glucose 6-phosphate dehydrogenase (G6PDH) (Buchanan, 1980; Cséke and Buchanan, 1986), a regulatory enzyme of the oxidative pentose phosphate pathway, a route of carbohydrate degradation alternate to glycolysis. The ferredoxin/thioredoxin system also functions in chloroplasts in regulating other enzymes such as NADP-malate dehydrogenase (NADP-MDH) and the 'coupling factor' (CF_1-ATPase) (Mills and Mitchell, 1982). The type of thioredoxin interacting with each of these chloroplast enzymes is shown in Fig. 17.2. Cyanobacteria, C3, C4 and Crassulacean acid metabolism (CAM) plants have been shown to utilize the ferredoxin/thioredoxin system in enzyme regulation for these processes.

The ferredoxin/thioredoxin system functions by changing the sulfhydryl status of target enzymes.

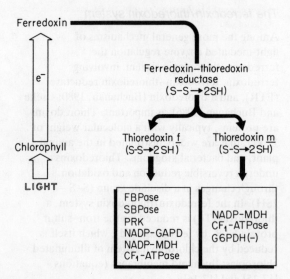

Fig. 17.2 Enzymes regulated by the ferredoxin/thioredoxin system. Role of thioredoxins in regulating phosphoglycerate kinase of C4 mesophyll cells is not indicated (cf. Cséke and Buchanan, 1986).

FBPase, which is a key regulatory enzyme of the RPP cycle, is activated by a net transfer of reducing equivalents (hydrogen) from reduced thioredoxin to enzyme disulfide (S–S) groups, thereby yielding oxidized thioredoxin f and reduced (SH) enzyme (Droux et al., 1987a). It is thought that deactivation of FBPase takes place through the oxidation (in the dark) of SH groups on reduced thioredoxin and the reduced (activated) enzyme. There is evidence that this light-dependent reduction mechanism also pertains to the activation of NADP-MDH (Droux et al., 1987b) and PRK (Porter et al., 1988).

In summary, current evidence is consistent with the view that the ferredoxin/thioredoxin system functions in photosynthetically diverse types of plants as a master switch to restrict the activity of degradatory enzymes and activate biosynthetic enzymes in the light. It is significant to note that, aside from NADP-G3PDH, enzymes controlled by the ferredoxin/thioredoxin system (FBPase, SBPase, and PRK) function in the regenerative phase of the reductive pentose phosphate cycle needed to sustain its continued operation, i.e. to regenerate the carbon dioxide acceptor, Rbu-1,5-P_2, from newly formed 3-PGA. It seems likely that one of these thioredoxin-linked enzymes limits the regeneration of Rbu-1,5-P_2.

Coordinate regulation of photosynthetic enzymes

Biochemical processes are generally regulated not by one, but by several interacting systems of regulation. From early work, it was concluded that the ferredoxin/thioredoxin system acts jointly with other light-actuated systems in achieving a particular regulatory effect, e.g. light-induced shifts in concentration of metabolite effectors and pH (Buchanan, 1980).

Since those early studies, results from a number of laboratories support such a coordinate function of the different regulatory systems. Noteworthy among the metabolite effector studies are the demonstration of the inhibition of thioredoxin-linked NADP-MDH activation by $NADP^+$, the inhibition of PRK by compounds such as 6-phosphogluconate, and the enhancement of thioredoxin-linked FBPase and SBPase activation by substrate (sugar bisphosphate) and divalent cations (Ca^{2+}, Mg^{2+}) (Cséke and Buchanan, 1986). In short, it appears that the ferredoxin/thioredoxin system functions jointly with mechanisms promoting light-dependent shifts in ions and metabolites in the regulation of a number of chloroplast enzymes (Hertig and Wolosiuk, 1983).

Considerable debate has centered on the question of whether it is the activity of Rubisco or the rate of regeneration of Rbu-1,5-P_2 (governed by the regulatory steps of the rest of the cycle, FBPase, SBPase, and PRK and by the supply of ATP and NADPH from the light reactions) that primarily sets the rate of the RPP cycle and CO_2 fixation *in vivo*. Recent results suggest that during rapid changes from high to low light, the rate of photosynthesis at subsaturating light intensity in certain plants is determined by the rate of Rbu-1,5-P_2 regeneration and not by the activity of Rubisco (Mott *et al.*, 1986). Subsequent variation in the activation state of Rubisco, however, occurs so as to match the Rbu-1,5-P_2 saturated rate of carboxylase activity to the rate of Rbu-1,5-P_2 regeneration. Most workers believe that under steady-state conditions Rubisco activity is the most important limitation on the rate of photosynthesis, even under saturating CO_2 (Dietz and Heber, 1984a,b).

Compartmentation and triose phosphate export

Because the RPP cycle is an exporter of fixed carbon, regulation of the cycle at several points may be insufficient to prevent the intermediates from being consumed by other metabolic processes. In addition to the biochemical controls discussed above, there is also compartmentation.

The chloroplast is encircled by a double membrane called the envelope. Of the two membranes, the inner is practically impermeable to hydrophilic compounds, such as P_i, phosphate esters, dicarboxylates, glucose and sucrose. Transport of certain of these metabolites is accomplished by carrier proteins, specific for groups of compounds. Individual carriers have been shown to facilitate the transport of P_i and phosphate esters, dicarboxylates, ATP and ADP, and glucose.

The carrier protein facilitating P_i and phosphate ester transport is of particular interest in leaves in connection with carbon processing, i.e. the synthesis, transport, and degradation of carbohydrate, all of which occur in the cytosol (Cséke *et al.*, 1984; Cséke and Buchanan, 1986). This metabolite carrier, called the phosphate translocator, is a polypeptide with a molecular weight of 29 000 and is a major component of the inner envelope membrane (Flügge and Heldt, 1984). The phosphate translocator mediates the countertransport of 3-PGA, DHAP and P_i. The rate of P_i transport alone is three orders of magnitude lower than with simultaneous DHAP or 3-PGA countertransport. Consequently, operation of the phosphate translocator keeps the total amount of esterified phosphate and P_i constant inside the chloroplast.

During photosynthesis, chloroplasts convert CO_2, water and P_i to triose phosphates that are exported to the cytosol. As phosphate is a substrate of this process, the continued operation of the RPP cycle is dependent on the utilization of triose phosphate for the synthesis of starch (in the chloroplast) and sucrose (in the cytosol). These synthetic processes release P_i, preventing the level of free P_i in the cell from falling to a concentration where photosynthesis may be limited by its availability. Such a limitation of photosynthesis is observed when mannose (which sequesters cytosolic P_i as mannose phosphate) is fed to a leaf (Robinson and Walker, 1981), and is suggested by the increase in CO_2 fixation detected on feeding P_i via the transpiration stream to a cut leaf (Sivak and Walker, 1986). It has long been known that isolated chloroplasts require a continuous supply of P_i in order to sustain photosynthesis.

A further ramification of the translocator-mediated exchange of exported triose phosphate and imported P_i pertains to starch synthesis. When cytosolic metabolism and P_i availability are limited, leading to a high 3-PGA/P_i ratio in the chloroplast, starch synthesis will be stimulated. The occurs because ADP-glucose pyrophosphorylase, the major regulatory enzyme in starch synthesis, is strongly activated by 3-PGA and inhibited by P_i (Preiss, 1982). As mentioned above, starch synthesis from triose phosphate will release P_i, relieving to some extent the P_i limitation of CO_2 fixation.

Coordination of CO_2 fixation and sucrose synthesis

The requirement of chloroplast photosynthesis for P_i and the release of P_i by sucrose synthesis in the cytosol mean that these two processes must be closely coordinated. This coordination is essential if photosynthesis is to continue, since a large fraction (five-sixths) of the triose phosphate produced in chloroplasts must be used to regenerate $Rbu-1,5-P_2$ to allow the continued function of the RPP cycle. Part of this coordination, as explained above, lies in the characteristics of the triose phosphate translocator and the supply of cytosolic P_i to the chloroplast (Robinson and Walker, 1981). Results obtained in the last few years have led to the identification of a regulatory metabolite that also serves this function. Fructose 2,6-bisphosphate ($Fru-2,6-P_2$) coordinates the metabolism of sucrose, starch and CO_2 fixation and, in so doing, links metabolic processes of the chloroplast with those of the cytosol.

Fructose 2,6-bisphosphate

$Fru-2,6-P_2$ occurs in plant tissues and exerts its effects on metabolism through the modulation of cytosolic FBPase and pyrophosphate, fructose 6-phosphate, 1-phosphotransferase (PFP) (see Chapters 5, 6 and 21). PFP catalyzes the reversible phosphorylation of fructose 6-phosphate (Fru-6-P) by pyrophosphate and is believed to be important in the regulation of glycolysis and gluconeogenesis (sucrose synthesis) in plant tissues.

Studies with spinach revealed that $Fru-2,6-P_2$ is present in the cytosolic fraction of photosynthetic (leaf parenchyma) cells; a PFP that is strongly activated by $Fru-2,6-P_2$ is present in the cytosol;

and that $Fru-2,6-P_2$ strongly inhibits cytosolic FBPase, an important regulatory enzyme of sucrose synthesis. $Fru-2,6-P_2$ can affect sucrose metabolism by inhibiting cytosolic FBPase and by activating PFP, an enzyme that, because of the reversibility of the reaction it catalyzes, can potentially function in both sucrose synthesis and breakdown. $Fru-2,6-P_2$ has little effect on phosphofructokinase (PKF) in plants, in contrast to its effect in animal cells. Instead, the cytosolic PFK of plants is regulated by changes in pivotal metabolites that alter enzyme activity either directly through activation or inhibition or indirectly through association or disassociation (Wong et al., 1987).

In the initial studies on plants (for review see Cséke et al., 1984; Cséke and Buchanan, 1986; Stitt, 1987), a substrate-specific fructose 6-phosphate, 2-kinase ($Fru-6-P,2K$) was identified in leaves, specifically in the cytosol fraction. Experiments revealed that the synthesis of $Fru-2,6-P_2$ by leaf $Fru-6-P,2K$ was regulated by metabolite effectors: P_i and Fru-6-P served as activators and 3-PGA and DHAP as inhibitors. Also, an enzyme was partially purified from spinach leaves that selectively hydrolyzed $Fru-2,6-P_2$ to Fru-6-P and P_i (Macdonald et al., 1987). The enzyme, designated fructose 2,6-bisphosphatase ($Fru-2,6-P_2ase$), was strongly inhibited by its products, Fru-6-P and P_i. Thus the regulation of $Fru-2,6-P_2ase$ by metabolites was found to be opposite to the regulation of $Fru-6-P,2K$ which, as noted above, is activated by the same metabolites (Fig. 17.3).

It should be noted that $Fru-6-P,2K$ and $Fru-2,6-P_2ase$ of animal tissues are regulated by phosphorylation via a cAMP-dependent protein kinase that, in turn, is regulated hormonally. So

Metabolite	Fru-6-P,2K	$Fru-2,6-P_2ase$
Pi	Activator	Inhibitor
Fru-6-P	Activator	Inhibitor
DHAP	Inhibitor	No effect
3-PGA	Inhibitor	No effect
PPi	Inhibitor	No effect
Mg^{2+}	Cofactor	Inhibitor

Fig. 17.3 Principal metabolites regulating Fru-6-P,2K and $Fru-2,6-P_2ase$ in leaves.

far, there is no conclusive evidence that Fru-6-P,2K and Fru-2,6-P_2ase in plants are regulated by phosphorylation physiologically. However, recent evidence suggests that plant Fru-6-P,2K is regulated covalently in addition to its regulation by metabolites (Stitt, 1987), perhaps through an ATP-dependent modification (Walker and Huber, 1987).

As discussed above, the P_i released in sucrose synthesis is recycled to the chloroplast via the phosphate translocator, in strict counter-exchange for triose phosphate (Flügge and Heldt, 1984). 3-Phosphoglycerate can also be transported by this same carrier, though its export from the chloroplast in the light is restricted by a pH-dependent change in the charge on the molecule.

It is thus obvious that the metabolites modulating Fru-6-P,2K and Fru-2,6-P_2ase occupy strategic positions in the pathway of sucrose synthesis in leaves. Extensive export of triose phosphates by chloroplasts into the cytosolic C3 pool would lower the Fru-2,6-P_2 concentration (by inhibiting Fru-6-P,2K) and thereby promote the use of photosynthate for sucrose synthesis by relieving the Fru-2,6-P_2-linked inhibition of cytosolic FBPase. On the other hand, elevated levels of P_i (e.g. in the dark) or Fru-6-P (e.g. as sucrose accumulated in the leaf) would tend to raise the Fru-2,6-P_2 concentration, and thus restrict sucrose synthesis or favor sucrose degradation.

The role of chloroplasts in controlling the content of Fru-2,6-P_2 in the cytosol via export and import of central metabolites is illustrated diagrammatically in Fig. 17.4.

Changes in DHAP concentration and the accompanying alteration of the Fru-2,6-P_2 concentration thus provide a mechanism to coordinate sucrose synthesis in the cytosol with the rate of carbon dioxide fixation in chloroplasts. This coordination is necessary to prevent the inhibition of photosynthesis that would result from either an excessive drain of metabolites from the chloroplast or from an inadequate release of P_i in the cytosol to sustain photophosphorylation in the chloroplast.

Other endproducts of CO_2 assimilation

Ultimately, all the fixed carbon which makes up plant cells is derived from CO_2 that is reduced by the RPP cycle. As stated above, the major endproducts of CO_2 fixation in a mature leaf are sucrose and starch. A number of other compounds, however, are derived so directly from intermediates of the cycle that their synthesis may be light dependent and these compounds may account for a significant percentage of the fixed carbon in developing tissues.

In unicellular green algae like *Chlorella* (Calvin and Bassham, 1962), as much as one-third of the carbon fixed during a one hour exposure to $^{14}CO_2$ is found in amino acids, and not in sucrose as is the case for higher plants (which produce sucrose as an easily transported carbohydrate). In the algae, the major amino acids labeled are glycine and serine. In addition, the label appears in alanine (from pyruvate) and aspartate, asparagine and glutamate, glutamine from oxaloacetate and 2-oxoglutarate of the tricarboxylic acid cycle. These compounds originate from the cytosolic metabolism of triose phosphates exported from the chloroplast. Though quantitatively insignificant in most C3 plants (with the exception of photorespiratory production of glycine and serine), such synthesis is clearly of great importance and is probably tightly controlled. There is evidence

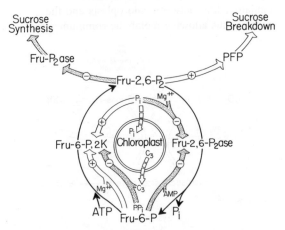

Fig. 17.4 Role of chloroplasts and effector metabolites in Fru-2,6-P_2-linked control of cytosolic sucrose transformations in spinach leaves. Fru-P_2ase is equivalent to FBPase. Regulation of Fru-6-P,2K and Fru-2,6-P_2ase is indicated by '+' for activation and '−' for inhibition.

of feedback inhibition of the pathways of amino acid production.

In addition to the cytosolic metabolism described above a number of products may be produced wholly within chloroplasts. These include the aromatic amino acids (tryptophan, tyrosine and phenylalanine) and plastoquinone derived from the shikimic acid pathway, the pyruvate-derived amino acids, fatty acids and terpenes produced from acetyl-CoA, and nucleic acids (Schulze-Siebert *et al.*, 1984). As with cytosolic metabolism, such products are quantitatively insignificant in comparison to export to the cytosol and starch production in the plastid, but are nevertheless of vital importance to the plant.

The existence in chloroplasts of complete pathways leading from the RPP cycle to the above mentioned compounds is a matter of debate and their synthesis at least in some plant species may require the cooperation of the cytosol and plastids. Pea chloroplasts, for example, appear not to contain phosphoglyceromutase activity which would be required for the production of PEP and pyruvate from photosynthetically generated PGA (ap Rees, 1985). Spinach chloroplasts, which apparently do contain a complete pathway leading from 3-PGA to PEP, pyruvate and acetyl-CoA, presumably face the problem of control so as to prevent metabolites being drained from the cycle during photosynthesis. Little is known of the control of this or other biosynthetic pathways in the chloroplast.

Concluding remarks

The mechanism of carbon dioxide assimilation by the RPP cycle has been known for three decades. During this interval, it has been established that light functions not only to produce ATP and $NADPH_2$ to drive the cycle, but also to regulate selected enzymes. In oxygen-evolving systems (chloroplasts and cyanobacteria), light absorbed by chlorophyll is converted to several different regulatory signals – changes in pH, cations, metabolite effectors, and sulfhydryl groups – that collectively interact to 'inform' selected enzymes that the light is on or off and that their activities should be altered accordingly. In the case of sulfhydryl changes, the light signal is carried from chlorophyll containing thylakoid membranes via ferredoxin to thioredoxins, which, through redox changes in their own sulfhydryl groups, bring about reversible changes in the sulfhydryl status of target enzymes, thereby altering key activities and directing major biosynthetic and degradatory pathways in the appropriate direction. With certain enzymes, the light-produced alkalization of the chloroplast stroma and increase in the concentration of cations and selected metabolite effectors enhance the sulfhydryl changes. By linking these regulatory changes to light, the cell is in command of its biosynthetic and degradatory capabilities at all times and can direct available resources to increase growth and survival under a wide range of environmental conditions. It is significant that photosynthetic bacteria (anaerobic photosynthetic organisms that lack the ability to evolve oxygen) seemingly do not regulate their metabolic processes in this manner.

In higher plants, which utilize photosynthetically fixed carbon to form transportable sugars such as sucrose, the photomodulation systems of chloroplasts interact with a newly discovered metabolite-directed system of enzyme regulation of the cytosol (Fig. 17.5). Here, Fru-2,6-P_2 plays a key role. In leaves, Fru-2,6-P_2 acts as a regulatory link between chloroplasts and the cytosol, thus allowing metabolic communication

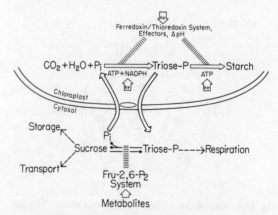

Fig. 17.5 Relation of carbon processing in the cytosol to photosynthetic carbon dioxide assimilation in chloroplasts. The dual function of light in supplying ATP and $NADPH_2$ and in regulation is shown.

between these compartments, and signalling changes in environmental conditions so that carbon processing, i.e. the synthesis, degradation and transport of carbohydrate in the cytosol, can be adjusted in accordance with needs of the plant. In performing its function, Fru-2,6-P_2 acts at several levels, i.e. sucrose synthesis (FBPase), sucrose degradation (PFP regulation), and the related process of carbon partitioning (accumulation of photosynthetically fixed carbon as sucrose versus starch).

The evidence at hand is thus in accord with the view that the Fru-2,6-P_2 system connects cytosolic carbohydrate metabolism with the light-directed regulatory mechanisms of chloroplasts, and with other regulatory signals significantly altering cytosolic metabolite status. This role of Fru-2,6-P_2 as an environmental sensor enables plants to make effective use of available energy for processes taking place either in leaves or in distal sink tissues.

Acknowledgments

Work from the authors' laboratory was supported by grants from the National Science Foundation, Competitive Research Grants office of the US Department of Agriculture, National Space and Aeronautics Administration, and Dow Chemical Company.

References

ap Rees, T. (1980). Integration of pathways of synthesis and degradation of hexose phosphates. In *The Biochemistry of Plants*, Vol. 3, ed. J. Preiss, Academic Press, New York, pp. 1–42.

ap Rees, T. (1985). The organization of glycolysis and the oxidative pentose phosphate pathway in plants. In *Encyclopedia of Plant Physiology*, Vol. 18, eds R. Douce and D. Day, Springer-Verlag, Berlin, pp. 391–417.

Barber, J. (1976). Cation control in photosynthesis. *Trends Biochem. Sci.* **1**, 33–6.

Bassham, J. A. and Buchanan, B. B. (1982). Carbon dioxide fixation pathways in plants and bacteria. In *Photosynthesis*, ed. Govindjee, Vol. II, Academic Press, New York, pp. 141–89.

Bassham, J. A. and Krause, G. H. (1969). Free energy changes and metabolic regulation in steady-state photosynthetic carbon reduction. *Biochim. Biophys. Acta* **189**, 207–21.

Berry, J. A., Lorimer, G. H., Pierce, J., Seeman, J. R., Meek, J. and Freas, S. (1986). Isolation, identification and synthesis of 2-carboxyarabinitol 1-phosphate, a diurnal regulator of ribulose-bisphosphate carboxylase activity. *Proc. Natl Acad. Sci. USA* **84**, 734–8.

Buchanan, B. B. (1980). Role of light in the regulation of chloroplast enzymes. *Ann. Rev. Plant Physiol.* **31**, 341–374.

Calvin, M. and Bassham, J. A. (1962). *The Photosynthesis of Carbon Compounds*, Benjamin, New York.

Cséke, C. and Buchanan, B. B. (1986). Regulation of the formation and utilization of photosynthate in leaves. *Biochim. Biophys. Acta* **853**, 43–64.

Cséke, C., Balogh, A., Wong, J. H., Buchanan, B. B., Stitt, M., Herzog, B. and Heldt, H. W. (1984). Fructose-2,6-bisphosphate: a regulator of carbon processing in leaves. *Trends Biochem. Sci.* **9**, 533–5.

Dietz, K. J. and Heber, U. (1984a). Rate-limiting factors in leaf photosynthesis 1. Carbon fluxes in the Calvin cycle. *Biochim. Biophys. Acta* **767**, 432–43.

Dietz, K. J. and Heber, U. (1984b). Rate-limiting factors in leaf photosynthesis 2. Electron transport. *Biochim. Biophys. Acta* **767**, 444–50.

Droux, M., Crawford, N. and Buchanan, B. B. (1987a). Mechanism of thioredoxin-linked activation of chloroplast fructose-1,6-bisphosphatase. *C. R. Acad. Sci. Paris, III* **305**, 335–41.

Droux, M., Miginiac-Maslow, M., Jacquot, J.-P., Gadal, P., Crawford, N. A., Kosower, N. S. and Buchanan, B. B. (1987b). Ferredoxin–thioredoxin reductase: a catalytically active dithiol group links photoreduced ferredoxin to thioredoxin functional in photosynthetic enzyme regulation. *Arch. Biochem. Biophys.* **256**, 372–80.

Flügge, U. I. and Heldt, H. W. (1984). The phosphate–triose phosphate–phosphoglycerate translocator of the chloroplast. *Trends Biochem. Sci.* **9**, 530–3.

Gutteridge, S., Parry, M. A. R., Burton, S., Keys, A. J., Mudd, A., Feeny, J., Servaites, J. C. and Pierce, J. (1986). A nocturnal inhibitor of carboxylation in leaves. *Nature* **324**, 274–6.

Hertig, C. M. and Wolosiuk, R. A. (1983). Studies on the hysteretic properties of chloroplast fructose-1,6-bisphosphatase. *J. Biol. Chem.* **258**, 984–9.

MacDonald, F. D., Cseke, C., Chou, Q. and Buchanan, B. B. (1987). Activities synthesizing and degrading fructose-2,6-bisphosphate in spinach leaves reside on different proteins. *Proc. Natl Acad. Sci. USA* **84**, 2742–6.

Mills, J. D. and Mitchell, P. (1982). Thiol modulation of

CF_0-CF_1 stimulates acid/base dependent phosphorylation of ADP by broken pea chloroplasts. *FEBS Lett.* **144**, 63–7.

Miziorko, H. M. and Lorimer, G. H. (1983). Ribulose-1,5-bisphosphate carboxylase-oxygenase. *Ann. Rev. Biochem.* **52**, 507–35.

Mott, K. A., Jensen, R. G. and Berry, J. A. (1986). Limitation of photosynthesis by RuBP regeneration rate. In *Biological Control of Photosynthesis*, eds R. Marcelle, H. Clijsters and M. Van Pouke, Martinus Nijhoff, Dordrecht, pp. 33–43.

Ogren, W., Salvucci, M. and Portis, A. (1986). The regulation of Rubisco activity. *Phil. Trans. R. Soc. London B* **313**, 337–46.

Perchorowicz, J. T., Raynes, D. A. and Jensen, R. G. (1981). Light limitation of photosynthesis and activation of ribulose bisphosphate carboxylase in wheat seedlings. *Proc. Natl Acad. Sci. USA* **78**, 2985–9.

Porter, M. A., Stringer, C. D. and Hartman, F. C. (1988). Characterization of the regulatory thioredoxin site of phosphoribulokinase. *J. Biol. Chem.* **263**, 123–9.

Preiss, J. (1982). Regulation of biosynthesis and degradation of starch. *Ann. Rev. Plant Physiol.* **33**, 431–54.

Robinson, S. P. and Walker, D. A. (1981). Photosynthetic carbon reduction cycle. In *The Biochemistry of Plants*, Vol. 8, Academic Press, New York, pp. 194–236.

Schulze-Siebert, D., Heinke, D., Scharf, H. and Schultz, G. (1984). Pyruvate-derived amino acids in spinach chloroplasts. *Plant Physiol.* **76**, 465–71.

Seeman, J. R. (1986). Mechanisms for the regulation of CO_2 fixation by ribulose-1,5-bisphosphate carboxylase. In *Biological Control of Photosynthesis*, eds R. Marcelle, H. Clijsters and M. Van Poucke, Martinus Nijhoff, Dordrecht, pp. 71–82.

Sivak, M. N. and Walker, D. A. (1986). Summing up: measuring photosynthesis *in vivo*. In *Biological Control of Photosynthesis*, eds R. Marcelle, H. Clijsters and M. Van Poucke, Martinus Nijhoff, Dordrecht, pp. 1–31.

Stitt, M. (1987). Fructose-2,6-bisphosphate and plant carbohydrate metabolism. *Plant Physiol.* **84**, 201–4.

Streusand, V. and Portis, A. (1987). Rubisco activase mediates ATP-dependent activation of ribulose bisphosphate carboxylase. *Plant Physiol.* **85**, 152–4.

Walker, G. and Huber, S. (1987). ATP-dependent activation of a new form of spinach leaf 6-phosphofructo-2-kinase/fructose-2,6-bisphosphatase. *Arch. Biochem. Biophys.* **258**, 58–64.

Wong, J., Yee. B. C. and Buchanan, B. B. (1987). A novel type of phosphofructokinase from plants. *J. Biol. Chem.* **262**, 3185–91.

18 Photorespiration and CO₂-concentrating mechanisms
David T. Canvin

Introduction

Photorespiration is light-dependent CO_2 evolution and O_2 uptake in green leaves (Fig. 18.1). In C3 plants, due to the existence of photorespiration, photosynthesis is inhibited by oxygen, a finite CO_2 compensation point of about 50 μl CO_2 l^{-1} exists and a burst of CO_2, called the post-illumination burst (PIB), is observed when the light is extinguished. Warburg observed the inhibition of photosynthesis by oxygen in 1920, the post-illumination burst was described in 1955, and the competition between O_2 and CO_2, as expressed in the direct dependence of the compensation point on O_2 concentration, was observed in 1966.

During the 1950s and 1960s, the extensive synthesis of glycolate by leaves in high O_2 concentrations or low CO_2 concentrations was demonstrated and the enzymatic pathway, now called the photosynthetic carbon oxidation (PCO) cycle, was elucidated. The origin of the glycolate was established and the gas exchange results were reconciled with the biochemistry of glycolate metabolism by the discovery (Bowes et al., 1971) that the CO_2 fixing enzyme, ribulose 1,5-bisphosphate carboxylase (Rubisco) also displays oxygenase activity. CO_2 and O_2 are competitive substrates for the enzyme (see Chapter 16). The addition of CO_2 to ribulose 1,5-bisphosphate (RuBP) by the enzyme yields two molecules of 3-phosphoglyceric acid (3-PGA), whereas the addition of O_2 to RuBP yields a molecule of 3-PGA and a molecule of phosphoglycolate. Glycolate is formed by the removal of the phosphate. Subsequent studies on

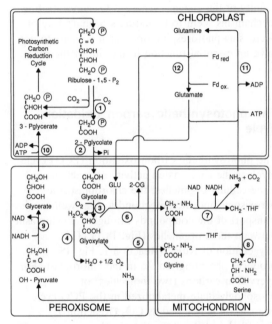

Fig. 18.1 The photosynthetic carbon oxidation cycle. The circled numbers refer to the enzyme or reaction: (1) ribulose 1,5-bisphosphate carboxylase/oxygenase, (2) phosphoglycolate phosphatase, (3) glycolate oxidase, (4) catalase, (5) serine:glyoxylate aminotransferase, (6) glutamate:glyoxylate aminotransferase, (7) glycine decarboxylase, (8) serine hydroxymethyltransferase, (9) hydroxypyruvate reductase, (10) glycerate kinase, (11) glutamine synthetase, (12) glutamate synthase.

the incorporation of oxygen into glycolate and other PCO cycle intermediates and the isolation of mutants deficient in PCO cycle enzymes definitively established the mechanics of photorespiration.

It was also apparent, however, that certain plants, such as CAM plants and C4 plants, had developed mechanisms to avoid oxygenase activity or the effect of oxygen on photosynthesis. In these plants elaborate biochemical and anatomical developments have provided a means for elevating the CO_2 concentration in the vicinity of Rubisco, thereby decreasing oxygenase activity. Microalgae and cyanobacteria have adopted a more direct approach by developing systems for the active transport of both CO_2 and HCO_3^- which also provide higher concentrations of CO_2 in the vicinity of the enzyme. These 'CO_2-concentrating' mechanisms seem to be nature's primary way of decreasing photorespiration, which appears to not have an essential function in the life of the plant.

The photosynthetic carbon oxidation cycle

The photosynthetic carbon oxidation (PCO) cycle involves the coordinated activity of reactions in three cellular organelles, the chloroplast, the peroxisome and the mitochondrion, as well as transport of the compounds through the cytoplasm between these organelles (Fig. 18.1). The oxygenation of RuBP (Reaction 1, Fig. 18.1) produces phosphoglycolate in the chloroplast. For every four carbons (two molecules) of phosphoglycolate metabolized in the PCO cycle, one carbon is released as CO_2 (Reaction 7, Fig. 18.1) and the other three carbons are returned to the photosynthetic carbon reduction (PCR) cycle (Reaction 10, Fig. 18.1). Hence, two molecules of O_2 are consumed in the oxygenase reaction for every CO_2 released, or the ratio between oxygenation and CO_2 evolution in photorespiration is 0.5. If total O_2 uptake, including that consumed by the oxidation of two glycolate molecules (Reaction 3, Fig. 18.1) is considered then three O_2 will be taken up for every CO_2 released.

This complex cycle, which can only function completely in intact cells, has been elucidated and formulated through a variety of approaches, including isolation and characterization of the various enzymes and isotope experiments with leaves (see Lorimer and Andrews, 1981). About 1950, glycolate was identified as an early product of photosynthesis and shown to be uniformly labeled during photosynthesis in $^{14}CO_2$. Uniform labeling is consistent with label distribution in the intermediates of the PCR cycle and the origin of glycolate from carbons 1 and 2 of RuBP through the oxygenation reaction. By using specifically labeled intermediates (e.g. 1-^{14}C-glycolate, 2-^{14}C glycolate, 2-^{14}C glycine, etc.) supplied to leaves in the light, it was shown that glycolate was directly converted to glycine. Carbon-1 of glycine was released as CO_2 and serine was derived from one complete molecule of glycine plus carbon-2 of another glycine molecule. When oxygenation occurs in the presence of $^{18}O_2$, one atom of ^{18}O ends up as water but one atom of ^{18}O is incorporated into the carboxyl group of phosphoglycolate and is retained in the glycine, serine and glycerate produced in subsequent reactions. By using intact isolated chloroplasts, the measured ^{18}O enrichment in glycolate was over 90% of the ^{18}O content supplied, showing that all, or essentially all, the glycolate was derived from the oxygenase reaction of Rubisco.

The initial reaction of photorespiration and the PCO cycle is the oxygenase reaction of Rubisco. The products of this reaction are 2-phosphoglycolate and 3-phosphoglycerate when O_2 is the substrate (Reaction 1, Fig. 18.1), whereas two molecules of 3-phosphoglycerate are produced when CO_2 is the substrate. CO_2 and O_2 are competitive substrates for the enzyme and the relative amounts of each that react will depend upon the relative concentrations of each gas at the active site of the enzyme and the kinetic properties of the enzyme. For higher C3 plant Rubisco, the K_m (CO_2) value is 10–20 μM and the K_m (O_2) value is 400–600 μM. A more detailed treatment of Rubisco can be found in Chapter 16 and more will be said about the rates of carboxylation and oxygenation later.

Phosphoglycolate, the first product of the oxygenase activity of Rubisco, is converted to glycolate and inorganic phosphate in the chloroplast by phosphoglycolate phosphatase (Reaction 2, Fig. 18.1). The enzyme has been found in a number of C3 and C4 plants and algae and has been purified from several sources (see

Husic et al., 1987). The molecular weight of the enzyme from spinach is 93 kD; it has a broad pH (6.0–8.5) optimum and requires a divalent cation and monovalent anion for activity. The enzyme has a high degree of specificity for phosphoglycolate and does not hydrolyze other phosphate esters that are normally present in the chloroplast. Definitive evidence for the importance of this enzyme in the PCO cycle was provided by Somerville and Ogren who isolated a mutant of *Arabidopsis thaliana* deficient in this enzyme (see Artus et al., 1986). Under non-oxygenating conditions (e.g. 1% CO_2) the mutant grew normally and the enzyme was not essential, but under oxygenating conditions (0.03% CO_2) phosphoglycolate accumulated. Photosynthesis was inhibited, presumably because phosphoglycolate inhibits triose phosphate isomerase.

The glycolate that is formed moves to the cytoplasm and enters the peroxisome. As the cell organelles move about in the living cell and as peroxisomes are frequently seen appressed to the chloroplasts, there may also be the possibility of direct transfer of glycolate across the appressed chloroplast and peroxisomal membranes without transit through the cytoplasm. There is now evidence for a glycolate transporter (which also transports glycerate) on the inner membrane of the chloroplast, but the mechanism by which glycolate crosses the peroxisomal membrane is not known, because it has been technically difficult to isolate intact peroxisomes for study.

In the peroxisome, glycolate is oxidized by a flavoprotein, glycolate oxidase (Reaction 3, Fig. 18.1). A molecule of O_2 is taken up and the products of the reaction are glyoxylate and H_2O_2. The H_2O_2 is broken down by catalase (Reaction 4, Fig. 18.1) to H_2O and 0.5 O_2 so that the net uptake of O_2 in the oxidation of one molecule of glycolate is 0.5 O_2. A catalase-deficient mutant of barley grew poorly in normal air with death of the older leaves, but grew well in 0.2% CO_2 (Kendall et al., 1983). The reported molecular weights of glycolate oxidase range from 93 000 to 199 000 from various sources, the pH optimum ranges from 7.8 to 8.5, the K_m (O_2) ranges from 76 to 130 μM and the K_m (glycolate) is around 0.25 to 0.4 mM. Glycolate oxidase contains flavin mononucleotide (FMN) which, in the reduced form, is oxidized by molecular O_2. The enzyme will also oxidize other α-hydroxy acids, including L-lactate. The enzyme is not inhibited by CN^- but is inhibited by a number of compounds, of which α-hydroxypyridine methane sulfonate (α-HPMS) and 2-hydroxy-3-butynoate (HBA) have been most widely used. The results of the use of these compounds for *in vivo* studies, however, should be interpreted with caution as they seem to inhibit a number of reactions.

Cyanobacteria and microalgae have, instead of the higher plant glycolate oxidase described above, a glycolate dehydrogenase that is inhibited by CN^- and oxidizes D-lactate (not L-lactate) in addition to glycolate. The natural electron acceptor is not known and the enzyme is assayed with artificial electron acceptors such as 2,6-dichlorophenol-indophenol (DCPIP) or phenazine methosulfate (PMS). In the green algae the enzyme appears to be localized in the mitochondria; in the cyanobacteria it appears to be associated with the thylakoids. The oxidation of glycolate in these organisms would appear to be linked to the respiratory electron transport chain.

The glyoxylate that is produced from the oxidation of glycolate is aminated by two aminotransferases, serine:glyoxylate aminotransferase (reaction 5, Fig. 18.1) and glutamate:glyoxylate aminotransferase (reaction 6, Fig. 18.1). The enzymes are essentially irreversible and both are necessary as two molecules of glycine must be produced to form one molecule of serine. From *in vitro* studies where both aminotransferases are present, serine is the preferred amino donor and this will ensure the production of hydroxypyruvate and the continued movement of carbon through the PCO cycle. The serine:glyoxylate aminotransferase is relatively specific for serine, although asparagine may function as a donor in pea. The glutamate:glyoxylate aminotransferase will also use alanine in addition to glutamate. In a Fd-GOGAT mutant of *Arabidopsis*, alanine is also used up along with glutamate under photorespiratory conditions, but the relative use of alanine versus glutamate is not known for normal conditions.

That the serine:glyoxylate aminotransferase is essential for the operation of the PCO cycle was demonstrated by the isolation of several mutants of *Arabidopsis* deficient in this enzyme. The plants grew under high CO_2 conditions but died under photorespiratory conditions. Glycine and serine accumulated as the glutamate:glyoxylate aminotransferase was still functional but little metabolism of serine could be demonstrated by a chase experiment of $^{12}CO_2$ after $^{14}CO_2$ fixation.

The aminotransferases are poorly inhibited with isonicotinyl hydrazide (INH) but strongly inhibited with hydroxylamine and aminooxyacetate (AOA). In algae inhibition with AOA results in glycolate excretion, but it has not been very useful for studies with higher plants.

In vitro, glyoxylate reacts non-enzymatically with H_2O_2 to produce formate and CO_2, and formate could be converted to CO_2 by an NAD-formate dehydrogenase in the mitochondria. It has been suggested that these reactions could produce at least part of the CO_2 observed in photorespiration. Indeed, a mutant of *Arabidopsis* deficient in serine transhydroxymethylase activity and hence unable to metabolize glycine did not accumulate glyoxylate but apparently metabolized it to CO_2 after all the amino donors in the cell were exhausted. However, if NH_3 (and serine) was supplied, the CO_2 evolution was prevented, which demonstrates that if amino donors are available the glyoxylate is aminated to glycine rather than being oxidized to CO_2. This view is further supported by results with catalase-deficient barley (Kendall *et al.*, 1983) where much higher concentrations of H_2O_2 than normal did not result in increased CO_2 production. Hence, with the possible exception of the situation in which there are no amino donors, glyoxylate is converted to glycine and is not oxidized and hence does not contribute to photorespiratory CO_2.

The glycine generated in the peroxisome moves to the mitochondria. The mechanism of uptake into the mitochondrion is not clear although there is some evidence for a glycine transporter in the inner mitochondrial membrane. In the mitochondria, glycine is metabolized by two enzymes, glycine decarboxylase (reaction 7, Fig. 18.1) and serine transhydroxymethylase (reaction 8, Fig. 18.1) in an overall reaction that converts two molecules of glycine to one molecule of serine and one molecule each of NADH, CO_2 and NH_3. There now seems to be little disagreement that this reaction produces the CO_2 that is observed in photorespiration. It also releases NH_3 that must be reassimilated for the continued operation of the PCO cycle.

Glycine decarboxylase activity has largely been studied with intact mitochondria, although now some results on the isolated enzyme complex have been reported. By analogy to the bacterial or vertebrate enzyme, the plant enzyme complex is thought to consist of four components. The first protein component decarboxylates glycine and also catalyzes the exchange of CO_2 with the carboxyl group of glycine. The second protein acts as a carrier for the methylamine and contains lipoic acid which acts as the electron acceptor for the electrons from glycine. The third component transfers the aminomethyl moiety to tetrahydrofolic acid (THF) with the release of NH_3 to form N^5,N^{10}-methylene THF. The fourth protein catalyzes the transfer of the electrons from reduced lipoic acid to NAD^+ to form NADH. The enzyme can be inhibited by INH and aminoacetonitrile (AAN), although AAN has also been reported to inhibit the uptake of glycine by the mitochondria.

Serine hydroxymethyltransferase activity results in the formation of serine through the transfer of the C_1 unit from N^5,N^{10}-methylene THF to another molecule of glycine. In addition to the mitochondrial enzyme, there is some evidence that an isozyme of the enzyme is present in the chloroplast. A serine hydroxymethyltransferase mutant of *Arabidopsis*, under non-photorespiratory conditions, did not require serine for growth or another source of C_1 units for other biosynthetic reactions. The enzyme is reversible but in leaves appears to function largely in the direction of serine synthesis.

Mutants deficient in glycine decarboxylase or serine hydroxymethyltransferase have been isolated in *Arabidopsis* and barley. In both cases, under photorespiratory conditions glycine accumulates and photosynthesis is inhibited for reasons which are not yet clear.

Only mitochondria from green plant tissues contain glycine decarboxylase and in *in vitro* mitochondria preferentially oxidize glycine as compared with other tricarboxylic acid cycle intermediates. In isolated mitochondria, in the absence of other electron acceptors, the NADH can be oxidized by the respiratory electron transport chain, with one molecule of O_2 taken up for every two molecules of CO_2 released. The oxidation of NADH involves all three coupling sites of the chain. An alternative mechanism of NADH oxidation results in the reduction of oxaloacetate (OAA) to malate in the mitochondria and the oxidation of this malate in the peroxisome or elsewhere in the cell will be discussed later. This latter mechanism would not result in any O_2 uptake or phosphorylation during the decarboxylation of glycine.

The serine that was produced in the mitochondria by the action of serine transhydroxymethylase moves to the peroxisome where the amino group is removed by the serine:glyoxylate aminotransferase (reaction 5, Fig. 18.1) and hydroxypyruvate is produced. The hydroxypyruvate is reduced by hydroxypyruvate reductase to glycerate using NADH as the electron donor (reaction 9, Fig. 18.1). Hydroxypyruvate reductase has been isolated from several plants and microalgae but no activity is found in cyanobacteria. The molecular weight of the enzyme is 91 000 to 97 000, the pH optimum is 6.0 to 7.4 and the K_m (hydroxypyruvate) for the spinach enzyme is 120 μM. The enzyme will also reduce glyoxylate.

Reducing equivalents must be imported into the peroxisome for the reduction of hydroxypyruvate, as a reaction generating NADH is not part of the major metabolic reactions of this organelle. NADH may be directly imported from the cytoplasm as peroxisomes may be permeable to nucleotides. However, peroxisomes do contain an isozyme of malate dehydrogenase, and import of reducing equivalents as malate is another possibility. Malate would be oxidized to oxaloacetate (OAA) with the production of NADH, and the OAA would leave the peroxisome.

The malate that is required for the above system could be generated in the mitochondria using the NADH produced from glycine decarboxylation to reduce OAA. Addition of oxaloacetate (OAA) to mitochondria metabolizing glycine severely inhibits oxygen uptake and greatly stimulates glycine oxidation (McC. Lilley *et al.*, 1987). The mitochondria possess a high affinity ($K_m = 7$ μM) transporter for OAA, and the observed effects are presumably due to the reduction of OAA to malate by malate dehydrogenase in the mitochondria and the reoxidation of NADH by that means rather than by the transfer of the electrons to O_2. The lowering of the NADH level relieves the inhibition of the glycine decarboxylase by this reduced nucleotide. It does not seem that the reoxidation of NADH *per se* by the respiratory electron transport chain is limiting, even in the presence of glycine oxidation, since the addition of another tricarboxylic acid cycle intermediate can stimulate O_2 uptake. The malate that is produced could leave the mitochondria and move through the cytosol to the peroxisome where it could be used to generate the NADH for the reduction of hydroxypyruvate to glycerate. This malate/OAA shuttle seems to be a feasible mechanism by which the production of NADH for the reduction of hydroxypyruvate is linked to the oxidation of glycine, another intermediate in the PCO cycle.

A shuttle involving the movement of OAA as aspartate (formed by transamination with glutamate) from the peroxisome to the mitochondria can also be formulated. This shuttle would also require the concurrent movement of 2-oxoglutarate and glutamate, but the rate of movement of 2-oxoglutarate into plant mitochondria is slow compared with oxaloacetate (Zoglowek *et al.*, 1988). Chloroplasts also possess a high affinity OAA transporter, and a malate/OAA shuttle between the chloroplast and the peroxisome is another means by which reducing equivalents could be supplied to the peroxisome. Such a reaction would, of course, affect the amount of NADH from glycine oxidation that could be reoxidized by the malate/OAA shuttle between the mitochondria and peroxisomes.

In vivo, it is not possible to determine the source of NADH for the reduction of hydroxypyruvate. The malate/OAA shuttle between the mitochondria and peroxisome conveniently balances the production and utilization of reducing equivalents in the cycle. It is possible that both the malate/OAA shuttle and oxidation through the respiratory electron transport chain contribute to the reoxidation of the NADH produced in glycine decarboxylation. If NADH from glycine decarboxylation is reoxidized by O_2, then NADH must be supplied to the peroxisome from some other source to satisfy the requirements for the reduction of hydroxypyruvate.

The glycerate that is produced leaves the peroxisome and is transported into the chloroplast via the glycerate (glycolate) transporter. In the chloroplast, glycerate kinase catalyzes the phosphorylation of glycerate (reaction 10, Fig. 18.1) to 3-phosphoglycerate, and the carbon re-enters the phosphorylated carbon pool of the photosynthetic carbon reduction cycle. Glycerate kinase is entirely located in the chloroplast and in spinach has a molecular weight of about 40 000, a pH optimum of 6.5 to 8.5 and a K_m (D-glycerate) of about 0.2 mM.

Photorespiratory nitrogen cycle

During the decarboxylation of glycine (reaction 7, Fig. 18.1), NH_3 is produced in stoichiometric amounts with the photorespiratory CO_2. As the rate of this NH_3 production in C3 plants can be 20 times the rate of primary nitrate assimilation, the plant could certainly not afford to lose this NH_3 as a volatile gas but must have an effective mechanism for its reassimilation. The reassimilation of this NH_3 occurs through the concerted action of glutamine synthetase (GS) (reaction 11, Fig. 18.1) and glutamate synthase (reaction 12, Fig. 18.1); the latter enzyme is also called glutamine:oxoglutarate aminotransferase (GOGAT) and the system is referred to as the GS/GOGAT system.

Two isozymes of GOGAT are exclusively located in the chloroplast. The major enzyme is a ferredoxin-linked GOGAT (Fd-GOGAT) and the minor enzyme, which occurs at 3% or less of the total activity of the former enzyme, uses NAD(P)H in place of ferredoxin as the electron donor. The GOGAT catalyzes the formation of two molecules of glutamate according to the following equation:

glutamine + 2-oxoglutarate + Fd_{red} → 2 glutamate + Fd_{ox}

The Fd-GOGAT from *Vicia faba* is reported to have a molecular weight of 150 kD, a pH optimum of 7.4, a K_m (Fd) of 2 μM and a K_m (glutamate) of 300 μM and a K_m (2-oxoglutarate) of 150 μM. Azaserine and 6-diazo-5-oxo-L-norvaline (DON), which are glutamine analogs, strongly inhibit the enzyme.

Fd-GOGAT is essential for the reassimilation of photorespiratory NH_3, as mutants of barley or *Arabidopsis* lacking this enzyme accumulate large quantities of NH_3 under photorespiratory conditions and eventually die. Under non-photorespiratory conditions the activity of the NAD(P)H GOGAT would seem to be sufficient for the assimilation of NH_3 produced from the reduction of nitrate.

Many plants also possess two isozymes of glutamine synthetase, one located in the chloroplast and one located in the cytoplasm. The chloroplast isozyme is the predominant activity and not all plants (e.g. spinach) possess the cytoplasmic form. Glutamine synthetase catalyzes the formation of glutamine according to the following equation:

gluamate + NH_3 + ATP $\xrightarrow{Mg^{2+}}$ glutamine + ADP + P_i

Glutamine synthetase has a molecular weight in the 350 000–400 000 range, a pH optimum in the region of 8, a K_m (glutamate) of 2–13 mM, a K_m (ATP) of 0.1–1.5 mM and a strikingly high affinity for ammonia, K_m (10–20 μM).

Assimilation of NH_3 into glutamine could occur in the cytoplasm (in those plants with the cytoplasmic isozyme) with the subsequent transport of glutamine to the chloroplast. But the transport rate of glutamine into the chloroplast is low compared with other dicarboxylic acids, and it seems likely that most of the NH_3 is assimilated by the chloroplast GS.

Mutants of *Arabidopsis* lacking glutamine synthetase activity have not been observed, but mutants of barley lacking the chloroplastic glutamine synthetase have been isolated. In barley, about 10–17% of the glutamine synthetase activity is found in the cytosol, with the remainder in the chloroplast. Studies with the barley mutants show that photosynthesis is inhibited in photorespiratory conditions and ammonia accumulates, indicating that the activity of the cytosolic glutamine synthetase is not sufficient to reassimilate the ammonia produced in photorespiration.

Although mitochondria possess glutamate dehydrogenase, this enzyme would appear to play no role in the assimilation of the NH_3. The enzyme has a low affinity for ammonia (K_m, 5–70 mM) and when 2-oxoglutarate is supplied to mitochondria that are actively decarboxylating glycine, little if any glutamate is produced. When GS is inhibited with L-methionine sulfoximine (MSO), little, if any, NH_3 is assimilated in the leaves.

The reassimilation of the NH_3 in the chloroplast requires the import of 2-oxoglutarate and the export of glutamate, the substrate for the glutamate:glyoxylate aminotransferase in the peroxisome. Recent evidence shows that there are two dicarboxylate transporters in the chloroplast membrane (Woo et al., 1987). The 2-oxoglutarate (2-OG) translocator transports 2-OG, malate, succinate, fumarate and glutarate. The other dicarboxylate transporter transports all these acids plus glutamate and aspartate. Yet another transporter for glutamine may be present. The two dicarboxylate transporters are thought to work in concert. The 2-OG translocater transports 2-OG into the chloroplast in exchange for malate and the dicarboxylate transporter transports glutamate out of the chloroplast in exchange for malate. Since the transporters are exchange transporters, the movement of malate in the opposite direction to 2-OG or glutamate is essential, but no net movement of malate into or out of the chloroplast occurs. The importance of the dicarboxylate transporter has been demonstrated by the isolation of a mutant of *Arabidopsis* defective in this protein. In non-photorespiratory conditions, growth of the mutant was normal, whereas the plants died under photorespiratory conditions.

Photorespiration and leaf photosynthesis

Because Rubisco uses either CO_2 or O_2 as substrate, higher O_2 concentrations produce striking inhibitions of photosynthesis in C3 leaves (Fig. 18.2A). Photosynthesis in C3 leaves is stimulated 40 to 60% when the oxygen concentration is decreased from 21% to 2% and photosynthesis is inhibited a similar amount when the oxygen concentration is increased to 50% (Fig. 18.2E).

Because photorespiration concurrently releases CO_2 during CO_2 fixation, a concentration of CO_2 is reached where release and fixation of CO_2 are equal. This CO_2 concentration is called the CO_2 compensation point and is normally 40–60 μl CO_2 l^{-1}. As O_2 fixation is required for CO_2 release, the compensation point is close to zero at 2% O_2 and increases linearly with the O_2 concentration in C3 leaves (Fig. 18.2C). Above the light compensation point (i.e. that light intensity where gross CO_2 fixation equals gross CO_2 release and net CO_2 exchange is zero) the CO_2 compensation point does not change with light intensity.

Just as photorespiration will vary with the O_2 concentration, it will also vary with the CO_2 concentration, and glycine formation is largely suppressed at CO_2 concentrations above 0.2% (Somerville and Somerville, 1983). O_2 uptake (which is largely due to oxygenase activity of Rubisco) in C3 leaves at low light intensities is also inhibited by increasing CO_2 concentration (Fig. 18.2F). More complex patterns of O_2 uptake occur as a function of CO_2 concentration at high light intensities (Badger and Canvin, 1981), but in all cases higher CO_2 concentrations inhibit O_2 uptake. In C3 leaves, the ratio of photorespiration to photosynthesis increases with leaf temperature. In part, this is due to the kinetics of Rubisco as increased temperature reduces the affinity of the enzyme for CO_2 and favors oxygenation. It is also a result of the increased solubility of O_2 and the decreased solubility of CO_2 that occurs at higher temperatures (see Chapter 16). As expected, the CO_2 compensation point of C3 plants increases with temperature (Fig. 18.2D) and the quantum yield decreases.

The increasing losses due to photorespiration with increasing temperature in C3 plants might be expected to limit the temperatures under which these plants can grow. In general, this is true, as the quantum yield in C3 plants is decreased by over 50% at temperatures above 35 °C. The quantum yield in C4 plants, which have developed a mechanism (see later) for suppressing

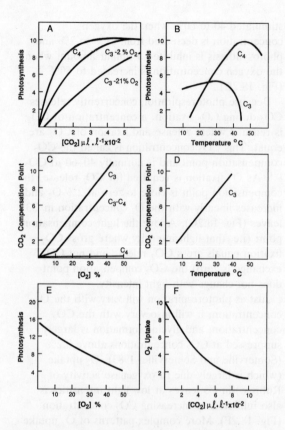

Fig. 18.2 Typical responses of some features of gas exchange in plant leaves. All rates are relative as detailed for individual panels. (A) Net photosynthesis of C3 leaves in 21% or 2% O_2 and C4 leaves in either 21% or 2% O_2 in response to the external CO_2 concentration. Rate of C4 leaf photosynthesis expressed relative to the maximum rate observed in the C4 leaf. Rate of C3 leaf photosynthesis expressed relative to the maximum rate observed in the C3 leaf in 2% O_2. Temperature about 25°C and about one-half full sunlight. (B) Effect of temperature on photosynthesis of C3 and C4 leaves. Rate of photosynthesis expressed relative to the maximum rate in the C4 leaf. Air and about one-half full sunlight. (C) Effect of O_2 concentration on the CO_2 compensation point of a C3 leaf, a C4 leaf and a C3–C4 intermediate leaf. Compensation points expressed relative to the compensation point in the C3 leaf at 20% O_2 being arbitrarily set to 5. Temperature about 25 °C and one-half full sunlight. (D) Effect of temperature on the CO_2 compensation point of a C3 leaf. Compensation point at 20 °C arbitrarily set equal to 5. Normal O_2 and about one-half full sunlight. (E) Net photosynthesis of a C3 leaf as a function of O_2 concentration. Rate of photosynthesis in 20% O_2 arbitrarily set equal to 10. Normal CO_2, temperature about 25 °C and one-half full sunlight. (F) The effect of CO_2 concentration on O_2 uptake of a C3 leaf. Maximum rate of O_2 uptake arbitrarily set equal to 10. Normal CO_2, temperature about 25 °C and one-quarter full sunlight.

photorespiration, is not affected by such temperatures. The optimal temperature for photosynthesis of C4 leaves is markedly higher than for C3 leaves (Fig. 18.2B), and results in C4 plants being found more frequently in environments with high temperatures.

Measurement and rate of photorespiration

Photorespiration, when measured as light-dependent O_2 uptake and CO_2 evolution, involves the same gases as photosynthesis but the fluxes are in the opposite direction. It is relatively easy to qualitatively detect photorespiration but, because photosynthesis and photorespiration occur simultaneously, it is exceedingly difficult to measure photorespiration quantitatively.

Ogren and his coworkers (Ogren, 1984) have developed the following equation to determine the ratio of the carboxylation to oxygenation reactions catalyzed by Rubisco:

$$v_c/v_o = (V_c K_o / V_o K_c)([CO_2]/[O_2])$$

where v_c and v_o are the velocity of carboxylation and oxygenation respectively, V_c and V_o the maximal velocities of these reactions, and K_c and K_o the K_m values of the enzyme for CO_2 and O_2. The term $(V_c K_o / V_o K_c)$ determines the relative rates at any given CO_2 or O_2 concentration and has been called the specificity factor (see chapter 16). The specificity factors for higher C3 and C4 plants are about 80. They decrease to about 50 to 60 for cyanobacteria and unicellular algae and are as low as 10 to 15 for some photosynthetic bacteria. Higher specificity factors indicate a greater affinity by the enzyme for CO_2.

As one CO_2 is released in photorespiration for every two O_2 fixed in the oxygenase reaction, the ratio of CO_2 evolution to oxygenation is 0.5 (see above). The velocity of the relative oxygenase activity ($V_o K_c$) can then be expressed in terms of CO_2 released as 0.5 $V_o K_c$. Hence, the specificity factor for carbon fixation in photosynthesis and carbon release in photorespiration is $V_c K_o/0.5 V_o K_c$. As Ogren states, this is the 'essence' of photorespiration (in terms of CO_2) as it will determine the relative rate of photorespiration in relation to photosynthesis.

With a specificity factor of 80 and assuming concentrations of CO_2 and O_2 at the enzyme site of 7.5 μM and 250 μM under normal ambient conditions, the ratio of carboxylation to oxygenation is about 2.5. Photorespiration as CO_2 release will be 0.5 or about 20% of the rate of photosynthesis. Since four carbons must traverse the PCO cycle for every CO_2 released in photorespiration, the rate of carbon flux through the PCO cycle will be 0.8 or 80% of the rate of CO_2 fixation or photosynthesis. But the actual concentrations of CO_2 and O_2 at the site of the enzyme are not known accurately and predictions should be verified by measurement.

In C3 plants, when the light is extinguished, there is a rapid evolution of CO_2 which greatly exceeds the rate of dark respiration and which has been called the post-illumination burst (PIB). The PIB is thought to arise from the continued metabolism of intermediates of the PCO cycle up to and including glycine that are present after the synthesis of phosphoglycolate has ceased. The PIB is then a qualitative indicator of photorespiration whose size depends on the amount of accumulated intermediates and the technical ability of the measuring system to measure the 'burst' with the minimum amount of dispersion. Computer treatment of the results are reported to improve the estimate of photorespiration obtained by this method (Peterson, 1987).

If the inward CO_2 flux of photosynthesis is reduced by depriving the leaf of the substrate (i.e. placing the leaf in CO_2-free air) the outward flux of CO_2 from photorespiration can be measured. This will be an underestimate because some of the CO_2 released in photorespiration will be refixed in photosynthesis before it can exit the leaf. The change in CO_2 concentration at the site of Rubisco could presumably also alter the amount of substrate produced for the PCO cycle. The method provides a quick and relatively easily obtained estimate of photorespiration, but it cannot be considered quantitative.

The measurement of opposite movements of the same component can be studied by altering the ratio of isotopes of the element in one compartment. For CO_2 uptake the external gas can be enriched in the carbon-14 isotope, and for O_2 uptake the gas can be enriched in the oxygen-18 isotope. Photorespiration as measured by the dual carbon ($^{14}CO_2/^{12}CO_2$) method yields rates of carbon flux through the PCO cycle of 0.65 to 0.9 the rate of CO_2 fixation and, hence, CO_2 evolution equal to 16–22% of the rate of photosynthesis under normal conditions. As some of the CO_2 that is produced from photorespiration must be refixed before it can exit the leaf, the obtained rates represent minimum estimates of photorespiration. The amount of refixation has been estimated (Gerbaud and Andre, 1987) as 16% under normal conditions, and application of this correction factor would yield rates of photorespiration as CO_2 evolution equal to 18–25% of the rate of photosynthesis.

Using oxygen-18 and the mass spectrometer the measurement of O_2 uptake is not complicated by the problem of extensive recycling (Gerbaud and Andre, 1987). It is, however, affected by O_2 uptake that occurs through reactions other than the Rubisco oxygenase reaction and hence tends to overestimate photorespiration. When allowance is made for oxygen uptake by other means the oxygenase/carboxylase ratio is about 0.55 (Badger and Canvin, 1981) under normal conditions. This would yield a rate of carbon flux through the PCO cycle that is 110% the rate of CO_2 fixation and result in a rate of photorespiration as CO_2 evolution that is 27% of the rate of photosynthesis.

The rate of photorespiration can also be determined by measuring the rate of carbon flux through glycine and serine, assuming a stoichiometry of one CO_2 released for every four carbons of traffic. These measurements indicate a flux of carbon through the PCO cycle about equal to the rate of apparent CO_2 fixation and a rate of photorespiration equal to 20–25% of the rate of photosynthesis.

The rate of photorespiration has also been

estimated by an elegant method using a mutant of *Arabidopsis* lacking in serine hydroxymethyl-transferase (Somerville and Somerville, 1983). If the mutant is supplied with serine and NH_3, the carbon that enters the PCO cycle accumulates as glycine. Under normal conditions 0.53 molecules of glycine accumulates for every CO_2 fixed. Hence, oxygenation would be 50% of the rate of carboxylation, carbon flow through the PCO cycle would be about equal to the rate of CO_2 fixation and photorespiration would be about 25% of the rate of photosynthesis.

Function and regulation of the PCO cycle

As described above, under normal conditions, the flux of carbon through the PCO cycle is 0.8 to 1.1 times the rate of CO_2 fixation, and CO_2 release in photorespiration is 20–27% of the CO_2 fixed. In addition to this direct loss of CO_2, the rate of CO_2 fixation is also decreased because of the competition between O_2 and CO_2 at the active site of Rubisco. It means that for every O_2 that is fixed by the oxygenase reaction a CO_2 cannot be fixed.

The inhibition of fixation of CO_2 by O_2 under normal conditions must be twice the rate of photorespiratory CO_2 release because two O_2 must be fixed to produce the two glycine required for the production of one CO_2. If photorespiratory CO_2 release is 25% of the true rate of CO_2 fixation, the true rate of CO_2 fixation could increase by 50% (equivalent to the two O_2) in the absence of O_2 and apparent photosynthesis (true CO_2 fixation minus photorespiratory CO_2 release) should theoretically increase by 75%. Experimentally, increases of 50 to 70% in both true and apparent CO_2 fixation rates are observed when the oxygen concentration is decreased from 21% to 1% (Fig. 18.2E). This is in the range of increase expected for true CO_2 fixation but is somewhat lower than that expected for apparent CO_2 fixation. The failure to observe the expected increase in CO_2 fixation when the O_2 concentration is decreased may be related to the reduction in photosynthetic electron transport or true photosynthetic capacity that occurs when the O_2 concentration is decreased (Canvin *et al.*, 1980).

It should now be apparent that the oxygenase activity and photorespiratory CO_2 loss substantially decreases CO_2 fixation and the productivity of plants growing in normal O_2 concentration. Indeed, soybean plants grown at 5% O_2 for 69 days produced twice the dry matter production of plants grown at 21% O_2 (Quebedeaux and Hardy, 1976). Experimentally, then, it has been possible to double plant productivity by reducing oxygenase activity and photorespiration by a reduction in the oxygen concentration around the plant.

The costs of oxygenase activity and photorespiration to the plant can also be determined from the energy requirements of the PCR and PCO cycles. These are shown in Table 18.1. The cost of fixing one molecule of CO_2 in the PCR cycle is nine ATP equivalents. The cost of fixing one molecule of O_2 in the oxygenase reaction is 9.5 ATP equivalents. However, 0.5 CO_2 is lost and the cost to fix this originally would have been 4.5 ATP equivalents. Hence the total cost of the oxygenase reaction would be 14 ATP per O_2 fixed or 28 ATP per CO_2 released. As oxygenase activity is approximately one-half the carboxylase activity under normal conditions, the utilization of all the energy for CO_2 fixation could support a rate of CO_2 fixation at least 50% higher.

All Rubisco's that have been studied, even those from chemolithotrophic bacteria and photoautotrophic bacteria such as *Rhodospirillum rubrum* that grow under anaerobic conditions, have oxygenase activity. The ubiquity of the oxygenase activity has led to the suggestion that oxygenase activity is an inevitable consequence of the reaction chemistry required for CO_2 fixation by the enzyme (see chapter 16). If this is indeed the case, the PCO cycle is a means whereby 75% of the carbon that is diverted to phosphoglycolate can be recovered into products useful to the plant. This recovery or 'salvage' role may be a sufficient and only function for the PCO cycle.

Because of the magnitude of carbon transit through the PCO cycle and the substantial expenditure of energy associated with it, however, one may continue to have a lingering feeling that there should be a more essential function for the cycle. Such a case can certainly not be demonstrated for the *Arabidopsis* or barley

Table 18.1 Energy requirements for CO_2 and O_2 fixation by Rubisco

CO_2 fixation
  3 RuBP + 3 CO_2 + 3 H_2O → 6 (3-PGA)
  6 (3-PGA) + 6 ATP + 6 (2 H^+ + 2 e^-) → 6 triose phosphate + 6 ADP + 6 P_i
  5 triose phosphate → 3 R-5-P + 2 P_i
  3 R-5-P + 3 ATP → 3 RuBP + 3 ADP

Sum 3 CO_2 + 3 H_2O + 9 ATP + 6 (2 H^+ + 2 e^-) → triose phosphate + 9 ADP + 8P_i
If (2 H^+ + 2 e^-) is assumed to be equivalent to 3 ATP, the cost for each CO_2 fixed would be 9 ATP equivalents.

O_2 fixation
  10 R-5-P + 10 ATP → 10 RuBP + 10 ADP
  10 RuBP + 10 O_2 → 10 (3-PGA) + 10 phosphoglycolate
  10 phosphoglycolate → 10 glycolate + 10 P_i
  10 glycolate + 5 O_2 → 10 glyoxylate + 10 H_2O
  10 glyoxylate + 10 glutamate → 10 glycine + 10 (2-oxoglutarate)
  10 glycine → 5 serine + 5 CO_2 + 5 NH_3 + 5 (2 H^+ + 2 e^-)
  5 serine + 5 (2-oxoglutarate) → 5 (hydroxypyruvate) + 5 glutamate
  5 (hydroxypyruvate) + 5 (2 H^+ + 2 e^-) → 5 glycerate
  5 glycerate + 5 ATP → 5 (3-PGA) + 5 ADP
  5 NH_3 + 5 ATP + 5 glutamate → 5 glutamine + 5 ADP + 5 P_i
  5 glutamine + 5 (2-oxoglutarate) + 5 (2 H^+ + 2 e^-) → 10 glutamate
  15 (3-PGA) + 15 ATP + 15 (2 H^+ + 2e^-) → 15 (triose phosphate) + 15 ADP + 15 P_i
  15 (triose phosphate) → 9 R-5-P + 6 P_i

Sum R-5-P + 1.5 O_2 + 3.5 ATP + 2 (2 H^+ + e^-) → 0.9 (R-5-P) + 3.5 ADP + 3.6 P_i + 0.5 CO_2 + H_2O
The cost for each O_2 fixed by oxygenase would be 9.5 ATP equivalents.

Adapted from Lorimer and Andrews, 1981.

mutants that lack essential enzymes of the PCO cycle, because they grow normally under conditions that prevent oxygenase activity and that in other respects are non-stressful. But is this also the case when plants are exposed to extreme stress, such as water deficiency or cold temperatures under high light conditions? In such situations, because CO_2 may be limiting when the stomata are closed or because the activity of carboxylation may be low, the capacity to dissipate photochemical energy may be limited. The high energy state that would result in the photochemical system leads to a damage of the system which has been called photoinhibition (Powles, 1984). Photoinhibition has been shown to occur when leaves are exposed to high light in the absence of both CO_2 and O_2, conditions where both the PCR and PCO cycles could not operate. With oxygen and oxygenase activity, CO_2 would be maintained at the compensation point and both O_2 and CO_2 fixation could contribute to using the photochemical energy. In C3 plants, either O_2 or CO_2 reduced photoinhibition to some degree, whereas in C4 plants only CO_2, and not O_2, was effective.

Any role of photorespiration in offering some protection against photoinhibition (Osmond, 1981) would seem to be entirely fortuitous. As noted above, all Rubisco's have oxygenase activity, so oxygenase activity was not fixed in the enzyme by evolutionary pressure in plants that routinely encountered stress conditions. There is no evidence that oxygenase activity is greater in plants that grow in regions where stress is frequent as compared with plants that grow in regions where stress is never encountered. C4 plants, in which oxygenase activity is reduced due to a 'CO$_2$-concentrating' mechanism and lower amounts of Rubisco, do not seem to be any more susceptible to photoinhibition than C3 plants. No doubt, this possible role of photorespiration will continue to be considered. One cannot escape from the fact, however, that if the carbon from phosphoglycolate is not returned to the PCR cycle, i.e. the 'salvage' function, the death of the plant follows, as shown in *Arabidopsis* and barley mutants deficient in PCO cycle enzymes.

Cyanobacteria and green algae, which are evolutionarily older than C3 terrestrial plants, do not seem to have a complete effectively functioning PCO cycle. They do, however, possess active transport systems for both CO_2 and HCO_3^- that function as a 'CO$_2$-concentrating' mechanism to elevate the internal CO_2 concentration and suppress oxygenase activity. Aquatic macrophytes seem to retain, at least in some plants, some 'CO$_2$-concentrating' activity, but they also have developed a PCO cycle. C3 terrestrial plants have apparently lost all 'CO$_2$-concentrating' capability. With the progressive loss of the 'CO$_2$-concentrating' capability it was imperative for the plants to develop an efficient PCO cycle to salvage the carbon that was diverted to phosphoglycolate and, hence, the PCO cycle is essential but the oxygenase reaction is merely unavoidable.

As a salvage pathway, it must have the capability of metabolizing all the substrate, i.e. the phosphoglycolate, that is produced. One should not expect to find that the operation of the pathway is regulated by any other factor than the availability of substrate. While all the enzymes of the PCO cycle have not been examined extensively, there is little evidence for regulation of them at the present time. The capacity of the total PCO cycle, however, may be able to moderate the activity of the PCR cycle. There is little doubt that in the extreme case where the PCO cycle cannot operate, such as in the PCO cycle enzyme-deficient mutants, one of the consequences is that the PCR cycle is inhibited. It is possible that where substrate supply exceeds the capacity of the PCO cycle, i.e. the PCO cycle is overloaded, the accumulation of intermediates would moderate the PCR cycle to again bring substrate production into line with the PCO cycle capacity.

C4 plants

In C4 plants two distinct types of photosynthetic cells are observed in the leaves. The mesophyll cells are arranged throughout the leaf lamina while the bundle sheath cells are arranged in a ring, called Kranz (wreath-like) anatomy, surrounding the vascular strands. The bundle sheath cells have thickened walls and prominent starch-filled chloroplasts, whereas the mesophyll cells have thinner walls and smaller chloroplasts.

The distinctive anatomy in C4 plants is accompanied by a specialized biochemistry of photosynthesis. Rubisco and the PCR cycle are confined to the bundle sheath. CO_2 fixation in the mesophyll cells consists of the addition of HCO_3^- to phosphoenolpyruvate (PEP) by PEP carboxylase to form oxaloacetic acid (OAA). The OAA is converted to either malate or aspartate and transported to the bundle sheath cells. In the bundle sheath cells the four carbon acid (malate or aspartate) is decarboxylated, depending on the plant, by NADP or NAD malic enzyme or PEP carboxykinase to provide CO_2 for Rubisco. The three carbon acid remaining (pyruvate or PEP) is transported back to the mesophyll cells where it serves as the acceptor for further CO_2 fixation (see Chapter 19).

In terms of whole-leaf CO_2 exchange, C4 photosynthesis is insensitive to the O_2 concentration (Fig. 18.2a) and the CO_2 compensation point is less than 5 μl CO_2 l^{-1} (Fig. 18.2c). O_2 uptake is observed in C4 plants but this O_2 uptake is largely insensitive to change in the CO_2 concentration. These characteristics of the O_2 and CO_2 exchange in C4 plants suggest that oxygenase activity and photorespiration are suppressed or absent in C4 plants.

As Rubisco is confined to the bundle sheath cells, oxygenase activity would be similarly limited and the

PCO cycle should also be confined to these cells. Photosynthesis of isolated bundle sheath cells is sensitive to O_2 concentration and while present in lower amounts than in C3 plants, many of the enzymes of the PCO cycle are present in the bundle sheath cells. Many of the enzymes, however, are also present in the mesophyll cells and, surprisingly, glycerate kinase seems to be localized in these cells. Peroxisomes are present in both cell types at 10 to 50% of the frequency observed in C3 plants, but they are usually several times more numerous in the bundle sheath cells. C4 leaves or bundle sheath strands metabolize glycolate or glycine to CO_2, and glycine and serine become labeled when C4 leaves photosynthesize in $^{14}CO_2$. The general picture that emerges is that oxygenase activity and PCO cycle activity occur under some circumstances in C4 leaves but these activities are much reduced. The operation of the PCO cycle may be more complex, involving both the bundle sheath and mesophyll cells, and any CO_2 that is released is effectively refixed in the mesophyll cells before it could exit to the surroundings.

The suppression of oxygenase activity and, hence, phosphoglycolate production in C4 plants is attributed to a high CO_2 concentration in the bundle sheath cells. This high CO_2 concentration is formed and maintained in the bundle sheath cells by the rapid decarboxylation of the C4 acids that were formed in the initial carboxylation in the mesophyll cells and transported to the bundle sheath cells. Only at elevated CO_2 concentrations will the carboxylation of RuBP become equal to the rates of decarboxylation to maintain the system in steady state. In general, the K_m (CO_2) of Rubisco from C4 plants is higher than that of the enzyme from C3 plants and the total Rubisco activity of C4 plants is lower than that in C3 plants.

Estimates of the inorganic carbon concentration in bundle sheath cells of six C4 grasses ranged from 160 to 990 μM (Furbank and Hatch, 1987). Bundle sheath cells do not contain carbonic anhydrase and it was estimated that 90% of this inorganic carbon would be in the form of CO_2. These concentrations of CO_2 would be 20 to 120 times the normal CO_2 concentration (c. 7.5 μM) in mesophyll cells of C3 plants and would be sufficient to largely eliminate oxygenase activity.

CAM plants

CAM plants are characterized by a marked diurnal variation in titratable acidity, mostly malate. This is due to extensive CO_2 fixation at night by PEP carboxylase to produce malate which is stored in the vacuole. Stomata are open at night when the evaporative demand is low and PEP is produced from stored carbohydrate. During the day, when evaporative demand is high, the stomata close and the malic acid produced during the preceding night is decarboxylated to produce CO_2. This CO_2 is then fixed by the normal PCR cycle. Decarboxylation of malic acid, depending on the plant, is catalyzed by either malic enzyme or PEP carboxykinase. The three carbon molecule that results from this decarboxylation is converted to carbohydrate in the light.

The stomata are closed during the day when decarboxylation is occurring and so the CO_2 concentration in the leaf can reach high levels. Sampling of the internal gas phase of a number of CAM plants revealed CO_2 concentrations ranging from 0.08% to as high as 2.5% (Cockburn et al., 1979; Spalding et al., 1979). These high internal CO_2 concentrations are sufficient to suppress or inhibit the oxygenase reaction and hence photorespiration.

At the end of the day when de-acidification is complete, CAM plants, growing under reasonable moisture levels, carry out normal C3 photosynthesis. During this phase, photosynthesis is sensitive to O_2 and a post-illumination burst can be observed. In these conditions, it is likely that photorespiration similar to that in C3 plants occurs in CAM plants. Hence, CAM plants would seem to be similar to C3 plants in the potential capacity for photorespiration, but this capacity is normally suppressed due to the elevated CO_2 concentrations that prevail in the light during photosynthesis.

C3–C4 intermediates

C3–C4 intermediates are characterized by a CO_2 compensation point of 7–28 μl CO_2 l^{-1} and a reduced sensitivity of photosynthesis to the O_2 concentration (Fig. 18.2c). The lower CO_2

compensation point in C3–C4 intermediates compared with C3 plants is consistent with reduced photorespiration in these plants.

C3–C4 intermediates also have a leaf anatomy intermediate between that of C3 plants and C4 plants. A bundle sheath can be observed, but it is not as well developed or defined as in C4 plants.

C3–C4 intermediates have been described in several genera, such as *Flaveria* (Asteraceae), *Panicum* and *Neurachne* (Gramineae), *Alternanthera* (Amaranthaceae) and *Molluga* (Aizoaceae) which contain both C3 and C4 species. However, other genera of C3–C4 intermediates such as *Moricandia* (Crucifereae) and *Parthenium* (Asteraceae) do not contain any C4 species. C4 plants are thought to have evolved from C3 plants and the C3–C4 intermediates may be a stage in that evolution or hybrids between existing C3 and C4 species.

C3–C4 intermediates have been intensively studied because they seem to represent natural instances of reduced photorespiration, and if the mechanism can be discovered, some guidance as to how photorespiration in C3 plants can be reduced may be provided. From these studies two proposals have emerged. The C4-like anatomy and the presence of a number of C4 photosynthesis enzymes has suggested that the C3–C4 intermediates are capable of a limited amount of C4 photosynthesis. Alternatively, the C3–C4 intermediates may possess a more efficient refixation system for any CO_2 that may be released. Support for this proposal rests largely on the fact that the CO_2 compensation point of the C3–C4 intermediates increases as the light intensity is decreased and at low light intensity becomes equal to closely related C3 plants. (In C3 plants, above the light compensation point, the CO_2 compensation point does not change with increasing light intensity.) Obviously, increased refixation is only possible at higher light intensities that would provide the energy for this process. Other explanations have been presented to account for reduced photorespiration in these plants, but none are entirely satisfactory (see Holaday and Chollet, 1984 or Monson et al., 1984).

Aquatic macrophytes

Many angiosperms grow in emergent or submerged conditions in aquatic habitats. The emergent leaves of these plants, which almost all have C3 metabolism have gas exchange characteristics similar to terrestrial C3 plants. The CO_2 compensation point is in the normal range and both the compensation point and photosynthesis show the normal magnitude of change in response to altered O_2 concentration. There is no reason to believe that such leaves differ substantially in photosynthesis or photorespiration from their terrestrial counterparts. When such plants are submerged, the rate of photosynthesis is low and much higher concentrations of dissolved inorganic carbon are required to saturate photosynthesis. This is primarily due to the high diffusion resistance of water to CO_2. Rubisco activity in these plants is low and the K_m (CO_2) of this enzyme is about double that of the enzyme from terrestrial leaves. Several of the enzymes of the PCO cycle have been demonstrated and photorespiration is likely to occur. It has been discovered, however, that the plants show two distinct photosynthetic forms – a high photorespiration (high PR) form and a low photorespiration (low PR) form. Under cooler temperatures and short days the plants exist in the high PR form. The CO_2 compensation point is 60–110 μl CO_2 l^{-1}, photosynthesis is inhibited 20–30% by 21% O_2, and significant CO_2 evolution or photorespiration can be detected. At higher temperatures and longer photoperiods the plants change to a low PR form. In this form the CO_2 compensation point is 10–25 μl CO_2 l^{-1}, the inhibition of photosynthesis by 21% O_2 ranges from 5 to 17%, and CO_2 evolution is greatly reduced. All evidence is consistent with reduced photorespiration in this form. The means by which this is accomplished is puzzling, as the change is not accompanied by changes in anatomy or by major changes in the activity of PCR or PCO cycle enzymes.

Among the aquatic genera that have been studied are *Hygrophila*, *Limnophila*, *Myriophyllum*, *Hydrilla* and the CAM plant *Isoetes*. In *Hydrilla*, when the low PR type develops, PEP carboxylase increases over two-fold. Other enzymes of C4 photosynthesis, including pyruvate P_i dikinase, also increase in amount. While both high PR and low PR plants incorporated $^{14}CO_2$ into malate and aspartate, only in the low PR plants does the label rapidly flush from these acids during a chase period. Evidence was also obtained for the fixation of CO_2 at night, similar to that in CAM plants. The low PR state is not yet clearly understood but it would seem to be due to limited intracellular operation of a C4 cycle with some possible contribution from CAM dark fixation.

In *Myriophyllum*, no evidence could be found for the development of C4 photosynthesis-like activity in the low PR plants. The low CO_2 compensation point of these plants increased to the high CO_2 compensation point as the light intensity was decreased, and a higher light intensity was required for saturating photosynthesis indicating a higher energy requirement for photosynthesis in the low PR plant. Carbonic anhydrase activity doubled in both *Myriophyllum* and *Hydrilla*. In *Myriophyllum*, the measurement of photosynthesis at pH 5.0 and pH 8.0 indicated an increased ability to use HCO_3^-. Ethoxyzolamide, an inhibitor of carbonic anhydrase, causes a rise in the compensation point and the K_m (CO_2) for photosynthesis. It has been suggested that an inorganic carbon concentrating mechanism involving active CO_2 and/or HCO_3^- transport is operating in the low PR plants of *Myriophyllum*. Such a system has been described for cyanobacteria and unicellular algae (see below). The active transport of inorganic carbon would require energy (hence the higher light requirement) and would result in higher internal CO_2 concentrations that could suppress oxygenase activity.

In the marine macroalgae there also seems to be a diversity of photosynthetic mechanisms, although, of course, the PCR cycle remains the basic means of net CO_2 assimilation. In *Sargassum* (a brown alga) and *Codium* (a green alga) the CO_2 compensation points are between 70 and 120 $\mu l\ CO_2\ l^{-1}$ and photosynthesis is inhibited about 30% by 21% O_2 compared with the rate in 1% O_2. Low PR states in these algae are not observed, and photosynthesis and photorespiration appear to be of the C3 type. In *Udotea* (a green alga), however, the CO_2 compensation point is 7–12 $\mu l\ CO_2\ l^{-1}$ and the inhibition of photosynthesis by 21% O_2 compared with 1% O_2 is 7–9%. An intracellular C4 photosynthetic-like mechanism may be responsible for the low PR state. It has been postulated that C4 acids are formed in the cytosol by PEP carboxykinase, transported to the chloroplast and decarboxylated by the same enzyme (since that was the only enzyme of C4 acid metabolism found). The resulting higher CO_2 in the chloroplast would suppress oxygenase activity (Reiskind *et al.*, 1988).

From the results that have been obtained with the small number of aquatic macrophytes investigated, it would appear that a number of systems that provide supplemental CO_2 for fixation in the PCR cycle have developed. At the moment these systems are not well understood.

Microalgae and cyanobacteria

The photosynthetic properties of cyanobacteria and microalgae vary with the dissolved inorganic carbon (DIC) concentration on which they are grown. Cyanobacteria, when grown on DIC concentrations ranging from 10 μM to 10 mM, develop $K_{\frac{1}{2}}$ (DIC) for photosynthesis ranging from about 5 μM to over 1 mM. Most experimental work, however, is done with cells grown either at high [DIC] (>1 mM) or at low [DIC] (<0.1 mM). It is well established that it is the total external DIC (CO_2 + HCO_3^- + CO_3^{2-}) concentration that is important and not just the CO_2 concentration that causes the adaptation that is observed in the cells. However, for convenience, high DIC cells are usually referred to as high CO_2 cells and low DIC cells are referred to as low CO_2 cells.

High CO_2 cells have a higher growth rate and maximum photosynthetic rate than low CO_2 cells. High CO_2 cells have a $K_{\frac{1}{2}}$ (DIC) that varies with the measured pH; it is low (< 50 μM) at pH 5.0 and rises to high values (> 5 mM) at pH 10.0, but is usually in the range of 1–2 mM at pH 8.0. High CO_2 unicellular algae do not have an external carbonic anhydrase; cyanobacteria do not have external carbonic anhydrase under any conditions.

Low CO_2 cells of unicellular algae and cyanobacteria have $K_{\frac{1}{2}}$ (DIC) for photosynthesis of less than 15 μM (Fig. 18.3A, B) and unicellular algae show substantial external carbonic anhydrase activity. The $K_{\frac{1}{2}}$ (DIC) is essentially independent of the pH. In low CO_2 cells, photosynthesis is not sensitive to the O_2 concentration, the CO_2 compensation point is less than 5 $\mu l\ CO_2\ l^{-1}$ and

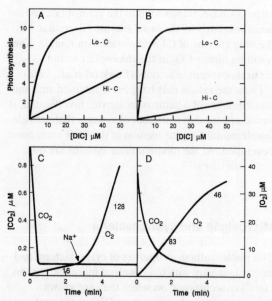

Fig. 18.3 Some properties of inorganic carbon assimilation in microalgae and cyanobacteria. (A) Photosynthesis of lo-C and hi-C cells of *Chlamydomonas reinhardtii* as a function of external dissolved inorganic carbon (DIC) concentration. Rates are relative to the maximum rate of photosynthesis in lo-C cells equal to 10. (Based on data from Berry *et al.*, 1976). (B) Photosynthesis of lo-C and hi-C *Synechococcus* as a function of DIC concentration. Rates are relative to maximum rate of photosynthesis in lo-C cells equal to 10. (Based on data from Miller *et al.*, 1984). (C) CO_2 uptake and O_2 evolution measured with a mass spectrometer in lo-C cells of *Synechococcus*. Light (600 μE m^{-2} s^{-1}) at time zero, 30 °C, pH 8.0 and total DIC 55 μM. At arrow 25 mM NaCl added to initiate HCO_3^- transport. Numbers next to O_2 trace show rates of photosynthesis in μmol mg^{-1} Chl h^{-1}. (Based on data of Espie *et al.*, 1988). (D) CO_2 uptake and O_2 evolution measured with a mass spectrometer in washed protoplasts (all external carbonic anhydrase removed) from lo-C *Chlamydomonas reinhardtii*. Light (220 μE m^{-2} s^{-1}) at time zero, 30 °C, pH 8.0 and total DIC 50 μM. Numbers next to O_2 trace show rates of photosynthesis in μmol mg^{-1} Chl h^{-1}. (Based on data of Sültemeyer *et al.*)

little photorespiration can be demonstrated.

Photosynthesis of high CO_2 cells, under high CO_2 conditions, is also not inhibited by O_2. When high CO_2 cells, however, are placed in low CO_2 conditions they are not able to effectively use the low DIC or CO_2 that is present and their photosynthetic rates are very low (Fig. 18.3A, B). Under conditions of low CO_2, photosynthesis of high CO_2 cells is sensitive to O_2, the CO_2 compensation point at pH 8.0 is usually greater than 60 μl CO_2 l^{-1}, and apparent photorespiration can be observed. Glycolate synthesis can be demonstrated and can be greatly increased if the incubation is done in the presence of aminooxyacetate, an inhibitor of the glyoxylate aminotransferase. While the cells have some ability to metabolize glycolate due to the presence of glycolate dehydrogenase, most of the glycolate is excreted into the medium.

Both cyanobacteria and unicellular algae fix CO_2 by the normal PCR cycle and are C3 plants in that respect. The Rubisco extracted from green unicellular algae has a K_m (CO_2) value in the range of 30–60 μM and the Rubisco of cyanobacteria 100–200 μM. There is thus quite a marked difference in the apparent ability of the primary CO_2-fixing enzyme to fix CO_2 and the ability of the low CO_2 cells to use the CO_2 in the medium.

This apparent contradiction was resolved when it was discovered that unicellular algae and cyanobacteria possess mechanisms for the active uptake of inorganic carbon. The active transport of inorganic carbon results in concentrations of DIC in the cells that can be in excess of 1000-fold the external concentrations. If the CO_2 and HCO_3^- are in equilibrium, the internal CO_2 concentration can be more than 17 000 times the external CO_2 concentration considering the internal pH. The rate of photosynthesis is directly correlated with the size of the internal pool and it is the high internal concentration of inorganic carbon that enhances photosynthesis and suppresses oxygenase activity. Initially, it was thought that low CO_2-grown cells possessed an inorganic carbon concentrating mechanism and high CO_2-grown cells lacked this mechanism. It is now apparent, at least in the cyanobacteria, that both high CO_2-grown and low CO_2-grown cells possess active inorganic carbon transport systems, but the transport systems differ substantially in their properties. Less is known about the transport systems in unicellular algae.

The cyanobacteria are considered to be simpler than the unicellular algae for the study of the transport systems because the cyanobacteria do not have organelles and the resulting compartmentation that exist in the algae. They also do not have an external carbonic anhydrase. Although cytosolic carbonic anhydrase levels in cyanobacteria are very low and in some cases not demonstrable, it is quite clear from using oxygen-18 labeled HCO_3^- and measuring the mass of the CO_2 produced, that in the light HCO_3^- and CO_2 are in rapid equilibrium (Badger et al., 1985).

Low CO_2-grown cyanobacteria possess active transport systems for both CO_2 and HCO_3^-. Transport of CO_2 requires micromolar concentrations of Na^+ and this action of Na^+ is not inhibited by Li^+. CO_2 transport is inhibited by carbon oxysulfide (COS), an analog of CO_2, and by H_2S. The mechanisms of these inhibitions are not known. HCO_3^- transport requires millimolar concentrations of Na^+ and the stimulation by Na^+ is competitively inhibited by Li^+. HCO_3^- transport is also inhibited by monensin, a compound that catalyzes the electroneutral exchange of Na^+ and H^+ across membranes and collapses Na^+ gradients. COS or H_2S seem to have little effect on HCO_3^- transport.

The differences in the properties of the CO_2 and HCO_3^- transport systems allows the study of the transport of one species with little or no contribution from the transport of the other species. If only micromolar amounts of Na^+ are supplied or the system is inhibited with Li^+, CO_2 transport can be studied with little or no HCO_3^- transport. On the other hand, HCO_3^- transport, without CO_2 transport, can be studied in the presence of the inhibitors COS or H_2S.

Low CO_2-grown cyanobacteria, in the presence of low levels of Na^+, light and DIC, will remove CO_2 from the medium and maintain it at a concentration near zero even though there is a substantial concentration of HCO_3^- (Fig. 18.3C). The maintenance of the CO_2/HCO_3^- system away from equilibrium and the high internal concentrations of inorganic carbon in the cells are strong evidence for an active CO_2 transport system (Badger and Andrews, 1982; Miller et al., 1988). The uptake of CO_2 does not occur in the dark and is inhibited by diethylstilbestrol, an ATPase inhibitor, and by the uncoupler, CCCP. Even in the presence of HCO_3^- transport, CO_2 is selectively removed from the medium. The $K_{\frac{1}{2}}$ for the CO_2 transporter has been estimated to be 0.4 μM. Nothing is known about the mechanism of transport, although a model which envisions a conversion of CO_2 to HCO_3^- at the transport site has been put forward (Volokita et al., 1984; Reinhold et al., 1987).

The active transport of HCO_3^- has been demonstrated in a number of cyanobacteria by showing that, at alkaline pH, the rate of photosynthesis is many-fold higher than that which could be sustained by the spontaneous formation and transport of CO_2. When millimolar Na^+ is added (Fig. 18.3c) the spontaneous rate of CO_2 production from HCO_3^- could not have supplied more than 12.5% of the CO_2 for the observed rate of photosynthesis. Again, little is known about the mechanism of transport, although some possibilities have been outlined (see Lucas and Berry, 1985).

High CO_2-grown cyanobacteria have an active transport system for CO_2 but appear unable to transport significant amounts of HCO_3^- (Badger and Gallagher, 1987; Miller and Canvin, 1987). Micromolar levels of Na^+ are required for transport but the amount of Na^+ required to give half-maximal rates of transport or photosynthesis varies with the DIC concentration. Higher concentrations are needed at low DIC concentrations and the amount required decreases as the DIC concentration is increased. At 500 μM DIC the $K_{\frac{1}{2}}$ (Na^+) for photosynthesis is about 18 μM. The effect is specific for Na^+ as other ions are inactive. Li^+ does not inhibit this CO_2 transport (similar to the CO_2 transport system in low CO_2 cells).

Several methods have been used to demonstrate that high CO_2 cells transport CO_2 and have only limited, if any, capacity to transport HCO_3^-. When photosynthesis is measured at pH values from 6.0 to 9.0, the $K_{\frac{1}{2}}$ (CO_2) remains relatively constant at about 15 μM whereas the $K_{\frac{1}{2}}$(DIC) changes from 29 μM at pH 6.0 to 5.6 mM at pH 9.0. In disequilibrium experiments (Espie and Colman, 1986) the inorganic carbon is supplied to the cells either in the form of CO_2 or HCO_3^-. When CO_2 is supplied, the internal pool of inorganic carbon rapidly increases in size and photosynthesis proceeds without any discernible lag. When HCO_3^- is supplied, the pool increases in size at a much lower rate and the size of the pool is much smaller. Consistent with the small internal pool, there is a lag in the initiation of photosynthesis and the photosynthetic rates are low. These results with

HCO_3^- are consistent with the spontaneous dehydration of HCO_3^- to CO_2 and the utilization of this CO_2 although the possibility exists that some HCO_3^- could be used. When rates of CO_2 production from HCO_3^- are calculated, though, it is apparent that the rate of photosynthesis of high CO_2 cells does not exceed the rate of spontaneous dehydration of HCO_3^- to CO_2.

Over a range of external DIC concentrations, the measured concentrations of DIC in the cell are about 20-fold higher; at 0.25 mM DIC in the medium the internal concentration of DIC is 70-fold higher. The high internal concentrations of DIC do not develop in the absence of Na^+ or in the dark; light is clearly required and the CO_2 is transported against a considerable concentration gradient. Hence, there is no doubt that high CO_2-grown cells are capable of active CO_2 transport but have little, if any, capacity to transport HCO_3^-.

It is not known if the transport system for CO_2 in high CO_2 cells is the same transport system for CO_2 which is found in low CO_2 cells. While the requirement for sodium is similar the apparent affinity of the transport system for CO_2 is quite different. In low CO_2 cells the $K_{\frac{1}{2}}(CO_2)$ for transport has been estimated to be about 0.4 μM CO_2. The response of transport to CO_2 concentration has not been directly determined in high CO_2 cells but an estimate of the $K_{\frac{1}{2}}(CO_2)$ for transport can be obtained from the $K_{\frac{1}{2}}(CO_2)$ for photosynthesis. Measurements of the $K_{\frac{1}{2}}(CO_2)$ for photosynthesis in high CO_2 cells range from 5 to 20 μM CO_2, indicating that the effectiveness of these cells to use CO_2 is 10 to 40 times less than low CO_2 cells.

There is, as yet, no direct evidence for an active inorganic carbon concentrating mechanism in high CO_2-grown unicellular algae. A comparison of the K_m of Rubisco and the $K_{\frac{1}{2}}$ of photosynthesis in *Chlamydomonas* indicates that little, if any, concentration of the CO_2 would be required, as the $K_m(CO_2)$ for Rubisco has been reported as 29–57 μm, whereas the $K_{\frac{1}{2}}(CO_2)$ for whole cell photosynthesis has been reported to be 20–30 μM. A low, if any, concentration of inorganic carbon would be required to reconcile the affinity of Rubisco for CO_2 with the apparent affinity of the cell for CO_2.

Low CO_2-grown cells of unicellular algae possess active transport systems for both CO_2 and HCO_3^-. In wild-type cells of unicellular algae, the transformation to the low CO_2 state is accompanied by a large increase in extracellular carbonic anhydrase activity. Intracellular concentrations of inorganic carbon more then 40 times the extracellular concentrations have been observed, so there is little doubt that the cells possess active inorganic carbon transport systems. The identification of the location of the transport systems and the form of inorganic carbon transported, however, has been complicated by the compartmentation that occurs in these cells and by the presence of the external carbonic anhydrase that keeps the CO_2/HCO_3^- system in equilibrium.

The locations of the inorganic carbon transport systems in unicellular algae have not been determined. Transport systems could be on the plasma membrane, on the chloroplast membrane (Moroney *et al.*, 1987), or on both membranes. The system that has been suggested as being located on the chloroplast membrane is a HCO_3^- transport system. Inorganic carbon would enter the cytoplasm by the diffusion of CO_2. The external carbonic anhydrase would be important to provide a continuous supply of CO_2 to the cell. While there is some evidence that chloroplasts from *Chlamydomonas* may be able to concentrate DIC, there is good evidence that CO_2 is actively transported into the cell as well as HCO_3^-, and it has been suggested that these transporters are on the plasma membrane (Williams and Turpin, 1987). Using protoplasts washed free of external carbonic anhydrase, mass spectrometer measurements of CO_2 have provided direct evidence for active CO_2 transport in *Chlamydomonas* (Fig. 18.3d). CO_2 is selectively removed from the medium and maintained at low levels even though considerable HCO_3^- is present. Active transport of HCO_3^-, however, must also occur as a substantial rate of photosynthesis is obtained when little CO_2 remains in the medium. Maximum photosynthetic rates are only achieved when both active CO_2 and active HCO_3^- transport occurs. The external carbonic anhydrase is important to ensure that CO_2 is available so that both transporters can operate. Mutants lacking the external carbonic anhydrase (see Aizawa and Miyachi, 1986) would be disadvantaged in photosynthesis and growth only under conditions where the spontaneous dehydration of HCO_3^- to CO_2 is inadequate to supply the substrate for the

CO_2 transporter. In conditions of suitable pH or DIC concentration, where CO_2 is available, little difference in photosynthetic rate or growth of these mutants compared to wild-type cells would be discernible.

Unicellular algae also have an internal carbonic anhydrase and this enzyme is essential for the supply of CO_2 to Rubisco. A mutant of *Chlamydomonas* deficient in internal carbonic anhydrase accumulates inorganic carbon to five times the amount of the wild-type cells but is only able to photosynthesize at less than 10% of the wild-type rate. This suggests that most of the DIC in the cell is in the form of HCO_3^- and carbonic anhydrase is necessary in the cell to catalyze the dehydration of HCO_3^- to maintain the CO_2/HCO_3^- equilibrium. Results such as these provide no evidence of the form of inorganic carbon transported, they merely indicate that the conditions (e.g. the pH) in the cell, or especially in the chloroplast, are such that the arriving inorganic carbon is maintained as, or is converted to, HCO_3^-.

In cyanobacteria, the transport systems can be studied in the absence of CO_2 fixation by inhibiting CO_2 fixation with iodoacetamide (IAA). In both IAA-inhibited cells or non-inhibited cells the provision of inorganic carbon causes a large increase in the photoreduction of oxygen. Photosynthetic phosphorylation, either through pseudocyclic or cyclic mechanisms, would seem to provide the energy for transport. Much work still needs to be done, however, to establish the mechanisms of transport and the distribution of such transport systems in the plant kingdom.

Concluding remarks

The majority of economically important plants are C3 plants. In such plants, photorespiration results in substantial losses of productivity and yield. Much research will continue to be devoted towards finding a means of suppressing or eliminating photorespiration as the perceived benefits of increased productivity are substantial. Whether or not this can be done remains an open question.

Photorespiration continues to cause perplexity as a process that results in considerable loss of fixed carbon and utilization of energy without any apparent beneficial function. Further research on its possible protective role in plants under severe stress is required, but this role has not been shown under natural conditions and oxygenase activity occurs in the Rubisco of all organisms, not just in higher plants that may experience stress. Oxygenase activity, the first reaction of photorespiration, may well be a chemical consequence of the carboxylation mechanism, and photorespiration is then a salvage operation. Support for this view is provided by the observation that evolution has not been able to eliminate oxygenase activity. Rather, nature has adopted other mechanisms to cope with the problem.

That is not to say that nature has not modified the carboxylase/oxygenase activities of Rubisco. The increased specificity of Rubisco for CO_2 in C3 higher plants compared with that of the Rubisco of microorganisms shows that the carboxylase/oxygenase ratio is not inflexible but can be modified. The substitution of Mn^{2+} for Mg^{2+} as the activating metal for Rubisco also alters the oxygenase/carboxylase ratio. But these changes, while they are no doubt significant, are relatively small compared with the changes that are observed from the other strategy nature has adopted to cope with oxygenase activity, namely, that of CO_2 concentration.

The CO_2-concentrating mechanisms, under normal conditions, virtually eliminate oxygenase activity as no effect of oxygen on photosynthesis can be observed. CAM plants and C4 plants have developed a supplementary system of fixation of CO_2 into C4 acids that acts as the front-end to the carboxylation reaction of Rubisco. In both plants, the decarboxylation of these C4 acids elevates the CO_2 in the vicinity of Rubisco to enhance carboxylation and suppress oxygenation. In CAM plants there is a temporal separation between the initial CO_2 fixation and the final CO_2 fixation. In C4 plants there is an anatomical separation between these events, with Rubisco specifically localized only in a certain portion of the chlorophyll-containing cells. In both types of plants an elaborate complex biochemical system has had to develop for the complete pathway of CO_2 fixation to operate. Microalgae and cyanobacteria seem to have taken a more direct approach to the problem by developing systems that directly concentrate CO_2. In both organisms, CO_2 as well as HCO_3^- are actively transported. In cyanobacteria these transport systems are located in the plasma membrane, but in the microalgae they may also be present in the chloroplast envelope in addition to the plasma membrane. There is no evidence for active CO_2 transport in the chloroplasts of higher plants. Why have these mechanisms for CO_2 concentration been lost in the evolution of algae into higher plants? This question and many others will continue to provide fertile ground for research for years to come, although the increasing CO_2 concentration of the atmosphere may make CO_2-concentrating mechanisms less important.

Further reading and references

Aizawa, K. and Miyachi, S. (1986). Carbonic anhydrase and CO_2 concentrating mechanisms in microalgae and cyanobacteria. *FEMS Microbiol. Rev.* **39**, 215–33.

Artus, N. N., Somerville, S. C. and Somerville, C. R. (1986). The biochemistry and cell biology of photorespiration. *CRC Crit. Rev. Plant Sci.* **4**, 121–47.

Badger, M. R. (1985). Photosynthetic oxygen exchange. *Ann. Rev. Plant Physiol.* **36**, 27–53.

Badger, M. R. and Andrews, T. J. (1982). Photosynthesis and inorganic carbon usage by the marine cyanobacterium *Synechococcus* sp. *Plant Physiol.* **70**, 517–23.

Badger, M. R. and Canvin, D. T. (1981). Oxygen uptake during photosynthesis in C3 and C4 plants. In *Photosynthesis IV. Regulation of Carbon Metabolism*, ed. G. Akoyunoglou, Balaban International Science Services, Philadelphia, pp. 151–61.

Badger, M. R. and Gallagher, A. (1987). Adaptation of photosynthetic CO_2 and HCO_3 accumulation by the cyanobacterium *Synechococcus* PCC6301 to growth at different inorganic carbon concentrations. *Aust. J. Plant Physiol.* **14**, 189–201.

Badger, M. R., Bassett, M. and Comins, H. N. (1985). A model for HCO_3 accumulation and photosynthesis in the cyanobacterium *Synechococcus* sp. *Plant Physiol.* **77**, 465–71.

Berry, J., Boynton, J., Kaplan, A. and Badger, M. (1976). Growth and photosynthesis of *Chlamydomonas reinhardtii* as a function of CO_2 concentration. *Carnegie Inst. Year Book* **75**, 423–32.

Blackwell, R. D., Murray, A. J. S., Lea, P. J., Kendall, A. C., Hall, N. P., Turner, J. C. and Wallsgrove, R. M. (1988). The value of mutants unable to carry out photorespiration. *Photosyn. Res.* **16**, 155–76.

Bowes, G. (1985). Pathways of CO_2 fixation by aquatic organisms. In *Inorganic Carbon Uptake by Aquatic Photosynthetic Organisms*, eds W. J. Lucas and J. A. Berry, American Society of Plant Physiologists, Rockville, MD, pp. 187–210.

Bowes, G., Ogren, W. L. and Hageman, R. H. (1971). Phosphoglycolate production catalyzed by ribulose diphosphate carboxylase. *Biochem. Biophys. Res. Commun.* **45**, 716–22.

Canvin, D. T. (1979). Photorespiration: Comparison between C3 and C4 plants. In *Encyclopedia of Plant Physiology*, Vol. 6, eds M. Gibbs and E. Latzko, Springer-Verlag, Berlin, pp. 368–96.

Canvin, D. T., Berry, J. A., Badger, M. R., Fock, H. and Osmond, C. B. (1980). Oxygen exchange in leaves. *Plant Physiol.* **66**, 302–7.

Chollet, R. and Ogren, W. L. (1975). Regulation of photorespiration in C3 and C4 species. *Bot. Rev.* **41**, 137–79.

Cockburn, W., Ting, I. P. and Sternberg, L. O. (1979). Relationships between stomatal behaviour and internal carbon dioxide concentration in Crassulacean acid metabolism plants. *Plant Physiol.* **63**, 1029–32.

Espie, G. S. and Colman, B. (1986). Inorganic carbon uptake during photosynthesis. I. A theoretical analysis using the isotope disequilibrium technique. *Plant Physiol* **80**, 863–9.

Espie, G. S., Miller, A. G., Birch, D. G. and Canvin, D. T. (1988). Simultaneous transport of CO_2 and HCO_3^- by the cyanobacterium *Synechococcus* UTEX 625. *Plant Physiol.* **87**, 551–4.

Flügge, I. V., Woo, K. C. and Heldt, H. W. (1988). Characteristics of 2-oxoglutarate and glutamate transport in spinach chloroplasts. *Planta* **174**, 534–41.

Furbank, R. T. and Hatch, M. D. (1987). Mechanism of C4 photosynthesis. The size and composition of the inorganic carbon pool in bundle sheath cells. *Plant Physiol.* **85**, 958–64.

Gerbaud, A. and Andre, M. (1987). An evaluation of the recycling in measurements of photorespiration. *Plant Physiol.* **83**, 933–7.

Holaday, A. S. and Chollet, R. (1984). Photosynthetic/photorespiratory characteristics of C3–C4 intermediate species. *Photosyn. Res.* **5**, 307–23.

Huang, A. H., Trelease, R. N. and Moore, T. S. (1983). *Plant Peroxisomes*, Academic Press, New York.

Husic, D. W., Husic, H. D. and Tolbert, N. E. (1987). The oxidative photosynthetic carbon cycle or C2 cycle. *CRC Crit. Rev. Plant Sci.* **5**, 45–100.

Kendall, A. C., Keys, A. J., Turner, J. C., Lea, P. J. and Miflin, B. J. (1983). The isolation and characterization of a catalase-deficient mutant of barley (*Hordeum vulgare* L.) *Planta* **159**, 505–11.

Kyle, D. J. and Lohad, I. (1986). The mechanisms of photoinhibition in higher plants and green algae. In *Encyclopedia of Plant Physiology*, Vol. 19, eds L. A. Staebelin and C. J. Arntzen, Springer-Verlag, Berlin, pp. 468–75.

Lorimer, G. (1981). The carboxylation and oxygenation of ribulose-1,5-bisphosphate: the primary events in photosynthesis and photorespiration. *Ann. Rev. Plant Physiol.* **32**, 349–83.

Lorimer, G. H. and Andrews, T. J. (1981). The C$_2$ chemo- and photorespiratory carbon oxidation cycle. In *The Biochemistry of Plants*, Vol. 8, eds M. D. Hatch and N. K. Boardman, Academic Press, New York, pp. 329–74.

Lucas, W. J. and Berry, J. A. (eds) (1985). *Inorganic Carbon Uptake by Aquatic Photosynthetic Organisms*, American Society of Plant Physiology, Rockville, MD.

McC. Lilley, R., Ebbinghausen, H. and Heldt, H. W. (1987). The simultaneous determination of carbon dioxide release and oxygen uptake in suspensions of plant leaf mitochondria oxidizing glycine. *Plant Physiol.* **83**, 349–53.

Miller, A. G. and Canvin, D. T. (1987). Na$^+$-stimulation of photosynthesis in the cyanobacterium *Synechococcus* UTEX 625 grown on high levels of inorganic carbon. *Plant Physiol.* **84**, 118–24.

Miller, A. G., Turpin, D. H. and Canvin, D. T. (1984). Growth and photosynthesis of the cyanobacterium *Synechococcus leopoliensis* in HCO$_3$-limited chemostats. *Plant Physiol.* **75**, 1064–70.

Miller, A. G., Espie, G. S. and Canvin, D. T. (1988). Active transport of CO$_2$ by the cyanobacterium *Synechococcus* UTEX 625. *Plant Physiol.* **86**, 677–83.

Monson, R. K., Edwards, G. E. and Ku, M. S. B. (1984). C3–C4 intermediate photosynthesis in plants. *Bioscience* **34**, 563–74.

Moroney, J. W., Kitayama, M., Togosaki, R. K. and Tolbert, N. E. (1987). Evidence for inorganic carbon transport by intact chloroplasts of *Chlamydomonas reinhardtii*. *Plant Physiol.* **83**, 460–3.

Ogren, W. L. (1984). Photorespiration: pathways, regulation, and modification. *Ann. Rev. Plant Physiol.* **35**, 415–42.

Osmond, C. B. (1981). Photorespiration and photoinhibition. Some implications for the energetics of photosynthesis. *Biochim. Biophys. Acta* **639**, 77–98.

Peterson, R. B. (1987). Quantitation of the O$_2$-dependent, CO$_2$ reversible component of the postillumination CO$_2$ exchange transient in tobacco and maize leaves. *Plant Physiol.* **84**, 862–7.

Pierce, J. and Omata, T. (1988). Uptake and utilization of inorganic carbon by cyanobacteria. *Photosyn. Res.* **16**, 141–54.

Powles, S. B. (1984). Photoinhibition of photosynthesis induced by visible light. *Ann. Rev. Plant Physiol.* **35**, 15–44.

Quebedeaux, B. and Hardy, R. W. F. (1976). Oxygen concentration: Regulation of crop growth and Productivity. In *CO$_2$ Metabolism and Plant Productivity*, eds R. H. Burris and C. C. Black, University Park Press, Baltimore, pp. 185–204.

Reinhold, L., Zviman, M. and Kaplan, A. (1987). Inorganic carbon fluxes and photosynthesis in cyanobacteria – a quantitative model. In *Progress in Photosynthesis Research*, Vol. IV, ed. J. Biggins, Martinus Nijhoff Publishers, Dordrecht, pp. 289–96.

Reiskind, J. B., Seamon, P. T. and Bowes, G. (1988). Alternative methods of photosynthetic carbon assimilation in marine macroalgae. *Plant Physiol.* **87**, 686–92.

Schnarrenberger, C. and Fock, H. (1976). Interactions among organelles involved in photorespiration. In *Encyclopedia of Plant Physiology*, Vol. 3, eds C. R. Stocking and U. Heber, Springer-Verlag, Berlin, pp. 185–234.

Somerville, C. R. (1986). Analysis of photosynthesis with mutants of higher plants and algae. *Ann. Rev. Plant Physiol.* **37**, 467–507.

Somerville, S. C. and Somerville, C. R. (1983). Effect of oxygen and carbon dioxide on photorespiratory flux determined from glycine accumulation in a mutant of *Arabidopsis thaliana*. *J. Expt. Bot.* **34**, 415–24.

Spalding, M. H., Stumpf, D. K., Ku, M. S. B., Burris, R. H. and Edwards, G. E. (1979). Crassulacean acid metabolism and diurnal variations of internal CO$_2$ and O$_2$ concentrations in *Sedum praealtum* DC. *Aust. J. Plant Physiol.* **6**, 557–67.

Sultemeyer, D. F., Miller, A. G., Espie, G. S., Fock, H. P. and Canvin, D. T. Active CO$_2$ transport by the green alga *Chlamydomonas reinhardtii*. *Plant Physiol.* **89**, 1213–19.

Ting, I. P. (1985). Crassulacean acid metabolism. *Ann. Rev. Plant Physiol.* **36**, 595–622.

Tolbert, N. E. (1979). Glycolate metabolism by higher plants and algae. In *Encyclopedia of Plant Physiology*, Vol. 6, eds M. Gibbs and E. Latzko, Springer-Verlag, Berlin, pp. 338–52.

Volokita, M., Zenvirth, D., Kaplan, A. and Reinhold, L. (1984). Nature of inorganic carbon species actively taken up by the cyanobacterium *Anabaena variabilis*. *Plant Physiol.* **76**, 599–602.

Wallsgrove, R. M., Keys, A. J., Lea, P. J and Miflin, B. J. (1983). Photosynthesis, photorespiration and nitrogen metabolism. *Plant Cell Environ.* **6**, 301–9.

Williams, T. G. and Turpin, D. H. (1987). The role of external carbonic, anhydrase in inorganic carbon acquisition by *Chlamydomonas reinhardtii* at alkaline pH. *Plant Physiol.* **83**, 92–6.

Woo, K. C., Flugge, V. I. and Heldt, N. W. (1987). A two-translocator model for the transport of 2-oxoglutarate and glutamate in chloroplasts during ammonia assimilation in the light. *Plant Physiol.* **84**, 624–32.

Zelitch, I. (1971). *Photosynthesis, Photorespiration and Plant Productivity*, Academic Press, New York.

Zoglowek, C., Kromer, S. and Heldt, H. W. (1988). Oxaloacetate and malate transport by plant mitochondria. *Plant Physiol.* **87**, 109–15.

19 The flux of metabolites in C4 and CAM plants
R. C. Leegood and C. B. Osmond

Introduction

Plants with the C4 dicarboxylic acid pathway of photosynthesis (C4 plants) and those with Crassulacean acid metabolism (CAM plants) can be distinguished from others by the initial products of $^{14}CO_2$ fixation and by leaf anatomy. These two features belie a complex of different metabolite transport processes which serve a common physiological function, a CO_2-concentrating mechanism which mitigates the oxygenase activity of Rubisco (Andrews and Lorimer, 1987). This potentially improves the efficiency of carboxylation, of water use and of nitrogen use in photosynthesis, with a number of important implications for plant performance and survival (Osmond et al., 1982; Edwards and Walker, 1983; Winter, 1985; Nobel, 1988).

Our appreciation of the significance of these leaf anatomical differences in relation to photosynthetic processes is quite recent. Although Haberlandt (1884) drew attention to the peculiar 'Kranz' (German for wreath) organization of photosynthetic cells which calls for intercellular metabolite transport in what we now know as C4 plants (Fig. 19.1a), it remained for Burr et al. (1957) and Karpilov (1960) to highlight the labeling of C4 acids as early products of $^{14}CO_2$ fixation in sugar cane and maize, respectively. The pathway was subsequently named after these initial products of carboxylation. Although the unusual capacity for dark CO_2 fixation (De Saussure, 1804) and nocturnal acidification and diurnal de-acidification (Heyne, 1815) in succulent plants was established in the 19th century, the importance of intracellular transport of malic acid into and out of the large vacuole (Fig. 19.1c) was not appreciated. The photosynthetic implications were not recognized until Kunitake and Saltman (1958) showed that malic acid labeled in the dark with $^{14}CO_2$ was conserved and converted to photosynthetic products in the light. Well before this, the metabolic pathway had been named after the family Crassulaceae, which contains many genera and species which show this acid metabolism. It is fair to say that interpretation of CAM as a photosynthetic process was stimulated by analogies with C4 photosynthesis (cf. Ranson and Thomas, 1960; Osmond, 1978).

The carboxylation and decarboxylation events which drive the CO_2-concentrating mechanisms of C4 and CAM plants are similar, but they operate on different anatomical, physiological and biochemical principles. They are initiated by the fixation of HCO_3^- via PEPcase, leading to the synthesis of malate and aspartate, and the subsequent decarboxylation of these C4 acids to generate CO_2 for Rubisco. In C4 plants the PEPcase and Rubisco are spatially separated in outer mesophyll and inner bundle sheath cells of the Kranz complex (Fig. 19.1a,b). Simultaneous bidirectional fluxes of C3 and C4 metabolites in the symplasm connecting these cells are required to sustain C4 photosynthesis, and the 4-C carboxyl carbon of malate and aspartate serves as a short-lived reservoir of CO_2 during transport. Small amounts of C4 acids (a few μmol per mg chlorophyll) are involved, and this transport pool turns over in about 10 s. In CAM plants, on the other hand, both PEPcase and Rubisco are present in all chloroplast-containing cells. The activities of these enzymes are thus regulated in

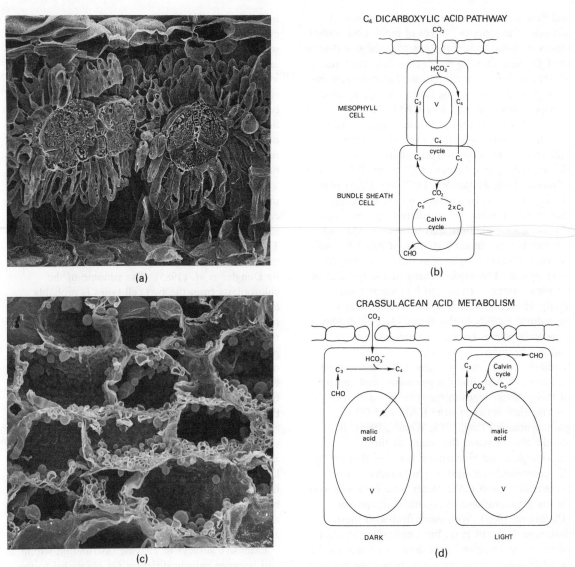

Fig. 19.1 Spatial (C4 plants) and temporal (CAM plants) separation of components of the C4 cycle and Calvin cycle which underlie metabolite transport in C4 and CAM plants. (a) SEM of a transverse section of the leaf of *Atriplex spongiosa*, a C4 plant, showing the radial arrangement of outer mesophyll (M) and inner bundle sheath cells (BS) (courtesy of J. H. Troughton). (b) A generalized schematic representation of the spatial separation of the carboxylation step in the C4 cycle from decarboxylation and the Calvin cycle in C4 plants. (c) SEM of part of a *Kalanchoe daigremontiana* leaf showing the large cell vacuoles (V) (courtesy R. A. Balsamo and E. G. Uribe). (d) A generalized schematic representation of the temporal separation of the carboxylation step of the C4 cycle from decarboxylation and the Calvin cycle in CAM plants.

time, rather than in space, with PEPcase active in the dark but inactive for much of the light period, and Rubisco active only in the light. In CAM plants the 4-C carboxyl of malate serves as a

longer-lived, much larger temporary store of CO_2. Large amounts of malic acid (several hundreds of μmoles per mg chlorophyll) may accumulate in the large cell vacuoles of CAM plants (Fig. 19.1c,d), and may have a turnover time of thousands of seconds. The tonoplast fluxes of malic acid, rather than symplastic transport, thus play the key role in the CO_2-concentrating mechanism of CAM, and also in the day–night regulation of enzyme activity.

In general, none of the reactions, regulations or transport processes in C4 or CAM plants is unique. However, they are coordinated in space and time in several unique ways which permit biochemical, physiological and anatomical differentiation of subtypes of C4 plants (Hatch and Osmond, 1976; Hattersley, 1987) and CAM plants (Osmond, 1978). In addition, it is now recognized that there are many intermediate forms between C4 and C3 metabolism, and between CAM and C3 metabolism. In early studies, artificial hybrids between C3 and C4 plants in the genus *Atriplex* were produced to explore many of the features of C4 photosynthesis (reviewed by Osmond *et al.*, 1980). However, it is now known that what may be natural hybrids in the genus *Flaveria* display genetically stable, but intermediate physiology and biochemistry (Edwards and Ku, 1987). Similarly, there are many genetically-determined variants of CAM, and the environmental control of this pathway leads to many manifestations intermediate between strict CAM and C3 photosynthesis (Ting, 1985; Winter, 1985). It should also be noted that many of the physiological and biochemical features (but rarely the anatomical features) of photosynthesis in terrestrial C4 and CAM plants have been observed during photosynthesis in aquatic macrophytes (Cockburn, 1985). We cannot deal with these variations in detail here, but emphasize the utility of these 'intermediate' organisms in evaluating the real functional significance, and hence the selective advantage and probable evolutionary relationships, of these photosynthetic pathways.

Intercellular and intracellular metabolite transport during C4 photosynthesis

The dramatic specialization of photosynthetic functions between the mesophyll and bundle sheath cells of C4 plants has few parallels. Perhaps the best analog is the cooperative function of heterocysts and vegetative cells in the N_2-fixing cyanobacteria (see Chapter 26). In its most sophisticated form, C4 photosynthesis in tropical grasses such as sugar cane and maize depends on mesophyll cell chloroplasts which lack most of the enzymes of the Calvin cycle, and bundle sheath cells with poorly developed PSII activities. This means that photosynthetic CO_2 fixation in C4 plants depends on inter- and intracellular regulation of metabolite transport within and between cells. Karpilov (1970) introduced the term 'cooperative photosynthesis' to describe the interdependencies of mesophyll and bundle sheath cells. There are many fascinating questions concerning the differentiation of the C4 system. The developmental anatomy of the 'Kranz' sheath in different types of C4 plants has been explored by Dengler *et al.* (1985). The genome of the chloroplasts from the two cell types is probably the same (Walbot, 1977) but transcriptional regulation of gene expression by light seems to be responsible for the differentiation of photosynthetic properties. Differential expression of the genes for the enzymes peculiar to the C4-cycle seems to involve light-modulation of transcript pools in the two cell types which parallel the development of etioplasts during greening, rather than the morphogenesis of the two cell types (Sheen and Bogorad, 1987). Greening is associated with the loss of mRNA for Rubisco large subunit in mesophyll plastids and a light-dependent arrest of its transcription. In bundle sheath plastids the expression of this gene, and of genes for the small subunit of Rubisco, is amplified by light (Sheen and Bogorad, 1985).

Extensive surveys of enzyme distribution within and between cells in different C4 plants led to the identification of three decarboxylase enzymes in the bundle sheath cells (Hatch and Osmond, 1976; Edwards and Walker, 1983). Apart from the different patterns of intercellular movement of C3 and C4 compounds that these subtypes of C4 photosynthesis imply, there are patterns of regulation in C4 plants which differ from those in C3 plants. One of the most striking examples is the way in which 3-PGA reduction is shared between the Calvin cycle in bundle sheath chloroplasts and the mesophyll chloroplasts in all subtypes of C4 plants.

Intercellular transport

In the majority of C4 plants the photosynthetic cells are organized in two concentric cylinders. Thin-walled mesophyll cells with large intercellular spaces radiate from the thick-walled bundle sheath cells (Fig. 19.1a). In some C4 plants, the cell walls between the bundle sheath and mesophyll contain suberin, but in all it seems likely that the thickened cell wall is itself a major barrier to the diffusion of solutes and gases (Hattersley, 1987). On the wall between mesophyll and bundle sheath cells there are extensive pit-fields with plasmodesmata which provide symplastic connections between cells (Botha and Evert, 1988). The plasmodesmata evidently exclude large molecules, such as the cytoplasmic enzyme PEPcase which is restricted to mesophyll cells, and the size exclusion limit seems to be about 900 daltons (Burnell, 1988). However, plasmodesmata generally permit the rapid exchange of solutes between cells and slower exchange of gases in solution.

Using a simple analogy based on symplastic transport of solutes across the concentric cylinders of cortical and vascular tissue in roots, Osmond (1971) concluded that metabolite transport in C4 photosynthesis could be sustained by diffusion, driven by gradients in the concentration of specific metabolites between the source cells and the sink cells. Thus, in the case of malate or aspartate, Hatch and Osmond (1976) estimated that a gradient with the concentration in mesophyll source cells 10–30 mM higher than in bundle sheath sink cells would be needed to sustain observed rates of photosynthesis in *Zea* and *Amaranthus*, respectively. *Zea*, which has centrifugally arranged bundle sheath chloroplasts and a short diffusion path, and *Amaranthus*, which has centripetally arranged chloroplasts and a longer diffusion path, are representative of a wide range of C4 plants (Fig. 19.2).

Many of the assumptions on which these conclusions were based have been supported by subsequent evidence. For example, Hatch and Osmond (1976) assumed an effective surface area of only 3% of the mesophyll/bundle sheath interface. Plasmodesmata are frequent only in primary pit-fields at the areas of contact between mesophyll and bundle sheath cells, and in the C4 grass *Themeda*, 57% of all the plasmodesmata in vascular bundles are at this interface (Botha and Evert, 1988). In several studies it has been found that the cross-sectional area of the sphincter in plasmodesmata, or of the desmotubule, occupies 1.5–3% of the pit-field of the surface between mesophyll and bundle sheath cells. Perhaps the greatest uncertainty stems from the need to sustain bidirectional metabolite fluxes, when there is a net flux of sucrose from mesophyll to bundle sheath coincident with water flux in the other direction.

Intercellular transport and leaf anatomy

The necessity for metabolite transport between the two types of cells requires intimate contact and therefore sets limits on the amount of mesophyll tissue which can be functionally associated with bundle sheath tissue. For this reason, the leaf thickness is limited in C4 plants, and there are significant differences between the various C4 decarboxylation types (Hattersley, 1982, 1987). For example, in C4 plants the maximum number of chlorenchymatous mesophyll cells which intervene between the photosynthetic bundle sheath cells is between two and four. Hence, the interveinal distance is usually smaller than in the leaves of C3 plants. A number of structural features, such as the position of chloroplasts within the bundle sheath cells, have been used to predict the different decarboxylation types (Fig. 19.2). Others include the presence of cells which intervene between the metaxylem vessel elements and the chlorenchymatous bundle sheath cells and the occurrence of an even or uneven outline to the bundle sheath cell walls, which influences the area of contact between the two cell types (Hattersley, 1987). However, the evidence suggests that such associations between biochemical type and structure may not be wholly reliable (e.g. PEPCK and NAD^+-ME type *Panicum* species are anatomically indistinguishable; Prendergast *et al.*, 1987). At the subcellular level, other differences that are related to metabolic pathways arise. In $NADP^+$-ME plants, the bundle sheath chloroplasts contain few grana and show a deficiency of PSII proteins and function (Hatch and Osmond, 1976;

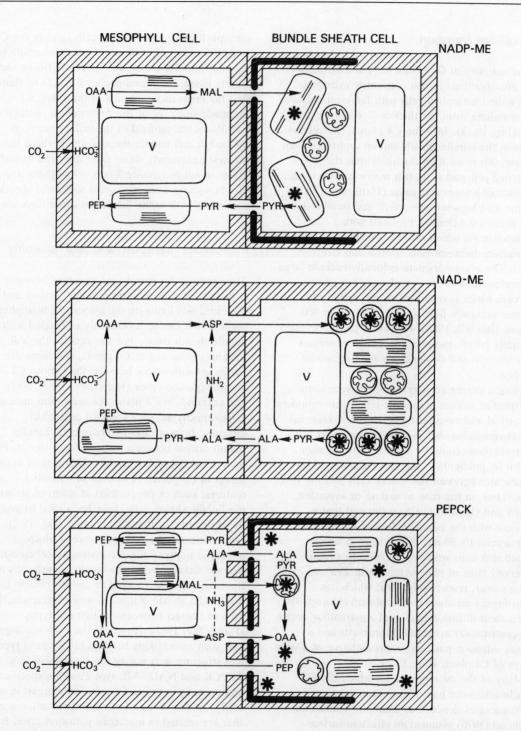

Fig. 19.2 Generalized schematic representation of biochemical and structural bases for metabolite transport and the differentiation of subtypes of C4 photosynthesis (based on Hatch and Osmond, 1976; Hattersley, 1987; Burnell and Hatch, 1988). The sites of C4 acid decarboxylation are indicated by the * and the suberin lamella by the solid black line.

Schuster et al., 1985). In NAD$^+$-ME and PEPCK plants which decarboxylate malate in the bundle sheath mitochondria, the mitochondrial frequency is between two and four times that in the bundle sheath of NADP$^+$-malic enzyme type C4 species (Fig. 19.2).

The chloroplast position and the presence of other cells between the mesophyll and the bundle sheath will affect the distance over which metabolites must be transported. These cells also need to be considered in relation to translocation of sucrose from the mesophyll cells to the vascular tissue. The occurrence of a suberized lamella varies; it is absent in dicotyledonous species, and in grasses is present only in species with either an uneven bundle sheath outline or with centrifugally located chloroplasts. In those species with uneven cell outlines the suberized lamella may be necessary to prevent CO_2 leakage through the high surface area of the bundle sheath/mesophyll interface (Hattersley, 1982).

Intercellular movement of C3 and C4 metabolites

Evidence of rapid metabolite movement may be inferred from the rapid transfer of ^{14}C from C4 acids (labeled in the mesophyll) to 3-PGA and products (labeled in the bundle sheath) which occurs when $^{14}CO_2$ is supplied to leaves of C4 plants (Hatch, 1971), and in the distribution of radioactivity in a leaf of *Atriplex spongiosa* in pulse-chase experiments, as revealed by microautoradiography (Osmond, 1971). After a 2 s pulse of $^{14}CO_2$ (when the majority of the label is in C4 acids), the cytoplasm of the mesophyll cells is clearly labeled, but a considerable amount of label is also found in the bundle sheath cells. The majority of the label is transferred to the bundle sheath during a 10 s chase. These data indicate very rapid movement of C4 acids, and also sizeable pools of these compounds within the bundle sheath cells.

Direct measurements of the gradients of metabolites in leaves of *Zea* are shown in Fig. 19.3 (Leegood, 1985; Stitt and Heldt, 1985b). The gradient of malate between the mesophyll and the bundle sheath is sufficient to generate a flux between the two cell types (7.2 mmol m^{-2} s^{-1}) which is much larger than the photosynthetic flux (0.5 mmol m^{-2} s^{-1}) estimated by Hatch and Osmond (1976). The gradient of malate in Fig. 19.3, and hence the flux, is likely to be a considerable overestimate, since much of the malate pool is inactive in photosynthesis (i.e. present in the vacuole or in non-photosynthetic cells). An anomalous feature shown in Fig. 19.3 is the predominance of pyruvate in the mesophyll compartment, so that the gradient of pyruvate appears to lie in the opposite direction to the expected flux, but this is a result of the intracellular transport of pyruvate within mesophyll cells (see below). The amounts of 3-PGA and triose-P in leaves of maize are typically 10–20 times higher than in the leaves of C3 plants. This reflects the opposing intercellular concentration gradients of these metabolites (Fig. 19.3) which are adequate to generate fluxes required by the assimilation rates.

Although most other enzymes of the Calvin cycle are absent from the mesophyll cells, all C4 species possess the enzymes for 3-PGA reduction in the mesophyll chloroplasts (Hatch and Osmond, 1976). The triose-P which is formed from 3-PGA in the mesophyll has two fates. The first is its conversion to sucrose or starch in the mesophyll cells (see below), while the second is its return to the bundle sheath to regenerate RuBP. In plants such as maize, in which 50% of the 3-PGA generated in the bundle sheath may be reduced in the mesophyll, a minimum of two-thirds of the triose-P must be returned to the bundle sheath to maintain pools of Calvin cycle intermediates. In NADP$^+$-ME plants this, and NADPH generation from malic enzyme demands less PSII activity in the bundle sheath and hence O_2 evolution is reduced. This favors the carboxylation of RuBP over its oxygenation. Another role for 3-PGA reduction in the mesophyll could be coordination of Calvin cycle and C4 cycle turnover (see below). Additionally, it may be a means of ensuring H$^+$ transport, and hence charge balance, between the two cell types. The reduction of 3-PGA to triose-P in the mesophyll consumes a proton, after which the triose-P is transported to the bundle sheath. The consumption of a proton in this reaction is necessary because a proton is released when CO_2

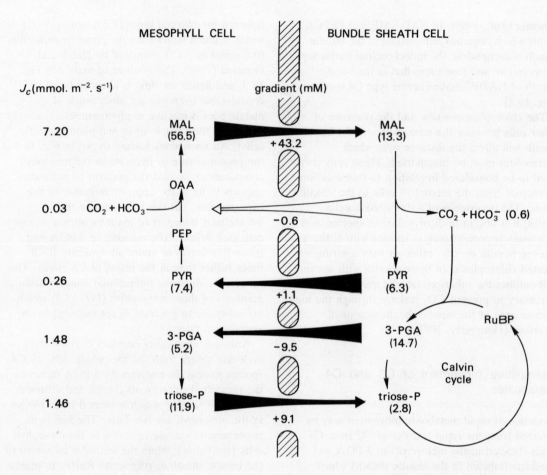

Fig. 19.3 Metabolite concentrations (mM) in mesophyll and bundle sheath cells which drive metabolite transport in *Zea mays* (mean values from Leegood (1985) and Stitt and Heldt (1985a) are shown in parentheses). The gradient in concentration (relative to the bundle sheath cells) is shown across the plasmodesmata, and the maximum flux which could be sustained by the gradient (J_c) is calculated using the assumptions of Hatch and Osmond (1976). These authors calculated that a flux of about 0.5 mmol m^{-2} s^{-1} would support the observed rate of photosynthesis.

is hydrated and the resulting HCO$_3^-$ is fixed by PEPcase.

Leaves of NAD$^+$-ME plants such as *Amaranthus edulis* contain amounts of triose-P and 3-PGA similar to those in maize, suggesting that extensive intercellular transport of these metabolites occurs. No direct measurements have been made of metabolite gradients in leaves of C4 plants other than maize. In leaves of *Amaranthus edulis*, an NAD$^+$-malic enzyme species which is believed to transfer aspartate from the mesophyll to the bundle sheath and return alanine to the mesophyll cells (Fig. 19.2), leaf pools of aspartate and alanine are sufficient to account for diffusion-driven transport of these compounds between the mesophyll and bundle sheath cells under many different flux conditions (Leegood and von Caemmerer, 1988). In PEPCK plants, transport is likely to be rather more complex, because both malate and aspartate are transferred from the mesophyll to the bundle sheath, and PEP and alanine return to the mesophyll (Fig. 19.2).

Intercellular movement of inorganic carbon

The symplastic pathway of metabolite movement also permits movement of inorganic carbon out of the bundle sheath. Direct measurements of the pool of HCO_3^- + CO_2 from radiotracer studies indicate that the total pool of inorganic carbon is about 0.6 mM in the bundle sheath of maize (Hatch, 1971). Analysis of CO_2-concentrating mechanisms in the three different decarboxylation types (Furbank and Hatch, 1987) indicates that the absence of carbonic anhydrase in bundle sheath cells is of critical importance. If the mesophyll inorganic carbon pool is assumed to be zero, then the flux of HCO_3^- via the plasmodesmata would be 0.03 mmol m^{-2} s^{-1}, which is less than 10% of the required flux for metabolites (Fig. 19.3). Thus leakage of HCO_3^- via the plasmodesmata is not likely to be a serious problem, nor is the leakage of CO_2, because the diffusion coefficients of gases in solution are 10^4 times less than in air. Measurements of the total pool of inorganic carbon in the bundle sheath of various C4 species show variation between the various decarboxylation types (Table 19.1). These differences could depend on the amount of O_2 evolved in bundle sheath chloroplasts and/or differences in the leakiness of the bundle sheath compartment. We might expect cells with higher inorganic carbon levels to be more leaky, and this appears to correlate with more negative $\delta^{13}C$ values (Hattersley 1982). The latter is an index of the extent to which Rubisco functions in a closed compartment, entirely dependent upon CO_2 previously fixed by PEPcase (O'Leary, 1981). The more leaky systems are less efficient, in that futile cycling of the energetically expensive C4 cycle leads to lower quantum yields (Edwards et al., 1985).

C3–C4 intermediates

Over twenty species of plants exhibit photosynthetic characteristics that are intermediate between C3 and C4 plants in that they show reduced rates of photorespiration and CO_2 compensation points in the range 7–15 $\mu l\ l^{-1}$ (Brown and Brown, 1975), compared with typical values in C3 plants of 50 $\mu l\ l^{-1}$ and in C4 plants of less than 5 $\mu l\ l^{-1}$. Although all show a degree of 'Kranz' anatomy, both mesophyll and bundle sheath cells contain Rubisco and amounts of C4 pathway enzymes are generally low. Two main types have been distinguished by Edwards and Ku (1987). In one type (e.g. *Panicum milioides* and *Moricandia arvensis*), there is little evidence for a functional C4 acid cycle which donates CO_2 from C4 acids to the Calvin cycle. The intermediate character with respect to CO_2 compensation point seems to be due to more efficient refixation of photorespiratory CO_2 in the bundle sheath cells. Like C4 plants, the mesophyll cell mitochondria of *Moricandia arvensis* and many other 'intermediate' plants have low activities of glycine decarboxylase. Thus photosynthesis in leaves of *M. arvensis* might involve shuttling of photorespiratory intermediates such as glycine from the mesophyll to the bundle sheath and the return of serine to the mesophyll (Hylton et al., 1988), although the operation of such a shuttle has yet to be demonstrated.

In the other type of 'intermediate' plants (such as *Flaveria anomala* and *Neurachne minor*), appreciable activities of the C4 pathway enzymes PEPcase, pyruvate P_i dikinase and $NADP^+$-malic enzyme are found. These plants show varying capacities to fix ^{14}C into C4 acids during

Table 19.1 Inorganic carbon pools in leaves of C4 plants from different subtypes and its correlation with ^{13}C values

C4 type	Inorganic carbon pool in bundle sheath (CO_2 and HCO_3^-)(mM)	$\delta^{13}C$ value (⁰/00)
$NADP^+$-ME type		−11.35
Echinochloa crusgalli	0.37	
Sorghum vulgare	0.16	
NAD^+-ME type		−11.95
Eleusine indica	0.77	
Panicum miliaceum	0.35	
PEPCK type		−12.7
Urochloa panicoides	0.99	
Panicum maximum	0.83	

Data from Hattersley (1982) and Furbank and Hatch (1987).

short-term exposure to $^{14}CO_2$, and to transfer this to products of the Calvin cycle, suggesting limited capacity for a C4 pathway. Comparison of quantum yields show that these 'intermediate' plants have higher energy requirements than C3 or C4 plants, presumably because the C4 pathway activity is largely futile (Edwards et al., 1985).

Intracellular transport in C4 plants

The primary carboxylase of C4 photosynthesis is a cytoplasmic enzyme, PEPcase, which draws on the light-dependent production of its substrate, PEP, in the chloroplast and in many cases the light-dependent reduction of its product, OAA, also occurs in the chloroplast (Fig. 19.2). Consequently, intracellular transport of metabolites between the cytoplasm and organelles is a key component of C4 photosynthesis in both mesophyll and bundle sheath cells. Leaf organelles in C4 plants share the translocators of organelles in leaves of C3 plants, but also contain translocators with unique or greatly altered kinetic properties.

Mesophyll chloroplasts

The observation that there is no overall gradient of pyruvate between the bundle sheath and mesophyll cells (Fig. 19.2) suggests that a gradient between the bundle sheath and the mesophyll might simply be obscured by accumulation of pyruvate within the mesophyll chloroplasts. Transport of pyruvate on a specific carrier (Huber and Edwards, 1977) occurs in chloroplasts of both C3 and C4 plants, but the translocator is much more active and is light-dependent in mesophyll chloroplasts of C4 plants (Flügge et al., 1985; Ohnishi and Kanai, 1987). Moreover, the light activation of pyruvate P_i dikinase seems to depend on maintenance of high pyruvate/PEP ratios (Burnell and Hatch, 1985), so it seems likely that pyruvate is maintained at high concentrations in the mesophyll chloroplasts.

The dicarboxylates, malate, oxaloacetate, 2-oxoglutarate, aspartate and glutamate are transported through the mesophyll chloroplast envelope in a carrier-mediated mode (Fig. 19.4). These compounds undergo counter-exchange on

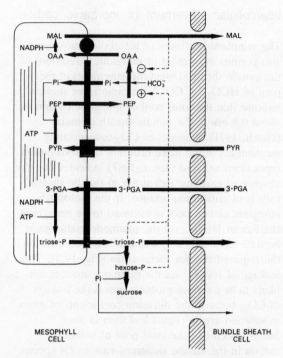

Fig. 19.4 Schematic representation of intercellular and mesophyll chloroplast metabolite transport in a $NADP^+$-ME-type C4 plant such as Zea mays. The dicarboxylate shuttle is shown by a circle, the pyruvate translocator by a square and the P_i translocator by the rectangular symbols on the chloroplast envelope. The equilibrium between 3-PGA and PEP in the cytoplasm is indicated by the dotted line. Feedback ($-$) and feedforward ($+$) modulation of cytoplasmic PEPcase by cytoplasmic malate and triose-P/hexose-P respectively, is indicated by the dashed lines.

the dicarboxylate translocator and the K_m for uptake of a particular dicarboxylic acid is similar to the K_i for the inhibition of uptake by other dicarboxylates. In maize, for example, the K_i (0.3 mM) for oxaloacetate inhibition of malate transport is comparable to the K_m (0.5 mM) for malate uptake (Day and Hatch, 1981a). Such a system is unsuitable for catalyzing oxaloacetate uptake where oxaloacetate concentrations are several orders of magnitude less than malate concentrations, as occurs in $NADP^+$-ME plants such as maize, in which OAA concentrations are probably less than 50 μM. Not surprisingly, Hatch et al. (1984) have provided evidence for a very active oxaloacetate carrier in maize (K_m (OAA) 45 μM) which is little affected by malate (K_i (malate) 7.5 mM).

One feature which distinguishes chloroplastic transport of phosphorylated intermediates in C4 plants from that in C3 plants is the direction of transport across the chloroplast envelope. During photosynthesis in C4 plants, the mesophyll chloroplasts *import* 3-PGA and *export* triose-P, and the bundle sheath chloroplasts *export* 3-PGA and *import* triose-P which has been reduced by the mesophyll chloroplasts (Fig. 19.4). In addition to the exchange of 3-PGA, triose-P and P_i, the mesophyll chloroplasts also catalyze the export of PEP, formed in the chloroplast by the action of pyruvate P_i dikinase, in exchange for P_i to sustain PEPcase in the cytoplasm. Although the phosphate translocator of C4 mesophyll chloroplasts has not been studied to the extent that it has in C3 chloroplasts (and has not yet been isolated), there is evidence which indicates that exchange of PEP/P_i and 3-PGA/triose-P occurs on a common translocator in the chloroplast envelope. In contrast, the phosphate translocator in spinach chloroplasts appears both to recognize, and transport, PEP poorly (Heldt and Flügge, 1986).

Studies on metabolite transport into C4 mesophyll chloroplasts show that 3-PGA uptake is competitively inhibited by PEP and P_i (Day and Hatch, 1981b). The translocator in C3 mesophyll chloroplasts can only recognize certain organic phosphates with phosphate in the C-3 position, while the translocator in C4 mesophyll chloroplasts of *Digitaria sanguinalis* can recognize certain organic phosphates with phosphate in the C-2 or C-3 position. Thus, in addition to PEP, phosphoglycolate and 2-PGA are effective inhibitors of 3-PGA-dependent O_2 evolution in C4 mesophyll chloroplasts but have no influence on 3-PGA-dependent O_2 evolution in barley chloroplasts (Rumpho et al., 1987). In terms of metabolic regulation, the practical consequences of having intermediates of the C3 and C4 pathways carried on a common translocator cannot easily be predicted since we are entirely ignorant of the subcellular compartmentation of these metabolites under varying environmental conditions.

Bundle sheath chloroplasts

Although transport processes across the envelope of C4 mesophyll chloroplasts are now adequately characterized, little is known of transport into the bundle sheath chloroplasts, largely because these have yet to be isolated intact and in appreciable quantities from any C4 plants. Bundle sheath chloroplasts are unusual in catalyzing 3-PGA export at high rates (Fig. 19.4). In C3 plants, 3-PGA is not exported by chloroplasts to any great extent because 3-PGA is transported by the translocator as the 3-PGA^{2-} ion, whereas 3-PGA^{3-} is the form which predominates at the pH which occurs in the illuminated stroma (Heldt and Flügge, 1986). It is not known whether the stromal pH is lower in bundle sheath chloroplasts, whether 3-PGA^{3-} is specifically exported, or whether 3-PGA within the chloroplast reaches such high concentrations that transport of 3-PGA^{2-} from the chloroplast is inevitable. Little is known about transport of dicarboxylates into bundle sheath chloroplasts, but aspartate has been shown to stimulate malate decarboxylation by bundle sheath chloroplasts of maize. It has therefore been suggested that a carrier specific for malate uptake is present which depends upon the presence of aspartate for maximum activity (Boag and Jenkins, 1985). In contrast to mesophyll chloroplasts, the bundle sheath chloroplasts of *Panicum miliaceum* show a slow carrier-mediated uptake of pyruvate which has very similar characteristics to the carrier in wheat and pea chloroplasts and which is not light-stimulated (Ohnishi and Kanai, 1987).

Leaf mitochondria and C4 photosynthesis

Since the mesophyll cell chloroplasts lack Rubisco, and because Rubisco in bundle sheath chloroplasts functions at high CO_2 concentrations, the flux of carbon through photorespiration is low in C4 plants. Leaf mitochondria in these plants contain low activities of the glycine decarboxylase complex, compared with C3 plants. However, in both the NAD$^+$-ME and PEPCK subtypes of the C4 pathway, bundle sheath cell mitochondria are the major sites of C4 acid decarboxylation (Fig. 19.3).

The carbon fluxes through the mitochondria in the above C4 sub-types are equivalent to the rate of photosynthesis, i.e. several-fold greater than those during photorespiration in C3 plants and 10 to 20-fold greater than respiratory carbon fluxes in leaves. However, even these fluxes are an order of magnitude slower than those involved in thermogenesis in the *Arum* spadix (see Chapter 9). Both in the *Arum* spadix and in NAD^+-ME-type C4 plants it seems that these high carbon fluxes are freed of respiratory control by engagement of the alternative, cyanide-insensitive pathway of respiration (Gardeström and Edwards, 1985). In PEPCK-type C4 plants, on the other hand, a much more complicated interaction between malate decarboxylation in mitochondria and OAA decarboxylation in the cytoplasm is indicated. Burnell and Hatch (1988) suggest that the ATP required for PEPCK in the cytoplasm may arise from the mitochondrial oxidation of malate and oxidative phosphorylation. These complex intracellular metabolite transport patterns are reminiscent of those which appear to operate during de-acidification in CAM plants. In none of these systems have the mitochondrial metabolite translocators been examined, chiefly because of difficulties in extraction of intact mitochondria, in quantity, from bundle sheath cells.

Regulation of C4 photosynthesis as a consequence of metabolite transport

The consequences of intercellular transport in C4 plants for the regulation of photosynthetic carbon metabolism and electron transport are far-reaching. For example, the export of reductant from the mesophyll as malate means that mitochondrial respiration may generate the ATP required for decarboxylation in the bundle sheath cytoplasm of PEPCK-type plants. The operation of the 3-PGA/triose-P shuttle has important consequences for the regulation of starch and sucrose synthesis, regulation of the C4 cycle and regulation of electron transport. Strict regulation of metabolism is also required if metabolite gradients (Fig. 19.3) are not to collapse. In maize, interchange of carbon between PEP and 3-PGA must be curtailed in the bundle sheath in order to prevent the collapse of the gradient of 3-PGA, as must the overall conversion of pyruvate to triose phosphate and of 3-PGA to malate in the mesophyll.

Coordination of the Calvin and C4 cycles

Coordination of the rate at which the Calvin and C4 cycles fix CO_2 is necessary if photosynthesis is to proceed efficiently under different environmental conditions. Coordination could occur in a variety of ways (Fig. 19.4), all of which are linked to metabolite fluxes between the bundle sheath and the mesophyll cells:

(1) In $NADP^+$-malic enzyme species such as maize, the C4 cycle is obligatorily coupled to the Calvin cycle, because NADPH generated by $NADP^+$-malic enzyme is reoxidized by the reduction of 3-PGA in bundle sheath chloroplasts.
(2) Electron transport in the mesophyll chloroplasts not only powers conversion of pyruvate to malate in the C4 cycle, but also drives 3-PGA reduction in all C4-types.
(3) Interconversion of 3-PGA and PEP, catalyzed by phosphoglycerate mutase and enolase in the mesophyll cytoplasm, provides metabolic communication between the C4 cycle and the Calvin cycle.
(4) Products of 3-PGA exported from the Calvin cycle, such as triose-P, and ultimately hexose-P, act as positive effectors of PEPcase, and relieve inhibition by malate.

The regulation of PEPcase in the mesophyll

Coordination of the C4 cycle with the Calvin cycle during photosynthesis is achieved by metabolite modulation of PEPcase in the mesophyll cytoplasm. This enzyme, and the enzyme which generates its substrate, pyruvate P_i dikinase, are known to be modulated by light (Burnell and Hatch, 1985) and by metabolites in the symplast. PEPcase from maize shows a light-dependent decrease in $S_{0.5}$ (PEP) (from 2.25 to 1.6 mM) and a decreased sensitivity to inhibition by malate ($I_{0.5}$ increases in the light from 5 to 12 mM) (Doncaster and Leegood, 1987) which may be a consequence

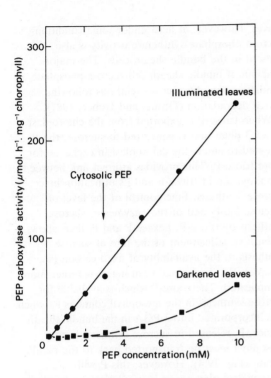

Fig. 19.5 Measured activity of PEPcase in extracts of illuminated and darkened leaves assayed in a reconstituted cytosol (20 mM malate, 15 mM triose-P) at different PEP concentrations, demonstrating the importance of interactions shown in Fig. 19.4. (Redrawn from Doncaster and Leegood, 1987.)

of phosphorylation of the enzyme (Nimmo et al., 1987b). Figure 19.5 shows how the two forms of PEPcase, from darkened and from illuminated leaves, respond to the concentration of PEP in the presence of concentrations of malate and triose-P which might be expected to occur in the mesophyll cytoplasm (Fig. 19.3). The enzyme from darkened leaves is virtually inactive at physiological concentrations of PEP (indicated by the arrow), whereas the enzyme from illuminated leaves shows a strong substrate dependence and a high activity.

Evidence of coordination between the C3 and C4 cycles is seen in the relationship between the assimilation rate and photosynthetic intermediates in *Amaranthus edulis* and maize in response to the intercellular concentration of CO_2 (Fig. 19.6). As the rate of photosynthesis increases with increasing CO_2, the RuBP pool falls, but the amounts of

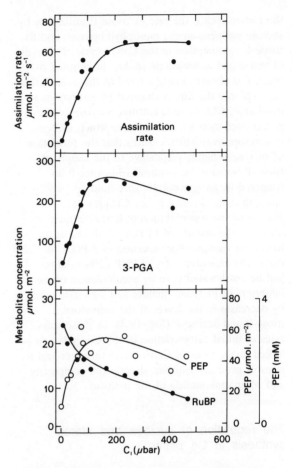

Fig. 19.6 Changes in steady-state photosynthesis and amounts of metabolites in leaves of *Zea mays* as a function of the intercellular partial pressure of CO_2 (C_i). The PEP concentration is estimated on the basis of a mesophyll cell cytoplasmic volume of 25 μl mg^{-1} chlorophyll. The arrow indicates C_i at ambient CO_2. (R. C. Leegood and S. von Caemmerer, unpublished.)

PEP and metabolites of the C4 cycle increase (Leegood and von Caemmerer, 1988). This behavior suggests that a feedback loop from the Calvin cycle is operative *in vivo*. The simplest explanation is that the amount of PEP declines when the assimilation rate falls because the amount of 3-PGA exported to the mesophyll also declines (Fig. 19.4).

Activation of PEPcase by triose-P and hexose-P is important in determining the response to the supply of 3-PGA (the metabolite 'message' from the Calvin cycle), the rate of triose-P utilization by sucrose synthesis in the mesophyll cytosol, and to triose-P consumption in the Calvin cycle. In leaves of maize and *Amaranthus edulis*, the amount of triose-P is always closely related to the assimilation rate whether the flux is changed by alterations in irradiance, CO_2 or temperature. Sucrose synthesis is regulated by fructose-2,6-P_2 (to which PEPcase is unresponsive) which ensures that the formation of sucrose is highly responsive to the supply of triose-P because the synthesis of hexose-P is triggered by an appropriate threshold concentration of triose-P (see Chapter 21). Increasing the concentration of hexose-P then increases the activity of PEPcase. On the other hand, the tendency for diversion of 3-PGA to malate by excessive rates of PEP carboxylation will be ameliorated both by accumulation of the inhibitors of PEPcase, malate and aspartate, and by decreases in the levels of the activators, triose-P and hexose-P (Fig. 19.4). In these ways rates of initial carboxylation in the C4 cycle can be linked to the rates of Calvin cycle turnover and to rates of product synthesis, despite the complexity and compartmentation of the pathway.

The regulation of sucrose and starch synthesis in C4 plants

Although in a species such as maize the synthesis of sucrose appears to occur largely in the mesophyll cells, while the synthesis of starch occurs mainly in the bundle sheath cells, there is considerable flexibility both within maize and between C4 plants in general. For example, while the mesophyll tissue of maize grown under normal conditions contains no detectable starch, growth of plants in continuous light or at low temperatures induces starch formation in the mesophyll. On the other hand, *Digitaria* spp., which are also $NADP^+$-ME plants, synthesize both sucrose and starch in the mesophyll compartment. During $^{14}CO_2$ fixation in maize, labeled sucrose appears first in the mesophyll cells (Furbank *et al.*, 1985). The majority of the sucrose-phosphate synthetase, fructose-6-P,2-kinase, fructose-2,6-P_2ase (Soll *et al.*, 1983) and fructose-2,6-P_2 itself (Stitt and Heldt, 1985a) is present in the mesophyll cells of maize leaves. However, at least under some conditions, sucrose-phosphate synthetase activity is also present in the bundle sheath cells. The major function of bundle sheath cell sucrose-phosphate synthetase may be sucrose synthesis following starch degradation (Ohsugi and Huber, 1987).

When triose-P is exported from the chloroplast of a C3 plant and is converted to sucrose, the P_i released re-enters the chloroplast in exchange for more triose-P. This provides a direct link between the provision of triose-P and its utilization in sucrose synthesis. Fine control of the level of fructose-2,6-P_2 and of the enzymes of sucrose synthesis by triose-P, hexose-P and P_i then allows a sensitive adjustment of the rate of sucrose synthesis to the availability of fixed carbon (see Chapter 21). The situation in maize is rather more complicated. The triose-P which is available for sucrose synthesis in the mesophyll contains P_i which was incorporated into 3-PGA in the bundle sheath. Hence, P_i released in sucrose synthesis in the mesophyll must be transported back to the bundle sheath (Fig. 19.4). However, this P_i will nevertheless play a part in regulating the transport of 3-PGA, triose-P and PEP across the mesophyll chloroplast envelope by the P_i translocator.

We have seen that triose-P is present in the mesophyll of C4 plants at far higher concentrations than in C3 plants. Consequently, the amount of fructose-1,6-P_2 formed through the action of aldolase is also higher in the mesophyll cytoplasm. In maize the cytosolic fructose 1,6-bisphosphatase (which is the first step in the synthesis of sucrose) shows a much higher K_m for FBP (20 μM in maize compared with 3 μM in spinach and, in the presence of fructose-2,6-P_2, 3–5 mM in maize and 20 μM in spinach; Stitt and Heldt, 1985a). These properties prevent discharge of the gradient of triose-P by sucrose synthesis. In addition, higher concentrations of triose-P and 3-PGA are needed to inhibit the fructose 6-phosphate,2-kinase from maize, and the enzyme is also inhibited by PEP and OAA (Soll *et al.*, 1983). Thus, fructose-2,6-P_2 will build up and inhibit sucrose synthesis when both C3 and C4 metabolites are low.

Studies have shown that the activities of starch synthase, of the branching enzyme and of ADP-glucose pyrophosphorylase are higher in the bundle sheath of maize than in the mesophyll. On the other hand, the enzymes of starch degradation, starch phosphorylase and amylase, are evenly distributed and may be slightly higher in the mesophyll (Spilatro and Preiss, 1987). The regulation of starch synthesis has also been modified in C4 plants. ADP-glucose pyrophosphorylase from spinach chloroplasts is activated by 3-PGA and inhibited by P_i, with ratios of 3-PGA/P_i for half-maximal activation typically being less than 1.5. ADP-glucose pyrophosphorylase from maize leaves requires a ratio of 3-PGA to P_i of between 7 and 10 for half-maximal activation in the bundle sheath and even higher ratios in the mesophyll (Spilatro and Preiss, 1987). The metabolite gradients which develop during photosynthesis lead to lower 3-PGA/P_i ratios in the mesophyll than in the bundle sheath which, together with the relatively low activities of enzymes of starch synthesis, would appear to limit synthesis of starch in the mesophyll relative to the bundle sheath.

Intracellular metabolite transport in CAM plants

The distinctive features of succulent plants that have CAM metabolism are the nocturnal fixation of CO_2 via PEPcase, accumulation of malic acid in the vacuole and subsequent diurnal de-acidification in which CO_2 released from malic acid is fixed via Rubisco during photosynthesis. This day/night pattern occurs within all chloroplast-containing cells of CAM plants, and the expression of its many components is very sensitive to biological and environmental factors. The essentials of intracellular transport of malic acid and the analogies in metabolic regulation between C4 and CAM plants are thus best discussed by dissecting the patterns of metabolism into four phases (Osmond, 1978; Fig. 19.7). Phase I embraces dark acidification, involving the stoichiometric conversion of carbohydrates (glucans or soluble sugars) to PEP, carboxylation via PEPcase,

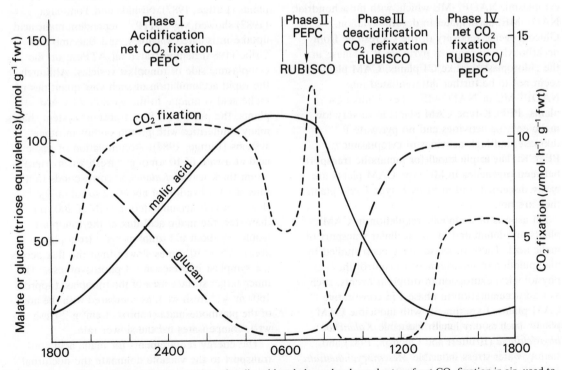

Fig. 19.7 Generalized schematic representation of malic acid and glucan levels, and rates of net CO_2 fixation in air, used to identify four phases of metabolism in ME-type CAM plants (redrawn from Osmond (1978) and other data).

reduction to malic acid and acid accumulation in the vacuole. Upon illumination, CO_2 fixation often increases transiently, involving both PEPcase and Rubisco, before malic acid efflux, de-acidification and stomatal closure. Designated phase II, the complex mechanism of this phase will not concern us further here. Phase III is the stoichiometric antithesis of phase I, in which de-acidification can be accomplished in 4 to 8 hours, depending on light and temperature. During phase III, the CO_2 which is released by decarboxylation of malic acid accumulates to high concentrations and is refixed by Rubisco. To sustain CAM it is essential that the C3 residue from malic acid decarboxylation should be conserved during phase III. In phase IV, stomata open again and CO_2 fixation by Rubisco will predominate, although some CO_2 fixation via PEPcase also occurs.

Like C4 plants, CAM plants can be divided into major subtypes on the basis of their decarboxylation pathways during phase III (Osmond, 1978). However, unlike C4 plants, the malic enzyme type (ME-CAM plants) have a cytoplasmic $NADP^+$-ME which, with mitochondrial NAD^+-ME, participates in decarboxylation. Gluconeogenic recovery of the C3 residue from decarboxylation requires pyruvate P_i dikinase in the chloroplast. Unlike C4 plants, CAM plants seem not to be further differentiated into $NADP^+$-ME or NAD^+-ME types. Unlike C4 plants, PEPCK-type CAM plants have very low malic enzyme activities and no pyruvate P_i dikinase, but high activities of cytoplasmic PEPCK. The implications for metabolic transport between organelles in ME-type CAM plants are easily discerned, but in PEPCK-type CAM plants they are not.

All aspects of metabolic regulation in CAM plants are dominated by intracellular transport of malic acid. These processes have been studied by manipulation of environment to control the physiological expression of different events, such as acid accumulation in phase I in constitutive CAM plants. Experiments with inducible CAM plants, such as day length inducible *Kalanchoe blossfeldiana* (Brulfert and Queiroz, 1982) and salinity/water stress inducible *Mesembryanthemum crystallinum* (Winter et al., 1982) have also been essential to the elucidation of these processes.

Induction of CAM in these plants involves the whole 'package' of enzyme activities and organelle translocators required to sustain CAM. This 'package' is expressed in pre-existing leaf cells of mature plants with C3 metabolism, making the phenomenon one of the most interesting cases of regulation of gene expression in response to physiological stress (Ostrem *et al.*, 1987; Vernon *et al.*, 1988).

Malic acid influx across the tonoplast

Simple considerations of volume, pH, and osmotic pressure imply that the massive amounts of malic acid that may be accumulated in CAM plants during phase I must be transported to the cell vacuole. Electrophysiological considerations indicate that active transport of malic acid to the vacuole must occur (Lüttge and Ball, 1979), and many studies have demonstrated both an ATP-dependent, and PP_i-dependent H^+ transport in tonoplast membranes and vesicles of CAM plants (Lüttge, 1987). Nishida and Tominaga (1987) showed Mg^{2+} ATP^{2-}-dependent malic acid uptake in isolated vacuoles, and Balsamo and Uribe (1988) demonstrated an ATPase on the cytoplasmic side of tonoplast vesicles. Although the rapid accumulation of such vast quantities of malic acid is unique to the vacuole of CAM plants, the proton-ATPase transport system shares many similarities with proton pumps in other systems (Lüttge, 1987). Accumulation of malic acid at a rate of 10 μmol g^{-1} fresh wt. h^{-1} in a 5 mm thick leaf of *Kalanchoe* corresponds to a rate of CO_2 fixation of about 12 μmol m^{-2} s^{-1} (leaf area). If accumulated in cells of 200 μm diameter, the malic acid flux at the tonoplast would be about 0.2 μmol m^{-2} s^{-1} (tonoplast area), 2.5×10^3 times slower than the flux across the symplast interface in C4 photosynthesis. The much larger surface area of the tonoplast (approx. 160 cm^2 g^{-1} fresh wt.), as compared with the area of the plasmodesmata (approx. 1 cm^2 g^{-1} fresh wt.), compensates for the slower rate.

The energy requirements for malic acid transport to the vacuole dominate the nocturnal respiratory activity of CAM plants. Malic acid synthesis is energetically self-contained (Osmond,

1978), and depending on the carbohydrate utilized, may even lead to net ATP production which can be used for transport. Thus, in ME-CAM plants such as *Kalanchoe*, glycolysis of glucans is initiated by hexose-P production via phosphorylase, and the overall equation for glycolysis, carboxylation and reduction to malate in these plants is

Glucan + P_i → hexose-P
Hexose-P + ADP + $2CO_2$ → 2 malate + ATP

In PEPCK-type CAM plants, including the bromeliad *Aechmea*, soluble hexoses are the principal carbohydrate sources for glycolysis. In these, the additional ATP required to synthesize hexose-P means that the overall equation for glycolysis, carboxylation and malate reduction is simply

Hexose + $2CO_2$ → 2 malate

Assuming a similar stoichiometry of the proton ATPase in both *Kalanchoe* and *Aechmea*, one can predict that the ATP demands on respiration for malic acid transport will be about twice as large in the latter as in the former. Lüttge and Ball (1987) confirmed these predictions, showing molar ratios of O_2 uptake to malic acid of about 6 in *Kalanchoe* and 12 in *Aechmea*. In pineapple, which can use both glucan and soluble sugar in glycolysis (Kenyon et al., 1985), the ratio was intermediate.

The stoichiometry of the proton ATPase, and the distinctly different enzyme, the proton PP_iase, have been explored by Lüttge (1987) and Bremberger et al. (1988) (Fig. 19.8). The electrochemical potential gradient across the tonoplast can be calculated from measurements of cell sap pH and the transmembrane electrical

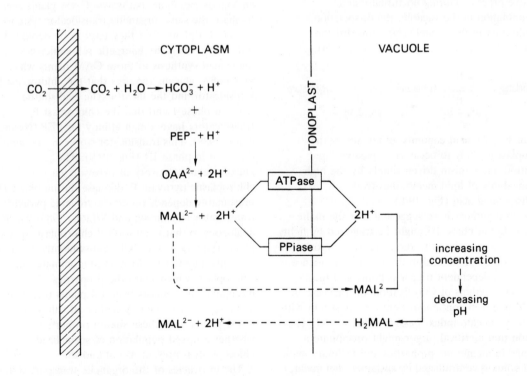

Fig. 19.8 Schematic representation of charge balance and malic acid transport during synthesis and accumulation of malate in CAM plants (redrawn from Lüttge, 1987). A tonoplast proton pump driven by a $2H^+$/ATPase and a pyrophosphatase are indicated (Bremberger et al., 1988). The malate anion is shown passively to follow the accumulation of protons, and efflux of the undissociated acid is also shown as a passive flux, denoted by broken lines.

potential at the tonoplast, assuming that malic acid transport at the tonoplast serves to maintain the cytosol at pH 7.5, when malate is fully dissociated (malate^{2-} 2H$^+$). In several CAM plants, at the highest malic acid concentrations, the calculated electrochemical potential gradient (which presumably drives the tonoplast ATPase) is equivalent to a Gibbs free energy of 20–30 kJ mol^{-1}. This corresponds to almost half the free energy of ATP hydrolysis, suggesting that the proton transport system involved in malic acid accumulation in CAM plants is a 2H$^+$/ATPase.

Malic acid efflux from the vacuole

In contrast to the active transport of protons and malate into the vacuole, the return passage of free malic acid during phase III is thought to be a passive process. During nocturnal malate accumulation in the vacuole, the dissociation equilibrium in the vacuole shifts towards the undissociated acid (Lüttge and Smith, 1984).

$$\text{malate}^{2-} \underset{-H^+}{\overset{+H^+}{\rightleftharpoons}} \text{H malate}^{1-} \underset{-H^+}{\overset{+H^+}{\rightleftharpoons}} \text{H}_2 \text{malate}$$
$$pK_a\ 4.25 \qquad\qquad pK_a\ 3.18$$

At high malic acid contents, efflux across the tonoplast is likely to occur via a passive 'lipid solution' mechanism driven simply by the high permeability of lipid membranes to the undissociated acid (Fig. 19.8).

Several authors have suggested that the malic acid efflux in phase III might be triggered by light, and may also respond to the increased turgor generated by malic acid in the vacuole. Evidence for a light-dependent trigger is poor, and the circadian rhythms of CO_2 fixation via PEPcase in CAM plants in continuous light are consistent with an active malate influx, passive efflux model. The notion that a critical turgor might precipitate a change in membrane properties and facilitate malic acid efflux is contradicted by sustained diel malate influx and efflux in wilted leaves (Rygol et al., 1987). Thus it seems likely that the malic acid storage capacity of the vacuole is set by the equilibrium between active influx and passive efflux. A corollary of this is that cytoplasmic malic acid levels also increase, and the rate of malic acid synthesis is impaired due to malate inhibition of PEPcase (Winter, 1985). Under natural conditions, this situation is reached at the end of a 10 to 14 hour dark period (Fig. 19.7). The efflux is presumably amplified following the activation of decarboxylase enzymes during illumination in phase II. Net efflux is sustained throughout phase III by malic acid decarboxylation in the cytoplasm or mitochondria, and the sustained malate inhibition of malate synthesis via PEPcase (Winter, 1985; Osmond et al., 1988).

Organelle transport processes in CAM plants

As might be expected on the basis of the analogous metabolic pathways, CAM plants seem to share the same organelle translocator systems as C3 and C4 plants, but they tend to be deployed differently. Thus the energetic self-sufficiency of malic acid synthesis in those CAM plants which metabolize glucans requires that the chloroplast P_i translocator and the dicarboxylic acid shuttle work in phase I and that the chloroplast P_i translocators have a high affinity for PEP (Neuhaus et al., 1988). This translocator works in the opposite direction in phase III (Fig. 19.9). The gluconeogenic recovery of pyruvate via chloroplastic pyruvate P_i dikinase during phase III presumably depends on an electrogenic pyruvate translocator (Demming and Winter, 1983) which is analogous to that in mesophyll chloroplasts of C4 plants (Figs 19.4 and 19.9). Subsequently, in phase IV, Calvin cycle CO_2 fixation predominates and chloroplasts which share the properties of mesophyll chloroplasts from C4 plants then behave as those of a C3 plant. Whether all chloroplasts in CAM plants have these shared properties, or whether a mixed population of specialized chloroplasts is present, is not known.

The intricacies of the organelle transport systems in other CAM plants which utilize soluble sugars, or decarboxylate malic acid via PEPCK rather than ME, remain to be explored. Little is known about metabolite exchanges in mitochondria of

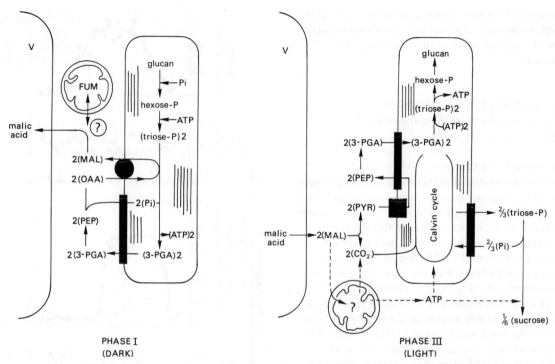

Fig. 19.9 Schematic representation of intracellular and chloroplast metabolite transport in a ME-CAM plant such as *Kalanchoe daigremontiana*. The organelle translocators are identified by the following symbols: dicarboxylate shuttle, a circle, the pyruvate translocator, a square, and the P_i translocator by rectangular symbols on the chloroplast envelope. During cytoplasmic malic acid synthesis in phase I, the translocators ensure that chloroplast glycolysis is energetically self-contained, but regulation of malic acid metabolism in mitochondria is not understood. In phase III the translocators ensure recovery of glucan in the chloroplast and sucrose synthesis in the cytosol, but once again the extent of mitochondrial metabolism is unclear.

CAM plants. However, a large part of the total malate flux in phase I equilibrates with fumarase, as indicated by randomization of C-1 and C-4 carboxyl carbons (Cockburn and McAuley, 1975), implying rapid exchange of C4 acids into and out of the mitochondria. The proportion of malic acid decarboxylated by NAD^+-ME in mitochondria of ME-type CAM plants during de-acidification is very much dependent on pH (Day, 1980). Moreover, it seems that mitochondrially-generated ATP can be used to make up the ATP requirements of pyruvate P_i dikinase during gluconeogenesis in phase III (Adams et al., 1986). This is analogous to mitochondrial ATP production during decarboxylation in PEPCK-type C4 plants (Burnell and Hatch, 1988).

Regulation of CAM as a consequence of metabolite transport

Because Rubisco, PEPcase and several decarboxylase enzymes are present at high activities in all chloroplast-containing cells of CAM plants, the regulation of these activities by metabolites is essential if CAM is to proceed. Thus for malic acid to accumulate in phase I, PEPcase must be active, and Rubisco and the decarboxylase systems must be inactive. If de-acidification and the Calvin cycle are to proceed in phase III, decarboxylases and Rubisco must be active and PEPcase inactive. At the same time, glycolysis and gluconeogenesis must be regulated to ensure substrate supply and restoration for the cyclical carbon fluxes in CAM. Not surprisingly, it seems that the above transport events mediate many of the key regulatory processes.

Regulation of futile carboxylation, decarboxylation, and competitive carboxylation in CAM

Regulation of PEPcase in CAM plants has been extensively studied and there is general agreement as to the response pattern, with some insights into mechanism. Although early studies implied coarse control, by means of changes in the amount of enzyme, this level of control is only observed during induction of CAM (Winter, 1982). The amount of PEPcase does not vary throughout the day–night cycle (Fig. 19.10). However, as in C3 and C4 plants, malic acid is a potent inhibitor of PEPcase and early models of CAM proposed·this to be the key regulator of the cycle of malic acid synthesis and consumption (Kluge, 1971). The PEPcase of CAM plants has a higher affinity for PEP, and a higher sensitivity to malic acid, than the enzyme from C4 plants. Hence, malic acid synthesis continues in phase I because this inhibitory endproduct of PEPcase is efficiently transported from cytoplasm to vacuole, and then declines as flux equilibrium is approached. In the light, at least during phase III, futile cycles of malic acid decarboxylation and resynthesis are prevented because PEPcase is inhibited by high malic acid concentration in the cytoplasm during de-acidification (Osmond et al., 1988).

This simple hypothesis has been augmented by evidence that PEPcase is sensitized to malate inhibition in the light, thus greatly amplifying the feedback control (Winter, 1982). When PEPcase is extracted from CAM plants in the dark, it shows a high affinity for PEP, is susceptible to glucose-6-P stimulation, and has a low sensitivity to malic acid. If extracted from the same tissues during the light period, the enzyme has a low affinity for its substrate, low sensitivity to effectors and is much more sensitive to malic acid (Fig. 19.10). Treatment of the enzyme isolated in the dark with malic acid *in vitro* suggests that malate is a slow binding inhibitor which sensitizes the enzyme to the

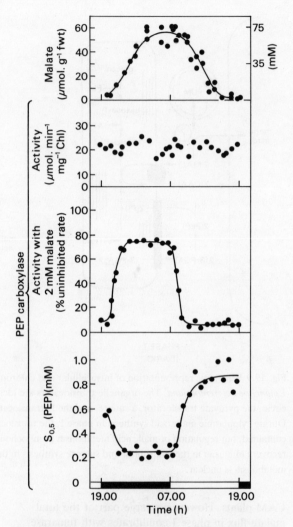

Fig. 19.10 Changes in the properties of PEPcase extracted from leaves of *Mesembryanthemum crystallinum* at different times of the diel cycle of CAM (redrawn from Winter, 1982). The concentration of malic acid in the vacuole was calculated assuming 85% water content. Although the total activity of PEP carboxylase remains unchanged the decreased affinity for PEP and the increased sensitivity to malate ensure that the enzyme is inhibited during de-acidification in the light.

inhibitor itself (Winter, 1982). Nimmo et al. (1987a) reported that the malic acid insensitive form of the enzyme isolated in the dark is phosphorylated, and that this covalent modification of the protein is responsible for changes in kinetic properties. Interestingly, an analogous phosphorylation of PEPcase from C4 plants in the light has been reported (Nimmo et al., 1987b). Direct evidence that PEPcase is inhibited during de-acidification in phase III, and that futile refixation of CO_2 into malic acid does not occur, comes from studies of ^{13}C label patterns in malate. The ^{13}C distribution among 4-^{13}C and 1-^{13}C labeled malate remains unchanged during de-acidification; futile cycling, if it were to occur, would yield doubly labeled malic acid (Osmond et al., 1988).

The regulation of decarboxylase enzymes in relation to the day–night cycle is not so well understood. It is often observed that the rate of CO_2 fixation in the dark declines as the malic acid content of the vacuole reaches a maximum level, towards the end of phase I (Fig. 19.7). This could reflect feedback inhibition due to malate, and/or the onset of decarboxylation. Control of the cytoplasmic decarboxylases (NADP-ME or PEPCK) in the dark is not easily explained, and whether futile cycles of carboxylation and decarboxylation occur in the dark probably depends on the residence time of malic acid in the cytosol, and on the balance of metabolite fluxes to the vacuole. Fumarase randomization of the (4-C) carboxyl implies that much of the total pool moves in and out of the mitochondria before entering the vacuole. There is evidence that mitochondrial NAD-ME is inactive in the dark in some CAM plants, but not in others (Wedding and Black, 1983; Artus and Edwards, 1985a, b). In these circumstances, inhibition of mitochondrial NAD-ME could prevent futile cycling.

It is now assumed that Rubisco is inactive during dark CO_2 fixation via PEPcase (cf. Ranson and Thomas, 1960), and there is evidence in some CAM plants that specific inhibitors, similar to those in some C3 plants, may be involved (Servaites et al., 1986). Both Rubisco and the decarboxylase enzymes are active in the light. After malic acid is consumed in photosynthesis, the relief of PEPcase inhibition permits complex parallel CO_2 fixation via both carboxylases in phase IV (Ritz et al., 1986; Osmond et al., 1988). However, because net CO_2 fixation in phase IV shows the O_2 sensitivity normally associated with Rubisco, it seems that PEPcase activity is quantitatively unimportant.

Regulation of carbohydrate metabolism

The pattern of malic acid synthesis and transport characteristic of CAM can only be sustained by stoichiometric degradation and resynthesis of carbohydrates as substrates and products of malic acid synthesis and breakdown. There is no analogy of this stoichiometry in other photosynthetic systems, and very little is known about the ways in which carbohydrate pools dedicated to malic acid synthesis (which are conserved during photosynthesis in phase III) are distinguished from photosynthetic endproducts produced for growth in phase IV. Yet there is tantalizing evidence for the isolation of the glycolytic and gluconeogenic pathways associated with CAM (Deleens and Quieroz, 1984).

Enzyme compartmentation and organelle metabolite translocators play key roles in the regulation of carbohydrate metabolism in CAM plants. These roles can be identified with some confidence in the glucan-utilizing ME-CAM plants, but are quite uncertain in PEPCK-CAM plants which convert soluble carbohydrates to malic acid and vice versa. In ME-CAM plants decarboxylation of malic acid produces pyruvate and CO_2. Gluconeogenic recovery of pyruvate can only take place in the chloroplast after conversion of pyruvate to PEP. Chloroplasts of CAM plants, like mesophyll chloroplasts of C4 plants, evidently contain a specific pyruvate translocator (Demming and Winter, 1983) which could ensure the stoichiometric transport of pyruvate into the chloroplast during de-acidification. However, gluconeogenic recovery of PEP involves further transport of PEP from, and 3-PGA into, the chloroplasts (Fig. 19.9). A specific form of the phosphate translocator in CAM chloroplasts which exchanges PEP and 3-PGA at high rates (Neuhaus et al., 1988) thus has a potentially significant role in restoring chloroplast glucan levels.

At the same time as these organelle metabolite transport events are taking place, CO_2 released from malic acid is fixed via the Calvin cycle and sucrose synthesis proceeds in the cytoplasm. This presumably involves metabolite and P_i exchanges across the chloroplast envelope analogous to those in C3 plants and similar regulatory interactions. Thus levels of fructose-2,6-P_2 are lowest during de-acidification in *Bryophyllum tubiflorum*, consistent with sustained cytoplasmic sucrose synthesis in the face of concurrent starch synthesis in the chloroplast (Fahrendorf et al., 1987). However, in a PEPCK-CAM plant (*Ananas comosus*) which derives only part of the carbon for nocturnal malic acid synthesis from glucans, these authors found 20 to 50-fold higher levels of fructose-2,6-P_2 throughout the 24 h cycle. Although these high levels of effector could inhibit sucrose synthesis in the cytosol and promote some glucan synthesis in the chloroplast during de-acidification, there is a paradox so far as sucrose accumulation is concerned. Fahrendorf et al. (1987) suggested that the high activity of pyrophosphate-dependent phosphofructokinase (PFP) in *Ananas*, which is promoted by fructose-2,6-P_2 (see Chapter 23), could serve as a path for cytoplasmic sucrose synthesis in the presence of high concentrations of fructose-2,6-P_2. The overall effect of high fructose-2,6-P_2 levels and high PFP activity may be to promote the recovery of both glucan and soluble carbohydrates during de-acidification. Osmond and Holtum (1981) deduced that, for many reasons, the ratio of 3-PGA to P_i is likely to be very high during de-acidification and hence to favor glucan synthesis via ADP-glucose pyrophosphorylase. Together with the extraordinary effect of 3-PGA on lowering the $S_{0.5}$ (glucose-1-P), and the fact that P_i does not completely inhibit the enzyme, ADP-glucose pyrophosphorylase from CAM plants seems uniquely suited to ensure glucan synthesis in the chloroplast (Singh et al., 1984).

Conclusions

Metabolic fluxes between and within chloroplast-containing cells underlie almost every aspect of photosynthetic metabolism in C4 and CAM plants. None of these processes is unique to these photosynthetic systems, but the activities and regulatory properties are markedly different from those in C3 plants. The broad distinction between the spatial and temporal separation of the carboxylation phase of the C4-cycle from the decarboxylation phase and the Calvin cycle, in C4 and CAM plants respectively, leads to further differentiation in respect of mitochondrial transport and regulation. Many details of the symplastic transport of photosynthetic metabolites in C4 plants (the most rapid of all symplast processes) remain to be resolved, but simple models based on concentration gradients and diffusion seem to confirm fundamental geometric, structural and biochemical properties of the system. Variants of the three major subtypes of C4 photosynthesis, and the C3–C4 intermediates, offer great scope for further testing of the principles of symplast transport in C4 photosynthesis. By the same token, progress in the evaluation of malic acid transport across the tonoplast of CAM plants, evidently driven by a $2H^+$ ATPase or pyrophosphatase, opens many avenues for the study of the most rapid cell membrane active transport processes in plants. The overwhelming influence of malic acid compartmentation on photosynthetic metabolism in CAM plants has yet to be explored in detail, especially in PEPCK-type CAM plants, and especially so far as leaf mitochondria are concerned. Understanding the regulation of carbohydrate metabolism in both C4 and CAM plants remains a major challenge.

Further reading and references

Adams, W. W. III, Nishida, K. and Osmond, C. B. (1986). Quantum yield of CAM plants measured by photosynthetic O_2 exchange. *Plant Physiol.* **81**, 297–300.

Andrews, T. J. and Lorimer, G. H. (1987). Rubisco: structure, mechanisms, and prospects for improvement. In *The Biochemistry of Plants*, Vol. 10, eds M. D. Hatch and N. K. Boardman, Academic Press, New York, pp. 131–218.

Artus, N. N. and Edwards, G. E. (1985a). NAD-malic enzyme from plants. *FEBS Lett.* **182**, 225–33.

Artus, N. N. and Edwards, G. E. (1985b). Properties of leaf NAD-malic enzyme from the inducible Crassulacean acid metabolism species *Mesembryanthemum crystallinum*. *Plant Cell Physiol.* **26**, 341–50.

Balsamo, R. A. and Uribe, E. G. (1988). Plasmalemma and tonoplast ATPase activity in mesophyll protoplasts, vacuoles and microsomes of the

Crassulacean-acid metabolism plant *Kalanchoe daigremontiana*. *Planta* **173**, 190–6.

Boag, S. and Jenkins C. L. D. (1985). CO_2 assimilation and malate decarboxylation by isolated bundle sheath chloroplasts from *Zea mays*. *Plant Physiol.* **79**, 165–70.

Botha, C. E. J. and Evert, R. F. (1988). Plasmodesmatal distribution and frequency in vascular bundles and contiguous tissues of the leaf of *Themeda triandra*. *Planta* **173**, 433–41.

Bremberger, C., Haschke, H.-P. and Lüttge, U. (1988). Separation and purification of the tonoplast ATPase and pyrophosphatase from plants with constitutive and inducible Crassulacean acid metabolism. *Planta* **175**, 465–70.

Brown, R. H. and Brown, W. V. (1975). Photosynthetic characteristics of *Panicum maximum*, a species with reduced photorespiration. *Crop Sci.* **15**, 681–5.

Brulfelt, J. and Quieroz, O. (1982). Photoperiodism and Crassulacean acid metabolism. III. Different characteristics of photoperiod-sensitive and nonsensitive isoforms of phosphoenolpyruvate carboxylase and Crassulacean acid metabolism operation. *Planta* **154**, 339–43.

Burnell, J. N. (1988). An enzymic method for measuring the molecular exclusion limit of plasmodesmata of bundle sheath cells of C4 plants. *J. Exp. Bot.* **39**, 1575–80.

Burnell, J. N. and Hatch, M. D. (1985). Light–dark regulation of pyruvate, P_i dikinase. *Trends Biochem. Sci.* **10**, 288–91.

Burnell, J. N. and Hatch, M. D. (1988). Photosynthesis in phosphoenolpyruvate carboxykinase-type C4 plants: pathways of C4 acid decarboxylation in bundle sheath cells of *Urochloa panicoides*. *Arch. Biochem. Biophys.* **260**, 187–99.

Burr, G. O., Hartt, C. E., Brodie, H. W., Tanimoto. T., Kortschak, H. P., Takahashi, D., Ashton, F. M. and Coleman, R. E. (1957). The sugar cane plant. *Ann. Rev. Plant Physiol.* **8**, 275–98.

Cockburn, W. (1985). Variation in photosynthetic acid metabolism in vascular plants: CAM and related phenomena. *New Phytol.* **101**, 3–25.

Cockburn, W. and McAuley, A. (1975). Pathway of dark CO_2 fixation in CAM plants. *Plant Physiol.* **55**, 87–9.

Day, D. A. (1980). Malate decarboxylation by *Kalanchoe daigremontiana* mitochondria and its role in crassulacean acid metabolism. *Plant Physiol.* **65**, 675–9.

Day, D. A. and Hatch, N. D. (1981a). Dicarboxylate transport in maize mesophyll chloroplasts. *Arch. Biochem. Biophys* **211**, 738–42.

Day, D. A. and Hatch, M. D. (1981b). Transport of 3-phosphoglyceric acid, phosphoenolpyruvate, and inorganic phosphate in maize mesophyll chloroplasts, and the effect of 3-phosphoglyceric acid on malate and phosphoenolpyruvate production. *Arch. Biochem. Biophys.* **211**, 743–9.

Deleens, E. and Quieroz, O. (1984). Effects of photoperiod and ageing on the carbon isotope composition of *Bryophyllum diagremontiana* Berger. *Plant Cell Environ.* **7**, 279–83.

Demmig, B. and Winter, K. (1983). Photosynthetic characteristics of chloroplasts isolated from *Mesembryanthemum crystallinum*; a halophyte plant capable of crassulacean acid metabolism. *Planta* **159**, 66–76.

Dengler, E. G., Dengler, R. F. and Hattersley, P. W. (1985). Differing ontogenetic origins of PCR ('Kranz') sheaths in leaf blades of C4 grasses (Poaceae). *Am. J. Bot.* **72**, 284–302.

De Saussure, T. (1804). *Recherches Chimiques sur la Vegetation*, Nyon, Paris.

Doncaster, H. D. and Leegood, R. C. (1987). Regulation of phosphoenolpyruvate carboxylase activity in maize leaves. *Plant Physiol.* **84**, 82–7.

Edwards, G. E. and Walker, D. A. (1983). C_3, C_4: *Mechanisms, and Cellular and Environmental Regulation, of Photosynthesis*, Blackwell, Oxford.

Edwards, G. E. and Ku, M. S. B. (1987). Biochemistry of C3–C4 intermediates. In *The Biochemistry of Plants*, Vol. 10, eds M. D. Hatch and N. K. Boardman, Academic Press, New York, pp. 275–325.

Edwards, G. E., Ku, M. S. B. and Monson, R. K. (1985). C4 photosynthesis and its regulation. In *Photosynthetic Mechanisms and the Environment*, eds J. Barber and N. R. Baker, Elsevier, Amsterdam, pp. 287–327.

Fahrendorf, T., Holtum J. A. M., Muckerjee, U. and Latzko, E. (1987). Fructose-2,6-bisphosphate, carbohydrate partitioning and crassulacean acid metabolism. *Plant Physiol.* **84**, 182–7.

Flügge, U. I., Stitt, M. and Heldt, H. W. (1985). Light-driven uptake of pyruvate into mesophyll chloroplasts from maize. *FEBS Lett.* **183**, 335–9.

Furbank, R. T. and Hatch, M. D. (1987). Mechanism of C4 photosynthesis. The size and composition of the inorganic carbon pool in bundle-sheath cells. *Plant Physiol.* **85**, 958–64.

Furbank, R. T., Stitt, M. and Foyer, C. H. (1985). Intercellular compartmentation of sucrose synthesis in leaves of *Zea mays*. *Planta* **164**, 172–8.

Gardeström, P. and Edwards, G. E. (1985). Leaf mitochondria (C3 + C4 + CAM). In *Encyclopedia of Plant Physiology*, New Series Vol. 18, eds R. Douce and D.A. Day, Springer-Verlag, Berlin, pp. 314–46.

Haberlandt, G. (1884). *Physiological Plant Anatomy*, (transl. M. Drummond), MacMillan, London.

Hatch, M. D. (1971). The C4 pathway of photosynthesis. Evidence for an intermediate pool of carbon dioxide and the identity of the donor C4 acid. *Biochem. J.* **125**, 425–32.

Hatch, M. D. and Osmond, C. B. (1976). Compartmentation and transport in C4 photosynthesis. In *Encyclopedia of Plant Physiology*, New Series Vol. 3, eds C. R. Stocking and U. Heber, Springer-Verlag, Berlin, pp. 144–84.

Hatch, M. D., Dröuscher, L., Flügge, U. I. and Heldt H. W. (1984). A specific translocator for oxaloacetate transport in chloroplasts. *FEBS Lett.* **178**, 15–19.

Hattersley, P. W. (1982). $\delta^{13}C$ values of C4 types in grasses. *Aust. J. Plant Physiol.* **9**, 139–54.

Hattersley, P. W. (1986). Variations in photosynthetic pathway. In *Grass Systematics and Evolution*, eds T. R. Sonderstron, K. W. Hilu, G. S. Campbell and M. E. Barkworth, Smithsonian Institute Press, Washington, DC, pp. 49–64.

Heldt, H. W. and Flügge, U.-I. (1986). Transport of metabolites accross the chloroplast envelope. *Methods Enzymol.* **125**, 705–16.

Heyne, B. (1815). On the deoxidation of the leaves of *Cotyledon calycina*. *Trans. Linn. Soc. Lond.* **11**, 213–15.

Holtum, J. A. M. and Osmond, C. B. (1981). The gluconeogenic metabolism of pyruvate during deacidification in plants with crassulacean acid metabolism. *Aust. J. Plant Physiol.* **8**, 31–44.

Huber, S. C. and Edwards, G. E. (1977). Transport in C4 mesophyll chloroplasts. Characterization of the pyruvate carrier. *Biochim. Biophys. Acta* **462**, 583–602.

Hylton, C. M., Rawsthorne, S., Smith, A. M., Jones, D. A. and Woolhouse, H. W. (1988). Glycine decarboxylase is confined to the bundle-sheath cells of C3–C4 intermediate species. *Planta* **175**, 452–9.

Karpilov, Y. S. (1960). The distribution of radioactivity of ^{14}C among the products of photosynthesis in maize. *Trans. Kazan Agric. Inst.* **41**, 15–24, (in Russian).

Karpilov, Y. S. (1970). Cooperative photosynthesis in xerophytes. *Proc. Moldavian Inst. Irrigation and Agric. Res.* **11**, 3–66, (in Russian).

Kenyon, W. H., Severson, R. F. and Black, C. C. (1985). Maintenance carbon cycle in Crassulacean acid metabolism plant leaves. Source and compartmentation of carbon for nocturnal malate synthesis. *Plant Physiol.* **77**, 183–9.

Kluge, M. (1971). Studies on CO_2 fixation by succulent plants in the light. In *Photosynthesis and Photorespiration*, eds M. D. Hatch, C. B. Osmond and R. O. Slatyer, Wiley Interscience, New York, pp. 283–7.

Kunitake, G. and Saltman, P. (1958). Dark CO_2 fixation by succulent leaves: conservation of the dark fixed CO_2 under diurnal conditions. *Plant Physiol.* **34**, 123–7.

Leegood, R. C. (1985). The intercellular compartmentation of metabolites in leaves of *Zea mays*. *Planta* **164**, 163–71.

Leegood, R. C. and von Caemmerer, S. (1988). The relationship between contents of photosynthetic intermediates and the rate of photosynthetic carbon assimilation in leaves of *Amaranthus edulis* L. *Planta* **174**, 253–62.

Lüttge, U. (1987a). Carbon dioxide and water demand: Crassulacean acid metabolism (CAM), a versatile ecological adaptation exemplifying the need for integration in ecophysiological work. *New Phytol.* **106**, 593–629

Lüttge, U. and Smith, J. A. C. (1984). Mechanism of passive malic-acid efflux from vacuoles of the CAM plant *Kalanchoe daigremontiana*. *J. Membrane Biol.* **81**, 149–58.

Lüttge, U. and Ball, E. (1987b). Dark respiration of CAM plants. *Plant Physiol. Biochem.* **25**, 3–10.

Neuhaus, H. E., Holtum, J. A. M. and Latzko, E. (1988). Transport of phosphoenolpyruvate by chloroplasts from *Mesembryanthemum crystallinum* L. exhibiting crassulacean acid metabolism. *Plant Physiol.* **87**, 64–8.

Nimmo, G. A., Wilkins, M. B., Fewson, C. A. and Nimmo, H. G. (1987a). Persistent circadian rhythms in the phosphorylation state of phosphoenolpyruvate carboxylase from *Bryophyllum feldtschenkoi* leaves and in its sensitivity to inhibition by malate. *Planta* **170**, 408–15.

Nimmo, G. A., McNaughton, G. A. L., Fewson, C. A., Wilkins, M. B. and Nimmo, H. G. (1987b). Changes in the kinetic properties and phosphorylation state of phosphoenolpyruvate carboxylase in *Zea mays* leaves in response to light and dark. *FEBS Lett.* **213**, 18–22.

Nishida, K. and Tominga, O. (1987). Energy-dependent uptake of malate into vacuoles isolated from a CAM plant, *Kalanchoe daigremontiana*. *J. Plant Physiol.* **127**, 385–93.

Nobel, P. S. (1988). *The Environmental Biology of Agaves and Cacti*, Cambridge University Press, New York.

Ohnishi J. and Kanai R. (1987). Pyruvate uptake by mesophyll and bundle-sheath chloroplasts of a C4 plant, *Panicum miliaceum* L. *Plant Cell Physiol.* **28**, 1–10.

Ohsugi, R. and Huber, S. C. (1987). Light modulation and localization of sucrose phosphate synthase activity between mesophyll cells and bundle sheath cells in C4 species. *Plant Physiol.* **84**, 1096–101.

O'Leary, M. H. (1981). Carbon isotope discrimination in plants. *Phytochemistry* **20**, 553–67.

Osmond, C. B. (1971). Metabolite transport in C4 photosynthesis. *Aust. J. Biol. Sci.* **24**, 159–63.

Osmond, C. B. (1978). Crassulacean acid metabolism, a curiosity in context. *Ann. Rev. Plant Physiol.* **29**, 379–414.

Osmond, C. B., Björkman, O. and Anderson, D. J. (1980). *Physiological Processes in Plant Ecology*, Ecological Studies, Vol. 36, Springer Verlag, Heidelberg.

Osmond, C. B. and Holtum, J. A. M. (1981). Crassulacean acid metabolism. In *The Biochemistry of Plants*, Vol. 8, eds M. D. Hatch and N. K. Boardman, Academic Press, New York, pp. 283–328.

Osmond, C. B., Holtum, J. A. M., O'Leary, M. H., Roeske, C., Summons, R. E., Wong, O. C. and Avadhani, P. N. (1988). Regulation of malic-acid metabolism in Crassulacean-acid-metabolism plants in the dark and light: In vitro evidence from ^{13}C-labelling patterns after $^{13}CO_2$ fixation. *Planta* **175**, 184–92.

Osmond, C. B., Winter, K. and Ziegler, H. (1982). Funtional significance of different pathways of CO_2 fixation in photosynthesis. In *Encyclopedia of Plant Physiology*, New Series Vol. 12B, eds O. L. Lange, P. S. Nobel, C. B. Osmond and H. Ziegler, Springer Verlag, Berlin, pp. 479–547.

Ostrem, J. A., Olson, S. W., Schmitt, J. M. and Bohnert, H. J. (1987). Salt stress increases the level of translatable mRNA for phosphoenolpyruvate carboxylase in *Mesembryanthemum crystallinum*. *Plant Physiol.* **84**, 1270–5.

Prendergast, H. D. V., Hattersley, P. W. and Stone, N. E. (1987). New structural/biochemical associations in leaf blades of C4 grasses (Poaceae). *Aust. J. Plant Physiol.* **14**, 403–20.

Ranson, S. L. and Thomas, M. (1960). Crassulacean acid metabolism. *Ann. Rev. Plant Physiol.* **11**, 81–110.

Ritz, D., Kluge, M. and Veith, H. J. (1986). Mass-spectrometric evidence for double-carboxylation pathway of malate synthesis in plants in light. *Planta* **167**, 284–91.

Rumpho, M. E., Wessinger M. E. and Edwards, G. E. (1987). Influence of organic phosphates on 3-phosphoglycerate dependent O_2 evolution in C3 and C4 mesophyll chloroplasts. *Plant Cell Physiol.* **28**, 805–13.

Rygol, J., Winter, K. and Zimmermann, U. (1987). The relationship between turgor pressure and titratable acidity in the mesophyll cells of intact leavels of a Crassulacean acid metabolism plant, *Kalanchoe daigremontiana* Hamet et Perr. *Planta* **172**, 487–93.

Schuster, G., Ohad, I., Martinean, B. and Taylor, W. C. (1985). Differentiation and development of bundle sheath and mesophyll thylakoids of maize. *J. Biol. Chem.* **260**, 1866–73.

Servaites, J. C., Parry, M. A. J., Gutteridge, S. and Keys, A. J. (1986). Species variation in the predawn inhibition of ribulose-1,5-bisphosphate carboxylase/oxygenase. *Plant Physiol.* **82**, 1161–3.

Sheen, J.-Y. and Bogorad, L. (1985). Differential expression of the ribulose bisphosphate carboxylase large subunit gene in bundle sheath and mesophyll cells of developing maize leaves is influenced by light. *Plant Physiol.* **79**, 1072–6.

Sheen, J.-Y. and Bogorad, L. (1987). Differential expression of C4 pathway genes in mesophyll and bundle sheath cells of greening maize leaves. *J. Biol. Chem.* **262**, 11726–30.

Singh, B. K., Greenberg, E. and Preiss, J. (1984). ADP glucose pyrophosphorylase from the CAM plants *Hoya carnosa* and *Xerosicyos danguyi*. *Plant Physiol.* **74**, 711–16.

Soll, J., Wörzer, C. and Buchanan, B. B. (1983). Fructose-2,6-bisphosphate and C4 plants. In *Advances in Photosynthesis Research*, Vol. 3, ed. C. Sybesma, Martinus Nijhoff, The Hague, pp. 485–8.

Spilatro, S. R. and Preiss, J. (1987). Regulation of starch synthesis in the bundle-sheath and mesophyll of *Zea mays* L. Intercellular compartmentation of enzymes of starch metabolism and the properties of the ADPglucose pyrophosphorylases. *Plant Physiol.* **83**, 621–7.

Stitt, M. and Heldt, H. W. (1985a). Control of photosynthetic sucrose synthesis by fructose-2,6-bisphosphate. Intercellular metabolite distribution and properties of the cytosolic fructose bisphosphatase in leaves of *Zea mays* L. *Planta* **164**, 179–88.

Stitt, M. and Heldt, H. W. (1985b). Generation and maintenance of concentration gradients between the mesophyll and bundle-sheath in maize leaves. *Biochim. Biophys. Acta* **808**, 400–14.

Stitt, M., Huber, S. C. and Kerr, P. (1987). Control of photosynthetic sucrose formation. In *The Biochemistry of Plants*, Vol. 10, eds M. D. Hatch and N. K. Boardman, Academic Press, New York, pp. 327–409.

Ting, I. P. (1985). Crassulacean acid metabolism. *Ann. Rev. Plant Physiol.* **36**, 595–622.

Vernon, D. M., Ostrem, J. A., Schmitt, J. M. and Bohnert, H. J. (1988). PEPcase transcript levels in *Mesembryanthemum crystallinum* decline rapidly upon relief from salt stress. *Plant Physiol.* **86**, 1002–4.

Walbot, V. (1977). The dimorphic chloroplasts of the C4 plant *Panicum maximum* contain identical genomes. *Cell* **11**, 729–37.

Wedding, R. T. and Black, C. C. (1983). Physical and kinetic properties and regulation of the NAD malic enzyme purified from leaves of *Crassula argentea*. *Plant Physiol.* **72**, 1021–8.

Winter, K. (1982). Properties of phosphoenolpyruvate carboxylase in rapidly prepared, desalted leaf extracts of the Crassulacean acid metabolism plant *Mesembryanthemum crystallinum* L. *Planta* **154**, 298–308.

Winter, K. (1985). Crassulacean acid metabolism. In *Photosynthetic Mechanisms and the Environment*. Topics in Photosynthesis, Vol. 6, eds J. Barber and N. R. Baker, Elsevier, Amsterdam, pp. 329–87.

Winter, K., Foster, J. G., Edwards, G. E. and Holtum, J. A. M. (1982). Intracellular localization of carbon metabolism in *Mesembryanthemum crystallinum* exhibiting photosynthetic characteristics of either a C3 or Crassulacean acid metabolism plant. *Plant Physiol.* **69**, 300–7.

Mesembryanthemum crystallinum L. *Planta* **154**, 298–308.

Winter, K., Foster, J. G., Edwards, G. E. and Holtum, J. A. M. (1982). Intracellular localization of carbon metabolism in *Mesembryanthemum crystallinum* exhibiting photosynthetic characteristics of either a C3 or Crassulacean acid metabolism plant. *Plant Physiol.* **69**, 300–7.

VI

Chloroplast–Cytosol Interactions

IV

Chloroplast–Cytosol Interactions

20 Transport of proteins into chloroplasts
Kenneth Keegstra

Introduction

Chloroplasts are functionally complex organelles that carry out a diverse array of metabolic processes in addition to their well-known role in photosynthesis. Consistent with their functional complexity, chloroplasts are structurally complex organelles. They possess three different lipid bilayer membranes enclosing three different aqueous compartments (Fig. 20.1). Each membrane and each aqueous compartment has a unique set of proteins and enzyme activities that reflects their respective functions. Because of the limited coding capacity of the plastid genome, most of these chloroplastic proteins are encoded by nuclear genes and synthesized in the cytoplasm as higher molecular weight precursors (Fig. 20.1). The extra region of peptide is present at the amino terminus of a precursor, and is called a transit peptide (Chua and Schmidt, 1979) because of its importance in directing the precursor into chloroplasts. Understanding how these precursor proteins are targeted from the cytoplasm to their proper location within chloroplasts is a major challenge, and progress in this effort is described in this chapter. This subject has been the topic of

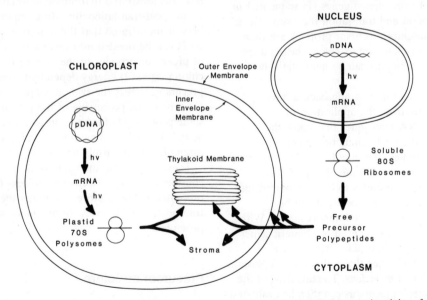

Fig. 20.1 Schematic diagram depicting the structural complexity of chloroplasts and the biogenetic origins of chloroplastic proteins. No arrow leads to the intermembrane space of the chloroplastic envelope because at present no proteins have been identified in this compartment.

several reviews in recent years, and readers should consult these reviews for further details (Schmidt and Mishkind, 1986; della-Cioppa et al., 1987; Szabo and Cashmore, 1987; Keegstra and Bauerle, 1988; Lubben et al. 1988).

Methods for studying protein transport into chloroplasts

Two different methods have been used to investigate the transport of proteins into chloroplasts. The first is *in vitro* assay in which the transport of radioactive precursor proteins into isolated intact chloroplasts is measured. The second is *in vivo* analysis where the intracellular localization of a protein is examined and information about the targeting process is deduced from the localization data. The *in vitro* procedure was first developed using a mixture of precursor proteins generated by *in vitro* translation of poly A^+ mRNA. Individual precursors and imported proteins were identified by immunoprecipitation. Recently, the utility of *in vitro* transport assays has been enhanced by the development of *in vitro* transcription systems. Precursor proteins can now be synthesized from cloned genes via sequential *in vitro* transcription and translation reactions. These radiochemically pure precursor proteins are then incubated with isolated intact chloroplast and the extent of protein import can be quantitatively measured.

Both *in vitro* and *in vivo* approaches have benefited greatly from the application of recombinant DNA techniques that allow the production of altered or chimeric precursor genes. With *in vitro* studies, the modified precursor genes are expressed *in vitro* and the ability of the modified precursor proteins to be transported into chloroplasts and properly localized can be examined using reconstituted import assays. With *in vivo* approaches, the altered or chimeric genes are introduced back into plant cells using a transformation technique that utilizes the Ti plasmid of *Agrobacterium tumefaciens*. The targeting of the altered or chimeric precursors can then be evaluated by examining the intracellular localization of the protein encoded by the introduced gene.

The *in vitro* reconstituted assays have allowed

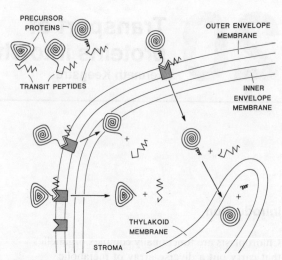

Fig. 20.2 Schematic representation of protein import into chloroplasts showing the steps of the import process. The binding step is represented as a dual interaction of the transit peptide with a receptor protein and of the transit peptide with the bilayer. Further details are explained in the text.

the import process to be dissected and analyzed in ways that are not possible *in vivo*. Such studies have demonstrated that import into chloroplasts occurs post-translationally. More importantly, they have demonstrated that the import process (Fig. 20.2) can be divided into several steps: (1) binding of precursor proteins to the surface of chloroplasts; (2) energy-dependent translocation of precursors across the two envelope membranes; (3) proteolytic processing of the precursor; and for some proteins (4) further translocation into or across the thylakoid membrane; and/or (5) assembly into multisubunit complexes.

These steps form the framework of the consideration of the transport process that is presented below. Before examining the details of transport, a brief review of precursor structure is presented.

Precursor structure

A major development in understanding protein targeting in eukaryotic cells was the demonstration that most translocated proteins are synthesized as higher molecular weight precursors. This was first demonstrated for secreted proteins, where the extra peptide was termed a signal peptide and the signal hypothesis was offered in 1975 to explain how this peptide leads to the cotranslational transport of secreted proteins into the lumen of the endoplasmic reticulum (Walter and Lingappa, 1986). In 1978 it was demonstrated that cytoplasmically synthesized chloroplastic proteins are made as higher molecular weight precursors (Chua and Schmidt, 1979). In this case the extra peptide was termed a transit peptide to denote its role in translocation, but still maintaining a distinction from signal peptides (Chua and Schmidt, 1979).

Experiments from several laboratories have demonstrated the importance of a transit peptide for the proper targeting of the precursor. The necessity of a transit peptide for transport was demonstrated by Mishkind et al. (1985) who showed that the precursor to the small subunit (prSS) of ribulose bisphosphate carboxylase could not be transported into chloroplasts if part or all of its transit peptide was removed. This has subsequently been confirmed using modified precursors generated by recombinant DNA techniques (Lubben et al., 1988). Precursors with a deletion in the transit peptide have either lost the ability to be imported or have impaired ability to be imported, depending upon the extent and location of the deletion.

Even more interesting is the recent demonstration that a transit peptide contains sufficient information for targeting 'foreign' proteins into chloroplasts. The first demonstrations came from the work of Van den Broeck et al. (1985) and Schreier et al. (1985), who used both in vivo and in vitro approaches to demonstrate that the prSS transit peptide from pea was capable of transporting a bacterial protein, neomycin phosphotransferase, into tobacco chloroplasts. This work has now been extended to include several other examples, using other transit peptides and other passenger proteins (Lubben et al., 1988). The conclusion derived from these types of studies is that it is possible to direct a number of different passenger proteins into chloroplasts simply by adding a transit sequence onto the amino terminus of the passenger.

Efforts to define essential regions within the primary structure of transit peptides have not been successful. Transit peptides have a distinctive amino acid composition, being rich in hydroxylated, basic, and small hydrophobic amino acids. However, sequence comparisons have failed to identify conserved regions possessed by all transit peptides. Given this variability among transit sequences, it seems unlikely that import functions can be correlated with conserved primary structures. Rather, it seems likely that the essential features lie within the secondary structure, as has been suggested for mitochondrial presequences (Roise and Schatz, 1988).

Organelle specificity

The similarities between transit peptides of chloroplastic precursors and the presequences of mitochondrial precursors raise an interesting question of how organelle specificity is determined in plant cells that contain both organelles. Boutry et al. (1987) examined this question with a series of in vivo experiments. They compared the targeting of chimeric precursors containing a chloroplastic transit sequence or a mitochondrial presequence fused to the same passenger protein, bacterial chloramphenicol acetyltransferase (CAT). These chimeric constructs were introduced back into plant cells by Ti-mediated transformation. Upon expression of the chimeric genes, the chloroplastic transit sequence directed CAT into chloroplasts, but not mitochondria. On the other hand, the mitochondrial presequence directed CAT to mitochondria, but not chloroplasts. These results provide convincing evidence that organelle specificity resides in the respective targeting sequences.

The import apparatus and the mechanism of protein translocation

Although the mechanisms by which proteins are transported into or across the various membranes in chloroplasts remain obscure, progress has been made in defining some of the steps in the import process. The first step during transport across the envelope membranes is binding of precursors to the surface of chloroplasts. Current evidence supports the hypothesis that this binding is mediated at least in part by a receptor protein, but it is also likely that lipid–protein interactions are important for binding (Fig. 20.2). Very little is known about the steps that result in protein translocation. It was established several years ago that protein translocation requires energy in the form of ATP (Schmidt and Mishkind, 1986); however, it is not known how the hydrolysis of ATP is connected to protein translocation. Once across the envelope membrane, a stromal processing protease removes the transit peptide. In the case of thylakoid lumen proteins, only part of the transit peptide is removed in the stroma. The remainder of the transit peptide is thought to direct these proteins across the thylakoid membrane, as discussed in more detail in the final section of this chapter. The final step of the import process requires that the imported protein be assembled into the proper macromolecular complex. For example, SS is assembled with chloroplast synthesized large subunit to yield holoenzyme. Each of these steps is described in more detail below.

Binding of precursor proteins

Binding of prSS to the surface of isolated chloroplasts can be observed if the translocation step is blocked (Cline et al., 1985). This binding is specific in that SS lacking a transit peptide does not bind, and is saturable at a level of 1000 to 3000 molecules per chloroplast (Friedman and Keegstra, submitted for publication). Earlier it was thought that binding did not require energy, but recent results demonstrate that precursor binding requires low levels of ATP (Olsen et al., submitted for publication). Because binding requires only 50–100 μM ATP whereas translocation requires 500–1000 μM ATP, it is possible to separate the binding and translocation steps.

Two different strategies have been utilized in efforts to identify envelope polypeptides that function as receptors to mediate precursor binding. Cornwell and Keegstra (1987) employed photoactivatable reagents to cross-link radioactive prSS to putative receptor proteins. They concluded that a 66 kD envelope protein is part of the receptor complex. Pain et al. (1988) generated anti-idiotypic antibodies directed against antibodies specific for the transit peptide of prSS. The anti-idiotypic antibodies mimic prSS, interact with the putative receptor and block import. They used these antibodies to identify a 30 kD envelope protein which they conclude is part of the receptor complex. Despite the apparent conflict between the results from the two groups, it is possible that both are correct and that the two different approaches are identifying different parts of the import apparatus. Much work is needed to clarify the function of each of these polypeptides. Another important question that needs to be resolved is whether the prSS receptor is specific for a single precursor or whether it is also involved in the transport of other precursors.

Protein translocation across the envelope membranes

Subsequent to precursor binding, it is not clear what steps are responsible for translocation of bound precursors across the two chloroplast envelope membranes. It has been established that at least one of these steps requires energy in the form of ATP. In the absence of ATP (or at least if ATP is below 100 μM) translocation is blocked and the precursors do not cross the envelope membranes. The translocation step is also blocked at low temperatures (e.g. 0 °C), whereas binding can occur at 0 °C. A disagreement exists on whether ATP is needed inside or outside the chloroplast envelope (Lubben et al., 1988). However, recent results from our laboratory support the conclusion suggested by Pain and Blobel (1987) that ATP is needed inside the chloroplast (Theg et al., manuscript in preparation). This conclusion concerning the location of ATP utilization has important implications with respect to a popular hypothesis for the mechanism of protein translocation. It has been postulated that during protein import into mitochondria, ATP is utilized outside the organelle to provide the energy needed to unfold precursors so that they can be threaded through the two mitochondrial membranes (Eilers and Schatz, 1988). If the ATP needed for protein import into chloroplasts is needed inside the organelle, this raises a question concerning the energy source for protein unfolding and raises the possibility the protein might be translocated without unfolding.

Another important question is whether the precursor proteins pass directly through the bilayer or whether they pass through a translocator protein. Singer et al. (1987) have postulated the existence of translocator proteins and suggest that they are involved whenever proteins are transported across membranes. However, translocator proteins have not yet been identified in any protein translocation system, and therefore remain hypothetical. Thus, with respect to protein transport into chloroplasts, at least two fundamental and related questions exist: (1) What is the conformation of the precursor during its translocation? (2) Are translocator proteins present in the envelope that mediate the movement of proteins across them?

Proteolytic processing and assembly

A processing protease capable of removing the transit peptide is located in the stromal space. The protease responsible for the conversion of prSS to SS has been extensively purified and partially characterized (Schmidt and Mishkind, 1986). The purified enzyme is capable of processing several different precursors. It has not yet been determined whether a single protease recognizes and processes all precursors or whether the purified preparation contains multiple proteases.

Many imported proteins contain prosthetic groups or are part of multisubunit complexes and must be modified or assembled after import. For example, ferredoxin contains iron and iron–sulfur centers and these are presumably added after import. The assembly of SS into holoenzyme has been most extensively studied. In this case it has been suggested that large subunit, which is synthesized inside chloroplasts, becomes associated with a large subunit binding protein (Hemmingsen et al., 1988). This binding protein has recently been demonstrated to be related to a ubiquitous class of heat shock proteins, and the term chaperonin has been suggested to describe this group of proteins (Hemmingsen et al., 1988). Gatenby et al. (1988) have recently provided evidence that SS associates transiently with the complex of large subunit and its binding protein before it appears in holoenzyme.

Targeting within chloroplasts

Although it is likely that cytoplasmically synthesized proteins are targeted to all chloroplastic compartments, few examples have been studied. Consequently, the targeting signals and pathways involved in transport to non-stromal locations are less well understood than targeting to the stroma. For example, no information is available regarding protein transport to the two envelope membranes or the intermembrane space between them (Fig. 20.1). Transport to the two envelope membranes can be viewed as being similar to transport to the two membranes of mitochondria, making it possible to formulate hypotheses for targeting to the chloroplastic envelope based on the information available for transport to the two mitochondrial membranes (Schatz, 1987). These hypotheses still need to be tested.

Mature chloroplasts contain an internal thylakoid membrane system which has no counterpart in mitochondria. This membrane is the site of photosynthetic electron transport and contains a specific complement of proteins which function in this process. Several of these proteins are encoded in the nucleus and synthesized in the cytoplasm. The transport and membrane insertion of one integral thylakoid membrane protein, the light-harvesting chlorophyll a/b protein of photosystem II (LHCP), has been reconstituted *in vitro*. The present results support the hypothesis that LHCP is imported into chloroplasts in a manner similar to stromal proteins and then is inserted into the thylakoid membrane from the stromal space. Cline (1986) has studied the membrane insertion step and demonstrated that the precursor to LHCP can integrate into isolated thylakoids in an energy-dependent manner. The question of how a protein can be translocated across two envelope membranes and then specifically incorporated in to the third (thylakoid) membrane which it encounters is still unanswered.

The targeting of cytoplasmically synthesized proteins to the thylakoid lumen is particularly interesting because these proteins must be transported across both the envelope membranes and the thylakoid membrane. Precursors of lumen proteins possess composite transit peptide with two distinct structural domains (Keegstra and Bauerle, 1988). The amino-terminal region (domain 1) resembles transit peptides of other imported proteins. The carboxyl-terminal region (domain 2) consists of a long hydrophobic stretch flanked by charged residues. Domain 2 is structurally similar to signal sequences, which mediate the transport of proteins across single membranes in both prokaryotes and eukaryotes. The second domain is unique to transit peptides of lumen proteins, and thus may be involved in targeting to that compartment.

Smeekens et al. (1986) have studied the transport pathway of plastocyanin (PC), a lumen protein involved in photosynthetic electron transport. Precursor to PC (prPC) from *Silene pratensis* was properly processed and localized by isolated pea chloroplasts. Interestingly, a form of PC (iPC), intermediate in size between prPC and mature PC, was also observed in the chloroplast soluble fraction. Based on the kinetics of appearance and disappearance of iPC, they proposed a two-step model for PC transport (Fig. 20.3). According to this model, prPC is initially directed into chloroplasts by domain 1 and is partially processed in the stroma to iPC. In the second step, domain 2 mediates transport of iPC into the lumen, where it is processed to its mature size.

The necessity of the two transit peptide domains for targeting to the thylakoid lumen has been supported by deletion analysis of the PC transit sequence. Deletion of the domain 1 coding region results in the production of a truncated precursor which is not able to be imported into chloroplasts *in vitro*. On the other hand, a precursor protein from which domain 2 has been deleted can still be imported into chloroplasts (J. Hageman and P. Weisbeek, personal communication). However, in this case the imported form is located in the stroma, indicating that lumen targeting information has been lost. These results support the conclusion that domain 2 is necessary for transport across the thylakoid membrane, and its deletion results in the passenger protein being stranded in the stroma.

To address the question of whether the PC transit peptide contains all the information necessary for targeting to the thylakoid lumen, chimeric proteins were constructed in which the transit sequence was fused to alternative passenger proteins. If all of the targeting information is present in the transit sequence, then these passenger proteins should be directed to the lumen. In one case, a stromal protein, ferredoxin, was used as the passenger (Smeekens et al., 1986). This chimeric precursor was imported into chloroplasts, but transport to the lumen did not occur. Rather, the imported ferredoxin was found in the soluble fraction as an intermediate-sized form. Interpretation of this result is complicated by the fact that ferredoxin interacts with several stromal components, which may hinder its transport across the thylakoid membrane. Therefore, a non-plastid protein, yeast mitochondrial superoxide dismutase, was used as the passenger (Smeekens et al., 1987). The results were the same: superoxide dismutase was imported but found only in the soluble fraction as

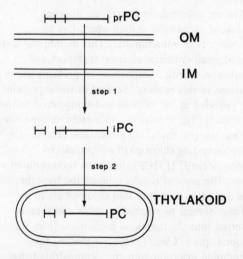

Fig. 20.3 Two-step model for transport of plastocyanin to the thylakoid lumen. Details are explained in the text.

an intermediate-sized form. If all of the results are considered, it may be concluded that the intact PC transit sequence is necessary for proper targeting to the lumen, but it alone is not sufficient to accomplish this targeting. It is intriguing that the requirements for an acceptable passenger protein are more stringent for thylakoid transport than for import across the two envelope membranes. Although this conclusion is based on a limited sampling of passenger proteins, it may have important implications for understanding the mechanisms involved in the two processes.

The mechanism for transport of proteins across the thylakoid membrane has received only limited study. A transport mechanism for chloroplast-encoded thylakoid membrane proteins is clearly implicated. These proteins are synthesized by thylakoid bound ribosomes and probably are inserted in a cotranslational manner. However, cytoplasmically synthesized proteins are transported across the thylakoid post-translationally (Fig. 20.3). Efforts to reconstitute post-translational thylakoid transport *in vitro* using prPC or iPC and isolated thylakoid membranes are ongoing in our laboratory. It is tempting to speculate that thylakoid transport is analogous to transport of proteins across bacterial membranes. Such a hypothesis is supported by the similarity between the putative thylakoid transfer domain (domain 2) and bacterial signal sequences. Moreover, chloroplast thylakoid membranes are closely related to thylakoid membranes of photosynthetic bacteria where a protein translocation system is presumably required. The ability to reconstitute post-translational thylakoid transport would allow the mechanism of this translocation process to be studied in detail.

Summary

Considerable progress has been made in the last decade in understanding how cytoplasmically synthesized chloroplastic proteins are directed to their proper location. These proteins are synthesized as larger precursors containing a transit peptide. The transit peptide has a crucial role in directing the precursor into chloroplasts. In the case of plastocyanin, the transit peptide also has a role in directing the protein to its proper place within the organelle. Despite these advances in our understanding of the targeting sequences and the transport pathways, we still have relatively little information concerning the mechanism whereby proteins are transported across biological membranes.

References

Boutry, M., Nagy, F., Poulsen, C., Aoyagi, K. and Chua, N.-H. (1987). Targeting of bacterial chloramphenicol acetyltransferase to mitochondria in transgenic plants. *Nature* **328**, 340–2.

Chua, N.-H. and Schmidt, G. W. (1979). Transport of proteins into mitochondria and chloroplasts. *J. Cell Biol.* **81**, 461–83.

Cline, K. (1986). Import of proteins into chloroplasts. Membrane integration of a thylakoid precursor protein reconstituted in chloroplast lysates. *J. Biol. Chem.* **261**, 14804–10.

Cline, K., Werner-Washburne, M., Lubben, T. H. and Keegstra, K. (1985). Precursors to two nuclear-encoded chloroplast proteins bind to the outer envelope membrane before being imported into chloroplasts. *J. Biol. Chem.* **260**, 3691–6.

Cornwell, K. L. and Keegstra, K. (1987). Evidence that a chloroplast surface protein is associated with a specific binding site for the precursor to the small subunit of ribulose-1,5-bisphosphate carboxylase. *Plant Physiol.* **85**, 780–5.

della-Cioppa, G., Kishore, G. M., Beachy, R. N and Fraley, R. T. (1987). Protein trafficking in plant cells. *Plant Physiol.* **84**, 965–8.

Douglas, M. G., McCammon, M. T. and Vassarottii, A. (1986). Targeting proteins into mitochondria. *Microbiol. Rev.* **50**, 166–78.

Eilers, M. and Schatz, G. (1988). Protein unfolding and the energetics of protein translocation across biological membranes. *Cell* **52**, 481–3.

Gatenby, A. A., Lubben, T. H., Ahlquist, P. and Keegstra, K. (1988). Imported large subunits of ribulose bisphosphate carboxylase/oxygenase, but not imported ATP synthase subunits, are assembled into holoenzyme in isolated chloroplasts. *EMBO J.* **7**, 1307–14.

Hemmingsen, S. M., Woolford, C., Van Der Vies, S. M., Tilly, T., Dennis, D. T., Georgopoulos, C. P., Hendrix, R. W. and Ellis, R. J. (1988). Homologous plant and bacterial proteins chaperone oligomeric protein assembly. *Nature* **333**, 330–4.

Keegstra, K. and Bauerle, K. (1988). Targeting of proteins into chloroplasts. *Bioessays* **9**, 15–19.

Lubben, T. H., Theg, S. M. and Keegstra, K. (1988). Transport of proteins into chloroplasts. *Phytosyn. Res.* **17**, 173–94.

Mishkind, M. L., Wessler, S. R. and Schmidt, G. W. (1985). Functional determinants in transit sequences: import and partial maturation by vascular plant chloroplasts of the ribulose-1,5-bisphosphate carboxylase small subunit of *Chlamydomonas*. *J. Cell Biol.* **100**, 226–34.

Pain, D. and Blobel, G. (1987). Protein import into chloroplasts requires a chloroplast ATPase. *Proc. Natl Acad. Sci. USA* **84**, 3288–92.

Pain, D., Kanwar, Y. S. and Blobel, G. (1988). Identification of a receptor for protein import into chloroplasts and its localization to envelope contact zones. *Nature* **331**, 232–7.

Roise, D. and Schatz, G. (1988). Mitochondrial presequences. *J. Biol. Chem.* **263**, 4509–11.

Schatz, G. (1987). Signals guiding proteins to their correct locations in mitochondria. *Eur. J. Biochem.* **165**, 1–6.

Schmidt, G. W. and Mishkind, M. L. (1986). The transport of proteins into chloroplasts. *Ann. Rev. Biochem.* **55**, 879–912.

Schreier, P. H., Seftor, E A., Schell, J. and Bohnert, H. J. (1985). The use of nuclear-encoded sequences to direct the light-regulated synthesis and transport of a foreign protein into plant chloroplasts. *EMBO J.* **4**, 25–32.

Singer, S. J., Maher, P. A. and Yaffe, M. P. (1987). On the translocation of proteins across membranes. *Proc. Natl Acad. Sci. USA* **84**, 1015–19.

Smeekens, S., Bauerle, C., Hageman, J., Keegstra, K. and Weisbeek, P. (1986). The role of the transit peptide in the routing of precursors toward different chloroplast compartments. *Cell* **46**, 365–75.

Smeekens, S., Van Steeg, H., Bauerle, C., Bettenbroek, H., Keegstra, K. and Weisbeek, P. (1987). Import into chloroplasts of a yeast mitochondrial protein directed by ferredoxin and plastocyanin transit peptides. *Plant Mol. Biol.* **9**, 377–88.

Szabo, L. J. and Cashmore, A. R. (1987). Targeting nuclear gene products into chloroplasts. In *Plant DNA Infectious Agents*, eds T. Hohn and J. Schell, Springer Verlag, Wein, New York, pp. 321–39.

Van Den Broeck, G., Timko, M. P., Kausch, A. P., Cashmore, A. R., Van Montagu, M. and Herrera-Estrella, L. (1985). Targeting of a foreign protein to chloroplasts by fusion to transit peptide from the small subunit of ribulose 1,5-bisphosphate carboxylase. *Nature* **313**, 358–63.

Walter, P. and Lingappa, V. P. (1986). Mechanism of protein translocation across the endoplasmic membrane. *Ann. Rev. Cell Biol.* **2**, 499–516.

21 The flux of carbon between the chloroplast and cytoplasm
Mark Stitt

Introduction

In plants, there is a compartmentation of metabolism between the cytosol and the plastid. The major storage carbohydrate, starch, is restricted to the plastid while the major transport carbohydrate, sucrose, is metabolized outside the plastid. This strict compartmentation is found in all plant tissues, from photosynthetic leaves to storage tissues like wheat grains or potato tubers. Clearly, processes which facilitate and control carbon movement across the plastid envelope membrane are crucial in determining how carbon is distributed between sucrose, starch and respiratory metabolism.

Furthering our understanding of carbon flow between the plastid and the cytosol is dependent upon advances in two areas. First, the transport proteins present in the envelope membrane must be identified and characterized. Obviously, until we know in what form carbon passes across the membrane, the biochemical pathways for starch and sucrose turnover will be unknown. Second, the mechanisms regulating the rate of transport must be elucidated. These might include direct regulation of the transport proteins, and could also involve regulation of metabolism in the interlinked subcellular compartments.

Our understanding of cytosol–plastid interactions is still incomplete, especially in non-photosynthetic tissues. In this chapter, one example of chloroplast–cytosol interactions will be discussed in detail, namely, the control of carbon fluxes across the chloroplast envelope membrane during photosynthetic starch and sucrose synthesis. Then the possible routes and regulation of carbon fluxes between starch and sucrose during respiratory metabolism will be introduced.

Transport in chloroplasts

Study of transport

Many early experiments attempted to characterize transport by incubating isolated organelles with a potential precursor (usually radioactive) and investigating its incorporation into a product within the organelle. At best, such studies only provide imprecise and indirect information, and can be very misleading for the following reasons. There is a strong possibility that the added compound is modified prior to uptake. The rate and kinetics of incorporation into products may be more dependent on the metabolism of the compound after it has been taken up, than on the transport step itself. For these reasons, the study of transport requires direct measurements of the initial rate of uptake into the organelle before other factors such as metabolism or back exchange can lead to an apparent change in the transport kinetics.

This has been achieved for isolated chloroplasts by the method of silicone oil centrifugation, pioneered by Heldt and coworkers (Heldt and Flügge 1984; Fig. 21.1). A suspension of isolated chloroplasts is pipetted into a 400 μl microcentrifuge tube which already contains 70 μl 10% (v/v) perchloric acid and, above this, 20 μl of silicone oil. Silicone oil is immiscible with aqueous solutions, and is used at a

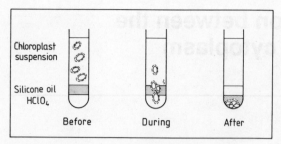

Fig. 21.1 Silicone oil centrifugation of chloroplasts.

density which is heavier than the suspension medium but is lighter than the chloroplasts themselves or the perchloric acid; hence, it provides a barrier between the suspension of chloroplasts and the perchloric acid. To measure transport, a small amount of radioactively-labeled substrate is mixed into the chloroplast suspension; a few seconds later uptake is terminated by centrifuging the chloroplasts out of the medium through the silicone oil into the perchloric acid. This serves to separate the chloroplasts from the incubation medium, and to stop any further uptake or metabolism. The amount of radioactivity in the perchloric acid is a measure of its uptake into the chloroplast, although (see below) control experiments also have to be carried out to correct for small amounts of medium which are carried through the silicone oil by the organelles.

This technique allows K_m and V_{max} values to be measured for the transport of a substrate, just as if the transporter were a soluble enzyme. Commonly, transporters mediate movement of more than one substrate. In this case, the alternative substrates may compete for binding and transport, and act as competitive inhibitors to one another. By comparing the K_m and K_i values of the various substrates, it is possible to decide which compounds are transported on the same protein (see, for example, Fliege et al., 1978). Further characterization and isolation of the protein usually depends upon finding a compound which will bind specifically and irreversibly to the protein, providing a 'tag' by which the protein can be identified. This is essential because the isolation of a protein requires that its presence during the various isolation steps can be monitored. The activity of a transport protein can obviously not be assayed once it has been removed from the membrane across which it performs the transport. The only exception to this are transport proteins that have a side reaction which is retained after solubilization, e.g. ATP-linked ion pumps that retain an ion-stimulated ATPase activity. Once a protein has been 'tagged', it can be solubilized with detergents, purified, and then reconstituted into artificial membranes or liposomes. Its kinetic properties and regulation can then be studied without other reactions or processes interfering.

The various transport systems found in the envelope membrane of chloroplasts will now be considered. It should be stressed that only one of these, the phosphate translocator, has been unambiguously characterized in the strict sense of having been isolated and reconstituted. The others have only been identified by measuring the uptake kinetics of substrates into isolated chloroplasts. So far, these studies have only been carried out on photosynthetic plastids. As yet, there is no direct evidence if these or other transporters are present in non-photosynthetic plastids although there is immunological evidence for a polypeptide in the envelope of non-photosynthetic plastids from sycamore cells and cauliflower florets which reacts with antibodies to the phosphate translocator (Nyernprasintsiri et al., 1988; J. Joyard, personal communication). For further details, see review by Heldt and Flügge (1987).

The inner membrane as the site of specific transport

The chloroplast envelope consists of two functionally and physically distinct membranes. Low molecular weight compounds can freely pass through the outer membrane, because it contains proteins called *porins*. Similar proteins, which are found in the outer membrane of mitochondria and Gram-negative bacteria, form pores in the membrane. In chloroplasts, these pores are large enough to allow free movement of molecules up to a molecular weight of 10 000 (Flügge and Benz, 1984).

The inner membrane is the site at which specific transport of metabolites occurs. This can be shown by incubating chloroplasts in increasing concentrations of a non-permeating osmoticum such as sorbitol. As the water is removed from the chloroplast the stromal volume decreases and the inner impermeable membrane contracts away from the outer membrane so that the volume of the space between these two membranes increases. This can be directly visualized in the electron microscope. The volume of the intermembrane space can be measured using the technique of silicone oil centrifugation. Chloroplasts are suspended in medium that contains tritiated water and ^{14}C-sorbitol. Tritiated water will readily permeate both envelope membranes but the inner membrane is impermeable to sorbitol, hence, the ^{14}C-sorbitol can only equilibrate with the intermembrane space. By comparing the amounts of ^3H and ^{14}C in the perchloric acid after centrifuging the chloroplasts through silicone oil, it is possible to quantitatively estimate the volumes of the stroma and intermembrane spaces. This approach also allows a correction to be made for the radioactivity which is carried through the silicone oil in the space between the two membranes when transport is being studied.

The phosphate translocator

The phosphate translocator is the major protein in the envelope of chloroplasts, comprising 15% of the total protein. This translocator has an essential role during photosynthesis. It was originally thought that chloroplasts could carry out photosynthesis autonomously. However, studies by Walker and coworkers in the 1960s showed that CO_2 fixation only occurred in isolated chloroplasts when they received a carefully controlled supply of P_i (Edwards and Walker, 1983). Studies by Heldt and coworkers demonstrated that the major products of photosynthesis in isolated chloroplasts were triose-P and glycerate-3-P, which accumulated in the medium (Lilley et al., 1976). Non-aqueous fractionation of leaves that had been exposed to $^{14}CO_2$ showed that the first radioactive metabolites in the cytosol were the triose-P and glycerate-3-P, rather than sucrose or its immediate precursors (Heber and Willenbrink, 1964). The evidence that the products of photosynthesis exported from the chloroplast are three carbon phosphoesters was complemented by the finding that the enzymes for sucrose synthesis are located outside the chloroplast (Bird et al., 1974). The discovery and characterization of the phosphate translocator by Heldt, Flügge and coworkers allowed

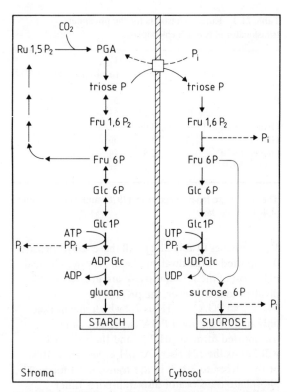

Fig. 21.2 Pathways of sucrose and starch synthesis in leaves of C3 plants.

the movement of these metabolites between the chloroplast and cytosol to be unambiguously defined. Photosynthate is exported from the chloroplast as three-carbon phosphoesters, and these are then converted to sucrose in the cytosol. The P_i which is released in the cytosol re-enters the chloroplast via the phosphate translocator, to allow further photosynthesis (Fig. 21.2). As will emerge later in this chapter, these findings provide a focal point for our current ideas about how the synthesis of endproducts is controlled during photosynthesis.

The phosphate translocator has been purified and incorporated into liposomes (Flügge and Heldt, 1984). The kinetic properties, which were first investigated in isolated chloroplasts (Fliege et al., 1978), could then be confirmed and extended in a purified system (Flügge and Heldt, 1984, 1986). The favored substrates are P_i, dihydroxyacetone-P, glyceraldehyde-3-P and glycerate-3-P, all of which have a K_m between 0.1–0.4 mM, and comparable

Table 21.1 Kinetic constants for the phosphate translocator of spinach chloroplasts

	K_m (mM)	V_{max} ($\mu mol\ mg\ Chl^{-1}\ h^{-1}$)
P_i	0.3	57
Dihydroxyacetone-P	0.13	51
Glyceraldehyde-P	0.08	41
Glycerate-3-P	0.14	36
Glycerate-2-P	2.8	24
Glycerol-1-P	1.1	59

The results are from Fliege *et al.* (1978) and were obtained at 4 °C in the dark.

V_{max} activities (Table 21.1). All these substrates are transported as the divalent anion. At physiological pH, most of the P_i and triose-P are present as the divalent ion. However, the pK_a for the reaction $PGA^{2-} \rightleftharpoons PGA^{3-} + H^+$ is 7.2 which means that only half of the total PGA is present as the transported form at pH 7.2, and the proportion will fall as the pH rises. At pH 8, less than 10% of the glycerate-3-P is in the transported form. The phosphate translocator also transports other compounds; erythrose-4-P, glycerol-1-P, phosphoenolpyruvate and glycerate-2-P are translocated with 2 to 5-fold lower affinities. Pentose-P and hexose-P are poor substrates, with K_m values of 40 mM or more.

Each substrate inhibits the transport of the other substrates competitively. Hence, a compound with a high K_m will be transported at low rates *in vivo*, if high-affinity substrates are present at comparable concentrations. The high-affinity substrates will also compete with each other for transport. For example, it is unlikely that glyceraldehyde-3-P will be important as a substrate for transport because its concentration is 10 to 20-fold lower than that of dihydroxyacetone-P concentration as a result of the equilibrium position of triose phosphate isomerase. The affinity of the phosphate translocator may vary in different species and tissues. For example, in C4 species the affinity for phosphoenolpyruvate may be increased in mesophyll chloroplasts where this metabolite has to be exchanged with P_i during photosynthesis.

The phosphate translocator catalyzes a *passive counterexchange* of the above substrates; that is, for each molecule transported in one direction, one molecule is transported in the opposite direction. This counterexchange is strictly coupled, the unidirectional transport of P_i being at least 1000-fold slower than the rate of counterexchange. This strict stoichiometry is important because it ensures that every molecule of phosphoester exported from the chloroplast is counterbalanced by the uptake into the chloroplast of a molecule of P_i. The molecular mechanism which ensures that transport only occurs by counterexchange is not yet known, but the mechanisms may be similar to that of the ADP/ATP translocator of mitochondria which has some similarities with the plastid phosphate translocator. Both these transport proteins are dimers, with subunit molecular weights of about 30 000, have only one binding site, and catalyze a strict counterexchange (Flügge, 1987). A 'gated' pore mechanism has been suggested for the ADP/ATP translocator by Klingenberg (1981). The binding site on this translocator is formed by the two subunits, and is directed to one side of the membrane. After a substrate is bound, the dimer undergoes a conformational change, the substrate is moved across the membrane and, simultaneously, the binding site is directed towards the other side of the membrane. In this mechanism, a strict coupling of transport in the two directions is ensured because the conformational change that reorientates the transporter can only occur when a ligand is bound in the pore.

The hexose transporter

The chloroplast envelope can also transport free sugars, including D-glucose, D-fructose and D-ribose (Schäfer *et al.*, 1977). The disaccharide maltose is also transported by this system (Beck, 1985). The uptake is selective (for example, L-glucose is not transported), shows saturation kinetics towards the substrates with K_m values of about 20 mM, and can be inhibited by phloretin which is known to inhibit hexose uptake in other systems, including mammalian red blood cells.

The maximum velocity of the hexose carrier (5 μmol hexose mg $Chl^{-1}\ h^{-1}$) is only one-tenth of the phosphate translocator in spinach chloroplasts. However, the fluxes of carbon across the envelope during respiratory metabolism are much lower than those needed during photosynthesis. As will be described later, hexose transport is rapid enough to cope with a significant part of the flux of carbon in leaves in the dark when starch is being degraded.

Organic acid transport

Chloroplasts resemble mitochondria in having several transport systems for dicarboxylates, carboxylates, and related organic acids. However, the actual transport systems differ between the two organelles. Transport of organic acids across the envelope plays an essential role in photorespiration, and in facilitating the flux of carbon that is associated with nitrogen metabolism. Many of these transporters are also involved in the accessory pathways of CO_2 fixation during C4 or CAM photosynthesis; indeed, they were often first described in these specialized systems. In addition, it is now becoming apparent that the transport of dicarboxylates may allow redox groups to be transferred between the chloroplast and the cytosol.

The dicarboxylate transporters

Chloroplasts transport L-malate, 2-oxoglutarate, L-aspartate, L-glutamate and L-glutamine. The transport is mediated by a counterexchange mechanism, with unidirectional transport being about 100-fold slower than exchange transport. The K_m values for organic acids are in the 1–2 mM range. It is likely that the transport is mediated by two translocators with different but overlapping substrate specificities. This is suggested by detailed studies of uptake kinetics in spinach chloroplasts (Lehner and Heldt, 1978) and by studies of *Arabidopsis* mutants, which revealed the presence of two separate translocators. One of them transported dicarboxylates but not glutamine, and the other transported glutamine in preference to dicarboxylates (Sommerville and Sommerville, 1985).

The oxaloacetate translocator

Oxaloacetate can also be transported by the dicarboxylate transporter but this route is unlikely to be of any significance *in vivo*. The dicarboxylate transporter has similar K_m values for malate and oxaloacetate but the *in vivo* concentration of oxaloacetate is 1000-fold lower than the malate concentration, because of the equilibrium position of the reaction catalyzed by malate dehydrogenase. Complex schemes in which oxaloacetate was aminated and deaminated to allow it to be moved across the membrane as aspartate were devised to overcome this problem. These schemes are no longer needed, because a high affinity ($K_m = 50$ μM) and very specific transporter for oxaloacetate has been identified in chloroplasts. This transporter has been found in maize mesophyll chloroplasts (Hatch *et al.*, 1984) where it is necessary for the carbon fluxes during C4 photosynthesis (see Chapter 19). It is also present in spinach chloroplasts, where it may cooperate with the malate transporter to export reducing equivalents from the chloroplast during photosynthesis (see Heldt and Flügge, 1987).

Pyruvate

The pyruvate transporter was also first discovered in mesophyll chloroplasts isolated from C4 plants. In species that use NADP-malic enzyme, the transporter is needed to facilitate the entry of the pyruvate returning from the bundle sheath into mesophyll chloroplasts where pyruvate P_i dikinase is located (see Chapter 19). The uptake of pyruvate is driven by a light-dependent cation gradient across the envelope (Flügge *et al.*, 1985). It allows accumulation to occur against a concentration gradient. The pyruvate transporter may be present at lower activities in other plastids, and could provide a source of pyruvate for the plastid pyruvate dehydrogenase complex. This implies that biosynthesis of fatty acids may not be entirely dependent on glycolysis within the plastid.

Glycerate and glycolate transport

During photosynthesis, glycolate has to be rapidly exported from, and glycerate returned to, the chloroplast. Two glycolate molecules are exported for every glycerate that returns. This transport occurs via a single transporter which binds and transports glycolate, glyoxylate, D-glycerate and D-lactate, all with a similar affinity (Howitz and McCarty, 1985). The transport is driven by either a proton symport or a hydroxyl antiport.

Exchange of ATP and ADP

Chloroplasts and mitochondria both contain a translocator in their inner membrane that exchanges ATP and ADP in their inner membrane, but they have very different properties (Heldt and Flügge, 1987). The mitochondrial ADP/ATP translocator is extremely active and catalyzes a strict counterexchange of ATP and ADP, which is energized by the electrical gradient across the inner membrane. This allows a selective uptake of ADP and export of ATP, and is obviously well suited to export energy to the cytosol so that a high phosphorylation potential can be maintained. In contrast, in chloroplasts the exchange of ATP and ADP is relatively slow, being 10 to 100-fold lower than that of the phosphate translocator. Chloroplasts also import ATP in preference to ADP. This, and the observation that the activity is higher in young immature pea leaves, suggests the ATP translocator operates to import ATP into chloroplasts in the dark.

Control of transport during photosynthesis

To study the regulation of transport, complementary approaches must be used. Although isolated chloroplasts, or preferably reconstituted systems in liposomes (Heldt and Flügge, 1984, 1987) can be used to study the regulatory properties of the transport proteins, it is also necessary to investigate how substrate concentrations and fluxes vary *in vivo*. This has become possible with the development of several techniques to measure the *sub*cellular metabolite distribution in leaves or in protoplasts (Stitt *et al*., in press). These measurements are dependent on the metabolism being rapidly quenched. They also depend upon adequate techniques being available for the extraction and assay of metabolites (ap Rees, 1980). The reliability of the data should therefore always be assessed by carrying out control experiments, in which the recovery of representative amounts of metabolites added to the plant material in the killing mixture is determined. Data presented without this kind of authentication have to be regarded as potentially unreliable.

Direct control of the phosphate translocator

The form in which photosynthate is exported can be regulated by light. As discussed previously, the phosphate translocator transports divalent anions. In the light, the pH of the chloroplast stroma increases and most of the glycerate-3-P is present in the trivalent form. This means glycerate-3-P is retained in the chloroplast in the light, where it is reduced to triose-P using the ATP and NADPH that are generated by photosynthetic electron transport and photophosphorylation in the thylakoids. The triose-P which remains as a divalent anion can preferentially be exported in exchange for P_i under these conditions.

The rate at which this triose-P/P_i exchange occurs must be under metabolic control. This can be illustrated by considering the influence of the P_i concentration in the medium on photosynthesis by isolated chloroplasts (Fig. 21.3). In this simplified system, the rate of exchange across the translocator can be varied by altering the concentration of P_i outside the chloroplast. When P_i is low, the phosphate translocator activity is restricted, which leads to an accumulation of metabolites in the chloroplast and a depletion of P_i in the stroma (Table 21.1). The low P_i concentration results in a decrease in ATP, which restricts Calvin cycle activity and, hence, inhibits photosynthesis. On the other hand, if too much P_i is present in the medium, photosynthesis is also inhibited. This results from the fact that CO_2 can only be fixed if the acceptor, Ru-1, 5-P_2 can be regenerated. For every three CO_2 molecules fixed, six molecules of triose-P are formed.

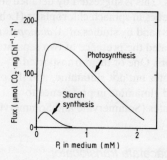

Fig. 21.3 Influence of the phosphate concentration in the medium on the rate of photosynthesis by isolated chloroplasts.

Table 21.2 Influence of the P_i concentration in the medium on photosynthesis, starch synthesis and stromal metabolites in isolated spinach chloroplasts

P_i in medium (mM)	Fluxes		Stromal metabolites				
	CO_2 fixation (μ mol mg Chl^{-1} h^{-1})	Starch synthesis (μ mol hexose mg Chl^{-1} h^{-1})	P_i	Glycerate-3-P	Triose-P (mM)	Hexose-P	Ru-1,5-P_2
2.0	10	–	16.0	tr	tr	0.1	0.06
0.5	107	1.4	7.0	6.0	0.33	4.1	0.28
0.1	52	7.9	2.2	8.9	0.25	4.2	0.57

tr = trace (50 μM).
The results are from Heldt et al. (1977) and Flügge et al. (1980).

One of the triose-P molecules can be exported, but the remaining five must be retained in the chloroplast and used to regenerate three molecules of Ru-1,5-P_2 again. If triose-P is withdrawn too rapidly, the Calvin cycle metabolite pools would be depleted, and photosynthesis would be inhibited. In other words, high rates of photosynthesis depend upon a controlled exchange of triose-P and P_i, so that there is a compromise between adequate P_i for ATP synthesis, and a sufficient concentration of stromal phosphorylated metabolites for the turnover of the Calvin cycle to regenerate Ru-1,5-P_2.

In a leaf, this balance must be achieved under more complex circumstances. The cytosol contains triose-P and glycerate-3-P, as well as P_i, all competing for entry into the chloroplast. This means the rate and direction of the resulting flux depends upon the relative concentrations of the substrates in the cytosol, and in the stroma. It also means that only a small fraction of the translocator capacity is actually involved in mediating the required export of triose-P and import of P_i. A considerable proportion will be involved in exchanges between other substrate pairs, or even so-called homologous exchanges between like partners. Hence, although the phosphate translocator is clearly in excess in isolated chloroplasts and has to be restrained by using a subsaturating P_i concentration in the suspension medium, there may not be a large excess of capacity in vivo. Non-aqueous fractionation of spinach leaves has actually shown that triose-P may not be equilibrated between the chloroplast and cytosol during photosynthesis (Gerhardt et al., 1987).

Another important difference between leaves and in vitro systems is that in leaves the rate of transport depends on metabolic events occurring in the cytosol. In leaves, the chloroplasts are not surrounded by a large volume of medium with a fixed P_i concentration, acting as an 'infinite' source and sink for P_i. Instead, triose-P is continually removed by being synthesized into sucrose. This pathway regenerates P_i which returns to the chloroplast (see Section II). The pools of metabolites in the cytosol are small, and turn over every few seconds (Stitt et al., 1988a). This means that photosynthesis will be inhibited unless the rate of sucrose synthesis and, hence, the supply of P_i is continually adjusted so that it is equal to the rate of CO_2 fixation. If sucrose synthesis is too rapid, CO_2 fixation will be inhibited because the high cytosolic P_i concentration will deplete the stromal metabolites and Ru-1,5-P_2 will no longer be regenerated at an adequate rate. If sucrose synthesis is too slow, the P_i concentration within the chloroplast will be too low, and ATP synthesis will be inhibited. How is this coordination of sucrose synthesis and CO_2 fixation achieved?

Fru-2,6-P_2 as a regulatory molecule to coordinate chloroplast and cytosol metabolism

The first irreversible reaction leading to sucrose synthesis is catalyzed by the cytosolic Fru-1,6-Pase. This enzyme has an extremely high affinity for Fru-1,6-P_2 ($K_m = 4$ μM). Until 1982, no metabolites were known that affected the activity of this enzyme when they were present at concentrations equivalent to those found in the cytosol. However, work in liver metabolism then led to the discovery of a regulator metabolite termed Fru-2,6-P_2.

In animals Fru-2,6-P_2 activates PFK and inhibits Fru-1,6-Pase. It plays an important role in regulating glycolysis and gluconeogenesis. Fru-2,6-P_2 is synthesized and degraded by specific enzymes termed Fru-6-P,2-kinase and Fru-2,6-P_2ase. In liver, both of these activities are present on a single bifunctional protein. They are regulated by a series of metabolites, often acting in an opposite manner on the synthesis and breakdown of Fru-2,6-P_2. They are also regulated in a reciprocal manner by cAMP-dependent protein kinase, which inactivates Fru-6-P,2-kinase and activates Fru-2,6-P_2ase. However, it is now apparent that other tissues differ from liver. In heart muscle and yeast, kinase and phosphatase activities can be separated and, in yeast, Fru-6-P,2-kinase is actually activated by phosphorylation.

Recent research has established that Fru-2,6-P_2 also plays an important role in controlling sucrose synthesis in plants. These results are reviewed in detail in Stitt et al. (1988a) and shorter accounts are to be found in Cseke et al. (1984), Stitt (1985, 1986), Stitt et al. (1987) and Huber (1986). Using non-aqueous fractionation, it was shown that Fru-2,6-P_2 is present in the cytosol of leaves, but not in the chloroplast. The concentration of this regulator is usually in the 1–10 μM range, although this could be an overestimate due to binding on the target proteins. Fru-2,6-P_2 is a potent inhibitor of the spinach leaf cytosolic Fru-1,6-Pase. It exerts its effect by decreasing the substrate affinity 100-fold and inducing sigmoidal substrate saturation kinetics. It also increases the enzyme's sensitivity to other inhibitors, such as AMP and P_i. As discussed in Section II, Fru-2,6-P_2, also activates pyrophosphate-dependent phosphofructokinase (PFP) which catalyzes a reversible conversion of Fru-6-P and PP_i to Fru-1,6-P_2 and P_i. Although PFP is present in leaves, its activity is usually much lower than the cytosolic Fru-1,6-Pase.

Many plant tissues, including leaves, have now been shown to contain Fru-6-P,2-kinase, and Fru-2,6-P_2ase. These activities are regulated by metabolites (Fig. 21.4). Fru-6-P,2-kinase is stimulated by Fru-6-P and P_i, and is inhibited by three-carbon phosphoesters such as glycerate-3-P and

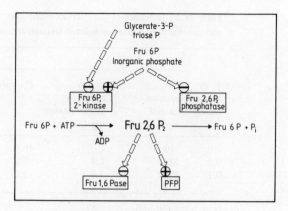

Fig. 21.4 The fructose 2, 6-bisphosphate system

dihydroxyacetone-P, the antagonistic interaction between glycerate-3-P and P_i being particularly important. In contrast, Fru-2,6-P_2ase is inhibited by Fru-6-P and P_i. Fru-6-P,2-kinase and Fru-2,6-P_2ase from plants can be separated, similar to the situation in heart muscle and yeast (MacDonald et al., 1989). At present, it is not known whether plant Fru-6-P,2-kinase is regulated by phosphorylation. Despite these uncertainties, it is already possible to propose the mechanism by which Fru-2,6-P_2 controls export from the chloroplast.

Feedforward control

When the rate of photosynthesis increases, the concentration of dihydroxyacetone-P will rise and, probably, also the glycerate-3-P:P_i ratio. As a result of this, there will be an inhibition of Fru-6-P,2-kinase, which will lead to a decrease of Fru-2,6-P_2. Hence, as the light intensity or the CO_2 concentration is raised, there will be a progressive decrease of Fru-2,6-P_2 coupled to an increase in photosynthesis (Fig. 21.5A). This decrease in Fru-2,6-P_2 is two to three-fold in spinach, but may be even more in other species such as barley or maize. The decrease of Fru-2,6-P_2 will relieve the inhibition of the cytosolic Fru-1,6-P_2ase, and sucrose synthesis will be stimulated. In this way, Fru-2,6-P_2 signals how much photosynthate is available to make sucrose, and whether the chloroplasts need more P_i.

The activity of the cytosolic Fru-1,6-Pase is *not* just affected by Fru-2,6-P_2. Like other important

The flux of carbon between the chloroplast and the cytosol

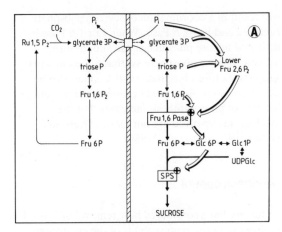

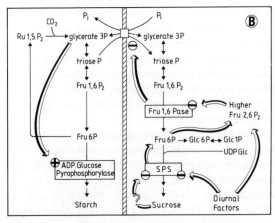

Fig. 21.5 Regulation of photosynthetic sucrose synthesis. (A) Feedforward control, (B) feedback control.

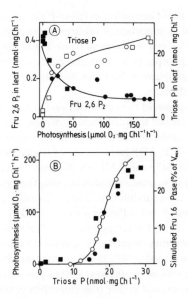

Fig. 21.6 Regulation of photosynthetic sucrose synthesis. (A) Changes of Fru-2,6-P_2 and triose-P as photosynthesis increases. (B) A model for the activation of the cytosolic Fru-1,6-Pase as availability of triose-P increases: (○) modeled Fru-1,6-Pase activity based on empirical measurements of Fru-2,6-P_2 and Fru-1,6-P_2 concentration estimated from the triose-P concentration assuming the reactions catalyzed by aldolase and triose phosphate isomerase are at equilibrium. The measured rate of photosynthesis is also shown (●, ■).

enzymes that are regulated, activity depends upon an interaction between several factors, e.g. the substrate concentration. As the rate of photosynthesis increases, triose-P is exported from the chloroplast and converted to Fru-1,6-P_2 by triose phosphate isomerase and aldolase. These reactions are near to equilibrium and, since two triose-P molecules are needed to produce one Fru-1,6-P_2 molecule, the Fru-1,6-P_2 concentration will increase as a square of the triose-P concentration. This means that there is a simultaneous increase of the substrate (Fru-1,6-P_2) and, because of the inhibition of Fru-6-P,2-kinase by three-carbon metabolites, a decrease of the concentration of the inhibitor (Fru-2,6-P_2) as photosynthesis increases (Fig. 21.6A). This interaction will exert a powerful regulation of the

cytosolic Fru-1,6-Pase in response to a changing rate of photosynthesis. An empirical model can be developed to illustrate how the cytosolic fluxes are regulated in response to the requirements of the chloroplast (Fig. 21.6B). This approach relates the estimated Fru-1,6-Pase activity and the measured rate of photosynthesis to the triose-P content of the leaf (see Stitt, 1986 or Stitt *et al.*, 1987 for details). Up to a threshold level of triose-P, the cytosolic Fru-1,6-Pase is effectively inactive. Obviously, when triose-P is low and there are inadequate levels of metabolites in the stroma for turnover of the Calvin cycle, the plant cannot afford to make sucrose. Once this 'threshold' is exceeded, there are adequate levels of metabolites for Calvin cycle turnover, and the 'surplus' may be removed for synthesis of sucrose. There is now a strong activation of the Fru-1,6-Pase in response to small increments of triose-P. This highly sensitive activation as triose-P rise above the 'threshold' is important: it ensures sucrose synthesis is stimulated and will remove triose-P and regenerate P_i before phosphorylated intermediates accumulate to the point where P_i becomes limiting for photosynthesis.

The response of the cytosolic Fru-1,6-Pase may also be modified by other factors. *Metabolites*, such as AMP, modulate this enzyme. If, for example, the cytosolic ATP/ADP ratio decreases, there would be a concomitant rise in AMP which would lead to an inhibition of sucrose synthesis because the cytosolic Fru-1,6-Pase is inhibited by AMP. The response of the cytosolic Fru-1,6-Pase is also modified by *temperature*. As the temperature is lowered this enzyme becomes increasingly sensitive to inhibition by Fru-2,6-P_2 and AMP. In effect, the 'threshold' for activating sucrose synthesis is shifted upwards at low temperatures, and higher concentrations of triose-P and other metabolites can be maintained. This may be an adaptation which promotes photosynthesis at low temperatures since the higher levels of metabolites in the chloroplast may partially compensate for a decreased turnover rate of the Calvin cycle enzymes. The response of the cytosolic Fru-1,6-Pase also depends on *genetic* factors. For example, cytosolic Fru-1,6-Pase from the mesophyll cells of maize has a 10-fold lower substrate affinity than the enzyme from C3 species. This change is functionally important. In maize, glycerate-3-P diffuses from the Calvin cycle in the bundle sheath to the mesophyll cells, where much of the glycerate-3-P reduction takes place. Triose-P has then to diffuse back to the bundle sheath cells. The low substrate affinity of the mesophyll Fru-1,6-Pase allows very high concentrations of triose-P to be maintained in the mesophyll. This provides a driving force for their diffusion back to the bundle sheath.

Hence, a variety of factors interact at the level of Fru-2,6-P_2 and the cytosolic Fru-1,6-Pase, and allow the removal of triose-P to be adjusted to changing rates of photosynthesis, to environmental conditions and to biochemical specializations. Of course, the cytosolic Fru-1,6-Pase is not the only site of control. Enzymes situated later in the pathway leading to sucrose are also regulated in response to the rate of photosynthesis. For example, sucrose phosphate synthase is regulated allosterically. It is activated by glucose-6-P and inhibited by P_i, and these properties will result in a coordination of sucrose phosphate synthase and Fru-1,6-Pase. When Fru-1,6-Pase is activated hexose phosphates rise and, in turn, activate sucrose phosphate synthase. Recent work suggests sucrose phosphate synthase may also be activated by covalent modification of the protein as the rate of photosynthesis increases, but the mechanisms involved are not known (Kerr and Huber 1987; Stitt *et al.*, 1988b).

Feedback control

So far, we have considered how removal of triose-P is adjusted to the requirements of the chloroplast. However, the export of triose-P from the chloroplast is also regulated by 'demand'. Although sucrose accumulates in leaves when photosynthesis is faster than the rate of sucrose export, in most plants, a point is reached where sucrose accumulation slows down or stops. At this point, the 'surplus' photosynthate is retained in the chloroplast and converted to starch.

Experimentally, the sucrose content of leaves can be increased by supplying sucrose exogenously, or by girdling or detaching leaves to prevent export. In spinach and soybean, these treatments lead to an increase of Fru-2,6-P_2 (Stitt *et al.*, 1984; Kerr and Huber, 1987). The increased Fru-2,6-P_2 inhibits Fru-1,6-Pase, which leads to an increase of triose-P and restricts the recycling of P_i to the chloroplast. An analogous situation can be obtained with isolated chloroplasts by lowering or omitting P_i from the medium. In both cases, starch synthesis is stimulated.

The stimulation of starch synthesis in isolated chloroplasts in low P_i results from the regulation of ADP-glucose pyrophosphorylase, the key enzyme in the pathway of starch synthesis. When P_i becomes limiting, ATP decreases and the rate at which glycerate-3-P is reduced to triose-P is lowered (Heldt *et al.*, 1977). Preiss and coworkers (Preiss, 1982) have shown that ADP-glucose pyrophosphorylase is activated by glycerate-3-P and inhibited by P_i. Heldt *et al.* (1977) demonstrated that glycerate-3-P accumulates in chloroplasts when P_i becomes limiting, probably because phosphoglycerate kinase is particularly sensitive to inhibition by the falling concentrations of ATP which are found in these

Table 21.3 Stimulation of starch synthesis by increased Fru-2, 6-P_2 in a mutant of *Clarkia xantiana* with reduced phosphoglucose isomerase activity

	Metabolite levels			Synthesis rates	
	Fru-6-P	Fru-2, 6-P_2	PGA	Sucrose	Starch
	(nmol mg Chl^{-1})			(μmol hexose mg Chl^{-1} h^{-1})	
Wild type	137	0.16	282	38	40
18% Mutant	176	0.25	360	34	56

The mutant line had 18% of the wild-type complement of phosphoglucose isomerase. The results were measured in leaves illuminated at 125 μmol m^{-2} s^{-1} in saturating CO_2, and are taken from Neuhaus et al. (1989).

conditions. An increase of the glycerate-3-P:P_i ratio is therefore a signal indicating that the supply of P_i has decreased. This results in a stimulation of starch synthesis. By allowing phosphorylated intermediates to be converted to starch within the chloroplast it facilitates the recycling of P_i within the stroma.

Mutants of *Clarkia xantiana*, have been used to confirm that increasing Fru-2,6-P_2 stimulates starch synthesis in this way in leaves. Genotypes of this species can be generated which have decreased cytosolic phosphoglucose isomerase activity. This enzyme is needed to convert Fru-6-P to Glc-6-P and, hence, these mutants contain more Fru-6-P than the wild type (Table 21.3). The higher Fru-6-P activates Fru-6-P,2-kinase and inhibits Fru-2,6-P_2ase which causes an increase of Fru-2,6-P_2. Compared with the wild type, the mutants have an increased level of glycerate-3-P and starch synthesis is found to be stimulated (Kruckeberg et al., 1989).

The mechanisms involved in the feedback control of sucrose synthesis itself are complex, and are not yet fully understood. It has often been suggested that sucrose directly inhibits the enzymes that are responsible for its synthesis, but the available evidence is not convincing. High levels of sucrose are needed to inhibit the enzymes involved in sucrose biosynthesis, and anyway most of the sucrose which accumulates in the leaf is located in the vacuole, not the cytosol (Gerhardt et al., 1987). An alternative explanation could be that sucrose acts indirectly by triggering protein modification or turnover. Studies of diurnal rhythms in leaves have provided some evidence for this possibility but the mechanisms involved still have to be clarified. However, it is already clear that there is a great deal of variability between species.

In soybean, there are endogenous rhythms of sucrose phosphate synthase activity, resulting from alterations in the amount of this protein (Kerr and Huber, 1987). Lower activity of this enzyme occurs concurrently with lower rates of sucrose synthesis, increases in Fru-2,6-P_2 and increased starch synthesis (Fig. 21.7). In spinach, there is a gradual shift towards starch synthesis during the day. This is again correlated with decreased sucrose phosphate synthase activity, but in this tissue it is not due to an endogenous rhythm (Stitt et al., 1988b; Fig. 21.7). The reduced sucrose phosphate synthase activity leads to an accumulation of Fru-6-P

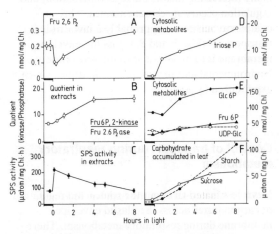

Fig. 21.7 Diurnal rhythms of enzyme activities and cytosolic metabolite levels in spinach leaves.

in the cytosol. This, in turn, would activate Fru-6-P,2-kinase and inhibit Fru-2,6-P_2ase, and result in an increase of Fru-2,6-P_2. The increased Fru-2,6-P_2 would then inhibit cytosolic Fru-1,6-Pase activity and result in an increase in the concentration of triose-P in the cytosol. The resulting restriction on the recycling of P_i to the chloroplast would then lead to a stimulation of starch accumulation.

Studies on soybean and spinach suggest there may also be a more direct control of the Fru-2,6-P_2 level during this feedback control (Stitt et al., 1987; Kerr and Huber, 1987). In both these species there is a diurnal rhythm of Fru-6-P,2-kinase activity in the leaves, which matches the alterations of Fru-2,6-P_2 and photosynthate partitioning. These changes have led to the hypothesis that Fru-6-P,2-kinase is regulated by protein modification analogous to the control of this protein by phosphorylation and dephosphorylation in mammalian tissues and yeast.

In summary, the rate of carbon export from the chloroplast depends on a balance between feedforward mechanisms which decrease Fru-2,6-P_2 and also activate sucrose phosphate synthase, and feedback mechanisms which deactivate sucrose phosphate synthase and increase Fru-2,6-P_2. A balance is achieved which allows the leaf to adjust the partitioning of photosynthate between various endproducts, while simultaneously maintaining levels of stromal metabolites and P_i which allow rapid photosynthesis to continue. The mechanisms involved in this regulation are complex, because many interacting factors make a contribution, and regulation occurs at several enzymes. This complexity is, however, probably unavoidable if the transport of triose-P and P_i across the envelope membrane is to be adjusted to the wide variety of environmental and physiological conditions which is experienced by a leaf.

Control of transport during respiratory metabolism

Much less is known about the routes and regulation of carbon fluxes across the plastid envelope membrane during respiratory metabolism. There have been no studies of transport in non-photosynthetic plastids. They are fragile and difficult to isolate in an intact and functional state. Plastids with large starch grains are especially difficult to prepare, because the starch grains disrupt the plastids during centrifugation. Study of metabolism is also more difficult in these tissues. For example, it is much easier to measure the flow of carbon from CO_2 to sucrose or starch than it is to measure the flux between them. It is also difficult to quantify the absolute rate of carbohydrate breakdown for respiration and biosynthesis. In addition, there is very little information about subcellular metabolite levels during respiratory metabolism.

In this section, starch degradation and carbon fluxes in leaves in the dark will be discussed first, since information is available about the transport properties of the plastid envelope membrane in this system. The relevance of this knowledge of transport in leaves to an elucidation of the pathways of starch–sucrose interconversions in non-photosynthetic tissues will then be considered. Further information on the enzymes involved in starch and sucrose turnover can be found in Chapter 5, and in reviews by Preiss (1982), ap Rees (1980), Stitt (1984) or Stitt and Steup (1985).

Starch mobilization in leaves in the dark

Intact chloroplasts containing large quantities of starch can be isolated from spinach leaves. When these chloroplasts are incubated in the dark, the starch is degraded and a mixture of glucose, maltose, triose-P and glycerate-3-P (Fig. 21.8), which are medium (Stitt and Heldt, 1981). The distribution between these products depends on the conditions during the incubation (Table 21.4).

When the concentration of P_i in the medium is high (over 1 mM) phosphorylated products and CO_2 account for up to half of the products. In these conditions, starch is degraded phosphorolytically, giving rise to hexose phosphates in the stroma. These are metabolized via the stromal phosphofructokinase and the oxidative pentose phosphate pathway to triose-P and glycerate-3-P (Fig. 21.8), which are exported via the phosphate translocator to the cytosol. This export occurs in counterexchange for P_i, which is required for the continued mobilization of starch by this route. The relation between metabolism in the chloroplast and cytosol during

Table 21.4 Influence of P_i in the medium on the products of starch breakdown in intact, starch-loaded spinach chloroplasts.

P_i in medium (mM)	Starch breakdown	Accumulation of products		
		Neutral sugars	Phosphorylated intermediates	CO_2
		(μatom C mg Chl^{-1} h^{-1})		
0.05	9.0	6.3	0.5	1.6
5.0	10.4	5.0	4.6	0.8

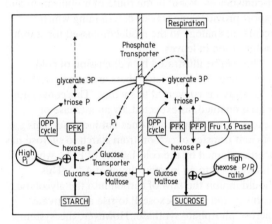

Fig. 21.8 Routes of starch mobilization in chloroplasts from spinach leaves. The possible fate of the starch degraded via the different routes is also indicated.

phosphorolytic starch breakdown could be very similar to the interaction that occurs during photosynthesis. It can be envisaged how increasing demand for respiratory substrates in the cytosol will be associated with a decrease of triose-P and glycerate-3-P, and an increase of P_i. This would increase the exchange of cytosolic P_i and chloroplast metabolites via the phosphate translocator. As a result, phosphorolysis of starch would be stimulated by the increased supply of P_i, while the products of starch mobilization would be removed from the chloroplast and made available for respiration.

However, this is not the only route that is available for starch mobilization in leaves (Fig. 21.8). Even when P_i is present, glucose and maltose account for half of the starch degradation in isolated chloroplasts. They are produced even more rapidly when P_i is omitted from the medium. Clearly, starch is being degraded hydrolytically, and maltose and glucose are then being released via the hexose transporter. It should also be mentioned that much of the α-glucan phosphorylase and endoamylase activity in leaves is actually in the cytosol (Stitt and Steup, 1985; Ziegler and Beck, 1987). It is not known whether glucans with a high degree of polymerization can also be exported from the chloroplasts to act as substrates for these enzymes.

Although more research is needed to clarify the precise routes of starch degradation, it is clear that export of hexose units could play a significant role during starch mobilization. This would allow export of carbon from the chloroplast when metabolite levels are high and P_i is low, and could be important for conversion of starch to sucrose. Sucrose phosphate synthase is stimulated by high concentrations of hexose phosphates and low P_i. Conditions of low P_i would also stimulate release of hexose from the chloroplast, but (see above) low P_i restricts phosphorolysis and release of carbon via the phosphate translocator. This suggests the major route from starch to sucrose might involve export of hexoses, rather than the phosphate translocator. This route would avoid having to convert triose-P back to hexose-P in the cytosol in the dark. As already discussed, the cytosolic Fru-1,6-Pase is inactive in the dark because leaves contain high Fru-2,6-P_2 and negligible Fru-1,6-P_2. There is, of course, a possibility that triose-P is converted to hexose-P in the dark via PFP. This enzyme would be active when Fru-2,6-P_2 is high and P_i is low (see Chapter 5).

Starch degradation in non-photosynthetic tissues

In principle, the route and regulation of carbon

fluxes during starch mobilization in non-photosynthetic plastids could be quite flexible and varied. As discussed in Chapter 6, there is convincing evidence that plastids from a wide variety of plant tissues contain PFK, and the enzymes needed for the oxidative pentose phosphate pathway and the oxidation of triose-P. They are usually able to convert glycerate-3-P to pyruvate, even though this section of glycolysis is always more active in the cytosol. In analogy with chloroplasts, they could contain transport systems allowing them to transport hexoses, three-carbon phosphoesters, pyruvate, or dicarboxylates. Detailed studies are needed to establish if these, or other, transport routes are actually present. Indeed, it is possible these could vary depending on the tissue, reflecting the different roles of starch in various plant organs.

In some tissues, starch is accumulated when sucrose import exceeds the immediate needs of the tissue. This starch is later degraded to support respiration and growth. Changes of P_i and metabolites could link the mobilization of starch and export from the plastid, as in leaves. Elegant evidence along these lines has been provided by Douce and coworkers (Rebeille et al., 1984), using ^{31}P-NMR to monitor the concentrations of P_i and metabolites in the cytoplasm of sycamore cell suspension cultures. When the supply of exogenous carbohydrate is removed, reserve carbohydrates in the cells are remobilized, a process that starts with sucrose. As sucrose is exhausted, there is a fall in the concentrations of phosphorylated metabolites, P_i increases, and then starch mobilization commences.

In other tissues, which we would normally designate as 'storage' tissues, starch is deposited in large quantities in specialized plastids (amyloplasts) as the seed or tuber develops. This starch is remobilized during germination or sprouting, when it provides a source of carbon from which sucrose can be synthesized for export. Two different strategies exist for moving carbon across the envelope membrane. In seeds in which starch is accumulated in an endosperm, the envelope and the remainder of the cell disintegrate when the seed matures. Upon germination, the starch is degraded by hydrolytic enzymes and the glucose is taken up into specialized cells where it is converted to sucrose. Many seeds and tubers, however, retain their plastid envelope during starch remobilization. These tissues also typically contain significant activities of α-glucan phosphorylase, as well as the hydrolytic enzymes. At least in tubers, starch can also be remobilized independently of sprouting. When the tuber is wounded, mobilization is linked to increased respiration and biosynthetic activity. Mobilization also occurs when tubers are transferred to the cold, when it leads to a quantitative accumulation of sucrose. Apparently, starch can be remobilized and selectively directed towards respiratory pathways, or towards sucrose synthesis, depending on the circumstances. Possibly the route of mobilization and export provides one way of determining which occurs, in analogy to the model discussed for starch mobilization in leaves.

This can be illustrated by a discussion of cold sweetening. When potato tubers are transferred to 4 °C, starch is converted to sucrose. This conversion is accompanied, as expected, by an increase of hexose-P, but respiration does not increase. This efficient transfer of carbon from starch to sucrose has been explained by the cold-lability of PFK, which dissociates into an inactive dimer at 4 °C. This should inhibit the flow of carbon into the glycolytic pathway and allow hexose-P to rise and activate sucrose phosphate synthase. However, this system will only conserve carbon if hexose units are being exported from the plastid.

Conversion of sucrose to starch

The presence of alternative pathways for the conversion of sucrose to starch can be postulated on the basis of the various transport systems known in chloroplasts (Fig. 21.9). The most direct would be the import of hexose units into the chloroplast. Alternatively, sucrose could be converted to triose-P and glycerate-3-P in the cytosol, imported to the plastid by the phosphate translocator and converted back to starch via a plastid Fru-1,6-Pase. It should be obvious that the regulation of starch accumulation cannot be studied until the contribution of these, or even of other, routes has been established. If the principal route from sucrose to starch is via hexoses, the major branch point between starch accumulation and respiration would be at the reactions catalyzed by PFK, PFP and the Fru-1,6-Pase. On the other hand, if there is rapid starch synthesis from the pools

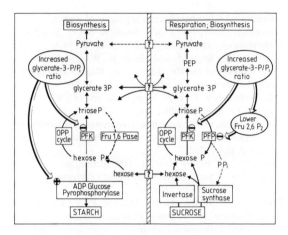

Fig. 21.9 Possible routes for starch accumulation in storage tissues.

of three-carbon phosphoesters, the major branch point between starch accumulation and respiration would involve pyruvate kinase and phosphoenolpyruvate carboxylase.

There is no direct evidence about transport, but four indirect approaches have been used to investigate which route is taken. One was to ask whether Fru-1,6-Pase is present in non-photosynthetic plastids. A low activity of Fru-1,6-Pase is found in plastids from soybean suspension cultures (MacDonald and ap Rees, 1983a), cauliflower florets (Journet and Douce, 1985) and maize endosperm (Escheveria et al., 1988) but not from wheat endosperm (Entwhistle and ap Rees, 1988) or guard cell protoplasts (Hedrich et al., 1985). A second approach involved supplying ^{14}C-glycerol to soybean suspension cells (MacDonald and ap Rees, 1983b). Some label was incorporated into starch, consistent with a route from glycerol to glycerol-3-P to dihydroxyacetone-P to Fru-1,6-P$_2$ to Fru-6-P to starch. However, this sequence could occur in the cytosol, or the chloroplast, and it is also possible to exchange radioactivity between triose-P and hexose-P via transketolase or transaldolase. Thus, the labeling of starch from ^{14}C-glycerol does not provide conclusive evidence that *net* starch synthesis proceeds from three-carbon units. A third approach involves supplying specifically labeled glucose, and then isolating the starch and investigating whether the radioactivity has been randomized. If the route occurs via the phosphate translocator, the radioactivity should be completely randomized between the top and bottom halves of the glucose molecule. In fact, over 60% of the radioactivity remains unrandomized during starch accumulation in wheat endosperm (Keeling et al., 1988) and in *Chenopodium* cell cultures (W. Hatzfeld and M. Stitt), in press suggesting that import of hexose units may make a major contribution to starch synthesis. The partial randomization could be due to a minor contribution from the phosphate translocator route, or could be due to randomization of the hexoses by other reactions. In a fourth approach, amyloplasts were isolated and incubated with different radioactively-labeled metabolites to investigate which allows the highest rate of starch synthesis. In this kind of experiment, it is important to demonstrate that the label is being converted to starch, and to work with amyloplast preparations which are carrying out starch synthesis at physiologically relevant rates (see also above for a discussion of further potential pitfalls in interpreting such experiments). The available evidence indicates that G-1-P may be the best precursor for starch synthesis by wheat amyloplasts (Tyson and ap Rees, 1988).

Taken together, these results suggest import of hexose units into the non-photosynthetic plastid may be important during starch accumulation. However, many more tissues need to be examined and direct studies of transport will be needed to show whether this import occurs as free sugars, or as a phosphorylated sugar via a transport system which is not present in chloroplasts. It also remains possible that the phosphate translocator is involved to some extent during starch accumulation. Even if it did not contribute to the net flow, it would still be important for regulation. The ADP-glucose pyrophosphorylase from storage tissues resembles the leaf enzyme in being activated by glycerate-3-P and inhibited by P_i. As discussed in Chapter 6, rising levels of glycerate-3-P and phosphoenolpyruvate and falling P_i provide a mechanism to inhibit PFK and restrict glycolysis in the cytosol when the supply of substrates exceeds the requirements of respiration. Rising glycerate-3-P : P_i ratios would also drive Fru-2,6-P$_2$ downwards and inactivate PFP (Chapter 6). It is plausible that the increasing glycerate-3-P : P_i ratio simultaneously stimulates starch synthesis, allowing storage of the excess carbohydrate. This depends upon the phosphate translocator being present in these non-photosynthetic plastids to transmit a rising cytosolic glycerate-3-P : P_i ratio to the plastid.

Concluding remarks

In this chapter the importance of understanding the transport of carbon between the chloroplast and cytosol has been emphasized. In leaves, considerable progress has been made in clarifying routes and regulation, and research in the future will probably be aimed at understanding protein modification and turnover, and in probing structure and function at a molecular level. This kind of information could open a way to a rational manipulation of export and storage strategies in leaves. However, plant productivity depends critically upon the distribution of photosynthate to growing tissues, and its efficient utilization or storage. It is a sad fact that we do not yet even know the pathway of starch accumulation in storage tissues, let alone how this storage is controlled. Our knowledge of the remobilization of starch is equally deficient. A considerable research effort at the biochemical and molecular level is needed to provide a basis for a rational approach to the physiology of these storage tissues.

Further reading and references

ap Rees, T. (1980). Integration of pathways of synthesis and degradation of hexose phosphates. In *The Biochemistry of Plants*, Vol 3, ed. J. Preiss, Academic Press, New York, pp. 1–42.

ap Rees, T. (1985). The organisation of glycolysis and the oxidative pentose phosphate pathway in plants. In *Encyclopedia of Plant Physiology*, Vol. 18, eds R. Douce and D. A. Day, Springer-Verlag, Berlin, pp. 347–90.

Beck, E. (1985). The degradation of transitory starch in chloroplasts. In *Regulation of Carbohydrate Partitioning in Photosynthetic Tissue*, eds R. L. Heath and J. Preiss, American Society of Plant Physiologists, Rockville, Maryland, pp. 27–40.

Bird, I. F., Cornelius, M. T., Dyer, T. A. and Keys, M. J. (1974). Intracellular site of sucrose synthesis in leaves. *Phytochemistry* **13**, 59–64.

Cseke, C., Balogh, A., Wong, J. H., Buchanan, B. B., Stitt, M., Herzog, B. and Heldt, H. W. (1984). Fructose-2,6-bisphosphate: a regulator of carbon processing in leaves. *Trends Biochem. Sci.* **9**, 533–5.

Ebbinghausen, M., Hatch, M. D., Lilley, R. McC., Krömer, S., Stitt, M. and Heldt, H. W. (1987). On the function of malate–oxaloacetate shuttles in a plant cell. In *Plant Mitochondria*, ed. A. L. Moore and R. B. Beechey, Plenum, New York, pp. 171–80.

Edwards G. E. and Walker, D. A. (1983). C_3 C_4: *Mechanisms, and cellular and environmental regulation of photosynthesis*, Blackwell Scientific Publication, Oxford, pp. 1–542.

Entwhistle, G. and ap Rees, T. (1988). Enzyme capacities of amyloplasts from wheat (*Triticum aestivum*) endosperm. *Biochem J.* **255**, 391–6.

Escheveria, E., Boyer, C. D., Thomas, P. A., Liu, K.-C. and Shannon, J. C. (1988). Enzyme activities associated with maize endosperm amyloplasts. *Plant Physiol.* **86**, 786–92.

Fliege, R., Flüge, U.-I., Werdan, K. and Heldt, H. W. (1978). Specific transport of inorganic phosphate, 3-phosphoglycerate and triose phosphate across the inner membrane of the envelope in spinach chloroplasts. *Biochem. Biophys. Acta* **502**, 232–47.

Flügge, U.-I. (1987). Physiological function and physical characteristics of the chloroplast phosphate translocator. In *Progress in Photosynthesis Research*, Vol. III, ed. J. Biggins, Martinus Nijhoff, Dordrecht, pp. 739–47.

Flügge, U.-I. and Benz, R. (1984). Pore forming activity in the outer membrane of the chloroplast envelope. *FEBS Lett.* **169**, 85–9.

Flügge, U.-I. and Heldt, H. W. (1984). The phosphate–triose phosphate–phosphoglycerate translocator of the chloroplast. *Trends Biochem. Sci.* **9**, 530–3.

Flügge, U.-I., Freisl, M. and Heldt, H. W. (1980). Balance between metabolite accumulation and transport on relation to photosynthesis by isolated spinach chloroplasts. *Plant Physiol.* **65**, 574–7.

Flüge, U.-I., Stitt, M. and Heldt, H. W. (1985). Light driven uptake of pyruvate in mesophyll chloroplasts from maize. *FEBS Lett.* **183**, 335–9.

Gerhardt, R., Stitt, M. and Heldt, H. W. (1987). Subcellular metabolite levels in spinach leaves. *Plant Physiol.* **87**, 399–407.

Hatch, M. D., Dröscher, L., Flügge, U.-I. and Heldt, H. W. (1984). A specific translocator for oxaloacetate transport in chloroplasts. *FEBS Lett.* **178**, 15–19.

Hatzfeld, W. D. and Stitt, M.(–). A study of the rate of recyling of triose-phophates in heterotrophic *Chenopodium rubrum* cells, potato tubers and maize endosperm. *Planta*. In press.

Heber, W. and Willenbrink, J. (1964). Sites of synthesis and transport of photosynthetic products within the leaf cell. *Z. Naturforschung* **25b**, 710–28.

Hedrich, R., Raschke, K. and Stitt, M. (1985). A role

for fructose-2,6- bisphosphate in regulating carbohydrate metabolism in guard cells. *Plant Physiol.* **79**, 977–82.

Heldt, H. W. and Flügge, U.-I. (1987). Subcellular transport of metabolites in a plant cell. In *The Biochemistry of Plants*, Vol 12, Academic Press, New York, pp. 49–85.

Heldt, H. W., Chon, C. J., Maronde, D., Herold, A., Stankovic, Z. S., Walker, B. A., Kraminer, A., Kirk, M. R. and Heber, U. (1977). Role of orthophosphate and other factors in the regulation of starch formation in leaves and isolated chloroplasts. *Plant Physiol.* **59**, 1146–55.

Howitz, K. T. and McCarty, R. E. (1985). Substrate specificity of the pea chloroplast glycolate transporter. *Biochemistry* **24**, 3645–50.

Huber, S. C. (1986). Fructose- 2,6-bisphosphate. *Ann. Rev. Plant Physiol.* **37**, 233–46.

Journet, E. and Douce, R. (1985). Enzyme capacities of purified cauliflower bud plastids for lipid synthesis and carbohydrate metabolism. *Plant Physiol.* **79**, 458–67.

Keeling, P. L., Wood, J. T., Tyson, R. H. and Bridges, I. G. (1988). Starch biosynthesis in developing wheat grain. Evidence against the direct involvement of triose phosphates in the metabolic pathway. *Plant Physiol.* **87**, 311–19.

Kerr, P. S. and Huber, S. C. (1987). Coordinate control of sucrose formation in soybean leaves by sucrose phosphate synthase and fructose-2,6-bisphosphate. *Planta* **170**, 197–204.

Klingenberg, M. (1981). The mitochondrial ATP/ADP translocator. *Nature* **290**, 449–54.

Kruckeberg, A. L., Neuhaus, E., Feil, R., Gottlieb, L. D. and Stitt, M. (1989). Reduced activity mutants of phosphoglucose isomerase in the chloroplast and cytosol of *Clarkia xantiana*. *Biochem. J.* **261**, 457–67.

Lehner, K. and Heldt, H. W. (1978). Dicarboxylate transport across the inner membrane of the chloroplast envelope. *Biochim. Biophys. Acta* **501**, 531–44.

Lilley, R. McC., Chon, C. J., Mosbach, A. M. and Heldt, H. W. (1976). The distribution of metabolites between spinach chloroplasts and medium during photosynthesis *in vitro*. *Biochim. Biophys. Acta* **460**, 259–72.

MacDonald, F. D. and ap Rees, T. (1983a). Enzymic properties of amyloplasts from suspension cultures of soybean. *Biochim. Biophys. Acta* **755**, 81–9.

MacDonald, F. D. and ap Rees, T. (1983b). Labelling of carbohydrate by ^{14}C glycerol supplied to suspension cultures of soybean. *Phytochemistry* **22**, 1141–3.

MacDonald, F. D., Chou, Q., Buchanan, B. B. and Stitt, M. (1989) Purification and characterisation of fructose-2,6-bisphosphate. A substrate-specific cytosolic enzyme from leaves. *J. Biol. Chem.* **264**, 5540–4

Neuhaus, H. E., Kruckeberg, A. L., Feil, R. and Stitt, M. (1989). Reduced activity mutants of phosphyoglucose isomerase in the cytosol and chloroplast of *Clarkia xantiana*. II Study of the mechanisms which regulate photosynthate partitioning. *Planta* **178**, 110–22.

Nyernprasintiri, J., Harinasut, P., Macheral, D., Strzalka, K ., Takebe, R., Akazawa, T. and Kojima, K. (1988). Isolation and characterization of the amyloplast envelope membrane from cultured white-wild cells of sycamore (*Acer pseudoplatanus*, L.). *Plant Physiol.* **87**, 371–8.

Preiss, J. (1982). Regulation of the synthesis and degradation of starch. *Ann. Rev. Plant Physiol.* **33**, 431–54.

Rebeille, F., Bligny, R., Martin, J.-P. and Douce, R. (1984). Effect of sucrose starvation on sycamore (*Acer pseudoplatanus*) cell carbohydrate and P_i status. *Biochem. J.* **226**, 679–84.

Schäfer, G., Heber, U. and Heldt, H. W. (1977). Glucose transport into spinach chloroplasts. *Plant Physiol.* **60**, 286–9.

Sommerville, S. C. and Sommerville, C. R. (1985). A mutant of *Arabidopsis* deficient in chloroplast dicarboxylase transport is missing an envelope protein. *Plant Sci. Lett.* **37**, 217–20.

Stitt, M. (1984). Degradation of starch in chloroplasts: a buffer to sucrose metabolism. In *Storage Carbohydrates in Vascular Plants*, ed. D. H. Lewis, Cambridge University Press, pp. 205–30.

Stitt, M. (1985). Fine control of sucrose synthesis by fructose-2,6-bisphosphate. In *Regulation of Carbon Partitioning in Photosynthesis Tissue*, eds R. L. Heath and J. Preiss, Waverley Press, Baltimore, pp. 199–214.

Stitt, M. (1986). Regulation of sucrose synthesis: integration, adaptation and limits. In *Proceedings of the 3rd International Congress on Phloem Transport*, ed. J. Cronshaw, Alan R. Liss, New York, pp. 331–46.

Stitt, M. (1987). Fructose-2,6-bisphosphate and plant carbohydrate metabolism. *Plant Physiol.* **87**, 201–4.

Stitt, M. and Heldt, H. W. (1981). Physiological rates of starch breakdown in isolated intact spinach chloroplasts. *Plant Physiol.* **68**, 755–61.

Stitt, M. and Steup, M. (1985). Starch and sucrose degradation. In *Encyclopedia of Plant Physiology*, Vol. 18, eds R. Douce and D. A. Day, Springer-Verlag, Heidelberg, pp. 347–90.

Stitt, M., Gerhardt, R., Wilke, I. and Heldt, H. W. (1987). The contribution of fructose- 2,6-bisphosphate to the regulation of sucrose synthesis during photosynthesis. *Physiol. Plantarum* **69**, 377–86.

Stitt, M., Huber, S. C. and Kerr, P. (1988a). Control of photosynthetic sucrose formation. In *The Biochemistry of Plants*, Vol. 8, eds M. D. Hatch and N. K. Boardman, Academic Press, New York, pp. 327–409.

Stitt, M., Huber, S. C. and Kerr, P. (1987). Control of photosynthetic sucrose synthesis. In *The Biochemistry of Plants*, Vol. 10, eds M. D. Hatch and N. K. Boardman, Academic Press, New York, pp. 327–409.

Stitt, M., Kurzel, B. and Heldt, H. W. (1984). Control of photosynthetic sucrose synthesis by fructose-2,6-bisphosphate. II Partitioning between sucrose and starch. *Plant Physiol.* **75**, 554–60.

Stitt, M., Lilley, R. McC., Gerhardt, R. and Heldt, H. W. (1988c). Determination of metabolite levels in specific cells and subcellular compartments of plant leaves. *Methods Enzymol.*, in press.

Stitt, M., Wilke, I., Gerhardt, R. and Heldt, H. W. (1988b). Coarse control of sucrose phosphate synthase in leaves. Alterations of the kinetic properties in response to the rate of photosynthesis and the accumulation of sucrose. *Planta*, **174**, 217–30.

Tyson, R. H. and ap Rees, T. (1988). Starch synthesis by isolated amyloplasts from wheat endosperm. *Planta* **175**, 33–8.

Ziegler, P. and Beck, E. (1986). Exoamylase activity in vacuoles isolated from pea and wheat leaf protoplasts. *Plant Physiol.* **82**, 1119–21.

The Formation and Breakdown of Lipids

VII

The Formation and Breakdown of Lipids

22 The structure and formation of microbodies
Claire Halpin and J. Michael Lord

Introduction

Microbodies were first described in mouse kidney cells in the early 1950s by Rhodin, a Swedish electron microscopist. They are small organelles about half a micrometer in diameter bounded by a single membrane and containing a fine granular matrix. Since morphologically similar organelles have now been discovered in a wide range of plant and animal tissue, in fungi and in unicellular organisms such as protozoa and yeasts, microbodies are considered to be ubiquitously present in eukaryotic cells. Almost without exception, microbodies can be biochemically characterized by the possession of a primitive respiratory chain involving hydrogen peroxide-producing oxidases and hydrogen peroxide-decomposing catalase, and as such are more commonly referred to as peroxisomes. In addition, an inducible fatty acid β-oxidation system is present in the peroxisomes of all organisms analyzed so far, although the activity of this system varies between tissues. In plants, β-oxidation enzymes are found exclusively in the peroxisome and are induced to high levels of activity during germination of fatty seeds, but show little activity during other developmental stages. In mammals, the bulk of fatty acid oxidation occurs in mitochondria, but enzymes present in peroxisomes are essential for the initial shortening of long chain fatty acids. In rat liver, high levels of activity of these enzymes can be induced by high fat diets or administration of drugs that are known to lower serum lipid levels.

Apart from these morphological and biochemical similarities, peroxisomes from different organisms and tissues can contain very different complements of enzymes reflecting the diverse metabolic functions that these organelles can perform. For example, one very specialized type of plant peroxisome contains all five enzymes of the glyoxylate cycle and has been given the special name glyoxysome. However, although peroxisomes in several algae and yeast strains metabolize certain specific compounds, they have neither been given special names nor have been classified separately.

Morphology and biochemistry of plant microbodies

Electron microscopic examination of microbodies has revealed them to be small organelles 0.5–1.5 μm in diameter, which exhibit spherical, elongate, or pleiomorphic profiles in section. They are bounded by a single membrane which encloses an amorphous or granular matrix. The matrix may also contain electron-dense, often semicrystalline, material known as a core or nucleoid. Microbodies do not contain any internal membranes, ribosomes, or nucleic acid. On sucrose density gradients, microbodies band at an equilibrium density of 1.23–1.26 g cm^{-3}.

Three specialized types of peroxisome have so far been identified in plant tissues: peroxisomes in photosynthetic leaves, glyoxysomes in the fat-storing cells of oily seeds, and peroxisomes found in the uninfected nodule cells of certain legumes. In addition, most other plant tissues contain some microbodies whose particular function is unknown, and these organelles are known as unspecialized peroxisomes.

Leaf peroxisomes

Peroxisomes in photosynthetic green leaves play an important role in photorespiration, a process by which carbon dioxide, newly fixed in the plant cell during photosynthesis, is liberated back into the atmosphere. Glycolate (produced in the chloroplast) is oxidized in the peroxisome to glyoxylate with the liberation of hydrogen peroxide, which is detoxified by catalase. Glyoxylate is then transaminated to yield glycine which is transported to the mitochondrion where it is used to produce serine and carbon dioxide. The serine can be converted in the peroxisome to glyceric acid which can be used for carbohydrate formation in the chloroplast. Thus leaf peroxisomes display a metabolic function common to most plant and protozoan peroxisomes; that is, they play a role in gluconeogenesis (the new formation of carbohydrate).

At an early stage of leaf development, as cotyledons are converted into green photosynthetic tissue, the enzymes associated with the photorespiratory glycolate pathway accumulate in leaf peroxisomes in a light-dependent manner. The leaves of plants with high photorespiration have much larger and more numerous peroxisomes than those of plants with low photorespiration, consistent with the difference in biochemical activities between the two types of plant (Newcomb, 1982). The peroxisomes may reach 1.5 μm in diameter and often contain large crystalline inclusions of the enzyme catalase (see Fig. 22.1).

Glyoxysomes

Glyoxysome is the name given to a specialized type of peroxisome that possesses enzymes of the glyoxylate cycle. In higher plants, glyoxysomes are found in the endosperm or cotyledons of oily or fatty seeds during the early stages of growth after germination, when conversion of stored fat to sucrose is the major metabolic process. This metabolic activity serves to provide developing shoots with energy in the form of carbohydrate, at the expense of the seed's stored fat, until the first green leaves appear and photosynthesis begins. The compartmentation of β-oxidation enzymes and those of the glyoxylate cycle together within the

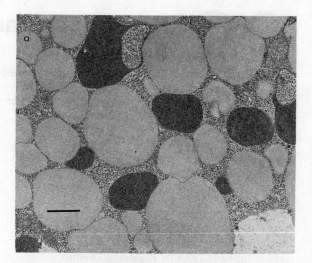

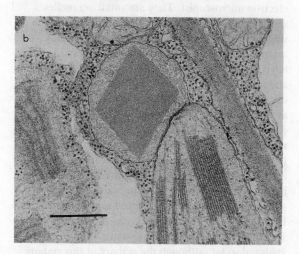

Fig. 22.1 Microbody morphology. (a) Glyoxysomes (in close association with spherosomes) in a tomato cotyledonary cell, (b) a crystalloid-containing spinach leaf peroxisome. Bar represents 1 μm (micrographs courtesy of Dr E. H. Newcomb).

glyoxysome is crucial to the efficiency of this process. Acetyl-CoA generated by β-oxidation of fatty acids avoids the oxidative decarboxylations of the mitochondrial citric acid cycle, and instead is converted exclusively to succinate via the glyoxylate cycle. Thus, up to 75% of the fatty acid carbon can be ultimately recovered in the cytoplasm as sucrose. Because of the magnitude of gluconeogenesis during this time, glyoxysomes may account for up to 20% of the total particulate protein in tissues such as castor bean endosperm when at the peak of their activity.

In oilseed species where fat reserves are stored in the cotyledons (e.g. cucumber, sunflower, pumpkin, cotton), relatively small, spherical glyoxysomes present at germination are thought to have been synthesized during seed formation (Trelease, 1984). Activities of glyoxysomal enzymes do not appear coordinately, but low levels of activity of certain enzymes notably catalase, β-oxidation enzymes, and the glyoxylate cycle enzyme malate synthase, have been detected during seed maturation. Following germination, the activity of all glyoxysomal enzymes increases dramatically during the first five days of growth (in seeds germinated at a constant 25–30 °C), peaks, and then declines again (Huang *et al.*, 1983). During this period of intense activity, the glyoxysomes themselves become more elongate and highly pleiomorphic, increasing in volume nearly seven-fold. The increase in enzyme activity has been shown to be due to massive *de novo* enzyme synthesis effected by a corresponding rise in the translatable messenger RNA (mRNA) for these enzymes, and suggesting that their synthesis is transcriptionally controlled (Weir *et al.*, 1980).

As the cotyledons emerge and are exposed to light (about day 3–4), the enzymes associated with the photorespiratory glycolate pathway, e.g. hydroxypyruvate reductase, serine:glycolate aminotransferase and glycolate oxidase, are induced and begin to accumulate in microbodies, while the glyoxylate cycle enzymes continue to decline. The nature of this transition from glyoxysomal to peroxisomal function in cotyledonary microbodies remains poorly understood and will be discussed later.

In oilseed species where fat reserves are stored in the endosperm (e.g. castor bean), this transition does not occur. Glyoxysomes and their constituent enzymes are rapidly synthesized during the first five days of growth (see Fig. 22.2). After this time the organelles, and indeed the whole endosperm tissue, senesces and effectively has disappeared by day nine of growth at 30 °C (Lord and Roberts, 1983).

Glyoxysomes are also found in algae, yeast and other fungi and protozoa which have been grown on acetate or compounds initially converted into acetyl units. In these cases, however, only two of the five glyoxylate cycle enzymes, malate synthase and isocitrate lyase, are present. For this reason some authors prefer to refer to these organelles as 'glyoxysome-like'.

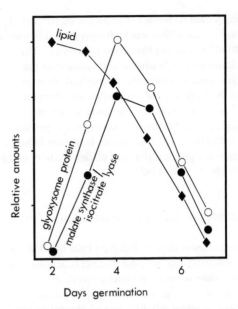

Fig. 22.2 Induction of glyoxysomes and glyoxylate cycle enzymes during germination of castor bean seeds.

Root nodule peroxisomes

A third specialized, though not very well characterized, type of plant peroxisome is found in certain cells of the root nodules of some legumes, e.g. soybean and cowpea. Early in nodule development, only a few relatively small peroxisomes are seen either in infected or uninfected cells. As the nodule develops, however, peroxisomes in the uninfected cells become larger and more numerous, and in soybean the induction of at least one peroxisomal enzyme, a uricase, has been demonstrated (Nguyen *et al.*, 1985). Such legumes principally produce and transport fixed nitrogen as the ureides allantoin and allantoic acid. These compounds are formed via purine biosynthesis and oxidation. It has been suggested that the enzymes involved in purine biosynthesis are located in the proplastids, whereas the oxidation of purines to ureides occurs in the peroxisomes. In this oxidation the purine xanthine is first converted to uric acid which is then converted to allantoin by uricase. This second reaction liberates hydrogen peroxide whose destruction requires the action of catalase. The localization of both uricase and catalase to the peroxisomes of uninfected nodule cells provides good evidence for the organelles' involvement in at least one reaction during the metabolism of compounds of recently fixed nitrogen. It has been suggested that such compounds may even induce peroxisome proliferation and uricase activity in this tissue (Newcomb, 1982).

Glyoxysome–peroxisome transition

Microbodies in the cotyledonary cells of oil seeds perform two successive but distinct functions. During germination when fat is being converted to carbohydrate, large numbers of glyoxysomes containing active β-oxidation and glyoxylate cycle enzymes are present; whereas later, after greening, the cotyledons contain abundant peroxisomes which carry out photorespiratory glycolate metabolism. During the transition between these states, the activity of the glyoxysomal glyoxylate cycle enzymes declines while that of the peroxisomal photorespiratory enzymes increases. Thus, for a short time during transition, both sets of enzyme activity are present in the same cell.

The nature of the glyoxysome–peroxisome transition has been investigated by a number of researchers over the past 10–15 years and a number of models have been proposed to explain it. Beevers (1979) suggested the two population model proposing that at transition two biochemically distinct sets of microbody exist, one containing the glyoxylate cycle enzymes and the other containing the glycolate pathway enzymes. In contrast, Trelease *et al.* (1971) proposed that the change in enzyme composition occurs within one population of homogeneous microbodies by gradual replacement of glyoxysomal enzymes by peroxisomal enzymes. A variation of this 'one population' model proposes that transition occurs by the continual synthesis and degradation of microbodies, which as development proceeds will contain changing patterns of enzymes (Schopfer *et al.*, 1976).

Only recently have the experimental techniques evolved which make it possible to differentiate between these models. Double-label immunoelectron microscopy has been employed by Titus and Becker (1985) to determine the localization of glyoxysomal and peroxisomal enzymes in transition-stage cotyledons of cucumber. Using two sizes of protein A-gold, these workers have unequivocally demonstrated the coexistence of a glyoxysomal marker enzyme (isocitrate lyase) and a peroxisomal marker enzyme (serine:glyoxylate aminotransferase) within the same microbody at transition, suggesting that only one population of microbodies exists which contains both sets of enzymes.

Biogenesis of microbody proteins

Elucidation of the pathway of microbody biogenesis has lagged behind that of nearly all other major cellular organelles. Over the past 15–20 years some controversy has developed between investigators supporting opposing models, and much conflicting and contradictory data has been produced, confusing the issue still further. Many technical difficulties have beset this area of investigation, making microbody biogenesis inherently more difficult to study than that of, say, chloroplasts or mitochondria. In recent years, however, the advent of modern molecular biology allied with improved biochemical techniques has overcome many of these problems, and all the recent evidence supports a simple post-translational mechanism for the incorporation of proteins into microbodies.

Microbody matrix proteins

The emerging model for the biogenesis of microbody matrix proteins can be summarized in a number of points.

(1) Microbody matrix proteins are synthesized on free polysomes. Classical fractionation techniques can be used to separate free cytosolic polysomes from membrane-bound polysomes, and these two fractions (or the mRNA derived from them) can then be translated *in vitro* in a heterologous cell-free system. In all cases so far investigated, mRNAs

encoding microbody matrix proteins have been found predominantly on free polysomes. In addition, studies employing radiolabeling followed by tissue fractionation have shown that *in vivo*, many microbody matrix proteins can be initially detected in the soluble cytosolic fraction before they accumulate in the microbody fraction. These data suggest that microbody matrix proteins are synthesized on free polysomes, released into the cytosol and post-translationally imported into microbodies. The endoplasmic reticulum (ER) appears to play no role in this process.

(2) Matrix proteins are post-translationally imported into microbodies. In addition to the evidence for post-translational import provided by the *in vivo* labeling studies mentioned above, some microbody matrix proteins have been successfully imported *in vitro* when cell-free translation products are incubated with isolated peroxisomes. Proteins such as cucumber malate synthase, *Neurospora crassa* isocitrate lyase, watermelon malate dehydrogenase, rat liver catalase and acyl-CoA oxidase have all been post-translationally imported *in vitro*, although with low efficiency.

(3) Most microbody matrix proteins are not made as precursors but are synthesized at their mature molecular weight. After SDS-gel electrophoresis, the initial products of isolated matrix protein mRNA translated *in vitro*, or newly synthesized matrix proteins specifically pulse-labeled *in vivo*, are not detectably different in size from the mature proteins isolated from microbodies. In the case of mammalian catalase, the N-terminal amino acid sequence of newly synthesized protein and of mature protein have been determined and found to be identical. Microbody proteins are unusual in this respect as endoplasmic reticulum, chloroplast and mitochondrial proteins are generally synthesized as larger molecular weight precursors with N-terminal 'signal' or 'targeting' sequences which direct the proteins to the correct organelle. These targeting sequences are cleaved during or after import. As microbody matrix proteins are not synthesized as precursors, the targeting signals that direct these proteins to microbodies must be present within the mature protein. To date nothing is known about these signals, and this remains one of the major areas where future work must be focused.

Although synthesis at the mature molecular weight is the rule for most microbody matrix proteins, five exceptions are known at present: (1) pumpkin catalase, (2) thiolase and (3) acyl-CoA oxidase from rat liver, (4) malate dehydrogenase from watermelon and cucumber cotyledons, and (5) carnitine acetyltransferase from *Candida tropicalis* which have all been shown to be initially synthesized as precursors that are 3–20 kD larger than their respective mature protein. Isocitrate lyase from cucumber has also been reported to be synthesized as a precursor, but this finding has been contradicted in later reports. The function of the extra sequences present in these precursors is unclear and there is no good evidence that they are involved in targeting or import, although this has often been suggested purely by analogy with the import mechanism of chloroplasts and mitochondria (for more detailed summary see Borst, 1986).

Microbody membrane proteins

The biosynthesis of only two microbody membrane proteins has been examined to date: the M_r 22 000 polypeptide of rat liver peroxisomes (Fujiki *et al.*, 1984), and a glyoxysomal alkaline lipase (M_r 62 000) from castor bean (Maeshima *et al.*, 1987). Both these proteins were found to be synthesized on free polysomes and must therefore enter the organelle post-translationally. The rat liver 22 000 M_r polypeptide is synthesized at its mature molecular weight, whereas the apparent molecular weight of the lipase synthesized *in vitro* is reported to be slightly higher than that of the mature enzyme. The pathway of biogenesis of at least these two membrane proteins is therefore similar to that described for microbody matrix proteins. There is no reason to postulate any involvement of the ER in their synthesis. This is important, as microbody membranes were originally proposed to arise directly from the ER membrane by budding (Goldman and Blobel, 1978). Similarities between the SDS-gel pattern of membrane proteins isolated from ER and those isolated from castor bean glyoxysomes were reported as supportive evidence. In addition, certain ER membrane enzymes were also reported to be present in rat liver peroxisomes and castor bean glyoxysomes, albeit in very low amounts. Improved techniques have shown that contamination of isolated microbody membranes with fragments of ER membranes may have been responsible for some of these results. The 22 000 M_r polypeptide and alkaline lipase are, however, major integral membrane proteins, unique to microbodies. The post-translational pathway deduced for their biogenesis with no contribution from ER is therefore more likely to reflect the true pathway of biosynthesis for all microbody membrane proteins.

Proliferation of microbodies

If both microbody matrix and integral membrane proteins are synthesized on free polysomes and enter the organelle post-translationally, where does the importing microbody originally come from? Do microbodies, like chloroplasts and mitochondria, arise by division of pre-existing organelles, or, conversely, might there still be a role for the ER in the production by vesiculation of a 'pre-microbody' vesicle into which matrix and membrane proteins might insert post-translationally? This question has still not been conclusively answered and various lines of evidence exist that would seem to support either theory.

ER vesiculation

In the early days of microbody research, it was expected that microbodies arose directly from the ER by vesiculation. Ultrastructural studies provided some evidence for this theory, as did early reports concerning the phospholipid composition of microbody membranes and the detection of glycoproteins in them. This theory, however, suggested that microbody proteins should be synthesized on bound polysomes and cotranslationally inserted into the ER, which conflicts with the growing body of biochemical and molecular evidence. A re-evaluation of the data has therefore been necessary which has left many of the earlier interpretations in doubt. In accordance with this, some investigators support a modified ER vesiculation model, i.e. that microbodies are formed by vesiculation from the ER and that matrix and possibly membrane proteins are added post-translationally during formation and/or following detachment from the ER (Trelease, 1984). The evidence supporting involvement of the ER in microbody biogenesis is as follows.

Proximity of the endoplasmic reticulum to microbodies

In ultrastructural studies, microbodies have often been seen in very close proximity to sections of the ER, sometimes being entirely surrounded by it. Regions where the ER and microbody are seen to touch have been interpreted as demonstrating direct membrane continuity between the two organelles (e.g. Novikoff and Shin, 1964). In many cases the membranes in these micrographs are broken or fuzzy around the putative connecting region. Unambiguous views of direct membrane and luminal connections between ER and microbodies are rare, if they exist at all. Several workers have looked for connections in rat liver, pneumocytes, and bean leaves, and failed to observe them. Other workers have found that connections are only apparent when the microsomal or ER membranes are sectioned tangentially. Connections could therefore be artefactual and no definitive evidence of their existence has been found.

Glycoproteins

Many workers have investigated the existence of glycoproteins in the peroxisomal matrix and membrane. As N-glycosylation is a cotranslational modification acquired during synthesis on the rough ER, identification of glycoproteins in peroxisomes would demonstrate ER involvement in their biogenesis. Efforts to detect such glycoproteins have been inconclusive. Malate synthase from castor bean was initially thought to be glycosylated but subsequent analyses have shown it to lack sugar residues. Reports conflict as to whether cucumber isocitrate lyase is glycosylated; the castor bean enzyme is not. Bergner and Tanner (1981) examined isolated glyoxysomes after incorporation of radioactive sugars, and found label in the membrane but not the matrix fraction. Membrane glycoproteins were also detected by Lord and Roberts (1983) in similar experiments. Lazarow and Fujiki (1985) suggest that the use of radioactive sugars in such experiments is dangerous because the method is so sensitive it permits the detection of trace contaminant glycoproteins in the cell fractions.

There has therefore been no totally convincing demonstration of glycoproteins in either the peroxisomal matrix or membrane.

Growth and division

As much of the evidence for the ER vesiculation theory is now in doubt, the model is being increasingly abandoned, and some investigators (e.g. Lazarow and Fujiki, 1985) have adopted an alternative one. As microbodies develop like mitochondria and chloroplasts by post-translational incorporation of proteins, might they not also be capable of division, as mitochondria and chloroplasts are? Some evidence supports this.

Studies of the morphological changes in the microbodies of yeasts following induction suggest that the rapid proliferation of microbodies that occurs is due to growth and division of pre-existing microbodies. When peroxisome proliferation is induced in *Candida tropicalis* by growth on alkanes, the size of the peroxisomal compartment increases twelve-fold (Osumi *et al.*, 1975). At the start of induction one or two very small peroxisomes (approximately 0.1 μm) can be seen per cell. By 6–8 h after induction both the number and the size of peroxisomes has increased five times and larger oblate peroxisomes, some containing septa which divide them unequally, are also obvious. After 46 h the cytoplasm is full of small peroxisomes 0.3–0.6 μm in diameter.

Similar changes in peroxisome morphology have been observed when *Hansenula polymorpha* cells, grown to stationary phase on methanol, are diluted into fresh methanol medium (Veenhuis *et al.*, 1978). After three hours the yeast cells begin to bud, and the peroxisomes inside them also appear to divide unequally, producing small new peroxisomes that apparently migrate into the yeast bud. After yeast division, small new yeast cells contain one or two small peroxisomes while the mother yeast cells are filled with large peroxisomes. During a second growth phase the small peroxisomes in the daughter cells themselves enlarge and divide.

Both these studies suggest that microbodies, in yeast at least, arise by division from pre-existing microbodies. Direct evidence for such division in higher plants is lacking, but glyoxysome and peroxisome 'preforms' have been documented that might be considered analogous to the 'small' peroxisomes found in yeast cells prior to induction. In suspension cultures of de-differentiated anise cells growing in sucrose, glyoxysome 'preforms', which contain three β-oxidation enzymes and catalase but which are less dense than mature organelles (equilibrating at 1.13 g ml^{-1} on sucrose gradients), have been described (Lutzenberger and Theimer, 1986). When glyoxysome proliferation is induced in these cells by growth on acetate for 48 h, six glyoxysomal marker enzymes accumulate in organelles equilibrating at 1.23 g ml^{-1} sucrose, indicating a development of functional glyoxysomes from the less dense organelles. The presence of small glyoxysomes in the cotyledonary cells of cotton and cucumber seeds that are detectable during seed development (inferred from appropriate enzyme activities) and that enlarge on germination, has already been mentioned. Might these organelles also be considered as 'pre' or progenitor glyoxysomes from which glyoxysome proliferation is effected on germination purely by growth and fission? The growth and division model is also in keeping with our recent understanding of the glyoxysome–peroxisome transition and explains the existence of small or pre-peroxisomes in cells when no specific microbody activity or proliferation is occurring. In fact, the growth and division model predicts that as microbodies never arise *de novo* but only through division of pre-existing organelles, most cells, especially germ cells, must contain at least one. In the case of yeast at least, this prediction has been confirmed, and yeast cells, grown on glucose which suppresses peroxisome formation, still contain one or two small peroxisomes.

Conclusions

The validity of the models for microbody proliferation, by ER vesiculation or by growth and division from pre-existing organelles, cannot yet be properly assessed. Much of the evidence suggesting ER involvement is now in doubt, although persistent claims that glycoproteins can be detected in the microbody membrane and matrix cannot be easily dismissed. On the other hand, evidence for the growth and division model is still largely circumstantial but is conceptually persuasive in the light of all the recent data which supports post-translational incorporation of proteins into microbodies. If the ER does play a role in microbody biogenesis, why is it that every well-characterized microbody protein investigated to date is synthesized on free ribosomes and post-translationally imported, while not a single microbody protein that will cotranslationally insert into ER microsomes has been documented.

The solution to this problem may soon be at hand. Using yeast cells, it may be possible to replace the gene for an essential peroxisome membrane component by a non-functional copy, resulting in a yeast strain devoid of peroxisomes. Reintroduction of the functional gene would then allow a critical test of whether or not microbodies can be made *de novo* (see Borst, 1986). The technology necessary for the performance of such an experiment is currently evolving and the first account of the expression of a peroxisomal protein in a foreign host has recently been published (Distel et al., 1987).

The simplest model in keeping with virtually all the recent evidence is that microbodies grow by the post-translational incorporation of new matrix and membrane proteins into pre-existing microbodies which then divide to form daughter microbodies. In this respect, microbodies are like mitochondria and chloroplasts. However, microbody proteins, unlike mitochondria and chloroplast proteins, are not initially synthesized as precursors. A role for the ER in this biogenetic process is still supported by many investigators and cannot be discounted, but as yet there is no unequivocal evidence for it.

Perspectives

Import into microbodies *in vitro*

A major tool in the investigation of post-translational protein import into mitochondria and chloroplasts has been the development of efficient *in vitro* import systems. Many workers have attempted to develop similar import systems for peroxisomes but have met with only limited success. This is due in part to the fragility of microbodies which makes their isolation and subsequent experimental manipulation difficult. In addition, peroxisomes represent a much smaller proportion of cell protein than mitochondria, in normal non-induced cells. It has also been suggested that the presence of high endogenous protease activity in some systems could influence the functional integrity of isolated peroxisomes. Successful *in vitro* import experiments have typically been inefficient, resulting in, at best, 20% protein uptake (Borst, 1986). Recently two more efficient import assays have been developed using rat liver (Fujiki and Lazarow, 1985) and yeast (Small et al., 1987) peroxisomes. The yeast paper in particular claims up to 70% association of peroxisomal proteins with the organelle *in vitro*. Internalization of these proteins into the peroxisome could not be assessed by protease protection experiments, however, as four protease inhibitors were present in the import mixture. Further experiments will therefore be necessary to fully characterize and optimize these systems. If this can be achieved, *in vitro* import systems will provide a powerful means of dissecting the exact mechanism of import. The possible involvement of cytosolic proteins, of receptors on the peroxisomal membrane, the nature of targeting signals on imported proteins, and the energy dependence of the import process, can all be investigated using this system.

Signals

Nothing is known about the nature of the topogenic signal (or signals) that directs microbody proteins to the correct organelle. The absence of signal removal upon import into microbodies provides no simple clues as to where the signal is located within these proteins. It has been suggested that the import of proteins into microbodies of *Trypanosoma brucei* is affected by a special configuration of basic residues on the surface of the molecules (Wierenga et al., 1987). Other workers have compared deduced amino acid sequences of three plant peroxisomal proteins and revealed a region of homology that may be involved in directing proteins to the peroxisome (Volokita and Somerville, 1987). Designation of such features as 'targeting signals' is completely speculative until tested by genetic manipulation. However, the lack of an efficient *in vitro* import assay makes the effect of genetically altering putative signal regions difficult to assess.

An alternative system has recently been developed by Distel *et al.* (1987), who have successfully introduced and expressed the *Hansenula polymorpha* gene for the peroxisomal protein alcohol oxidase into *Saccharomyces cerevisiae* which normally lacks this protein. The introduced protein can be immunocytochemically localized to the peroxisome. As alcohol oxidase is a non-essential protein in *Saccharomyces cerevisiae* it can be altered at liberty and the effect of such alterations on import into peroxisomes can be assessed. This *in vivo* system may well provide an alternative to *in vitro* import assays that have proved so problematical.

References

Beevers, H. (1979). Microbodies in higher plants. *Ann. Rev. Plant Physiol.* **30**, 159–93.
Bergner, U. and Tanner, W. (1981). Occurrence of several glycoproteins in glyoxysomal membranes of castor beans. *FEBS Lett.* **131**, 68–72.
Borst, P. (1986). How proteins get into microbodies (peroxisomes, glyoxysomes, glycosomes). *Biochim. Biophys. Acta* **866**, 179–204.
De Duve, C. (1983). Microbodies in the living cell. *Sci. Amer.* **248**, 74–84.
Distel, B., Veenhuis, M. and Tabak, H. F. (1987). Import of alcohol oxidase into peroxisomes of *Saccharomyces cerevisiae*. *EMBO J.* **6**, 3111–16.
Fujiki, Y. and Lazarow, P. B. (1985). Post-translational import of fatty acyl CoA oxidase and catalase into peroxisomes of rat liver *in vitro*. *J. Biol. Chem.* **260**, 5603–9.
Fujiki, Y., Rachubinski, R. A. and Lazarow, P. B. (1984). Synthesis of a major integral membrane polypeptide of rat liver peroxisomes on free polysomes. *Proc. Natl Acad. Sci. USA* **81**, 7127–31.
Goldman, B. M. and Blobel, G. (1978). Biogenesis of peroxisomes: Intracellular site of synthesis of catalase and uricase. *Proc. Natl Acad. Sci. USA* **75**, 5066–70.
Huang, A. H. C., Trelease, R. N. and Moore, T. S. Jr. (1983). *Plant Peroxisomes*, Academic Press, New York.
Lazarow, P. B. and Fujiki, Y. (1985). Biogenesis of peroxisomes. *Ann. Rev. Cell Biol.* **1**, 489–530.
Lord, J. M. and Roberts, M. R. (1983). Formation of glyoxysomes. In *Aspects of Cell Regulation*, ed. J. F. Danielli, International Review of Cytology Suppl. 15, pp. 115–56.
Lutzenberger, A. and Theimer, R. R. (1986). Nutrient dependent induction of formation and degradation of glyoxysomal isocitrate lyase in cell suspension cultures of anise (*Pinpinella anisum* L.). *Eur. J. Cell Biol.* **41**, 28–33.
Maeshima, M., Takeuchi, A. and Asahi, T. (1987). Cell-free synthesis of alkaline lipase, a glyoxysomal membrane protein, from castor bean endosperm, *FEBS Lett.* **220**, 23–6.
Newcomb, E. H. (1982). Ultrastructure and cytochemistry of plant peroxisomes and glyoxysomes. In *Peroxisomes and Glyoxysomes*, eds H. Kindl and P. B. Lazarow, Annals of the New York Academy of Sciences, Vol. 386, pp. 228–41.
Nguyen, T., Zelechowska, M., Foster, V., Bergmann, H. and Verma, D. P. S. (1985). Primary structure of the soybean nodulin-35 gene encoding uricase 11 localized in the peroxisomes of uninfected cells of nodules. *Proc. Natl Acad. Sci. USA* **82**, 5040–4.
Novikoff, A. B. and Shin, W.-Y. (1964). The endoplasmic reticulum in the Golgi zone and its relations to microbodies, Golgi apparatus and autophagic vacuoles in rat liver cells. *J. Micros. Oxford* **3**, 187–206.
Osumi, M., Fukuzumi, T., Teranishi, Y., Tanaka, A., and Fukui, S. (1975). Development of microbodies in *Candida tropicalis* during incubation in *n*-alkane medium. *Arch. Microbiol.* **103**, 1–11.
Schopfer, P., Bajracharya, D., Bergfeld, R. and Falk, H. (1976). Phytochrome-mediated transformation of glyoxysomes into peroxisomes in the cotyledons of mustard seedlings. *Planta* **133**, 73–80.
Small, G. M., Imanaka, T., Shio, H. and Lazarow, P. B. (1987). Efficient association of *in vitro* translation products with purified, stable *Candida tropicalis* peroxisomes. *Mol. Cell. Biol.* **7**, 1848–55.
Titus, D. E. and Becker, W. M. (1985). Investigation of the glyoxysome–peroxisome transition in germinating cucumber cotyledons using double-label immunoelectron microscopy. *J. Cell Biol.* **101**, 1288–99.
Trelease, R. N. (1984). Biogenesis of glyoxysomes. *Ann. Rev. Plant Physiol.* **35**, 321–47.
Trelease, R. N., Becker, W. M., Gruber, P. J. and Newcomb, E. H. (1971). Microbodies (glyoxysomes and peroxisomes) in cucumber cotyledons. Correlative biochemical and ultrastructural study in light- and dark-grown seedlings. *Plant Physiol.* **48**, 461–75.
Veenhuis, M., Van Dijken, J. P., Pilon, S. A. F. and Harder, W. (1978). Development of crystalline peroxisome in methanol-grown cells of the yeast *Hansenula polymorpha* and its relation to environmental conditions. *Arch. Microbiol.* **117**, 153–63.
Volokita, M. and Somerville, C. R. (1978). The primary structure of spinach glycolate oxidase deduced from

the DNA sequence of a cDNA clone. *J. Biol. Chem.* **262**, 15825–8.

Weir, E. M., Riezman, H., Grienenberger, J.-M., Becker, W. M. and Leaver, C. J. (1980). Regulation of glyoxysomal enzymes during germination of cucumber. *Eur. J. Biochem.* **112**, 469–77.

Wierenga, R. K., Swinkels, B., Michels, P. A. M., Osinga, K., Misset, O., Van Beeumen, J., Gibson, W. C., Postma, J. P. M., Borst, P., Opperdoes, F. R. and Hol, W. G. J. (1987). Common elements on the surface of glycolytic enzymes from *Trypanosoma brucei* may serve as topogenic signals for import into glycosomes. *EMBO J.* **6**, 215–21.

23 Fatty acid and lipid biosynthesis and degradation

Jaen Andrews and John Ohlrogge

Structures and functions of plant lipids

Lipids are a diverse group of chemicals which perform several major functions in plants. Phospholipids, galactolipids and sterol esters form the central hydrophobic barrier of cell membranes (Fig. 23.1). Cuticular lipids and wax esters form a coating on the aerial surface of plants which serves to prevent water loss and as a protection from environmental and biological stress. An analogous material, suberin, is formed by underground organs or in response to wounding. In most seeds, triacylglycerol is a major form of carbon storage (see Fig. 23.6). Plant triacylglycerols from seeds of soybean, sunflower, maize, etc. are a major source of calories for human consumption. Lipids also may serve a variety of less well defined functions such as hormones, second messengers, insect attractants, and defense chemicals (phytoalexins).

Most plant membrane lipids contain two long chain fatty acids esterified to the sn 1 and 2 positions of glycerol. These fatty acids are almost always 16 or 18 carbons in length and contain from 0 to 3 cis double bonds. (Fig. 23.1) Attached to the third position on the glycerol backbone is a polar head group. The combination of the non-polar fatty acyl chains and the polar head group leads to the amphipathic properties of membrane lipids. Most such lipids spontaneously form bilayer or micellar structures when mixed with water, and in biological systems they are organized into the classic fluid bilayer.

Table 23.1 Glycerolipid composition of plant cell membranes

	Chloroplast[a]			ER[b]	Mitochondria[c]	
	Thylakoid	IEM[d]	OEM[e]		IM[f]	OM[g]
MGDG	56.0	52.9	1.4	1.2	ND	ND
DGDG	31.6	31.0	40.4	2.5	ND	ND
SL	3.5	3.5	3.7	–	ND	ND
PC	0	0	42.2	28.9	41	42
PE	0.2	0.5	0.4	17.0	37	24
PG	7.8	10.6	5.5	4.2	3	10
PI	0.6	0.1	4.4	6.1	5	21
PS	–	–	–	3.0	–	
CL	–	–	–	ND	14 (PGP?)	3 (PGP?)

[a]J. Andrews, PhD thesis, pea leaf chloroplasts, Mol%; [b]Musgrave, A. *et al.* (1976) *Phytochemistry* **15**, 1219, pea shoot microsomes, Mol%; [c]Moreau *et al.* (1974) *Biochim. Biophys. Acta* **345**, 294, potato tuber mitochondria, wt%; [d]IEM = inner envelope membrane; [e]OEM = outer envelope membrane; [f]IM = inner membrane; [g]OM = outer membrane. ND = not detected.

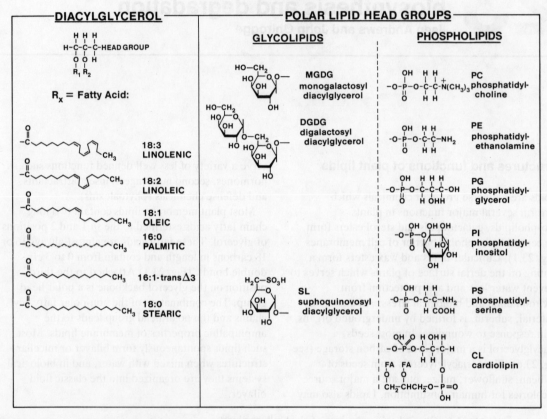

Fig. 23.1 Glycerolipids of plant cell membranes. Note that the fatty acids are referred to by the number of carbon atoms followed by the number of double bonds after the colon.

The composition of plant membrane lipids depends on a number of factors including the tissue type, subcellular localization and environmental influences (Harwood, 1980). The lipid composition of plant leaves is dominated by the photosynthetic thylakoid membranes of the chloroplast. These membranes contain high levels of MGDG and DGDG. These galactolipids are absent or present in low levels in other subcellular organelles or in non-photosynthetic tissues (Table 23.1). The composition of non-photosynthetic membranes in plants is not substantially different from that of other eukaryotic organisms in that phosphatidylcholine, phosphatidylethanolamine and phosphatidylinositol are major components.

Fatty acid biosynthesis

The hydrophobic properties of lipids are determined by their fatty acid constituents. The synthesis of the long chain non-polar fatty acids is conducted through sequential two carbon additions which are catalyzed by a group of soluble enzymes. Because soluble proteins are easier to purify and study, much more is known about the biochemistry of fatty acid synthesis than about the subsequent membrane-bound reactions of fatty acid desaturation and glycerolipid assembly. Most of the enzymes for the early steps of plant lipid metabolism have been purified, and some are now yielding to efforts to obtain cDNA and genomic clones.

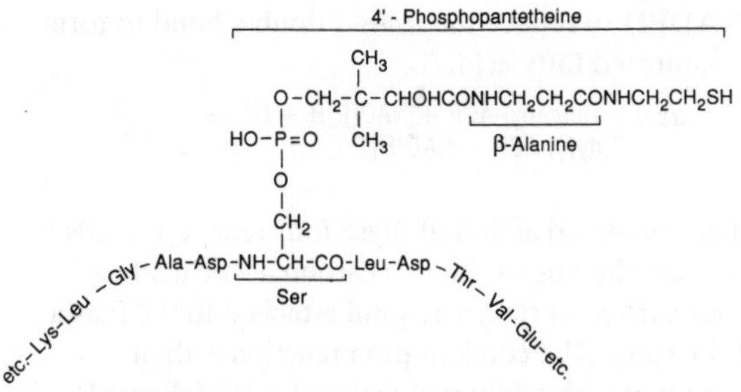

Fig. 23.2 Prosthetic group structure of acyl carrier protein. The 4'-phosphopantetheine group is attached to a serine residue near the middle of the polypeptide chain.

Role of acyl carrier protein (ACP)

ACP is a small acidic protein that has a phosphopantetheine prosthetic group attached to a serine residue near the middle of the protein (Fig. 23.2) (Ohlrogge, 1987). The prosthetic group is similar to the structure of coenzyme A and serves a similar function. A sulfhydryl group at the end of the pantetheine can be joined with the carboxyl carbon of a fatty acid to form a thioester. Formation of the thioester bond requires energy (supplied by ATP) and results in the activation of the carbonyl carbon of the acyl group so that it can participate in several key reactions of lipid metabolism. In general, acyl groups are not directly esterified to ACP but are first linked to coenzyme A and then transferred from coenzyme A to ACP by the action of a transacylase.

Subcellular localization

In animals, yeast and many other eukaryotic organisms, fatty acid synthesis occurs in the cytoplasm. In plants, however, there is a fundamentally different subcellular organization. It had been known since 1960 that isolated chloroplasts have the capacity to synthesize fatty acids. However, it was expected that additional FAS activity would occur in the cytoplasm (similar to animals) to provide fatty acids needed for membrane synthesis outside the chloroplasts. However, it now appears that *de novo* fatty acid synthesis is confined to the chloroplasts of plant mesophyll cells: therefore, this organelle must supply not only its own needs but also must export fatty acids to supply the needs of membrane synthesis outside the plastid. The plastid also is known to be an important site of fatty acid synthesis in several other types of plant cells, including developing seeds. Unfortunately, because of difficulties in subcellular fractionation of these other tissues, it is not yet certain if the plastid is the sole site of fatty acid synthesis in all tissues.

The enzymes of fatty acid synthesis

The fatty acid synthesis pathway uses acetyl-CoA as the building block for assembly of long chain (C16 and C18) fatty acids. Acetyl-CoA is supplied to the pathway by the action of the pyruvate dehydrogenase reaction. Although this enzyme occurs both in the mitochondria and in plastids, it is now believed that the plastid isozyme has sufficient activity to account for *in vivo* rates of fatty acid synthesis. In non-photosynthesizing tissues, pyruvate is provided from the glycolytic pathway. In chloroplasts, pyruvate can be derived from 3-phosphoglycerate produced by the Calvin cycle reactions. Plastids also contain a very active acetyl-CoA synthetase, and isolated chloroplasts or developing seed plastids rapidly incorporate radiolabeled acetate into fatty acids. Therefore, it appears that acetyl-CoA for fatty acid biosynthesis can derive either from the pyruvate dehydrogenase reaction inside the plastid or from extraplastidial production of acetate followed by its activation inside the plastid by acetyl-CoA synthetase.

The assembly of an eighteen carbon fatty acid requires the condensation of nine two-carbon units. All of these units are derived from acetyl-CoA. However, their actual assembly first requires that the acetyl group be further 'activated'. This activation is accomplished when acetyl-CoA is carboxylated to form malonyl-CoA by the action of acetyl-CoA carboxylase (ACC).

$$ATP + HCO_3^- + BCCP \xrightarrow{Mg^{2+}} ADP + P_i + BCCP\text{-}CO_2 \quad [23.1]$$

Biotin carboxylase

$$BCCP\text{-}CO_2 + \text{Acetyl-CoA} \rightarrow BCCP + \text{Malonyl-CoA} \quad [23.2]$$

Trans carboxylase

This enzyme appears to be a multifunctional polypeptide with three functional domains. Biotin is attached covalently to the biotin carboxyl carrier protein domain. CO_2 is activated and attached to biotin by the action of the biotin carboxylase domain. After formation of the carboxyl-biotin, the CO_2 is transferred to acetyl-CoA by the acetyl-CoA: malonyl-CoA transcarboxylase domain.

In animals, ACC appears to be the rate-limiting step for fatty acid synthesis. However, because malonyl-CoA is used in several additional pathways in plants, this is unlikely to be the case for plants. ACC has been localized to the chloroplasts of leaf tissue and the plastids of developing seeds. It seems likely, however, that isozymes of this enzyme may exist to provide malonyl-CoA for diverse pathways.

All the subsequent steps of plant fatty acid synthesis involve reactions which require acyl carrier protein (ACP). Malonyl-CoA produced by aceytyl-CoA carboxylase is transferred to ACP by the action of a transacylase.

Malonyl-CoA + ACP → Malonyl-ACP + CoA

The next step in fatty acid synthesis involves the condensation of acetyl and malonyl groups to form the four carbon intermediate, acetoacetyl-ACP.

Acetyl-CoA + Malonyl-ACP → Acetoacetyl-ACP + CO_2

This key reaction results in the release of the CO_2 which was added by the acetyl-CoA carboxylase reaction. The removal of CO_2 in this reaction helps to drive this reaction in the forward direction, making it essentially irreversible.

The acetoacetyl-ACP is next reduced at the carbonyl group by the enzyme 3-ketoacyl-ACP reductase which uses NADPH as the electron donor.

Acetoacetyl-ACP + NADPH + H^+ →
D-3-Hydroxybutyryl-ACP + $NADP^+$

The third reaction in the fatty acid synthesis cycle is the dehydration of hydroxy-acyl-ACP to yield trans-2-acyl-ACP which is catalyzed by the enzyme 3-hydroxyacyl-ACP dehydratase.

D-3-Hydroxybutyryl-ACP → trans-2-Butenoyl-ACP + H_2O

One round of fatty acid synthesis is completed by the enzyme enoyl-ACP reductase which uses NADH or NADPH to reduce the trans-2 double bond to form a saturated fatty acid.

trans-2-Butenoyl-ACP + NAD(P)H + H^+ →
Butyryl-ACP + $NAD(P)^+$

The combined action of these four reactions leads to the lengthening of the two carbon acetic acid to the four carbon butyric acid (still attached to ACP as a thioester). The condensation reaction is then repeated with additional malonyl-ACP followed again by the keto reduction, dehydration, and enoyl reduction steps. This cycle continues until the fatty acid chain length is 16 carbons long. The 16 carbon palmitoyl-ACP may then be acted upon by three different sets of enzymes; it may be elongated to an 18 carbon fatty acid, it may be used in glycerolipid synthesis in the plastid itself or it may be hydrolyzed to a free fatty acid for eventual use outside the plastid. The elongation of palmitoyl-ACP requires a separate condensing enzyme, but otherwise uses the same three enzymes described earlier for fatty acid synthesis; the resulting product is the 18 carbon stearoyl-ACP. This saturated fatty acid is the substrate for the introduction of the first cis double bond in plant fatty acids. Almost all aerobic fatty acid desaturation in nature is catalyzed by membrane-bound enzymes. However, in plants, the enzyme stearoyl-ACP desaturase is a soluble component of the chloroplast stroma. This enzyme requires O_2 and an electron donor such as reduced ferredoxin. A cis double bond is introduced exactly in the middle of the acyl chain between carbons 9 and 10.

Both palmitoyl-ACP and oleoyl-ACP produced by the stearoyl-ACP desaturase are potential substrates for two additional branch point enzymes in chloroplast fatty acid metabolism. These fatty acids may be transferred from ACP to glycerolipids or they may be released as free fatty acids from ACP by the action of a soluble acyl-ACP thioesterase. If the fatty acids are used in glycerolipid synthesis, they remain in the plastid. On the other hand, if they are released as free fatty acids they may then be exported outside the plastid for use by other cellular membranes. Thus, the different fates of oleate and palmitate in these two reactions determine the allocation of fatty acids between retention in or export from the chloroplast.

Source of NADPH and ATP

The ATP and reducing power needed for the fatty acid biosynthesis pathway is provided differently in different tissues. In leaves, the photosynthetic electron transport chain provides ATP and also electrons which reduce $NADP^+$ (with ferredoxin as an intermediate electron donor). In developing seeds, energy must be derived from sucrose imported from the leaves. Therefore, the pentose phosphate pathway in the plastid functions specifically to provide reducing power for fatty acid synthesis. ATP required for acetyl-CoA carboxylase can be derived from the glycolytic pathway which occurs both in plastids and the cytosol of developing seeds (Dennis and Miernyk, 1982).

Location of genes

Although the fatty acid synthesis pathway is localized inside the plastid, it is now clear that the genes coding for at least some of the pathway members are in the nuclear genome. This was first suggested by observing that mutants which lack plastid protein synthesis are still capable of producing characteristic plastid lipids. More recently, cDNA clones for acyl carrier protein have been found to code for a 50–55 amino acid transit peptide extension (Scherer and Knauf, 1987). Thus, ACP (and presumably the other pathway members) are first synthesized in the cytoplasm as precursors and are then taken up into plastids for processing to their mature size.

Occurrence of isozymes

In at least some cases, the proteins involved in plant fatty acid synthesis occur in different forms which are expressed differently in different tissues. The situation with ACP is best understood. In spinach leaves, ACP exists in two major forms which have been shown to have different amino acid sequences (Ohlrogge, 1987). Both forms are localized in the chloroplast, but their relative expression is different in light versus dark-grown leaves. In addition, the minor form found in spinach leaf appears to be the major form in seeds or roots. A similar tissue-specific expression of isozymes has been found for the malonyl-CoA : ACP transacylase in leaves and seeds of soybean.

Why do plants have different forms of the same fatty acid synthetase proteins? In the case of ACP, there is evidence that the different forms have different activity in the oleoyl-ACP hydrolase and oleoyl-ACP : glycerol 3-phosphate acyltransferase reactions. This difference may afford a means for the cell to use ACP isoform expression to control the proportions of oleate exported or retained within the plastid.

The presence of multiple genes for fatty acid synthetase components may also provide a mechanism for regulation of fatty acid synthesis under a greater variety of cellular demands. Fatty acids serve a 'housekeeping function' needed in all cells for membrane biosynthesis. In addition, fatty acids are produced in developing seeds for triacylglycerol storage and in epidermal cells for cuticular lipids. The presence of multiple genes with different promoters and regulatory sequences may provide mechanisms so that these diverse functions of fatty acid synthesis can be distinctly controlled.

Membrane glycerolipid synthesis

The original concept that, in plant cells, membrane lipids were first synthesized in the ER and subsequently exported to other cellular membranes has undergone considerable revision. This is due to observations that other cellular compartments, such as the chloroplast and mitochondria, are also capable, to varying extents, of lipid synthesis. The emerging scheme thus envisions a greater interaction and interdependence among various plant cell compartments in lipid synthesis, with a resulting rather brisk traffic of membrane lipids and their component parts.

Much of the research in plant lipid metabolism has been done with leaf tissue, due in large part to the relative ease of obtaining the material and in preparing tissue homogenate and cell fractions. Considerable work has also been directed toward elucidating biosynthetic pathways in oilseeds due to their economic and agricultural importance.

Unfortunately, much less is known about lipid metabolism in other parts of the plant. In fact, there are still very large and serious gaps in our knowledge of how lipids are synthesized, and questions of how this synthesis is regulated are just now being approached.

Two pathways

The assembly of membrane glycerolipids may be considered to occur in two fairly distinct stages. The first is the sequential transfer of fatty acids onto positions 1 and 2 of glycerol 3-phosphate. This provides the diacylglycerol portion of the lipid. The second stage is the addition of the head group to position 3 of the glycerol moiety. The head group thus defines the class (or type) to which the lipid belongs (Table 23.1).

As mentioned earlier, the fatty acids synthesized by the plastid may be used for glycerolipid synthesis either by the plastid itself, or they may be exported for use by other cell membranes. Although both the plastid and the extraplastidial membranes (such as the ER) are able to synthesize glycerolipids, the nature of the diacylglycerol moieties synthesized in the two compartments is dissimilar, and the difference lies in the type of fatty acid esterified to position 2 of the glycerol backbone (Roughan and Slack, 1982).

In lipids synthesized in the plastid, position 2 of the glycerol backbone is occupied almost exclusively by 16 carbon fatty acids (Fig. 23.3). Both 16 and 18 carbon fatty acids are found at position 1. This fatty acid distribution is characteristic of the lipids found in photosynthetic prokaryotic organisms such as cyanobacteria, and hence is designated 'prokaryotic'. The lipids synthesized by the ER are so-called 'eukaryotic' in nature, in that position 2 is occupied primarily by 18 carbon fatty acids (Fig. 23.3). In these lipids as well, position 1 may contain either 18 carbon or 16 carbon fatty acids. Lipids synthesized by other plant cell organelles are probably also eukaryotic, although this remains to be demonstrated. The positional distribution of the fatty acids is established by the activities of the enzymes which transfer the fatty acids onto the glycerol backbone of the lipid, and this specificity thus varies in the different compartments.

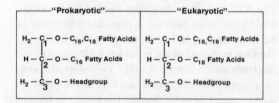

Fig. 23.3 Glycerolipid classes based on fatty acid distribution. The 'prokaryotic' and 'eukaryotic' lipid classes are based on the distribution of the fatty acids within the diacylglycerol moiety; the major difference occurs at position 2 of the glycerol backbone.

Glycerol 3-phosphate is synthesized by the enzyme dihydroxyacetone-phosphate reductase. The substrate for this enzyme is derived from glyceraldehyde 3-phosphate. At least two isozymes of the reductase occur, one in the chloroplast and one in the cytoplasm. Thus, both the plastidial and the extraplastidial compartments can produce the backbone for glycerolipid synthesis.

Plastidial lipid synthesis

In the chloroplast, the fatty acid substrates for glycerolipid synthesis are the endproducts of fatty acid synthesis, namely 16:0- and 18:1-ACP. The first acyltransferase transfers predominantly 18:1 to position 1 on glycerol 3-P, forming lysophosphatidic acid, or LPA (Fig. 23.4) (Frentzen, 1986). This is a soluble enzyme, found in the chloroplast stroma. The second acyltransferase is located in the inner membrane of the chloroplast envelope, and transfers almost exclusively 16:0 to position 2 of LPA, resulting in the formation of phosphatidic acid, or PA (Fig. 23.4) (Frentzen, 1986). As a result of the specificities of the acyltransferases, this newly-synthesized PA is prokaryotic in nature.

PA is a central intermediate in glycerolipid synthesis. It may be directed into any one of the lipids which the chloroplast synthesizes by first entering one of two branching reactions (Fig. 23.4) (Mudd et al., 1985). In one, PA is dephosphorylated to yield diacylglycerol (DG), which is a precursor

Fatty acid and lipid biosynthesis and degradation

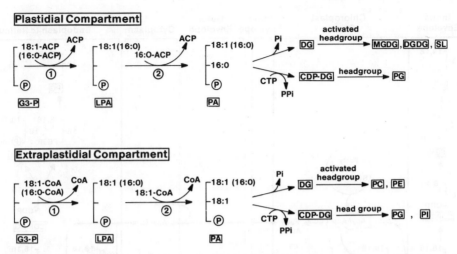

Fig. 23.4 Acyltransferases involved in glycerolipid synthesis. The sequential acylation of glycerol 3-P (G3P) by (1) acyl-ACP (or acyl-CoA): glycerol 3-P acyltransferase and (2) acyl-ACP (or acyl-CoA): lysophosphatidic acid acyltransferase gives rise to PA in both the plastid and ER. PA may then be either dephosphorylated to DG by (3) phosphatase or (4) activated by the addition of CTP to CDP-DG by phosphatidate cytidyltransferase in both compartments.

for those lipids which are synthesized by the addition of an activated head group (i.e. one to which a high energy phosphate bond has been added). In the other, the DG portion of PA is itself activated by the addition of CTP to form CDP-DG. The lipid head group is then added to this high-energy intermediate.

The phosphatidylglycerol (PG) which is found in plastid membranes appears to be universally synthesized by the plastid, as it is in every case so far examined prokaryotic in nature. PG is formed from CDP-DG by the sequential action of two enzymes (Figs 23.4 and 23.5) (Mudd et al., 1987). The first adds glycerol 3-P to CDP-DG to form phosphatidylglycerol 3-P (PGP); and PGP is then rapidly dephosphorylated to PG by the second enzyme. In fact, the only way to detect the intermediate PGP is to inhibit the activity of the second enzyme. All three enzymes are located in the inner membrane of the chloroplast envelope.

DG is the precursor for those glycolipids which the chloroplast synthesizes (Figs 23.4 and 23.5) (Mudd et al., 1985). In this case, PA is first dephosphorylated to diacylglycerol (DG) by the action of a phosphatase, which again appears to be located in the inner envelope membrane. The DG is galactosylated from UDP-galactose by a galactosyltransferase to form monogalactosyl diacylglycerol (MGDG) (Joyard and Douce, 1987). MGDG is a major glycerolipid in the chloroplast membranes (Table 23.1). The synthesis of the other major glycerolipid, digalactosyl diacylglycerol (DGDG), is unclear (Joyard and Douce, 1987). A galactose unit may be donated from one MGDG to another to form DGDG by the action of an intergalactolipid galactosyltransferase, at least *in vitro*. However, the physiological significance of this reaction is unclear, as it does not appear to be the route of synthesis *in vivo*. Finally, DG may also be incorporated in sulfolipid (SL), probably in a manner similar to the synthesis of MGDG, although the nature of the head group donor and the means of its synthesis are still unknown (Mudd and Kleppinger-Sparace, 1987).

Glycerolipid synthesis outside the chloroplast nonetheless utilizes fatty acids which are synthesized inside the plastid. These exported fatty acids are first hydrolyzed by a soluble thioesterase to form free fatty acids in the stroma (Fig. 23.5) (Mudd et al., 1985). These are then added to CoA to form acyl-CoA's by the action of an acyl-CoA synthetase which is located in the outer envelope membrane (Fig. 23.5) (Mudd et al., 1985). However, the high energy acyl-CoA's are water-soluble forms by which the

346 The Formation and Breakdown of Lipids

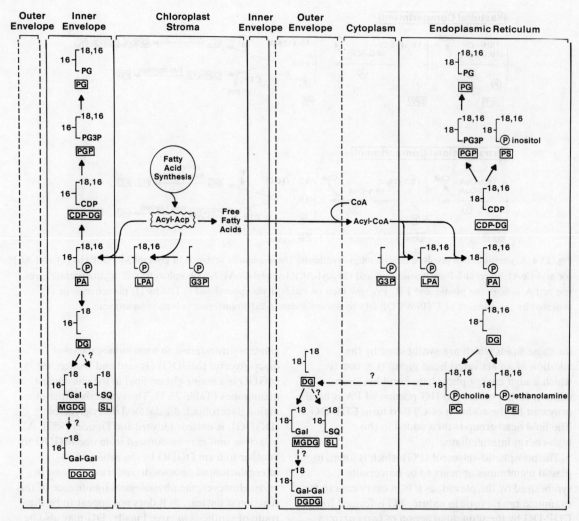

Fig. 23.5 Summary of membrane glycerolipid synthesis. Lipid synthesis in the plastid and ER is interdependent for the supply of fatty acids (ER) and the supply of DG moieties (plastid). Within the plastid, compartmentalization of lipid synthesis suggests that lipids must be moved from their site of synthesis in the envelope to the thylakoids.

fatty acids may be transported to other cell membranes and used directly for lipid synthesis.

Extraplastidial lipid synthesis: endoplasmic reticulum

The assembly of phospholipids is known to occur in both the endoplasmic reticulum (ER) and the mitochondria. Whether it also occurs in other cell membranes is under investigation. Currently, it is thought that the phospholipids synthesized in the ER are not only used in constructing its own membranes but that these lipids are also exported to other cellular membranes. These other membranes may then assemble those lipids which are unique to them.

In the ER, glycerolipid synthesis begins in much the same manner as it does in the chloroplast, with the sequential acylation of glycerol 3-P to form PA (Fig. 23.4) (Mudd, 1980; Roughan and Slack, 1982; Frentzen, 1986). There are two major differences between the ER and the plastid. The first is that the fatty acid donors for the acyltransferases in the ER are acyl-CoA's, rather than the acyl-ACPs used in the plastid. The second is that while the first acyltransferase will use either 16:0-CoA or 18:1-CoA, the second enzyme utilizes almost exclusively 18:1-CoA. This results in newly-synthesized PA with a eukaryotic fatty acid pattern.

As in the plastid, PA is the precursor for all the phospholipids assembled in the ER. It may be directed into one of two major sets of pathways (Figs 23.4 and 23.5) (Mudd, 1980). Dephosphorylation of PA yields DG, which serves as the precursor to phosphatidylcholine (PC) or phosphatidyl-ethanolamine (PE). The head groups for these lipids are first activated by the addition of CTP, resulting in CDP-choline (for PC) or CDP-ethanolamine (for PE). They are then added to DG to form PC or PE. Thus, the synthesis of these lipids is analogous to the synthesis of the glycolipids in the plastid. In fact, PC and PE are the major lipids found in the ER membranes (Table 23.1).

Alternatively, the diacylglycerol portion of PA may be itself activated by the addition of CTP, which forms CDP-DG. This serves as the precursor for either PG or PI, both of which are relatively minor components of the ER membrane (Table 23.1). PG is synthesized by the same steps which occur in the plastid: glycerol 3-P is added to CDP-DG to form PGP, which is then dephosphorylated to PG. The addition of inositol to CDP-DG results in the formation of PI.

The plastidial–endoplasmic reticulum connection

The only phospholipid synthesized by the plastid appears to be PG. Yet the plastid membranes also contain PC, and in much smaller amounts PS and PI as well (Table 23.1), and these lipids possess eukaryotic fatty acid distributions. Therefore, it is thought that these lipids are imported from their site of synthesis in the ER to the plastid. In fact, PC may play a special role. Not only is it present as a structural component in the plastid membranes, but it may also serve as a precursor to some of the plastidial glycolipids as well.

As mentioned earlier, the chloroplast can synthesize the glycolipids MGDG, DGDG and SL, which should then be prokaryotic in nature. However, in some plants, only about half of these lipids are prokaryotic, while in other plants almost none of them are; instead, they are eukaryotic in nature. The results of many experiments suggest that PC which is synthesized in the ER is the precursor for these eukaryotic glycolipids. After its assembly in the ER, at least the diacylglycerol portion from PC is transported to the plastid (Fig. 23.5) (Frentzen, 1986). It may be that the entire PC molecule is moved by a phospholipid transport protein, although there is no direct evidence to support this idea. At some point, the phosphocholine head group is removed. The resulting DG can then serve as a precursor for the glycolipids which are subsequently synthesized in the plastid by the same steps described earlier.

What determines the proportion of prokaryotic to eukaryotic glycolipids that a particular plant will synthesize? One obvious control point are the enzymes which control the fate of PA synthesized in the plastid. The prokaryotic glycolipids are synthesized from DG which is derived from the PA; if PA is not dephosphorylated, DG is no longer available to form the glycolipids. In fact, it has been shown that those plants with high proportions of prokaryotic glycolipids have high levels of phosphatase, while plants with low proportions of prokaryotic glycolipids have low levels of phosphatase activity (Mudd *et al.*, 1985).

Additional regulation may be exerted at the point at which fatty acids are either used by the plastid for lipid synthesis, or hydrolyzed for export to other cell membranes. In those plants with low levels of prokaryotic glycolipids, the predominant lipid synthesized in the plastid is PG, which is a minor lipid component of the plastid membranes. Therefore, it seems reasonable that the plastids in these plants would synthesize relatively less PA, which is directed only into PG, and that they would export relatively more fatty acids, which then return in the form of PC to be used in the synthesis of the glycolipids, which are a major component of the plastid membranes. Recent research indicates that this may in fact be the case.

Thus it can be seen that lipid metabolism may be regulated at several points within one organelle, and that it must be coordinated among several different organelles. Many other factors will come into play, such as the age of the leaf (a rapidly expanding leaf will require increased lipid synthesis to keep pace with rapid membrane biogenesis) and environmental conditions.

Extraplastidial lipid synthesis: mitochondria

Of the other cellular compartments, only the mitochondria have been shown to assemble glycerolipids, and these activities have been less well characterized than those in the chloroplast and the ER membranes. This is due to several reasons. One is the difficulty in obtaining very pure cellular subfractions, which are not contaminated with other membranes which may also be able to synthesize lipids. Another is preparing subfractions which are still active, and yet another is determining the appropriate conditions under which to investigate lipid synthetic activities.

Plant mitochondria are able to synthesize PA by the sequential acylation of glycerol 3-P in much the same manner as the ER (Mudd, 1980). This synthesis appears to occur in both the inner and outer membranes of the mitochondria, and the resulting PA is eukaryotic in nature. It may be used in the synthesis of PG, which apparently occurs only in the inner mitochondrial membrane, by the same set of steps described earlier. PA may also be used to synthesize cardiolipin, a lipid unique to the mitochondria, although there is no evidence of this as yet. No other lipids are known to be synthesized in the mitochondria, although new information may change this statement.

Further modification of lipids

Newly-synthesized membrane glycerolipids contain varying amounts of the fatty acids 16:0 and 18:1. However, the lipids actually found in plant membranes may contain other fatty acids as well; MGDG contains 18:3 (and 16:3 in those plants which synthesize prokaryotic lipids), while the phospholipids may contain 18:2 and 18:3. Thus, it appears that except for the desaturation of 18:0 to 18:1, fatty acids are desaturated after their incorporation into lipids (Roughan and Slack, 1982; Frentzen, 1986). In fact, desaturation probably occurs only in 'completely assembled' lipids which contain two fatty acids and the head group. It is known to occur in both the chloroplast and the ER, although it may occur in other cell organelles as well.

In the ER, desaturation requires both NADH and O_2. Although there is no evidence as yet, it is thought that electrons are passed from the NADH to the desaturase by two membrane-bound electron transport components, in a manner analogous to that observed in animal systems. The 18 carbon fatty acids at either position of the glycerol backbone may be desaturated, first to 18:2 and then to 18:3. 16:0 does not appear to be desaturated by the ER desaturates.

Much less is known about desaturation in the chloroplast, although O_2 again appears to be required. In those plants which synthesize prokaryotic glycolipids, the fatty acids in MGDG are desaturated to 18:3 and 16:3 (hence the occasional designation of these plants as '16:3' plants), while in DGDG only 18:1 is desaturated to 18:3. In SL, only 18:1 may be desaturated to 18:2 and 18:3. In PG, 18:1 may be desaturated to 18:2 and 18:3; in addition, 16:0 may be desaturated to an unusual 16:1 3-*trans*. This fatty acid is found predominantly in the plastidial PG, although it is reported to occur in some mitochondrial cardiolipin and in the seed oil of certain plants.

The fatty acid composition of a particular lipid class (i.e. the type and proportion of the constituent fatty acids) appears to be a characteristic of that lipid class, as well as perhaps of the membrane in which the lipid is found. Thus, desaturation must be somewhat specific with respect to lipid class and membrane in order to reach the steady-state levels of fatty acid compositions observed in the plant cell. In addition, fatty acid compositions change in response to changing environmental conditions, and perhaps during growth and development as well.

There are two means by which the fatty acid composition of membrane lipids may be changed. One is by *de novo* synthesis of new lipids, which may then be added to a membrane, in addition to or in replacement of pre-existing lipids. Such a mechanism could account for slow changes observed during growth. However, some changes occur too rapidly to be accounted for by new lipid synthesis. In this case, the fatty acid composition may be modified by removing fatty acids from lipids while they remain in the membrane, and replacing them with new fatty acids. This type of 'acyl exchange' may be accomplished by a combination of the reverse and forward reactions of acyltransferases, where the reverse reaction removes a fatty acid by attaching it to CoA and the forward reaction replaces it with a new fatty acid from another acyl-CoA (see Fig. 23.8) (Stymne and Stobart, 1987). Very little is known about these acyltransferases, although they appear to differ from those involved in *de novo* lipid synthesis. Another possibility is to remove fatty acids through the actions of specific lipases, and to then insert new ones via acyltransferases. Even less is known about such lipases, although there are recent reports that at least some of these activities may in fact exist. In any case, since most of the changes appear to involve a degree of fatty acid desaturation, the activity of desaturases may be modified by environmental conditions.

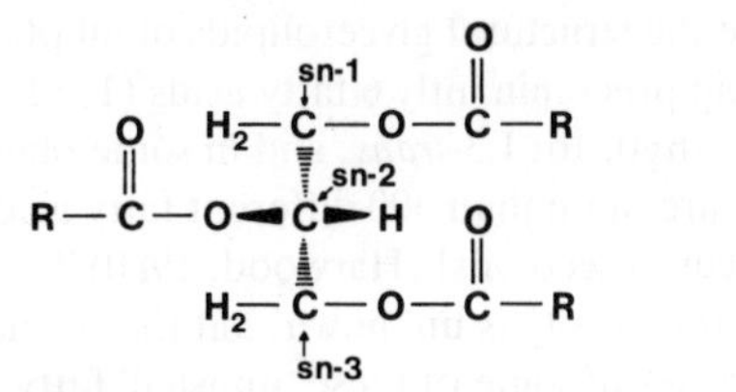

Fig. 23.6 Structure of triacylglycerol.

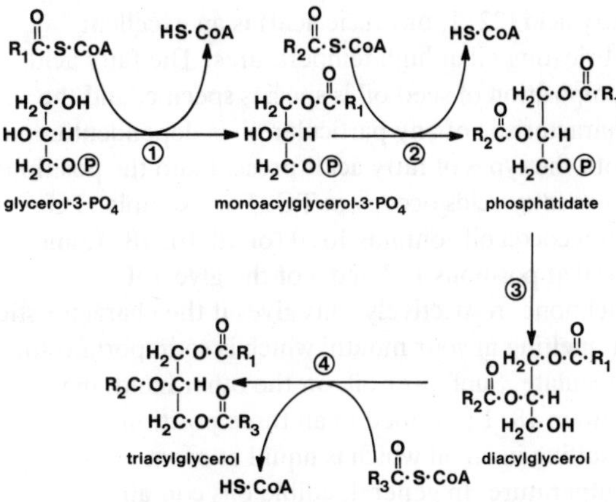

Fig. 23.7 Synthesis of triacylglycerol. The DG moiety of TG may be derived from either PA by the action of phosphatase (1) or PC by the reverse reaction of CDP-choline:DG choline transferase (2).

Seed lipid metabolism

Triacylglycerol synthesis

Lipid metabolism in the developing seed is geared toward the synthesis of triacylglycerols (TG), which are storage oils. The amount of this oil in different species may vary, from as little as 1–2% to as much as 60% of the total dry weight of the seed. Unlike the glycerolipids found in membranes, triacylglycerols do not perform a structural role but instead serve primarily as a storage form of carbon. Plants which produce seeds containing oil of economic importance for either food or industrial use include sunflower, safflower, soybean, cotton, peanut, rapeseed, coconut and palm. Although there is a great deal of information available concerning the composition of seed oils, there is much less known about how these oils are synthesized, and even less about the regulation of oil synthesis. This is due to the difficulty in procuring enough seed material for experiments; as mature seeds are relatively inactive in lipid metabolism, it is necessary to use developing seeds. In fact, very rapid lipid synthesis usually occurs during a limited period when the seed is rapidly gaining weight, and it essentially stops as the seed begins to mature and dehisce.

Triacylglycerol consists of a glycerol backbone with fatty acids esterified to all 3 carbons (Fig. 23.6). Its synthesis is relatively straightforward (Fig. 23.7); as in the leaf tissue, fatty acids are sequentially transferred to positions 1 and 2 of glycerol 3-P, resulting in the formation of LPA and PA respectively. PA is dephosphorylated to DG, and a third fatty acid is transferred to position 3. However, in certain seeds, at least some of the DG used in TG synthesis is derived from PC (see below). The proportions that each type of DG contributes to the final oil is unclear (Slack and Browse, 1984; Stymne and Stobert, 1987).

Fatty acid modifications in seeds

The fatty acid composition of storage oils varies much more than that of membrane glycerolipids.

While the structural glycerolipids of all plants contain predominantly 6 fatty acids (18:1, 18:2, 18:3, 16:0, 16:1 3-*trans*, and in some plants 16:3), there are more than 300 different fatty acids known to occur in seed oils! (Harwood, 1980) The reason for this diversity is unknown, but the special properties of some of these 'unusual' fatty acids are utilized commercially. The short chain fatty acids (12:0 or lauric acid) derived from coconut or palm kernel oil are used as detergents in shampoos and toothpastes, while the longer chain monounsaturated fatty acid (22:1, or erucic acid) is an excellent lubricating oil at high temperatures. The fatty acid composition of seed oil is species specific, and the characteristic of any particular oil is dependent upon both the types of fatty acids present and the positions these fatty acids occupy in TG. For example, TG from cocoa oil contains 16:0 (or 18:0), 18:1, and 18:0 at positions 1, 2 and 3 of the glycerol backbone, respectively; this gives it the characteristic of 'melting in your mouth' which is so important for chocolate. Sunflower oil, on the other hand, may contain 18:1 esterified to all three positions, resulting in an oil which is liquid at room temperature. In general, edible oils contain predominantly the saturated fatty acids 16:0 and 18:0 and the unsaturated fatty acids 18:1, 18:2 and 18:3.

The variations observed in fatty acids may be roughly characterized as those occurring in chain length and those occurring through modification within the fatty acid itself, such as the addition of double bonds or chemical groups (Harwood, 1980). From the little that is known, it appears as though modification within fatty acids occur while they are esterified to position 2 of PC (Stymne and Stobart, 1987). These fatty acids may then become available for triacylglycerol synthesis by one of two mechanisms.

In the first, the modified fatty acid may be removed from PC and replaced by a fatty acid to be modified. Such an 'acyl exchange' probably occurs by the combined reverse and forward reactions of an acyl-CoA : PC acyltransferase (Fig. 23.8) (Stymne and Stobart, 1987). The resulting acyl-CoA with the modified fatty acid may then be used as an acyl donor in triacylglycerol synthesis. It thus appears that the acyltransferases involved in TG synthesis are not as 'restricted' as those involved in membrane

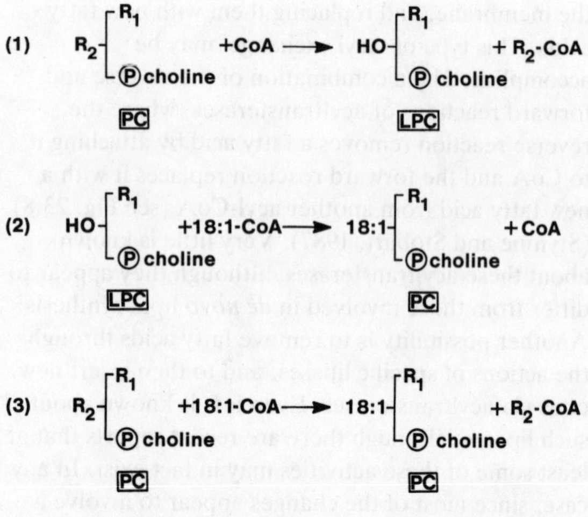

Fig. 23.8 Acyl exchange in PC. The combined reverse (1) and forward (2) reactions of acyl-CoA : lysophosphatidylcholine acyltransferase will result in acyl exchange between acyl-CoA and phosphatidylcholine (3).

glycerolipid synthesis, as many fatty acids other than 16:0 and 18:1 may be used as acyl donors in TG synthesis. In fact, the final composition of the TG probably depends on both the selectivities and specificities of the three acyltransferases (i.e. which acyl-CoA's the enzymes preferentially select to transfer, and how fast they transfer any particular acyl-CoA) and what is available to use as substrates (i.e. the types and amounts of the acyl-CoA's available).

The second mechanism by which modified fatty acids are incorporated into TG is via the entire DG portion of the PC itself (Slack and Browse, 1984; Stymne and Stobart, 1987). In some plants, the synthesis of PC from DG and CDP-choline appears to be rapidly reversible. This would allow the DG moiety of PC which contains the modified fatty acid to become available for TG synthesis.

Variations in chain length occur as fatty acids longer than 18:1 (e.g. 20:1 and 22:1) or shorter than 16:0 (such as 12:0). Elongation of fatty acids, from 18:1 to 20:1 or 22:1 (or even longer) occurs outside the plastid in the ER with the fatty acid esterified to CoA (Slack and Browse, 1984; Stymne and Stobart, 1987). Two carbon units are sequentially transferred from malonyl-CoA to 18:1-CoA in much the same manner by which the fatty acids are synthesized. How shorter chain fatty acids are synthesized is unknown. It seems a reasonable assumption that fatty acid synthesis would be terminated before a chain length of 16 carbons was reached, probably by the action of the thioesterase. However, investigation of thioesterase activity in seeds with shorter fatty acids demonstrates that the enzyme preferentially hydrolyzes 18:1. In addition, when incubated with labeled acetate, homogenate from these developing seeds synthesize predominantly 18:1 and 16:0.

Very little is known about the regulation of the fatty acid content of TG. In some seeds, the proportion of unsaturated fatty acids is decreased as the temperature during seed development is increased (Stymne and Stobart, 1987). However, fatty acid desaturases exhibit normal enzyme kinetics in the test tube, in that the activity increases as the temperature is raised. Some other factors must then regulate the final proportion of unsaturated fatty acid, such as possibly the relative rates of fatty acid biosynthesis and desaturation which may respond differently to increased temperature.

Storage of triacylglycerols in oil bodies

In the mature seed, TG is stored in densely packed oil bodies, which are roughly spherical in shape with an average diameter of 1 μm (Slack and Browse, 1984; Stymne and Stobart, 1987). This size does not change during seed development, and accumulation of oil is accompanied by an increase in the number of oil bodies. Evidence suggests that oil bodies are discrete organelles surrounded by a type of membrane which may be only half of a normal bilayer membrane. The membrane probably contains phospholipids and perhaps several low molecular weight proteins, as these components are associated with isolated oil bodies.

The synthesis of TG is located in the ER, as these membranes contain all three acyltransferase activities. In addition, the reactions by which fatty acids are modified (such as desaturation) or elongated also appear to occur in the ER. How triacylglycerols are moved from their site of synthesis in the ER to oil bodies is unknown. In fact, how oil bodies are generated is also poorly understood. *In vitro* experiments indicate that TG synthesized in the ER may first accumulate between the membrane bilayers and then be released as a 'naked' oil droplet (Stymne and Stobart, 1987). Formation of the membrane-type of structure may occur afterwards in the cytoplasm by association of phospholipids and proteins with the surface of the droplet.

Fatty acid utilization during germination

When oil storing seeds germinate, a massive breakdown of the triacylglycerol reserves is initiated. This process is begun by the action of lipases which lyze the hydrolysis of fatty acids from the glycerol backbone (Huang, 1987). In most cases, lipase activity is absent in ungerminated seeds and thus its synthesis appears to be turned on rapidly early after imbibition. The fatty acids released by lipase activity are further metabolized in the glyoxysomes (Beevers, 1980). Glyoxysomes are a specialized form of microbody which contain enzymes for both fatty acid oxidation and the glyoxylate cycle.

The fatty acid oxidation pathway results in the breakdown of long chain fatty acids to acetyl-CoA. In animals, the acetyl-CoA would be further completely oxidized via the TCA cycle to release CO_2, H_2O and energy. In germinating seeds, however, the acetyl-CoA is instead used as a source of carbon for carbohydrate synthesis. This conversion of lipid to carbohydrate requires the two key enzymes, isocitrate lyase and malate synthase, which are characteristic of glyoxysomes in germinating seeds. Together with citrate synthetase, aconitase and malate dehydrogenase these five enzymes constitute the glyoxylate cycle shown in Fig. 23.9.

Succinate, which is produced by the isocitrate lyase reaction in the glyoxysome, is further metabolized in the mitochondria by TCA cycle enzymes to produce oxaloacetate. Oxaloacetate then leaves the mitochondria where in the cytosol PEP is produced and formation of sucrose occurs by way of gluconeogenesis. Thus the conversion of triacylglycerol in oil bodies to sucrose in the cytosol is a dramatic example of the interrelationships of several compartments in plant metabolism. It appears that the function of this multi-organelle and multi-pathway process is to convert the carbon stored as insoluble triacylglycerol into sucrose so that

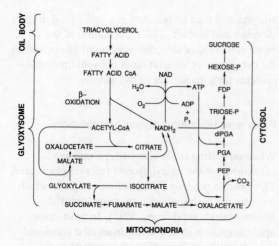

Fig. 23.9 Summary of reactions in the pathway of conversion of fatty acid to sucrose, showing cellular compartmentation of the four major parts of the sequence.

it can be easily exported to support new growth by other cells in the germinating seedling.

Further reading and references

Beevers, H. (1980). The role of the glyoxylate cycle. In *The Biochemistry of Plants*, Vol. 4, ed. P. K. Stumpf, Academic Press, New York, pp. 117–29.

Dennis, D. T. and Miernyk, J. A. (1982). Compartmentation of nonphotosynthetic carbohydrate metabolism. *Ann. Rev. Plant Physiol.* **33**, 27–50.

Frentzen, M. (1986). Biosynthesis and desaturation of the different diacylglycerol moieties in higher plants. *J. Plant Physiol.* **124**, 193–209.

Harwood, J. L. (1980). Plant acyl lipids: Structure, distribution, and analysis. In *The Biochemistry of Plants*, Vol. 4, ed. P. K. Stumpf, Academic Press, New York, pp. 2–56.

Huang, A. H. C. (1987). Lipases. In *The Biochemistry of Plants*, Vol. 9, eds P. K. Stumpf and E. E. Conn, Academic Press, New York, pp. 91–116.

Jaworski, J. G. (1987). Biosynthesis of monoenoic and polyenoic fatty acids. In *The Biochemistry of Plants*, Vol. 9, eds P. K. Stumpf and E. E. Conn, Academic Press, New York, pp. 159–73.

Joyard, J. and Douce, R. (1987). Galactolipid synthesis. In *The Biochemistry of Plants*, Vol. 9, eds P. K. Stumpf and E. E. Conn, Academic Press, New York, pp. 215–75.

Kolattukudy, P. E. (1984). Biochemistry and function of cutin and suberin. *Can. J. Bot.* **62**, 2918–33.

Mudd, J. B. (1980). Phospholipid biosynthesis. In *The Biochemistry of Plants*, Vol. 4, ed. P. K. Stumpf, Academic Press, New York, pp. 249–81.

Mudd, J. B. and Kleppinger-Sparace, K. (1987). Sulpholipids. In *The Biochemistry of Plants*, Vol. 9, eds P. K. Stumpf and E. E. Conn, Academic Press, New York, pp. 275–90.

Mudd, J. B., Andrews, J. E., Sanchez, J., Sparace, S. A. and Kleppinger-Sparace, K. F. (1985). Biosynthesis of chloroplast lipids. In *Frontiers of Membrane Research in Agriculture*, eds J. B. St John, E. Berlin and P. C. Jackson, Rowman and Allanheld, Ottawa, pp. 35–53.

Mudd, J. B., Andrews, J. E. and Sparace, S. A. (1987). Phosphatidylglycerol synthesis in chloroplast membranes. *Methods Enzymol.* **148**, 338–45.

Ohlrogge, J. B. (1987). Biochemistry of plant acyl carrier proteins. In *The Biochemistry of Plants*, Vol. 9, eds P. K. Stumpf and E. E. Conn, Academic Press, New York, pp. 137–49.

Roughan, P. G. and Slack, C. R. (1982). Cellular organization of glycerolipid metabolism. *Ann. Rev. Plant Physiol.* **33**, 97–132.

Scherer, D. E. and Knauf, V. C. (1987). Isolation of a cDNA clone for acyl carrier protein I of spinach. *Plant Mol. Biol.* **9**, 127–34.

Slack, C. R. and Browse, J. A. (1984). Synthesis of storage lipids in developing seeds. In *Seed Physiology*, Vol. 1, ed. D. R. Murray, Academic Press, Australia, pp. 209–43.

Stumpf, P. K. (1987). The biosynthesis of saturated fatty acids. In *The Biochemistry of Plants*, Vol. 9, eds P. K. Stumpf and E. E. Conn, Academic Press, New York, pp. 121–34.

Stymne, S. and Stobart, A. K. (1987). Triacylglycerol biosynthesis. In *The Biochemistry of Plants*, Vol. 9, eds P. K. Stumpf and E. E. Conn, Academic Press, New York, pp. 175–214.

Vick, B. A. and Zimmerman, D. C. (1987). Oxidative systems for modification of fatty acids: the lipoxygenase pathway. In *The Biochemistry of Plants*, Vol. 9, eds P. K. Stumpf and E. E. Conn, Academic Press, New York, pp. 54–85.

24 Terpene biosynthesis and metabolism
Charles West

Introduction

It was recognized early that the structures of a large, diverse array of natural products could be rationalized as covalently linked, branched C_5 units related to isoprene (methylbutadiene). The name terpene was derived from one group of these so-called isoprenoid substances that are derived from terpentine and shown to contain two C_5 units. In time, terpene became a generic name for all classes of substances composed of isopentenoid units irrespective of the number present per molecule. The major classes of terpenes are: monoterpenes (C_{10}), sesquiterpenes (C_{15}), diterpenes (C_{20}), sesterterpenes (C_{25}), triterpenes (C_{30}), tetraterpenes (C_{40}) and polyterpenes ($>C_{40}$). Higher plants produce members of all these classes of terpenes. In addition, natural products are known in which a terpenoid moiety is covalently linked with another moiety of different biogenetic origin. Examples of such mixed terpenoids from higher plants include chlorophyll, with its diterpenoid phytyl side chain esterified to the cyclic tetrapyrrole nucleus; and plastoquinone, with a polyprenyl side chain substituted by alkylation on the benzoquinone ring.

The existence of common structural features in all terpenoid substances implies that their biosynthetic origins also share common features. Ruzicka (1953) formulated the 'biogenetic isoprene rule' which predicted the biosynthetic relationships between the various classes of terpenes. Subsequent investigations of the biosynthetic pathways have established the essential correctness of these ideas and have led to the generalized scheme illustrated in Fig. 24.1. A common pathway leads from acetyl-CoA via mevalonate to the central intermediate isopentenyl pyrophosphate (IPP). IPP is isomerized to dimethylallyl pyrophosphate (DMAPP) and also participates in the series of chain elongation steps to generate the acyclic precursors of the various classes of terpenes, geranyl pyrophosphate (GPP) (C_{10}), farnesyl pyrophosphate (FPP) (C_{15}) and geranylgeranyl pyrophosphate (GGPP) (C_{20}). These acyclic precursors (prenyl pyrophosphates) are converted to the various members of the monoterpene, sesquiterpene and diterpene classes by an initial cyclization, often followed by further, mostly oxidative, transformations of the cyclic hydrocarbons. In some instances, rearrangements occur and carbons are lost or gained, resulting in structures in which the regular isoprenoid skeleton is no longer intact. Reductive coupling of two FPP molecules in a 'head-to-head' fashion generates the triterpene, squalene, that in turn can be modified by cyclization and further transformations to the family of triterpenes and related sterols. In analogous fashion, GGPP is the precursor of the tetraterpenes that include the carotenes and xanthophylls. The polyterpenes result from the further elongation of GGPP, or in some cases FPP or GPP more directly, by the addition of units from IPP. Based on considerations evident from Fig. 24.1, terpenes can be defined as those substances that are derived biosynthetically from IPP.

Higher plants produce a diverse array of isoprenoid substances. Some found in most or all plants have a role in fundamental processes of energy metabolism and macromolecular assemblages required for growth. Sterols, which

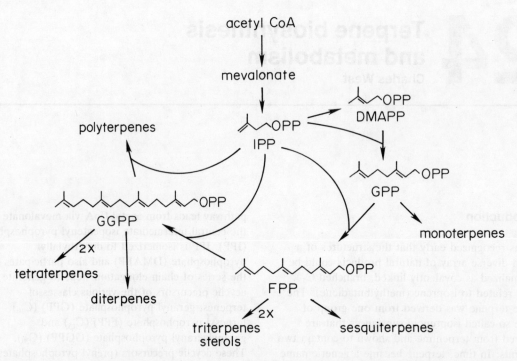

Fig. 24.1 Biosynthetic relationships among the various classes of terpenes. IPP = isopentenyl pyrophosphate; DMAPP = dimethylallyl pyrophosphate; GPP = geranyl pyrophosphate; FPP = farnesyl pyrophosphate; GGPP = geranylgeranyl pyrophosphate. PP = pyrophosphate or diphosphate; both names are used in the literature.

are required membrane components, and carotenes, which serve as accessory photosynthetic pigments, are examples of such primary terpenoid metabolites. Many other isoprenoid substances are of a more specialized nature in that they are found in only a limited range of plants. In cases where the function of such secondary metabolites is known or suspected, they serve a regulatory role in permitting the plant that produces them to adapt to changes in the environment or biological stresses. For example, the sesquiterpenoid and diterpenoid phytoalexins produced by some plants are thought to participate in resistance to infection by microbial pathogens. Isoprenoid growth regulators, such as the gibberellins and abscisic acid, fall somewhere in between these two groups in that they are produced by most or all higher plants and function as regulatory agents for normal growth and development as well as in adaptation to stresses. Many isoprenoid substances in plants, like other secondary metabolites, are of unknown function.

Table 24.1 lists the major classes of higher plant terpenes and some of their general characteristics.

The biosynthesis of isoprenoid substances is much better understood than their degradative metabolism, although it is clear that many isoprenoid substances do turn over in plant tissues. Also, we know that the metabolism of terpenoid substances in plants must be subject to regulation at the enzyme level, at the level of gene expression, and by features of compartmentation; however, our understanding of these features of terpene biosynthesis and catabolism are still very fragmentary. This chapter will attempt to summarize with selected examples the current status of understanding of these problems, and will try to show how terpene metabolism is integrated with other aspects of plant metabolism.

Table 24.1 Characteristics of major classes of higher plant terpenes

Class	Precursor	Sites of synthesis	Some suggested functions	General references
Cyclic monoterpenes	GPP	Epidermal oil glands Plastids[a]	Attractants Defensive agents C and energy source	Croteau (1987) Croteau (1981)
Cyclic sesquiterpenes	FPP	Epidermal oil glands	Defensive agents Phytoalexins Anti-feeding agents	Cane (1981)
Cyclic diterpenes	GGPP	Cytoplasm/ER[b] Plastids	Defensive agents Phytoalexins Plant hormones–gibberellins	West (1981)
Cyclic triterpenes (sterols)	2 × FPP	Cytoplasm/ER[b]	Membrane components Defensive agents Plant hormones–brassinosteroids	Goodwin (1981)
Tetraterpenes (carotenes, xanthophylls)	2 × GGPP	Plastids	Photosynthesis–photoreception and photoprotection Source of plant hormone–abscisic acid	Spurgeon and Porter (1983)
Prenylated pigments Ubiquinones	FPP or GGPP	Mitochondria	Mitochondrial electron transport	Pennock and Threlfall (1983)
Plastoquinones Tocopherols Phylloquinone Chlorophyll	GGPP	Plastids	Photosynthesis Photosynthetic electron transport	Pennock and Threlfall (1983)
Polyprenols Dolichols	FPP or GGPP	Mitochondria, Golgi	Biosynthesis of glycoconjugates	Hemming (1983)
Rubber	FPP or GGPP	–	–	Benedict (1983)

[a]Mettal *et al.* (1988).
[b]ER = endoplasmic reticulum.

Biosynthesis of isopentenyl pyrophosphate

Pathway

The pathway for the conversion of acetyl-CoA to IPP is illustrated in Fig. 24.2. This pathway and the participating enzymes have been studied in detail from yeast and mammalian tissues (Qureshi and Porter, 1981). The pathway is the same in plants, and the catalytic properties of the participating enzymes are similar, although they have not been so extensively investigated from higher plant sources. Two molecules of acetyl-CoA are condensed to form acetoacetyl-CoA in a reaction catalyzed by acetoacetyl-CoA thiolase (step 1). This thermodynamically unfavorable reaction is driven to completion through its coupling with the strongly favorable synthesis of 3-hydroxy-3-methylglutaryl-CoA from acetyl-CoA and acetoacetyl-CoA catalyzed by the condensing enzyme (step 2). Mevalonate is then formed from 3-hydroxy-3-methylglutaryl-CoA in the essentially irreversible reaction involving successive NADPH-dependent reductions at the same catalytic site of 3-hydroxy-3-methylglutaryl-CoA reductase (step 3). Two successive phosphoryl transfers from ATP catalyzed by separate kinases

356 The Formation and Breakdown of Lipids

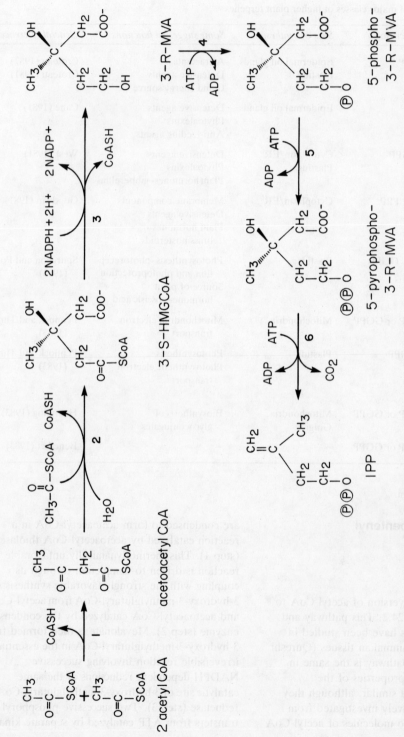

Fig. 24.2 Pathway for the conversion of acetyl-CoA to isopentenyl pyrophosphate. Catalysts are: step 1, acetoacetyl-CoA thiolase [acetyl-CoA acetyltransferase (EC 2.3.1.9)]; step 2, 3-hydroxy-3-methylglutaryl-CoA synthase (EC 4.1.2.5); step 3, 3-hydroxy-3-methylglutaryl-CoA reductase [mevalonate: NADP$^+$ oxidoreductase acetylating CoA (EC 1.1.1.34)]; step 4, mevalonate kinase [ATP: mevalonate 5-phosphotransferase (EC 2.7.1.36)]; step 5, phosphomevalonate kinase [ATP: 5-phosphomevalonate phosphotransferase (EC 2.7.4.2)]; step 6, pyrophosphomevalonate decarboxylase [ATP: 5-pyrophosphomevalonate decarboxylase (dehydrating) (EC 4.1.1.33)]; P = phosphoryl group.

(steps 4 and 5) generates
5-pyrophosphomevalonate. And finally, IPP is generated from 5-pyrophosphomevalonate by a decarboxylation coupled with dehydration and the obligate hydrolysis of ATP (step 6). The overall stoichiometry of this sequence emphasizes the substrates that must be supplied for IPP synthesis:

3acetyl-CoA + 2NADPH + 2H$^+$ + 3ATP + H$_2$O →
IPP + CO$_2$ + 3CoASH + 2NADP$^+$ + 3ADP + P$_i$

Compartmentation of the pathway in plants

Early experiments of Goodwin and his associates suggested the possibility that the biosynthesis of IPP and isoprenoid products formed from it were localized in at least two separate compartments in plant cells (Goodwin and Mercer, 1963). Dark-grown maize seedlings excised from their roots were fed DL-[2-^{14}C]-mevalonate and illuminated for 24 h. Radioactivity was readily detected in squalene, phytosterols, β-amyrin (a pentacyclic triterpene synthesized in plants) and ubiquinone, but not in β-carotene, the phytyl side-chain of chlorophyll or plastoquinone, all of which are plastid constituents being rapidly synthesized under these conditions. Conversely, when the precursor supplied in these experiments was ^{14}CO$_2$ instead of [2-^{14}C]-mevalonate, the plastid pigments were readily labeled while squalene, sterol precursors and β-amyrin were not. These and other results were interpreted to suggest the existence of two separate isoprenoid biosynthetic pathways, one in the chloroplast and the other in the cytoplasm/endoplasmic reticulum where sterol synthesis was thought to occur. They speculated that exogenously supplied mevalonate failed to penetrate the chloroplast membranes.

Several lines of evidence developed during the ensuing years to support the existence of an autonomous pathway for IPP synthesis and utilization in plastid compartments separate from the cytoplasm/endoplasmic reticulum pathway. (See Leidvogel (1986) for a summary and references.) Firstly, isolated plastids were reported to contain mevalonate kinase and 3-hydroxy-3-methylglutaryl-CoA reductase activities. The existence of a plastid mevalonate kinase was subsequently challenged, but several laboratories have reported 3-hydroxy-3-methylglutaryl-CoA reductase activity in plastids. A recent report (Reddy and Das, 1987) identifies the presence of five of the six enzymes (Fig. 24.2) required for IPP synthesis in lysates of freshly harvested *Parthenium argentatum* and *Phaseolus vulgaris* leaves. The presence of 3-hydroxy-3-methylglutaryl-CoA synthase was not tested for. These authors argue for the autonomy of chloroplasts in the biosynthesis of IPP. Secondly, isolated chloroplasts incorporate early pathway intermediates into isoprenoid endproducts of the plastid. For example, Grumbach and Forn (1980) demonstrated significant incorporations of radioactivity from ^{14}CO$_2$, [^{14}C]-3-phosphoglycerate, [^{14}C]-acetate and [^{14}C]-mevalonate into β-carotene and plastoquinone-9 in isolated and seemingly intact chloroplasts from spinach leaves. Thirdly, in agreement with the experiments from Goodwin's laboratory, protoplasts and suspension cells were able to incorporate [^{14}C]-acetate and [2-^{14}C]-mevalonate into sterols, but not into the isoprenoid lipids of plastids. These results imply that IPP generated in the cytoplasmic compartment does not participate in plastid lipid synthesis in these cells. However, these *in vivo* results differ from those of Grumbach and Forn (1980) who found that mevalonate served as a substrate for prenyl lipid biosynthesis in isolated chloroplasts. Finally, mevinolin, a potent inhibitor of 3-hydroxy-3-methylglutaryl-CoA reductase, showed differential effects on the accumulation of prenyl lipids in intact radish seedlings. Whereas accumulation of the sterol fraction was inhibited in a mevinolin concentration-dependent manner, plastidic prenyl lipid accumulation was unaffected by even the highest concentrations of mevinolin tested. It was proposed that a separate plastid pool of 3-hydroxy-3-methylglutaryl-CoA reductase that is not readily accessible to mevinolin must exist.

In spite of this body of evidence that favors an autonomous pathway in plastids for IPP synthesis, Kleinig and his associates have reached a different conclusion from their investigations that IPP produced in the cytoplasmic/endoplasmic reticulum compartment is utilized for the synthesis of prenyl lipids in plastids and ubiquinone in mitochondria, as well as for sterol and triterpene biosynthesis in the cytoplasm/endoplasmic reticulum (Kreuz and Kleinig, 1981; Luetke-Brinkhaus et al., 1984). In their work, seemingly intact chloroplasts from spinach or chromoplasts from daffodil readily utilize exogenously supplied IPP for plastid prenyl lipid synthesis, whereas these same organelle preparations cannot use mevalonate, 5-phosphomevalonate or 5-pyrophosphomevalonate unless a cytoplasmic source of IPP-synthesis enzymes and ATP are also provided. In more recent work, similar results were obtained with isolated intact or disrupted etioplasts from mustard seedlings grown under continuous far-red light (Luetke-Brinkhaus and Kleinig, 1987). They further demonstrated that none of the enzymes, mevalonate kinase, 5-phosphomevalonate kinase and 5-pyrophosphomevalonate decarboxylase, were detectable in isolated etioplasts, whereas these enzymes were present in the cytoplasm of this tissue. This group has also reported that [^{14}C]-acetate is not incorporated into plastid prenyl lipids, whereas it is readily incorporated into fatty acids in isolated plastid preparations. These latter results indicate that supplied acetate gets into the plastid and is activated, but cannot be incorporated into prenyl lipids because a system for IPP biosynthesis is not available there.

It is not possible at present to reconcile these seemingly contradictory sets of observations. One explanation could be that the requisite enzymes for IPP synthesis are present in the plastids from some plant tissues, but are absent or inactive in plastids of other tissues. The absence of active IPP synthesis in the latter cases might represent the physiological state of the tissue, or might result from artifactual inactivation of IPP synthesis during cell disruption or organelle isolation. Further work will be necessary before this situation can be clarified.

Sources of precursors for isopentenyl pyrophosphate biosynthesis.

The summary equation for IPP biosynthesis indicates that three substrates – acetyl-CoA, NADPH and ATP – must be available in any cellular compartment where this process is occurring. Mechanisms for maintaining NADPH and ATP levels in the cytoplasm, chloroplasts, and mitochondria, are well known. However, the source of acetyl-CoA for use in the cytoplasm/endoplasmic reticulum and in plastids, the two compartments considered the most likely sites for IPP synthesis, is less well established. Reviews by Givan (1983) and Leidvogel (1986) discuss this problem and summarize the evidence from the original literature.

The source of acetyl-CoA for use as a substrate for IPP synthesis in the cytoplasm is unclear. Pyruvate dehydrogenase complexes that generate acetyl-CoA appear to be restricted to organelles – mitochondria and in some cases chloroplasts. Since organellar membranes are thought not to be permeable to acetyl-CoA, an indirect transport mechanism would presumably be necessary to provide acetyl-CoA in the cytoplasm from that formed by pyruvate dehydrogenase action within the organelles. There is some evidence for a cytosolic ATP-dependent citrate lyase (Kaethner and ap Rees, 1985). The presence of this enzyme in the cytoplasm would present the possibility for citrate generated from acetyl-CoA in mitochondria and transported to the cytoplasm to serve as a source of cytoplasmic acetyl-CoA. An alternative to this possibility is to activate acetate to form acetyl-CoA in the cytoplasm. The incorporation of [^{14}C]-acetate into sterols has been reported; however, the pathway involved is not known. Acetyl-CoA synthase, which catalyzes the synthesis of acetyl-CoA from acetate, is present in spinach leaves, but is restricted to the stromal compartment of the chloroplast (Kuhn et al., 1981). A mechanism for transporting acetyl-CoA from the stroma to the cytoplasm is not known at present. Thus, the source of acetyl-CoA required in the cytoplasmic compartment of plant cells remains unclear.

Two sources of acetyl-CoA for the plastid compartment have been considered most likely. The first route was proposed by Stumpf (1984). A number of investigators had shown that [^{14}C]-acetate is efficiently incorporated into fatty acids in the chloroplast. Stumpf and his associates demonstrated acetyl-CoA hydrolase activity in mitochondria and an acetyl-CoA synthase activity

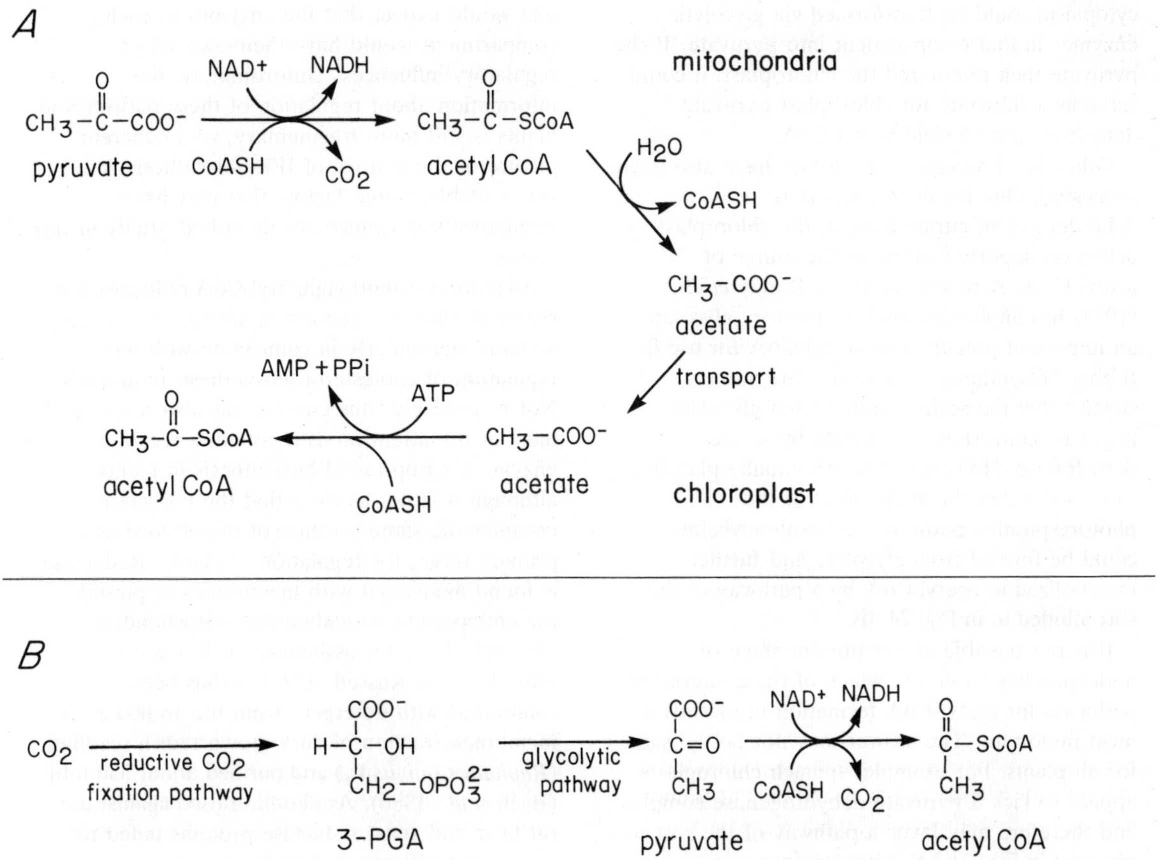

Fig. 24.3 Proposed pathways for acetyl-CoA formation in chloroplasts. (A) Pathway via acetate. (B) Pathway utilizing pyruvate dehydrogenase.

in the stromal compartment of the chloroplast. The pathway they suggested is illustrated in Fig. 24.3A. Acetyl-CoA is produced from pyruvate in mitochondria, and then hydrolyzed to free acetate. The acetate is then transported into the stromal compartment of the plastid where it is reactivated to provide acetyl-CoA.

The second proposal (Fig. 24.3B) relates acetyl-CoA formation to photosynthetic CO_2 fixation. As pictured, 3-phosphoglycerate formed by CO_2 fixation in the Calvin–Benson–Bassham cycle is directed through a series of glycolytic steps to pyruvate, which then undergoes oxidative decarboxylation through the action of the stromal pyruvate dehydrogenase to yield acetyl-CoA. Evidence has been presented for the presence of all of the required glycolytic enzymes in the chloroplast except for phosphoglyceromutase, which has not been detected to date. Also, it should be noted that the pyruvate dehydrogenase complex has been found in chloroplasts from a number of plants including pea, but efforts to detect this enzyme in spinach leaf chloroplasts have failed. Obviously, the absence of either of these enzymes in chloroplasts would preclude the operation of the intraplastidic pathway shown in Fig. 24.3B. The very low rates of incorporation of $^{14}CO_2$ into fatty acids in isolated chloroplasts (Givan, 1983) might suggest that this pathway operates to only a limited extent as shown, if at all. An alternative to the scheme shown in Fig. 24.3B might be more likely in view of the questions that have been raised. Triose phosphate that is transported from the chloroplast to the cytoplasm could be transformed via glycolytic enzymes in that compartment into pyruvate. If the pyruvate then re-entered the chloroplast, it could serve as a substrate for chloroplast pyruvate dehydrogenase to yield acetyl-CoA.

Other less investigated pathways have also been suggested. One involves a reported ATP-dependent citrate lyase in the chloroplast acting on imported citrate as the source of acetyl-CoA. A recent report by Reddy et al., (1987) has implicated photorespiratory glycolate as an important precursor of acetyl-CoA for use in rubber biosynthesis in guayule. The authors suggest that the serine formed from glycolate might be converted to pyruvate by serine dehydratase. However, it seems equally plausible to consider that the endproduct of the photorespiration pathway, 3-phosphoglycerate, could be formed from glycolate and further metabolized to acetyl-CoA by a pathway of the sort alluded to in Fig. 24.3B.

It is not possible at our present stage of understanding to decide which of these suggested pathways for acetyl-CoA formation in plastids is most important. The answer may not be the same for all plants. For example, spinach chloroplasts appear to lack a pyruvate dehydrogenase complex and therefore may favor a pathway of the sort pictured in Fig. 24.3A, whereas since pea chloroplasts do possess pyruvate dehydrogenase activity the pathway shown in Fig. 24.3B may assume more importance. It is also conceivable that a given plant has the potential to produce acetyl-CoA in the chloroplast by more than one pathway, with the relative importance of the pathways a function of the conditions.

Regulation of isopentenyl pyrophosphate biosynthesis

Since the pathway pictured in Fig. 24.2 is the starting point for the biosynthesis of all isoprenoid endproducts, one might expect to find regulation of both the activity and levels of regulatory enzymes of the pathway in response to the plants' momentary needs for various endproducts. Furthermore, if the pathway operates in different subcellular compartments, as has been suggested, one would expect that the enzymes in each compartment would have their own set of regulatory influences. Unfortunately, the information about regulation of these pathways in plants is still quite fragmentary, so a coherent picture of the control of IPP biosynthesis is not yet available. Some factors that may have regulatory significance are described briefly in this section.

3-Hydroxy-3-methylglutaryl-CoA reductase has received a lot of attention in animal and fungal systems, particularly in connection with the regulation of cholesterol biosynthesis in animals. Not surprisingly, this enzyme has also received most of the attention as a potential regulatory enzyme for isoprenoid biosynthesis in plants, although it is not as clear that the reductase occupies the same position of importance as a primary target for regulation in plants. Reductase is found associated with membranes of plastids, the endoplasmic reticulum and mitochondria, although the latter assignment is less certain (Brooker and Russell, 1975). It has been solubilized with detergent from the 16 000 g membrane fraction of dark-grown radish seedlings (*Raphanus sativus* L.) and purified about 350-fold (Bach et al., 1986). Antibodies raised against the rat liver and yeast reductase proteins failed to cross-react with the radish enzyme.

Russell (1985) reported no feedback inhibition by isoprenoid products, including gibberellins and abscisic acid, with reductase preparations from either microsomal or chloroplast membranes of pea seedlings. Reductase preparations from both pea seedlings (Russell, 1985) and the latex of *Hevea brasiliensis* (Sipat, 1985) were inactivated by treatment with ATP and reactivated by phosphatase in a manner suggesting phosphorylation–dephosphorylation reminiscent of that seen with rat liver reductase. However, this phenomenon has not been examined in much detail, nor has its physiological significance been evaluated. Preliminary results also suggest that the pea seedling and radish reductases are under phytochrome control *in vivo* in a manner that is modulated by plant growth regulators. Together, these results indicate a regulatory role for the reductase *in vivo*, but a coherent picture of the regulation has not yet emerged.

Two observations point to 5-pyrophosphomevalonate decarboxylase (Fig. 24.2, step 6) as a possible site of regulation between mevalonate and IPP. In one study, the activity of this enzyme was found to be rate limiting and seen to increase most markedly after induction of furanosesquiterpenoid phytoalexin synthesis in sweet potato root tissue by treatment with a fungus or toxic chemical (Oba et al., 1976). In another study, adenylate energy charge was found to regulate the activity of this enzyme in a cell-free enzyme preparation from immature *Marah macrocarpus* seeds that was catalyzing the conversion of mevalonate to the diterpene hydrocarbon *ent*-kaurene (Knotz et al., 1977). The response was typical of that seen for regulatory enzymes in a biosynthetic sequence. No other step was affected by adenylate energy charge. These indications that 5-pyrophosphomevalonate decarboxylase is rate limiting and may be regulated are surprising because of its position in the middle of the pathway removed from a branch point.

Utilization of isopentenyl pyrophosphate for the synthesis of prenyl pyrophosphates

Isomerase and prenyl transferases

The scheme portrayed in Fig. 24.1 emphasizes the central importance of IPP as the precursor of all classes of terpenes. Figure 24.4 illustrates the reactions responsible for the conversion of IPP to the family of four prenyl pyrophosphates that serve in turn as precursors of all of the major classes of terpenes.

Dimethylallyl pyrophosphate is formed from IPP by an isomerization reaction (Fig. 24.4, reaction 1). This involves the addition of a proton to C-4 of IPP coupled with the stereospecific elimination of proton from C-2 and the resulting shift of the double bond from the 3–4 to the 2–3 position. Isomerase activity is readily detected in plant tissues, but the enzyme has not been very thoroughly characterized from these sources.

Prenyl transferase activities (Fig. 24.4, reactions 2, 3, 4) are responsible for transferring the prenyl unit of a prenyl pyrophosphate donor to pyrophosphate to generate a new prenyl pyrophosphate donor with one additional

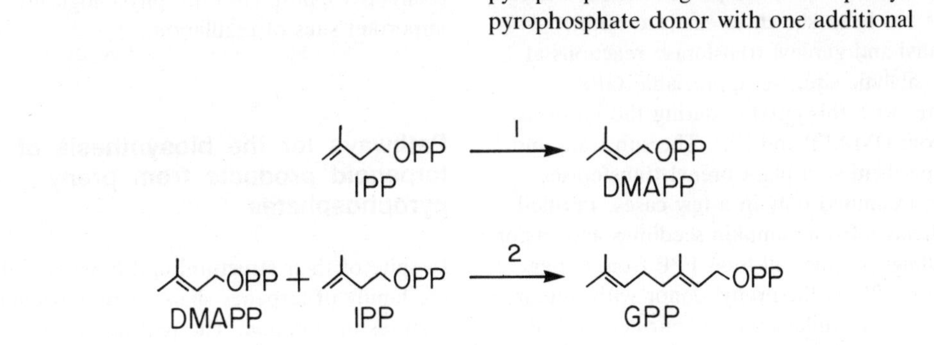

Fig. 24.4 Reactions utilizing IPP in the formation of prenyl pyrophosphates. Catalysts: Reaction 1, isopentenyl pyrophosphate:dimethylallyl pyrophosphate isomerase; Reaction 2, dimethylallyl transferase; Reaction 3, geranyl transferase; Reaction 4, farnesyl transferase.

C_5-prenyl unit in the chain. The reaction involves the electrophilic addition of a carbonium ion generated at C-1 of the prenyl donor by elimination of pyrophosphate to the electron-rich C-4 position of the acceptor IPP molecule. A proton elimination from the adduct establishes the 2–3 double bond of the new prenyl pyrophosphate product to complete the process.

A detailed mechanism that serves as a model for all prenyl transferase reactions has been developed by Poulter and Rilling (1981) on the basis of extensive supporting evidence from studies of farnesyl pyrophosphate synthetase.

GPP synthetase accepts only DMAPP as a prenyl donor at physiological concentrations and thus possesses only a single prenyl transferase activity. FPP synthetase and GGPP synthetase, on the other hand, may possess more than one prenyl transferase activity. For example, liver FPP synthetase has been shown to catalyze both dimethylallyl and geranyl transferase reactions at the same catalytic site. No appreciable GPP accumulates with this enzyme during the synthesis of FPP from DMAPP and IPP. The substrate and product specificities of plant prenyl transferases have been examined only in a few cases. Purified FPP synthetases from pumpkin seedlings and castor bean seedlings produce all-*trans*-FPP from either DMAPP or GPP as the prenyl donor with similar efficiencies. In a similar manner, purified GGPP synthetases from carrot root, pumpkin seedlings and tomato fruit plastids produce all-*trans*-GGPP from DMAPP, GPP or FPP with similar efficiencies. However, the purified farnesyl transferase from fungally infected castor bean seedlings used only FPP as a prenyl donor at physiological substrate concentrations (Dudley *et al.*, 1986). It was therefore proposed that this enzyme must function in conjunction with isomerase and FPP synthetase to produce GGPP from IPP *in vivo*.

Compartmentation and regulation

The FPP synthetases in plant cells are believed to be localized in the cytoplasm/endoplasmic reticulum region where sterol biosynthesis occurs. GGPP synthetase, on the other hand, appears to reside predominantly in plastids. This is consistent with the observed production of the major diterpene-containing pigment, chlorophyll, as well as carotenes and xanthophylls in plastid compartments. It is possible that smaller amounts of these prenyl transferases are present in other compartments as well. Localization of prenyl transferases in subcellular compartments is doubtless an important feature in the regulation of terpene biosynthesis in higher plants.

There is little at present to suggest that the enzymes participating in the conversion of IPP to prenyl pyrophosphates are physiologically important sites of regulation.

Pathways for the biosynthesis of terpenoid products from prenyl pyrophosphates

In spite of their structural and functional diversity, the family of terpenes share many biogenetic features in common. The pathway from acetyl-CoA to IPP is common to all classes of terpenes. The next stage involves a combination of isomerase and prenyl transferase to generate four prenyl pyrophosphate substrates – DMAPP (C_5), GPP (C_{10}), FPP (C_{15}) and DMAPP (C_{20}). It now seems likely that each of the diverse array of terpenoid substances that are found in nature is derived from one of these four prenyl pyrophosphates plus, in the case of polyprenyl compounds, additional IPP.

The primary reactions for the utilization of

prenyl pyrophosphates for the synthesis of this array of terpenoid products are of four general types.

Cyclization reactions

In the case of the monoterpenes, sesquiterpenes and diterpenes, most of which are carbocyclic compounds, the primary reaction is a cyclization using the appropriate prenyl pyrophosphate as a substrate.

Coupling reactions

The primary step in the biosynthesis of triterpenes and tetraterpenes involves the 'head-to-head' covalent coupling of two molecules of the appropriate prenyl pyrophosphate. The reductive coupling of two FPP molecules yields squalene, the acyclic C_{30} precursor of the triterpenes and sterols. The non-redox coupling of two GGPP molecules yields phytoene, the acyclic C_{40} precursor of the family of carotenes and xanthophylls.

Prenyl transfer reactions

The synthesis of long chain polyprenyl compounds involves a chain elongating prenyl transferase that catalyzes the processive addition of IPP units to a prenyl pyrophosphate prenyl-donating primer that can be either GPP or FPP or GGPP depending on the case.

Prenylation (alkylation) reactions

This type of reaction is involved in the synthesis of mixed terpenoids in which a prenyl chain is linked to a moiety of different biosynthetic origin, often an aromatic or heterocyclic ring compound. In these cases, a prenyl pyrophosphate acts as a prenyl donor to alkylate (prenylate) the acceptor. These reactions resemble a prenyl transferase reaction: a prenyl carbocation is first formed from the ionization of the prenyl pyrophosphate, followed by electrophilic attack of the prenyl carbocation on an electron-rich aromatic ring or heteroatom. Ubiquinone, plastoquinone, phylloquinone, tocopherol and cytokinins are examples of mixed terpenoid compounds which require a prenylation reaction in their biosynthesis.

It is noteworthy that none of the reactions involved in the biosynthesis of terpenes from acetyl-CoA through the primary steps of prenyl pyrophosphate utilization described above are oxidative in nature or require molecular oxygen as a substrate. On the other hand, the endproducts of terpene biosynthetic pathways frequently contain oxygenated functional groups and other structural features not represented in the primary products of prenyl pyrophosphate utilization. It is clear that a secondary phase of most terpene biosynthetic pathways is required to introduce these structural features. The reactions involved are frequently oxidative in nature, including those catalyzed by mono- and dioxygenases requiring molecular oxygen for the more hydrophobic substrates, and others catalyzed by dehydrogenases. Other secondary reactions include cyclizations and rearrangements of the carbon skeletons, desaturation, isomerization and alkylation reactions.

A great deal is now known about the pathways leading to terpenoid products. In some cases, the nature of the enzyme catalysts involved has been investigated, although much remains to be done in this area. A detailed consideration of this complex subject is not possible here. The following sections will elaborate somewhat on the very general summary just presented with a few selected examples. It will be necessary to consult the references and monographs cited for a more detailed consideration of the biosynthetic pathways of the major classes of terpenes.

Cyclic diterpene biosynthesis

The outline of the biosynthetic pathway leading to a plant growth regulating gibberellin is shown in Fig. 24.5 to illustrate the general features of biosynthesis of a polycyclic diterpenoid compound.

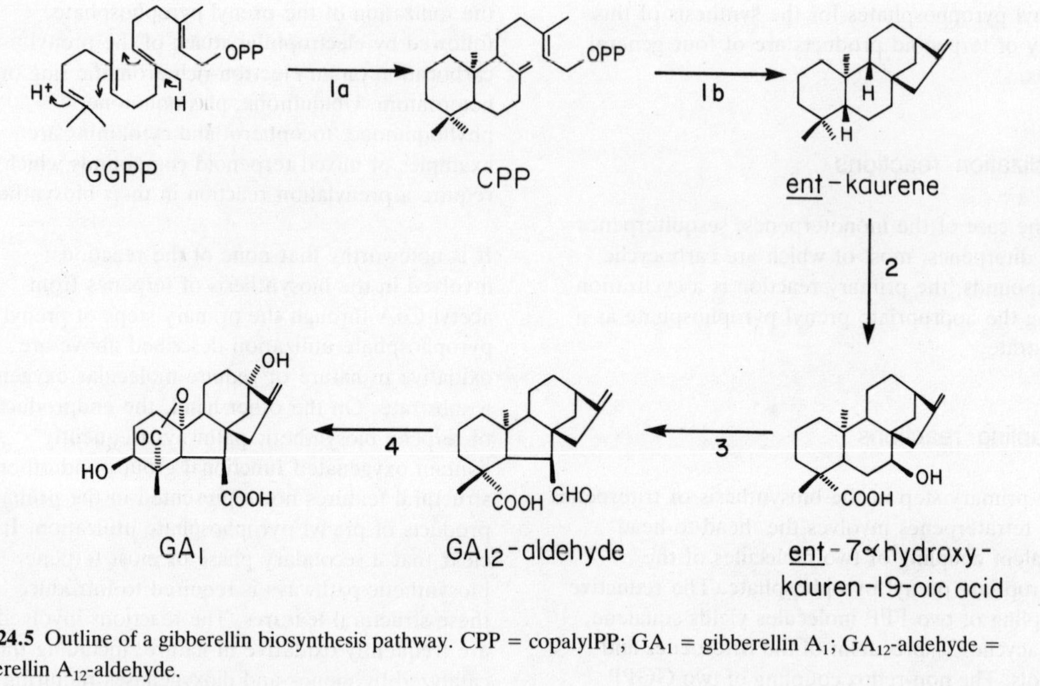

Fig. 24.5 Outline of a gibberellin biosynthesis pathway. CPP = copalylPP; GA_1 = gibberellin A_1; GA_{12}-aldehyde = gibberellin A_{12}-aldehyde.

A more detailed consideration of the characteristics of biosynthesis of the gibberellins will be found in the chapter by Jones and MacMillan (1984). All-*trans*-GGPP is the acyclic precursor of the entire family of diterpenes including the gibberellins. GGPP is cyclized to *ent*-kaurene (steps 1a and 1b) by the successive action of two catalysts which are known collectively as kaurene synthetase. Step 1a involves the proton-initiated cyclization of GGPP to copalyl pyrophosphate (CPP), as shown by the curved arrows that indicate the direction of electron migrations to establish new bonds (Fig. 24.5). CPP is further cyclized in step 1b to form *ent*-kaurene; this reaction involves the elimination of pyrophosphate, coupled with a further cyclization and a rearrangement of the carbon skeleton to generate the C and D rings of *ent*-kaurene. Kaurene synthetase from the endosperm of immature *Marah macrocarpus* fruit is composed of two separable enzymes which must function as a complex to catalyze the efficient accumulation of *ent*-kaurene from GGPP without significant accumulation of free CPP (Duncan and West,

1981). This enzyme is specific for the production of *ent*-kaurene among the diterpene hydrocarbons. However, CPP can serve with enzymes from other plant sources as a precursor of additional polycyclic diterpenes with different carbon skeletons.

The further transformation of *ent*-kaurene to *ent*-7α-hydroxykauren-19-oic acid (Fig. 24.5, step 2) is catalyzed by a series of specific cytochrome P_{450}-dependent, membrane-bound mixed function monooxygenases requiring NADPH and O_2 as cosubstrates. The interesting contraction of the B ring (step 3) producing gibberellin A_{12}-aldehyde, the first intermediate with a gibberellane skeleton, is a reaction of the same type. However, investigations of the oxygenation enzymes catalyzing the interconversions of the gibberellins as part of the complex group of reactions represented by step 4 indicate that these are soluble dioxygenases requiring Fe^{2+} and 2-oxoglutarate as cosubstrates with O_2.

This sequence is generally characteristic of the biosynthesis of cyclic monoterpenes, sesquiterpenes and diterpenes where an initial intramolecular cyclization of an acyclic prenyl pyrophosphate is followed by secondary transformations with an emphasis on oxygenation steps.

Biosynthesis of carotenes

The initial step utilizing prenyl pryrophosphates in the synthesis of triterpenes and tetraterpenes is the coupling of two molecules of either FPP or GGPP, respectively, to form a symmetrical coupled precursor. The reaction involved in tetraterpene synthesis is illustrated as the first step in the outline of a tetraterpene biosynthetic pathway in a higher plant (Fig. 24.6). The coupling reaction involves the initial formation of a discrete cyclopropyl carbinyl pyrophosphate intermediate, called prephytoene pyrophosphate, which is further rearranged to form phytoene. The initial product of rearrangement is *cis*-phytoene in which

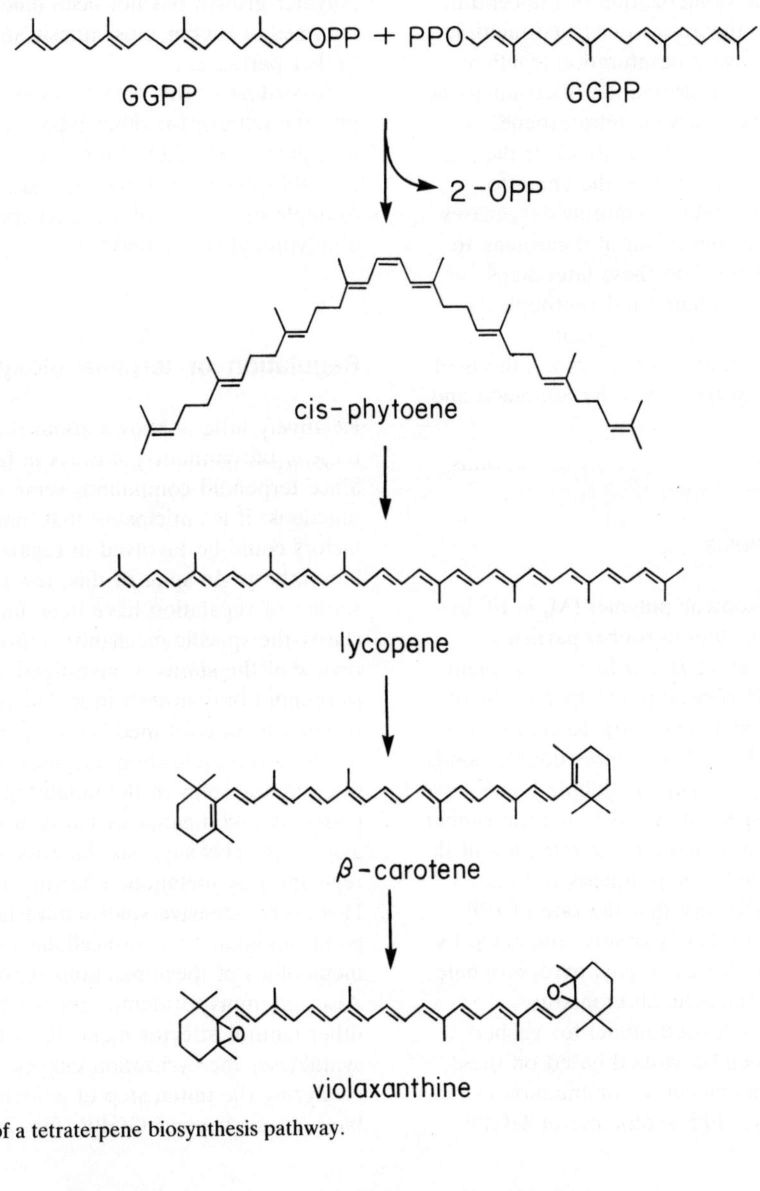

Fig. 24.6 Outline of a tetraterpene biosynthesis pathway.

the newly synthesized central double bond has the *cis* configuration. The coupling reaction between two FPP molecules that is a part of triterpene synthesis occurs in an analogous manner, except that the intermediate presqualene pyrophosphate undergoes reductive rearrangement in the presence of NADPH to form squalene in which the central double bond is lacking.

Further transformations of *cis*-phytoene lead to the tetraterpenes that accumulate in higher plants, including β-carotene and violaxanthin. A series of steps, including the isomerization of the central *cis* double bond to the *trans* isomer and multiple dehydrogenase-catalyzed desaturation reactions, are involved in the transformation of *cis*-phytoene to the fully conjugated acyclic tetraterpene, lycopene. Cyclization enzymes introduce the carbocyclic rings at the ends of the chain to produce β-carotene, and O_2-requiring oxygenases catalyze the further conversion of β-carotene to violaxanthin. Variations on these later steps produce the other carotenes and xanthophylls present in the plastids of higher plants.

A more detailed account of the biosynthesis of carotenes is given in the review by Spurgeon and Porter (1983).

Rubber biosynthesis

Rubber is *cis*-polyisoprene polymer ($M_r = 10^5$ to 4×10^6) that accumulates in rubber particles dispersed in the latex of *Hevea brasiliensis* plants or in stem and root cortical parenchyma cells of guayule (*Parthenium argentatum*). Recently, the presence of a small number of *trans* double bonds along with the predominant *cis* double bonds have been detected by spectral means in natural rubber. This observation prompted a re-examination of the requirements for rubber biosynthesis in isolated systems and the discovery that the rate of IPP incorporation into rubber is greatly stimulated by the inclusion of an all-*trans* prenyl pyrophosphate, FPP or GGPP, in the incubation mixture (Benedict, 1986). A revised model for rubber biosynthesis has been developed based on these observations. In this model, a combination of isomerase and either FPP synthetase or GGPP synthetase generates a pool of 'initiator' all-*trans* prenyl pyrophosphate from IPP. Then, another prenyl transferase catalyzes a reaction in which the 'initiator' prenyl pyrophosphate donates its prenyl moiety to IPP to generate a product with a new prenyl unit added in the *cis* configuration. This prenyl transferase continues to add more IPP molecules in processive fashion until a long *cis*-polymer is generated, with the '*trans*' double bonds of the initiator molecule incorporated at the distal end. The mechanism for termination of polymer growth has not been elucidated. These enzymes of rubber biosynthesis are associated with rubber particles.

According to this view, rubber biosynthesis fits into the pattern for other types of terpenes in that it requires one of the four standard prenyl pyrophosphates to initiate the synthesis. This is an example of the use of a prenyl transferase to form a polyprenyl compound.

Regulation of terpene biosynthesis

Relatively little is known about the regulation of terpene biosynthetic pathways in higher plants. Since terpenoid compounds serve a wide variety of functions, it is anticipated that many different factors could be involved in regulating terpene biosynthesis. In spite of this, few systematic studies of regulation have been undertaken to clarify the specific mechanisms involved. A critical review of the status of investigations of control of isoprenoid biosynthesis in higher plants has recently been published (Gray, 1987).

The initial cyclization enzymes appear to be rate-limiting steps in the monoterpene biosynthetic pathways investigated by Croteau and his associates. This suggests the possibility of regulation by metabolic effectors at these steps. However, extensive studies have failed to identify good candidates for intracellular modifier metabolites of these reactions (Croteau, 1987). Also, attempts to identify feedback inhibitors or other natural effector metabolites for kaurene synthetase, the cyclization enzyme complex catalyzing the initial step of gibberellin biosynthesis beyond GGPP (Fig. 24.5), were similarly negative (Frost and West, 1977). Although these are negative results and limited to two systems, they suggest that regulation of performed enzyme by intracellular effectors may not be of such general importance in these types of terpene biosynthetic pathways.

On the other hand, transcriptional activation of genes may be a more important regulatory feature for some terpene biosynthetic pathways. Casbene, a macrocyclic diterpene synthesized from GGPP through the action of a single enzyme, casbene synthetase, is produced in appreciable quantities in castor bean seedling extracts only after the seedlings have been in contact with microbial pathogens or elicitor substances derived from them. Casbene has antimicrobial properties and thus has been considered to serve the castor bean plant as a possible phytoalexin. Transient increases in hybridizable and translatable casbene synthetase mRNA precede the accumulation of active enzyme after treatment of the seedlings with an elicitor (Moesta and West, 1986). These findings, coupled with other unpublished observations, strongly support the idea that the appearance of active casbene synthetase, and the potential for casbene biosynthesis, is regulated by elicitor at the level of transcription of the casbene synthetase gene. It seems likely that activation of gene transcription and translation will prove to be important means of regulation of pathways for the production of secondary terpenoid metabolites like casbene that are produced in response to environmental or biological stresses.

The activity of some terpene biosynthetic pathways is regulated by environmental factors. For example, the synthesis of prenyl lipids in chloroplasts and other plastids – chlorophyll, carotenoids, plastoquinone, α-tocopherol and phylloquinone (vitamin K_1) – are regulated by phytochrome and light treatment (Mohr, 1981). Also, it has been observed that periodic low temperature treatment at night in field-grown guayule plants strikingly stimulates the formation of rubber while inhibiting the accumulation of oleoresins (Goss et al., 1984). In neither of these cases has the regulatory mechanism been elucidated at the molecular level. Clearly much remains to be done to develop understanding of the regulation of terpene metabolism.

Catabolism of terpenes

It is evident that at least some endproducts of terpene biosynthesis do undergo further metabolic transformations in the plant that produces them. Carotenoids have been shown to undergo chemical changes during leaf senescence, some of which appear to be enzyme-catalyzed (Spurgeon and Porter, 1983). Physiologically active gibberellins are converted to physiologically inactive forms by hydroxylation in the 2β-position, and the 2β-hydroxy derivatives are then subject to further oxidative degradations to unidentified catabolites (Jones and MacMillan, 1984). In many cases, the further metabolic fates of terpenes accumulated in the plant have not been examined. One notable exception involves the cyclic monoterpenes that accumulate in oil glands. Croteau (1987) has shown that these monoterpenes are in a state of metabolic flux, and has undertaken a detailed examination of the catabolic pathways involved in *Salvia* and *Mentha* species. This long-term turnover results in a net decrease in the monoterpene content at a late stage in development during which the oil gland is undergoing ultrastructural changes characteristic of senescence. (−)-Menthone, the major terpene component in the leaf, is converted by reduction to (−)-menthol and (+)-neomenthol. These two products are conjugated to form (−)-menthyl acetate and (+)-neomenthyl-β-D-glucoside in the leaf. It was further demonstrated that the glucoside is transported in large part to the rhizome where it undergoes catabolism. This catabolic pathway in the rhizome has been elucidated. The glycoside is first hydrolyzed by a glucosidase, the resulting (+)-neomenthol is oxidized to (−)-menthone, and the latter is oxidized to menthone lactone which is subjected to a β-oxidation scheme yielding acetyl-CoA as the endproduct. This acetyl-CoA is available for reincorporation into fatty acids and lipids, or for oxidative catabolism to yield energy and other biosynthetic intermediates. Thus, this might be viewed as a salvage pathway during leaf senescence, in which a portion of the carbon that had been stored in abundant monoterpenes of the leaf can be recovered in a form for recycling by a non-senescing tissue. An analogous pathway seems to operate for the recovery of camphor carbon from oil glands in sage.

References

Bach, T. J., Rogers, D. H. and Rudney, H. (1986). Detergent-solubilization, purification, and characterization of membrane-bound-3-methylglutaryl-coenzyme A reductase from radish seedlings. *Eur. J. Biochem.* **154**, 103–11.

Benedict, C. R. (1983). Biosynthesis of rubber. In *Biosynthesis of Isoprenoid Compounds,* Vol. 2, eds J. W. Porter and S. L. Spurgeon, John Wiley and Sons, New York, pp. 355–69.

Benedict, C. R. (ed) (1986). *Biochemistry and Regulation of cis-Polyisoprene in Plants,* NSF Sponsored Workshop, Texas A & M University, College Station, Texas.

Brooker, J. D. and Russell, D. W. (1975). Subcellular localization of 3-hydroxy-3-methylglutaryl Coenzyme A reductase in *Pisum sativum* seedlings. *Arch. Biochem. Biophys.* **167**, 730–7.

Cane, D. E. (1981). Biosynthesis of sesquiterpenes. In *Biosynthesis of Isoprenoid Compounds,* Vol. 1, eds J. W. Porter and S. Spurgeon, John Wiley and Sons, New York, pp. 283–374.

Croteau, R. (1981). Biosynthesis of monoterpenes. In *Biosynthesis of Isoprenoid Compounds,* Vol. 1, eds J. W. Porter and S. Spurgeon, John Wiley and Sons, New York, pp. 225–82.

Croteau, R. (1987). Biosynthesis and catabolism of monoterpenoids. *Chem. Rev.* **87**, 929–54.

Dudley, M. W., Green, T. R. and West, C. A. (1986). Biosynthesis of the macrocyclic diterpene casbene in castor bean (*Ricinus communis* L.) seedlings. The purification and properties of farnesyl transferase from elicited seedlings. *Plant Physiol.* **81**, 343–8.

Duncan, J. D. and West, C. A. (1981). Properties of kaurene synthetase from *Marah macrocarpus* endosperm: evidence for the participation of separate but interacting enzymes. *Plant Physiol.* **68**, 1128–34.

Frost, R. G. and West, C. A. (1977). Properties of kaurene synthetase from *Marah macrocarpus. Plant Physiol.* **59**, 22–9.

Givan, C. V. (1983). The source of acetyl coenzyme A in higher plants. *Physiol. Plant.* **57**, 311–16.

Goodwin, T. W. (1981). Biosynthesis of plant sterols and other triterpenoids. In *Biosynthesis of Isoprenoid Compounds,* Vol. 1, eds J. W. Porter and S. Spurgeon, John Wiley and Sons, New York, pp. 443–80.

Goodwin, T. W. and Mercer, E. I. (1963). The regulation of sterol and carotene metabolism in germinating seedlings. *Biochem. Soc. Symp.* **24**, 37–41.

Goss, R. A., Benedict, C. R., Keithley, J. H., Nessler, C. L., Madhavan, S. and Stipanovic, R. D. (1984). cis-Polyisoprene synthesis in guayule (*Parthenium argentatum* Gray) exposed to low, non-freezing temperatures. *Plant Physiol.* **74**, 534–7.

Gray, J. C. (1987). Control of isoprenoid biosynthesis in higher plants. *Adv. Botan. Res.* **14**, 25–91.

Grumbach, K. H. and Forn, B. (1980). Chloroplast autonomy in acetyl-Coenzyme-A-formation and terpenoid biosynthesis, *Z. Naturforsch. C.* **35**, 645–8.

Hemming, F. W. (1983). Biosynthesis of dolichols and related compounds. In *Biosynthesis of Isoprenoid Compounds,* Vol. 2, eds J. W. Porter and S. L. Spurgeon, John Wiley and Sons, New York, pp. 305–54.

Jones, R. L. and MacMillan, J. (1984). Gibberellins. In *Advanced Plant Physiology,* ed. M. Wilkins, Pitman Publishing Ltd, London, pp. 21–52.

Kaethner, T. M. and ap Rees, T. (1985). Intracellular location of ATP citrate lyase in leaves of *Pisum sativum* L. *Planta* **163**, 290–4.

Knotz, J., Coolbaugh, R. C. and West, C. A. (1977). Regulation of the biosynthesis of *ent*-kaurene from mevalonate in the endosperm of immature *Marah macrocarpus* seeds by adenylate energy charge. *Plant Physiol.* **60**, 81–5.

Kreuz, K. and Kleinig, H. (1981). On the compartmentation of isopentenyl diphosphate synthesis and utilization in plant cells. *Planta* **153**, 578–81.

Kuhn, D. N., Knauf, M. J. and Stumpf, P. K. (1981). Subcellular localization of acetyl CoA synthesis in leaf protoplasts of *Spinacia oleracea. Arch. Biochem. Biophys.* **209**, 441–50.

Liedvogel, B. (1986). Acetyl Coenzyme A and isopentenyl pyrophosphate as lipid precursors in plant cells – biosynthesis and compartmentation. *J. Plant Physiol.* **124**, 211–22.

Luetke-Brinkhaus, F. and Kleinig, H. (1987). Formation of isopentenyl diphosphate via mevalonate does not occur within etioplasts and etiochloroplasts of mustard (*Sinapis alba* L.) seedlings. *Planta* **171**, 406–11.

Luetke-Brinkhaus, F., Leidvogel, B. and Kleinig, H. (1984). On the biosynthesis of ubiquinones in plant mitochondria. *Eur. J. Biochem.* **141**, 537–41.

Mettal, U., Boland, W., Beyer, P. and Kleinig, H. (1988). Biosynthesis of monoterpene hydrocarbons by isolated chromoplasts. *Eur. J. Biochem.* **170**, 613–16.

Moesta, P. and West, C. A. (1986). Casbene synthetase: regulation of phytoalexin biosynthesis in *Ricinus communis* L. seedlings. Purification of casbene synthetase and regulation of its biosynthesis during elicitation. *Arch. Biochem. Biophys,* **238**, 325–8.

Mohr, H. (1981). Control of chloroplast development by light – some recent aspects. In *Photosynthesis V.*

Chloroplast Development, ed. G. Akoyunglow, Balaban International Science Services, Philadelphia, pp. 869–83.

Oba, K., Tatematsu, H., Yamashita, K. and Uritani, I. (1976). Induction of furano-terpene production and formation of the enzyme system from mevalonate to isopentenyl pyrophosphate in sweet potato tissue injured by *Ceratocystis fimbriata* and by toxic chemicals. *Plant Physiol.* **58**, 51–6.

Pennock, J. F. and Threlfall, D. R. (1983). Biosynthesis of ubiquinone and related compounds. In *Biosynthesis of Isoprenoid Compounds*, Vol. 2, eds J. W. Porter and S. L. Spurgeon, John Wiley and Sons, New York, pp. 191–303.

Poulter, C. D. and Rilling, H. C. (1981). Prenyl transferases and isomerase. In *Biosynthesis of Isoprenoid Compounds*, Vol. 1, eds J. W. Porter and S. L. Spurgeon, John Wiley and Sons, New York, pp. 161–224.

Qureshi, N. and Porter, J. W. (1981). Conversion of acetyl-Coenzyme A to isopentenyl pyrophosphate. In *Biosynthesis of Isoprenoid Compounds*, Vol. 1, eds. J. W. Porter and S. L. Spurgeon, John Wiley and Sons, New York, pp. 47–94.

Reddy, R. A. and Das, V. S. R. (1987). Chloroplast autonomy for the biosynthesis of isopentenyl diphosphate in guayule (*Parthenium argentatum* Gray). *New Phytol.* **106**, 457–64.

Reddy, A. R., Suhasini, M. and Das, V. S. R. (1987). Impairment of photorespiratory carbon flow into rubber by the inhibition of the glycolate path in guayule (*Parthenium argentatum* Gray). *Plant Physiol.* **84**, 1447–50.

Russell, D. W. (1985). 3-Hydroxy-3-methylglutaryl-CoA reductases from pea seedlings. *Methods Enzymol.* **110**, 26–40.

Ruzicka, L. (1953). The isoprene rule and the biogenesis of terpenic compounds. *Experientia* **9**, 357–67.

Sipat, A. (1985). 3-Hydroxy-3-methyl-CoA reductase in the latex of *Hevea brasiliensis*. *Methods Enzymol.* **110**, 40–51.

Spurgeon, S. L. and Porter, J. W. (1983). Biosynthesis of carotenoids. In *Biosynthesis of Isoprenoid Compounds*, Vol 2, eds, J. W. Porter and S. L. Spurgeon, John Wiley and Sons, New York, pp. 1–122.

Stumpf, P. K. (1984). Fatty acid biosynthesis in higher plants. In *Fatty Acid Metabolism and Its Regulation*, ed. S. Numa, Elsevier Science Publishers, Amsterdam, New York, Oxford, pp. 155–79.

West, C. A. (1981). Biosynthesis of diterpenes. In *Biosynthesis of Isoprenoid Compounds*, Vol. 1, eds. J. W. Porter and S. L. Spurgeon, John Wiley and Sons, New York, pp. 375–411.

VIII

Nitrogen Metabolism

25 The molecular biology of N metabolism*
C. P. Vance and S. M. Griffith

Introduction

Nitrogen is the major limiting nutrient for most plant species. Acquisition and assimilation of nitrogen is second in importance only to photosynthetic carbon assimilation for plant growth and development. Seed viability and germination are directly related to nitrogen content. Production of high-quality, protein-rich food is extremely dependent on availability of sufficient nitrogen. Clearly, the crucial role that nitrogen plays in plant growth requires that physiologists understand the biochemical and molecular events that regulate nitrogen metabolism.

Plants acquire nitrogen from two principal sources: (1) the soil, through commercial fertilizer, manure, and/or mineralization of indigenous organic matter; and (2) the atmosphere, through symbiotic N_2 fixation. Soil-derived nitrogen, generally in the form of nitrate (NO_3^-) and atmospheric N_2 must be reduced to ammonia (NH_4^+) to become available for amino acid and protein synthesis (Chapter 26). Nitrate is reduced to NH_4^+ by the plant enzymes nitrate reductase (NR; EC 1.6.6.1) and nitrite reductase (NiR; EC 1.6.6.4), while atmospheric N_2 is reduced by the microbial enzyme nitrogenase (EC 1.18.6.1).

$$NO_3^- + NAD(P)H + H^+ \rightarrow NO_2^- + NAD^+ + H_2O \quad [25.1]$$
(Nitrate reductase)

$$NO_2^- + 6e^- + 8H^+ \rightarrow NH_4^+ - 2H_2O \quad [25.2]$$
(Nitrite reductase)

$$N_2 + 16ATP + 8e^- + 10H^+ \rightarrow 2NH_4^+ + 16ADP + 16P_i + H_2O \quad [25.3]$$
(Nitrogenase)

Since NH_3 is toxic, it must be rapidly assimilated into non-toxic metabolites so it does not accumulate. Over the past 20 years ammonia assimilation has been the focus of intense study. It is now generally agreed that glutamine synthetase (GS; EC 6.3.1.2) and glutamate synthase (GOGAT; EC 1.4.1.14) catalyze the initial steps in ammonia assimilation (Fig. 25.1). Further assimilation of NH_4^+ into aspartate and asparagine is mediated by aspartate aminotransferase (AAT; EC 2.6.1.1) and asparagine synthetase (AS; EC 6.3.5.4) with phosphoenolpyruvate carboxylase (PEPC; EC 4.1.1.31) providing a portion of the carbon skeleton of these amino acids (see Chapters 26 and 27).

This chapter will attempt to familiarize the reader with molecular aspects relating to nitrogen metabolism. Our aim is not to present a comprehensive review, but more to provide a working knowledge of the current status of the topic. Because some steps in nitrogen metabolism have received more attention than others, this chapter will not devote equal space to each reaction.

Nitrate reductase

Characteristics

Reduction of NO_3^- to NH_4^+ is a two step process (eqns [25.1] and [25.2]). Nitrite reductase activity

---

*Joint contribution of the United States Department of Agriculture-Agricultural Research Service and the Minnesota Agricultural Experiment Station. This work was supported in part by United States Department of Agriculture CRGO grant 87-CRCR-1-2588. Paper 16046 Scientific Journal Series, Minnesota Agricultural Experiment Station.

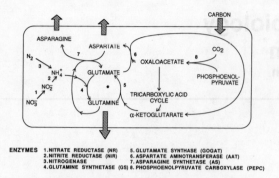

ENZYMES
1. NITRATE REDUCTASE (NR)
2. NITRITE REDUCTASE (NiR)
3. NITROGENASE
4. GLUTAMINE SYNTHETASE (GS)
5. GLUTAMATE SYNTHASE (GOGAT)
6. ASPARTATE AMINOTRANSFERASE (AAT)
7. ASPARAGINE SYNTHETASE (AS)
8. PHOSPHOENOLPYRUVATE CARBOXYLASE (PEPC)

Fig. 25.1 The general scheme of nitrogen assimilation in higher plants and the enzymes involved. In this particular scheme, glutamine, asparagine, and aspartate are the primary amino acids transported to other cells and plant organs. Photosynthate is used via the TCA cycle to generate carbon skeletons for amino acid biosynthesis. Carbon for amino acids can also be derived from non-photosynthetic CO_2 fixation of respired and/or atmospheric CO_2.

(eqn[25.2]) is generally several-fold greater than NR activity (eqn [25.1]). This suggests that NR is the rate-limiting enzyme and the key to regulating NO_3^- assimilation (Campbell, 1985; Kleinhofs et al., 1985).

The NR enzyme is a large, complex protein containing FAD, cytochrome b_{557} and molybdenum as prosthetic groups. Both fungal and plant NRs have been purified and have characteristics in common. The best characterized plant NRs have been isolated from barley (*Hordeum vulgare* L.) and squash (*Curcurbita maxima* L.) The enzyme is a homodimer with a native molecular weight of 200 000–230 000 and a subunit size of 100 000–115 000. Squash NR appears to contain two molecules each of FAD, cytochrome b_{557} and molybdenum cofactor (MoCo per native enzyme (Fig. 25.2). Barley NR has been reported to contain only one MoCo per native enzyme, thus either the MoCo content may vary with species or the estimates may need to be revised.

The NR mechanism is proposed to be a two site ping-pong model (Campbell, 1985). Reduced pyridine nucleotide donates two electrons to the initial acceptor, FAD. Electrons then flow through the cytochrome b_{557} site to the MoCo sites, where they are donated to NO_3^- reducing it to NO_2^-. The

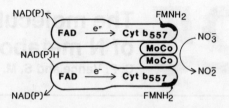

Fig. 25.2 Model of plant nitrate reductase structure. Electrons from NAD(P)H are accepted by FAD and transferred directly to cyt.b_{557} and then used at the molybdenum cofactor (MoCo) site to reduce NO_3^-. $FMNH_2$ can also donate electrons for the direct reduction of NO_3^-. The enzyme is composed of two identical subunits of approximately 110 000 MW, two FAD, two cyt. b_{557}, and two MoCo.

enzyme can function as a diaphorase and can catalyze NAD(P)H-dependent reduction of many electron acceptors including cytochrome c, ferricyanide, dichlorophenolindophenol, and methylene blue.

The MoCo is required for NR activity. Mutants defective in MoCo lack NR activity. Activity from such mutants can be restored through *in vitro* complementation with MoCo from other sources. For example, when barley *nar2* mutants, which are defective in MoCo and lack NR activity, are used as the apoprotein source and complemented *in vitro* with MoCo from xanthine oxidase, over 70% of the NR activity is restored (Warner et al.,1984). All molybdoenzymes tested so far, except nitrogenase, have MoCos that are active in the *in vitro* complementation test.

Genetic regulation

Expression and genetic regulation of NR have been evaluated through: (1) selection of plant mutants lacking NR activity; (2) immunological characterization of the NR polypeptides; and (3) preparation of cDNAs to NR mRNAs. Each approach provides unique information about NR and how the enzyme is regulated.

Nitrate reductase-deficient mutants have been isolated in *Hordeum, Pisum, Glycine, Arabidopsis, Datura, Rosa, Petunia*, and *Hyoscyamus*. Kleinhofs et al. (1985) have reviewed and described the significance of these NR mutants. Both apoprotein and MoCo mutants have been isolated.

In barley, the apoprotein mutants characterized to date belong to a single complementation group designated *nar1* (Warner et al., 1984; Kleinhofs et al., 1985) This locus codes for the NADH-specific NR structural gene. The *nar1* alleles are co-dominant with the wild-type allele and represent a single gene. Surprisingly, many *nar1* mutants are capable of growth, comparable with the parent, when grown on NO_3^-. This discrepancy was resolved when it was shown that *nar1* mutants contain a unique NAD(P)H bispecific NR which appears to be responsible for NO_3^- assimilation.

The MoCo function in barley appears to be under the control of at least three loci designated *nar2*, *nar3*, and *nar4* (Warner et al., 1984; Kleinhofs et al., 1985). Segregation patterns indicate that all three are controlled by a single gene. The *nar2* and *nar3* alleles are recessive. Mutants in *nar2* and *nar3* contain very low levels of functional MoCo and can be complemented *in vitro* (restored to NR activity) by exogenous MoCo from milk xanthine oxidase. The *nar4* locus has not been well characterized. Three loci for the MoCo have also been identified in *Nicotiana*, and up to five loci have been shown to control MoCo in fungi. This suggests that other MoCo loci may be identified in plants.

Single genes, each regulating NO_3^- induction and NH_4^+ repression of NR, have been identified in fungi. However, similar genes have not yet been identified in higher plants. Support for additional genes involved in NR activity and regulation is provided by studies which suggest that NR activity in *Arabidopsis thaliana* may be under the control of seven loci.

In all plants, NR activity increases dramatically in response to applied NO_3^- (Campbell, 1985). As NO_3^- is depleted or removed NR activity diminishes. Light is also required for the maintenance of high levels of NR in leaves of higher plants. The turnover in NR activity could result from: (1) NO_3^--induced *de novo* synthesis of NR protein; (2) NO_3^--dependent activation of inactive NR protein; and (3) light-regulated protein synthesis and degradation. Antibodies to barley, spinach, and squash NR have been used to ascertain how NR expression is regulated.

Western blotting and rocket immunoelectrophoresis have shown that barley proteins which cross-react with NR antibodies are not detectable in roots or shoots of seedlings grown without NO_3^- (Somers et al., 1983). Polypeptides which cross-react with NR antibodies appear within 10 hours after exposure of plants to NO_3^-, and increased NR activity is directly related to the appearance of NR polypeptides. As NO_3^- is depleted or removed, NR activity declines and is accompanied by a reduction in NR polypeptides. Plants grown in the presence of NO_3^- and maintained in the dark had low NR activity and low NR polypeptides. Similarly, changes in NR activity of squash leaves and cultured spinach cells were also directly related to changes in NR polypeptides. Taken inclusively, these data indicate that NR activity in plants is primarily regulated by the *de novo* synthesis and degradation of NR proteins.

Polyclonal antibodies raised against squash, barley and maize (*Zea mays* L.) have been extremely useful in evaluating whether NR mRNAs increase during NO_3^--induced increases in NR activity and in identifying recombinant NR cDNAs that were cloned into the expression vector λgt11 (Cheng et al., 1986; Crawford et al., 1986; Calza et al., 1987). Barley seedlings grown with and without NO_3^- were labeled *in vivo* with [^{35}S]-methionine. RNA was also extracted from seedlings grown with and without NO_3^- and translated *in vitro*. Examination of both *in vivo* and *in vitro* immunoprecipitates showed a striking increase in the 110 kD NR band in NO_3^- induced as compared with uninduced seedlings. These data showed that the induction of the NR protein was attributable to an increase in translatable NR mRNAs.

Further studies on the molecular basis of NR induction in barley have focused on the isolation of a cDNA clone for barley NR (Cheng et al., 1986). A population of polyA$^+$ RNAs enriched for NR was isolated by sucrose density gradient centrifugation. These NR-enriched polyA$^+$ RNAs were used to construct a cDNA library in the expression vector λgt11. Under proper conditions, λgt11 recombinants will express the polypeptides encoded by the cDNA inserts. A total of 25 000 recombinants were screened with barley antibodies and two NR clones identified. These two NR clones were then used to screen a second barley cDNA library prepared from NR-enriched mRNAs isolated by methylmercuric hydroxide agarose gel electrophoresis. From this second library ten NR cDNA clones were identified. Hybrid-select translation with one of the cDNAs confirmed that it corresponded to NR. Northern blots of polyA$^+$ RNAs isolated from NO_3^--induced and non-induced seedlings, probed with NR cDNA, showed a dramatic induction of a single 3.5 kbp transcript corresponding to NR mRNAs. Thus, increases in steady-state levels of NR mRNA accompany or precede NO_3^--induced NR activity.

Using a similar approach and antibodies to maize NR, eight recombinant clones expressing NR polypeptides were isolated from a cDNA library prepared from tobacco NR-enriched polyA$^+$ RNAs (Calza et al., 1987). Two clones were shown by Western blot analysis to express NR polypeptides. One of the clones, 13–29, was used for further study. Monoclonal antibodies to tobacco NR recognized the polypeptide expressed by 13–29 and antibodies that bound to the polypeptide expressed by 13–29 were inhibitory to in vitro NR activity. This evidence confirmed that the polypeptide expressed by 13–29 was tobacco NR. The 13–29 cDNA sequence shares strong homology at the amino acid level with the heme-binding domain of the cytochrome b_5 gene superfamily. All thirteen conserved residues of the cytochrome b_5 superfamily are present in the 13–29 tobacco NR cDNA sequence. The data confirmed that the cloned sequences corresponded to fragments of NR mRNA and defined the heme-binding domain of tobacco NR. Another portion of the 13–29 cloned sequence is homologous to the gene for human flavoprotein cytochrome b_5 and it was suggested that it probably corresponds to the NADH/FAD domain of tobacco NR. Sequence and homology information indicates that the heme-binding site and reducing flavoprotein domain lie side by side in the tobacco NR polypeptide chain.

Southern blot analysis of tobacco genomic DNA digested with EcoRI and probed with 13–29 shows a 3.4 kbp and a 5.0 kbp fragment and a weak signal corresponding to a 2.9 kbp fragment. These data support the assumption that there are two functional structural NR genes present in the N. tabacum genome. The two genes are found because N. tabacum is an amphidiploid derived from hybridization of N. sylvestris and N. tomentosiformis (Kleinhofs et al., 1985; Calza et al., 1987). Therefore, one of the NR genes is proposed to originate from N. sylvestris and the other from N. tomentosiformis.

A squash NR cDNA clone has also been selected by screening a cDNA library prepared with polyA$^+$RNA isolated from cotyledons of NO_3^--treated seedlings (Crawford et al., 1986). The 1.2 kbp NR cDNA hybridized to a 3.2 kbp mRNA that appeared long enough to contain the coding sequence for the NR enzyme (3.0 kbp). Nitrate reductase mRNA was 120-fold more abundant in NO_3^--induced cotyledons as compared with uninduced tissue. Induction of squash NR activity involves de novo synthesis of the enzyme from newly synthesized message.

The availability of purified NR, antibodies to NR (both polyclonal and monoclonal) and the cloned NR gene places plant physiologists and molecular biologists on the edge of developing a comprehensive understanding of plant NR from the whole plant to the subcellular level.

Nitrogenase

Characteristics

The second major process by which plants acquire nitrogen is through the microbial enzyme nitrogenase expressed by Rhizobium and Bradyrhizobium during symbiotic N_2 fixation (eqn [25.3]). It should be noted that nitrogenase is also expressed in certain free-living bacteria such as Azotobacter, Clostridium, Klebsiella, and some cyanobacteria (e.g. Anabaena). Amino acid analyses have shown that the enzyme is highly conserved between symbiotic and free-living bacteria. Therefore, the characteristics described in this section are derived from experiments with both types of bacteria. For molecular analysis of nif genes the Klebsiella pneumoniae system will be described and then, superimposed, the Rhizobium system will be considered.

Nitrogenase is comprised of two easily separable proteins designated the iron protein (Fe protein) and the molybdenum–iron protein (MoFe protein). The Fe protein (encoded by the *nifH* gene) is a homodimer with a native molecular weight of 60 000–64 000 and a subunit molecular weight of 30 000–32 000 (Burgess, 1984). The Fe protein contains approximately 4 g-atoms of Fe and S per mole of preparation. The Fe and S form a single [4Fe–4S] cluster which is bound between the subunits. The Fe protein has two MgATP binding sites. As ATP binds to these sites the potential of electrons present at the [4Fe–4S] cluster becomes more negative allowing the Fe protein to donate electrons to the MoFe protein.

The MoFe protein is a tetramer ($\alpha_2 \beta_2$) of approximately 220 000 molecular weight (Burgess, 1984). The α subunit has a molecular weight of about 56 000 and is encoded by the *nifD* gene while the β subunit has a molecular weight of approximately 60 000 and is encoded by the *nifK* gene. The MoFe protein contains two atoms of Mo and 24 to 32 atoms of Fe and S per molecule. There appear to be four [4Fe–4S] clusters and two [MoFe$_6$S$_8$] clusters. The two [MoFe$_6$S$_8$] clusters comprise the MoFe cofactor. The role of the MoFe protein is to transfer electrons to N_2 and H^+.

Genetic regulation

The genes required for nitrogen fixation have been most clearly defined in the free-living bacterium *K. pneumoniae*. The organization of the nitrogen fixing (*nif*) gene cluster of this bacterum is shown in Fig. 25.3. Some 17 genes are transcribed in eight adjacent operons which occupy 23 kbp of the genome (Ow and Ausubel, 1983; Dixon, 1984; Brookes *et al.*, 1985). The *nif* gene functions can be grouped into several categories: (1) *nifH*, *nifD*, *nifK*: structural proteins for nitrogenase; (2) *nifB*, *nifV*, *nifN*, *nifE*, *nifQ*: proteins involved in the synthesis of the MoFe cofactor; (3) *nifF*, *nifJ*: flavodoxin electron transport proteins; (4) *nifM*, *nifS*: proteins involved in processing of nitrogenase; (5) *nifU*, *nifX*, *nifY*: unknown; and (6) *nifA*, *nifL*: regulatory proteins affecting the other *nif* genes. All 17 *nif* genes have been cloned

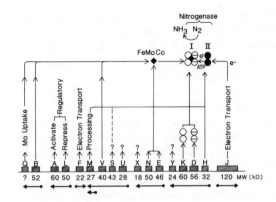

Fig. 25.3 Map of the *nif* gene cluster in *Klebsiella pneumoniae*. The function of each gene is denoted by arrows above the gene letter designation and the molecular weight of the gene product is given below each gene. The size of the individual operons and the direction of transcription are noted by arrows below the gene product MW (modified from Brookes *et al.*, 1985 and Dixon, 1984).

and the molecular weight determined for most of the gene products. Regulation of the *nif* genes in *K. pneumoniae* is complex and at present not completely understood. Only the salient features of this regulation will be covered. For more complete reviews the reader should consult Brookes *et al.* (1985), Dixon (1984), and Ow and Ausubel (1983).

Nitrogenase is synthesized when *K. pneumoniae* is grown under anaerobic, nitrogen-limiting conditions (Ow and Ausubel, 1983; Dixon, 1984). This is not surprising since nitrogenase is irreversibly denatured in the presence of oxygen and is not required when alternative sources of reduced nitrogen are available. Regulation of nitrogenase is controlled by the *nifA* and *nifL* proteins in the *nifLA* operon. The *nifLA* operon is functionally regulated by several genes designated *ntr*, which control many operons involved in nitrogen assimilation. The primary components of the *ntr* system are *ntrA*, *ntrB*, and *ntrC*. Under anaerobic, nitrogen-limiting conditions, and *ntrA* and *ntrC* gene products activate transcription of the *nifLA* operon. The *nifA* and *ntrA* gene products then activate transcription of all other *nif* operons. Since activation of the *nif* operons requires gene products in common, it seems reasonable that the promoter regions of the *nif* genes would be similar, which is in fact the case. The promoter regions of all *K. pneumoniae nif* genes contain conserved regions of homology at -24 and -12 with respect to the transcription start sites (Ow and Ausubel, 1983; Dixon, 1984). Activation of these common regulatory regions under appropriate environmental and nutritional conditions results in a cascade effect leading to assembly of functional nitrogenase.

Since *nif* gene expression is positively controlled by transcriptional activators and requires the *ntrA* gene product, repression of nitrogenase synthesis in the presence of excess nitrogen and/or oxygen involves inactivation of these positive controlling elements. In the presence of oxygen and/or excess nitrogen, the *nifL* gene product is altered. The altered *nifL* gene product in some, as yet unknown, fashion inactivates the *nifA* gene product resulting in repression of the other *nif* operons. The *ntrB* gene product is also thought to be involved in sensing of excess nitrogen and repression of *nif* genes. Evidence for the involvement of *nifL* and *ntrB* in repression of nitrogenase has been obtained through *nifL* and *ntrB* mutants which synthesize nitrogenase in the presence of oxygen and/or excess nitrogen. Understanding the organization and regulation of the *K. pneumoniae nif* genes has provided the framework and tools to more fully ascertain how symbiotic N_2 fixation is controlled.

Symbiotic N_2 fixation

Rhizobium–legume symbiosis

Symbiotic N_2 fixation results from the complex interaction between host plant and microorganism (Verma *et al.*, 1986; Govers *et al.*, 1987). The host plant provides the microorganism with a source of energy for growth, metabolism and a specialized ecological niche. The microorganism fixes atmospheric N_2 and provides the plant with a source of reduced nitrogen. While the *Rhizobium*–legume symbiosis typifies such an interaction, symbiotic N_2 fixation also occurs in non-legumes such as plants of the family Eleagnaceae in association with the actinomycete *Frankia* and between the water fern *Azolla* and cyanobacteria. Since genetic regulation of the *Rhizobium*-legume symbiosis has been the most fully characterized, that system will be covered in some detail.

The ecological niche for *Rhizobium*–legume symbioses is the root nodule. Root nodules are highly organized, hyperplastic tissue masses derived from root cortical cells (Vance *et al.*, 1988). Based on agricultural significance, nodules are generally divided into two major groupings characterized by shape, meristematic activity, and fixed N transport products: (1) nodules that are elongate–cylindrical with indeterminate apical meristematic activity that transport fixed N as amides such as alfalfa (*Medicago sativa* L.), pea (*Pisum sativum* L.) and clover (*Trifolium*); and (2) nodules that are spherical with determinate internal meristematic activity that transport fixed N as ureides, such as soybean (*Glycine max* L. Merr) and common bean (*Phaseolus vulgaris* L.). The complex series of events leading from bacterial colonization of the legume rhizosphere to fixation of N_2 and export of that fixed N requires controlled coordinated expression of both bacterial and host plant genes. The contribution of these genes to symbiosis can be grouped into several functions, including recognition, root hair invasion, infection thread growth, nodule differentiation, carbon assimilation, organic acid metabolism, ammonia assimilation and, possibly, suppression of host plant defense responses. In the last 10 years, substantial progress has been made in understanding molecular aspects of both the bacterial and host plant role in symbiosis.

Rhizobium symbiotic genes

Nod genes

Symbiotic genes of rhizobia are categorized as those affecting nodulation (*nod* genes), those controlling nitrogenase based on their homology to *K. pneumoniae* (*nif* genes), and others that affect symbiotic N_2 fixation, but share no homology to known *K. pneumoniae* genes (*fix* genes) (Earl *et al.*, 1987; Johnston *et al.*, 1987). For the most

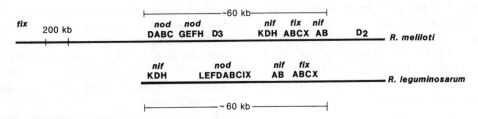

Fig. 25.4 Identification and organization of the symbiotic genes on plasmids of *Rhizobium meliloti* and *Rhizobium leguminosarum*. Note the difference in proximity of *nifK,D,H* to *fixA,B,C,X* in *R. meliloti* as compared to *R. leguminosarum*. Other features of interest include three copies of the *nodD* regulatory gene found in *R. meliloti*, and a second *fix* region located some 200 kbp away from the 60 kbp symbiotic gene cluster of *R. meliloti*.

part, these genes are highly conserved in all rhizobia; however, their location in the genome may vary. For example, the symbiotic genes of *R. meliloti* and *R. leguminosarum* are clustered within an approximately 60 kbp region on indigenous plasmids (Fig. 25.4), while those in *Bradyrhizobium* are more scattered and located on the chromosome. Irrespective of the organism and the location in the genome, the organization and regulation of symbiotic genes is similar. The symbiotic genes of *R. meliloti* and *R. leguminosarum* have been analyzed in great detail and serve as the foundation of our understanding of the molecular control of N_2 fixation in the microbial symbiont.

A region about 15 kbp in size containing about eight genes comprises the *nod* gene cluster of the *R. leguminosarum* symbiotic plasmid (Johnston et al., 1987). Two small clusters 6 kbp apart make up the analogous *nod* gene region of the *R. meliloti* symbiotic (sym) plasmid. Transfer of these regions to *Agrobacterium* and other rhizobia cured of their sym plasmid confers the ability to nodulate pea or alfalfa. Those nodules are, however, ineffective (non-N_2 fixing), indicating that, although the heterologous genes carry the information for nodule development, other genes are required for effective N_2 fixation.

Mutations in the *nodA,B,C* genes block root hair curling, infection, and nodulation indicating that these genes are involved in the earliest steps of the symbiotic interaction (Kondorosi and Kondorosi, 1986). The *nodD,A,B,C* genes are conserved in most rhizobia and are functionally identical, as shown in experiments where the *nodD,A,B,C* genes from one *Rhizobium* or *Bradyrhizobium* species could complement mutations in homologous genes of other species. The *nodD,A,B,C* genes are frequently referred to as the 'common *nod* genes'.

Unique features of *nodD*, however, suggest that this gene is not simply another common *nod* gene (Johnston et al., 1987). Mutations of *nodD* in *R. leguminosarum* prevent nodulation. However, mutations in *nodD* in *R. meliloti* only delay nodulation because *R. meliloti* has three functional copies of *nodD*. The *nodD* gene of the wide host range siratro-type *Rhizobium* when transferred to *R. meliloti* confers upon the transconjugants the capacity to nodulate siratro. Lastly, and discussed later, *nodD* appears to play a regulatory role in activating transcription of other *nod* genes.

Nodulation by rhizobial species is very host specific. For example, *R. meliloti* nodulates alfalfa, but not clover or soybean. This host-specific nodulation characteristic is regulated by the *nodE,F,G*, and *H* genes in *R. meliloti* and by *nodE* and *F* in *R. leguminosarum* (Kondorosi and Kondorosi, 1986; Johnston et al., 1987). Mutations in these genes cannot be complemented by similar genes from other species. In fact, host range specificity can be changed by modifying these genes or transferring them to other species. Other interesting host-specific genes occur within strains of some *Rhizobium* species. The *R. leguminosarum* strain, designated TOM, nodulates primitive pea plants from Afghanistan, but not adapted European pea cultivars. A gene designated *nodX* has been shown to be required for *R. leguminosarum* TOM to nodulate primitive pea. European strains of *R. leguminosarum* have no homologous *nodX* gene.

Several of the *nod* gene products have been isolated and characterized (Kondorosi and Kondorosi, 1986; Johnston *et al.*, 1987). Many appear to be membrane bound; one is a transport protein and another is similar to acyl carrier protein.

Regulation of the *nod* genes has been a particularly exciting area of study in recent years because experiments have demonstrated that activation of the *nod* genes is controlled by secondary plant products found in root exudates (Peters *et al.*, 1986; Kosslak *et al.*, 1987). This control is affected by the interaction of the *nodD* gene product with plant root exudates. The *nodD* gene is transcribed constitutively in free-living cultures of *Rhizobium* and *Bradyrhizobium*. However, the other *nod* genes are transcribed little, if any, except in the presence of root exudates. The induction of *nodA,B,C,E,F,G*, and *H* is dependent upon the presence of the *nodD* gene product and factors in legume root exudates. The *nodD* gene product is modified by the active factor in root exudates. The *nodD*–root exudate product binds to the promoter regions of the other *nod* operons and thus activates transcription. The promoter regions of the *nod* operons contain a 35 to 45 bp conserved sequence (called the *nod* box) to which the *nodD*–root exudate product binds, thus promoting transcription of the *nod* operons.

The compounds in root exudates responsible for induction of *nod* gene transcription are flavones, flavanones, and isoflavones (Fig. 25.5). These compounds are derived from the condensation of phenolic cinnamic acid derivatives and malonate units. The primary inducing compounds from alfalfa, pea, clover, and soybean root exudates are luteolin, hesperitin, 7,4′-dihydroxyflavon, and 4′,7-dihydroxyisoflavone (daidzein), respectively (Peters *et al.*, 1986; Johnston *et al.*, 1987; Kosslak *et al.*, 1987). The most effective inducer compounds have hydroxyl groups substituted at the 3′ or 4′ position on the B ring and a hydroxyl or glucoside linkage at position 7 of the A ring (Johnston *et al.*, 1987). While isoflavonoid compounds induce the expression of *nod* genes in *B. japonicum* and *R. fredii* (strains that nodulate soybean), these compounds act as antagonists of *nod* gene expression in *R. meliloti*, *R. trifolii* and *R. leguminosarum*. Isoflavonoids not only play a role in legume–*Rhizobium* symbiosis, but also in plant disease resistance in legumes. Microbial infection of legumes frequently induces the accumulation of certain isoflavonoids that act as antibiotics (phytoalexins) which limit the growth of invading organisms (Vance *et al.*, 1988). Therefore, subtle differences in secondary plant products may regulate whether an interaction results in symbiosis or pathogenesis.

The *nodD* gene product is constitutively expressed and acts as a positive regulator for transcription of other *nod* operons. The *nodD* gene initially was thought to be a common *nod* gene because mutations in *nodD* of one *Rhizobium* sp. could be complemented, in part, by the *nodD* gene of another species. In addition, coding regions of *nodD* genes of various *Rhizobium* sp. share about 75% homology. However, studies showing that isogenic strains of *Rhizobium* which vary only in the source of their *nodD* gene differed in response to a variety of flavonoid inducers and root exudates suggest that the *nodD* gene products of various species are not common and may mediate host-specific nodulation (Johnston *et al.*, 1987). Further support for *nodD* involvement in species-specific nodulation is demonstrated by the fact *nodD* genes of *R.*

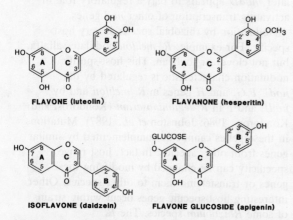

Fig. 25.5 Flavonoid and isoflavonoid compounds exuded from legume roots that activate and/or inhibit transcription of the *nod* genes in *Rhizobium* and *Bradyrhizobium*. The B ring is derived from phenylalanine, while the A and C rings are derived from malonate.

leguminosarum and *R. meliloti* can not completely complement *R. trifolii nodD* mutants. Replacement of *R. trifolii nodD* with the *nodD* from other species results in a narrowing of the host range of *R. trifolii*. While there is selective induction of the *nod* genes by the interaction of the *nodD* product with different flavonoids and isoflavonoids, legume root exudates do not appear to be the sole regulators of legume–*Rhizobium* host range specificity. For example, root exudates of clover and alfalfa will activate transcription of pea rhizobia *nod* genes. Since all plants contain flavonoids and isoflavonoids and decomposition of plant debris would release such compounds in the field, specificity of nodulation must involve a finely tuned balance between inducers and antagonists and probably other as yet unidentified genes.

Nif *and* fix *genes*

Using *K. pneumoniae nif* genes as probes, DNA sequences homologous to *nifK,D,H,A*, and *B* have been identified in all *Rhizobium* and *Bradyrhizobium* species (Earl *et al.*, 1987; Johnston *et al.*, 1987). These corresponding genes are located on plasmids in fast growing *R. meliloti*, *R. trifolii*, and *R. leguminosarum* and on the chromosome in slow growing *Bradyrhizobium*. The *nifK,D,H,A*, and *B* genes have the same functions in rhizobia as in *Klebsiella*. Although a *nifL* gene has not yet been identified in *Rhizobium*, regulation of the *nif* operons is similar to that in *Klebsiella*. The *nifA* gene product is a transcriptional activator for other *nif* operons. Mutations in *nifA* block symbiotic N_2 fixation and such mutants do not synthesize *nif* polypeptides. In addition, the promoter sequences of *Rhizobium nif* genes are similar to those in *Klebsiella*.

In addition to the *nif* genes, another group of genes essential for symbiotic N_2 fixation, designated *fix* genes, have been identified in *Rhizobium*. Mutation in *fix* genes result in nodules with a Fix⁻ phenotype. In *R. meliloti* four genes, *fixA, fixB, fixC*, and *fixX*, have been identified and characterized (Earl *et al.*, 1987). They are closely linked to the *nif* clusters on the symbiotic plasmid. Polypeptides encoded by *fixA,B,C* and *X* have molecular weights of 31 000, 38 000, 47 000 and 11 000 respectively. All *Rhizobium* and *Bradyrhizobium* species examined have DNA sequences homologous to the *R. meliloti fixA,B,C* and *X*. However, only one free-living diazotroph, *Azotobacter vinelandii*, has DNA homologous to these *fix* genes. The *fixX* gene of *R. meliloti* is highly homologous to an *A. vinelandii* ferredoxin gene which donates electrons to nitrogenase. The *fixA,B,C,X* genes constitute a single operon which contains a promoter region that is very similar to other *nifA*-activated promoters.

Lastly, there are other *nod* and *fix* genes in rhizobia that have yet to be mapped and characterized. In fast growing *R. meliloti, R. trifolii*, and *R. leguminosarum*, these genes affecting symbiotic N_2 fixation are located on both plasmids and the chromosome. In *Bradyrhizobium* such genes have been shown to be on the chromosome. In addition, a second large plasmid having genes involved in nodulation has recently been identified in *R. meliloti*. As our knowledge of the biochemistry and physiology of symbiosis grows, undoubtedly more bacterial genes affecting symbiosis will be identified.

Plant genes involved in symbiosis

A complete discussion of host plant genes involved in symbiosis is beyond the scope of this chapter. However, a brief overview of these genes will provide some perspective of the progress in this area. A combination of approaches utilizing immunology, *in vitro* translation of nodule mRNAs, and nodule cDNA cloning have shown some 25 plant genes or plant gene products are either induced or enhanced during root nodule development and function (Verma *et al.*, 1986; Govers *et al.*, 1987). Of the ten nodule-specific or enhanced gene products (nodulins) identified to date, five are involved in assimilation of ammonia and carbon (Table 25.1), while the others are involved in nodule morphogenesis and control of nodule oxygen concentration. To date, genes that encode leghemoglobin, sucrose synthase, uricase, glutamine synthetase, nodulins 23 and 24, and hydroxyproline-rich glycoproteins (HPRG) have been isolated.

Although we do not understand how host plant

Table 25.1 Nodule specific/enhanced host genes and their products

Nodulin	Gene isolated	Subcellular location	MW(10^3)	Function
Leghemoglobin	Yes	Infected cell (cytoplasm)	16	Oxygen carrier
Sucrose synthase	Yes	Unknown	90–100	Carbon metabolism
Uricase	Yes	Uninfected cell (peroxisome)	35	N assimilation
Glutamine synthetase	Yes	Infected cell (cytoplasm, plastids)	37, 38, 44	N assimilation
Choline kinase	No	Infected cell (peribacteroid membrane)	60	Membrane structure
Phosphoenolpyruvate carboxylase	No	Infected cell (cytoplasm)	101	N assimilation and organic acids
Glutamate synthase (NADH)	No	Infected cell (cytoplasm)	200	N assimilation
Aspartate aminotransferase	No	Infected cell (cytoplasm, plastids, peribacteroid membrane)	37–40	N assimilation, carbon metabolism
Nodulin 23 and 24	Yes	Infected cell (peribacteroid membrane)	23–30	Unknown
(Hydroxy) proline rich glycoprotein	Yes	Unknown	75	Early nodule development

nodule genes are regulated, expression can be influenced by: (1) stage of nodule development; (2) nodule effectiveness; and (3) presence of bacteria in nodules (Govers et al., 1987; Vance et al., 1988). For most nodule genes, transcription is induced either just prior to or concomitant with the appearance of visible nodules and nitrogenase activity. Some genes, however, are expressed early in development, such as HPRG, while others are expressed after the onset of the N_2 fixation. Proteolytic enzyme activity increases late in nodule development and some genes expressed during this period may code for these enzymes. Transcription of nodule genes in effective nodules is 3 to 10-fold greater than that in ineffective nodules. In tumor-like ineffective nodules lacking bacteria, few if any of the genes specific for nodules are transcribed and translated. Therefore, not only are bacteria required for nodule gene expression, but maximum expression requires a product associated with effective bacteria. It is not clear whether ammonia production by effective bacteria is the inducer of maximum expression of plant genes in nodules. In soybean nodules, treatments that inhibit nitrogenase and ammonia production inhibit the accumulation of nodule GS mRNAs. In contrast, alfalfa nodules containing ineffective bacteroids accumulate nodule GS mRNAs.

Since many nodule gene products appear simultaneously, they probably have similar regulatory sequences. Sequences in the 5' untranslated regions of genes control initiation of transcription. Therefore, genes which are coordinately expressed may have conserved sequences in this region (Verma et al., 1986; Stougaard et al., 1987). The soybean leghemoglobin, nodulin 23 and nodulin 24 genes share significantly conserved sequences in the 5' region. Verma's laboratory (Verma et al., 1986) has demonstrated that a *trans*-acting factor(s) from soybean nodule extracts specifically interacts with the 5' region of these genes and activates transcription. Ineffective nodules may lack such factors or they may be altered to be less effective.

Marcker's group (Stougaard et al., 1987) have also demonstrated strong positive regulatory sequences in the 5' region of soybean nodulin genes. In addition, they showed that organ-specific expression of *Lb* genes may be due to *cis* regulatory elements located between positions −139 and −102 in the 5' region of the genes. They demonstrated nodule-specific expression of soybean nodulin genes in transgenic *Lotus corniculatus* and *Trifolium repens* indicating that the promoter regions of soybean nodulin genes are recognized in nodules of other species. Thus, the molecular mechanisms responsible for activation of some nodulin genes are conserved in various *Rhizobium*–legume associations.

Some plant-controlled gene products are also involved in the regulation of expression of *Rhizobium* nitrogenase. Several plant mutants from a number of legume species have been identified that block nitrogenase expression or function (Vance *et al.*, 1988). These mutant plants form ineffective nodules when inoculated with strains of *Rhizobium* that normally form effective nodules on wild-type plants. Plant control of ineffective nodulation is usually controlled by a single recessive gene. This implies that the mutants lack a gene product required for effective nodulation. Nodules from plant-controlled ineffective mutants have reduced leghemoglobin, lowered capacity to assimilate ammonia, and altered peribacteroid membranes surrounding the root nodule bacteria. Whether any of these changes are the cause or effect of reduced nitrogenase expression is not known. As our knowledge of the biochemistry and genetics of symbiosis advances, it becomes increasingly clear that not only do the plant and microbe contribute to the nutritional status of each other, but each partner exerts a degree of gene regulation to the other.

Control of ammonia assimilation

Overview

In higher plants, the reduced form of nitrogen ultimately available for direct assimilation is NH_4^+ (Wallsgrove *et al.*, 1983). The degree of NH_4^+ assimilation in tissues varies as a function of organ development, environmental conditions, nutritional status, and genotype or species. Since NH_4^+ is generally toxic to plant cells at high concentrations, NH_4^+ pools are maintained at low levels through rapid assimilation into amino acids via the GS/GOGAT pathway (Fig. 25.1). The enzyme glutamate dehydrogenase (GDH; EC 1.4.1.3):

$$\alpha\text{-ketoglutarate} + NH_4^+ + NAD(P)H \rightarrow \text{glutamate } NAD(P) + H_2O \quad [25.4]$$

offers an alternative route for assimilation (eqn [25.4]).

However, it now appears that this ubiquitous enzyme functions in glutamate catabolism during growth and/or senescence rather than NH_4^+ assimilation. Although the physiology and biochemistry of NH_4^+ assimilation is considered elsewhere in this text (Chapter 26), a brief overview here will suffice as a reference for further discussion of molecular aspects of NH_4^+ assimilation.

Ammonia entering the GS/GOGAT pathway can come from NO_3^- reduction, symbiotic N_2 fixation, and photorespiration (Wallsgrove *et al.*, 1983). Irrespective of the source, the initial reaction in NH_4^+ assimilation involves the ATP-dependent amination of glutamate by GS to yield glutamine. The next step, catalyzed by GOGAT, involves the reductive transfer of the amide-amino group of glutamine to α-ketoglutarate to yield two glutamates. At this point, glutamate can either be used to replenish the GS–glutamate pool or may donate its N to form other nitrogenous compounds such as amino acids, nucleic acids, alkaloids, and polyamines. A second tier of control for NH_4^+ assimilation is represented by the enzyme aspartate aminotransferase, which controls the flow of carbon between amino and organic acid biosynthesis and regulates the formation of the key amino acids, aspartate and asparagine. During the last 10 years numerous studies have documented physical properties, tissue specificity, changes in activity due to environment and ontogeny, and selection for mutations in enzymes of NH_4^+ assimilation. However, the molecular processes controlling the expression of these enzymes is still poorly understood.

Glutamine synthetase

What is known of molecular mechanisms regulating NH_4^+ assimilation has been gleaned primarily from studies of GS (Cullimore *et al.*,

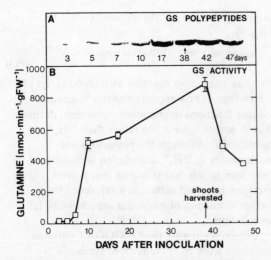

Fig. 25.6 The induction of (A) GS polypeptides and (B) GS activity during alfalfa (*Medicago sativa* L.) root nodule development. (A) Increased enzyme activity is accompanied by a striking increase in the 39 kD GS polypeptide and the appearance of a new GS polypeptide at 44 kD. Reduced enzyme activity beyond 38 days after inoculation (DAI) is accompanied by a decrease in the amount of both GS polypeptides. (B) Enzyme activity is very low until seven to ten days after inoculation (DAI). Activity then increases until 38 DAI. At 38 DAI, plant shoots were removed to mimic forage grazing. Removal of plant shoots caused a rapid decline in GS activity.

1984; Gebhardt *et al.*, 1986; Tingey *et al.*, 1987), which has been studied in a number of species. The enzyme can comprise 1 to 2% of the total soluble protein in organs actively assimilating NH_4^+. Glutamine synthetase activity increases dramatically during legume root nodule development (Fig. 25.6), in etiolated leaves exposed to light, and in leaves and roots of plants grown in NO_3^-. Changes in GS activity during organ development are attributed to differential expression of isozymes (Lara *et al.*, 1983; Cullimore and Bennett, 1988). Individual isozymes of a single species can be separated by ion exchange chromatography and native-PAGE, but not by exclusion chromatography. This indicates that GS holoenzymes are generally equivalent in size, but vary in charge. In higher plants, native GS ranges in molecular weight from 300 000 to 370 000 and is comprised of eight identical subunits of 38 000 to 46 000.

Leaves generally contain two forms of GS, one cytosolic (GS_1) and one of chloroplast origin (GS_2) (McNally and Hirel, 1983). Roots contain a single form of GS that is similar in characteristics to GS_1. The plastid GS_2 form comprises the major portion of leaf GS, although in some species GS_1 may account for as much as 40% of the total leaf activity. Antibodies prepared to GS_2 recognize the plastid form of GS from other species, but only partially recognize GS_1. Antibodies prepared to barley leaf GS_1 recognize cytosolic GS forms from leaves, roots, and root nodules of other species, but only poorly recognize plastid forms of GS. Thus, the antigenic determinants of GS_1 and GS_2 seem to be conserved across a number of species indicating that the various forms of GS are similar across species. Greening of etiolated rice leaves is accompanied by a five-fold increase in chloroplast GS_2 and a two-fold decrease in cytosolic GS_1. Immunological and deuterium labeling studies showed that the increase in GS_2 activity is due to increased synthesis of GS_2 protein. Recent studies of barley mutants lacking GS_2 suggest that plastid GS is controlled by a single nuclear gene and is involved primarily in reassimilation of NH_4^+ released during photorespiration. The above evidence indicates that leaf GS is controlled by at least two genes.

Several GS genes have been invoked in expression of root nodule GS activity. Root nodules of all legumes tested to date contain multiple GS isozymes. *Phaseolus vulgaris* L. root nodules contain two cytosolic forms of GS (Lara *et al.*, 1983). One is specific to nodules (GS_{n1}), while the other is nearly identical to cytosolic root GS. The increase in nodule GS activity is due to the appearance of the nodule-specific GS_{n1} isozyme. Increased GS activity during alfalfa root nodule development involves the enhanced expression of a form of GS found at low levels in roots and also the appearance of a new GS polypeptide (Fig. 25.6). By comparison, root nodules of pea (*Pisum sativum* L.) contain both plastid and cytosolic forms of GS, similar to those in leaves and roots (Tingey *et al.*, 1987). The large increase in GS during pea nodule development is due to an increase in cytosolic GS and there does not appear to be a nodule-specific GS in peas. The story is less clear in soybean nodules, where conflicting reports exist regarding the existence of a nodule-specific form of GS. Studies of the regulation of *Phaseolus* and pea nodule GS genes have provided our greatest insights into molecular control of NH_4^+ assimilation.

Glutamine synthetase in *Phaseolus vulgaris* occurs as five forms, two leaf, one root, and two nodule (one of which is nodule-specific GS_{n1}) (Lara et al., 1984). The leaf chloroplast GS is composed of four 45 kD polypeptides, while the cytosolic forms of roots and leaves are composed of two 43 kD polypeptides. The two 43 kD GS polypeptides common to root and leaf cytosol have been designated as the α and β subunits. The nodule-specific form of GS is also composed of two 43 kD polypeptides, one of which corresponds to the 43 kD β subunit of leaves and roots, while the other 43 kD subunit found only in nodules is designated as γ. The ratios of the various subunits vary with development and environments, suggesting that seven independent GS genes occur in *Phaseolus*.

Antibodies prepared to GS_{n1} have been used to study expression of GS_{n1} mRNAs and to isolate the GS_{n1} cDNA (Cullimore and Miflin, 1983; Cullimore et al., 1984). Immunoprecipitation of *in vitro* translation products with anti-GS_{n1} showed that the increased activity of nodule GS is due to increased GS_{n1} mRNAs. Immunopurified GS_{n1} mRNAs have been used to probe a *Phaseolus* nodule cDNA library. A single GS_{n1} cDNA was isolated and characterized. Hybrid-select translation verified that the clone actually coded for nodule-specific GS_{n1}. Southern blots of restriction digested pea genomic DNA probed with the GS_{n1} cDNA showed that four to six genes code for the enzyme, further support that GS is a multigene family.

The GS_{n1} cDNA clone was used to isolate two full-length GS cDNAs (R-1 and R-2) from a *Phaseolus* root cDNA library (Gebhardt et al., 1986). Sequence analysis of R-1 and R-2 compared to GS_{n1} showed 85% and 90% nucleotide and amino acid homology, respectively. However, the 5'- and 3'-untranslated regions of these clones were highly divergent, indicating that they are related yet distinct. The three cDNAs were used as probes to examine the expression of GS_{n1}, R-1, and R-2 during root nodule development. The genes were differentially expressed. The GS_{n1} gene was expressed only in nodules, while R-1 was expressed in both roots and nodules and R-2 was expressed only in roots. This suggests that GS_{n1} codes for the γ-subunit of nodule specific GS, R-2 codes for the α-subunit of root and leaf cytosolic GS, and R-1 codes for the β-subunit of GS present in all *Phaseolus* cytosolic GS.

A rather different picture has emerged from studies of GS in pea, where five distinct GS polypeptides are differentially expressed in leaves, roots, and nodules (Tingey et al., 1987). The predominant leaf GS is composed of a 44 kD polypeptide localized to chloroplast stroma. Leaves also express, at a much lower level, a cytosolic 38 kD GS polypeptide. This 38 kD polypeptide is the major form of GS in root cytosol. Roots and leaves also express, at a very reduced level, three 37 kD GS polypeptides. The high GS activity that occurs as root nodules develop is accompanied by a striking increase in the three 37 kD polypeptides. Nodules also express at low levels the 44 and 38 kD GS polypeptides. Similar to *Phaseolus*, pea GS appears to be under multigene control. However, there does not appear to be a solely nodule-specific form of GS in pea as there is in *Phaseolus*.

Using a heterologous GS probe, three distinct GS cDNAs (pGS197, pGS134, and pGS341) were isolated from pea leaf, root, and nodule cDNA libraries (Tingey et al., 1987). Sequence analyses showed that the coding regions of the 3 cDNAs were 72–74% homologous at the nucleotide level. The pea GS cDNAs also shared greater than 80% nucleotide homology with GS cDNA isolated from tobacco, *Phaseolus*, and alfalfa. The three pea cDNAs, however, were divergent in the 3'- and 5'-non-coding regions.

Hybrid-select translation was used to confirm that the clones were GS clones and to determine the size of the primary translation products encoded by their corresponding mRNAs. The pGS197 cDNA clone selected mRNAs that encode a primary product of 49 kD, 11–12 kD larger than the pGS341 and 134 clones, and larger by 5 kD than the *in vivo* chloroplast GS polypeptide. The 49 kD translation product represents the precursor to the 44 kD chloroplast GS_2 polypeptide and it contains a 5 kD transit peptide required for import into the chloroplasts. Northern analysis showed that pGS197 mRNAs were most highly expressed in green leaves.

The primary product translated from root RNA, selected by pGS341, was the 38 kD GS polypeptide. This same clone selected mRNAs from nodules that translated into both 38 and 37 kD GS polypeptides. Northern blot analysis with pGS341 and pGS134 inserts detected GS mRNAs whose steady-state levels in leaves and roots were low, but increased 10 to 20-fold during nodule formation. Pea cytosolic forms of GS appear to be encoded by pGS341 and 134 cDNA inserts. These data verify that three different mRNAs translate into three different sized GS polypeptides, additional evidence for separate GS genes. Southern blots of pea genomic DNA probed with the three pea GS cDNAs confirm that there are at least three distinct GS genes in pea.

Lastly, the herbicide L-phosphinothricin (L-PPT), a mixed competitive inhibitor of GS, has been used to select alfalfa altered in GS activity. Resistant lines have three to seven-fold higher GS activity than the wild type (Donn et al., 1984). The L-PPT enhanced GS had a subunit molecular weight of 39 000. A *Phaseolus* GS cDNA was used to screen a cDNA library prepared to mRNA from L-PPT resistant alfalfa callus, in order to obtain the L-PPT resistant alfalfa GS cDNA. Northern blot analysis of mRNAs from resistant and wild-type calli indicated that the increased GS activity in L-PPT resistant lines was associated with increased steady-state levels of a specific GS mRNA. Southern blot analysis of DNA from resistant and wild-type lines showed that resistance was due to the amplification of one member of the alfalfa GS gene family (Tischer et al., 1986). A genomic GS clone corresponding to the amplified gene was subsequently isolated and sequenced. The GS gene amplified in L-PPT resistant lines has two transcription start sites. These may be involved in gene regulation in response to nitrogen source as in bacterial GS. The amplified GS gene was used to complement an *Escherichia coli* GS-deficient mutant, proving that the gene encoded a complete, enzymatically active GS protein (Das Sarma et al., 1986). This complementation experiment was the first example of expression of a functional plant gene in a bacterium. The interrelatedness of bacterial GS and eukaryotic GS was further shown when Carlson and Chelm (1986) demonstrated that *B. japonicum* GS II was probably of plant origin. Thus, it would appear that not only can bacteria donate genes to plants, as in the case of *Agrobacterium*, but plants may donate genes to bacteria.

While we know that GS belongs to a multigene family in many plant species, little is known of how expression of those genes is controlled. Further understanding of how the GS genes are regulated awaits the isolation and characterization of more full-length genomic GS clones.

Other plant enzymes of NH_4^+ assimilation

Although most of the other plant enzymes involved in NH_4^+ assimilation have received attention and have in some cases been purified and characterized, knowledge of genetic mechanisms regulating their activity is sparse. Other than for maize PEPcase, cDNA and/or genomic clones for the NH_4^+ assimilating enzymes are not presently available. Because of the importance of these enzymes to plant productivity, isolation and characterization of genes for NH_4^+ assimilation will be the focus of intense research in the near future. Several approaches are currently being utilized to address the research needs in this area including: (1) generation and characterization of plant mutants in NH_4^+ assimilation; (2) isolation and characterization of enzymes; (3) production of antibodies to NH_4^+ assimilating enzymes; (4) *in vitro* translation and immunoprecipitation of primary translation products; (5) Western blot analysis of *in vivo* soluble proteins with antibodies to the various enzymes; and (6) isolation of the corresponding cDNAs for enzymes of NH_4^+ assimilation. As is evident from the foregoing list, a multidisciplinary approach is required. Rapid progress will be made through the cooperative efforts of physiologists, biochemists, breeders, geneticists and molecular biologists.

References

Brookes, S. J., Imperial, J. and Brill, W. J. (1985). Biochemical genetics of nitrogen fixation in *Klebsiella pneumoniae*. In *Nitrogen Fixation and CO_2 Metabolism*, eds P. W. Ludden and J. E. Burris, Elsevier, New York, pp. 66–74.

Burgess, B. K. (1984). Structure and reactivity of nitrogenase – an overview. In *Advances in Nitrogen Fixation Research*, eds C.Veeger and W. E. Newton, Nijhoff/Junk, Pudoc, Wageningen, pp. 103–14.

Calza, R., Huttner, E., Vincentz, M., Rouze, P., Galangau, F., Vaucheret. H., Cherel, I., Meyer, C., Kronenberger, J. and Caboche. M. (1987). Cloning of DNA fragments complementary to tobacco nitrate reductase mRNA encoding epitopes common to nitrate reductases from higher plants. *Mol. Gen. Genet.* **209**, 552–62.

Campbell, W. H. (1985). The biochemistry of higher plant nitrate reductase. In *Nitrogen Fixation and CO_2 Metabolism*, eds P.W. Ludden and J.E. Burris, Elsevier, New York, pp. 143–51.

Carlson, T. A. and Chelm, B. K. (1986). Apparent eukaryotic origin of glutamine synthetase II from the bacterium *Bradyrhizobium japonicum*. *Nature* **322**, 568–70.

Cheng, C. L., Dewdney, J., Kleinhofs, A. and Goodman, H. M. (1986). Cloning and nitrate induction of nitrate reductase mRNA. *Proc. Natl Acad. Sci. USA* **83**, 6825–8.

Crawford, N. M., Campbell, W. H. and Davis, R. W. (1986). Nitrate reductase from squash: cDNA cloning and nitrate regulation. *Proc. Natl Acad. Sci. USA* **83**, 8073–6.

Cullimore, J. V. and Bennett, M. J. (1988). The molecular biology and biochemistry of plant glutamine synthetase from root nodules of *Phaseolus vulgaris* L. and other legumes. *J. Plant Physiol.* **132**, 387–93.

Cullimore, J. V. and Miflin, B. J. (1983). Glutamine synthetase from the plant fraction of *Phaseolus* root nodules. Purification of the mRNA and *in vitro* synthesis of the enzyme. *FEBS Lett.* **158**, 107–12.

Cullimore, J. V., Gebhardt, C., Saarelainen, R., Miflin, B. J., Idler, K. B. and Barker, R. F. (1984). Glutamine synthetase of *Phaseolus vulgaris* L: 7 Organ-specific expression of a multigene family. *J. Mol. Appl. Genet.* **2**, 589–99.

Das Sarma, S., Tischer, E. and Goodman, H. M. (1986). Plant glutamine synthetase complements a *glnA* mutation in *Escherichia coli*. *Science* **232**, 1242–44.

Dixon, R. A. (1984). The genetic complexity of nitrogen fixation. *J. Gen. Microbiol.* **130**, 2745–55.

Donn, G., Tischer, E., Smith, J. A. and Goodman, H. M. (1984). Herbicide resistant alfalfa cells: an example of gene amplification in plants. *J. Mol. Appl. Genet.* **2**, 621–35.

Earl, C. D., Ronson, C. W. and Ausubel, F. M. (1987). Genetic and structural analysis of the *Rhizobium meliloti fix A, fix B, fix C* and *fix X* genes. *J. Bacteriol.* **169**, 1127–36.

Gebhardt, C., Oliver, J. E., Forde, B. G., Saarelainen, R. and Miflin, B. J. (1986). Primary structure and differential expression of glutamine synthetase genes in nodules, roots and leaves of *Phaseolus vulgaris*. *EMBO J.* **5**, 1429–35.

Govers, F., Nap, J. P., Van Kammen, A. and Bisseling, T. (1987). Nodulins in the developing root nodule. *Plant Physiol. Biochem.* **25**, 309–22.

Johnston, A. W. B., Downie, J. A., Rossen, L., Shearman, C. A., Firmin, J. L., Borthakur, D., Wood, E. A., Bradley, D. and Brewin, N. J. (1987). Molecular analysis of the *Rhizobium* genes involved in the induction of nitrogen-fixing nodules in legumes. *Phil. Trans. R. Soc. London B.* **317**, 193–207.

Kleinhofs, A., Warner, R. L. and Navayanan, K. R. (1985). Current progress toward an understanding of the genetics and molecular biology of nitrate reductase in higher plants. In *Oxford Surveys of Plant Molecular and Cell Biology*, Vol. 2, ed. B. J. Miflin, Oxford University Press, Oxford, pp. 91–121.

Kondorosi, E. and Kondorosi, A. (1986). Nodule induction on plant roots by *Rhizobium*. *Trends Biochem. Sci.* **11**, 296–9.

Kosslak, R. M., Bookland, R., Barkei, J., Paaren, H. E. and Appelbaum, E. R. (1987). Induction of *Bradyrhizobium japonicum* common *nod* genes by isoflavones from *Glycine max*. *Proc. Natl Acad. Sci. USA* **84**, 7428–32.

Lara, M., Cullimore, J. V., Lea, P. L., Miflin, B. J., Johnston, A. W. B. and Lamb, J. W. (1983). Appearance of a novel form of plant glutamine synthetase during nodule development in *Phaseolus vulgaris* L. *Planta* **157**, 254–8.

Lara, M., Porta, H., Padilla, J., Folch, J. and Sanchez, F. (1984). Heterogeneity of glutamine synthetase polypeptides in *Phaseolus vulgaris* L. *Plant Physiol.* **76**, 1019–23.

McNally, S. and Hirel, B. (1983). Glutamine synthetase isoforms in higher plants. *Physiol. Veg.* **21**, 761–74.

Ow, D. W. and Ausubel, F. A. (1983). Regulation of nitrogen metabolism genes by *nif A* gene product in *Klebsiella pneumoniae*. *Nature* **301**, 307–13.

Peters, N. K., Frost, J. W. and Long, S. R. (1986). A plant flavone, luteolin, induces expression of *Rhizobium meliloti* nodulation genes. *Science* **233**, 977–80.

Somers, D. A., Kuo, T. M., Kleinhofs, A., Warner, R. L. and Oaks, A. (1983). Synthesis and degradation of barley nitrate reductase. *Plant Physiol.* **72**, 949–52.

Stougaard, J., Sandal, N. N., Gron, A., Kuhle, A. and

Marcker, K. A. (1987). 5' Analysis of the soybean leghemoglobin lbc$_3$ gene: Regulatory elements required for promoter activity and organ specificity. *EMBO J.* **6**, 3565–9.

Tingey, S. V., Walker, E. L. and Coruzzi, G. M. (1987). Glutamine synthetase genes of pea encode distinct polypeptides which are differentially expressed in leaves, roots, and nodules. *EMBO J.* **6**, 1–9.

Tischer, E., Das Sarma, S and Goodman, H. M. (1986). Nucleotide sequence of an alfalfa glutamine synthetase gene. *Mol. Gen. Genet.* **203**, 221–9.

Vance, C. P., Egli, M. A., Griffith, S. M. and Miller, S. S. (1988). Plant regulated aspects of nodulation and N$_2$ fixation. *Plant Cell. Environ.* **11**, 413–27.

Verma, D. P. S., Fortin, M. G., Stanley, J., Mauro, V. P., Purohit, S. and Morrison, N. (1986). Nodulins and nodulin genes of *Glycine max*. A perspective. *Plant Mol. Biol.* **7**, 51–61.

Wallsgrove. R. M., Keys. A. J., Lea. P. and Miflin, B. J. (1983). Photosynthesis, photorespiration, and nitrogen metabolism. *Plant Cell. Environ.* **6**, 301–9.

Warner, R. L., Kleinhofs, A. and Narayanan, K. R. (1984). Genetics, biochemistry and physiology of nitrate reductase-deficient mutants in barley. In *Exploitation of Physiological and Genetic Variability to Enhance Crop Productivity*, eds J. E. Harper, L. E. Schrader and R. W. Howell, American Society of Plant Physiologists, Rockville, Maryland, pp. 23–30.

26 N_2 fixation, NO_3^- reduction and NH_4^+ assimilation

David B. Layzell

Introduction

Elemental nitrogen is one of the most important nutrients required for plant growth. It is a key constituent of protein, nucleic acids and other cellular components and its availability in the environment frequently limits the growth and yield of crop plants. Most plants acquire N from the soil solution as either nitrate (NO_3^-) or ammonium (NH_4^+) ions. In addition, some plants are able to utilize the atmospheric N_2 pool through symbiotic associations with species of bacteria, cyanobacteria or actinomycetes that contain the N_2 fixing enzyme, nitrogenase.

This chapter will summarize some of what is known of the structure, localization, catalytic mechanism and regulation of those enzymes which are responsible for the assimilation of inorganic N. The pathway of N assimilation will be considered to the point of glutamate formation. Glutamate is one of the first organic forms of N synthesized in plants, and its subsequent metabolism is central to the synthesis of virtually all N compounds within the plant (Miflin et al., 1981). Finally, the relative energy cost of assimilating N_2, NO_3^- and NH_4^+ will be considered briefly, along with a discussion of how theoretical costings such as these are done.

N_2 fixation

Nitrogenase-catalyzed reactions and the electron allocation coefficient

Nitrogenase is a prokaryotic enzyme which is known to catalyze the transfer of electrons to a number of substrates. In an atmosphere containing N_2 gas, nitrogenase catalyzes the reduction of both N_2 to NH_4^+ (N_2 fixation, eqn [26.1]) and protons (H^+) to H_2 gas (H_2 production, eqn [26.2]):

$$N_2 + 12ATP + 6e^- + 8H^+ \rightarrow 2NH_4^+ + 12ADP + 12P_i$$
(N_2 fixation) [26.1a]

$$4ATP + 2e^- + 2H^+ \rightarrow H_2 + 4ADP + 4P_i$$
(H_2 production) [26.1b]

As will be discussed later, the production of H_2 seems to be associated with the binding of N_2 to the enzyme; hence a ratio of at least $1H_2$ produced per N_2 fixed is observed (Simpson, 1987). In many N_2 fixing systems, rates of H_2 production occur which are in excess of the 1:1 ratio with N_2 fixation and the term *electron allocation coefficient* (EAC) is used to describe the proportion of total electron flow through the enzyme which is allocated to N_2 fixation rather than H_2 evolution. Since N_2 fixation consumes three electron pairs and H_2 production only one electron pair (eqn [26.1a,b]), the maximum EAC (i.e. at $1H_2/N_2$ fixed) is 0.75. In legume symbioses, values for EAC vary from 0.40 to 0.75, with most values around 0.60. The reasons for this variation are not fully understood. Methods for measuring EAC will be discussed in more detail below.

In vivo, electrons are passed to nitrogenase, one at a time, by either ferredoxin or flavodoxin, with ferredoxin being the most common electron donor. *In vitro*, sodium dithionite ($Na_2S_2O_4$) is generally used as an artificial electron donor since it is far more convenient to use than the natural compounds. Besides electrons, the nitrogenase reaction requires the hydrolysis of at least two MgATP for every electron passed, or about 16

MgATP to fix one N_2 and produce one H_2. A possible role for ATP and a mechanism for electron transfer to N_2 will be proposed later.

In the absence of N_2 gas (e.g. in an atmosphere where N_2 is replaced by Ar or He), all electron flow through nitrogenase is used to reduce protons to H_2 as shown in eqn [26.2].

$$16ATP + 8e^- + 8H^+ \rightarrow 4H_2 + 16ADP + 16P_i$$
(H_2 production in Ar) [26.2]

A stoichiometry showing the production of $4H_2$ has been chosen here so that the total electron flow through the enzyme ($8e^-$ or 4 electron pairs) will be similar to that in the combined eqns of [26.1a] and [26.1b].

Similarly, in the presence of saturating concentrations of acetylene (e.g. 10 kPa C_2H_2), virtually all of the electron flow through nitrogenase is used in the reduction of C_2H_2 to ethylene (C_2H_4) (eqn [26.3]) and few, if any of the electrons are allocated to N_2 fixation or H_2 production.

$$4C_2H_2 + 16ATP + 8e^- + 8H^+ \rightarrow 4C_2H_4 + 16ADP + 16P_i$$
(C_2H_2 reduction) [26.3]

C_2H_2 is a non-competitive inhibitor of N_2 fixation and H_2 production, acting at a site which is separate from the N_2 binding site, but competing for electrons from the same pool as N_2 fixation and H_2 production (Ludden and Burris, 1986).

Carbon monoxide (CO) is another non-competitive inhibitor of N_2 fixation. However, unlike C_2H_2, CO only binds to the enzyme and is not reduced. At saturating levels of CO, N_2 fixation by purified nitrogenase is blocked and all electron flow through nitrogenase is diverted to H^+ reduction to H_2.

The rate of total electron flow through nitrogenase (*total nitrogenase activity, TNA*) is routinely measured as either H_2 evolution in $Ar:O_2$ (79:21) or C_2H_4 production in an atmosphere of 10 kPa C_2H_2. The former approach can only be used in symbioses in which all of the H_2 produced by nitrogenase is lost to the environment. This may not always be the case since some N_2 fixers have an enzyme called an uptake hydrogenase (HUP) which can recover some or all of the H_2 produced.

Uptake hydrogenase activity and relative efficiency

A membrane-bound, unidirectional uptake hydrogenase (HUP) may be found in many N_2 fixers. Like other hydrogenases, it is a Fe–S protein, and in rhizobia species and some other bacteria, Ni is an integral component of the active enzyme. HUP catalyzes the oxidation of H_2 and, in most cases, passes the electrons to a respiratory pathway to produce ATP and consume O_2 (Maier, 1986). Presumably, the ATP produced can be utilized by the bacteria, thereby improving the efficiency of energy use by the bacteria or symbiotic association. Some evidence exists for increased N_2 fixation and plant productivity in symbioses which express HUP activity (Eisbrenner and Evans, 1983).

While some N_2 fixing bacteria lack the gene that codes for HUP, in those with the gene, HUP expression is regulated by a variety of factors. For example, H_2 is required for the induction of HUP while O_2 and C substrates repress its expression. Studies with a range of *Bradyrhizobium japonicum* mutants have lead to the suggestion (Maier, 1986) that the regulation by O_2 and C substrates could be related to the redox state of the pyridine nucleotide pool and/or the energy charge of the cells. It is interesting to note that various workers have reported a host effect on the expression of HUP activity (Phillips *et al.*, 1985).

The term '*relative efficiency*' (RE) may be defined as the proportion of electron flow through nitrogenase which is not lost as H_2 evolution to the environment. Hence, RE differs from EAC in that it takes into consideration the net H_2 exchange resulting from the activities of both nitrogenase and HUP (Saari and Ludden, 1987). In a N_2 fixing system lacking uptake hydrogenase activity, RE = EAC.

Measurement of nitrogenase activity, EAC and RE

A number of methods have been developed to measure N_2 fixation and electron flow through nitrogenase (Layzell *et al.*, 1984; Reporter, 1985; Rennie 1986). In general, techniques involving the direct measurement of N_2 fixation are expensive and time consuming, and are difficult to perform in an assay which is not destructive to the plant material being analyzed. More commonly, nitrogenase activity and the associated EAC or RE are measured as rates of H_2 production (in the presence or absence of N_2) or as C_2H_2 reduction activity. In a symbiosis lacking HUP activity (HUP$^-$ symbiosis), EAC is routinely estimated as one of the two formulas of eqn [26.4].

$$EAC = 1 - \frac{\text{rate of } H_2 \text{ prod'n in } N_2 : O_2}{\text{rate of } H_2 \text{ prod'n in Ar} : O_2}$$

$$= 1 - \frac{\text{rate of } H_2 \text{ prod'n in } N_2 : O_2}{\text{rate of } C_2H_2 \text{ reduction}} \quad [26.4]$$

In symbioses which express HUP activity (HUP$^+$ symbioses), H_2 evolution cannot be used in the determination of EAC, therefore necessitating the use of one of the following formulas:

$$EAC = \frac{\text{rate of } ^{15}N_2 \text{ fixation} \times 3}{\text{rate of } C_2H_2 \text{ reduction}}$$

$$= \frac{\text{rate of } NH_3 \text{ production} \times 1.5}{\text{rate of } C_2H_2 \text{ reduction}} \quad [26.5]$$

where N_2 fixation and NH_3 production are assumed to consume 3 and 1.6 times, respectively, the number of electrons used in the reduction of C_2H_2 to C_2H_4. ^{15}N is a stable isotope of N which is found in air at a natural abundance of 0.365%. Analysis of ^{15}N content of plant material following a period when the nodules are exposed to an atmosphere enriched in ^{15}N provides a measure of the N_2 fixation rate. ^{13}N, a radioactive isotope of N, could also be used in such experiments but its short half-life (c. 10 min) limits its usefulness.

As mentioned previously, in HUP$^-$ symbioses, EAC = RE. In HUP$^+$ symbioses, the latter formula in eqn [26.4] may be used to determine RE.

Nitrogenase structure

Nitrogenase is a prokaryotic enzyme which catalyzes the reduction of N_2 to NH_3. It is made up of two component proteins. The Mo–Fe protein (also called dinitrogenase or component 1), has a molecular weight of about 200 000–250 000 and is known to contain the active site of the enzyme (Burgess, 1985). It has 2Mo and about 32Fe and 32 acid-labile S per molecule. When dissociated, the MoFe protein separates into 4 subunits; 2 each of two types (i.e. A_2B_2). The Fe protein (also called dinitrogenase reductase or component 2) has a total molecular weight of 60 000, and can be dissociated into 2 similar subunits. Each Fe protein contains an active center with four Fe and four acid-labile S atoms. As will be discussed below, the Fe protein is thought to accept electrons from ferredoxin and pass these to the Mo–Fe protein.

In recent years, 'alternative' nitrogenases have been discovered in some N_2 fixing prokaryotes. In one of these, component 1 is a vanadium–iron (V–Fe) protein (Robson et al., 1986), and both the Fe protein and V–Fe protein of this nitrogenase are coded for in a region of the bacterial genome distinct from the region which codes for the standard Mo–Fe and Fe protein described in the previous paragraph. More recently, another nitrogenase has been identified in *Azotobacter* in which the component 2 lacks either Mo or V (Bishop et al., 1988). There have been some suggestions that these alternative nitrogenases may be remnants of primitive, evolutionary forms of the nitrogenase enzyme. Since they have yet to be identified in symbiotic bacteria, and relatively little is known of the alternative nitrogenases, all subsequent discussion will be in reference to the Mo-containing nitrogenase.

Nitrogenase regulation

Nitrogenase activity *in vivo* is almost certainly regulated at a variety of levels, including transcription, translation, substrate supply, covalent modification and allosteric effectors. In addition, the importance of any one regulatory process seems to be dependent upon the N_2 fixing species and the symbiosis being considered. Therefore, to date, no widely accepted mechanism has been proposed to describe how nitrogenase is regulated in a free-living N_2 fixing bacteria, let alone in a legume symbiosis. Recently, reviews have been written of nitrogenase regulation at the level of transcription and translation (Ludden and Burris, 1986) and the subject is also covered in

Chapter 25. Other aspects of regulation will be briefly considered here.

NH_4^+ and NO_3^- as inhibitors

Although NH_4^+ is the first stable product of nitrogenase, there is no evidence of direct feedback regulation of the enzyme by this compound. However, in some organisms (*Azotobacter, Rhizobium* sp., cyanobacteria), high concentrations of NH_4^+ can disrupt membrane potential and thereby indirectly cause a loss of *in vivo* nitrogenase activity (Ludden and Burris, 1986). In addition, it is well known that in legume symbioses, NO_3^- or NH_4^+ additions result in a decline of nitrogenase activity in existing nodules and an inhibition of both new nodule formation and nodule development. While various hypotheses have been proposed to account for these phenomena, no single regulatory mechanism is widely accepted (Streeter, 1988, Vessey *et al.*, 1988b).

Role of MgADP and MgATP

MgADP, another product of nitrogenase activity, is a potent inhibitor of the purified enzyme, having a $K_i(MgADP)$ of 20 μM. Considering the cellular ADP concentration, the ADP/ATP ratio *in vivo* and the relatively high $K_m(MgATP)$ for nitrogenase (80–200 μM), it has been estimated that nitrogenase activity *in vivo* should be inhibited by 90% or more (Ludden and Burris, 1986). This is a paradox. Not only do the cells show little, if any, evidence of an ADP inhibition, but *in vivo* nitrogenase activity seems to be relatively independent of energy charge.

One possible explanation is that the proton gradient across the bacterial membrane may regulate both electron flow to nitrogenase and the Mg^{2+} level in the cell. Since ATP binds Mg^{2+} more tightly than ADP, when Mg^{2+} is limiting, the effective ratio of MgATP/MgADP operating in the cell would be much higher than the pool measurements may indicate (Ludden and Burris, 1986).

Role of O_2

O_2 is an irreversible inhibitor of nitrogenase. *In vitro* studies have shown the Fe protein to be the most O_2 labile, with a half-life in air of only seconds to minutes. The paradox is that O_2 is also required as a substrate for oxidative phosphorylation, enabling the production of ATP. This has necessitated the evolution of highly specialized mechanisms to provide the microsymbiont with a high flux of O_2 to support ATP production, yet at a low O_2 concentration so as not to inhibit nitrogenase. Recent modeling studies with legumes have proposed the existence of a variable barrier to O_2 diffusion in the nodule inner cortex (Hunt *et al.*, 1988) which works in concert with nodule respiration to maintain the central zone of the nodule at a low, but stable O_2 concentration (c. 5–60 nM in the cytosol). The high flux of O_2 through the infected cell to the bacteroids is achieved by facilitated diffusion of O_2 by oxygenated leghemoglobin. Leghemoglobin is an O_2-binding protein which is present at high concentration (c. 0.7 mM; Bergersen, 1982) in the cytosol of the infected cells.

Studies with legumes have shown that treatments involving a decrease in photosynthate supply to nodules are associated with an increase in the resistance of the diffusion barrier and an increase in the O_2 limitation of nodule metabolism and, therefore, nitrogenase activity (Vessey *et al.*, 1988a). It has been suggested that O_2 supply to the bacteroids may limit oxidative phosphorylation such that ATP availability regulates nitrogenase activity *in vivo*. It is interesting to note that NO_3^- treatment of legume roots also causes an increase in the O_2 limitation of respiration and nitrogenase activities (Vessey *et al.*, 1988b), an observation which is consistent with NO_3^- inhibiting N_2 fixation by causing a decline in photosynthate supply to nodules.

In filamentous cyanobacteria, nitrogenase is localized within heterocysts and glycolipid layers external to the heterocyst wall are thought to provide a barrier to O_2 diffusion (Wolk, 1982). The O_2 that does diffuse into the heterocyst is consumed in respiration (electrons derived from carbohydrate metabolism or uptake hydrogenase activity). Since heterocysts lack Photosystem II, they do not produce O_2 and can only carry out cyclic photophosphorylation. Therefore, the light reactions of photosynthesis can provide some or all of the ATP requirements of nitrogenase while its reductant requirements seem to be provided by the metabolism (pentose phosphate pathway) of a disaccharide formed in the adjacent vegetative cells by photosynthesis (Wolk, 1982).

In *Azotobacter*, high O_2 results in the reversible binding of a protein to nitrogenase. This alters the conformation state of nitrogenase and protects it from irreversible inactivation. If conditions of low pO_2 are restored, the bound enzyme is released and nitrogenase activity is recovered. Hence, O_2 acts as an uncompetitive inhibitor in *Azotobacter*. No direct evidence exists for a similar mechanism in any symbiotic association.

H_2 as an inhibitor

If, as the evidence suggests, the cortical barrier to gas diffusion in legume nodules is a physical barrier, then O_2 and N_2 diffusion into the nodules would be restricted, as would H_2 and CO_2 diffusion out of the nodule. Under normal atmospheric conditions, there is no evidence that N_2 or CO_2 concentration in the infected cells limits nitrogenase activity. However, it has been suggested that H_2 may build up to concentrations which are inhibitory to N_2 fixation, especially in nodules which lack uptake hydrogenase activity (Hunt et al., 1988).

H_2 is both a product of nitrogenase activity and a specific, competitive inhibitor of the N_2 fixation reaction (Burris, 1985). Studies with D_2, the stable isotope of H_2, have shown that in the presence of N_2, H_2 (or D_2) acts at the active site of nitrogenase to inhibit N_2 fixation and to induce the production of two H_2 (or two HD) for every H_2 (or D_2) reacting at the active site. To date, little work has been done to determine whether H_2 inhibits N_2 fixation *in vivo*, although some theoretical considerations indicate that the H_2 concentration in legume nodules may be near the $K_i(H_2)$ values for N_2 fixation (80–150 μM, Hunt et al., 1988).

The nitrogenase mechanism

Thorneley and Lowe (1985) recently proposed a model which attempts to describe the mechanism of nitrogenase action in the fixation of N_2 and the production of H_2. This model, which is consistent

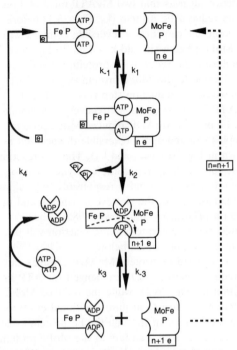

Fig. 26.1 Model of the Fe protein cycle of nitrogenase as proposed by Thorneley and Lowe (1985). The binding sites for the MgATP (ATP) and MgADP (ADP) molecules are shown to be localized between the Fe protein (Fe P) and MoFe protein (MoFe P) components. In fact, little is known of the precise location of the MgATP binding sites. The MoFe component shown here represents one of the two independently-functioning halves of the tetrameric MoFe protein. The MoFe component receives from the Fe protein one of the eight electrons (e) that it requires in order to reduce N_2 or protons to NH_4^+ or H_2, respectively. As each electron is passed to the MoFe component, the number of electrons it holds (n) is incremented by one (i.e. n = n + 1). Note that the protein dissociation step (reaction k_{-3}) is the slow, rate-determining step in the cycle. The MoFe protein cycle of Fig. 26.2 shows how the 8 cycles of the Fe protein cycle may be organized to bring about N_2 fixation and H_2 production.

with virtually all experimental results collected to date, separates the mechanism into two cycles, the Fe protein cycle, and the MoFe protein cycle. In the Fe protein cycle (Fig. 26.1), electrons are passed first to the Fe protein, and then from the

Fe protein to the MoFe protein. Experimental evidence suggests that two MgATP must be bound to the reduced Fe protein (k_4, Fig. 26.1) before this molecule can form a reversible complex with the MoFe protein (k_1 and k_{-1}, Fig. 26.1). Once complexed, the reduced Fe protein can transfer the electron to the MoFe protein (k_2, Fig. 26.1). This oxidation–reduction step is coupled to the MgATP hydrolysis, and is effectively irreversible. The Fe protein:MgADP:reduced MoFe protein complex can undergo a reversible dissociation as shown in Fig. 26.1 (k_{-3} and k_3). The k_{-3} reaction is the rate-limiting step in the catalytic cycle for substrate reduction, and is responsible for making nitrogenase one of the slowest enzymes found in bacteria (Thorneley and Lowe, 1985). The significance of this rate-limiting reaction will be discussed in more detail below.

The oxidized Fe protein:MgADP complex can be reduced and the ADP exchanged for ATP in reaction k_4 (Fig. 26.1), while the reduced MoFe protein can return to acquire additional electrons from the Fe protein through the Fe protein cycle. In total, 8 electron transfers to the MoFe protein are required to reduce $1N_2$ and produce $1H_2$, and between each electron transfer the Fe protein:MoFe protein complex dissociates completely. This cycle of 8 electron transfers has been called the MoFe protein cycle (Thorneley and Lowe, 1985) and is summarized in Fig. 26.2. Note that the Fe protein cycle (Fig. 26.1) forms the basic unit for a MoFe cycle.

When the MoFe protein has received 3 or 4 electrons (i.e. E_3H_3 or E_4H_4), it is thought to be in a state which is receptive to N_2 binding (k_{10} and k_{11}, Fig. 26.2). The binding of N_2 to the enzyme is thought to result in a concomitant release of H_2 gas, thereby accounting for the prerequisite one H_2 evolved per N_2 fixed, discussed previously. In this model, additional H_2 can be evolved from the dissociation of the E_2H_2, E_3H_3 and E_4H_4 intermediates (reactions k_7, k_8 and k_9, Fig. 26.2). Therefore, these intermediates are the substrates for the reactions leading to both N_2 binding/fixation and wasteful H_2 production. Based on a kinetic analysis of the competing reactions, Thorneley and Lowe (1985) predict that minimal rates of H_2 production by nitrogenase will result if (1) the *in vivo* concentration of nitrogenase is as

The MoFe Protein Cycle

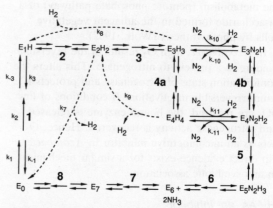

Fig. 26.2 Model of the MoFe protein cycle of nitrogenase as proposed by Thorneley and Lowe (1985). Note that reactions k_1 to k_3 (and k_{-1} to k_{-3}) of the Fe protein cycle (Fig. 26.1) forms the basic unit for the MoFe cycle. The large numbers (1 to 8) associated with each set of arrows represents the sequential transfer of electrons from the Fe protein to the MoFe protein. The species E_n is equivalent to the MoFe protein of Fig. 26.1, where the subscript n represents the total number of electrons which have been passed to that protein. The H_n and N_n associated with E refers to the number (n) of H and N atoms bound to the enzyme, respectively. N_2 binding to the enzyme and the H_2 production associated with that binding is depicted in reactions k_{10} and k_{11} while reactions k_{-10} and k_{-11} show a possible mechanism for the competitive inhibition of H_2 on N_2 fixation. The dashed lines (k_7 to k_9) show possible pathways of H_2 production not associated directly with N_2 binding to the enzyme. The protons involved in the reaction are assumed to be derived from pools within the bacterial cytosol, and the precise stage at which ammonia is released from the enzyme is not known.

high as possible, and (2) the pool sizes of the E_2H_2, E_3H_3 and E_4H_4 intermediates are very low. Herein lies explanations both for the high *in vivo* concentration of nitrogenase and for the slow rate of catalysis of reaction k_{-3} (Fig. 26.1), thereby making it the rate-determining step in the nitrogenase mechanism. If the dissociation of the Fe protein:MgADP:MoFe protein complex is slow, pool sizes of the reduced MoFe protein will be very low, and N_2 binding will be favored over H_2 production. In the absence of N_2 gas (i.e. eqn

[26.2]), reactions k_7, k_8 and k_9 (Fig. 26.2) would be favored and, in effect, all electron flow through nitrogenase would go to H_2 production.

The binding of N_2 and release of H_2 seems to be a reversible reaction, such that H_2 can displace N_2 from the active site. This portion of the model can account for the observation that H_2 is a competitive inhibitor of N_2 fixation and that, in the presence of D_2, N_2 fixation is inhibited and HD is produced (presumably through dissociation of intermediates E_3HD_2 and $E_4H_2D_2$ in Fig. 26.2).

Nitrate assimilation

NO_3^- uptake, transport and the site of assimilation

In well-aerated, non-acidic soils, the activity of nitrifying bacteria ensure that virtually all of the available N is present as NO_3^-. Most plants are very efficient at extracting NO_3^- from their root environment, and are able to maintain tissue or xylem sap concentrations at levels which are much higher than that in the soil solution. Unfortunately, to date the mechanism and regulation of NO_3^- uptake by the root are not well understood. Deane-Drummond and Thayer (1986) have provided evidence that the plant regulates both uptake and efflux, and the net uptake rate is the difference between the two processes. Whatever the mechanism, the uptake of the nitrate anion must be balanced by either the uptake of a cation (e.g. K^+, Mg^{2+}, etc.) or the excretion of another anion (e.g. OH^- or HCO_3^-) to maintain electroneutrality within the plant. These concepts are discussed in detail in a recent review (Jackson et al., 1986).

The NO_3^- which is taken up may be either reduced in the roots, stored in the vacuole of root cells or transported to the shoots in the xylem. Nitrate received by the shoot can either be stored or reduced by stem or leaf organs. In general, nitrate is thought not to be phloem-mobile. The relative importance of leaves, stems and roots in contributing to whole-plant NO_3^- assimilation varies with species and the rate of NO_3^- uptake (Andrews, 1986). In most tropical or subtropical plants, shoot NO_3^- reduction seems to be predominate. However, in many temperate plants, especially legumes, roots rather than shoots reduce most of the NO_3^-, particularly at low NO_3^- concentrations. Andrews (1986) has suggested that in these temperate species, the lower proportion of shoot NO_3^- assimilation may contribute to improved tolerance to low temperature.

NO_3^- reduction to ammonia

The reduction of NO_3^- to ammonia is a two-step, two enzyme process, with nitrite (NO_2^-) being the intermediate species. Nitrate reductase (NR) catalyzes the two-electron reduction of nitrate to nitrite:

$$2H^+ + NO_3^- + 2e^- \rightarrow NO_2^- + H_2O \qquad [26.6]$$

while nitrite reductase (NiR) passes 6 electrons to nitrite and produces ammonium:

$$8H^+ + NO_2^- + 6e^- \rightarrow NH_4^+ + 2H_2O \qquad [26.7]$$

In general, NR is considered to be a cytosolic enzyme, but some evidence suggests that it is located on the outer membranes of either the chloroplasts in the leaf or the plastids in the root. NiR is a chloroplast or plastid enzyme in the leaf or roots, respectively.

Numerous pieces of evidence support the conclusion that NR is the rate-determining enzyme in the pathway of NO_3^- assimilation. For example, as the first enzyme in the pathway, it is a logical place for regulation, and nitrite – the toxic product of the NR reaction – is rarely found in high concentrations in plant tissues. Also, NR is a relatively unstable enzyme, and its activity is substrate inducible. Finally, the maximal activity (per gram dry weight of tissue) for NiR is not only 5-fold (or more) higher than that for NR, but its $K_m(NO_2^-)$ is much lower than the $K_m(NO_3^-)$ for NR. These characteristics will tend to ensure that any NO_2^- that is produced will be quickly reduced to NH_4^+. Because of the apparent role of NR in regulating nitrate assimilation, much more is known of its structure, properties and mechanism than that of NiR. The following discussion will consider only NR; further details on both of these enzymes may be found in the review by Beevers and Hageman (1980).

Structure and mechanism of nitrate reductase

Like most reduction/oxidation enzymes, NR can be classified according to the nature of the electron donor. In cyanobacteria and other prokaryotes, NR has a specific requirement for reduced ferredoxin, while in eukaryotes, more than one NR may be present, but all types require pyridine nucleotides (NADH, NADPH) as electron donors. In soybean, for example, there are two constitutive NR which use NADPH as an electron donor, but differ in their $K_m(NO_3^-)$. A third NR is NO_3^- inducible and requires NADH (Harper, 1987). The presence of one or more constitutive enzymes in soybean makes it unusual among higher plants, and more work is needed to determine the significance and role of these isozymes in the N metabolism of this plant.

Much of what is known about NR is from work done on the NADH-dependent, NO_3^-- inducible form of the enzyme which has been isolated from various eukaryotes. This enzyme is composed of two subunit proteins of approximately 115 kD each (Fig. 26.3A), and containing three 40 kD 'domains' which house the individual components of a self-contained electron transport chain. The first domain contains a flavin adenine dinucleotide (FAD domain), the second, cytochrome b_{557} (cyt. b domain), and the third, molybdenum (Mo domain). In the 'complete' reaction of nitrate reduction to nitrite (eqn[26.6]), electrons are passed from NADH to the FAD domain and then to the cyt. b and Mo domains before being passed to NO_3^- (Fig. 26.3B). The active site for NO_3^- binding and reduction is within the Mo domain.

Alternative substrates such as chlorate, bromate and iodate can be reduced to chlorite, bromite and iodite, respectively. While of little significance in nature, chlorate treatment of mutagenized plants has been used in the selection for NR mutants (Campbell and Smarrelli, 1986). The chlorite produced in the reaction is not metabolized further by the plant and since it is toxic, only plants lacking NR activity will survive the chlorate treatment.

NR will also catalyze a number of 'partial' reactions in which electrons are passed through only one or two of the three domains of the

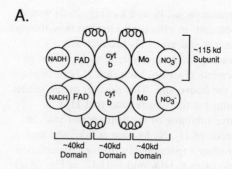

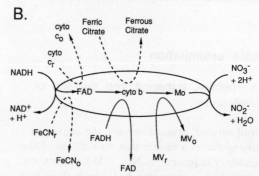

Fig. 26.3 The structure and reactions of nitrate reductase. (A) Structure of nitrate reductase showing the three 40 kD domains of the dimeric protein and the location of binding sites for NADH and NO_3^-. (B) Summary of some of the complete (solid lines) and partial reactions of nitrate reductase. Where the partial reactions deviate from the complete reaction, they are identified as having either dehydrogenase (dashed lines) and reductase (dotted lines) types of activities. Other terminal electron acceptors besides nitrate for the complete and partial reductase activities include chlorate, bromate or iodate. MV, methyl viologen. After Campbell and Smarrelli (1986).

enzyme. While virtually all of the partial reactions are non-physiological, they have been extremely valuable in elucidating the various properties and the structure of the enzyme. There are two types of partial reactions (Fig. 26.3B): (1) the *dehydrogenase* reactions receive electrons from NADH at the FAD domain, and pass them directly (or through the cyt. b domain) to reduce cytochrome c, ferricyanide, methylene blue or ferric citrate. Since ferrous citrate may be used as a substrate in porphyrin synthesis, this is one partial reaction which may have physiological relevance; (2) *the reductase* reactions use electron donors such as FADH or reduced methyl viologen to pass electrons to the cyt. *b* or Mo domain to reduce nitrate (or other substrates) at the active site on the Mo domain.

Regulation of nitrate assimilation

To elucidate the physiological and biochemical mechanism(s) responsible for regulating NR activity in plants, numerous studies have examined the influence of the following factors on the synthesis, degradation and activity of NR: light, mineral nutrition, N source, water stress, temperature, plant growth regulators, plant age, metabolic inhibitors, and energy sources. The results of these studies have been discussed in detail elsewhere (Beevers and Hageman, 1980; Naik *et al.*, 1982; Campbell and Smarrelli, 1986), and will not be dealt with here. However, it is important to note that NR activity in photosynthetic tissues generally varies on a diurnal basis, with peak rates during the light period and the lowest rates at the end of the dark period. Unfortunately, at the present time, no comprehensive, well-accepted explanation exists to describe the mechanism of regulation in these tissues. Certainly, the activity of NR is likely to be a function of a complex set of factors, and these may vary in importance in different plants, or in different organs within the same plant (Campbell and Smarrelli, 1986). Some important pieces in the 'puzzle' of NR regulation are discussed below.

NR turnover

The steady-state activity of NR in a plant tissue seems largely to be a function of the simultaneous rates of synthesis and degradation of active NR. Studies by A. Oaks, W. Wallace, T. Yamaya and coworkers (see Campbell and Smarrelli (1986) for references) have provided evidence for the existence of a protease with a high specificity for NR. Much work remains to be done in elucidating the factors responsible for regulating the *de novo* synthesis of NR and its specific protease.

NR affectors

Various workers have discovered certain proteins in plant tissues that can bind reversibly to NR and inhibit NR activity upon binding (Cambell and Smarrelli, 1986). Such proteins have been reported in rice cell cultures, rice roots and soybean leaves. In soybean, the inhibitor was found to be present in the dark at higher concentrations than in the light, thereby accounting for the diurnal fluctuations observed in NR activity. Further work is required to ascertain the physiological significance of these protein inhibitors. While present in some tissues, it is possible that they are localized in a cellular compartment separate from NR and therefore bind with and inhibit NR activity only when the tissues are ground for analysis (Campbell and Smarrelli 1986).

Substrate availability

As mentioned previously, the activity of the NADH-dependent form of NR increases dramatically in the presence of nitrate. This is one of the few higher plant examples of substrate induction, a phenomenon that is common in bacterial enzymes. Indeed, in the majority of higher plants that have been studied, NR activity is most greatly affected by nitrate availability, and feedback regulation of NH_4^+ on NR does not seem to be an important aspect of regulatory control.

If nitrate is the only oxidant available to NR *in vivo*, then its concentration may control the catalytic turnover rate of the enzyme. This is because NR is more susceptible to degradation when it becomes over-reduced, and becoming over-reduced would be likely to occur under conditions of low nitrate availability (Beevers and Hageman, 1980). Another factor which may be important in regulating enzyme degradation is the concentration of Fe chelators, since they may also oxidize NR and thereby maintain some degree of catalytic turnover, even in the absence of nitrate.

The increase in NR activity following the transfer of leaves from dark to light may also be a function of nitrate concentration at the site of the enzyme. In the light, transpiration increases and nitrate enters the leaf in the xylem stream. Alternatively, light may cause the release of nitrate from the vacuolar storage pools within the leaf.

Reductant in the form of NADH, NADPH or ferredoxin is another important substrate for NR and NiR which may be involved in the regulation of nitrate assimilation in the light and dark. In photosynthetic tissues in the light, the light reactions in the choroplasts are probably the major source of electrons to nitrate assimilation. As a chloroplastic enzyme, NiR can receive its electrons directly from reduced ferredoxin. The cytosolic NR may receive its electron requirement by way of a triose phosphate shuttle (Fig. 26.4) or malate/aspartate shuttle. In the dark, rates of nitrate assimilation are generally much lower than in the light, and reducing power must be provided through either mitochondrial activity or through the extramitochondrial oxidation of carbohydrate reserves stored in the light (e.g. malate; Lee, 1980). Similarly, in roots and in other non-photosynthetic organs, phloem-supplied carbohydrates are oxidized in mitochondrial or extramitochondrial pathways to provide reducing power to NR and NiR (Lee, 1980). Therefore, nitrate assimilation must compete with other cellular processes of reductant, and the availability of reducing power is more likely to be limiting in the dark or in non-photosynthetic tissues than in conditions of high light in photosynthetic tissues. In addition to its potential role as a limiting substrate, NADH may activate NR by causing the removal of the protein inhibitor previously described (Campbell and Smarrelli, 1986).

Ammonium assimilation

NH_4^+ uptake and its production *in vivo*

In general, acidic soils are inhibitory to nitrifying bacteria and NH_4^+, rather than NO_3^-, is the predominant form of inorganic N present in the soil solution. Although high concentrations of NH_4^+ are toxic to most plants, many (e.g. some forest tree species) are adapted to acidic soils and prefer NH_4^+ over NO_3^- as a N source (Haynes, 1986).

Little is known of the mechanism by which

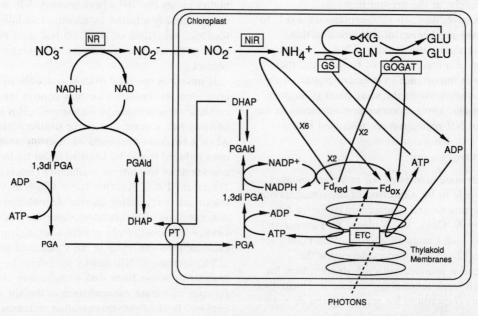

Fig. 26.4 Proposed pathway of nitrate reduction to glutamate in a photosynthetic, eukaryotic cell in the light. Note that the light reactions of photosynthesis provide the reductant and ATP for both the cytosolic nitrate reductase (NR) and the chloroplastic nitrite reductase (NiR), glutamine synthetase (GS) and glutamate synthase (GOGAT) activities. PT, phosphate translocator; PGA, 3-phosphoglycerate; PGAld, phosphoglyceraldehyde; DHAP, dihydroxyacetone phosphate; ETC, electron transport chain.

NH_4^+ is absorbed from the soil solution. It is known to be an active process with a $K_m(NH_4^+)$ of 10–70 μM, a range similar to the concentration of NH_4^+ in an arable, non-acidic soil. Some studies (but not all) indicate that NH_4^+ uptake may be similar to K^+ uptake, and it has been suggested that the two cations may share a common uptake system. Studies of K^+ uptake suggest the presence of a carrier which simultaneously exports protons and is coupled to a membrane-bound ATPase (Haynes, 1986).

In addition to NH_4^+ uptake from the soil, the NH_4^+ pools within the plant can originate from the pathways of NO_3^- assimilation, N_2 fixation or photorespiratory N metabolism. Photorespiration occurs in the light in tissues of plants with C3-type photosynthesis. In C3 photosynthesis, O_2 competes with CO_2 for the active site of ribulose bisphosphate carboxylase/oxygenase (Rubisco), the enzyme responsible for CO_2 fixation. The oxygenase activity of Rubisco forms 2-phosphoglycolate and this enters the glycolate pathway for conversion to 3-phosphoglyceric acid. In this pathway, CO_2 and NH_4^+ are released. In leaves of a C3 plant under normal atmospheric conditions in the light, NH_4^+ production from photorespiration can be 10–20 times the rate of primary NH_4^+ assimilation (Canvin, 1981; Chapter 18).

In general, NH_4^+ taken up from the soil is assimilated rapidly into amino acids or other nitrogenous compounds for subsequent transport to the shoot. Compared with NO_3^-, little NH_4^+ is stored in root tissue, or transported to the shoot in the xylem. The primary pathway of NH_4^+ assimilation in plants consists of two enzymes: glutamine synthetase (GS) and glutamate synthase (also known as glutamine oxoglutarate aminotransferase or GOGAT). These two enzymes will provide the focus for this discussion and are referred to as the GS/GOGAT system. Another NH_4^+ assimilating enzyme (glutamate dehydrogenase or GDH) may be of importance under certain conditions, and will be discussed briefly. There are other plant enzymes which are known to react with NH_4^+ (Durzan and Steward, 1983), but these reactions are generally thought to be of minor significance in primary N assimilation and will not be discussed here.

Glutamine synthetase (GS)

This enzyme is found throughout the plant, bacterial and animal kingdoms and catalyzes the conversion of the amino acid, glutamate, into the amide, glutamine:

$$NH_4^+ + \text{L-glutamate} + ATP \xrightarrow{Me^{2+}} \text{L-glutamine} + ADP + P_i \quad [26.8]$$

The reaction requires ammonium, ATP and a divalent cation (Me^{2+}) such as Mg^{2+}, Co^{2+} or Mn^{2+} as a cofactor. The pH optimum and kinetics of the enzyme are dependent upon the cation present. With Mg^{2+} as a cofactor, the K_m (NH_4^+) is about 10–50 μM (Miflin et al., 1981).

There are a number of isoforms of GS which have been identified on the basis of differences in ion-exchange, isoelectric focusing or immunological properties. In leaves of many higher plants GS_1 (so named because it elutes first from DEAE columns) has been shown to be of cytosolic origin, whereas GS_2 originates in the chloroplasts. Non-photosynthetic tissues such as roots contain a cytosolic form of GS (GS_r) which seems to be different from the GS_1 found in leaves (Oaks and Hirel, 1985). The relative importance of GS_1 and GS_2 in leaves is species dependent. In general, achlorophylous, parasitic plants have only GS_1, plants with a C3 photosynthetic system have 60–100% of their total GS activity as GS_2, and GS_1 is more predominant than GS_2 in most plants with C4 or CAM-type photosynthetic systems. These results suggest that the chloroplastic form of GS (i.e. GS_2) is responsible for reassimilating the ammonia lost in photorespiration (Oaks and Hirel, 1985; Chapter 18).

In addition to the physiological reaction resulting in the production of glutamine (eqn [26.8]), GS also catalyzes the following two reactions:

The synthetase reaction

$$NH_2OH + \text{L-glutamate} + ATP \xrightarrow{Me^{2+}} \gamma\text{-glutamyl hydroxamate} + ADP + P_i \quad [26.9]$$

The transferase reaction

$$NH_2OH + \text{Glutamine} \xrightarrow{ADP, P_i, Me^{2+}} \gamma\text{-glutamyl hydroxamate} + NH_3 \quad [26.10]$$

Since hydroxamate production is relatively easy to quantify, these reactions are sometimes used to estimate GS activity in enzyme assays. However, the transferase assay is particularly susceptible to artifacts, and its use is not recommended (Miflin et al., 1981).

Studies of the catalytic mechanism of GS indicate that L-glutamate binds to the active site first, the hydroxyl group is then phosphorylated to form the intermediate, γ-glutamyl phosphate (Fig. 26.5), and finally NH_4^+ displaces the phosphate. A number of structural analogs of glutamate are effective inhibitors of the reaction. The most widely used inhibitor for physiological and biochemical studies is methionine sulfoximine (MSO) (Fig. 26.5). In the presence of ATP and metal ions, the MSO binds to the enzyme and becomes phosphorylated. However, unlike γ-glutamyl phosphate, the MSO-phosphate is tightly bound and thereby inhibits further activity. While MSO inhibition is generally considered to be specific to GS activity, it is known to have additional effects such as the inhibition of NH_4^+ transport in cyanobacteria or methionine uptake in plant tissues (Miflin and Lea, 1980).

Glutamate synthase (GOGAT)

Glutamate synthase was first discovered as a bacterial enzyme in 1970, and later found to be virtually ubiquitous in the plant kingdom. Subsequent work has shown that this enzyme plays a key role in the assimilation of NH_4^+ into amino compounds (Miflin et al., 1981). It catalyzes the reductive transfer of the amide-amino group of the glutamine formed by GS to the 2-oxo position of 2-oxoglutarate to form 2 molecules of glutamate:

Glutamine + 2-oxoglutarate + 2e$^-$ → 2 Glutamate [26.11]

When one of the glutamate molecules is cycled back as a substrate for the GS reaction, the net reaction of GS and GOGAT is:

NH_4^+ + 2-oxoglutarate + 2e$^-$ + ATP → Glutamate + ADP + P$_1$ [26.12]

There are two types of GOGAT enzyme, and they differ in the nature of the electron donor. Ferredoxin is the electron donor in green algae, cyanobacteria and photosynthetic tissues of higher plants, while in bacteria and in the non-photosynthetic tissues of higher plants, electrons are donated from reduced pyridine nucleotides, generally NADPH in bacteria and NADH in higher plants. A NADH-dependent GOGAT has also been found in leaf tissues of some plants, but generally at much lower activities than the Fd-dependent enzyme.

The reaction mechanism of the NADH-GOGAT from root nodules of lupin has been studied and involves the binding of NADH to the enzyme followed by a random binding of either glutamine or oxoglutarate before the amide group is transferred. Azaserine (AZA) is a structural analog of glutamine (Fig. 26.5) and a widely used

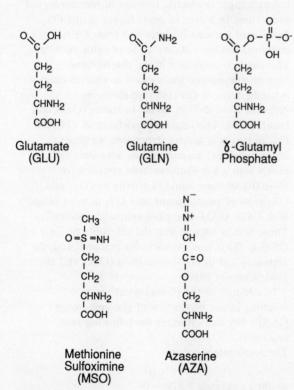

Fig. 26.5 Structures of the amino acid, glutamate (GLU); the amide, glutamine (GLN); the intermediate in glutamine synthetase (GS) activity, γ-glutamyl phosphate; the GS inhibitor, methionine sulfoximine (MSO) and the glutamate synthase inhibitor, azaserine (AZA). Note the structural similarity of MSO to GLU, and AZA to GLN.

inhibitor of the GOGAT reaction and other reactions involving the transfer of the amide-amino group of glutamine.

Glutamate dehydrogenase (GDH)

GDH catalyzes the conversion of 2-oxoglutarate to glutamate in the presence of reductant and NH_4^+:

$$NH_4^+ + \text{2-oxoglutarate} + 2e^- \rightarrow \text{glutamate} \qquad [26.13]$$

In roots or leaves, the enzyme is localized in the mitochondria and the electron donor may be either NADH or NADPH. In photosynthetic tissues of higher plants, a chloroplastic form of the enzyme is found and it has a specific requirement for NADPH.

Prior to the discovery of GOGAT in 1970, GDH was considered to be the key enzyme responsible for glutamate synthesis in plants. However, since that time, the majority of evidence indicates that NH_4^+ assimilation and glutamate production is predominantly via the GS/GOGAT pathway rather than GDH. While some of this evidence will be summarized below, a more detailed and critical analysis may be found in Miflin and Lea (1980) and Yamaya and Oaks (1987).

(1) The K_m (NH_4^+) for GDH (3–80 mM) is much higher than that for GS (10–50 μM), and therefore GDH would presumably have difficulty in competing with GS for the available NH_4^+. As will be discussed below, this argument does not consider the possibility that NH_4^+ concentrations may vary widely between subcellular compartments, thereby permitting both enzymes to operate *in vivo*. Also, K_m values measured *in vitro*, may have little relevance to those operating *in vivo* (Yamaya and Oaks, 1987).

(2) The GS inhibitor, MSO, and the GOGAT inhibitor, AZA, are not inhibitory to GDH activity and therefore are useful in determining the path of ammonia assimilation. When tissues are treated with these inhibitors, NH_4^+ accumulates and $^{15}NH_4^+$ incorporation into amino acids is blocked.

(3) ^{13}N is a short-lived (half-life = 10 min), radioactive isotope of N which has been used to identify the first products of $^{13}NH_4^+$ assimilation.

Studies with cyanobacteria and soybean nodules have shown that $^{13}N_2$ is first incorporated into glutamine, and subsequently into glutamate; a pattern characteristic of the GS/GOGAT pathway of N assimilation.

(4) Chemical or site-specific mutations giving rise to plants or fungi which lack GDH activity are generally not lethal; however, plant mutants of the Fd-GOGAT are lethal unless they are maintained under non-photorespiratory conditions (e.g. high pCO_2 or low pO_2 in shoot atmosphere).

In summary, convincing evidence exists to support the conclusion that the GS/GOGAT pathway is the major port of entry of NH_4^+ in most plants, and especially in the reassimilation of NH_4^+ released during photorespiration. However, the evidence which exists cannot eliminate a possible role for GDH. Indeed, Yamaya and Oaks (1987) have shown that the NH_4^+ concentration in mitochondria is sufficiently high to permit GDH to function at reasonable rates. They have suggested that GDH may have an anaplerotic role in replacing the amino acids which are either incorporated into protein or exported from root or shoot organs in the xylem or phloem. It is also possible that GDH may play an *in vivo* role in catalyzing the reverse reaction: glutamate breakdown to NH_4^+ and 2-oxoglutarate. This reaction may be important in replenishing Krebs cycle intermediates during protein catabolism.

Regulation of NH_4^+ assimilation

In land plants there is probably little negative regulation of NH_4^+ assimilation: NH_4^+ taken up from the environment or produced within the tissues will be rapidly assimilated into amino acids (Miflin and Lea, 1980). Such a strategy would help to ensure that the NH_4^+ does not accumulate to toxic levels in tissues.

In aquatic eukaryotic plants and in yeast there is good evidence for regulatory control of NH_4^+ assimilation, especially through feedback control and product inhibition of GS and other N assimilating enzymes (Miflin and Lea, 1980).

Energetics of nitrogen assimilation

The preceding sections have considered some of the biochemical and physiological aspects associated with the uptake and assimilation of elemental N as N_2, NO_3^- or NH_4^+ In a recent review, Haynes (1986) discussed a number of factors which may be important in determining why plant growth and development in some species is greater under one N source than another. For example, one advantage of NO_3^- over NH_4^+ assimilation may be related to a difficulty in some plants to regulate NH_4^+ pool size within the tissues or to the effects of lower pH (associated with NH_4^+ nutrition) on the levels of toxic metals in the soil. On the other hand, plants assimilating NH_4^+ may have an advantage over those assimilating NO_3^- due to the absence of a requirement for NR synthesis, an inability to regulate the uptake of cations associated with nitrate uptake, or a lower energy cost associated with NH_4^+ versus NO_3^- assimilation (Haynes, 1986). A difference in energy cost has also been used to account for the observation that most legumes or cyanobacteria will assimilate NO_3^- or NH_4^+ in preference to N_2 gas. The remainder of this section will consider the approach and some of the conclusions of studies which have attempted to estimate the relative energy costs associated with the assimilation of one or more of these three N sources (Schubert, 1982; Saari and Ludden, 1987; Layzell et al., 1988).

The approach: normalizing energy cost to a common 'currency'

The energy cost of biochemical reactions or transport processes are known quantitatively in units of ATP hydrolyzed (or synthesized) or reductant consumed (or produced) per unit of compound metabolized or transported. Most theoretical estimates of energy cost involve a normalization of these units to 'ATP equivalents', where one ATP equivalent is the energy required to phosphorylate an ADP to make ATP. Therefore, AMP phosphorylation to ATP is analogous to two ATP equivalents.

The relationship between the number of reductant produced or consumed and the number of ATP equivalents is defined by the P:O ratio of the tissue in which the reaction occurs. The P:O ratio can be defined as the number of high energy phosphate bonds which are formed (i.e. number of ADP $\rightarrow$ATP) for every pair of electrons moving through an electron transport chain to consume $\frac{1}{2} O_2$. It is important to remember that ferredoxin and flavodoxin are one-electron carriers, so oxidation of two of these electron carriers is equivalent to the oxidation of one NADH or one NADPH. In aerobic mitochondria of higher plants, the P:O ratio is probably about 3.0 (3ATP being produced per electron pair passed to O_2 from 2Fd, NADH or NADPH). One exception is if the electron pair originates with FADH: then a P:O ratio of 2 is generally used (see Chapter 9). Similarly, in prokaryotes, Hinkle and McCarty (1978) have estimated that a P:O ratio of 2 is more realistic, regardless of the electron donor. Also, a P:O ratio of 3.0 in plant tissues assumes that the cytochrome pathway is the only electron transport chain which is operational. A number of studies (Moller et al., 1988) have provided evidence that tissues from many plants display significant alternative chain activity, and for this component of respiration, the P:O ratio would be approximately 1.0 (Chapter 9).

Some aspects of the energy cost of N assimilation do not involve a specific biochemical reaction or the movement of a substance across a membrane. Examples include the respiratory cost of producing the plant structure responsible for assimilating the N (i.e. growth cost), and the respiratory cost of maintaining the ion gradients, protein turnover, etc. of this tissue or organ (i.e. maintenance cost). Empirical and theoretical estimates have been made of the growth and maintenance costs associated with producing plant tissues having various chemical compositions (Amthor, 1984; Layzell et al., 1988). These values are generally expressed in units of moles of sugar consumed or moles of CO_2 evolved per gram dry weight produced (growth cost) or per gram dry weight maintained per hour (maintenance cost). To normalize these values with the biochemical and transport costs, it is necessary to define the biomass and growth rate of the structure which is produced for the assimilation of N, and then estimate the number of ATP equivalents which would be produced per mole of sugar oxidized to CO_2 and H_2O in support of the growth and maintenance of this structure. Assuming that the sugar is sucrose, and that it is fully oxidized either through glycolysis and the Krebs cycle or through the pentose phosphate pathway, approximately 72 ATP equivalents (assumes P:O = 3) and $12CO_2$ will be produced per sucrose consumed. This value permits one to express an estimate of energy cost in units of moles of sucrose consumed, ATP equivalents hydrolyzed or CO_2 evolved per N assimilated. The latter of these units has been referred to as the respiratory cost of N assimilation and it can be compared with empirically obtained values (Schubert, 1982).

Finally, if the cost of establishing the physical structure (e.g. nodule or root) is also to be considered as part of the energy cost associated with N assimilation, then it is necessary to know the relationship between the biomass of the structure and the number of moles of sucrose equivalents incorporated into that biomass. Assuming that all the C within the plant tissue originates from phloem-supplied sucrose is a simple, but realistic approach. The C content of sucrose is 42%, while that of plant dry matter ranges from 38% to 60% C, depending on tissue composition. Nodules are typically 46% C and roots about 40% C in dry matter. If this component is to be considered in the cost estimate, then the overall, theoretical energy cost of N assimilation will be in units of sucrose consumed per N assimilated. Examples of this basic approach to estimating the cost of N_2 assimilation may be found in Schubert (1982), Saari and Ludden (1987) and Layzell et al. (1988).

Conclusions relating to the relative energy costs of N_2, NO_3^- and NH_4^+ assimilation

Theoretical estimates of the respiratory cost of N assimilation have compared favorably with empirical values available in the literature (Schubert, 1982; Saari and Ludden, 1987; Layzell et al., 1988). This has given researchers the confidence that the major components of cost have been considered, and that the cost estimates which have been made are realistic, at least for the major components. These studies have also identified key areas of N assimilation in which it may be possible to develop plants with improved energy use efficiency. Some of the observations and conclusions from these studies are summarized below.

The importance of the site of N assimilation in the energy cost

When comparing the theoretical energy cost of N assimilation, it is important to consider physiological aspects of the organ or tissue where the N is being assimilated. In roots or in N_2 fixing nodules, phloem-delivered sucrose supplies virtually all of the C and energy requirements; therefore the energy cost may be considered in terms of sucrose consumed or CO_2 evolved per N assimilated. However, in photosynthetic organs, NO_3^- reduction and NH_4^+ assimilation (from NO_3^- reduction or photorespiration) can be supported by the light reactions of photosynthesis. At low light, the energy requirements for N assimilation may compete with CO_2 fixation for the available reductant and ATP, while at high light, when the light reactions produce excess reducing power, the actual cost of N assimilation may be negligible to the plant. A lower apparent cost for NO_3^- assimilation in shoots than roots may account for the predominance of shoot NO_3^- assimilation in many tropical and subtropical species (Andrews, 1986). Little work has been done on the energetic implications of the photosynthetic cortical tissues in stem nodules of some tropical legumes.

H_2 exchange and the energy cost of N_2 fixation

In theory, the energetic cost of N_2 fixation would be lowest per N fixed in nodules which display both an optimal EAC (i.e. minimal rates of H_2 production) and an efficient HUP activity to recover any H_2 that is produced by nitrogenase. HUP activity may also be a factor in determining EAC. By maintaining low concentrations of H_2 within the infected cells, HUP activity may minimize the inhibitory effects that H_2 may have on N_2 fixation in vivo. Certainly, further information is required on the mechanism of EAC regulation *in vivo*, and the possible role that the HUP enzyme may play in this regulation.

Ammonia assimilation and the apparent respiratory cost of N assimilation

The transported endproduct of N assimilation varies among species and with N source within the same species. For example, N_2 fixing soybeans transport most of their N as ureides (allantoin and allantoic acid), while NO_3^--fed roots transport mostly asparagine. N_2 fixing lupin nodules also produce asparagine for N transport. Although the GS/GOGAT pathway of NH_4^+ assimilation is common to all, the pathways responsible for the production of ureides and amino acids/amides are very different, and special attention is frequently given to the presence of a CO_2 fixation site in the pathway of amino acids and amide synthesis. Various workers have speculated on the effect that this CO_2 fixation might have on the measured rates of respiration per N assimilated (Schubert, 1982; Saari and Ludden, 1987). Recently, Layzell and coworkers (1988) calculated the theoretical sucrose cost and the CO_2 exchange rate associated with the synthesis of ureides and amides from sucrose and NH_4^+. On a whole nodule basis, it was predicted that ureide-producing nodules would consume 8% less sucrose per N fixed than asparagine-producing nodules, but would display an apparent respiratory cost which would be 5% higher than that in asparagine-producing nodules. These predictions suggest that measured values of respiratory loss per N fixed must be treated with caution if used as an indicator of the relative energy cost of N assimilation.

Estimating structural costs

In N_2 fixing cyanobacteria and higher plant symbioses, N assimilation is associated with distinct structures such as heterocysts or root nodules. The cost of producing and maintaining these specialized, N_2 fixing structures represents a significant energy cost to the organism or symbiosis (Layzell *et al.*, 1984, 1988; Turpin *et al.*, 1985).

Estimating the structural costs associated with NO_3^- or NH_4^+ assimilation is difficult since these processes are integrated into plant tissues which have many other functions. Evidence that additional structural costs are associated with NO_3^- or NH_4^+ assimilation includes the observation that plants grown under limited NO_3^- supply support a larger root biomass than plants grown under non-limiting NO_3^- (Vessey and Layzell, 1987). Therefore, the more limiting the N, the higher the structural costs associated with assimilating that N.

What are the absolute and relative energy costs of N assimilation?

Attempts to measure the energy cost of N assimilation have generally involved an assessment of nodule respiration in N_2 fixing legume symbioses. Although many of the methods that have been used are open to criticism, reasonable cost estimates generally fall between 2.4 and 7.0 g C respired per g N fixed (Saari and Ludden, 1987). These values are similar to those derived from a theoretical assessment of the component cost of N_2 fixation, and the range of empirical values in the literature is consistent with the variation which is known to occur between nodules in physiological processes such as specific nitrogenase activity, EAC and HUP activity (Layzell *et al.*, 1988).

The question of whether the cost of N_2 fixation is greater than the costs of NO_3^- or NH_4^+ assimilation is difficult to answer empirically. On a theoretical basis – when only the direct and the obvious indirect costs are considered – it seems likely that the relative costs would be in the following order $NH_4^+ < NO_3^- < N_2$. However, we have much to learn about the complex interrelationships which occur in plants, and the actual costs to a plant assimilating a particular form of N in a specific environment may be very different from what may be predicted on the basis of our limited understanding of these complexities.

Acknowledgments

I wish to thank Dr Stephen Hunt for his useful comments on this manuscript.

Rererences

Amthor, J. S. (1984). The role of maintenance respiration in plant growth. *Plant Cell Environ.* **7**, 561–9.

Andrews, M. (1986). The partitioning of nitrate assimilation between root and shoot of higher plants. *Plant Cell Environ.* **9**, 511–19.

Beevers, L. and Hageman, R. H. (1980). Nitrate and nitrite reduction, In *The Biochemistry of Plants*, Vol. 5, ed. B. J. Miflin, Academic Press, New York, pp. 115–68.

Bergersen, F. J. (1982). *Root Nodules of Legumes, Structure and Function*, Research Studies Press, J. Wiley and Sons Ltd, New York, 164 pp.

Bishop, P. E., :Premakumar, R., Joerger, R. D., Jacobsen, M. R., Dalton, D. A., Chisnell, J. R. and Wolnger, E. D. (1988). Alternative nitrogen fixation systems in *Azotobacter vinelandii*. In *Nitrogen Fixation: Hundred Years After*, eds H. Bothe, F. J. de Bruijn and W. E. Newton, Proc. 7th International Congress on Nitrogen Fixation, Cologne, FRG, March 13–20 1988, pp. 71–80.

Burgess, B. A. (1985). Substrate reactions of nitrogenase. In *Molybdenum Enzymes*, ed. T. G. Spiro, J Wiley and Sons, New York, pp. 161–219.

Burris, R. H. (1985). H_2 as an inhibitor of N_2 fixation. *Physiol. Veg.* **23**, 843–8.

Campbell, W. R. and Smarrelli, J. (1986). Nitrate reductases: biochemistry and regulation. In *Biochemical Basis of Plant Breeding*, Vol II, ed. C. A. Neyra, CRC Press, Boca Raton, pp. 1–39.

Canvin, D. T. (1981). Photorespiration and nitrogen metabolism. In *Nitrogen and Carbon Metabolism*, ed. J. D. Bewley, Martinus Nijhoff/Dr W. Junk Publishers, The Hague, pp. 178–94.

Deane-Drummond, C. E. and Thayer, J. R. (1986). A substrate cycling model for NO_3^- uptake by *Pisum sativum* seedlings: a key to sensitivity of response of net efflux to substrate and effectors? *Plant Soil* **91**, 307–11.

Durzan, D. J. and Steward, F. C. (1983). Nitrogen metabolism. In *Plant Physiology, A Treatise*, Vol. VIII eds F. C. Steward and R. G. S. Bidwell, Academic Press, Toronto, pp. 55–261.

Eisbrenner, G. and Evans, H. J. (1983). Aspects of H_2 metabolism in N_2-fixing legumes and other plant microbe associations. *Ann. Rev. Plant Physiol.* **34**, 105–36.

Harper, J. E. (1987). Nitrogen metabolism. In *Soybeans: Improvement, Production and Uses*, 2nd edn, Agronomy Monograph no. 16, pp. 497–533.

Haynes, R. J. (1986). Uptake and assimilation of mineral nitrogen by plants. In *Mineral Nitrogen in the Plant–Soil System*, ed. R. J. Haynes, Academic Press, Toronto, pp. 303–78.

Hinkle, P. C. and McCarty, R. E. (1978). How cells make ATP. *Sci. Amer.* **238**, 104–23.

Hunt, S., Gaito, S. T. and Layzell, D. B. (1988). Model of gas exchange and diffusion in legume nodules. II. Characterisation of the diffusion barrier and estimation of the concentrations of CO_2, H_2, and N_2 in the infected cells. *Planta* **173**, 128–41.

Jackson, W. A., Pan, W. L., Moll, R. H. and Kamprath, E. J. (1986). Uptake, translocation and reduction of nitrate. In *Biochemical Basis of Plant Breeding*, Vol. II, ed. C. A Neyra, CRC Press, Boca Raton, pp. 73–108.

Layzell, D. B., Weagle, G. E. and Canvin, D. T. (1984). A highly sensitive flow through H_2 gas analyzer for use in nitrogen fixation studies. *Plant Physiol.* **75**, 582–5.

Layzell, D. B., Gaito, S. T. and Hunt, S. (1988). Model of gas exchange and diffusion in legume nodules, 1. Calculation of gas exchange rates and the energy cost of N_2 fixation. *Planta* **173**, 117–27.

Lee, R. B. (1980). Sources of reductant for nitrate assimilation in non-photosynthetic tissue: a review. *Plant Cell Environ.* **3**, 65.

Ludden, P. K. and Burris, R. H (1986). Nitrogenase: properties and regulation. In *Biochemical Basis of Plant Breeding*, Vol. II, ed C. A. Neyra, CRC Press, Boca Raton, pp. 41–58.

Maier, R. J. (1986). Biochemistry, regulation and genetics of hydrogen oxidation in *Rhizobium*. *CRC Crit. Rev. Biotechnol.* **3**, 17–38.

Miflin, B. J. and Lea, P. J. (1980). Ammonia assimilation. In *The Biochemistry of Plants*, Vol. 5, ed. B. J. Miflin, Academic Press, Toronto, pp. 169–202.

Miflin, B. J., Wallsgrove, R. M. and Lea, P. J. (1981). Glutamine metabolism in higher plants. *Curr. Topics Cell. Reg.* **20**, 1–43.

Moller, I. M., Berczi, A., Van Der Plas, L. J. W. and Lambers, H. (1988). Measurement of the activity and capacity of the alternative pathway in intact plant tissues; Identification of problems and possible solutions. *Physiol. Plant.* **72**, 642–9.

Naik, M. S., Arrol, Y. P., Nair, T. V. R. and Ramaroa, C. S. (1982). Nitrate assimilation – its regulation and relationship to reduced nitrogen in higher plants. *Phytochemistry* **21**, 495–504.

Oaks, A. and Hirel, B. (1985). Nitrogen metabolism in roots. *Ann. Rev. Plant Physiol.* **36**, 345–65.

Phillips, D. A., Bedmar, E. J., Qualset, C. O. and Teuber, L. R. (1985). Host legume control of *Rhizobium* function. In *Nitrogen Fixation and CO_2 Metabolism*, eds P. W. Ludden and J. E. Burris, Elsevier Publ. Co., New York, pp. 203–12.

Rennie, R. J. (1986). Comparison of methods of enriching a soil with N-15 to estimate dinitrogen fixation by isotope dilution. *Agron. J.* **78**, 158–63.

Reporter, M. (1985). Nitrogen fixation. In *Techniques in Bioproductivity and Photosynthesis*, 2nd edn, eds J. Coombs, D. O. Hall, S. P. Long and J. M. O. Scurlock, Pergamon Press, Toronto, pp. 158–64.

Robson, R. L., Eady, R. R., Richardson, T. H., Miller, R. W., Hawkins, M. and Postgate, J. R. (1986). The alternative nitrogenase of *Azotobacter chroococcum* is a vanadium enzyme. *Nature* **322**, 388–90.

Saari, L. L. and Ludden, P. W. (1987). The energetics and energy cost of symbiotic nitrogen fixation. In *Plant Microbe Interactions*, Vol. 2, eds T Kosuge and E. W. Nester, Macmillan Publ. Co., New York, pp. 147–93.

Schubert, K. R. (1982). The energetics of biological nitrogen fixation. Workshop Summaries I, American Society of Plant Physiologists, Rockville, Md, pp 1–30.

Simpson, F. B. (1987). The hydrogen reactions of nitrogenase. *Physiol. Plant.* **69**, 187–90.

Streeter, J. (1988). Inhibition of legume nodule formation and N_2 fixation by nitrate. *CRC Crit. Rev. Plant Sci.* **7**, 1–23.

Thorneley, R. N. F. and Lowe, D. J. (1985). Kinetics and mechanisms of the nitrogenase enzyme system. In *Molybdenum Enzymes*, ed. T. G. Spiro, J. Wiley and Sons, New York, pp. 220–84.

Turpin, D. H., Layzell, D. B. and Elrifi, I. R. (1985). Modelling the C economy of *Anabaena flos-aquae*. Estimates of establishment, maintenance and active costs associated with growth on NH_4^+, NO_3^-, and N_2. *Plant Physiol.* **78**, 746–52.

Vessey, J. K. and Layzell, D. B. (1987). Regulation of assimilate partitioning in soybean. Initial effects following change in nitrate supply. *Plant Physiol.* **83**, 341–8.

Vessey, J. K., Walsh, K. B. and Layzell, D. B. (1988a). O_2 limitation of N_2 fixation in stem-girdled and nitrate-treated soybean. *Physiol. Plant.* **73**, 113–21.

Vessey, J. K., Walsh, K. B. and Layzell, D. B. (1988b). Can a limitation in phloem supply to nodules account for the inhibitory effect of nitrate on nitrogenase activity in soybean? *Physiol. Plant.* **74**, 137–46.

Wolk, C. P. (1982). Heterocysts. In *The Biology of Cyanobacteria*, eds N. G. Carr and B. A. Whitton, University of California Press, Berkeley, pp. 359–86.

Yamaya, T. and Oaks, A. (1987). Synthesis of glutamate by mitochondria – an anaplerotic function for glutamate dehydrogenase. *Physiol. Plant.* **70**, 749–56.

27 Amino acid and ureide biosynthesis
Robert Ireland

Introduction

Apart from being the structural units of proteins, amino acids have a wide range of other functions in plants. They are involved in the transporting of nitrogen between roots, leaves, fruits, etc. and are precursors in the syntheses of chlorophyll and many other nitrogen-containing compounds, such as the enzyme cofactors biotin (from aspartic acid), thiamine pyrophosphate (from alanine and methionine) and coenzyme A (from valine, aspartate and cysteine). Amino acids also serve as the carbon and nitrogen source for the production of most 'secondary', or 'natural' products, such as alkaloids, phenolic acids and cyanogenic compounds. This latter role involves them in the complex realm of 'chemical ecology' since the majority of the compounds responsible for plant color, taste, smell and toxicity are derived from amino acids. Despite their obvious importance and the fact that they are usually present in plant tissues at far higher concentrations than most other metabolites, we know much less about amino acid metabolism than we do, for example, about the glycolytic or other carbon pathways. In several cases the pathway generally accepted for the synthesis of a particular amino acid has been assumed to be the same as that in bacterial or animal systems, and it is only relatively recently that these have been confirmed or novel pathways demonstrated. A number of amino acids can be made by more than one synthetic pathway, depending on, for example, the tissue concerned, time of day, stage of growth, etc. The synthesis of some amino acids in plants remains unclear, as is the case with histidine and many of the non-protein amino acids.

The branched and interwoven nature of the pathways for amino acid synthesis requires strict and precise control. The nitrogen and carbon flowing through, for example, aspartate, does not all end up in a single amino acid, but is distributed to several, in a defined manner, in response to the needs of the cell at that particular time. The consumption of amino acids for synthetic processes requires the establishment and maintenance of pools. For example, the conversion of phenylalanine to secondary products requires the activity of the shikimate pathway to restore phenylalanine levels, without affecting tyrosine levels. Other metabolic conditions may require the synthesis of phenylalanine and tyrosine simultaneously, and thus these pathways have to be precisely controlled. Enzymes are regulated to different degrees: at one end of the spectrum are enzymes like the transaminases, which do not exhibit any regulatory properties, and with activities solely dependent on pH and substrate/product concentrations. At the other end of the scale are enzymes like aspartate kinase, that are highly regulated by several metabolites. As with many metabolic pathways, endproduct inhibition appears to be the principal mechanism by which amino acid synthesis is regulated.

Amino acids can be grouped together into 'families', each of which are derived from a single 'head' amino acid. For example, the 'aspartate family' is comprised of asparagine, homoserine, threonine, isoleucine and lysine, all synthesized from aspartic acid. Because more than one amino acid may be involved in the synthesis of another, a single amino acid may be assigned to more than one family, and it is not unusual for different authors to differ in their assignments.

This chapter will largely be confined to the 20 (or so) 'protein' amino acids, so-called because they are those commonly found in proteins. Plants do, however, synthesize hundreds of other amino acids, the 'non-protein' amino acids, which will be dealt with briefly towards the end of the chapter.

Ammonia assimilation: glutamine, glutamate and asparagine

As described in the previous chapter, it appears that nearly all plant nitrogen is first assimilated as glutamine, due to the high affinity of glutamine synthetase for ammonia (Miflin and Lea, 1980). This enzyme works in concert with glutamate synthase in what is generally referred to as the glutamate synthase cycle (see Chapter 26). The net effect of this is the amination of α-ketoglutarate with the concurrent hydrolysis of one molecule of ATP and the oxidation of one molecule of NAD(P)H or equivalent. A similar result is achieved by the enzyme, glutamate dehydrogenase, which also produces glutamate from α-ketoglutarate and ammonia, but this time with only the oxidation of one molecule of NADH (see Chapter 26). There has been considerable discussion over the role of glutamate dehydrogenase in ammonia assimilation (e.g. Yamaya and Oaks, 1987), and its precise function is still unclear. Other pathways have also been implicated in ammonia assimilation, including alanine dehydrogenase, aspartate dehydrogenase, aspartase and asparagine synthetase. Alanine dehydrogenase, which has been purified from the blue–green alga, *Anabaena* (Rowell and Stewart, 1976), catalyzes the amination of pyruvate to alanine:

$$\begin{array}{c} CH_3 \\ | \\ C=O \\ | \\ COOH \end{array} + NH_3 + NAD(P)H + H^+ \rightarrow \begin{array}{c} CH_3 \\ | \\ HCNH_2 \\ | \\ COOH \end{array} + H_2O + NAD(P)^+$$

pyruvate → alanine

In a similar reaction, aspartate dehydrogenase aminates oxaloacetate to aspartate:

$$\begin{array}{c} COOH \\ | \\ CH_2 \\ | \\ C=O \\ | \\ COOH \end{array} + NH_3 + NAD(P)H + H^+ \rightarrow \begin{array}{c} COOH \\ | \\ CH_2 \\ | \\ HCNH_2 \\ | \\ COOH \end{array} + H_2O + NAD(P)^+$$

oxaloacetate → aspartate

Aspartase is another ammonia-assimilating enzyme which catalyzes the addition of ammonia to fumarate, yielding aspartic acid:

$$\begin{array}{c} COOH \\ | \\ CH \\ || \\ CH \\ | \\ COOH \end{array} + NH_3 \rightarrow \begin{array}{c} COOH \\ | \\ CH_2 \\ | \\ HCNH_2 \\ | \\ COOH \end{array}$$

fumarate → aspartate

These three enzymes have been detected in a few plants, but are not thought to make significant contributions to the assimilation of ammonia in plant tissues (Miflin and Lea, 1982). Asparagine synthetase, on the other hand, is present in many plants:

$$\begin{array}{c} COOH \\ | \\ CH_2 \\ | \\ HCNH_2 \\ | \\ COOH \end{array} + NH_3 \xrightarrow[Mg^{2+}]{ATP \quad AMP + PP_i} \begin{array}{c} CONH_2 \\ | \\ CH_2 \\ | \\ HCNH_2 \\ | \\ COOH \end{array}$$

aspartate → asparagine

but in most cases it appears that the *in vivo* substrate for this enzyme is actually glutamine, not ammonia:

$$\begin{array}{cc} CONH_2 & COOH \\ | & | \\ CH_2 & CH_2 \\ | & | \\ CH_2 & HCNH_2 \\ | & | \\ HCNH_2 & COOH \\ | \\ COOH \end{array} \xrightarrow[Mg^{2+}]{ATP \quad AMP + PP_i} \begin{array}{cc} COOH & CONH_2 \\ | & | \\ CH_2 & CH_2 \\ | & | \\ CH_2 & HCNH_2 \\ | & | \\ HCNH_2 & COOH \\ | \\ COOH \end{array}$$

glutamine aspartate → glutamate asparagine

and hence asparagine synthetase does not usually constitute a route for ammonia assimilation.

Recent work has shown that, at least in some tissues, ammonia is the preferred substrate, and asparagine synthetase may indeed represent an entry point for ammonia into organic metabolism in some plants (Oaks and Hirel, 1985). Asparagine may also be synthesized from the hydrolysis of β-cyanoalanine, which is formed from hydrogen cyanide and cysteine:

$$\underset{\text{cysteine}}{\begin{array}{c}\text{SH}\\|\\\text{CH}_2\\|\\\text{HCNH}_2\\|\\\text{COOH}\end{array}} \xrightarrow[\text{H}_2\text{O}]{\text{HCN}} \underset{\text{β-cyanoalanine}}{\begin{array}{c}\text{C}\equiv\text{N}\\|\\\text{CH}_2\\|\\\text{HCNH}_2\\|\\\text{COOH}\end{array}} \xrightarrow{\text{H}_2\text{O}} \underset{\text{asparagine}}{\begin{array}{c}\text{CONH}_2\\|\\\text{CH}_2\\|\\\text{HCNH}_2\\|\\\text{COOH}\end{array}}$$

This pathway has been demonstrated in a number of plants, including lupins, sorghum, sweet pea and asparagus, but relies on the supply of cyanide, which, due to its toxicity, is thought to be limited in most plants. However, recent work on the synthesis of methionine and ethylene suggests that HCN synthesis may be much more widespread than is currently thought, and the β-cyanoalanine pathway of HCN detoxification may be functioning in many, if not most plants. A third route for asparagine synthesis involves the transamination of α-ketosuccinamate (the β-amide of oxaloacetate) with a suitable amino donor, but this reaction proceeds far more favorably in the reverse direction and is thus thought to be more significant in asparagine catabolism (Sieciechowicz et al., 1988).

Asparagine synthesis is particularly important in the root nodules of legumes, where much of the nitrogen fixed by the bacteria is rapidly transferred to asparagine through the joint activities of glutamine synthetase and asparagine synthetase. Thus much of the nitrogen transported in the xylem, away from a nodule, is in the form of asparagine (Sieciechowicz et al., 1988). Asparagine levels in plant tissues often increase under stress conditions, such as mineral deficiencies, salt stress, or drought. An example of this can be seen in tomatoes, where zinc deficiency can lead to a 50-fold increase in the concentration of asparagine (Stewart and Larher, 1980). The significance of such increases has not been established, but may be a means of storing nitrogen when protein synthesis is inhibited because of stress.

For most plants, the data are in agreement with ammonia being assimilated via the actions of the glutamate synthase cycle, with perhaps some contribution from glutamate dehydrogenase. When nitrate labeled with the heavy isotope of nitrogen, ^{15}N, is fed to plants, the label rapidly appears in both glutamine and glutamate, which then act as the precursors for the synthesis of all of the other amino acids. Nitrate reduction is not the only source of ammonia production in plant tissues; in fact most of the ammonia assimilated by glutamine synthetase in photosynthetic tissues appears to come from photorespiration (see Chapter 18). Other sources of ammonia include reactions catalyzed by enzymes such as asparaginase and arginase (Miflin and Lea, 1980).

As well as being the entry point for nitrogen into plant metabolism, glutamine also serves as a nitrogen transport compound, as does asparagine. These two amides differ only in chain length, but exhibit considerable differences in their biochemical activities and roles in the cell: asparagine is more soluble and less reactive than glutamine and is thus more suited to its role as a transport and storage compound (Sieciechowicz et al., 1988).

Nitrogen flow

The route that nitrogen flow takes in a plant cell depends very much on the tissue and plant species involved. Developing wheat leaves, for example, receive most of their nitrogen as nitrate, transported in the xylem from the roots, and thus their starting point for amino acid biosynthesis is glutamine/glutamate, as described above. In many other plants, such as the legumes, much of the nitrate is reduced and converted to organic form in the roots prior to transport in the xylem, and thus the nitrogen arrives at the leaves in a different form, often as asparagine and glutamine. Similarly, developing seeds or fruits, which are very active in amino acid biosynthesis, will receive most of their nitrogen in the form of amino acids supplied by the phloem. In both cases, these transport compounds are subsequently metabolized to the other amino acids (Lea and Miflin, 1980). The age of the tissue also affects nitrogen flow: young leaves consume all of the incoming nitrogen for growth, mature leaves re-export (in the phloem) much of the nitrogen they receive to the growing apex or developing fruit, as do senescing leaves, which also convert a lot of their proteins and other nitrogenous molecules to transport compounds for export. Diurnal variations in flow are also seen (see below).

Carbon flow

There are three metabolites in glycolysis and the citric acid cycle which serve as the major withdrawal points for organic carbon in the

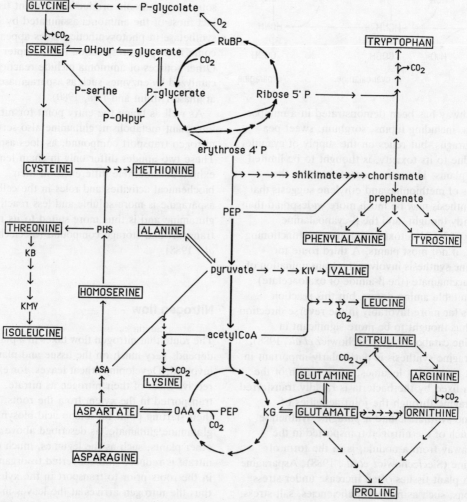

Fig. 27.1 Carbon flow in amino acid biosynthesis. Each arrow represents a separate reaction: for clarity some reactions in the oxidative/reductive pentose phosphate pathway have been omitted, as have the links between it and glycolysis. The intermediates indicated are those discussed in the text: ketoisovalerate (KIV), ketobutyrate (KB), ketomethylvalerate (KMV), aspartate semialdehyde (ASA).

syntheses of amino acids: pyruvate, oxaloacetate, and α-ketoglutarate (Fig. 27.1). Significant amounts of phosphoenolpyruvate and acetyl coenzyme A are also consumed for amino acid synthesis, and the pentose phosphate pathways (reductive and oxidative) yield ribose 5-phosphate and erythrose 4-phosphate for the synthesis of the aromatic amino acids. The Calvin cycle provides the carbon skeletons for the photorespiratory synthesis of glycine and serine, following the oxidation of ribulose 1,5-bisphosphate to phosphoglycolate (and 3-phosphoglycerate). Considerable quantities of α-ketoglutarate are required by the ammonia-assimilating activities of the glutamate synthase cycle, and this represents the greatest drain on these carbon pathways. Consumption of amino acids for protein or other syntheses requires that carbon compounds be withdrawn from these pathways to replenish the amino acids and maintain their concentrations in the 'amino acid pools'. Increased glycolytic and anaplerotic activities (such as PEP carboxylase) are then required to restore the levels of these carbon compounds (see Chapters 8 and 28).

Demand for carbon skeletons varies considerably according to many factors including plant species, tissue, age, time of day, stress level, etc. In wounded or infected plant tissues, there is an increase in the activity of glycolytic and pentose phosphate pathway enzymes, in order to increase production of carbon skeletons for the synthesis of the aromatic amino acids. These are the precursors to so-called 'secondary' metabolites, such as phenolic acids and flavonoids, which are used in defense reactions. The relationship between carbon and nitrogen flow can clearly be seen when nitrogen-deficient plants are transferred to a nitrogen-rich environment. This results in an increase in cellular respiration, because of the increased demand placed on respiratory pathways to supply carbon skeletons needed for the incorporation of nitrogen into organic form (see Chapter 28). The consumption of α-ketoglutarate by the nitrogen-assimilating activities of the glutamate synthase cycle causes a depletion of citric acid cycle intermediates. Responses to this include increased glycolytic activity which serves to restore the levels of citric acid cycle metabolites, as does increased PEP carboxylase activity. Similar responses are seen when oxaloacetate is withdrawn for aspartate synthesis and when pyruvate is withdrawn from glycolysis for transamination to alanine (Beevers, 1976).

Transamination

It is difficult to discuss the biosynthesis of amino acids without considering their breakdown, since the synthesis of one amino acid often involves the degradation of another. Such is the case in transamination reactions, which are central to amino acid metabolism since these reactions result in the redistribution of nitrogen from glutamate to other amino acids in all cells and most subcellular compartments. Transaminases catalyze the transfer of an amino group from the α-carbon of an amino acid to the α-carbon of a keto acid, producing a new amino and a new keto acid. Less commonly, amines may act as amino donors and aldehydes as amino acceptors in these reactions.

$$\underset{\text{amino acid A}}{\overset{R_1}{\underset{\text{COOH}}{\text{HCNH}_2}}} + \underset{\text{keto acid B}}{\overset{R_2}{\underset{\text{COOH}}{\text{C=O}}}} \rightleftharpoons \underset{\text{keto acid A}}{\overset{R_1}{\underset{\text{COOH}}{\text{C=O}}}} + \underset{\text{amino acid B}}{\overset{R_2}{\underset{\text{COOH}}{\text{HCNH}_2}}}$$

Transaminases, also known as aminotransferases, each have a tightly-bound coenzyme; pyridoxal 5-phosphate. This coenzyme accepts an amino group from the amino acid substrate (A), becoming aminated to pyridoxamine phosphate. The keto acid thus produced (A) is released and the aminated form of the coenzyme then undergoes a reversal of the process, giving up its newly-acquired amino group to the keto acid substrate (B) to produce the amino acid product (B).

Enzymes have been isolated from plant tissues that can catalyze the transamination of all of the common, or protein amino acids, except proline, which is in fact not an amino acid but an imino acid and thus has no primary amino group available for transamination. These data indicate that most amino acids are transaminated.

Transaminases play a key role in the synthesis of amino acids. Since they are usually freely reversible, it is theoretically possible for all amino acids to be synthesized by transamination, but in some cases, the only source of the required keto acid appears to be the transamination reaction itself, so no net synthesis of the amino acid can occur by this route. Where there is another pathway able to provide the keto acid, transamination can contribute significantly to amino acid synthesis, and in fact transaminases catalyze the final step in the synthesis of many amino acids. An example of this is seen in the complex pathway leading to the synthesis of phenylalanine, where the final step is the transamination of phenylpyruvate. Similarly, tyrosine and leucine are produced by the amination of hydroxyphenylpyruvate and ketoisocaproate, respectively (Ireland and Joy, 1985).

Unlike other metabolites, many amino acids are often present in millimolar concentrations in plant cells and occur not only in the vacuole, which seems to act as a repository for high concentrations of many compounds, but also in the cytosol, chloroplasts and other organelles. For example, corn leaf chloroplasts have been shown to contain 16 mM asparagine, 5 mM aspartate, 10 mM serine and alanine, and 5 mM glycine (Chapman and Leech, 1979). Not surprisingly, the kinetic constants for the enzymes metabolizing these compounds also seem to be high when compared with, for example, the enzymes of the glycolytic pathway. For example, it is common to see a K_m for the amino acid substrate in the 2–6 mM range in the case of transaminases. It appears that transaminases are not highly specific, and will react with a number of amino acid or keto acid substrates. For example, a transaminase may be designated 'aspartate transaminase', but also be able to use glutamate as the amino donor, and either α-ketoglutarate or oxaloacetate as keto acid substrates (Givan, 1980). Given that nearly all plant nitrogen has to pass through glutamate, it is not surprising that many transaminases can use glutamate or, to a lesser extent, aspartate as amino donors in the synthesis of a wide range of amino acids such as glycine, phenylalanine, tyrosine, serine, etc. The extensive use of glutamate as an amino donor results in the production of large quantities of α-ketoglutarate, which is probably re-aminated by the glutamate synthase cycle during ammonia assimilation.

Following the assimilation of nitrogen into glutamine and glutamate, transaminases serve to redistribute the nitrogen to a range of other amino acids and they contribute to the maintenance of relatively stable amino acid pools. Several transaminases have functions that are unique to plants, such as those involved in carbon assimilation and other processes requiring the 'shuttling' of metabolites. Carbon shuttling in C4 plants makes extensive use of transaminases to produce aspartate and alanine, which are used as transport compounds for the movement of fixed carbon between the mesophyll and bundle sheath cells (Hatch, 1976) (see Chapter 19). Organelle membranes are largely impermeable to pyridine nucleotides, and so shuttle mechanisms involving transaminases can be used to transfer reducing power across chloroplast and mitochondrial membranes (see Chapters 15 and 22).

Aspartate and alanine

After incorporation into glutamate, nitrogen is quickly distributed to other amino acids, much of it being transferred to alanine and aspartate via transamination:

```
COOH              COOH              COOH              COOH
|                 |                 |                 |
CH2               COOH              CH2               COOH
|                 |                 |                 |
CH2        +      CH2        ⇌      CH2        +      CH2
|                 |                 |                 |
HCNH2             C=O               C=O               HCNH2
|                 |                 |                 |
COOH              COOH              COOH              COOH
glutamate         OAA               α-ketoglutarate   aspartate

COOH                                COOH
|                                   |
CH2               CH3               CH2               CH3
|                 |                 |                 |
CH2        +      C=O        ⇌      CH2        +      HCNH2
|                 |                 |                 |
HCNH2             COOH              C=O               COOH
|                                   |
COOH                                COOH
glutamate         pyruvate          α-ketoglutarate   alanine
```

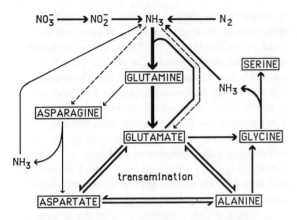

Fig. 27.2 Primary routes of nitrogen flow in amino acid synthesis. Ammonia is assimilated into the amino acids shown, which then serve as precursors and amino donors in the synthesis of the other amino acids.

In both of these reactions, α-ketoglutarate is produced, which can be reused in the glutamate synthase cycle. A third transamination linking aspartate and alanine completes the aspartate–alanine–glutamate triangle seen in Fig. 27.2, transferring amino groups between alanine and aspartate:

$$
\begin{array}{c}
\text{CH}_3 \\
| \\
\text{HCNH}_2 \\
| \\
\text{COOH} \\
\text{alanine}
\end{array}
+
\begin{array}{c}
\text{COOH} \\
| \\
\text{CH}_2 \\
| \\
\text{C=O} \\
| \\
\text{COOH} \\
\text{OAA}
\end{array}
\rightleftharpoons
\begin{array}{c}
\text{CH}_3 \\
| \\
\text{C=O} \\
| \\
\text{COOH} \\
\text{pyruvate}
\end{array}
+
\begin{array}{c}
\text{COOH} \\
| \\
\text{CH}_2 \\
| \\
\text{HCNH}_2 \\
| \\
\text{COOH} \\
\text{aspartate}
\end{array}
$$

These three enzymes are all present in both the chloroplast (Kirk and Leech, 1972) and the cytosol, and the first two are also found in mitochondria (Ireland and Joy, 1985).

Another enzyme responsible for aspartate synthesis is asparaginase, a cytosolic enzyme that catalyzes the hydrolysis of asparagine to aspartate and ammonia. Asparagine is a major source of aspartate in developing legume leaves and fruits/seeds. These tissues often receive much of their nitrogen in the form of asparagine (from the xylem or phloem). Asparaginase activity varies during the development of legume seeds and leaves. It also varies diurnally, increasing in light and decreasing in darkness. It appears that the diurnal variation is due to a relatively rapid turnover of the enzyme, with synthesis occurring in the light and degradation by proteolysis in the dark. Since the reaction releases ammonia, which can be harmful to plant tissues, this regulation may have metabolic significance. The enzyme is only functional in the light when there is adequate ATP and reducing power (from photosynthetic electron transport) to allow the glutamate synthase cycle to assimilate the ammonia produced. In the dark, ATP is produced from the respiration of storage carbohydrates. Hence, reducing ammonia production by decreasing the level of asparaginase activity will serve to alleviate the demand on this limited ATP supply. Furthermore, in darkness protein synthesis is reduced because of the lower concentration of ATP and hence there will be a reduced demand for aspartate which acts as a precursor in the synthesis of many of the other amino acids. (Sieciechowicz *et al.*, 1988).

Regardless of the form in which nitrogen arrives at a tissue, be it nitrate, asparagine, glutamate or other amino acid, it is quickly distributed within pools of amino acids in different subcellular compartments, within which aspartate, glutamate and alanine are often the major components.

Serine and glycine: photorespiration

The synthesis of glycine and serine during photorespiration involves four subcellular compartments. Glycolate, produced in the chloroplast by the oxygenase activity of Rubisco (see Chapter 18; Fig. 18.1), is transported via the cytosol, into the peroxisome where it is oxidized to glyoxylate. It is then transaminated to glycine, using either glutamate, alanine, serine or asparagine as the amino donor (Ta *et al.*, 1985). Following transport into the mitochondrion, two glycine molecules react to produce serine (see Chapter 16; Fig. 16.1 and Chapter 18; Fig. 18.1). The ammonia produced in this reaction probably diffuses out of the mitochondrion and is reassimilated by glutamine synthetase in the cytosol or, more likely, in the chloroplast. The serine enters the amino acid pools where some is used for protein synthesis, some for the synthesis of other amino acids, but most of it is transaminated with glyoxylate (see above) in the peroxisome with the formation of glycine and the keto analog of serine, hydroxypyruvate. The latter is reduced to glycerate and subsequently converted to sugars (Miflin and Lea, 1982).

Serine can also be synthesized by two other routes not involving glycine. Both these pathways use phosphoglycerate (diverted from either the Calvin cycle or the glycolytic pathway) as the starting point. Phosphoglycerate phosphatase and glycerate dehydrogenase can convert phosphoglycerate first to glycerate and then to hydroxypyruvate, which can be transaminated to serine. These enzymes are present in plant tissues but it is probable that they are more involved in serine degradation than its synthesis. The other route from phosphoglycerate is via phosphohydroxypyruvate and phosphoserine (Fig. 27.1). The enzymes for this pathway have been found in plants, but the contribution it makes to serine synthesis has not been determined (Miflin and Lea, 1982).

Regulation of photorespiratory production of glycine and serine is largely dependent on the relative concentrations of CO_2 and O_2, which compete for the active site of Rubisco. Regulation by metabolites has not been established (see Chapter 18).

Lysine, isoleucine and threonine

The relationship amongst the amino acids in the 'aspartate family' is shown in Fig. 27.1. The key reactions in the synthesis of these amino acids are the two involved in the synthesis of aspartic semialdehyde (ASA) from aspartate, catalyzed by the enzymes aspartate kinase (i) and aspartate semialdehyde dehydrogenase (ii) (see below). Aspartate semialdehyde is at an important branch point in amino acid synthesis, since it can either be reduced to homoserine by homoserine dehydrogenase (iii), or condensed (iv) with pyruvate to give dihydropicolinic acid, which subsequently undergoes a series of reactions to produce lysine (Fig. 27.1). The synthesis of lysine involves six steps, including acylation (from acetyl-CoA), transamination and decarboxylation but few of the enzymes responsible for these reactions have been isolated from plant tissues. The pathway has been deduced from feeding studies, isolation of intermediates, and by comparison with the bacterial pathway. Homoserine is an amino acid not found in proteins and hence is not usually present in appreciable concentrations in plants with the exception of peas, where it can constitute 70% of the soluble nitrogen in 1-week old seedlings (Mitchell and Bidwell, 1970). In most plants nearly all of the homoserine is phosphorylated to phosphohomoserine. This metabolite is also at a metabolic branch point since it can be converted to methionine (see below) or rearranged in a single step to threonine by the enzyme threonine synthase. Threonine can either be used for protein synthesis or metabolized to isoleucine, the synthesis of which begins with the deamination of threonine to α-ketobutyrate. This is then converted by a series of reactions to α-ketomethylvalerate, which is subsequently transaminated to isoleucine (Bryan, 1980).

The branched nature of the pathway requires that the regulation of the synthesis of the aspartate family of amino acids must occur at several points. The first enzyme showing regulatory properties is aspartate kinase, which occurs in most plants as two distinct isozymes. Several metabolites affect the activity of these isozymes, but the most effective are lysine and threonine, each of which inhibits only one of the isozymes. Thus there are lysine-sensitive and threonine-sensitive aspartate kinase isozymes in plant cells. Hence, the presence of high levels of only one of these compounds is insufficient to shut the pathway down since the other isozyme will still allow the pathway to operate. Lysine is also a major regulator of another key enzyme, dihydropicolinate synthase, which catalyzes the first step in the sequence of

$$ASP \xrightarrow[Mg^{2+}]{\underset{ATP\ ADP}{(i)}} ASP\text{-}P \xrightarrow[H_2PO_4]{\underset{NADPH+H^+\ NADP^+}{(ii)}} ASA \xrightarrow[]{\underset{NADPH+H^+\ \ \ NADP^+}{(iii)}} homoserine$$
$$ASA \xrightarrow[pyruvate]{(iv)} dihydropicolinic\ acid$$

reactions leading from aspartate semialdehyde to lysine. Most of this inhibition data comes from work on isolated enzymes, but is supported by feeding studies. For example, when lysine was fed to isolated pea leaf chloroplasts, the synthesis of both lysine and threonine was inhibited (Mills *et al.*, 1980).

The metabolism of phosphohomoserine is also regulated, as would be expected since it is the common precursor of threonine, isoleucine and methionine. In barley, the methionine derivative, S-adenosylmethionine, has been shown to stimulate the activity of threonine synthase up to 20-fold. This would divert carbon and nitrogen towards threonine synthesis when levels of methionine are high. In radish and sugar beet, threonine synthesis is inhibited by cysteine. Cysteine reacts with phosphohomoserine as the first step in methionine synthesis, so that when cysteine levels are high, threonine synthesis is inhibited so that the concentration of phosphohomoserine will increase and be available to react with cysteine. The first step in isoleucine synthesis, catalyzed by threonine dehydratase, is subject to feedback inhibition by isoleucine (Bryan, 1980; Miflin and Lea, 1982).

Most of the enzymes involved in aspartate family synthesis are found in the chloroplast, and it is likely that the whole pathway occurs in this organelle. Illuminated chloroplasts will convert ^{14}C-aspartate or malate to lysine, threonine and isoleucine (Mills *et al.*, 1980).

Valine and leucine

Pyruvate and acetyl-CoA provide the carbon skeletons for these two 'branched-chain' amino acids. Isoleucine is also a branched chain amino acid, similar in structure to leucine, but derives its carbon from aspartate (see above). These three amino acids are often grouped together, not only because of structural considerations, but also because they share several common enzymes in their synthesis. The same enzymes that convert α-ketobutyrate to isoleucine also convert pyruvate to valine in a parallel but distinct pathway, with no sharing of intermediates. A branch point in the valine pathway is at α-ketoisovalerate, which can be considered equivalent to α-ketobutyrate in the isoleucine pathway. α-Ketoisovalerate can either be transaminated directly to valine, or condensed with the methyl group of acetyl coenzyme A to give isopropylmalate, which is isomerized and decarboxylated to α-ketoisocaproate (see Fig. 27.1), then transaminated to leucine (Bryan, 1980). As with other branched pathways, close regulation is necessary to produce the desired levels of the different products, and again feedback inhibition is in effect. Valine and leucine both inhibit the second step in the pathway from pyruvate, and leucine inhibits the synthesis of isopropylmalate from ketoisovalerate (Miflin and Lea, 1982)

These pathways appear to be located in the chloroplast since isolated chloroplasts can synthesize valine from $^{14}CO_2$, and several enzymes of the pathway have been found in isolated chloroplasts (Miflin and Lea, 1982).

The sulfur amino acids: cysteine and methionine

Sulfur is usually taken up by plants as sulfate, which is transported in the xylem to leaf tissue and reduced by a series of reactions to protein-bound sulfide, which is then incorporated into amino acids. Nearly all sulfate reduction occurs in photosynthetic tissues, specifically in chloroplasts, where it accounts for a significant amount of ATP and reduced ferredoxin consumed. Most of the sulfide is incorporated into cysteine by reaction with *o*-acetylserine, also in the chloroplast:

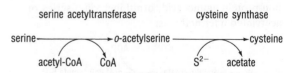

The control mechanisms in cysteine synthesis appear to involve feedback inhibition by cysteine at several steps in the reduction of sulfate to sulfide, and at sulfate uptake itself. There has been some discussion, but little conclusive evidence, about the regulation of cysteine synthase and serine acetyltransferase (Giovanelli et al., 1980).

Methionine can be considered to be a member of the aspartate family, since one of its precursors is phosphohomoserine. However, it is often put in a separate group with the other sulfur amino acid, cysteine, which is another precursor in its synthesis.

Cysteine and phosphohomoserine react (cystathionine synthase) to produce cystathionine, which is subsequently cleaved by cystathionine lyase to give homocysteine. The final step in methionine synthesis is the methylation of homocysteine, which involves the polyglutamyl derivative of folic acid (Cossins, 1980).

cysteine + PHS → cystathionine → homocysteine → methionine

These reactions have been demonstrated in many plants, but the specific nature and location of the enzymes involved has not been established. There is some evidence that cystathionine synthase occurs in the chloroplast, and that the subsequent reactions leading to methionine occur in the mitochondria.

As well as being used in protein and ethylene synthesis, methionine is also withdrawn from amino acid pools for the synthesis of, amongst other compounds, the enzyme cofactors spermine and spermidine. Methionine is known to inhibit its own synthesis, but the nature of this control is not understood (Giovanelli et al., 1980).

Arginine and proline

Glutamate is the precursor of the amino acids, glutamine, arginine and proline. In the synthesis of arginine, glutamate is first metabolized to the non-protein amino acid, ornithine, via a series of acetylated intermediates:

The subsequent metabolism of ornithine appears to be by the same series of reactions that occur in animal tissues, the ornithine cycle:

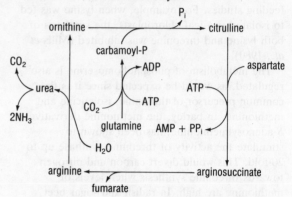

Little is known about the subcellular localization of these reactions in plant cells, but some of the reactions have been shown to be chloroplastic (Taylor and Stewart, 1981). The use of acetylated intermediates in this pathway probably serves to prevent competition between the syntheses of arginine and proline, which are similar in several ways. Arginine has been shown to inhibit the phosphorylation of acetylglutamate, another example of an endproduct inhibiting an early reaction in its synthesis.

In the first step of proline synthesis, glutamate is reduced to glutamyl 5-semialdehyde, via an enzyme-bound intermediate, glutamyl 5-phosphate (similar to the production of aspartate semialdehyde from aspartate). The semialdehyde spontaneously cyclizes to give pyrroline 5-carboxylic acid, which is then reduced to proline (Thompson, 1980). (See diagram.)

Again, not much is known about the regulation of this pathway, although it has been reported that proline feeds back to inhibit its own synthesis. Proline synthesis probably occurs in the cytosol, but the same series of reactions apparently occurs in the reverse direction in the mitochondrion, providing another route for glutamate synthesis. This pathway is not of great significance, but its importance may increase under stress conditions when it also appears that proline and arginine metabolism become linked: ornithine can be converted to glutamate semialdehyde either directly by transamination, or through a series of reactions via proline and the reverse of the proline

Amino acid and ureide biosynthesis

```
COOH                              CHO
|                                 |
CH₂    ATP   ADP   NADPH + H⁺ NADP⁺ CH₂
|         ↘ ↙         ↘    ↙       |
CH₂       Mg²⁺              ↘      CH₂
|                           Pi     |
HCNH₂                              HCNH₂
|                                  |
COOH                               COOH
glutamate                          glutamyl semialdehyde

         NADP⁺  NADPH + H⁺
H₂C——CH₂   ↖   ↙            H₂C——CH₂
|    |       ←              |    |
H₂C  HCCOOH                 HC   HCCOOH
  \ /                         \\ //
   NH                           N
proline                      pyrroline 5'-carboxylate
```

synthetic pathway. These complex interconversions have been demonstrated by feeding experiments with labeled intermediates: radioactive label introduced into plant tissue via ornithine or arginine is transferred to proline, and label from proline is transferred to pyrroline 5-carboxylate, glutamyl 5-semialdehyde and glutamate (with prolonged feedings also to other amino acids and citric acid cycle intermediates) (Thompson, 1980). Proline levels are often observed to increase dramatically (up to 200-fold) under stress conditions such as increased salinity, drought, or high temperatures (Stewart and Larher, 1980).

The aromatic amino acids: phenylalanine, tyrosine and tryptophan

These three amino acids appear to be synthesized exclusively by the 'shikimic acid pathway', the first step of which is the condensation of erythrose 4-phosphate (derived from the oxidative pentose phosphate pathway or the Calvin cycle) with phosphoenolpyruvate (from glycolysis) to produce 3-deoxy D-arabino heptulosonic acid 7-phosphate (DAHP). This undergoes a series of reactions, including condensation with another molecule of PEP to give chorismic acid (Fig. 27.1). Chorismic acid is at a branch point in this pathway, and can undergo two different reactions, one leading to tryptophan, and the other to phenylalanine and tyrosine.

The synthesis of tryptophan from chorismate begins with the reaction of chorismate with the amide group of glutamine to produce anthranilic acid, which subsequently condenses with phosphoribosyl pyrophosphate (derived from ribose 5-phosphate) to give phosphoribosyl anthranilate. This molecule undergoes a further series of reactions to produce indole, which then reacts with serine to produce tryptophan (catalyzed by tryptophan synthetase) (Gilchrist and Kosuge, 1980).

The synthesis of phenylalanine and tyrosine starts with the rearrangement of chorismate by chorismate mutase to prephenic acid, whose further metabolism has recently been the subject of some debate (Fig. 27.3). For some time the synthesis of phenylalanine and tyrosine from prephenate in plants was assumed to be the same as in bacteria, where the prephenate is either dehydrated to phenylpyruvate (prephenate

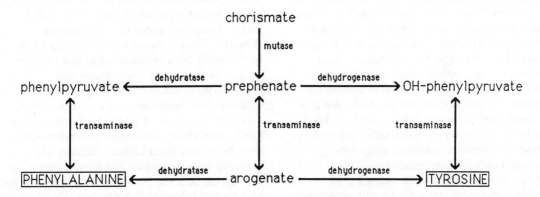

Fig. 27.3 Two routes for the synthesis of phenylalanine and tyrosine from chorismic acid.

dehydratase) or oxidatively decarboxylated to hydroxyphenylpyruvate (prephenate dehydrogenase). Both of these keto acids are subsequently aminated by transaminases, the former to phenylalanine and the latter to tyrosine. Although some of the enzymes involved in this route have been found in plants, there is a growing body of evidence which suggests that another route is either also, or in some plants solely, in operation (see Rubin and Jensen, 1979). This 'alternative' pathway, formerly called the 'pretyrosine' pathway, is now generally referred to as the 'arogenate pathway', and involves the transamination of prephenate to arogenate which is then directly converted to either phenylalanine (arogenate dehydratase) or tyrosine (arogenate dehydrogenase). Arogenate dehydratase has been purified from sorghum and its activity shown to be inhibited by phenylalanine and stimulated by tyrosine, as might be expected from its position in the pathway (Siehl and Conn, 1988). Arogenate dehydrogenase has also been purified from sorghum and characterized, and is strongly inhibited by tyrosine, but unaffected by phenylalanine or other metabolites of the pathway (Connelly and Conn, 1985). There is good evidence that the arogenate pathway is functioning in a number of plants, but further work is required to determine the contribution made by the two pathways to aromatic amino acid synthesis in other plants.

Many of the enzymes of tryptophan synthesis have been found in the chloroplast, and labeling studies with $^{14}CO_2$ have shown that chloroplasts contain the complete pathways for the synthesis of the aromatic amino acids. It is believed that these pathways also exist in the cytosol, and perhaps other subcellular compartments, but this has not been proven. As might be expected, feedback inhibition by tryptophan affects the synthesis of anthranilate from chorismate. Phenylalanine and tyrosine also inhibit their own synthesis, but it is not clear how this occurs (Miflin and Lea, 1982).

When plant tissues are wounded or infected by parasites, respiration increases, along with associated carbohydrate metabolism. This is principally to provide the precursors for the synthesis of defense and wound repair compounds, such as phenolic acids, suberin and lignin. Most of these compounds are synthesized from phenylalanine, tyrosine or components of the shikimate pathway, all of which are synthesized from phosphoenolpyruvate (from glycolysis) and erythrose 4-phosphate (form the pentose phosphate pathways). Under such circumstances, many of the enzymes of glycolysis, such as phosphofructokinase and pyruvate kinase, show increased activity, and in some cases, such as pyruvate kinase, there is increased synthesis of the enzyme. The net result is a dramatic increase in glycolytic capacity. Pentose phosphate pathway activity increases even more than glycolysis in response to wounding or infection, largely as a result of increased synthesis of glucose 6-phosphate dehydrogenase and 6-phosphogluconate dehydrogenase. Thus the supply of shikimate pathway precursors is enhanced, facilitating an increase in aromatic amino acid production, which is the basis for the synthesis of most of the defense or wound repair compounds (Uritani and Asahi, 1980).

Non-protein amino acids

Many plants channel large amounts of nitrogen into amino acids that are not usually constituents of proteins. These non-protein amino acids comprise a very diverse and often complex group of compounds: several hundred of them have been found in plants, usually in seeds, where they can accumulate to high levels. They are found in all plant tissues as intermediates in the synthesis of protein amino acids, and in a more restricted range of plants as metabolic 'endproducts' (Fowden, 1980). Examples of the former category have already been encountered in this chapter: homoserine is a non-protein amino acid intermediate in the synthesis of threonine, isoleucine and methionine, as is diaminopimelic acid in the synthesis of lysine. In the case of the non-protein amino acids which are endproducts, their function is often unclear, but they are generally toxic to animals and found in seeds, so they appear to serve both as a storage reserve and as a feeding deterrent to herbivores. One non-protein amino acid fairly common in legumes is canavanine, which can account for up to 6% of the fresh weight of the seeds of the jackbean (*Canavalia ensiformis*). It is very similar in structure to arginine and is thus able to interfere with arginine metabolism in animals that ingest the seeds.

$$\underset{\text{canavanine}}{\begin{array}{c}NH\\\|\\CNH_2\\|\\NH\\|\\O\\|\\CH_2\\|\\CH_2\\|\\HCNH_2\\|\\COOH\end{array}} \qquad \underset{\text{arginine}}{\begin{array}{c}NH\\\|\\CNH_2\\|\\NH\\|\\CH_2\\|\\CH_2\\|\\CH_2\\|\\HCNH_2\\|\\COOH\end{array}}$$

As with most non-protein amino acids, the pathway by which canavanine is synthesized is not clear, but seems to be similar to the ornithine cycle which produces arginine (Rosenthal, 1982). Non-protein amino acids often increase under stress conditions; for example, the levels of γ-aminobutyric acid, the decarboxylation product of glutamic acid, often increase in response to stresses such as anoxia (Stewart and Larher, 1980).

Ureides

Ureides are compounds such as citrulline, allantoin and allantoic acid, which are used for nitrogen transport in a variety of plants, including many legumes. Citrulline synthesis has been described above, and so this section will focus on the synthesis of the nitrogen-transport compounds, allantoin and allantoic acid. A more detailed discussion of this subject can be found in the review by Schubert (1986).

Many legumes which normally use the amides glutamine and asparagine for nitrogen transport from their roots switch to ureide synthesis when they are nodulated. These plants are referred to as the 'tropical' legumes, and include soybeans, cowpeas and mungbeans. The 'temperate' legumes such as peas, lupins and alfalfa, continue to synthesize amides regardless of whether they are nodulated or not. The rationale for this switch from amide to ureide synthesis in the tropical legumes involves the economy of carbon use: the relative metabolic costs of using ureides or amides has been the subject of some debate, but it does appear that the ureide producers use less organic carbon to transport the same amount of nitrogen as do the amides (Pate *et al.*, 1981). Catabolic costs are difficult to estimate, since the pathways have not been completely elucidated, but catabolism of ureides appears to be less efficient than catabolism of amides, since some of the ureide carbon is lost as CO_2. This may not be particularly significant in the light since there is presumably an excess of ATP under such conditions (Schubert, 1986).

Ureides are synthesized by the oxidation of purines, which are themselves derived from glutamine, glycine, aspartate and ribose 5-phosphate. Root nodules contain both infected (with the nitrogen-fixing bacteroids) and uninfected cells, both of which are involved in ureide synthesis. In the bacteroids, dinitrogen gas is reduced to ammonia, which is excreted into the cytosol of the infected host cell, then assimilated into glutamine by glutamine synthetase. Purine synthesis from this glutamine then appears to occur in the plastids of these infected cells (Fig. 27.4). Feeding experiments with ^{14}C-glycine have shown that *de novo* synthesis is far more important than the salvage pathways from the nucleic acids in providing purines for ureide synthesis.

The purine, xanthine, is exported to the cytosol where it is oxidized to uric acid by xanthine dehydrogenase (Fig. 27.5). There has been some debate about the location of this reaction in the

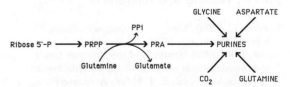

Fig. 27.4 Biosynthesis of purines in root nodules: phosphoribosylpyrophosphate (PRPP), phosphoribosylamine (PRA).

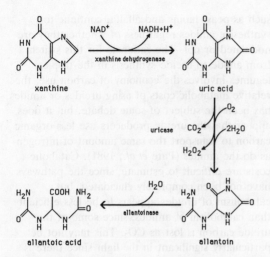

Fig. 27.5 Synthesis of allantoin and allantoic acid from the purine, xanthine.

nodule. The use of sucrose density gradients indicated that xanthine dehydrogenase is in the cytosol of both infected and uninfected cells (Shelp et al., 1983), whereas immunochemical studies have assigned it solely to infected cells (Triplett, 1985), or to the plastids of uninfected cells (Nguyen, 1986). Further work is needed to sort out the cellular and subcellular localization of these reactions. The next step occurs in the peroxisomes, apparently in the uninfected cells, and involves the action of uricase on uric acid, producing allantoin, CO_2 and H_2O_2. Depending on the plant species concerned, differing proportions of the allantoin are used directly for transport in the xylem, or first converted to allantoic acid in the endoplasmic reticulum by the enzyme allantoinase.

Further reading and references

Beevers, L. (1976). *Nitrogen Metabolism in Plants*, Edward Arnold, pp. 28–58.

Bryan, J. K. (1980). Synthesis of the aspartate family and branched-chain amino acids. In *The Biochemistry of Plants*, Vol. 5, ed. B. J. Miflin, Academic Press, New York, pp. 403–52.

Chapman, D. J. and Leech, R. M. (1979). Changes in pool sizes of free amino acids and amides in leaves and plastids of *Zea mays* during leaf development. *Plant Physiol.* **63**, 567–72.

Connelly, J. A. and Conn, E. E. (1986). Tyrosine biosynthesis in *Sorghum bicolor*. Isolation and regulatory properties of arogenate dehydrogenase. *Z. Naturforsch.* **41c**, 69–78.

Cossins, E. A. (1980). One carbon metabolism. In *The Biochemistry of Plants*, Vol. 2, ed. D. D. Davies, Academic Press, New York, pp. 366–418.

Fowden, L. (1981). Nonprotein amino acids. In *The Biochemistry of Plants*, Vol. 7, ed. E. E. Conn, Academic Press, New York, pp. 215–47.

Gilchrist, G. D. and Kosuge, T. (1980). Aromatic amino acid biosynthesis and its regulation. In *The Biochemistry of Plants*, Vol. 5, ed. B. J. Miflin, Academic Press, New York, pp. 507–531.

Giovanelli, J., Mudd, S. H. and Datko, A. H. (1980). Sulfur amino acids in plants. In *The Biochemistry of Plants*, Vol. 5, ed. B. J. Miflin, Academic Press, New York, pp. 453–505.

Givan, C. V. (1980). Aminotransferases in higher plants. In *The Biochemistry of Plants*, Vol. 5, ed. B. J. Miflin, Academic Press, New York, pp. 329–57.

Hatch, M. D. (1976). In *CO_2 Metabolism and Plant Productivity*, eds R. M. Burris and C. C. Black, University Park Press, Baltimore, pp. 59–81.

Ireland, R. J. and Joy, K. W. (1985). Plant transaminases. In *Transaminases*, eds P. Christen and D. E. Metzler, John Wiley & Sons, New York, pp. 376–84.

Kirk, P. R. and Leech, R. M. (1972). Amino acid biosynthesis by isolated chloroplasts during photosynthesis. *Plant Physiol.* **50**, 228–34.

Kushad, M. M., Richardson, D. G. and Ferro, A. J. (1983). Intermediates in the recycling of 5-methylribose to methionine in fruits. *Plant Physiol.* **73**, 257–61.

Lea, P. J. and Miflin, B. J. (1980). Transport and metabolism of asparagine and other nitrogen compounds within the plant. In *The Biochemistry of Plants*, Vol. 5, ed. B. J. Miflin, Academic Press, New York, pp. 569–607.

Miflin, B. J. and Lea, P. J. (1980). Ammonia assimilation. In *The Biochemistry of Plants*, Vol. 5, ed. B. J. Miflin, Academic Press, New York, pp. 169–202.

Miflin, B. J. and Lea, P. J. (1982). Ammonia assimilation and amino acid metabolism. In *Encyclopedia of Plant Physiology*, New Series Vol. 14A, eds D. Boulter and D. Parthier, Springer-Verlag, Berlin, pp. 5–64.

Mills, W. R., Lea, P. J. and Miflin, B. J. (1980) Photosynthetic formation of the aspartate family of amino acids in isolated chloroplasts. *Plant Physiol.* **65**, 1166–72.

Mitchell, D. J. and Bidwell, R. G. S. (1970). Compartments of organic acids in the synthesis of

asparagine and homoserine in pea roots. *Can. J. Bot.* **48**, 2001–7.
Nguyen, J., Machal, L., Vidal, J., Perrot-Rechenmann, C. and Gadal, P. (1986). Immunochemical studies on xanthine dehydrogenase of soybean root nodules. *Planta* **167**, 190–5.
Oaks, A. and Hirel, B. (1985). Nitrogen metabolism in roots. *Ann. Rev. Plant Physiol.* **36**, 345–65.
Pate, J. S., Atkins, C. A. and Rainbird, R. M. (1981). Theoretical and experimental costing of nitrogen fixation and related processes in nodules of legumes. In *Current Perspectives in Nitrogen Fixation*, eds A. H. Gibson and W. E. Newton, Australian Academy of Sciences, Canberra, pp. 105–16.
Rosenthal, G. A. (1982). *Plant Nonprotein Amino and Imino Acids*, Academic Press, New York.
Rowell, P. and Stewart, W. D. P. (1976). Alanine dehydrogenase of the N_2-fixing blue–green alga *Anabaena cylindrica*. *Arch. Microbiol.* **107**, 115–24.
Rubin, J. L. and Jensen, R. A. (1979). Enzymology of L-tyrosine biosynthesis in mung bean (*Vigna radiata* [L.] Wilczek). *Plant Physiol.* **64**, 727–34.
Schubert, K. R. (1986). Products of biological nitrogen fixation in higher plants: synthesis, transport and metabolism. *Ann. Rev. Plant Physiol.* **37**, 539–74.
Shelp, B. J., Atkins, C. A., Storer, P. J. and Canvin, D. T. (1983). Cellular and subcellular organization of pathways of ammonia assimilation and ureide synthesis in nodules of cowpea (*Vigna unguiculata* L. Walp.). *Arch. Biochem. Biophys.* **224**, 429–41.
Sieciechowicz, K. A., Joy, K. W. and Ireland, R. J. (1988). The metabolism of asparagine in plants. *Phytochemistry* **27**, 663–71.
Siehl, D. L. and Conn, E. E. (1988). Kinetic and regulatory properties of arogenate dehydratase in seedlings of *Sorghum bicolor* (L.) Moench. *Arch. Biochem. Biophys.* **260**, 822–9.
Stewart, G. R. and Larher, F. (1980). Accumulation of amino acids and related compounds in relation to environmental stress. In *The Biochemistry of Plants*, Vol. 5, ed. B. J. Miflin, Academic Press, New York, pp. 609–30.
Ta, T. C., Joy, K. W. and Ireland, R. J. (1985). The role of asparagine in the photorespiratory nitrogen metabolism of pea leaves. *Plant Physiol.* **78**, 334–7.
Taylor, A. A. and Stewart, G. G. (1981). Tissue and subcellular localization of enzymes of arginine metabolism in *Pisum sativum*. *Biochem. Biophys. Res. Commun.* **101**, 1281–9.
Thompson, J. F. (1980). Arginine synthesis, proline synthesis, and related processes. In *The Biochemistry of Plants*, Vol. 5, ed. B. J. Miflin, Academic Press, New York, pp. 375–402.
Triplett, E. W. (1985). Intercellular nodule localization and nodule specificity of xanthine dehydrogenase in soybean. *Plant Physiol.* **77**, 1004–9.
Uritani, I. and Asahi, T. (1980). Respiration and related metabolic activity in wounded and infected tissues. In *The Biochemistry of Plants*, Vol. 2, ed. D. D. Davies, Academic Press, New York, pp. 463–85.
Yamaya, T. and Oaks, A. (1987). Synthesis of glutamate by mitochondria – an anaplerotic function for glutamate dehydrogenase. *Physiol. Plant.* **70**, 749–56.

28 Interactions between photosynthesis, respiration and nitrogen assimilation

David H. Turpin and Harold G. Weger

Introduction

Photosynthesis, respiration and N assimilation are interrelated processes (Fig. 28.1) (Turpin et al., 1988). Many of the interactions between these processes have been addressed separately in considerable detail in other chapters of this book. The purpose of this chapter is not to reiterate detail covered elsewhere but to integrate these processes by providing a simplified overview of the interactions involved.

In a simplistic sense photosynthesis and respiration are fundamentally opposed processes. Photosynthesis involves the light-driven oxidation of water resulting in the production of O_2. This oxidation is coupled to the reduction of CO_2 or other physiological electron acceptors (NO_2^-, SO_4^-, O_2). The transfer of electrons between H_2O and $NADP^+$ is also coupled to ATP production. Subsequent electron transfer to CO_2 is also associated with ATP consumption. Conversely, respiration is the oxidation of reduced carbon compounds with the production of CO_2 and the transfer of electrons to O_2 resulting in its reduction to H_2O. Similar to photosynthesis, the electron transfer reactions may be coupled to ATP production. Fig. 28.2 compares and contrasts these processes at this fundamental level. Given the opposing characteristics of photosynthesis and respiration one would expect some form of coordinate regulation between them in order to minimize the futile cycling of CO_2 to carbohydrate and back to CO_2 during photosynthesis. There is little evidence for futile cycling but nonetheless, mitochondrial respiration plays a significant role in the growth and metabolism of photosynthetic organisms in the light.

The effects of photosynthesis on respiration

Given the importance of both photosynthesis and respiration in plant metabolism it is surprising that the occurrence and magnitude of mitochondrial respiration during photosynthesis remains controversial. Mitochondrial respiration consists of two distinct but interrelated processes. The first is the oxidation of organic carbon by glycolysis and the TCA cycle resulting in the production of CO_2, reducing equivalents (NADH and $FADH_2$) and carbon skeletons for use in biosynthesis or in continued TCA cycle oxidation.

The second process is the oxidation of TCA cycle-generated reductant by the mitochondrial electron transport chain (ETC), resulting in the reduction of O_2 to H_2O. Although TCA cycle activity and mitochondrial electron transport are usually considered interdependent, these processes can operate in isolation of one another. For example, maintenance of TCA cycle carbon flow requires that the NADH and $FADH_2$ produced be continually oxidized to ensure a continuing supply of NAD^+ and FAD. Although we commonly think of the mitochondrial electron transport chain serving this role, many other reactions use NADH and produce NAD^+ (e.g. OAA reduction to malate, NO_3^- reduction to NO_2^-). Likewise, the mitochondrial electron transport chain may oxidize

Interactions between photosynthesis, respiration and nitrogen assimilation

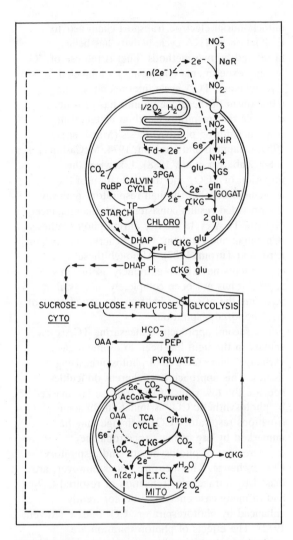

Fig. 28.1 Interactions between photosynthesis, respiration and N assimilation during photosynthetic suppression in response to N pulsing in N-limited *S. minutum*. The stoichiometry is not balanced and provides a simplified representation of N assimilation into glutamate. α-KG, α-ketoglutarate; ETC, electron transport chain; Fd, ferredoxin; glu, glutamate; OAA, oxaloacetate; PEP, phosphoenolpyruvate; PGA, 3-phosphoglycerate; RuBP, ribulose bisphosphate; TP, triose phosphate; NaR, nitrate reductase; NiR, nitrite reductase; MITO, mitochondrion; CYTO, cytoplasm; CHLORO, chloroplast; DHAP, dihydroxyacetone phosphate. Many other pathways of carbon flow are involved in the synthesis of other amino acids (Chapter 27) (adapted from Turpin *et al.*, 1988).

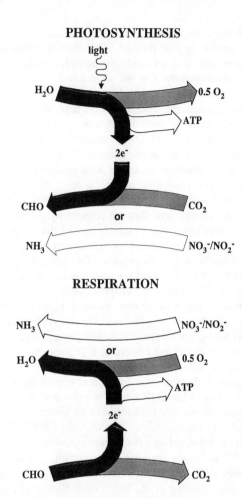

Fig. 28.2 A simplified representation of photosynthesis and respiration illustrating the opposing nature of these two processes. Photosynthesis is the oxidation of water and the reduction of some physiological electron acceptor such as CO_2 or NO_3^-/NO_2^-. Respiration is the oxidation of carbohydrate (CHO) and the reduction of O_2 to H_2O.

NADH produced in reactions other than the TCA cycle. For example, the NADH produced by glycine decarboxylase during photorespiration may also serve as substrate for the mitochondrial electron transport chain (see Chapter 18). Consequently, when discussing interactions between respiration and other metabolic processes it is important to deal with both the TCA cycle and mitochondrial electron transport separately.

In the light, cellular needs for ATP can presumably be met through photophosphorylation, without the need for oxidative phosphorylation in the mitochondrion. On the other hand, photoautotrophs maintain high rates of biosynthesis in the light. Consequently, there is a large demand for carbon skeletons, particularly keto acids (e.g. αKG, OAA, succinate and pyruvate) from glycolysis and the TCA cycle for use in the synthesis of amino acids and other compounds. Therefore, maintenance of the biosynthetic processes during photosynthesis requires carbon flow through respiratory pathways. Not surprisingly, it is generally agreed that at least some 'respiratory activity' is maintained during photosynthesis for this purpose (Graham, 1980; Raven, 1984). The case for maintenance of mitochondrial electron transport during photosynthesis is more controversial. Until recently it was believed that an increase in the ATP/ADP ratio during photosynthesis would prevent operation of the mitochondrial ETC, thereby removing one potential sink for TCA cycle generated electrons. Several recent reports have shown that ATP/ADP ratios, as influenced by photosynthesis, are probably not an important regulatory mechanism for mitochondrial ETC activity. First, cellular (especially cytosolic) ATP/ADP ratios are relatively constant, and not greatly affected by light/dark transitions (Goller et al., 1982; Hampp et al., 1982, 1985; Stitt et al., 1982). Second, in vitro evidence has shown that extremely high ATP/ADP is necessary to inhibit mitochondrial ETC activity. (Dry and Wiskich, 1982) and such conditions are unlikely during steady-state photosynthesis. Indeed, in vivo ETC activity is probably much more sensitive to the absolute ADP concentration. Therefore, any biosynthetic process which consumes ATP and produces ADP could conceivably stimulate ETC activity. Finally, the existence of alternative pathway respiration would minimize ATP/ADP effects on ETC activity, especially if electron flow through the alternative pathway is completely uncoupled from ATP production through utilization of the rotenone-resistant bypass (Chapter 8).

Measurements of mitochondrial respiration in the light have focused on either TCA cycle or mitochondrial electron transport chain activity. Evaluation of TCA cycle activity has been facilitated by two methods. First is the use of ^{14}C feeding experiments employing radiolabeled glycolytic and TCA cycle intermediates. The metabolism of these compounds has provided evidence that respiratory carbon metabolism continues during photosynthesis (Marsh et al., 1965; Chapman and Graham, 1974; McCashin et al., 1988; Zabkova et al., 1988). Likewise, the provision of $^{14}CO_2$ and the evaluation of the movement of label through respiratory pathway has provided results consistent with maintenance of some TCA cycle activity during photosynthesis (Chapman and Graham, 1974; Scherer et al., 1984; Elrifi and Turpin, 1987). Although these experiments have been extremely useful, this approach has not been completely successful in resolving the effects of light on the actual rates of mitochondrial respiration.

The second approach to measuring TCA cycle activity in the light has been to examine gas exchange characteristics of photosynthesizing tissues. This approach has inherent difficulties because the low rates of CO_2 release are masked by photosynthetic CO_2 assimilation. In some situations respiratory CO_2 exchange may be unmasked by use of mass spectrometric measurement of photosynthetic and respiratory CO_2 exchange. The measured rates however, are biased by intracellular refixation of respired CO_2 and in many cases may also be apparently enhanced by photorespiratory CO_2 release (Raven, 1972). The effects of photorespiration are minimized in studies with unicellular algae which possess a CO_2-concentrating mechanism. In such cases, TCA cycle CO_2 release has been observed during photosynthesis; however, the problem of photosynthetic refixation has minimized the utility of this approach in the absolute quantification of TCA cycle activity in the light (Weger et al., 1988).

Mass spectrometry has also been used to evaluate mitochondrial electron transport during photosynthesis by measuring the rates of O_2 consumption in the light. Although there are numerous studies of O_2 exchange in the light there are also a variety of interpretations of these results. A clear appreciation of the role of mitochondrial electron transport during photosynthesis has yet to emerge. In part this may be due to the number of potential mechanisms for O_2 consumption during photosynthesis including photorespiration, O_2 photoreduction (the Mehler reaction) and mitochondrial respiration (Badger, 1985). Under conditions where photorespiration is suppressed (e.g. high CO_2), a high rate of O_2 consumption at high irradiance has often been demonstrated and attributed to photoreduction (Radmer and Kok, 1976; Marsho et al., 1979; Furbank et al., 1982; Ishii and Schmid, 1982; Sültemeyer and Fock, 1986, 1987; Shiraiwa et al., 1988). Many of the latter experiments have also suggested that illumination results in a decrease in mitochondrial O_2 consumption. Such a decline in mitochondrial electron transport chain activity may provide an explanation for the higher apparent quantum yield at low light (Scherer et al., 1984). On the other hand, other experiments have indicated that O_2 consumption is decreased in the light, and does not respond to increasing irradiance (Bate et al., 1988). Yet other work has demonstrated that O_2 consumption in the light occurs at rates comparable to those in the dark (Gerbaud and Andre, 1980; Peltier and Thibault, 1985; Weger et al., 1988).

One technique which has been used to separate O_2 consumption by the Mehler reaction and photorespiration from mitochondrial respiration is the addition of DCMU. DCMU inhibits non-cyclic electron flow so that maintenance of O_2 consumption in the presence of DCMU rules out photorespiration and the Mehler reaction. In the green alga *Selenastrum minutum* O_2 consumption in the light continued at high rates and was completely unaffected by DCMU. This indicated little change in mitochondrial ETC activity between the light and the dark and suggested that, in this case, rates of both photorespiratory and Mehler O_2 consumption are low during photosynthesis (Weger et al., 1988). However, in a Rubisco-deficient mutant of *Chlamydomonas reinhardtii*, transfer to light decreased the ADP concentration, thereby limiting glycolytic carbon flow and TCA cycle activity (Gans and Rebeille, 1988). In contrast, in wild-type *Chlamydomonas*, transitions from dark to light do not result in this marked inhibition of mitochondrial respiration (Weger and Turpin, unpublished).

Finally, there is some evidence for enhanced mitochondrial respiration following periods of illumination. This is presumably due to increased substrate levels in recently darkened cells (Azcon-Bieto and Osmond 1983; Weger et al., 1989), which may include triose phosphate from the Calvin cycle or from the breakdown of recently synthesized starch. Some authors have suggested that the enhanced rate of post-illumination mitochondrial respiration is representative of the rate in the preceding light period, and may sometimes be associated with the alternative pathway (Azcon-Bieto et al., 1983).

Another issue of potential relevance to the debate about the magnitude of mitochondrial respiration in the light is the potential for chloroplast respiration ('chlororespiration') (Chapter 14). To date, the existence of chlororespiration has been postulated only for green algae, but some estimates of the rate suggest that it may account for up to 20% of dark O_2 consumption (Bennoun, 1982), although others have been unable to demonstrate O_2 consumption by isolated green algal chloroplasts (Klock et al., 1989).

Clearly there exist many possible relationships between the rates of respiration and photosynthesis. Based on work to date it is unlikely that a single point of view will be valid under all conditions in every organism. We should expect there to be conditions where photosynthesis enhances respiration, others in which it causes its inhibition, and still others where there is little effect.

The effects of respiration on photosynthesis

Photosynthesis is usually viewed as affecting respiration. Recently it has been suggested that mitochondrial respiration may play a role in photosynthesis. Kromer et al. (1988) have shown that by selectively inhibiting mitochondrial ETC activity with oligomycin, the photosynthetic rate of protoplasts, but not chloroplasts, declines. They suggest that the mitochondria oxidize chloroplast reductant to provide for cytosolic ATP demands, which is a much more efficient way to produce ATP than by cyclic photosynthetic electron flow. In another example, a *Chlamydomonas* mutant deficient in chloroplast ATP synthase can grow photoautotrophically using ATP generated in the mitochondria to support CO_2 fixation by the Calvin cycle (Lemaire et al., 1988).

Interactions between N assimilation and photosynthesis

The light reactions

The reductive assimilation of inorganic N into amino acids can be considered a photosynthetic process. The reduction of NO_3^- to NH_4^+ and its assimilation into glutamate requires 5 electron pairs ($2e^-$). Consequently, this process can support the photosynthetic evolution of $2\frac{1}{2}$ moles of O_2 per NO_3^- assimilated (Fig. 28.3) (Grant and Turner, 1969; Ullrich and Eisele, 1977; Larsson et al., 1982).

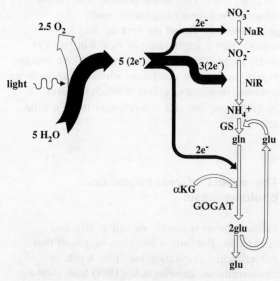

Fig. 28.3 A diagrammatic representation of the use of photosynthetic electron flow in the reduction and assimilation of NO_3^- to glutamate. NaR, nitrate reductase; NiR, nitrite reductase; GS, glutamine synthetase; GOGAT, glutamine : oxoglutarate aminotransferase.

Although other reductive processes also make use of photogenerated electrons, the reduction of NO_3^- is the most significant process after CO_2 (Syrett, 1981). Many plants are also capable of using mitochondrial reductant for the assimilation of NO_3^- which is not possible for CO_2 reduction (Sawhney et al., 1978; Woo et al., 1980). This is clearly the case during NO_3^- assimilation by roots and during the assimilation of NO_3^- in the dark by photosynthetic tissues (see Chapter 23; Schrader and Thomas, 1981; Weger and Turpin, 1989). It has also been shown that, in certain cases, mitochondrial reductant is used for NO_3^- and NO_2^- reduction during photosynthesis (Weger and Turpin, 1989).

The simultaneous reduction of both NO_3^-/NO_2^- and CO_2 in photosynthesizing tissues leads to the potential for competition between these two processes for photogenerated reductant. Recent work suggests that this does not occur (Robinson, 1988). This may result from the photosynthetic electron transport having a capacity that is in excess of that required for carbon fixation (Stitt, 1986) or it may be due to the dedicated association of specific photosystems with nitrite reductase (Robinson, 1988).

Photosynthetic carbon fixation

Nitrate or ammonia addition to nitrogen-sufficient higher plant cells, protoplasts, or microalgae generally results in little change or in some cases a slight enhancement of photosynthetic carbon fixation (see review by Turpin et al., 1988). In at least one reported case, a slight suppression of carbon fixation has been observed (Larsson et al., 1982). These authors claim that this suppression results from a competition for reducing power between CO_2 and N assimilation. However, similar effects were observed during both NH_4^+ and NO_3^- assimilation, suggesting other factors may also be involved. Although the effects of N assimilation on CO_2 fixation by N-sufficient tissue are relatively minor, there is a striking effect of CO_2 availability on the ability of photosynthetic cells to assimilate NO_3^- and NH_4^+ (Larsson et al., 1985; Lara et al., 1987). Most authors attribute this to a need for recent photosynthate to provide carbon skeletons for amino acid biosynthesis. There is also a possibility that CO_2 removal causes HCO_3^- limitation of PEPcase. This would decrease the ability of PEPcase to replenish TCA cycle intermediates. This would cause amino acid biosynthesis to stop because of a lack of carbon intermediates that ultimately would inhibit N assimilation. It is therefore difficult to determine if the CO_2 requirement for N assimilation is mediated entirely through Rubisco or if PEPcase plays an important role. This role of PEPcase in the provision of carbon for amino acid synthesis will be explored in more detail later. Kramer et al. (1988) have also proposed a direct effect of CO_2 on NO_2^- transport into Chlorella chloroplasts.

Although the effects of N assimilation on photosynthetic carbon fixation in N-sufficient tissues are relatively minor, a very different situation occurs when algae are grown under N deficiency. Under these conditions, their capacity for N assimilation increases dramatically. The effect of N resupply on carbon metabolism in N-limited cells is greatly increased and in many cases leads to a suppression in photosynthetic carbon fixation (see Turpin et al., 1988). Photosynthetic carbon fixation is controlled by many factors, including the quantity and activation of Rubisco and the availability of its substrates: CO_2, O_2, and ribulose bisphosphate (RuBP) (Jones, 1973; Walker, 1976; Collatz, 1977; Collatz, et al., 1979; Perchorowicz et al., 1981; von Caemmerer et al., 1984; Sharkey, 1985; Salvucci et al., 1987). At light intensities saturating to carbon fixation, ammonium or nitrate addition to N-limited S. minutum results in a major suppression in photosynthetic carbon fixation, which coincides with a decrease in RuBP (Elrifi and Turpin, 1986). The concentration of RuBP remains low until the added N is assimilated, at which time both RuBP and carbon fixation increase. Measurement of the RuBP binding site density with [^{14}C]-carboxyarabinitol bisphosphate has shown that during N assimilation RuBP decreases below this value indicating that the decrease in photosynthetic carbon fixation is due, at least in part, to the limitation of Rubisco by RuBP (Elrifi et al., 1988).

In contrast to these results, ammonium assimilation by N-limited Chlorella pyrenoidosa did not affect carbon fixation at saturating light levels. Although there is a decrease in the cellular RuBP concentration, it remains above the binding site density of Rubisco. Therefore, light-saturated photosynthetic carbon fixation remains RuBP saturated during ammonium assimilation by N-limited Chlorella. This implies that the rate of RuBP regeneration by Chlorella is high enough to ensure a saturating pool of RuBP during ammonium assimilation. If the rate of RuBP regeneration is experimentally lowered by decreasing the light intensity, the addition of ammonium results in a large suppression in photosynthetic carbon fixation. As with S. minutum the suppression of carbon fixation is accompanied by a decrease in the RuBP concentration below the RuBP binding site density of Rubisco. The mechanism by which N assimilation causes a decrease in RuBP regeneration is uncertain. However, measurements of Calvin cycle metabolite changes immediately following NH_4^+ addition to N-limited S. minutum shows a rapid increase in both triose phosphate and FBP (Smith et al., 1989). This suggests that the regulatory step(s) is downstream of FBPase and upstream of phosphoribulokinase.

These N-induced changes in Calvin cycle activity which occur during N assimilation by N-limited algae should not be viewed in isolation from the rest of cell metabolism. During this transient suppression of Calvin cycle activity the cellular demand for carbon skeletons in amino acid synthesis increases dramatically. This requires an increase in carbon export from the chloroplast to provide substrate for respiratory production of carbon skeletons used in the synthesis of amino acids.

Interactions between N assimilation and respiration

The supply of carbon skeletons for amino acid biosynthesis

In C3 plants the largest quantity of nitrogen assimilated results from the NH_4^+ released during photorespiration. As has been previously discussed

(Chapter 18), the assimilation of photorespiratory NH_4^+ occurs via the photorespiratory nitrogen cycle. Although the rates of N cycling through this pathway are extremely high there is no net production of amino nitrogen. However, during primary N assimilation inorganic nitrogen obtained from the environment is assimilated resulting in the net production of amino acids. The net synthesis of amino acids requires carbon skeletons in the form of keto acids.

Figures 28.1 and 28.4 illustrate the role of glycolysis and the TCA cycle in providing α-ketoglutarate for net glutamate synthesis. The synthesis of the other amino acids also require the provision of carbon skeletons, most of which are also intermediates in respiratory pathways (Chapter 27). The implication is that the assimilation of new nitrogen must be met by an increase in the supply of keto acids which requires an increase in carbon flow through respiratory pathways. A simple demonstration of this requirement is the stimulation of TCA cycle CO_2 efflux in both the dark and during photosynthesis by the supply of inorganic nitrogen (Fig. 28.5).

The utilization of TCA cycle intermediates in amino acid synthesis requires replenishment of these intermediates. This is because an organic acid removed from the cycle is unavailable for the regeneration of OAA. As OAA is a substrate of citrate synthase, the enzyme responsible for incorporation of acetyl-CoA into the TCA cycle, depletion of OAA would cause a slowdown in the TCA cycle. The only way in which TCA cycle intermediates may be consumed in biosynthetic reactions is if there is a continuous replenishment of intermediates (Chapter 8). The most common reaction serving this anaplerotic function is the

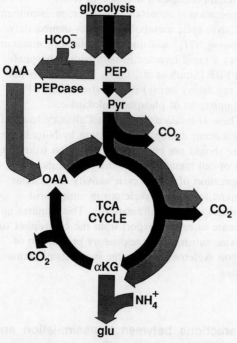

Fig. 28.4 A diagrammatic representation of the stoichiometry required to maintain both catabolic TCA cycle activity (black arrows) or anabolic TCA cycle activity (light arrows) for the net synthesis of glutamate. The widths of the arrows are proportional to the rates of carbon flow. This figure illustrates the requirement for anaplerotic reactions (e.g. PEPcase) to replenish OAA during biosynthetic consumption of TCA cycle intermediates.

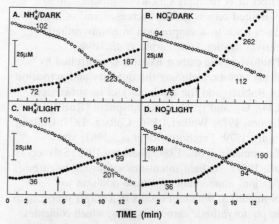

Fig. 28.5 The interactions between respiration and nitrogen assimilation in both the light and dark in the N-limited green alga *Selenastrum minutum*. Respiratory O_2 consumption (○) and CO_2 release (●) were measured in quadruple isotope experiments with mass-spectrometry which enables the determination of respiratory gas exchange during photosynthesis. The rates reported are in units of μmol (O_2 or CO_2) $mg^{-1}Chl\ h^{-1}$. (A) The effects of the onset of ammonium assimilation on CO_2 release (●) and O_2 consumption (○) in the dark; (B) the effects of NO_3^- in the dark; (C) the effects of NH_4^+ during photosynthesis; (D) the effects of NO_3^- during photosynthesis. Arrows indicate the addition of nitrogen.

carboxylation of PEP resulting in the production of OAA by the enzyme PEPcase. This enzyme is located in the cytosol and uses PEP produced in glycolysis. The OAA produced is rapidly reduced to malate via cytosolic malate dehydrogenase (MDH) and imported into the mitochondrion where it enters the TCA cycle. The significance of this process has been dramatically demonstrated in the N-limited green alga *S. minutum*. NH_4^+ addition to these algae causes a 20–40 fold increase in the rate of PEP carboxylation in both the light and dark in support of α-KG export for amino acid synthesis (Elrifi and Turpin, 1986; Guy *et al.*, 1989). Other reactions serving this anaplerotic function are dealt with in Chapter 8. Figure 28.4 illustrates the simplified stoichiometry of carbon flow which must occur during simultaneous operation of the TCA cycle for both catabolic and anabolic (biosynthetic) processes, assuming that glutamate is the only amino acid being produced. On balance the removal of one molecule of α-ketoglutarate for use in net glutamate synthesis would require the production of one OAA via PEP carboxylation and the entry of one additional molecule of acetyl-CoA. Hence, net glutamate synthesis requires two additional molecules of PEP to enter the TCA cycle; one via PEPcase or another anaplerotic route, the other via pyruvate kinase and the pyruvate dehydrogenase complex (Fig. 28.4).

Physiological analysis has shown that both PEPcase and PK are key regulatory enzymes governing the increase in anaplerotic carbon flow to the TCA cycle during N assimilation (Smith *et al.*, 1989). In *S. minutum* the initiation of N assimilation causes a large increase in the Pyr/PEP and malate/PEP ratios. If such changes occurred alone they would serve to decrease the rate of carbon flow through these enzymes. Given that carbon flow actually increases during N assimilation, these enzymes must be activated. Examination of the kinetic properties of cytosolic PK from the green alga *S. minutum* show it is stimulated by DHAP and inhibited by glutamate and P_i (Lin *et al.*, 1989; see Chapter 6; Fig. 28.4). Initiation of N assimilation in this organism causes a transient decrease in glutamate due to the action of GS as well as an increase in DHAP (Smith *et al.*, 1989). An increase in the rate of DHAP

export from the chloroplast would also serve to decrease cytosolic P_i due to operation of the phosphate translocater. As a result N assimilation would be expected to yield a rapid activation of cytosolic PK. The regulatory mechanisms involved in the control of the anaplerotic functioning of PEPcase have yet to be investigated in any detail.

The source of respiratory substrate for these anaplerotic reactions depends in part on whether or not N assimilation is occurring in photosynthesizing or non-photosynthesizing tissues. In the absence of photosynthesis the immediate source of carbon substrate is either starch or sucrose. Stimulation of N assimilation in the dark has been shown to stimulate the mobilization of these carbon sources (Vanlerberghe and Turpin, unpublished). In the light under N-sufficient conditions, recent photosynthate is most likely the primary source of carbon for steady-state amino acid synthesis. In some cases, however, high rates of N assimilation during photosynthesis may actually result in net starch breakdown during photosynthesis (Smith *et al.*, 1989). Our understanding of the regulatory mechanisms controlling carbon partitioning in response to N assimilation is incomplete, but many of the mechanisms already encountered in controlling photosynthate partitioning between starch and sucrose and the flow of carbon through glycolysis and the TCA cycle must also play a major role in assimilate partitioning during N assimilation (see Chapters 5, 6, 8 and 21).

Oxidation of TCA cycle reductant during N assimilation

Maintenance of enhanced TCA cycle carbon flow during N assimilation requires continued oxidation of NADH and $FADH_2$. In the case of NH_4^+ assimilation this apparently occurs via the mitochondrial electron transport chain resulting in enhanced rates of O_2 consumption. During NH_4^+ assimilation by *S. minutum* such a situation is apparent in both the light and dark (Fig. 28.5) (Weger *et al.*, 1988; Weger and Turpin, 1989). Studies with respiratory inhibitors suggest that most of the mitochondrial electron transport chain activity is a result of the cytochrome pathway and

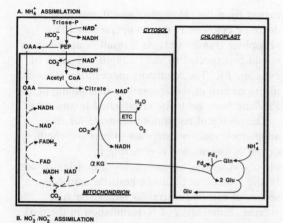

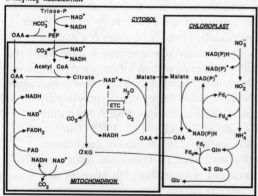

Fig. 28.6 Proposed pathways of TCA cycle electron and carbon flow during transient NH_4^+ assimilation (A) and transient NO_3^-/NO_2^- assimilation (B) in both the light and the dark by the N-limited green alga *Selenastrum minutum*. Cytosolic triose phosphate would be provided by the chloroplast, from the Calvin cycle and from starch breakdown in the light. The actual mechanism of reductant shuttling during inorganic N assimilation is unknown, and has been represented in (B) as malate/oxaloacetate exchange across the mitochondrial and chloroplast envelopes. The light reactions of photosynthesis would also provide photogenerated reductant (NADPH and Fd_r), and therefore contribute to NO_3^-/NO_2^- reduction in the light. This is not illustrated. Higher plant nitrate reductase is located in the cytosol but some recent work suggests that in green algae it may be plastidic. This figure represents the latter possibility for the sake of diagrammatic simplicity. Abbreviations: KG, α-ketoglutarate; OAA, oxaloacetate; Fd_r, reduced ferredoxin; Fd_o, oxidized ferredoxin (adapted from Weger and Turpin, 1989).

is presumably coupled to oxidative phosphorylation. The situation during NO_3^- assimilation is quite different. Although NO_3^- assimilation increases TCA cycle carbon flow, in many cases to a greater extent than NH_4^+, there is a much smaller increase in mitochondrial O_2 consumption both in the light and the dark (Fig. 28.5; Weger and Turpin, 1989). Apparently the TCA cycle is capable of supplying both carbon skeletons for amino acid synthesis and reducing power for NO_3^- and NO_2^- reduction not only in the dark but during photosynthesis (Weger and Turpin, 1989). In essence these oxidized N sources may serve as the oxidizing agent for TCA cycle-generated reductant. This contrast between NO_3^- and NH_4^+ assimilation is illustrated diagramatically in Fig. 28.6. The observation that reducing power may be transported from the mitochondrion to the chloroplast for use in NO_2^- reduction during photosynthesis serves to further highlight the potential complexity of the interactions between photosynthesis, respiration and nitrogen assimilation (Weger and Turpin, 1989).

Summary

The processes of photosynthesis, respiration and nitrogen assimilation interact in profound and dramatic ways. To view these processes as isolated pathways is to ignore the integrated metabolism upon which natural selection has acted. Clearly the time is right for the launching of major research initiatives designed to unlock the regulatory mechanisms which serve to integrate and control photosynthesis, respiration and N assimilation.

References

Azcón-Bieto, J. and Osmond, C. B. (1983). Relationship between photosynthesis and respiration: the effect of carbohydrate status on the rate of CO_2 production by respiration in darkened and illuminated wheat leaves. *Plant Physiol.* **71**, 574–81.

Azcón-Bieto, J., Lambers, H. and Day, D. A. (1983).

Effect of photosynthesis and carbohydrate status on respiratory rates and the involvement of the alternative pathway in leaf respiration. *Plant Physiol.* **72**, 598–603.

Badger, M. R. (1985). Photosynthetic oxygen exchange. *Ann. Rev. Plant Physiol.* **36**, 27–53.

Bate, G. C., Sültemeyer, D. F. and Fock H. P. (1988). $^{16}O_2/^{18}O_2$ analysis of oxygen exchange in *Dunaliella tertiolecta*. Evidence for the inhibition of mitochondrial respiration in the light. *Photosyn. Res.* **16**, 219–31.

Bennoun, P. (1982). Evidence for a respiratory chain in the chloroplast. *Proc. Natl Acad. Sci. USA* **79**, 4352–6.

Chapman, E. A. and Graham, D. (1974). The effects of light on the tricarboxylic acid cycle in green leaves. I. Relative rates of the cycle in the dark and the light. *Plant Physiol.* **53**, 879–85.

Collatz, G. J. (1977). The interaction between photosynthesis and ribulose-P_2 concentration – effects of light, CO_2 and O_2. *Carnegie Inst. Washington Year Book, 1977*, 248–51.

Collatz, G. J., Badger, M., Smith, C. and Berry, J. A. (1979). A radioimmune assay for RuP_2 carboxylase protein. *Carnegie Inst. Washington Year Book, 1979*, 171–5.

Dry, I. B. and Wiskich, J. T. (1982). Role of external adenosine triphosphate/adenosine diphosphate ratio in the control of plant mitochondrial respiration. *Arch. Biochem. Biophys.* **217**, 72–9.

Elrifi, I. R. and Turpin, D. H. (1986). Nitrate and ammonium induced photosynthetic suppression in N-limited *Selenastrum minutum*. *Plant Physiol.* **81**, 273–9.

Elrifi, I. R. and Turpin, D. H. (1987). The path of carbon flow during NO_3^- induced photosynthetic suppression in N-limited *Selenastrum minutum*. *Plant Physiol.* **83**, 97–104.

Elrifi, I. R., Holmes, J. J., Weger, H. P. Mayo, W. P. and Turpin, D. H. (1988). RuBP limitation of photosynthetic carbon fixation during NH_3 assimilation: Interaction between photosynthesis, respiration and ammonium assimilation in N-limited green algae. *Plant Physiol.* **87**, 395–401.

Furbank, R. T., Badger, M. R. and Osmond, C. B. (1982). Photosynthetic oxygen exchange in isolated cells and chloroplasts of C_3 plants. *Plant Physiol.* **70**, 927–31.

Gans, P. and Rebeille, F. (1988). Light inhibition of mitochondrial respiration in a mutant of *Chlamydomonas reinhardtii* devoid of ribulose-1,5 bisphosphate carboxylase/oxygenase activity. *Arch. Biochem. Biophys.* **260**, 109–17.

Gerbaud, A. and Andre, M. (1980). Effect of CO_2, O_2 and light on photosynthesis and photorespiration in wheat. *Plant Physiol.* **66**, 1032–6.

Goller, M., Hampp, R. and Ziegler, H. (1982). Regulation of the cytosolic adenylate ratio as determined by rapid fractionation of mesophyll photoplasts of oat. Effect of electron transport inhibitors and uncouplers. *Planta* **156**, 255–63.

Graham, D. (1980). Effects of light on 'dark' respiration. In *The Biochemistry of Plants*, Vol. 2, ed. D. D. Davies, Academic Press, New York, pp. 525–79.

Grant, B. R. and Turner, I. M. (1969). Light-stimulated nitrate and nitrite assimilation in several species of algae. *Comp. Biochem. Physiol. A* **29**, 995–1004.

Guy, R. D., Vanlerberghe, G. C. and Turpin, D. H. (1989). Significance of phosphoenolypyruvate carboxylase during ammonium assimilation: Carbon isotope discrimination in photosynthesis and respiration by the N-limited green alga *Selenastrum minutum*. *Plant Physiol.* **89**, 1150–7.

Hampp, R., Goller, M. and Ziegler, H. (1982). Adenylate levels, energy charge, and phosphorylation potential during dark/light and light/dark transition in chloroplasts, mitochondria, and cytosol of mesophyll protoplasts from *Avena sativa* L. *Plant Physiol.* **69**, 448–55.

Hampp, R., Goller, M., Fullgraf, H. and Eberle, I. (1985). Pyridine and adenine nucleotide status, and pool sizes for a range of metabolites in chloroplasts, mitochondria, and the cytosol/vacuole of *Avena* mesophyll protoplasts during dark/light transition: Effect of pyridoxal phosphate. *Plant Cell Physiol.* **26**, 99–108.

Hoch, G. and Owens, O. H. (1963). Photosynthesis and respiration. *Arch. Biochem. Biophys.* **101**, 171–80.

Ishii, R. and Schmid, G. H (1982). Studies on $^{18}O_2$ uptake in the light by entire plants of different tobacco mutants. *Z. Naturforsch.* **37c**, 93–101.

Jones, H. G. (1973). Limiting factors in photosynthesis. *New Phytol.* **62**, 1089–94.

Klock, G., Sültemeyer, D. F., Fock, H. P. and Kreuzberg, K. (1989). Gas exchange in intact isolated chloroplasts from *Chlamydomonas reinhardtii* during starch degradation in the dark. *Physiol. Plant.* **75**, 109–13.

Kramer, E., Tischner, R. and Schmidt, A. (1988). Regulation of assimilatory nitrate reduction at the level of nitrite in *Chlorella fusca*. *Planta* **176**, 28–35.

Kromer, S., Stitt, M. and Heldt, H. W. (1988). Mitochondrial oxidative phosphorylationn participating in photosynthetic metabolism of a leaf cell. *FEBS Lett* **226**, 352–6.

Lara, C., Romero, J. M., Coronil, T. and Guerrero, M. G. (1987). Interactions between photosynthetic nitrate assimilation and CO_2 fixation in cyanobacteria. In *Inorganic Nitrogen Metabolism*, eds W. R. Ullrich, P. J. Aparicio, P. J. Syrett and F. Castillo, Springer-Verlag, New York, pp. 45–52.

Larsson, M., Ingemarsson, B. and Larsson, C-M. (1982). Photosynthetic energy supply for NO_3^- assimilation in *Scenedesmus*. *Physiol. Plant.* **55**, 301–8.

Larsson, M., Olsson, T. and Larsson, C-M. (1985). Distribution of reducing power between photosynthetic carbon and nitrogen assimilation in *Scenedesmus*. *Planta* **164**, 246–53.

Lemaire, C., Wollman, F-A. and Bennoun, P. (1988). Restoration of phototrophic growth in a mutant of *Chlamydomonas reinhardtii* in which the chloroplast at *pB* gene of ATP synthase has a deletion: An example of mitochondria-dependent photosynthesis. *Proc. Natl Acad. Sci. USA* **85**, 1344–8.

Lin, M., Turpin, D. H. and Plaxton, W. (1989). Pyruvate kinase isozymes from the green alga *Selenastrum minutum* II. Kinetic and regulatory properties. *Arch. Biochem. Biophys.* **269**, 228–38.

Marsh, H. V. Jr., Galmiche, J. M. and Gibbs, M. (1965). Effect of light on tricarboxylic acid cycle in *Scenedesmus*. *Plant Physiol.* **40**, 1913–22.

Marsho, T. V., Behrens, P. W. and Radmer, R. J. (1979). Photosythetic oxygen reduction in isolated intact chloroplasts and cells from spinach. *Plant Physiol.* **64**, 656–9.

McCashin, B. G., Cossin, E. A. and Canvin, D. T. (1988). Dark respiration during photosynthesis in wheat leaf slices. *Plant Physiol.* **87**, 155–61.

Peltier, G. and Thibault, P. (1985). O_2 uptake in the light in *Chlamydomonas*: evidence for persistent mitochondrial respiration. *Plant Physiol.* **79**, 225–30.

Perchorowicz, J. T., Raynes D. A. and Jensen, R. G. (1981). Light limitation of photosynthesis and activation of ribulose bisphosphate carboxylase in wheat seedlings. *Proc. Natl Acad. Sci. USA* **78**, 2895–989

Radmer, R. J. and Kok, B. (1976). Photoreduction of O_2 primes and replaces CO_2 assimilation. *Plant Physiol.* **58**, 336–40.

Raven, J. A. (1972). Endogenous inorganic carbon sources in plant photosynthesis II. Comparison of total CO_2 production in the light with measured CO_2 evolution in the light. *New Phytol.* **71**, 995–1014.

Raven, J. A. (1984). *Energetics and Transport in Aquatic Plants*. Alan R. Liss, Inc., New York, 597 pp.

Rebeille, F. and Gans, P. (1988). Interaction between chloroplasts and mitochondria in microalgae: role of glycolysis. *Plant Physiol.* **88**, 973–5.

Robinson, J. M. (1988). Spinach leaf chloroplast CO_2 and NO_2^- photoassimilations do not compete for photogenerated reductant: Manipulation of reductant levels by quantum flux density titrations. *Plant Physiol.* **88**, 1373–80.

Salvucci, M. E., Werneke, J. M., Ogren, W. L. and Portis, A. R. (1987). Purification and species distribution of Rubisco activase. *Plant Physiol.* **84**, 930–6.

Sawhney, S. K., Naik, M. S. and Nicholas, D. J. D. (1978). Regulation of NADH supply for nitrate reduction in green plants via photosynthesis and mitochondrial respiration. *Biochem. Biophys. Res. Comm.* **81**, 1209–16.

Scherer, S., Sturzel, E. and Boger, P. (1984). Photoinhibition of respiratory CO_2 release in the green alga *Scenedesmu. Physiol. Plant.* **60**, 557–60.

Schrader, L. E. and Thomas, R. J. (1981). Nitrate uptake, reduction and transport in the whole plant. In *Nitrogen and Carbon Metabolism*, ed. J. D. Bewley, Junk, The Hague, pp.

Sharkey, T. D. (1985). Photosynthesis in intact leaves of C_3 plants: physics, physiology and rate limitations. *Bot. Rev.* **51**, 53–105.

Shiraiwa, Y., Bader, K. P. and Schmid, G. H. (1988). Mass spectrometric analysis of oxygen exchange in high and low-CO_2 cells of *Chlorella vulgaris*. *Z. Naturforsch.* **43C**, 709–16.

Smith, R. G., Vanlerberghe, G. C., Stitt, M. and Turpin, D. H. (1989). Short-term metabolite changes during transient ammonium assimilation by the N-limited green alga *Selenastrum minutum*. *Plant Physiol.*, **91**, 749–55.

Stitt, M. (1986). Limitation of photosynthesis by carbon metabolism I. Evidence for excess electron transport capacity in leaves carrying out photosynthesis in saturating light and CO_2. *Plant Physiol.* **81**, 1115–22.

Stitt, M., Lilley, R. McC. and Heldt, H. W. (1982). Adenine nucleotide levels in the cytosol, chloroplasts and mitochondria of wheat leaf protoplast. *Plant Physiol.* **70**, 971–7.

Sültemeyer, D. F., and Fock, H. P. (1986). Effect of photon fluence rate on oxygen evolution and uptake by *Chlamydomonas reinhardtii* suspensions grown in ambient and CO_2 enriched air. *Plant Physiol.* **81**, 372–5.

Sültemeyer, D. F., and Fock, H. P. (1987). Effect of dissolved inorganic carbon on oxygen evolution and uptake by *Chlamydomonas reinhardtii* suspensions adapted to ambient and CO_2 enriched air. *Photosyn. Res.* **12**, 25–33.

Syrett, P. J. (1981). Nitrogen metabolism of microalgae. *Can. Bull. Fish. Aquat. Sci.* **219**, 182–210.

Turpin, D. H., Elrifi, I. R., Birch, D. G., Weger, H. G. and Holmes, J. J. (1988). Interactions between photosynthesis, respiration, and nitrogen assimilation in microalgae. *Can. J. Bot.* **66**, 2083–97.

Ullrich, W. R. and Eisele, R. (1977). Relations between nitrate uptake and nitrate reduction in *Ankistrodesmus braunii*. In *Transmembrane Ionic Exchange in Plants*, eds. M. Thellier, A. Monnier, M. Demarty and J. Dainty, *Colloquedu CNRS, Rouen*, pp 307–13.

von Caemmerer, S., Coleman, J. R. and Berry, J. A. (1984). Control of photosynthesis by RuBP concentration: studies with high- and low-CO_2 adapted cells of *Chlamydomonas reinhardtii*. *Carnegie Inst. Washington Year Book, 1982*, 91–5.

Walker, D. A. (1976). Regulatory mechanisms in photosynthetic carbon metabolism. *Curr. Top. Cell. Regul.* **11**, 203–41.

Weger, H. G. and Turpin, D. H. (1989). Mitochondrial respiration can support NO_3^- and NO_2^- reduction during photosynthesis: Interactions between photosynthesis, respiration, and N assimilation in the N-limited green alga *Selenastrum minutum*. *Plant Physiol.* **89**, 409–15.

Weger, H. G., Birch, D. G., Elrifi, I. R. and Turpin, D. H. (1988). Ammonium assimilation requires mitochondrial respiration in the light: A study with the green alga *Selenastrum minutum*. *Plant Physiol.* **86**, 688–92.

Weger, H. G., Herzig, R., Falkowski, P. G. and Turpin, D. H. (1989). Respiratory losses in the light in a marine diatom: Measurements by short-term mass-spectrometry. *Limnol. Oceanogr.*, in press.

Woo, K. C., Jokinen, M. and Canvin, D. T. (1980). Reduction of nitrate via a dicarboxylate shuttle in a reconstituted system of supernatant and mitochondria from spinach leaves. *Plant Physiol.* **65**, 433–6.

Zubkova, E. K., Filippova, I. S., Manushina, N. S. and Chupakhina, G. N. (1988). Effects of light on dark respiration of albino and green regions of the barley leaf. *Fiziologiya Rastenii.* **35**, 254–9.

29 Long-distance transport of nitrogen and carbon from sources to sinks in higher plants

Mark B. Peoples and Roger M. Gifford

Introduction

A poorly understood aspect of plant function is the regulation of the interactions between sources (organs which supply assimilates) and sinks (sites of storage or growth) for nutrients, and the role that such interactions play in determining assimilate partitioning between plant parts. Nonetheless, overall partitioning of nitrogen (N) and carbon (C) assimilates over a growing season is a major determinant of crop yield and product quality (Gifford and Evans, 1981; Gifford, 1986). Thus, it is important to consider the mechanisms governing assimilate export, translocation, import and utilization, and factors which may influence N and C allocation and distribution within the plant, on both a whole-season basis and in the short term. The broad overview presented here draws largely on reviews devoted to aspects of N (Raven and Smith 1976; Pate, 1980; Hocking et al., 1984; Simpson, 1986) and C assimilation and transport (Giaquinta, 1983; Delrot and Bonnemain, 1985) and partitioning (Gifford and Evans, 1981; Pate et al., 1988), or assimilate unloading in sink organs (Thorne, 1985; Wolswinkel, 1985; Murray, 1987).

Translocation of N and C

Requirements, characteristics and constraints

The specific requirements of each plant part for C and N compounds change throughout development. With fixed C originating in leaves and N compounds (and minerals) coming from roots, higher plants have faced a challenging task to move these substrates simultaneously in opposite directions over long distances to meet these demands.

To achieve upwards transport of nutrients from the roots, plants capitalize on the large flow of water from root to shoot which is an inevitable consequence of the open stomata required to ventilate the leaves with CO_2 for photosynthesis. The driving force for upward movement of solutes in the xylem is therefore principally 'suction' generated by transpiration. It is a wick effect, relying on capillary continuity in the xylem vessel network, a 'super-apoplast of dead cells within the living plant axis' (Raven, 1977). 'Root pressure', a positive hydrostatic pressure in the xylem (when transpiration rate is very low), arising from osmotic uptake of soil water following active accumulation of ions (or in special cases, organic solutes), plays only a minor part by day. It may, however, serve to re-establish capillary continuity at night after this has been lost in some xylem vessels by cavitation due to excessive tension during rapid daytime transpiration.

Export of C-compounds from the leaves to areas of growth and storage needs a different mechanism since the flow is against an overall water potential gradient throughout most of the plant. The C-transport system needs (1) to use minimal water, (2) to be compartmentalized away from the xylem flow, and to minimize uncontrolled leakage of C-compounds into the fast-flowing dilute super-apoplast, and (3) to have an active driving force in contrast to the 'free-rider' method for xylem solutes. The solution to this problem has been that the phloem's sieve tubes, forming a network of 'super-symplastic' tubes (Raven, 1977) throughout the plant, are (together with companion cells) powerful active scavengers of sugars from the neighbouring apoplast. The translocated sugars, usually sucrose, having favorable solubility concentration/viscosity relationships (Lang, 1978), can accumulate to concentrations as high as 500–1000 mM and still move by mass solution flow. In contrast, xylem sap concentration is only approximately 1–50 mM depending on transpiration rate.

The most favored hypothesis for the phloem transport mechanism – that of mass flow in the sieve tube lumen ('Münch' pressure flow) – requires a hydrostatic pressure gradient along the lumen of the sieve tube. This is achieved by active loading of carbohydrates at source, causing water uptake osmotically, and unloading of the solutes and water at the sink. Assuming open sieve plate pores, the required gradient in osmolarity to drive the observed flow is only a small part of the total osmolarity of sieve tube sap.

While the apoplastic xylem sap flows under tension, requiring vessel reinforcement by encircling lignin bands to prevent collapse (Raven, 1977), the symplastic sieve tube sap is driven under pressure. These two conduits are in close proximity in vascular strands throughout the plant. This may seem to provide a risk of inefficiency by 'short-circuiting'. In practice, it offers an opportunity for improved efficiency via regulated solute exchanges between xylem and phloem at strategic sites. The need for such regulated exchange arises in part because N in the xylem is predominantly transported directly to the sites of transpiration, not to the growing points. On the other hand, solutes and water transported to growing points in phloem sap may not all be utilized (or transpired) by the growing points. The unused molecules can then be carried away in the xylem back to the leaves. Whereas loading and unloading of carbohydrates induces the very driving force needed to move this material to where it is required from where is produced, N-compounds have to 'hitchhike' first to where the transpiration stream takes them, and then (after phloem loading in the leaves) to where the carbohydrate stream takes them. Transfer of N-compounds between xylem and phloem *en route* could potentially adjust the sieve tube sap composition to avoid delivery of unwanted compounds to final sinks.

There is a substantial C requirement in the roots for growth and respiration. It would therefore seem, at first sight, most efficient for N to move out of the roots in the form in which it is absorbed from the soil (as NH_4^+ and/or NO_3^-), not as organic N-compounds, and for phloem not to transport N-compounds into the root system. This rarely happens. Linked constraints of pH regulation, charge balance, avoidance of osmotic stress and energetics at cell, tissue and whole system levels influence the molecules transported and cause close coordination between xylem and phloem transport (Raven and Smith, 1976).

When NO_3^- is assimilated to the redox level of protein, excess OH^- is produced in the cytoplasm. When NH_4^+ is assimilated, excess H^+ is produced. When N_2 is converted to protein by nodulated legumes, a small H^+ surplus is produced. Movement of such excess protons and hydroxyls into and out of vacuoles and apoplast, in association with organic or inorganic anions and cations, would allow short-term buffering of cytoplasmic pH. However, this would involve substantial pH and osmotic shifts in the vacuole and cell walls. In the long term, overall plant pH stability can be achieved only by three methods: (1) uptake and assimilation of NH_4^+, NO_3^- (and N_2 in legumes) in such ratios that OH^- production everywhere balances H^+ production; (2) assimilation of NH_4^+ and NO_3^- solely in the roots so that H^+ and OH^- (respectively) can be excreted into the soil water; or (3) if N assimilation occurs in the shoot, utilization of a biochemical 'pH-stat' and phloem transport to shift H^+ or OH^- equivalents to the roots for excretion. Method 1 (pH-balancing NH_4^+/NO_3^- uptake) would be excessively restrictive, since the proportions of these ions in the soil varies widely. In practice, free NH_4^+ is never found in more than trace amounts in the xylem. All the assimilation of both NO_3^- and NH_4^+ to organic N would have to occur entirely in the roots to maintain pH constant: this would require extra C import by the roots. If NH_4^+ and NO_3^- were assimilated solely in the roots then the availability of the soil water as a sink for excreted H^+ and/or OH^- would mean that the two N-forms did not have to be taken up in a pH-balancing ratio anyway (i.e. method 2 could be used). However, the import of carbohydrate into the roots would still be required to provide C-skeletons for organic-N export to the shoot. A further cost of NO_3^- reduction in the roots, instead of in the chloroplasts, is that it must be energized wholly by respiration of phloem-transported carbohydrate. The opportunity to utilize photophosphorylation energy surplus to the needs of CO_2 reduction in the chloroplast is then foregone.

To take advantage of photophosphorylation energy for NO_3^- assimilation in the shoot (involving method 3) requires the operation of a biochemical 'pH-stat' to neutralize the surplus OH^-. This is considered to involve the carboxylation of neutral photosynthetic precursors to organic acids like malic, succinic, or citric acids. Salts of these organic acids (e.g. K^+ malate) must then either be stored in shoot vacuoles or be phloem-translocated to the root where they can be decarboxylated, releasing CO_2 and OH^- to the soil in exchange for NO_3^- uptake. This suggested route has the disadvantage of transporting K^+ against its net flux out of the root system. But the products of decarboxylation, such as pyruvate, would be readily available for root respiration to energize further NO_3^- uptake, and to provide C-skeletons to carry N assimilated from NH_4^+ up the xylem. Thus, where NO_3^- is available, its uptake and assimilation in the shoot appears to confer overall advantages for plant function and efficiency.

In summary, the constraints of pH, charge balance, and osmotic regulation, together with the 'hitchhiker' mechanism of main channel N-transport, result in the apparent inefficiencies of transporting C out of the roots and N and minerals into the roots. However, the circulation of materials around the plant so generated provides flexibility for growing points to extract what solutes they need when they need them from the circulating mixed composition saps. Different species have adopted different ways of coping with the various conflicting requirements as we next document. A common feature, however, is that N-compounds transported are often those with relatively low C:N ratio and large number of N atoms per molecule.

Identity of transported solutes

It is relatively easy to examine the solutes of xylem fluids by vacuum extracting 'tracheal' sap from freshly harvested shoot segments, by collecting sap exuding from the cut xylem of roots or stems passively under root pressure, or by applying pneumatic pressure to the root system. However, the delicacy of phloem tissue and the ease with which sieve tubes become blocked with slime and callose create difficulties in collecting phloem contents. Phloem sap exudes spontaneously from shallow incisions for only relatively few woody and non-woody species. Techniques involving severing the stylets of phloem-feeding insects (Fisher and Frame, 1984), the use of aqueous solutions of chelating agents to maintain incision exudation (Simpson and Dalling, 1981), or 'cryopuncturing' vascular strands of legume pods (Peoples et al., 1985b) have been employed to sample phloem translocates.

Analyses indicate that xylem saps are mildly acidic (e.g. pH 6–6.5), while phloem saps are alkaline (pH 7–8.5). Nitrogenous compounds frequently comprise the main solute component of xylem, while carbohydrates are the major constituent of phloem sap (although high levels of N-solutes are also found). K^+ is the main inorganic ion in phloem. The relative importance of N and C in the two transport streams is reflected in their C:N weight ratios which can range between 1.5 to 6 in xylem exudate, and 10 to 200 in phloem sap. For comparison, a typical range in C:N ratio of whole plants is 15 to 45.

In plants such as *Xanthium*, where virtually no nitrate is reduced in the root, over 95% of xylem soluble-N can consist of free nitrate. The roots of most other plants, however, exhibit some capacity to reduce nitrate, and a proportion of xylem N is transported in an organic form (Fig. 29.1). Nitrate is essentially absent from phloem saps (Fig. 29.1). Generally one N-rich molecule, characteristic of the species, dominates the spectrum of organic N compounds present in xylem and phloem. While the advantage of N being transported in N-rich

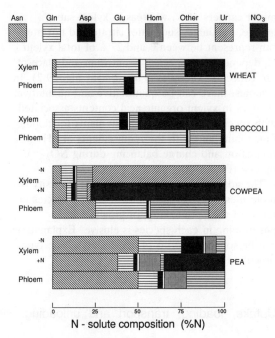

Fig. 29.1 Composition of nitrogenous fractions of xylem exudates and phloem saps collected from wheat (*Triticum aestivum* L.), broccoli (*Brassica oleracea* var. *italica*), cowpea (*Vigna unguiculata* (L.) Walp.), and pea (*Pisum sativum* L.). The major N-solutes in xylem exudates from the legumes cowpea and pea are depicted for plants grown under two different N-regimes. −N: plants totally dependent upon an effective symbiosis, or +N: plants fed with culture solution containing nitrate. Phloem sap constituents were similar for each legume regardless of whether grown under −N or +N regimes. Data for wheat were derived from Simpson and Dalling (1981) and Simpson *et al.* (1982); broccoli from Shelp (1987); cowpea from Peoples *et al.* (1985a); and pea from Peoples *et al.* (1987) and unpublished data. Asn: asparagine; Gln: glutamine; Asp: aspartate; Glu: glutamate; Hom: homoserine; Other: all other amino compounds detailed in above references; Ur: ureides (allantoin plus allantoic acid); NO$_3$: nitrate.

molecules in the xylem is evident, it is not so clear why this should be so in phloem. The amide glutamine for instance, is a major N-compound commonly detected in both xylem and phloem saps collected from non-leguminous crops (Hocking *et al.*, 1984; Fig. 29.1). In legumes, glutamine is only rarely the main N-solute transported from the nodule in xylem (Peoples *et al.*, 1987), despite its synthesis as the initial product of ammonia assimilation during N$_2$ fixation. Most nodulated legumes tend to export either the ureides, allantoin and allantoic acid (tropical species, e.g. cowpea, soybean, mungbean), or the amide asparagine (mainly temperate species, e.g. lupin, pea, clovers) as the principal forms of symbiotically fixed N in xylem exudate (Fig. 29.1). The ureides, having four N atoms per molecule (C:N = 1:1), have an advantage over asparagine with only two N atoms per molecule (C:N = 2:1) in terms of N export and efficient C use. On the other hand, the low solubility of ureides may limit the amount of N that can be transported in this form in temperate conditions (Sprent, 1980). Both tropical and temperate legumes appear to translocate a mixture of asparagine and glutamine in phloem (Fig. 29.1).

In ureide-producing species, the ureides dominate xylem composition only so long as the legume derives a high proportion of its N requirements for growth from atmospheric N$_2$. With amide-producing legumes, asparagine is the principal exported organic product of both root nitrate reduction and N$_2$ fixation, and continues to be the major form of reduced-N despite changes in N nutrition (Fig. 29.1). The principal N-solutes of broccoli (a non-legume), cowpea and pea phloem exudates, depicted in Fig. 29.1, have also been shown to remain relatively unchanged regardless of whether the roots are assimilating soil NO$_3^-$ or NH$_4^+$ or fixing atmospheric N$_2$. In view of the different constraints relating to the regulation of pH imposed by the assimilation of varying forms of N, this invariance is perplexing. Other compounds such as arginine can be important xylem components in some deciduous perennial trees, and high levels of citrulline have been found in *Alnus* xylem fluid. Significant amounts of N in certain species are carried as alkaloids, while the non-protein amino acids homoserine and δ-methylene glutamine carry N in other species (Pate, 1980; Fig. 29.1).

The C-skeletons of the main organic N constituents often form the bulk of xylem-C (Fig. 29.2). Sugars and sugar alcohols are generally not the major C component in xylem of herbaceous plants (e.g. Fig. 29.2), but they may occur in

438 Nitrogen Metabolism

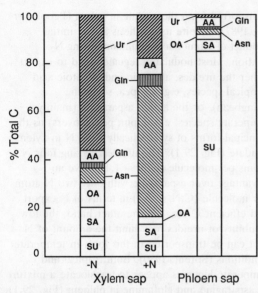

Fig. 29.2 The relative contribution of sugars (SU), sugar alcohols (SA), organic acids (OA), the amides asparagine (Asn) and glutamine (Gln), amino acids (AA), and the ureides allantoin and allantoic acid (Ur) to the carbon composition of xylem and phloem exudates collected from effectively nodulated (−N) or nitrate fed (+N) soybean (*Glycine max* [L.] Merrill.). Data derived from Layzell and LaRue (1982). The major sugars were fructose, glucose and sucrose in xylem saps, but over 90% of the sugar was as sucrose in phloem. Pinitol was the principal sugar alcohol detected. Malonate and malate were the main organic acids in −N xylem and phloem exudates, while malonate and citrate dominated +N xylem sap.

quantity in deciduous trees such as maple in early spring or apple during winter dormancy. These are times of year when the lack of transpiring leaves means that the mechanism of xylem transport must be analogous to that of phloem – osmotically driven pressure flow rather than suction flow, i.e. 'root pressure'. In the phloem, on the other hand, soluble carbohydrates commonly account for over 80% of total C (Fig. 29.2). Sucrose is most commonly the predominant phloem component, especially in herbaceous species. The oligosaccharides raffinose, stachyose and verbascose are present in phloem of some species in a few genera and families, while sugar alcohols like sorbitol and mannitol occur in others. Various organic acids detected in phloem translocates are minor carriers of C, whereas in xylem saps, malate, malonate, succinate, citrate and tartrate can represent between 3 and 27% of total xylem C, depending on species, N-nutrition and place of sap collection (Peoples *et al.*, 1985b; Fig. 29.2). In legumes, there can be quantitative and qualitative changes in xylem organic acid contents associated with specific organic acid requirements for nodule functioning and with the requirements of pH regulation and charge balancing during NO_3^- uptake (Fig. 29.2).

The adaptive rationale for the wide diversity of molecules in xylem and phloem among species in terms of the optimization of growth and partitioning in each species is elusive. Explanations can be sought in terms of constraints of the type described above.

Uptake, loading, transport and unloading

Nitrogen

After exhaustion of the seed protein reserves, plants are solely dependent upon the roots for a continued N-input for growth and development. Crop plants may differ in their N source preferences but most grow faster on NO_3^-. This is believed to be the commonest form of N taken up from agricultural soils, although this may not be true for species adapted to acid soils or anaerobic conditions (e.g. paddy rice). NO_3^- uptake by root cells is the net result of an active influx causing concentrations several hundred-fold greater in the cells than in the soil solution, and of a passive efflux (leakage) of nitrate back into the rooting medium. Nitrate influx is thought to include ATPase-facilitated transport across the plasmalemma and from cytoplasmic nitrate pools within the root to other pools (e.g. the vacuole). Such energy-driven transport creates a C demand and, indeed, roots have a high respiration rate on account of the energy needs for ion uptake. As discussed, owing to the requirements of pH regulation and maintenance of electroneutrality, the uptake of soil mineral N by roots also requires C import for the parallel transformation of organic acids (mostly malate) and the excretion of hydroxyl or bicarbonate ions.

The root cortex can have high NO_3^- reductase activity and concentration of enzymes involved in glutamine synthesis. The tissue appears to be the main zone for N assimilation in many plants, although further N metabolism can occur in the vascular bundle region of the stele. Transport of NO_3^- and newly-synthesized amino compounds across the root cortex to xylem is probably symplastic, but there is little information on the mechanism of release into the xylem. There is also little known about the route or mechanisms in legumes involved in transfer of N-solutes from their site of synthesis to the vascular elements which carry them out of the nodule to the host plant. However, specialized pericycle transfer cells, which could facilitate vein loading of a symplastic flux of N-solutes, have been found adjacent to the vascular bundles in nodules of amide-exporting legumes (Pate, 1980).

While movement of N-solutes in the xylem is by mass flow in the transpiration stream, the rate of movement may also be influenced to some extent by adsorption to the walls of the vessels and exchange with the surrounding apoplast, by xylem to phloem transfer *en route*, or by diffusion down a potential gradient within the flowing system. The importance of transpirational flow for export of N from roots is especially evident in legume nodules. Nitrogenous solutes tend to accumulate within nodules during the night, but in the early morning this pool is rapidly depleted as the night's backlog of fixation products is swept from the nodule with the onset of transpiration.

Since xylem flow is towards the points of evaporative water loss, most of the xylem-borne N enters the leaf apoplast close to those sites. Phloem transport is then required to redistribute N-compounds to weakly transpiring growth centers. Nevertheless, transfer cells in leaf vascular tissue may also assist in regulating the supply of N. Some N-solutes (e.g. glutamine, asparagine) may be transferred preferentially from the xylem directly into the phloem, whereas several amino acids (e.g. glutamate, aspartate, glycine, arginine), nitrate, and ureides require metabolic conversion to specific amino compounds in mesophyll tissue. These are then exported in the phloem. Loading of amino acids into the phloem may be accomplished by cotransport with protons in a manner akin to that suggested for sucrose transport in phloem (see below). Translocation speeds of amino compounds in phloem have been found to be similar to those of sugars, probably because they flow with the sugars.

Carbon

The regulation of C movement from sources to sinks involves different processes on various time-scales. Seasonally, phenological and developmental processes are of paramount importance to understanding the cumulative partitioning of C into the various plant parts. On a diurnal scale, the regulation of temporary storage in leaf and stem assumes significance, as does the daily photosynthetic integral, the daily sink growth and the feedback interaction between sources and sinks. On a minute or second time-scale, regulation can be considered in terms of enzymic and carrier-mediated transport processes, transmembrane movement, phloem loading in the leaves, and phloem unloading in the sinks. Most of the partitioning literature relates to this very short-term regulation (principally within the leaf) and is difficult or impossible to relate to long-term C partitioning between organs.

In the leaf, the complex fine regulation of partitioning of photosynthetically produced triose phosphate between chloroplast starch formation and sucrose synthesis in the cytosol or stored in the mesophyll vacuoles is described in Chapters 5 and 21. The mechanism involves the regulatory compound fructose 2, 6-bisphosphate which may also play a role in relating photosynthesis and sucrose synthesis to sink demand for assimilate. The net effect of this fine metabolic control is that over a diurnal cycle or longer, approximately all the photosynthetic assimilate produced by a fully expanded leaf is loaded into phloem and exported while high photosynthetic efficiency is sustained. The starch and vacuolar pools act as buffers smoothing out sucrose availability for loading during the light/dark cycle. Under some circumstances the level of sink demand can feed back through translocation rate, loading, and sucrose synthesis, hence modifying photosynthesis rate to some extent. Just how important such feedback is in field photosynthesis is still unclear, as are the details of its mechanism. It is often said that the amount of sucrose available for export is determined by partitioning between pools within the leaves. Since the sizes of these pools do not grow or shrink continually, it seems more likely that on longer time-scales the amount available is usually the amount of photosynthetic C fixed according to environmental conditions, but modified by sink demand feedback under extreme circumstances.

The mechanism of sucrose movement from mesophyll to sieve tube is not definitively established and is keenly debated. The favored hypothesis is that sucrose diffuses symplastically from cell to cell towards the phloem. Any leakage *en route* would presumably flow away from the vascular strand in the apoplastic transpiration drift between the mesophyll cells and be reabsorbed by the 'general [sucrose] accumulating system' (Maynard and Lucas, 1982) of those cells. Sucrose unloaded or leaked from mesophyll cells in close proximity to phloem – in a 'backwater' from the transpirational drift – would be retrieved by the powerful retrieval system of companion cells and/or sieve tubes. The loading from acidic apoplast to alkaline sieve tube is considered to be by a proton-cotransport mechanism driven by a plasmalemma-bound ATPase extruding protons into the apoplast (Giaquinta, 1983). No mechanism has been demonstrated whereby purely symplastic transport from mesophyll through plasmodesmata to sieve tubes could operate against the very high concentration gradient that exists between them, viz. 1–20 mol m^{-3} in mesophyll cytoplasm to 500–1000 mol m^{-3} in the sieve tubes of the fine leaf veins. The high osmotic concentrations in the minor veins of source leaves generates, by osmotic water uptake, the high turgor pressure necessary to drive mass flow towards sinks where sucrose is unloaded.

At the sink, solutes are unloaded from the sieve tubes. Pertinent reviews include those by Thorne (1985), Wolswinkel (1985) and Murray (1987). Transfer of sugars (and amino acids) from parent plant to developing embryo or endosperm certainly involves apoplastic transport, since embryo and endosperm do not have symplastic continuity with the parent plant. But, in roots and expanding leaves, phloem unloading can be wholly symplastic and directly into the growing cells. In no tissue (meristem, storage tissue or seed) is the mechanism of phloem unloading clearly understood, but most progress has been made with legume seeds because their embryos can be removed and replaced with solutions into which phloem contents continue to be released. These solutions can subsequently be collected and analyzed.

In developing seeds, transport from sieve tube–companion cell complex to the apoplastic interface with the filial sink is down a steep diffusion gradient through parenchymatous maternal cells of the seed coat (in legume seeds) or funicular-chalazal and nucellar tissue (in temperate cereals). There is bi-directional equilibration of sugars between symplast and apoplast of these transfer tissues (in soybean at least; Gifford, 1986), so it is not necessarily meaningful to ask where exactly translocated solutes leave the symplast and enter the apoplast – net transfer to apoplast could be dispersed along the parenchymatous path from sieve tube to sink. However, in fruits where water imported through the phloem is partly removed in the xylem (Peoples *et al.*, 1985b), there is presumably a zone around the xylem where efflux of solutes from the seed coat symplast is minimal, otherwise they would appear in the xylem export stream.

During the passage of translocated substances across the seed coat they can be transformed. For example, a range of amino acids are synthesized in seed coats, ureides are withdrawn, and sucrose may be partially hydrolyzed. The compounds secreted from the seed coat inner surface are therefore derived from, but not identical to, the material unloaded from the sieve tubes into the seed coat tissue. The mechanisms of control over sieve tube unloading and over secretion from the maternal seed coat or pericarp tissue seem different, having different responses to inhibitors and other exogenous agents, but details are unclear. Despite the steep solute gradients from sieve tubes to final sinks, the transfer processes appear to involve respiration and to be metabolically controlled, possibly involving phloem turgor relations.

By taking up the unloaded solutes, sink cells maintain the sucrose concentration gradient needed to sustain continued unloading by conversion to other soluble products (e.g. hydrolysis of sucrose), vacuolar storage (e.g. in sugar beet roots), or conversion to insoluble products (e.g. starch in seeds). Most of the source-to-sink gradient in solute concentration appears to occur across the unloading zone. This places phloem unloading strategically for involvement in the regulation of sink growth and partitioning.

Redistribution and utilization of N and C

Nitrogen

Partitioning of N between the roots and shoot is influenced by environmental factors and by the physiological and nutritional status of the plant (e.g. plants tend to invest more N in roots relative to the shoot when growing in soils of low N status). Roots may preferentially utilize the N they absorb to satisfy their own requirements before translocating the remainder to the shoot. However, roots have also been reported to be partly dependent for their growth on N cycled through the shoot. In some cases, the movement of N to the roots via the phloem is in excess of the root's total N requirements. In this case some N is re-exported in the xylem (Simpson et al., 1982), possibly undergoing metabolic conversion before release from phloem to xylem.

Nitrogen translocated from roots in the xylem stream is not partitioned to leaves and growing points of the shoot in strict proportion to their transpirational activity. Older leave acquire less N per volume of water transpired than do expanding leaves, apices, fruits and stems. It has been suggested that the stem plays a significant role in partitioning N to leaves and apices, partially through storage in stem parenchyma cells and partially through a succession of xylem-to-xylem and xylem-to-phloem transfers of N-solutes (Pate, 1980; Simpson, 1986). It is proposed that there is lateral withdrawal by stem nodal tissue N, initially directed in the transpiration stream towards lower leaves and then a subsequent re-routing of this N to xylem streams moving further up the shoot and a transfer of amino acids and amides to the phloem. The importance of such N exchanges are illustrated in the models of N flow prepared for the legume, lupin (Pate et al., 1988). Empirical models of whole lupin plant partitioning of N indicate that of the total N requirements of the developing lateral apices and terminal inflorescence: (1) 41% was attracted directly through xylem in accordance with transpiration loss; (2) 40% was donated through phloem from leaves; and (3) 19% was obtained through xylem to phloem transfer in upper regions of the stem.

Bearing in mind that these three sources of N all involved in xylem-to-xylem transfer lower down the shoot, and that (2) and (3) were implemented by xylem-to-phloem transfer, the importance of active short distance exchanges between conducting elements in coping with the limitations imposed by whole plant pH regulation, and the passive mechanism of long-distance N transport in both xylem and phloem, is apparent.

The N content of many dry land crops reaches a maximum during early reproductive phase. This is usually because of limitations to further N uptake caused by water stress, by depletion of available N in the root zone or, in annual legumes, by nodule senescence. Seed filling in these circumstances is usually dependent on extensive redistribution of N assimilated and invested in the vegetative organs prior to anthesis (Peoples and Dalling, 1988). The various vegetative organs differ significantly from one another in the extent to which they contribute by N mobilization to the final yield of grain-N. The differences are a reflection of the absolute N content of the organ at anthesis and the extent of N removal from the organ. Mobilization of N from leaves usually contributes most to the seeds' N requirements (from 16–40% of seed-N), but fruit parts (e.g. glumes in cereals and grasses, legume pod walls, the burr in cotton, or the husk and cob of maize) are also capable of N redistribution to seeds (Hocking et al., 1984; Peoples and Dalling, 1988). In general, leaves, stem and fruit parts are characterized by a high efficiency of N removal ($> 65\%$). Roots reallocate N less readily ($< 30\%$). It has been suggested that these differences are a reflection of an 'ordered priority system' within the plant whereby there is a need to balance integrity of organ function during senescence against any inherent tendency towards a high efficiency of N removal. The 'cost' of this compromise is less N redistribution than would otherwise occur, especially in the roots, which seem to be the last organ to senesce fully, as befits their role of keeping the shoot hydrated and physically supported.

Although some crops accumulate large amounts of soluble N in storage organs or stems, the majority of the N redistributed within the plant arises from the enzymatic degradation of protein by peptide hydrolases and the metabolic interconversion of breakdown products. It is likely that, through turnover, much of the N entering the plant early in its life passes through several age groups and types of vegetative structures, and is incorporated into several generations of proteins and other compounds before being finally stored in developing seed. In species having the C_3 photosynthetic pathway, the chloroplast enzyme ribulose-1,5-bisphosphate carboxylase is likely to represent the single largest remobilizable reserve of protein-N since it generally accounts for 30 to 60% of the total soluble protein and between 20 to 30% of total leaf N (Peoples and Dalling, 1988).

The reallocation of N is regarded as occurring principally via the phloem, but there is evidence that N can also be redistributed via the xylem if phloem transport is disrupted. In wheat, less than 50% of the N redistributed from leaves may be exported directly to the grain via the phloem. The rest appears to be transported to the roots and re-exported to the shoot via the xylem, and transferred again to the phloem for export to grain (see Hocking et al., 1984). Studies with legumes suggest that plants have the ability to manipulate solute composition via metabolic conversions and by xylem-to-xylem and xylem-to-phloem transfers in transit to reproductive structures so that fruits receive a uniform spectrum of N-compounds in the xylem or phloem regardless of whether they are utilizing current assimilates from the roots or N redistributed from senescing tissue for seed growth (Peoples et al., 1985a).

Carbon

During vegetative development, photoassimilate tends to be translocated preferentially to the sink closest to the assimilating leaf. The uppermost fully expanded leaf, for instance, exports C predominantly to the shoot apex and newly emerging leaves, while leaves below it direct most photoassimilate downwards to the roots (and tillers). In legumes, N_2 fixation by nodules may consume up to 18% of the net photosynthate produced, with total phloem translocation of C to the nodulated root representing over half of the shoot's photoassimilate supply. Some 16% of the C directed to the nodulated roots may be returned to the shoot in the xylem stream as the C-skeletons of N-transport compounds and as organic acids, and in some legumes as much as 53% can be lost through respiration (Pate et al., 1988). This large respiratory release of C, however, is likely to represent only a small net respiratory loss, since part of the CO_2 respired will have been refixed by the active phosphoenolpyruvate carboxylase (PEPcase) enzymes in legume nodules and roots. Such non-photosynthetic fixation of CO_2 by below-ground parts appears to play a key role in the synthesis of C-skeletons for amino acid biosynthesis, respiratory substrates, pH and ion balance regulation, and in the *de novo* synthesis of purines for ureide biogenesis in legumes such as soybean. Dark CO_2 fixation rates of the nodulated roots of alfalfa have been estimated as averaging 26% of the gross respiratory rate, with nodule PEPcase contributing as much as 25% of the C required for the assimilation of fixed N (Anderson et al., 1987). In contrast to legumes, relatively less of the C allocated to the roots of cereals appears to be lost through root respiration (32% in young wheat plants), but more (39% of the total C translocated to roots) may be cycled through the roots to be exported back to the shoot associated with amino acids (Lambers et al., 1982).

There appears to be some capacity for C to be exchanged between translocation streams within the stem by mechanisms similar to those described above for N. It seems that the phloem transport pathway is to some extent leaky, with passive unloading into, and active reloading from, the apoplast of legume stems. Continual phloem unloading (leakage) and reloading within the stem could provide a sucrose pool in the apoplast buffering the sieve tubes against sudden changes in phloem sucrose concentration. The significance of phloem-to-phloem or phloem-to-xylem C-transfer to shoot nutrition has not been fully evaluated (Minchin and McNaughton, 1987).

Towards the end of the vegetative phase of growth, the upper leaves transport progressively more of their photoassimilate to the stem and the developing inflorescence until anthesis and early flowering. There is commonly, thereafter, an abrupt decline in the flow of photosynthate to below-ground parts (and tillers), a gradual decrease in source leaf photosynthetic rate, and an increasing monopolization of photosynthate by reproductive organs. Temporary storage of non-structural carbohydrates in upper stem internodes of cereals during this phase, and subsequent movement to the ear during grain filling, has been reported in many studies. Fructan levels can decline in wheat from around 20% of the dry weight of the second internode below the ear after ear emergence to zero at the end of grain filling. Quantitatively, however, such mobilization of C reserves in temperate cereals contributes less than 10% to the final ear weight, except when the crop becomes droughted during grain filling. In this respect the time-course of C assimilation and partitioning appears very different from that of N. Whereas a major portion of the seed N yield is derived from the reallocation of N from vegetative organs to the seed, more than 75% of the final seed C usually comes from direct transfer of current photosynthate from source leaves. Even in cowpea, a legume noted for its rapid fruit growth and early senescence of leaves, reallocation of C from vegetative parts contributes less than 20% to the plant's post-anthesis C requirements, even though up to 50% of leaflet, stem, petiole and peduncle C may be mobilized during the period of fruit development. Appreciable amounts of extractable starch and sugars can be lost from vegetative tissues between anthesis and seed maturity in cowpea; however, these alone could not account for more than 75% of the total C loss from any one organ, indicating that C from structural components may also be mobilized in addition to non-structural carbohydrates to help satisfy the C demand of fruiting plants (Pate et al., 1983).

N and C interrelationships of individual organs

The following case studies for the leaves and fruits of legumes exemplify the dynamics of the interrelationships between xylem and phloem transport in the distribution of C and N during growth and development.

The leaf

The net turnover of C and N in a white lupin leaf during its life is summarized in Fig. 29.3a, and the time-course of exchanges made with the rest of the plant through xylem and phloem is shown in Fig. 29.3b and c.

Initially, both C and N were imported by the very young leaf mainly via the phloem (Fig. 29.3b, c). As the developing leaf began to transpire, the xylem became an increasingly important conduit for the import of N, contributing more than 80% of the N utilized for leaf growth by the time the leaf began to export C (i.e. at between one-third to one-half of its fully expanded size), and 94% of the leaf's final dry matter N requirements (Fig. 29.3a). Rapid leaf growth and accumulation of N continued in the period after the leaf's transition from C sink to C source until maximum N content was reached just prior to full leaf expansion. Except for the early phloem importing phase, the leaf continuously cycled N (i.e. simultaneously receiving and exporting N) by xylem to phloem transfer in the minor leaf veins (Fig. 29.3a). Of the N that was received through the xylem during the leaf's life, two-thirds were cycled back to the plant more or less immediately by xylem to phloem exchange, and a further one-sixth was translocated out of the leaf, following protein breakdown at the approach of leaf senescence. The time-courses of these transfers are depicted in Fig. 29.3c.

The budget for C (Fig. 29.3a) records a net photosynthetic input of 807 mg C and phloem export of 766 mg C compared with 65 mg C left behind in the leaf as non-retrievable dry matter, and a loss of 52 mg C in night respiration. Figure 29.3b shows a sharp peak in export of C shortly after the leaf achieved maximum size and N

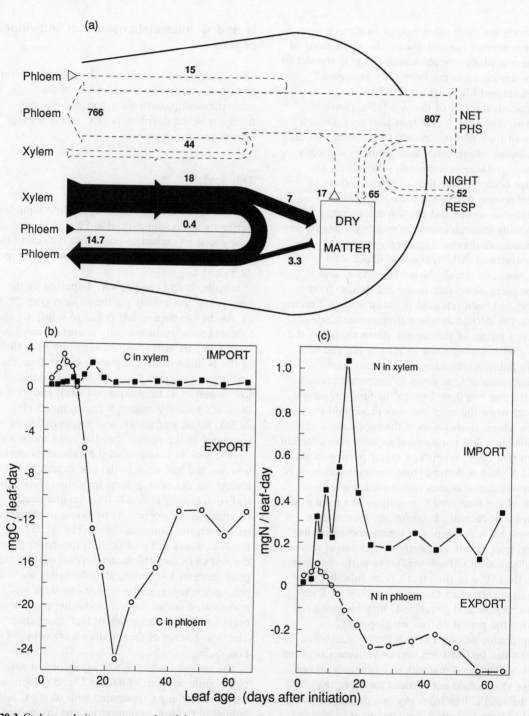

Fig. 29.3 Carbon and nitrogen economies of the uppermost main stem leaf of white lupin (*Lupinus albus* L.). Data are expressed as (a) net total exchanges of C and N (solid arrows) between plant and leaf (mg C and N) and as daily rates of export and import of (b) C and (c) N by the leaf through phloem (○) and xylem (■). Adapted from Pate *et al.*, 1988. NET PHS = net photosynthetic C input; NIGHT RESP = night-time respiratory C loss.

content, and a gradual lowering of the C:N ratio of the phloem export stream due to declining photosynthesis relative to N cycling and, later, to increasing mobilization of N from the leaf.

The legume fruit

The fruits on a modern cultivar of grain legume contain up to 40% of the plant's final content of C and 60 to 85% of its final N. To meet this especially large demand for N and C during seed filling, a heavy premium is placed on mechanisms to allocate these to the developing fruit. This applies particularly to fast maturing, determinant cultivars in which near-synchronous maturation of fruits generates a climacteric in demand for C and N during the final weeks of plant growth. The following discussion compares the strategies developed by a legume with a long reproductive phase (lupin) with one characterized by rapid rates of seed filling (cowpea) to effectively supply C and N for fruit growth.

Budgets for C and N have been provided for both the lupin fruit (Fig. 29.4a) and the cowpea fruit (Fig. 29.4b). Both species conserved C by refixing some or all CO_2 respired by day in green tissues of the pod wall (cowpea) and pod and seed (lupin). Mobilization of pod N during seed filling provided a significant proportion (17% in lupin, 9% in cowpea) of the seeds' N requirements (Fig. 29.4).

In lupin the C, N and H_2O budgets for the fruit balanced closely over its 12-week period of growth. Phloem supplied 89% of the N and 98% of the C (Fig. 29.4a). Transpiration and tissue water accumulation matched water input via xylem and phloem. While the phloem supplied a similar proportion of the C (97%) and slightly less N (72%) in the cowpea fruit, the water accompanying the C and N inputs in xylem and phloem vastly exceeded the fruit's tissue requirements and transpiration losses during the 3 weeks between anthesis and maturation. It appears that xylem intake by the cowpea fruit was mostly at night. However, during the day intake of phloem-borne water associated with the high rates of dry matter gain by pod and seed exceeded fruit

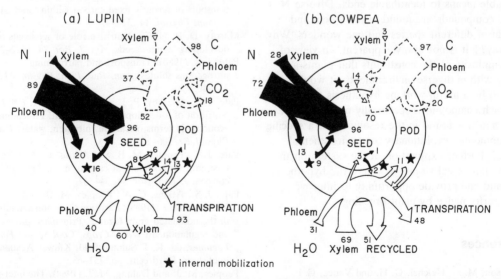

Fig. 29.4 Proportional intakes of C, N and water through xylem and phloem during development of (a) white lupin (*Lupinus albus* L.) and (b) cowpea (*Vigna unguiculata* [L.] Walp.). All components of a fruit's budget are expressed relative to a net intake of 100 units of C, N or water through xylem and phloem. Ratios of absolute amounts by weight of C, N and water consumed by lupin fruits are given in Pate *et al*. (1977) and by cowpea in Peoples *et al*. (1985b) from which the developmental budgets were derived. The hypothetical 'recycled' component of cowpea's water budget is shown in (b). Values for mobilization of C, N, or water are indicated by a star (★).

transpiration. It is surmised that the water potential gradient must have been in the direction for water to flow out of the pod in the xylem when there is this surplus of phloem-delivered water. This would account for the discrepancy in the water budget. Whether such diurnally reversible xylem exchanges of water carry solutes, including 'fruit-induced leaf senescence' hormones, out of the seed is worthy of investigation (see Pate et al., 1988).

Conclusions

Higher plants employ a wide range of strategies to cope with the conflicting and ever-changing boundary conditions involved in taking up N from the soil and C from the air and moving them to where they are needed for integrated operation of the whole growing plant. In achieving this there are apparent inefficiencies, such as the transfer of C from the roots to the leaves and N from the shoots to the roots. But the closer one looks, the more one realizes that the inefficiencies may be inevitable means to identifiable ends. Diverse N and C compounds are found in the xylem and phloem of different species and one wonders 'Why so many?'. It is not always apparent. In view of the complexity of the constraints that plants are coping with in diverse environments, it would seem to be a useful working hypothesis to assume that each compound is specifically needed where and when it is found in the resolution of a specific developmental, evolutionary, or environmental conflict. Further exploration of the system on that basis might reveal situations where the hypothesis is invalid and provide opportunity to improve plant performance by plant breeding.

References

Anderson, M. P., Heichel, G. H. and Vance, C. P. (1987). Nonphotosynthetic CO_2 fixation by alfalfa (*Medicago sativa* L.) roots and nodules. *Plant Physiol.* **85**, 283–9.

Delrot, S. and Bonnemain, J. L. (1985). Mechanism and control of phloem transport. *Physiol. Vég*, **23**, 199–220.

Fisher, D. B. and Frame, J. M. (1984). A guide to the use of the exuding-stylet technique in phloem physiology. *Planta* **161**, 385–93.

Giaquinta, R. T. (1983). Phloem loading of sucrose. *Ann. Rev. Plant Physiol.* **34**, 347–87.

Gifford, R. M. (1986). Partitioning of photoassimilate in the development of crop yield. In *Phloem Transport*, eds J. Cronshaw, W. J. Lucas and R. T. Giaquinta, Alan R. Liss, New York, pp. 535–49.

Gifford, R. M. and Evans, L. T. (1981). Photosynthesis, carbon partitioning, and yield. *Ann. Rev. Plant Physiol.* **32**, 485–509.

Hocking, P. J., Steer, B. T. and Pearson, C. J. (1984). Nitrogen nutrition of non-leguminous crops: A review. Part 1. *Field Crop Abst.* **37**, 625–36.

Lambers, H., Simpson, R. J., Beilharz, V. C. and Dalling, M. J. (1982). Translocation and utilization of carbon in wheat (*Triticum aestivum*). *Physiol. Plant.* **56**, 18–22.

Lang, A. (1978). A model of mass flow in phloem. *Aust. J. Plant Physiol.* **5**, 535–46.

Layzell, D. B. and LaRue, T. A. (1982). Modelling C and N transport to developing soybean fruits. *Plant Physiol.* **70**, 1290–8.

Maynard, J. W. and Lucas, W. (1982). Sucrose and glucose uptake into *Beta vulgaris* leaf tissue: a case for general (apoplastic) retrieval systems. *Plant Physiol.* **70**, 1436–43.

Minchin, P. E. H. and McNaughton, G. S. (1987). Xylem transport of recently fixed carbon within lupin. *Aust. J. Plant Physiol.* **14**, 325–9.

Murray, D. R. (1987). Nutritive role of seedcoats in developing legume seeds. *Am. J. Bot.* **74**, 1122–37.

Pate, J. S. (1980). Transport and partitioning of nitrogenous solutes. *Ann. Rev. Plant Physiol.* **31**, 313–40.

Pate, J. S., Sharkey, P. J. and Atkins, C. A. (1977). Nutrition of a developing legume fruit. Functional economy in terms of carbon, nitrogen, water. *Plant Physiol.* **59**, 506–10.

Pate, J. S., Peoples, M. B. and Atkins, C. A. (1983). Post-anthesis economy of carbon in a cultivar of cowpea. *J. Exp. Bot.* **34**, 544–62.

Pate, J. S., Atkins, C. A., Peoples, M. B. and Herridge, D. F. (1988). Partition of carbon and nitrogen in the nodulated grain legume: Principles, processes and regulation. In *World Crops: Cool Season Food Legumes*, ed. R. J. Summerfield, Kluwer Academic Publishers, Dordrecht, pp. 751–65.

Peoples, M. J. and Dalling, M. J. (1988). The inter-play between proteolysis and amino acid metabolism during senescence and nitrogen re-allocation. In *Senescence and Aging in Plants*, eds L. D. Noodén and A. C. Leopold, Academic Press, New York, pp. 181–217.

Peoples, M. B., Pate, J. S. and Atkins, C. A. (1985a). The effect of nitrogen source on transport and metabolism of nitrogen in fruiting plants of cowpea (*Vigna unguiculata* (L.) Walp.). *J. Exp. Bot.* **36**, 567–82.

Peoples, M. B., Pate, J. S., Atkins, C. A. and Murray, D. R. (1985b). Economy of water, carbon, and nitrogen in the developing cowpea fruit. *Plant Physiol.* **77**, 142–7.

Peoples, M. B., Sudin, M. N. and Herridge, D. F. (1987). Translocation of nitrogenous compounds in symbiotic and nitrate-fed amide-exporting legumes. *J. Exp. Bot.* **38**, 567–79.

Raven, J. A. (1977). The evolution of vascular land plants in relation to supracellular transport processes. *Adv. Bot. Res.* **5**, 154–219.

Raven, J. A. and Smith, F. A. (1976). Nitrogen assimilation and transport in vascular land plants in relation to intracellular pH regulation. *New Phytol.* **76**, 415–31.

Shelp, B. J. (1987). The composition of phloem exudate and xylem sap from broccoli (*Brassica oleracea* var. italica) supplied with NH_4^+, NO_3^- or NH_4NO_3. *J. Exp. Bot.* **38**, 1619–36.

Simpson, R. J. (1986). Translocation and metabolism of nitrogen: whole plant aspects. In *Fundamental, Ecological and Agricultural Aspects of Nitrogen Metabolism in Higher Plants*, eds H. Lambers, J. J. Neetsson and I. Stulen, Martinus Nijhoff Publ., Dordrecht, pp. 71–96.

Simpson, R. J. and Dalling, M. J. (1981). Nitrogen redistribution during grain growth in wheat (*Triticum aestivum* L.) III Enzymology and transport of amino acids from senescing flag leaves. *Planta* **151**, 447–56.

Simpson, R. J., Lambers, H. and Dalling, M. J. (1982). Translocation of nitrogen in a vegetative wheat plant. *Physiol. Plant.* **56**, 11–17.

Sprent, J. I. (1980). Root nodule anatomy, type of export product and evolutionary origin in some Leguminosae. *Plant Cell Environ.* **3**, 35–43.

Thorne, J. H. (1985). Phloem unloading of C and N assimilates in developing seeds. *Ann. Rev. Plant Physiol.* **36**, 317–43.

Wolswinkel, P. (1985). Phloem unloading and turgor-sensitive transport: Factors involved in sink control of assimilate partitioning. *Physiol. Plant.* **65**, 331–9.

30 Protein degradation
Peggy M. Hatfield and Richard D. Vierstra

Introduction

Protein turnover can be defined as the flow of amino acids from existing protein into newly synthesized protein. As such, both protein degradation and synthesis are important components of this process. The composition of newly synthesized proteins may be different from those that existed previously, underscoring a critical role for turnover in altering cellular constituents during the life of a plant cell. Such alterations of protein content are essential for normal development and for responding to environmental conditions.

Most proteins have a lifetime that is less than that of a cell, and so are continuously degraded and, if necessary, resynthesized. The turnover time of total soluble protein has been estimated at between 3–7 days in leaves of actively growing plants of *Lemna minor* (Davies, 1982). This observation implies that the metabolic flow of nitrogen and amino acids is both cyclic and continuous. Thus, excluding the special case of seed storage proteins, proteins cannot be viewed simply as a storage form of nitrogen. Nonetheless, rates of protein turnover can be altered during times of amino acid starvation by changing the rate of degradation relative to synthesis, thereby converting less essential proteins into more essential proteins.

Protein turnover serves a significant role in the regulation of basic cellular metabolism. Goldberg and St John (1976) compared the half-lives of various rat liver proteins and found that proteins with the shortest half-lives occupy critical control points in major metabolic pathways. Because of this, they proposed that short half-lives allow the cell to change the concentration of such proteins quickly. Activity of the major metabolic pathways would then be regulated by the turnover of key enzymes, thus altering metabolic activity rapidly in response to environmental or developmental stimuli.

The intent of this chapter is to focus on the role of degradation in protein turnover. Specifically, the manner in which various plant proteases function in degradation and possible levels of their regulation will be examined. As will be seen, protein degradation is a complex process, with many questions still unanswered.

Plant proteases

Plants contain a variety of hydrolytic activities capable of degrading proteins (for reviews see Ryan, 1973; Davies, 1982; Matile, 1982; Storey, 1986). In fact, some of the common commercially available proteases are from plants, including papain (papaya), ficin (fig), and bromelain (pineapple). They are generally characterized as endopeptidases (cleaving internal peptide bonds, thus generating a range of peptide fragments) or exopeptidases (cleaving peptide bonds progressively from either the C-terminus (carboxypeptidase) or N-terminus (aminopeptidase), liberating free amino acids) (Barrett, 1986). Proteases have been found within the vacuole, the cytoplasm, and chloroplasts. The existence of proteases in the mitochondrion of plant cells has not been documented, but is inferred from their existence in animal mitochondria (Goldberg and St John, 1976). The function of specific plant proteases *in vivo* is largely unknown though, as most have been characterized *in vitro* using model substrates (e.g. hemoglobin, casein). Thus, a major challenge remains to define the exact functions of the myriad of plant proteases.

The plant vacuole contains a variety of acid hydrolases, including several proteases (e.g. endopeptidases, carboxypeptidases) (Boller, 1986; Boller and Wiemken, 1986; Mikola and Mikola, 1986; Storey, 1986). Most vacuolar proteases function optimally *in vitro* under acidic conditions (pH 3 to 6) (Ryan, 1973), conditions that exist in the vacuole *in vivo* (Roberts *et al.*, 1980). Because the vacuole contains most of the proteolytic activities measured *in vitro*, it was proposed that the vacuole was the major site of *in vivo* degradation for most cellular constituents, including proteins from the cytoplasm and chloroplast (Matile, 1982). However, it now seems that each subcellular compartment has one or more proteolytic enzymes of its own (Goldberg and St John, 1976; Matile, 1982; Liu and Jagendorf, 1984, 1986; Malek *et al.*, 1984; Boller and Wiemken, 1986; Vierstra, 1987). Thus, vacuolar proteases probably have a minor role in endogenous protein degradation (except for protein bodies, see below) until the final stages of senescence when the tonoplast membrane has ruptured (Boller, 1986). They may serve an alternative role in defense of the plant against pathogens, parasites, or herbivores (Boller, 1986).

Exposure to adverse environments, though, can result in the transport of specific proteins from other cellular compartments into the vacuole (Boller and Wiemken, 1986; Canut *et al.*, 1986). For example, protoplasts incubated with amino acid analogs synthesize proteins with abnormal conformations. These abnormal proteins subsequently accumulate in the vacuole, where they are degraded rapidly (Canut *et al.*, 1986).

Protein bodies of plant seeds are part of the same membrane system as the central vacuole, and, in legumes, are formed by fragmentation of the vacuole itself (Boller and Wiemken, 1986; Harris, 1986). Protein bodies have been described as specialized vacuoles, serving to contain and, during seedling development, to degrade enclosed storage proteins. The proteases associated with these structures have activities similar to those described for the vacuole, exhibiting acidic pH optima, and consisting mainly of acid endopeptidases and carboxypeptidases (Ryan, 1973; Matile, 1982; Mikola and Mikola, 1986).

Proteases have been isolated from extracts of germinating tissues, mature leaves, and fruits that have optimal activity under more basic conditions (pH 6–9) than are found in the vacuolar system, indicating these may function in the more basic environment of the cytoplasm (Ryan, 1973; Matile, 1982). A proteolytic pathway has been elucidated that degrades cytoplasmic proteins with the involvement of the small protein ubiquitin. The pathway was described initially in animals (Hershko and Ciechanover, 1986; Rechsteiner, 1987) but is also functional in plants (Vierstra, 1987) (Fig. 30.1). This pathway is unique in that

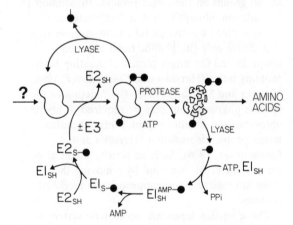

Fig. 30.1 Diagram of ubiquitin conjugate formation and subsequent degradation or disassembly. Ubiquitin (−●) is activated via adenylation by the enzyme E1 (or ubiquitin activating enzyme), and transferred to one of many small carrier proteins, E2's, as a reactive thiolester. Ubiquitin is ligated to a primary amino group on the target protein with or without the help of another enzyme, E3 (or ubiquitin–protein ligase). The modified target protein is either degraded by a specific ATP-dependent protease, or the ubiquitin molecule is clipped off by ubiquitin–protein lyase, yielding both ubiquitin and target protein intact. During target protein degradation, it is likely that ubiquitin–protein lyase removes ubiquitin from protein fragments, allowing ubiquitin to re-enter the pathway.

ubiquitin mediates protein degradation, not by degrading proteins itself, but by becoming covalently ligated to selected target proteins. Target proteins, so modified, are rapidly degraded by ATP-dependent protease(s) specific for ubiquitin conjugates, without damaging the ubiquitin moiety (Rechsteiner, 1987). In this way, ubiquitin serves as a reusable signal for selective proteolysis (Fig. 30.1). A conjugate-specific protease has been purified recently from mammalian tissues, and is composed of several polypeptides (8–10) forming a large, multimeric complex (MW 1 000 000) (Rechsteiner, 1987).

Ligation of ubiquitin to target proteins requires ATP and is accomplished via the sequential action of two, and sometimes three, enzymes (Fig. 30.1) (Hershko and Ciechanover, 1986; Rechsteiner, 1987). It occurs through the formation of an unusual peptide bond between the carboxyl group of the C-terminal glycine of ubiquitin and primary amino groups on the target protein. In addition to degradation, ubiquitin–protein conjugates can be disassembled by activities (ubiquitin–protein lyases) that cleave only the peptide bond between ubiquitin and the target protein, liberating both proteins intact (Hershko and Ciechanover, 1986; Vierstra and Sullivan, 1988). The function of the lyase is unknown, but may be required to liberate ubiquitin from smaller peptides formed during target protein degradation (Hershko and Ciechanover, 1986). Such an activity might also serve a corrective function, by removing the ubiquitin moiety from inappropriately modified proteins.

The ubiquitin-dependent proteolytic system is extremely conserved among eukaryotes as diverse as humans, yeast, and higher plants (Hershko and Ciechanover, 1986; Rechsteiner, 1987; Vierstra, 1987). This conservation emphasizes the importance of ubiquitin conjugation to eukaryotic cell biology. The function(s) of ubiquitin conjugation in plant cell physiology are not completely resolved. By comparison with its roles in animal systems, the ubiquitin proteolytic pathway likely mediates the turnover of many abnormal or short-lived proteins in plants (Rechsteiner, 1987). As mentioned above, many significant regulatory proteins have relatively short half-lives, implying that ubiquitin would have an important role in modulating cell metabolism (Goldberg and St John, 1976).

Ubiquitin may be important in the action of phytochrome which is a key regulatory protein in plants. It is a developmental photoreceptor that triggers a multitude of morphogenic responses after photoconversion from the biologically inactive, red light absorbing (Pr) form to the biologically active, far-red absorbing form (Pfr) (Vierstra and Quail, 1986). The Pr form of phytochrome is quite stable in plants (Table 30.1), whereas Pfr is degraded rapidly. Evidence suggests that the rapid degradation of Pfr occurs through ubiquitinated intermediates (Shanklin et al., 1987).

Protein degradation has also been detected within the chloroplast (Liu and Jagendorf, 1984, 1986; Malek et al., 1984; Greenberg et al., 1987). Proteolysis within this compartment serves a variety of purposes, including cleaving signal peptides, regulating the stoichiometric accumulation of subunits, and hydrolyzing damaged proteins. Chloroplast proteins that are encoded in the nuclear genome are synthesized with N-terminal amino acid extensions responsible for correct transport of these proteins into this organelle. The transit peptides are removed and rapidly degraded during import of the protein into the chloroplast. In addition, the stoichiometric accumulation of large and small subunits of Rubisco in the chloroplast is controlled in part by degradation of excess subunits. Similarly, the apoprotein of the light-harvesting chlorophyll a/b binding protein and plastocyanin are unstable without their respective cofactors, chlorophyll and Cu^{2+}. By their nature, the photochemical reactions of the chloroplast generate free radicals harmful to protein integrity. The 32 kD herbicide-binding protein, which acts on the reducing side of PSII, seems to be particularly sensitive to these free radicals, and is rapidly degraded under intense light.

Both ATP-dependent and ATP-independent proteolytic activities have been identified in isolated, intact chloroplasts (Liu and Jagendorf, 1984, 1986; Malek et al., 1984). The ATP-dependent protease of pea chloroplasts has been tentatively localized to the thylakoid

Table 30.1 Half-lives of proteins from plant or animal sources

Protein	Source	Half-life (h)	Reference
Plant			
Nitrate reductase (+ NO$_3$)	Tobacco, corn, cotton	1.5–6	Wallace and Oaks, 1986
LHCP	Duckweed	10	Dalling and Nettleton, 1986
Phytochrome (Pfr)	Oat	1–2	Shanklin et al., 1987
Phytochrome (Pr)	Oat	≥100	Shanklin et al., 1987
Rubisco	Corn	144–156	Davies, 1982
Animal			
Ornithine decarboxylase	Mouse kidney	0.2–1.3	Rechsteiner et al., 1987
RNA polymerase I	Rat liver	1.3	Goldberg and St John, 1976
HSP 70 (heat shock protein)	Fruit fly	1–2	Rechsteiner et al., 1987
PEP carboxykinase	Rat liver	5.0	Goldberg and St John, 1976
Glucose -6- phosphate dehydrogenase	Rat liver	15.0	Goldberg and John, 1976
Acetyl-CoA carboxylase	Rat liver	48.0	Rechsteiner et al., 1987
Histone H2A, H2B	–	>200	Rechsteiner et al., 1987

membrane, and may degrade normal but unassembled plastid subunits (Liu and Jagendorf, 1984). The ATP-independent protease may be responsible for the removal of abnormal or prematurely terminated proteins (Liu and Jagendorf, 1986). One endopeptidase and three aminopeptidase activities have also been detected, though their relationship to the ATP-dependent and ATP-independent activities has not been established (Liu and Jagendorf, 1986).

Protein degradation within the chloroplast seems to be influenced by factors from the cytoplasm. For example, during leaf senescence, degradation of proteins within the chloroplast requires the presence of proteins synthesized by the nuclear genome (Dalling and Nettleton, 1986). Furthermore, a mutation in fescue displaying classical Mendelian inheritance (indicative of a nuclear gene) prevents the breakdown of certain thylakoid membrane components during senescence (Stoddart and Thomas, 1982; Dalling and Nettleton, 1986).

Regulation of protein degradation

Individual proteins differ substantially in the rate at which they are degraded *in vivo*. This selectivity allows proteins to coexist in the same compartment with half-lives ranging from minutes to several days. The range of protein half-lives is strikingly demonstrated by comparing degradation rates of different plant and animal proteins (Table 30.1). This poses a question as to how the proteolytic systems within a single cellular compartment are able to selectively recognize and degrade proteins at different rates. The mechanism(s) responsible for this selectivity are largely unknown, though it is likely that selective degradation is accomplished by a combination of factors. These include inherent physicochemical characteristics of the protein to be degraded, *de novo* synthesis or activation of proteases, and changing the compartmentation of either the protease or the substrate.

Substrate characteristics

Early studies determined that the conformation of proteins significantly influenced their rates of degradation. In plants, the rapid degradation of phytochrome is initiated by an undetermined conformational difference between the unstable Pfr form and the stable Pr form (Vierstra and Quail, 1986). How the cell recognizes this conformational change is unknown. Experiments with mammalian proteins revealed that proteins with short half-lives *in vivo* were usually more susceptible to *in vitro* proteolysis as well (Goldberg and St John, 1976). However, a single common structural element could not be identified; rather, several basic structural features may influence degradative rates. Among these, primary sequence, molecular mass, and charge may affect protein stability either singly, or, more likely, in concert with other factors. Likewise, situations that alter normal protein structure (e.g. mutation, denaturation, incorporation of amino acid analogs) substantially increase rates of degradation *in vivo*.

Primary sequence

The primary sequence of a protein may affect its catabolic stability (Rechsteiner, 1987). One important domain appears to be the N-terminus. Using novel β-galactosidase gene constructs, Bachmair *et al.* (1986) found a significant correlation between the N-terminal amino acid residue and the rate of galactosidase degradation. When the N-terminal amino acid was Met, Ala, or Val, the protein was stabilized. Conversely, the protein was destabilized when Arg, Lys, Leu, and Asp residues were at the N-terminus. A survey of half-lives of proteins with known N-termini revealed that non-compartmentalized, intracellular proteins tended to have stabilizing amino acids. Compartmentalized proteins, such as secretory proteins, usually had destabilizing amino acids at the N-terminus, possibly insuring that these proteins are removed quickly if released inadvertently into the cytoplasm. Proteins with destabilizing amino residues, however, may be modified (e.g. acetylation of the N-terminal residue) to render the proteins less susceptible to breakdown.

Internal amino acid sequences may influence degradative rates as well. Peptide domains enriched in certain amino acids, e.g. Pro, Glu, Ser, Thr, or containing specific sequences, e.g. Lys–Phe–Glu–Arg–Gln, are correlated with a greater susceptibility to degradation *in vivo* (Rechsteiner, 1987). In addition, oxidative modification of specific residues may render a protein more vulnerable to degradation. Generally, though, amino acids sensitive to this type of modification (e.g. His, Tyr, Trp, Met, Cys) are less abundant in proteins.

Molecular mass

In general, large proteins have shorter half-lives than small proteins. In particular, a correlation has been noted between rate of degradation and subunit molecular mass of both plant and animal proteins (Goldberg and St John, 1976; Davies, 1982). Davies (1982) postulated that large proteins are less stable because they have potentially more sites sensitive to proteolytic cleavage than smaller proteins. It is also possible that large proteins are simply more susceptible to chemical damage (i.e. oxidative damage or denaturation), which precedes proteolysis.

Charge

A relationship has been identified between degradative rate and protein charge, such that proteins with acidic isoelectric points usually have higher rates of turnover than proteins with neutral or basic isoelectric points (Goldberg and St John, 1976; Davies, 1982). Both mammalian and plant proteins follow this general postulate. Moreover, the effects of protein size and charge are interrelated, such that small, basic proteins tend to be the most stable (Davies, 1982).

Protein conformation

Proteins with abnormal conformations are degraded rapidly, though it is not known how these abnormal conformations are detected (Goldberg and St John, 1976; Canut *et al.*, 1986). Incorrect conformation can be induced either by incorporation of amino acid analogs, premature translational termination, biosynthetic errors, or oxidative damage. Enzymes devoid of cofactors or ligands likewise may assume alternate, unstable conformations. Because non-dividing cells are incapable of removing aberrant proteins by dilution through cell division, efficient removal by proteolysis is necessary to avoid accumulation of these proteins to toxic levels.

Other factors

Other structural characteristics have been correlated with degradative rates in plants and animals (Goldberg and St John, 1976; Davies, 1982). These include: hydrophobicity, glycosylation, thermostability, and assembly of subunits into multimeric complexes.

Regulation of protease activity

Most plant proteases, while demonstrating narrow specificity with regard to cleavage site, display broad specificity with regard to protein substrate. Thus, they are generally not specific for a single protein or class of proteins. Consequently, proteases must be tightly regulated to insure that proteolysis is confined to appropriate substrates. Possible mechanisms for regulation include: (1) altering the concentration of the protease by changes in synthesis and/or degradation; (2) conversion of the protease from an inactive zymogen into an active enzyme; (3) changes in the concentration of protease inhibitors; or (4) changes in compartmentation of the protease and or substrate. The following observations about regulation of proteolysis have been obtained mainly from germinating tissue, and thus may or may not apply to mature or senescing tissues.

Rate of protease turnover

Degradation of specific proteins may be regulated by changing the concentration of the corresponding protease. For example, in the aleurone layer of barley seeds, endopeptidase activity increases during germination as a consequence of *de novo* synthesis, induced by the plant hormone gibberellic acid (Ryan, 1973; Bewley, 1982; Davies, 1982). Eight other proteinases exhibiting increased activity during germination have been identified in barley seeds, some localized to specific tissues (Ryan, 1973).

Another example of regulation of proteinase activity by *de novo* synthesis is in mung bean. The major storage protein in mung bean seeds, vicilin, can be completely digested in *vitro* through the combined action of a specific endopeptidase and a carboxypeptidase (Matile, 1982; Mikola and Mikola, 1986). The endopeptidase, vicilin peptidohydrolase, is absent in ungerminated seeds, but is synthesized within three days of the onset of germination (Davies, 1982; Matile, 1982). The enzyme is synthesized in the cytoplasm and transferred via vesicles from the endoplasmic reticulum to the protein bodies containing vicilin (Davies, 1982; Matile, 1982). In this instance, then, mobilization of storage protein amino acids is regulated both by synthesis of vicilin peptidohydrolase and its compartmentation with the substrate.

Activation of zymogens

In a few examples, the increase in proteolytic activity associated with germination results from the activation of a precursor. In lettuce seeds, a protease was identified whose activity increased upon germination. However, the enzyme was not synthesized *de novo*, but was formed by proteolytic processing of a larger precursor (Ryan, 1973; Davies, 1982).

Protease inhibitors

Protease inhibitors have been identified in tubers and seeds of various plant species. Several inhibit trypsin-like proteases from animals and bacteria (Ryan, 1973; Huffaker and Peterson, 1974; Davies, 1982). Some of these inhibitors affect endogenous plant proteases, but most do not. It is possible that such inhibitors prevent premature breakdown of storage proteins. They may also protect cytoplasmic constituents from inadvertent contact during transport of the proteases to protein bodies (Davies, 1982). Alternatively, they may function in defense of the plant during attack by herbivores or pathogens by inactivating the proteases of the pest (Ryan, 1973; Davies, 1982).

Compartmentation of proteolytic activities

Protein degradation may be controlled by confining proteases to certain cellular compartments, thereby limiting access to inappropriate substrates. The primary examples of this in plants are the vacuolar proteases. The vacuole contains several hydrolytic activities with relatively broad substrate specificities that must be separated from the contents of other cellular compartments. As mentioned above, active proteases are confined to protein bodies during seedling development. Localization of proteases in the membranes of the chloroplast, mitochondrion, or plasmalemma also physically restricts the interaction of potential substrates with active proteases.

The increase in protein degradation apparent with the onset of senescence is not necessarily accompanied by an increase in total proteolytic activity (Davies, 1982; Stoddart and Thomas, 1982; Sacher, 1983). This suggests that senescence is not triggered by the synthesis of new proteases. If so, then before senescence is initiated, either the intracellular proteases are inactivated or inhibited, or else their substrates are inaccessible. Proteases may also be activated by a shift in the physical environment, such as a change in the pH of the chloroplast upon cessation of electron transport (Stoddart and Thomas, 1982). In addition, plant hormones, especially cytokinins and abscisic acid, influence the rate of protein degradation during leaf senescence, though their mode of action is not understood (Sacher, 1983).

Conclusions

The manner in which plants degrade their proteins is largely unresolved. Many questions remain to be answered regarding substrate selection and the mechanisms used to degrade target proteins. It is already evident that multiple degradative systems operate within the plant cell, perhaps delineated by compartmental boundaries. Moreover, several factors seem to influence protein stability, and these can change in response to environmental stimuli and developmental state. A renewed interest in protein degradation in plants may result in the discovery of answers to these basic questions. The answers will undoubtedly be complex.

References

Bachmair, A., Finley, D. and Varshavsky, A. (1986). *In vivo* half-life of a protein is a function of its amino-terminal residue. *Science* **234**, 179–186.

Barrett, A. J. (1986). The classes of proteolytic enzymes. In *Plant Proteolytic Enzymes*, Vol. I, ed. M. J. Dalling, CRC Press, Boca Raton, pp. 1–16.

Bewley, J. D. (1982). Protein and nucleic acid synthesis during seed germination and early seedling growth. In *Encyclopedia of Plant Physiology*, Vol. 14A, eds D. Boulter and B. Parthier, Springer-Verlag, Berlin, pp. 559–91.

Boller, T. (1986). Roles of proteolytic enzymes in interactions of plants with other organisms. In *Plant Proteolytic Enzymes*, Vol. I, ed. M. J.Dalling, CRC Press, Boca Raton, pp. 67–96.

Boller, T. and Wiemken, A. (1986). Dynamics of vacuolar compartmentation. *Ann. Rev. Plant Physiol.* **37**, 137–64.

Canut, H., Alibert, G., Carrasco, A. and Boudet, A. M. (1986). Rapid degradation of abnormal proteins in vacuoles from *Acer psuedoplatanus* L. cells. *Plant Physiol.* **81**, 460–3.

Dalling, M. J. and Nettleton, A. M. (1986). Chloroplast senescence and proteolytic enzymes. In *Plant Proteolytic Enzymes*, Vol. II, ed. M. J. Dalling, CRC Press, Boca Raton, pp. 125–53.

Davies, D. D. (1982). Physiological aspects of protein turnover. In *Encyclopedia of Plant Physiology*, Vol. 14A, eds D. Boulter and B. Parthier, Springer-Verlag, Berlin, pp. 189–228.

Goldberg, A. L. and St John, A. C. (1976). Intracellular protein degradation in mammalian and bacterial cells: Part 2. *Ann. Rev. Biochem.* **45**, 747–803.

Greenberg, B. M., Gaba, C., Mattoo, A. K. and Edelman, M. (1987). Identification of a primary *in vivo* degradation product of the rapidly-turning over 32 kd protein of photosystem II. *EMBO J.* **6**, 2865–9.

Harris, N. (1986). Organization of the endomembrane system. *Ann. Rev. Plant Physiol.* **37**, 73–92.

Hershko, A. and Ciechanover, A. (1986). The ubiquitin pathway for the degradation of intracellular proteins. *Prog. Nucl. Acid Res. Mol. Biol.* **33**, 19–56.

Huffaker, R. C. and Peterson, L. W. (1974). Protein turnover in plants and possible means of its regulation. *Ann. Rev. Plant Physiol.* **25**, 363–92.

Liu, X.-Q. and Jagendorf, A. T. (1984). ATP-dependent proteolysis in pea chloroplasts. *FEBS Lett.* **166**, 248–52.

Liu, X. Q. and Jagendorf, A. T. (1986). Neutral peptidases in the stroma of pea chloroplasts. *Plant Physiol.* **81**, 603–8.

Malek, L., Bogorad, L., Ayers, A. R. and Goldberg, A. L. (1984). Newly synthesized proteins are degraded by an ATP-stimulated proteolytic process in isolated pea chloroplasts. *FEBS Lett.* **166**, 253–7.

Matile, P. H. (1982). Protein degradation. In *Encyclopedia of Plant Physiology*, Vol. 14A, eds D. Boulter and B. Parthier, Springer-Verlag, Berlin, pp. 169–88.

Mikola, L. and Mikola, J. (1986). Occurrence and properties of different types of peptidases in higher plants. In *Plant Proteolytic Enzymes*, Vol. I, ed. M. J. Dalling, CRC Press, Boca Raton, pp. 97–117.

Rechsteiner, M. (1987). Ubiquitin-mediated pathways for intracellular proteolysis. *Ann. Rev. Cell Biol.* **3**, 1–30.

Rechsteiner, M., Rogers, S. and Rote, K. (1987). Protein structure and intracellular stability. *Trends Biochem. Sci.* **12**, 390–4.

Roberts, J. K., Ray, P. M., Wade-Jardetzky, N. and Jardetzky, O. (1980). Estimation of cytoplasmic and vacuolar pH in higher plants by ^{31}P NMR. *Nature* **283**, 870–2.

Ryan, C. A. (1973). Proteolytic enzymes and their inhibitors in plants. *Ann. Rev. Plant Physiol.* **24** 173–96.

Sacher, J. A. (1983). Abscisic acid in leaf senescence. In *Abscisic Acid*, ed. F. T. Addicott, Praeger Publishers, New York, pp. 479–522.

Shanklin, J., Jabben, M. and Vierstra, R. D. (1987). Red light-induced formation of ubiquitin–phytochrome conjugates: Identification of possible intermediates of phytochrome degradation. *Proc. Natl Acad. Sci. USA* **84**, 359–63.

Stoddart, J. L. and Thomas, H. (1982). Leaf senescence. In *Encyclopedia of Plant Physiology*, Vol. 14A, eds D. Boulter and B. Parthier, Springer-Verlag, Berlin, pp. 592–636.

Storey, R. D. (1986). Plant endopeptidases. In *Plant Proteolytic Enzymes*, Vol. I, ed. M. J. Dalling, CRC Press, Boca Raton, pp. 119–40.

Vierstra, R. D. (1987). Demonstration of ATP-dependent, ubiquitin-conjugating activities in higher plants. *Plant Physiol.* **84**, 332–6.

Vierstra, R. D. and Quail, P. H. (1986). Phytochrome the protein. In *Photomorphogenesis in Plants*, eds R. E. Kendrick and G. H. M. Kronenberg, Martinus-Nijhoff, Boston, pp. 35–60.

Vierstra, R. D. and Sullivan, M. L. (1988). Hemin inhibits ubiquitin-dependent proteolysis in both a higher plant and yeast. *Biochemistry* **27**, 3290–5.

Wallace, W. and Oaks, A. (1986). Role of proteinases in the regulation of nitrate reductase. In *Plant Proteolytic Enzymes*, Vol. II, ed. M. J. Dalling, CRC Press, Boca Raton, pp. 81–9.

31 Protein storage and utilization in seeds
J. Derek Bewley and John S. Greenwood

Introduction

During development, seeds characteristically synthesize relatively large quantities of food reserves which are sequestered in storage tissues such as the cotyledons or endosperm. These reserves are mobilized following germination, and their catabolites are used to support the growth of the seedling until it can establish itself as a photosynthesizing, autotrophic plant. The major reserves are commonly laid down in discrete storage organelles: protein in protein bodies (often in conjunction with the minor reserve phytin, a source of phosphate and micronutrients), lipid in lipid bodies, and starch in starch grains (amyloplasts). Since approximately 70% of all food for human consumption comes directly from seeds, and a large proportion of the remainder is derived from animals that are fed on seeds, it is not surprising that there has been considerable research into seed nutritional qualities and the events that are involved in food reserve deposition. Part of this research has incorporated the synthesis, packaging and mobilization of seed proteins, largely in cultivated plants, and especially in the legumes and cereals. Consequently, in this chapter we will concentrate on protein reserve anabolism and catabolism in the seeds of these two major groups. A fuller account of both events can be found in Bewley and Black (1985) and Murray (1984 a,b).

Protein types and their composition

The classification of seed proteins has remained unchanged since the end of the last century, and while it is less than adequate for all proteins known today, it is a useful guide to their general properties. According to Osborne, who developed the classification, proteins are divisible into four classes in relation to their solubility: (1) albumins – soluble in water and dilute buffers at neutral pHs; (2) globulins – soluble in salt solutions but insoluble in water; (3) glutelins – soluble in dilute acids and alkalis; and (4) prolamins – soluble in 70–90% aqueous ethanol. The latter two protein types are found exclusively in cereal seeds, while the globulins are predominant in legumes and oats (Table 31.1).

Characteristically, storage proteins are oligomeric, the holoprotein being comprised of two to many subunits. The subunits may contain two to several polypeptide chains which are joined by hydrogen bonding, or by disulfide bonds between cysteine residues. The latter may be separated into component polypeptide chains by reducing agents such as β-mercaptoethanol, and the former by sodium dodecyl sulfate (SDS). The structure and composition of many of the major storage proteins is very complex, and only a few simplified examples will be outlined here: for more details consult Casey *et al*. (1986). The globulins of the legumes fall into two classes of holoprotein, separable on the basis of their sedimentation coefficients (S), determined by ultracentrifugation. The larger of the globulin types (11S or legumin proteins) has a MW of some 320 000 to 360 000, made up of six subunits which range in MW from 35 000 to 80 000, depending upon the species of seed. The subunits of legumin usually contain acidic and basic polypeptides in equimolar amounts. In pea and soybean most acidic polypeptides are approximately

Table 31.1 The percent protein composition of some cereals and legumes, with the common name of the protein in brackets

Species	Albumin	Globulin	Prolamin	Glutelin
Wheat	9	5	40 (gliadin)	46 (glutenin)
Maize	4	2	55 (zein)	39
Rice	5	10	5 (oryzin)	80 (oryzenin)
Oat	11	56	9 (avenin)	23
Pea	40	60 (vicilin, convicilin, legumin)	0	0
Soybean	30	70 (glycinin, β-conglycinin)	0	0

MW 40 000 in size and each is linked by a disulfide bond to a basic polypeptide of approximately MW 20 000. Because of variations in the amino acid composition of the constituent polypeptides, and variation in their association to form subunits, the storage legumins of any particular species can be quite heterogeneous. Similarly, the smaller of the globulin types (7S or vicilin proteins) is heterogeneous, although vicilins do not incorporate sulfur-containing amino acids and there is no disulfide bridging between subunits, only hydrogen bonding. The complexity of storage proteins is illustrated most dramatically in the prolamins of cereals. Zein, from maize, consists of two major and two minor classes of subunits, made up of combinations of about 30 polypeptides; in wheat there are present at least 46 polypeptides in gliadin.

In some seeds, the low MW globulins are the more important and predominant storage proteins, e.g. the 2–3S protein in rape seed. In others, albumins are present in large quantities (over 60% of the sunflower storage proteins). Some seeds contain unusual storage albumins (e.g. urease in jack bean) or those which are nutritionally undesirable (proteolytic inhibitors in legumes), toxic (ricin D in castor bean) or have interesting and spurious biological properties (lectins, many of which can agglutinate erythrocytes). A number of storage globulins and albumins are glycoproteins, i.e. they contain carbohydrate moieties such as mannose, glucose and N-acetylglucosamine (Murray, 1984a). Vicilins, but not legumins, are glycosylated, as are the lectins. Differences in the extent of glycosylation of the subunits which make up the storage vicilins also contribute to the variability in MW of the holoprotein.

Protein bodies

The subcellular repositories for the majority of storage proteins in reserve tissues of mature seeds are the protein bodies. These organelles are unit membrane-bound and commonly spherical in shape, having a diameter of 0.1–25 μm. Both size and the presence of inclusions, such as phytin-containing globoids or proteinaceous crystalloids (see Fig. 31.1), within the proteinaceous matrix of protein bodies are species and tissue dependent (Lott, 1980). In the following sections we will review the structure and formation of protein bodies in seeds of the legumes and cereals.

Storage protein synthesis

The biosynthesis of storage proteins in seeds occurs during the last two-thirds of their development, commencing after cell division and formation of the embryo body is completed, and ceasing during the latter stages of maturation drying (Higgins, 1984). The timing, rate, and extent of synthesis of the various storage proteins and their subunits in any seed may be different,

Nitrogen Metabolism

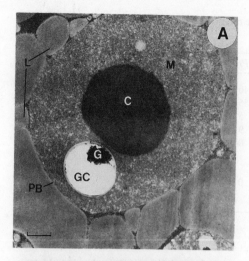

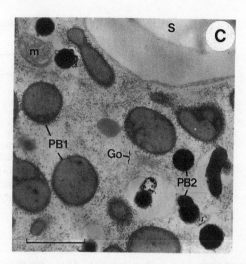

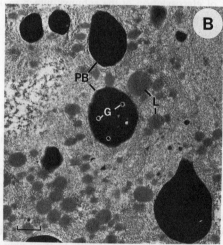

Fig. 31.1 Types of protein bodies found in storage parenchyma cells of seeds. (A) A protein body with protein crystalloid (C) and phytin-containing globoid (G) inclusions embedded in the amorphous proteinaceous matrix (M). A portion of the globoid has been lost due to sectioning (GC) (*Ricinus communis* endosperm). (B) Protein bodies typical of those found in mature legume seed cotyledon cells. Crystalloid inclusions are absent, numerous small globoids may be found embedded in the matrix protein, as illustrated here (*Medicago sativa* cotyledon). (C) Protein bodies of cereal grains. Spherical (PB1) and crystalline (PB2) protein bodies, as found in the cells of the starchy endosperm, are illustrated. Matrix protein in both types of protein bodies is usually free of inclusions (*Zea mays* starchy endosperm). L, lipid; m, mitochondrion; Go, Golgi; S, starch.

although these variables are genetically regulated, with the environment playing a modulating role (see later).

The accumulation of any protein is a reflection of the equilibrium established between its rate of synthesis and degradation. Since storage proteins in cotyledons and endosperms undergo little degradation during their deposition, their levels are determined almost exclusively by their rate of synthesis and processing. The latter plays only a minor regulatory role, and the quantitatively limiting role in the synthesis of any protein appears to be the amount of its mRNA which is present in the cells of the seed storage tissue. A quantitative analysis of mRNA levels during seed development, as illustrated in Fig. 31.2 for soybean cotyledons, shows that during the early stages the messages for storage albumins and globulins are in low amounts, rising to a peak at the mid-maturation stage, the time of maximum protein deposition. At this time there are in excess of 30 000 molecules of glycinin (legumin) mRNA per cell, constituting nearly 10% of the total message population. As the seed begins to mature, and dry out, the amount of all mRNAs declines, and storage protein synthesis ceases. For some proteins, the relationship between mRNA levels and protein synthesis is not so clear-cut. For example, in soybean the mRNA for the β-subunit of the vicilin β-conglycinin is present in the cells

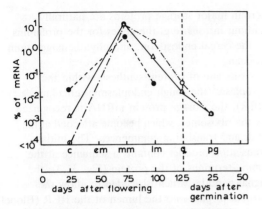

Fig. 31.2 Changes in the messenger RNA content for storage proteins in *Glycine max* (soybean) cotyledons during development and after germination. △, mRNA for α and α' subunits of the vicilin storage protein (β-conglycinin); ○, mRNA for the legumin storage protein (glycinin); ●, mRNA for trypsin inhibitor (albumin). Developmental stages represented are c, cotyledon; em, early maturation; lm, late maturation; q, quiescent; pg, post-germination. After Goldberg *et al.* (1981).

of the cotyledons for some 7–10 days prior to the appearance of the protein. This is suggestive of a control of synthesis of the β-subunit at the level of translation, although some synthesis might occur at earlier times, with the product being subjected to hydrolysis by proteolytic enzymes.

Different tissues within the same seed often accumulate different storage proteins, an observation which is indicative of tissue-specific gene regulation. The soybean axis, for example, produces very little of the legumin storage protein (and none of certain of its subunits), or of the subunit of β-conglycinin, both of which are present in the cotyledons. Moreover, an α-subunit (α') is present, which is not synthesized in the cotyledons. The reduced synthesis of the α and α'-subunits of β-conglycinin in the axis is related to the lower presence of their mRNAs, which could be due to reduced transcription, stability, or translatability. The mRNA for the β-subunit is 40-fold less in amount in the axis than in the cotyledons, although any subunit which is synthesized is immediately degraded by proteolysis. Hence, in the axis, both mRNA levels and protein turnover can contribute to the regulation of protein deposition. In some mutants of maize, which synthesize relatively minor amounts of zein, there are high levels of ribonuclease. These appear to control zein levels through the post-transcriptional degradation of its mRNA.

In cereals, the prolamins and glutelins of the starchy endosperm are never synthesized in the surrounding aleurone layer, which sequesters a unique set of proteins. Likewise, the embryo contains unique storage proteins, including certain lectins (e.g. wheat germ agglutinin), but these are never found in either the starchy endosperm or aleurone layer.

The regulation of expression of storage protein genes, both in a tissue-specific manner and in temporal manner during seed development is poorly understood at the present time. An attractive hypothesis, which is rapidly gaining support, is that 'upstream' of the gene for a particular storage protein is to be found certain regions which are involved in the regulation of gene expression (Gatehouse *et al.*, 1986; Goldberg, 1986). These so-called *cis*-acting control elements are activated by a specific *trans*-acting factor, a protein, which binds to the DNA and by interacting with the *cis*-element triggers transcription of the gene under their regulatory influence. Thus, each seed protein gene might contain a hierarchy of *cis*-control sequences that allows it to respond to developmental situations in a tissue-specific and temporal manner.

The production of a storage protein commences with transcription of its gene within the nucleus, and the resultant mRNA then is transported to the protein synthesizing site within the cytoplasm (Muntz, 1987). The gene for a particular protein may contain regions which are transcribed and are present as part of the mature mRNA (exons) and regions which are transcribed but are spliced out before maturation of the message (introns). This is illustrated in Fig. 31.3, which shows the synthesis and processing of a legumin (glycinin) subunit in soybean. The role of the introns in storage protein synthesis is unclear, since by using sophisticated molecular techniques, genes of legume storage proteins have been made intron-less and then inserted into other seeds, where they are transcribed normally. Moreover, the genes of

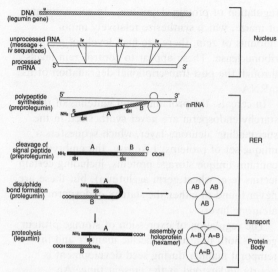

Fig. 31.3 A schematic representation of the synthesis of the legumin (glycinin) in soybean cotyledons. The gene for legumin possesses an upstream sequence (u) of several hundred nucleotides which controls its tissue-specific and temporal expression. The first transcriptional product is an unprocessed mRNA which contains non-translated intron regions (iv), about 1150 bases altogether, and which are spliced out to produce the mature processed legumin mRNA, which is 1455 bases long. (The mRNA is polyadenylated at the 3' end.) This mRNA is transported from the nucleus to the rough endoplasmic reticulum (RER) where it is translated. The mRNAs for the acidic (A) and basic (B) subunits are joined by codons for a 4-amino acid linker sequence (1) on the primary translation product (pre-prolegumin). The mRNA also contains a code for the signal peptide (s) at the amino-terminal end and a pentapeptide (c) at the carboxy-terminal end. The primary product (prolegumin) is processed to yield the A and B subunits joined by the linker sequence (1) and disulfide bonds (S–S); segments s and c are cleaved off. In the final step of processing the mature protein is formed as 1 is removed and the acidic and basic subunits are joined only by the disulfide bonding. The mature acidic subunit contains 278 amino acids (approx. 40 kD) and the basic subunit 180 amino acids (approx. 20 kD). Assembly of the subunits within the protein bodies yields the mature hexameric legumin holoprotein. Based on Krochko and Bewley, 1989.

certain major storage proteins are naturally without introns, e.g. the genes for the prolamins of the cereal endosperm, and phytohemagglutinin in bean.

At its site of protein synthesis within the cytoplasm (the rough endoplasmic reticulum, RER), the storage protein mRNA is recognized by the ribosomes, which become attached at the 5' end, and translation commences. The initial translation product contains a sequence at the amino-terminus of the protein (Fig. 31.3), the signal peptide, which is needed to allow the protein to pass into the lumen of the RER (Blobel et al., 1979). Once the protein has passed through a specific site within the membrane the signal peptide is enzymatically removed.

The subunits of glycinin, which are comprised of both an acidic and a basic polypeptide, are synthesized sequentially on the same mRNA template. The primary translation product (pre-prolegumin) contains the signal peptide at the amino-terminal end, the acidic and basic polypeptides joined by a linker sequence, and a short (penta-) peptide at the carboxy-terminus (Fig. 31.3). Legumins in other seed species are produced in a similar manner, although those in peas and broad bean contain neither the linker sequence nor the carboxy-terminal peptide. The next step in the processing of glycinin (in association with the RER) is the cleaving off of the peptides at both ends of the molecule, and the joining together of the SH groups of the cysteine residues to form a single sulfhydryl link. The result is the production of prolegumin (Fig. 31.3). The final stage in the maturation of the legumin, the separation of the acidic polypeptide from the basic, occurs in the protein body. For glycinin, this involves the removal of the four amino acid linker sequence by a thiol-containing proteinase. The separated acidic and basic polypeptides remain joined by the single disulfide bond as the subunit. The six subunits which comprise the legumin holoprotein become associated by surface charge interactions.

The synthesis of vicilin in legume storage tissues also requires a number of steps, some of which are different from those for legumin. The signal peptide is cleaved off cotranslationally on the ER, as for legumins, but other co- and post-translational modifications, the latter involving the Golgi apparatus, occur; these are dealt with in more detail in the following section. Within the protein body, at least some of the vicilin may be partially hydrolyzed ('nicked') at one or more points along the chain to yield shorter polypeptides (subfragments) which remain associated due to their surface charges.

Synthesis and processing of the major cereal storage proteins is generally less complex, for while the precursor form contains a signal peptide which is removed after passage into the lumen of the ER, there is no glycosylation and usually no post-translational hydrolysis (Shewry and Miflin, 1983). Rice glutelins appear to be an exception, for they are processed in a manner similar to the legumin-type globulins in legumes (i.e. cleavage of a 57 kD proglutelin to 22 and 38 kD polypeptides, that are then joined by disulfide bonds) with which there is considerable homology.

Protein body formation in legume seeds

Protein bodies in the storage parenchyma cells of legume seeds are generally 1–10 μm in diameter and are relatively homogeneous in structure, the proteinaceous matrix containing only a few small globoid inclusions (Lott, 1980). The formation of protein bodies in developing legume seeds has centered most recently around two topics: the origin of the organelles themselves, and the mechanisms and organelles involved in the synthesis, transport and ultimate sequestering of the storage proteins within the protein bodies. The development of legume protein bodies is outlined in Fig. 31.4.

At the onset of storage protein accumulation, deposits of protein collect against the luminal side of the tonoplast, the limiting membrane of the central vacuole. The membrane appears to evaginate around these deposits and virtually mature protein bodies are pinched off. With continued development, the vacuole itself divides and is gradually replaced by a large number (e.g. up to 150 000 per cell in *Pisum sativum*) of smaller vacuoles and protein bodies (Craig *et al.*, 1980).

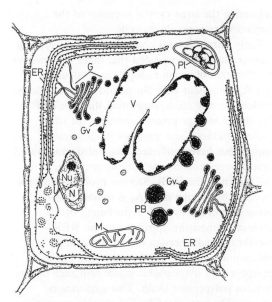

Fig. 31.4 Diagrammatic representation of the events involved in protein body formation in storage parenchyma cells of legume seeds. Storage proteins are synthesized on the rough endoplasmic reticulum and transported to the Golgi apparatus via tubular smooth ER connections. The storage proteins are sorted and packaged into Golgi-derived vesicles and transported to the vacuole/protein body compartment. Fusion of the vesicles with the tonoplast, illustrated on the left of the central vacuole, discharges the storage protein into the vacuolar lumen. Evagination of the tonoplast around concentrations of storage proteins, seen on the right of the vacuole, results in the formation of virtually mature protein bodies. Continued vacuolar subdivision gives rise to numerous protein bodies. ER, endoplasmic reticulum; G, Golgi apparatus; Gv, Golgi-derived vesicle; M, mitochondrion; N, nucleus; Nu, nucleolus; Pl, plastid; V, vacuole.

Filling of the smaller vacuoles with storage protein continues until the maturation–drying phase of seed development. In addition to the storage proteins (and in many seed tissues, phytic acid), protein bodies contain a number of hydrolases commonly found in the vacuoles of plant cells. Thus, the protein bodies must be considered as small vacuoles temporarily specialized for protein storage (Chrispeels, 1985). As will be discussed later, after germination of the seed the storage proteins are mobilized and the protein bodies fuse, reforming the large central vacuoles of the cotyledonary parenchyma cells.

Early ultrastructural observations of developing legume cotyledons revealed that there is a massive proliferation of rough endoplasmic reticulum (RER) coincident with the period of rapid storage protein accumulation, suggesting that the RER is the site of storage protein synthesis. *In vitro* protein synthesis and immunocytochemical approaches have confirmed this and demonstrated that storage proteins are initially sequestered in the lumen of the RER. Storage protein synthesis, directed by specific mRNA, is initiated on ribosomes in the cytosol. Attachment of the protein synthesizing apparatus to the RER membrane and insertion of the elongating polypeptide into the ER lumen is due to the presence of a hydrophobic signal sequence on the nascent polypeptide chain. This sequence is presumably recognized by ER specific receptors. Once through the ER membrane, the signal sequence is cotranslationally removed by a specific peptidase on the luminal side; elongation and insertion of the polypeptide chain into the lumen continues until the chain is completed (Higgins, 1984). A second cotranslational event, glycosylation, may occur if the proper amino acid sequence is present on the elongating chain (see later).

The site of storage protein synthesis (the RER) and sites of accumulation (the vacuolar/protein body compartment) are spatially separated within the legume storage parenchyma cells. Again, earlier ultrastructural studies implicated the involvement of the Golgi apparatus and Golgi-derived vesicles in the transport of storage proteins from the ER to the protein bodies (Dieckert and Dieckert, 1976) but, of course, it was impossible to demonstrate this dynamic process using a series of static micrographs. More recently, pulse-chase labeling and cell fractionation procedures have been used to confirm the involvement of the Golgi apparatus in the transport of storage proteins to the protein bodies. Enzymes responsible for the modification of oligosaccharide side-chains on glycosylated storage proteins reside in the Golgi apparatus. In addition, storage proteins have been identified in both the Golgi apparatuses and Golgi-derived vesicles of developing legume seed storage parenchyma cells (Chrispeels, 1985).

Several ultrastructural studies have demonstrated that direct tubular ER connections exist between the RER and Golgi apparatus cisternae. Transport of storage proteins to the Golgi apparatus most likely takes place via similar connections (Harris, 1986). The storage proteins are sorted and oligosaccharide side-chains are modified within the cisternae of the Golgi apparatus. The proteins are then packaged into vesicles for transport to the vacuolar/protein body compartment. Fusion of vesicular membranes with the tonoplast/protein body membrane discharges the storage proteins into the lumen of the vacuole or protein body. Concomitant with protein body formation, there is a 100-fold increase in the membrane surface area of the vacuolar/protein body compartment. The fusion of vesicles with the existing compartment membranes explains this increase.

Protein body formation in cereal grains

The structure of protein bodies in the cells of the major storage organ of cereal grains, the endosperm, is tissue specific and reflects differences in storage protein content. Protein bodies in the cells of the aleurone layer (as well as those in embryonic tissues) are commonly spherical, 2–5 μm in diameter, and often contain globoid, and occasionally crystalloid, inclusions. Cells of the starchy endosperm, however, contain as many as three structurally distinct types of protein bodies: large (1–3 μm) and small (0.1–1 μm diameter) spherical protein bodies and angular, 1–5 μm diameter 'crystalline' protein bodies. Globoids and crystalloids are absent from the proteinaceous matrix of these protein bodies.

There are two mechanisms of protein body formation in developing cereal grains. The first is most likely an adaptation to allow the sequestering of the highly insoluble prolamin storage proteins. The small and large spherical protein bodies of the starchy endosperm are the sites of prolamin accumulation, as evidenced by immunocytochemical localization. During development, these protein bodies are continuous with the RER and polysomes are attached to the outer surface of the protein body membrane. Isolated spherical protein bodies with attached polysomes are able to direct the synthesis of prolamin storage proteins *in vitro*. Thus the spherical protein bodies form as localized dilations of the RER (see Higgins, 1984).

In addition to the prolamins, cereal grains also sequester other storage proteins (glutelins or globulins depending on species) in the cells of the starchy endosperm. The 'crystalline' protein bodies are the sites of accumulation of the non-prolamin storage proteins. Recent ultrastructural and immunocytochemical evidence strongly suggests that their formation in cells of the starchy endosperm, as well as in the aleurone layer and embryonic tissues, occurs in the same manner as for protein bodies in the storage parenchyma cells of legume seeds. The non-prolamin storage proteins are synthesized on the RER, transported to the Golgi apparatus and, from there, to the protein bodies via Golgi-derived vesicles (Krishnan et al., 1986).

Co-translational glycosylation of proteins and post-translational modification of oligosaccharide side-chains

Many of the storage proteins and lectins in mature seeds are glycoproteins; one or more oligosaccharide side-chains are covalently linked to asparagine (Asn) residues in the constituent polypeptides. Examples include the vicilins of legume seeds (e.g. β-conglycinin of soybean) and lectins of common bean (phytohemagglutinin) and soybean (soyin). The oligosaccharide side-chains present on the storage proteins fall into two major categories: (1) simple or high-mannose oligosaccharides, composed exclusively of mannose (Man) and *N*-acetylglucosamine (GlcNAc) residues, usually in a 5–9:2 ratio, and (2) complex, or modified, oligosaccharides which, although often rich in Man, have sugar residues other than GlcNAc and Man. In seed proteins these additional sugar residues include fucose (Fuc), xylose (Xyl) and galactose (Gal) (Fig. 31.5).

Fig. 31.5 Probable structures of typical high-mannose and complex oligosaccharide side-chains of seed storage glycoproteins.

The initial glycosylation of storage protein polypeptides is a cotranslational event. High-mannose chains [$Glc_3Man_9(GlcNAc)_2$] are synthesized on carrier molecules of dolichol pyrophosphate (a lipid) via ER-associated glycosyltransferases and are then transferred *en bloc* to Asn residues occurring in the sequence –Asn–X–threonine (or serine) on nascent polypeptide chains. The glucose moieties and one to four Man are subsequently cleaved from the side-chain(s), leaving mature high-mannose chains attached to the polypeptide (Chrispeels, 1984).

Complex oligosaccharide side-chains on glycoproteins are derived from the high-Man chains following the covalent linkage to the polypeptide. Although there is considerable variation in the structure and sugar residues that comprise complex side-chains, depending on the glycoprotein, the mechanism of modification of these oligosaccharides commonly involves a somewhat perplexing series of steps where, sequentially, terminal Man residues are removed and other sugar residues are added.

The post-translational modification of high-mannose chains to complex chains, as it is postulated to occur for the storage glycoproteins phaseolin and phytohemagglutinin of *Phaseolus vulgaris* seed, is outlined in Fig. 31.6. The production of the modified oligosaccharide side-chains begins with the obligatory trimming of the $Man_9(GlcNAc)_2Asn$ to $Man_5(GlcNAc)_2Asn$ by the action of α-mannosidases, specific for α1, 2-linked Man (see Figs 31.5 and 31.6), located in

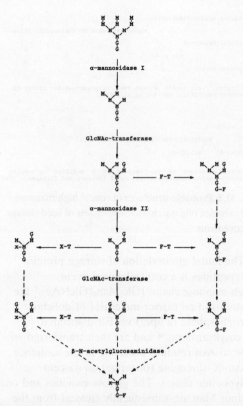

Fig. 31.6 Steps involved in the processing of certain asparagine-linked oligosaccharide side-chains of *Phaseolus vulgaris* storage proteins. Slashed arrows indicate probable, but as yet unconfirmed, steps in the processing. Dotted connections in the final conformation indicate that either xylose or fucose, or both, are found on the modified side-chain. M, mannose; G, *N*-acetylglucosamine; X, xylose; F, fucose; F-T, fucosyl transferase; X-T, xylosyl transferase. Modified from Johnson and Chrispeels, 1987.

the ER and Golgi apparatus. For further processing to occur, an additional GlcNAc must be transferred, by the action of a Golgi-localized GlcNAc transferase, to the terminal Man that is linked α1, 3 to the β-linked Man. In the case of the complex side-chain of phytohemagglutinin, Fuc may be added to the β1, N-linked core GlcNAc at any stage following the addition of this terminal GlcNAc. Further processing, involving the removal of two further terminal Man and addition of a second terminal GlcNAc, can occur with the Fuc attached (Fig. 31.6). However, addition of Xyl to the β-linked Man (occurring in both phytohemagglutinin and phaseolin) requires the prior removal of two additional terminal Man from the GlcNAcMan$_5$(GlcNAc)$_2$Asn (Fig. 31.6). It seems that the attachment of the second terminal GlcNAc can occur before or after the addition of Xyl. Addition of Fuc, Xyl and the terminal GlcNAc occurs in the Golgi apparatus prior to transport of the glycoproteins to the protein bodies. The two terminal GlcNAc are removed from the complex side-chains after the glycoproteins are deposited in the protein bodies (Johnson and Chrispeels, 1987).

Processing of high-Man side-chains to complex side-chains does not always occur. Some glycoproteins have only high-Man side-chains and, indeed, many glycosylated storage proteins may have both high-Man and complex side-chains on the same polypeptides. The lack of processing of high-Man side-chains appears to be a function of the accessibility of the oligosaccharides to the Golgi localized α-mannosidases and glycosyltransferases. If the high-Man side-chain is hidden within the tertiary structure of the polypeptide then it will not be modified (Faye *et al.*, 1986).

Finally, although we have a good understanding of the sequence of events involved in glycosylation and modification of glycoprotein oligosaccharide side-chains, the necessity for glycosylation is not at all clear. Tunicamycin, a drug that prevents protein glycosylation, has no effect upon subunit association, the binding ability of lectins or the transport and targeting of storage protein polypeptides to the protein bodies (Chrispeels, 1985).

Sorting and targeting of storage proteins to the protein bodies

One of the more interesting, but least understood, facets of protein transport in cells of developing and germinated seeds is targeting. How does a cell recognize that a particular protein, synthesized on and deposited within the lumen of the RER, is to be transported to the protein bodies? In the case of the prolamins in cereal grains, the process of targeting is fairly simple. The signal sequence on the nascent polypeptide chain promotes its insertion into the lumen of the RER, where it remains: since prolamin-containing protein bodies are derived directly from the RER, no further targeting mechanism is necessary. This is not so for the non-prolamin protein bodies in the starchy endosperm (or protein bodies in cells of the aleurone layer in cereals), nor for protein bodies in the storage parenchyma cells of dicot seeds. In all of these cases, sites of accumulation of storage protein are spatially separated from the site of synthesis. Transport from the RER to the protein bodies, to the best of our knowledge, is mediated by the Golgi apparatus.

In animal cells the Golgi apparatus is responsible for sorting and directing the transport of enzymes to the lysosomal compartments. Receptors in the membranes of the Golgi recognize a specific signal, a terminal Man phosphorylated in the sixth position, on the glycan side-chains of hydrolases destined for transport to lysosomes. If the terminal Man-6-P is not present, the proteins are transported to the plasmalemma via Golgi-derived vesicles and secreted. Although the vacuoles and protein bodies of seeds are considered to be lysosomal, the Man-6-P signal–receptor mechanism of protein targeting does not operate, for Man-6-P residues do not occur on storage proteins (Gaudreault and Beevers, 1984). Oligosaccharide side-chains cannot play any significant role in the targeting of proteins to the protein bodies either since many storage proteins, legumins being prime examples, are never glycosylated. Furthermore, tunicamycin, which inhibits glycosylation, does not prevent proper sequestering of the affected proteins. However, the possibility still remains that the Golgi apparatus plays a role in targeting proteins to protein bodies (Chrispeels, 1985).

Instead of the Man-6-P signal–receptor mechanism, targeting of proteins to the vacuole/protein body compartment in plant cells may depend on common regions of primary sequence homology or on common areas of secondary or tertiary structure of the polypeptides. Studies of protein targeting in yeast cells have indicated that the first 30 N-terminal amino acids following the signal peptide are important for correct transport of a protease to the vacuole (Johnson et al., 1987). More recently, Tague and Chrispeels have found that a small post-signal sequence region of the N-terminus of PHA, the bean seed storage lectin, was recognized as the vacuolar targeting signal by transformed yeast cells. Fusion of the region of the PHA gene coding for the targeting signal to a truncated yeast invertase gene caused invertase, which is normally secreted, to be transported to the vacuole in the transformed yeast cells. The lack of homology in the amino acids in the targeting region of PHA and yeast protease suggests that the secondary or tertiary structure, rather than the primary structure, of the N-terminus of the polypeptide is important for targeting. It has been demonstrated that the bean storage protein, phaseolin, is properly targeted to the protein bodies in transgenic tobacco seeds (Greenwood and Chrispeels, 1985). This, and the recent data of Tague and Chrispeels, indicates that the mechanism for targeting proteins to the vacuole/protein body compartment is similar between unrelated species.

An environmental effect on protein deposition

Environmental factors such as temperature and plant nutrition can modulate storage protein accumulation during development. Increased nitrogen supply may or may not increase storage protein synthesis, depending upon the species or time of application during development. The most dramatic effects of nutritional factors have been observed in developing legume and cereal seeds deprived of sulfur, potassium or phosphorus. In relation to the first of these elements, the synthesis of proteins which are normally rich in S-containing amino acids is severely affected by sulfur deficiency, as illustrated for the legumin and low MW albumin storage proteins in pea (Higgins, 1984). In sulfur-deficient pea seeds, total protein is reduced by 20%, but legumin declines almost completely and the total albumin fraction is reduced by 35%. In contrast, the relative level of vicilin, which does not contain S amino acids, increases by some 40%. The reduced accumulation of legumin is not caused by increased degradation, but by reduced synthesis of the protein, which can be directly correlated with a reduction in the level of its mRNA; vicilin mRNA levels are somewhat enhanced under the same S-deficient conditions. Studies on the mRNA of the low MW albumin have shown that the reduction in its level in the cotyledons of developing seeds low in S is not due to the reduced transcription, but rather due to its degradation, presumably in the cytoplasm after release from the nucleus. Hence, the control exerted by S-deficiency is at the post-transcriptional level of organization.

The switch from protein deposition to protein mobilization

One of the striking differences between the metabolic events occurring during seed development and those during post-germinative seedling growth is that during the former there is an extensive deposition of reserves, and during the latter, their mobilization. What then, causes this switch in direction of metabolism from an anabolic to a catabolic mode? The terminal event in seed development is maturation drying, during which protein synthesis gradually declines, the ribosomes become dissociated from the mRNAs, and the latter is degraded so as to be present in only low amounts in the dry seed (Fig. 31.2). Upon subsequent rehydration of the seed, as germination proceeds, the residual mRNAs for storage proteins continue to be degraded; in addition, their synthesis is completely suppressed. A new set of mRNAs do arise specifically after drying, however, which include proteins essential for germination and for the mobilization of the stored reserves, including proteolytic enzymes. Thus drying is an important controlling event for suppressing the synthesis of developmentally and related proteins, and activating the synthesis of those essential for germination and growth (Kermode and Bewley, 1986). The mode of action of drying in causing the switch is not known.

Mobilization of stored protein reserves

The hydrolysis of storage proteins to their constituent amino acids requires the presence of a series of proteinases each with their own substrate specificity, some broad ranging and some narrow (Shutov and Vaintraub, 1987). It is likely that the storage tissues of germinated seeds contain several to many proteinases, but the full range of these enzymes in any one seed remains to be elucidated. Initially, the large oligomeric storage proteins, which are frequently insoluble, are subjected to limit proteolysis within the protein bodies by endopeptidases to produce soluble polypeptide fragments. These, in turn, are subjected to further degradation by exopeptidases (mostly carboxypeptidases, which sequentially cleave the terminal amino acid from the carboxy-terminal end of the polypeptide chain) and endopeptidases which expose new carboxy-terminal sequences for exopeptidase attack. The final stages of degradation, from small oligopeptides to amino acids may occur outside of the protein bodies, and involve aminopeptidases (exopeptidases) and peptide hydrolases, e.g. dipeptidases (Muntz et al., 1985). For those proteins stored within the cytoplasm, the whole enzyme complement is presumably present therein. The liberated amino acids may be reutilized for protein synthesis or be deaminated to provide carbon skeletons for respiration. The sequence of events can be outlined as follows:

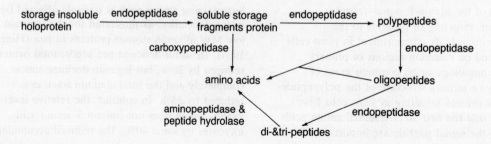

Mobilization in cereals

Reserve proteins are stored in two distinct regions in the cereal grain: in the aleurone grains of the aleurone layer (up to 30% of total seed protein) and in the protein bodies (or remnants thereof, e.g. barley and wheat) of the starchy endosperm. Mobilization of the aleurone layer proteins is the result of the activity of *de novo* synthesized proteinases, mostly unidentified, which arise within the aleurone layer itself, sometimes in response to hormonal (gibberellic acid) stimulation. The amino acids produced may be recycled to produce more hydrolytic enzymes, which are released into the non-living starchy endosperm. Since protein hydrolysis and protein (enzyme) synthesis occur concurrently in the same cells of the aleurone layer, it stands to reason that the two processes must be separated. Hydrolysis of the proteins occurs within the aleurone grains (protein bodies) and the resultant peptides and amino acids are released to the protein synthetic sites in the cytoplasm. Excess amino acids diffuse into the embryonic axis for use by the developing seedling. The cells of the starchy endosperm are non-living at maturity, and hence the supply of enzymes for protein mobilization comes either from the aleurone layer, following *de novo* synthesis, or by activation of enzymes already present within the dry seed. This activation may be simply by hydration of the enzymes due to imbibition, or be due to their hydrolytic activation from a latent or sequestered precursor within the starchy endosperm.

The catabolites of protein hydrolysis, i.e. the amino acids, dipeptides and small oligopeptides, are taken up into the growing embryonic axis via the scutellum, the absorptive region of the embryo which lies against the starchy endosperm. The uptake mechanisms are active, and selective, and some amino acids can enter the embryo more readily than others. Within the growing axis they are utilized for the resynthesis of proteins essential for sustained growth.

Mobilization in legumes

Cotyledons of mature dry legume seeds contain few proteolytic enzymes, but their levels increase markedly soon after germination is completed. Not

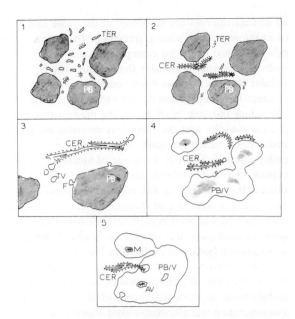

Fig. 31.7 A diagrammatic representation of part of a cell from a cotyledon of *Vigna radiata* (mung bean) to illustrate changes undergone by the protein body and ER during reserve hydrolysis and cell autolysis. (1) Dry state and during germination. Intact protein bodies and tubular endoplasmic reticulum (TER) present with few ribosomes attached. (2) Starting about the time germination is completed, the TER is dismantled and cisternal endoplasmic reticulum (CER) synthesis commences, to which ribosomes become attached. No change to protein bodies. (3) Three to five days after start of imbibition, when seedlings are growing. Vicilin peptidohydrolase is synthesized on polyribosomes attached to CER and inserted into the lumen. Dilations (D) of the cisternae form which contain the enzyme; these break off as transport vesicles (TV) and carry the peptidohydrolase to the protein bodies, with which they fuse (F). The protein commences hydrolysis of the vicilin. Other hydrolytic enzymes (e.g. ribonuclease, phosphatases) are targeted to the protein body and inserted. (4) As storage protein is hydrolyzed the protein bodies coalesce to form larger vacuoles (PB/V) and other hydrolytic enzymes are inserted. (5) Autophagic vacuoles (AV) are formed, engulfing cell contents such as the CER and mitochondria (M). More protein bodies fuse to form a large central vacuole, containing autolytic enzymes. Based on studies of Chrispeels and coworkers. See Chrispeels and Jones, 1980/81 and Herman *et al.*, 1981.

all storage reserves are mobilized at the same rate, or at the same time after germination. In garden pea and broad bean, for example, legumin is hydrolyzed earlier and faster than vicilin. The cellular changes associated with proteolysis in legumes have been studied most extensively in the cotyledons of mung bean. Here, the major storage protein is vicilin and the enzyme responsible for its hydrolysis is vicilin peptidohydrolase, an endopeptidase. The ER is the site of vicilin peptidohydrolase synthesis, and associated with this event is a change in ER type from tubular to cisternal. The sequence of changes the cell undergoes during protein mobilization, with the eventual formation of a vacuole by fusion of the spent protein bodies, is outlined in Fig. 31.7. Hydrolytic enzymes other than peptidohydrolase are inserted into the protein bodies during protein mobilization, including enzymes such as α-mannosidase and glucosaminidase which remove the glycosidic components from the vicilin glycoprotein. An array of non-proteolytic enzymes are also acquired by the vacuoles following their synthesis in the cytoplasm, including ribonuclease, acid phosphatase, phosphodiesterase and phospholipase D. Thus the vacuole assumes lysosome-like properties and is ultimately involved in the total autolysis and death of the storage cells. The products of hydrolysis, including the amino acids, are redistributed to the growing seedling via the vascular system which is connected to the cotyledons. Since the favored form of transportation of amino acid nitrogen is in the form of glutamine and asparagine, enzymes which catalyze their synthesis frequently arise in storage tissues at the time of storage protein mobilization.

References

Bewley, J. D. and Black, M. (1985). *Seeds. Physiology of Development and Germination*, Plenum Press, New York.

Blobel, G., Walker, P., Chang, C. N., Goldman, B. M., Erickson, A. H. and Lingappa, V. R. (1979). Translocation of proteins across membranes: The signal hypothesis and beyond. In *Secretory Mechanisms*, The Society of Experimental Biology Symposium 33, eds C. R. Hopkins and C. J. Duncan, Cambridge University Press, Cambridge, pp. 9–36.

Casey, R., Domoney, C. and Ellis, N. (1986). Legume storage proteins and their genes. *Oxford Surv. Plant Mol. Cell Biol.* **3**, 1–95.

Chrispeels, M. J. (1984). Biosynthesis, processing and transport of storage proteins and lectins in cotyledons of developing legume seeds. *Phil. Trans. Royal Soc. London B* **304**, 309–22.

Chrispeels, M. J. (1985). The role of the Golgi apparatus in the transport and post-translational modification of vacuolar (protein body) proteins. *Oxford Surv. Plant Mol. Cell Biol.* **2**, 43–68.

Chrispeels, M. J. and Jones, R. L. (1980/81). The role of the endoplasmic reticulum in the mobilization of reserve macromolecules during seedling growth. *Israel J. Bot.* **29**, 225–45.

Craig. S., Goodchild, D. J. and Miller, C. (1980). Structural aspects of protein accumulation in developing pea cotyledons. II. Three-dimensional reconstructions of vacuoles and protein bodies from serial sections. *Aust. J. Plant Physiol.* **7**, 329–37.

Dieckert, J. W. and Dieckert, M. C. (1976). The chemistry and cell biology of the vacuolar proteins of seeds. *J. Food Science* **41**, 475–82.

Dorland, L., Van Halbeek, H., Vliegenthart, J. F. G., Lis, H. and Sharon, N. (1981). Primary structure of the carbohydrate chain of soybean. A reinvestigation by high resolution ^{1}H NMR spectroscopy. *J. Biol. Chem.* **256**, 7708–11.

Faye, L., Johnson, K. D. and Chrispeels, M. J. (1986). Oligosaccharide side chains of glycoproteins that remain in the high-mannose form are not accessible to glycosidases. *Plant Physiol.* **81**, 206–11.

Gatehouse, J. A., Evans, I. M., Croy, R. R. D. and Boulter, D. (1986). Differential expression of genes during legume seed development. *Phil. Trans. Royal Soc. London. B* **314**, 367–84.

Gaudreault, P. R. and Beevers, L. (1984). Protein bodies and vacuoles as lysosomes. *Plant Physiol.* **76**, 228–32.

Goldberg, R. B. (1986). Regulators of plant gene expression. *Phil. Trans. Royal Soc. London. B* **314**, 343–53.

Goldberg, R. B., Hoschek, G., Ditta, G. S. and Breidenbach, R. W. (1981). Developmental regulation of cloned superabundant mRNA in soybean. *Dev. Biol.* **83**, 218–31.

Greenwood, J. S. and Chrispeels, M. J. (1985). Correct targeting of the bean storage protein phaseolin in the seeds of transformed tobacco. *Plant Physiol.* **79**, 65–71.

Harris, N. (1986). Organization of the endomembrane system. *Ann. Rev. Plant Physiol.* **37**, 73–92.

Herman, E. M., Baumgartner, B. and Chrispeels, M. J. (1981). Uptake and apparent digestion of cytoplasmic organelles by protein bodies (protein storage vacuoles) in mung bean cotyledons. *Eur. J. Cell Biol.* **24**, 226–35.

Higgins, T. J. V. (1984). Synthesis and regulation of major proteins in seeds. *Ann. Rev. Plant Physiol.* **35**, 191–221.

Johnson, K. D. and Chrispeels, M. J. (1987). Substrate specificities of *N*-acetylglucosaminyl-, fucosyl-, and xylosyltransferases that modify glycoproteins in the Golgi apparatus of bean cotyledons. *Plant Physiol.* **84**, 1301–8.

Johnson, L. M., Bankaitis, V. A. and Emr, S. A. (1987). Distinct sequence determinants direct intracellular sorting and modification of a yeast vacuolar protease. *Cell* **48**, 875–85.

Kermode, A. R. and Bewley, J. D. (1986). Alteration of genetically related syntheses in seeds by desiccation. In *Membranes, Metabolism and Dry Organisms*, ed. A. C. Leopold, Cornell University Press, Ithaca, pp. 59–84.

Krishnan, H. B., Franceschi, V. R. and Okita, T. W. (1986). Immunochemical studies on the role of the Golgi complex in protein body formation in rice seeds. *Planta* **169**, 471–80.

Krochko, J. E. and Bewley, J. D. (1989). Use of electrophoretic techniques in determining the composition of seed storage proteins. *Electrophoresis* **9**, 751–63.

Lott, J. N. A. (1980). Protein bodies. In *The Biochemistry of Plants*, Vol. 1, ed. N. E. Tolbert, Academic Press, New York, pp. 589–623.

Muntz, K. (1987). Developmental control of storage protein formation and its modulation by some internal and external factors during embryogenesis in plant seeds. *Biochem. Physiol. Pflanzen* **182**, 93–116.

Muntz, K., Bassuner, R., Lichtenfeld, C., Scholz, G. and Weber, E. (1985). Proteolytic cleavage of storage proteins during embryogenesis and germination of legume seeds. *Physiol. Veg.* **23**, 75–94.

Murray, D. R. (1984a). Accumulation of seed reserves of nitrogen. In *Seed Physiology. I. Development*, ed. D. R. Murray, Academic Press, Sydney, pp. 83–137.

Murray, D. R. (ed) (1984b). *Seed Physiology. II. Germination and Reserve Mobilization.* Academic Press, Sydney.

Shewry, P. R. and Miflin, B. J. (1983). Characterization and synthesis of barley seed proteins. In *Seed Proteins*, eds W. Gottschalk and H. P. Muller, Martinus Nijhoff/Junk, The Hague, pp. 143–205.

Shutov, A. D. and Vaintraub, I. A. (1987). Degradation of storage proteins in germinating seeds. *Phytochemistry* **26**, 1557–66.

Prospects for Plant Improvement

IX

Prospects for Plant Improvement

32 Fundamentals of gene transfer in plants
Brian L. A. Miki and V. N. Iyer

Introduction

The ability has been developed to isolate and clone genes from plant cells or their organelles, to experimentally alter them and to assess the effect of such manipulation *in vitro*. However, these technologies will be incomplete unless they are complemented by dependable techniques that will enable these manipulated genes to be introduced back into the organism. It is by this method that the experimental alterations to the gene can best be correlated with specific biochemical and physiological responses in the different cells of the whole plant, which is even more important in cases where the gene to be installed in the plant is derived from a foreign source. In this chapter, several methods that have been developed recently for such transfers and their relevant advantages will be described and compared. The present limitations and promise will also be discussed.

Initially, the growth and differentiation of plant cells in culture and the use of marker genes to identify transformed tissues will be discussed. It is essential to have genes which can be used for the selection of those few cells in a population that have been genetically transformed. This initial discussion is followed by sections describing transfer vectors and modes of transfer. The chapter concludes with sections considering the demonstrated or potential utility of these techniques for increasing our understanding of plant gene regulation and in obtaining transgenic plants with potential for the agricultural industry.

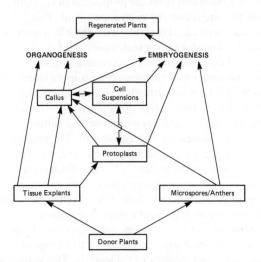

Fig. 32.1 Potential pathways for the growth and differentiation of plant cells and tissues in culture. The regeneration of plantlets in culture may occur by organogenesis or embryogenesis. This diagram illustrates some of the manipulations that have been performed with plant cells in culture. The experimental capabilities that exist for a given plant species vary greatly.

Plant cell and tissue culture

A great number of experimental opportunities exist for the genetic manipulation of plant cells and tissues in culture. Plants are unique among higher organisms in having the capacity to regenerate whole organisms from a variety of cell types as well as from a fertilized ovum. Genetic transformation or the successful introduction and integration of exogenous DNA into these cells can give rise to transgenic plants. Figure 32.1 illustrates potential pathways for growth and differentiation of plant cells and tissues in culture. The regeneration of plantlets can occur by two different processes. Organogenesis usually involves the differentiation of shoots followed by roots. Embryogenesis in culture follows stages of maturation similar to zygotic embryos in seeds but originates from different cells. Culture conditions have been determined for most of these steps for species such as tobacco, which is frequently used as an experimental plant and as a model system for studying organogenesis in culture. Similar manipulations can be performed with some crop species such as alfalfa or rapeseed, but with lower efficiencies. However, these latter two species are excellent for studying the process of embryogenesis in culture. The regeneration of cereals from protoplasts is difficult, but it has recently been achieved with rice suggesting that it is possible with monocotyledons. Cells at any of the stages shown in Fig. 32.1 can be used as recipients for transformation with isolated DNA and successful transformation of most of them has been demonstrated using a variety of DNA transfer techniques.

Marker genes

A critical step in the development and evaluation of transformation strategies for plants was the construction of vectors with genes that could act as dominant selectable markers. Under various selection pressures, these genes provide a growth advantage to cells that integrate the vectors and express the marker gene. An example is *nptII* from the bacterial transposon Tn5 which codes for the enzyme neomycin phosphotransferase. This enzyme catalyzes the transfer of a phosphate moiety from ATP to a number of aminoglycoside antibiotics, including kanamycin (Km), thereby detoxifying them. In the presence of kanamycin, transformed cells can grow and differentiate into plantlets; however, normal cells and tissues of most plant species cannot. Among recipient cells, transformation events may occur at very low frequencies such as one in a hundred thousand cells. Without an efficient selection system, the transformants would be difficult to detect and could not be recovered or separated from the

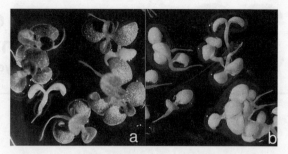

Fig. 32.2 Segregation of kanamycin resistance among seedlings of transgenic (a) and normal (b) tobacco plants. Seedlings were germinated on media with 200 μg ml^{-1} kanamycin. Under these conditions seedlings of normal plants (b) were white and eventually died. In self-pollinated transgenic plants, kanamycin resistance segregated as a single dominant marker, i.e. approximately three green resistant seedlings were recovered for every white sensitive seedling. (Photographs courtesy of Jérôme Gabard and Pierre Charest, Agriculture Canada and Carleton University.)

untransformed cells. The transformation frequencies, or proportion of cells that are transformed, may be roughly estimated from the proportion of resistant cells that are recovered in an experiment. This provides a standard for comparing the efficiencies of different transformation techniques and for manipulating conditions to optimize a system. Genetic analyses of transgenic plants using selectable markers is relatively straightforward. Figure 32.2 illustrates the segregation of kanamycin resistance among tobacco seedlings germinated on culture media supplemented with kanamycin. The ratio of resistant, i.e. green seedlings to sensitive, i.e. white seedlings is approximately 3:1 in this illustration. This is expected for the segregation of a dominant marker among progeny derived from self-fertilization of a transgenic plant with a single integration.

A number of selectable marker genes have been developed which provide resistance to a range of antibiotics and chemicals (Table 32.1). This variety is important because plant species and particular tissues of a plant may differ widely in their sensitivities to different selective agents. Furthermore, the choice of several markers increases the range of genetic manipulations and analyses that can be performed with transgenic plants.

Most of the selectable markers are not of plant

Table 32.1 Examples of selectable markers

Chimeric gene	Source	Selective agent	
nos-nptII-nos	Tn5	kanamycin	Herrera-Estrella et al., 1983
ocs-aphIV-nos	E. coli	hygromycin	Waldron et al., 1985
35S-ble-nos	Tn5	bleomycin	Hille et al., 1986
35S-dhfr-nos	mouse	methotrexate	Eichholtz et al., 1987

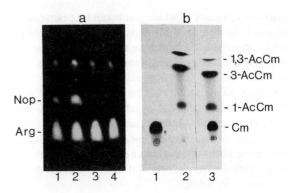

Fig. 32.3 (a) Detection of nopaline synthase activity in broccoli crown gall tumor tissues transformed with nopaline Ti plasmid C58. Extracts from transformed tissues in culture revealed small amounts of nopaline (lane 1) whereas normal tissues did not (lane 4). Following standard assay protocols, nopaline was synthesized in cell-free extracts of transformed tissues (lane 2) but not with normal tissues (lane 3). Nopaline (Nop) was separated from the substrate arginine (Arg) by paper electrophoresis and detected by staining of guanidines with a fluorescent dye. (Reproduced from Holbrook et al., 1986 with permission.) (b) Detection of chloramphenicol acetyltransferase (cat) activity in transgenic tobacco. [^{14}C]-chloramphenicol (Cm) was acetylated (Ac) in cell-free extracts; separated into 1-AcCm, 3-AcCm and 1,3-AcCm forms by thin-layer chromatography; and detected by autoradiography. Cat activity can be detected in transgenic tobacco (lane 2) but not in normal tobacco (lane 1). High levels of cat activity are normally present in *Brassica* species (lane 3). (Photograph courtesy of Pierre Charest, Carleton University.)

origin. To achieve constitutive expression in culture and in the various tissues of the plants, the coding regions of the genes for the marker have been fused to promoters or other regulatory sequences known to function in plants such as those from the nopaline synthase (*nos*) or octopine synthase (*ocs*) genes of the Ti plasmids and from the cauliflower mosaic virus (CaMV) 35S or 19S transcript. The 35S promoter ensures high levels of transcription. Although various 3' termination signals have been used, that of *nos* is the most common.

Reporter genes are genes that cannot be used for selection but that code for enzymes that can be easily detected with great sensitivity. These genes have been important for monitoring transformation or for the detection of promoter activity. The opine synthase genes of the Ti plasmids, *nos* and *ocs*, and the chloramphenicol acetyltransferase (*cat*) gene from bacteria are good examples of reporter genes. *Ocs* and *nos* catalyze the reductive condensation of arginine with α-ketoglutarate and pyruvate, respectively. These reactions can be performed with cell-free extracts of transformed tissues and the opines detected by staining or fluorography after paper electrophoresis (Fig. 32.3a). In many early studies, this method provided clear evidence for transformation with Ti plasmids of *Agrobacterium tumefaciens* (Kemp, 1982). This provides a generally useful detection system; however, comigrating compounds and loss of activity may interfere with analyses of samples. The acetylation of chloramphenicol catalyzed by cat can be detected by autoradiography (Fig. 32.3b). This system is valuable for monitoring transient expression and the regulation of transcription by promoters in transgenic plants (Hauptmann et al., 1987). The main problem is that some species such as those of the genus *Brassica*, have a very high level of endogenous plant cat activity (Fig. 32.3b) and also have inhibitors of bacterial cat which obscure the activity derived from the transformed bacterial gene.

Recently, more versatile reporter genes have been developed. The firefly luciferase gene (Ow et al., 1986) and the *E. coli* β-glucuronidase gene (*uid A* locus; Jefferson et al., 1987), in particular, allow great sensitivity of detection. Luciferase catalyzes the oxidative decarboxylation of luciferin which results in the emission of light which can be detected in cell-free extracts by luminometry. Alternatively, the activities in tissues or whole plants can be examined by exposing them to X-ray films (autoluminography). β-Glucuronidase is a hydrolase that cleaves a variety of β-glucuronides including fluorogenic substrates that can be detected by fluorometry. β-Glucuronidase activity can be detected histochemically in cells using substrates such as X-glu (5-bromo-4-chloro-3-indolyl glucuronide). The histochemical detection of gene expression in transgenic plants offers greater precision in identifying the specific cells and tissues in which the promoters are active.

Agrobacterium-mediated gene transfer

Bacteria of the genus *Agrobacterium* are free-living but opportunistic soil bacteria that have evolved the unique capacity to interact genetically with susceptible plants. This interaction results in the stable insertion of part of the genome of the bacterium into the genome of the plant. It is this natural ability of the bacterium to genetically transform the plant tissue that is the basis of all gene transfer technology that exploits this system. However, our current understanding of the details of this transfer process that integrates the gene is far from complete. Nevertheless, sufficient is known to encourage speculation about the potential and to permit some initial exploitation. The molecular biology of the *Agrobacterium*–plant interaction and its potential have been reviewed (e.g. Nester *et al.*, 1984; Gheyson *et al.*, 1985; Rogers and Klee, 1987).

Natural history and molecular basis

The definition of 'species' within the genus *Agrobacterium* is a matter of convenience and has no phylogenetic implications. All isolates have been recognized on the basis of the symptoms they produce on susceptible plants. Thus, *A. tumefaciens* induces tumors and *A. rhizogenes* roots at the site of infection on susceptible plants. These symptoms do not reflect basic differences in the gene transfer process, but rather the genetic determinants contained in the DNA segment that is transferred to the plant. This DNA segment, called the T-DNA (Transferred DNA), is initially present in the bacterium as part of a bacterial plasmid which is transmissible by inter-bacterial mating and called the Ti (Tumor inducing) plasmid. Thus it is feasible to interconvert one 'species' of *Agrobacterium* into another simply by substituting the Ti plasmid.

Figure 32.4 is a diagrammatic summary of what is known about the infection and transformation process as it is thought to occur in nature. Infection is initiated by accidental wounding of plant tissue. The wound site releases a mixture of several compounds, the nature of which varies between plants, and which are often phenolic in nature; e.g. 4-acetyl-2,6-dimethoxyphenol (acetosyringone) and α-hydroxyacetosyringone from tobacco wounds. These compounds diffuse from the wound, and, at low concentrations, some of them exert positive chemotaxis on the bacterium (Fig. 32.4). Chemotaxis is presumably one factor that contributes to the colonization of the wound site. There are several other factors that may also contribute to this colonization. They include (1) the extracellular formation of cellulose microfibrils by the bacteria upon contact or close exposure with plant cells, and (2) the synthesis and release by the transformed plant tissue of a class of amino acid-derived metabolites called the opines. These can be selectively or preferentially catabolized by the infecting bacterium and can also induce the inter-bacterial mating process. The molecular genetics of these different pathways to wound colonization have yet to be examined in detail. In laboratory experiments colonization steps can apparently be bypassed by artificially infecting wound sites. The colonization pathways are therefore probably important in nature but not essential to the transformation process. One role of wound exudate compounds is to induce certain bacterial genes, the expression of which is essential for T-DNA transfer.

The process of T-DNA transfer from the Ti plasmid to the plant nuclear DNA has been difficult to examine directly. Nevertheless, indirect approaches have led to some useful working hypotheses. There are only two regions on the Ti plasmid that are essential to this transmission, the T-DNA region and another region called the *vir* (virulence) region. Although these two regions are linked in all native plasmids, they will also function

Fundamentals of gene transfer in plants 477

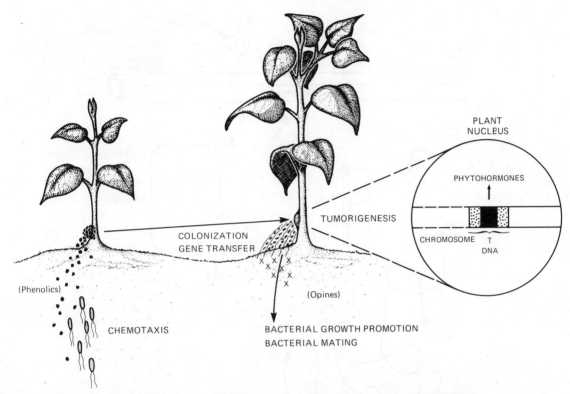

Fig. 32.4 Outline of events in the genetic transformation of susceptible plant tissue by *Agrobacterium*.

Phenolic compounds diffusing from a plant wound site serve as chemotactic attractants for *Agrobacterium*. At higher concentrations, such compounds also induce the expression of several bacterial genes some of which are essential for plant cell transformation (see Fig. 32.5).

Bacteria attach to and colonize cells in the wound site. This colonization is aided by the induction of a mesh of cellulose fibrils by the colonizing bacteria.

A region of the Ti plasmid DNA defined and bordered by 25 bp directly repeated sequences is transferred and inserted into the plant nuclear genome. The molecular intermediates and mechanisms of these processes are not yet firmly established.

Genes within the T-DNA are expressed by the recipient plant cells including (1) genes for the synthesis of rare amino acids (opines) that are a second colonizing factor as they promote both selective bacterial growth and inter-bacterial conjugation, and (2) genes for novel and unregulated pathways of biosynthesis of auxin and/or cytokinin that results in the localized tumors. Not shown in this diagram is that substantial parts of the T-DNA (except one or both borders) can be deleted and replaced by other desired and functional genes and that regenerant transgenic plants can be obtained from transformed tissue.

when they are separated. Most of the T-DNA region can be deleted or inactivated without affecting its transmission; the only parts of the T-DNA that are essential for this to occur are 25 bp of directly repeated sequences that border the T-DNA region. The *vir* region has not been found in transformed plant cells. It can function in the bacterium when it is *in trans* to the T-DNA border regions. This suggests that *vir* genes and their products function primarily or exclusively in the bacterium. One or both T-DNA border regions provide the sites for the recognition of gene product(s) from the *vir* region. In addition, sites (called overdrive) outside the border regions may enhance T-DNA transmission in some cases. In

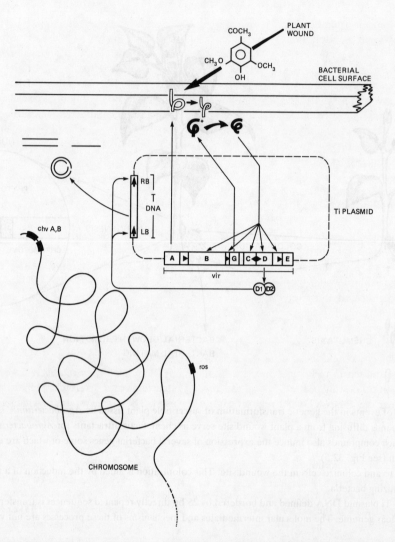

Fig. 32.5 A diagram that illustrates relevant genes and events that occur in the *Agrobacterium* during the transformation process.

There are three genetic regions on the chromosome that have been found to be necessary or helpful. Of these, *chvA* and *chvB* are needed for stable binding to plant cells and *ros* influences the efficiency of expression of some of the genes in the *vir* region.

The *vir*A and *vir*G genes and their products constitute a two-component system that allows *Agrobacterium* to respond to plant wound exudate via VirA (which is a transmembrane protein) and VirG which when modified is a positive regulator of the other *vir* genes including *virD*.

Part of the VirD1 protein is able to nick DNA specifically at one or both the 25 bp directly-repeated borders of T-DNA. This has resulted either directly or indirectly in the three forms of DNA shown in the figure and in the speculation that the nick site at one of these borders provides a leading end for DNA transfer into the plant cell. A molecular understanding of the functions of the other *vir* genes is not yet known.

artificial circumstances, when both border regions are deleted, functional pseudoborders have been revealed.

Genes in the *vir* region must play a major role in T-DNA transmission. Recent analyses of their structure and regulation have provided useful insights, but do not as yet provide definitive information on their role in the transfer process. In large part, this results from the limitation that, experimentally, only events in the bacterium can be analyzed. These events are summarized in Fig. 32.5. Transcription of most of the known loci of the *vir* region is coordinately and positively regulated by the products of the autoregulated *vir*A and *vir*G genes and the phenolic compounds present in plant wound exudates. It is likely that *vir*A and *vir*G together constitute a regulatory unit, with *vir*A being responsive to the wound exudate compounds at concentrations higher than those required for chemotaxis, and in turn interacting with the *vir*G product to activate the transcription of the remaining *vir* loci including *vir*D. The diversity and mode of action of inducing compounds is an active area of research. Gene products of *vir*D have been shown to be involved in the processing of the DNA of the Ti plasmid at or near the T-DNA border regions while the plasmid is in the bacterium. Several different kinds of plasmid DNA structures have been detected as a result of this processing. These include single-stranded T-DNA and double-stranded linear or circular T-DNA, the formation of which are all dependent on the function of *vir*D. They are derived from DNA strand-nicks at the T-DNA border sequence. It seems unlikely that all such structures are formed in a single pathway that is causally related to T-DNA transfer. Several models have been considered and summarized recently by Koukolikova-Nicola *et al.* (1987). The form of molecules actually transferred to plant cells remains to be conclusively identified. At present a model of single strand T-DNA transfer initiated by a single-strand nick at a border is attractive. However, the model is derived largely by analogy to events that occur in the donor during the inter-bacterial transfer of conjugative plasmids, events that have themselves not been completely described.

As indicated earlier, the T-DNA is the region from Ti plasmids that is found to be inserted into the nuclear genome of susceptible plant cells. However, it is important to bear in mind that inferences concerning its structure are not usually based on an analysis of the primary products of the insertion event itself. Rather, they are based on comparative analyses of T-DNA and of T-DNA–plant DNA junctions from populations of cells of transformed tissue that have usually been subject to selective pressures. Secondary rearrangement of the primary recombinant structure is therefore a distinct possibility and has been observed. Nevertheless, there is sufficient evidence to indicate that insertion into the plant genome varies only within 10 bp of the right border. Figure 32.6 illustrates the kinds of inserted structures that have been found for representatives of different kinds of plasmids. In several, but not all, instances, the DNA that has been transferred is found in two structures (called TL and TR) that are genetically and physically distinguishable from one another and separated by plant DNA sequences. Markers that determine the tumor phenotype or the ability to produce specific opines can be found in one or both of these structures.

There have been relatively fewer analyses of the structures that result when binary vector systems are used. These are systems in which a continuous T-DNA region is linked to a cloning vector that is maintained in a stable state in *Agrobacterium*, with another stable plasmid providing the *vir* functions. With such systems, it appears that either or both of the DNA segments enclosed by the border regions can be frequently transferred to the plant genome.

In all naturally occurring Ti (or Ri) plasmids, the genes present in their respective T-DNA have evolved to function in plants and not in the bacterium (Figs 32.5 and 32.6). Prominent among these functions are those that result in the production of auxins, cytokinins or both. This results in the disease symptoms of tumorigenesis and of the phytohormone independence of transformed cell growth in culture. This change in phenotype results in a loss of regenerative capacity in the transformed tissue. In early studies, a correlation was observed between the ability of *A. tumefaciens*-transformed tobacco cell cultures to regenerate into plants and the loss of a central portion of the T-DNA. This portion coded for auxin and cytokinin biosynthesis pathways novel to the plant. Since an important objective in using the *Agrobacterium* system is to

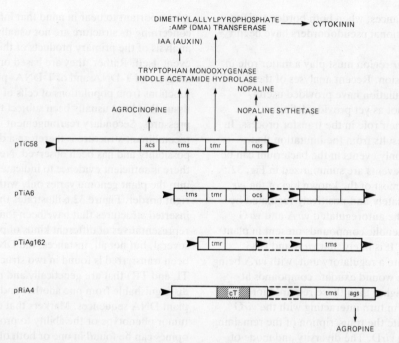

Fig. 32.6 Examples of the structure of T-DNA in plant tissue transformed by *Agrobacterium* strains carrying different Ti (or Ri) plasmids. pTiC58 and pTiA6 are the intensively studied plasmids that incite tumors producing nopaline and octopine, respectively. pTiAg162 is a plasmid from a grapevine isolate. It confers a limited host range. pRiA4 is a plasmid that incites tumors with a profusion of roots (hairy root disease). Shaded arrows indicate the border regions that signal the extent of DNA that is transferred. cT is a DNA region that has been found to hydridize with the genomic DNA of some untransformed plant species.

obtain transgenic plants, it is the general practice to 'disarm' natural or derived Ti plasmid systems by deleting from their T-DNA these genes that interfere with normal plant morphogenesis. Examples of such 'disarmed' vectors will be considered in the next section.

Gene vectors

All systems in which *Agrobacterium* is employed for plant genetic engineering rely on strategies for manipulating the bacterium and its plasmid(s). These in turn engineer the plant cells. In the decade since the first demonstration of the usefulness of this system, these strategies have become progressively simpler. The development of expression and selection systems for plant cells has also kept pace with these developments. In this section these systems are summarized in a practical rather than in

an historical context. In general, an *Agrobacterium* strain carrying a disarmed and engineered Ti plasmid, or one that carries the binary system composed of a deleted Ti plasmid providing the *vir* functions and a separate vector providing the engineered T-DNA, are used. The basic attributes of *Agrobacterium* strains that provide either of these two systems is illustrated in Fig. 32.7. In binary systems, the *vir* region is *in trans*, i.e. located on a separate plasmid to the T-DNA region, and is usually provided as a deleted derivative of a Ti plasmid. This acts as a helper plasmid for the transmission of T-DNA. Two such helper plasmids are pAL4404, which is derived from the octopine-inducing plasmid pTiAch5 (closely related to pTiA6), and pMP90RK, derived from the nopaline-inducing plasmid pTiC58. Although each of the *vir* genes from one of the parental plasmids can be replaced by the corresponding gene from the other plasmid, there is evidence that the efficiencies

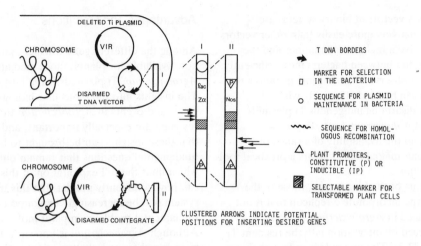

Fig. 32.7 General attributes of *Agrobacterium* plasmid constructs that are used in the genetic engineering of plants. The top figure of the bacterium illustrates a binary system, one in which the *vir* functions are provided *in trans* to the plasmid carrying the engineered T-DNA. In the bottom figure, these two essential regions are on the same molecule, usually the modified Ti plasmid that is referred to as the *cointegrate* because of the way in which it is usually constructed. In either situation, at least one border 25 bp sequence is essential (shown as thick shaded arrows). I and II illustrate two of several possible kinds of cassettes that can be engineered to be contained within the border sequence (and therefore within the T-DNA). The various symbols used are as follows. Thick shaded arrows, T-DNA border sequences; small open rectangle, marker that permits selection for stable maintenance of the construct in the bacterium (often *E. coli* and *Agrobacterium* species); open circle, sequence that permits replication of the plasmid in the bacterium. Note that in the binary system (top figure), this sequence is shown as part of the cassette I, while in the cointegrate system (bottom figure) it is not within cassette II but still within the T-DNA borders. Both such systems can be used. The inclusion of such sequences within the T-DNA borders can allow recovery and recloning in bacteria of these regions from the transformed plant cells; wavy line, region of sequence similarity that permits cointegrate formation by homologous recombination; open triangles, promoters functional in plants (P, constitutive; IP, inducible); hatched rectangle, dominant marker that is selectable for the transformed plant tissue; clustered arrows, positions for inserting a desired gene (by *in vitro* methods); *lac*Z, portion of the β-galactosidase gene that permits assay by alpha-complementation; NOS, nopaline synthase (assayable in transformed plant tissue).

of expression of each *vir* region are different. It may therefore be advantageous to use a particular helper for a specific purpose.

A large number of existing plasmid vectors provide the T-DNA component as part of a binary system. As a minimum, these vectors contain (1) one or both the right and left borders of the T region, (2) a selectable dominant marker under the control of plant transcription signals, (3) a single replicon and markers for maintenance and selection in *Agrobacterium* and *E. coli*, and (4) restriction sites suitably positioned to allow the cloning of desired plant sequences and their detection. Additionally, it is useful to incorporate *ori*T regions (origins of transfer) into these vectors. These allow the facile transfer of the vectors across bacterial species boundaries. Lambda *cos* (cohesive end) sites are also useful by enabling the packaging of lambda DNA into lambda virions. The replicons in these vectors are usually derived from the Incompatibility (Inc)P or IncQ plasmid groups of *E. coli* which have a broad bacterial host range. Virtually any plasmid marker conferring resistance to a bacterial antibiotic or inhibitor can be used as a selection marker to ensure plasmid maintenance in *Agrobacterium* before they are used for transformation of plants.

Several resistance markers have been used successfully to select plants that have been transformed. As indicated above, the most appropriate marker may be different for each plant species. T-DNA vectors of binary systems are manipulated relatively more easily than other vectors because they have a higher copy number and they can be transmitted between bacteria more efficiently. It has also been observed that in some instances the binary system can be used effectively without eliminating or disarming indigenous Ti plasmids native to the *Agrobacterium* host. These observations and the availability of a range of selection systems make binary vectors a practical tool at the present time.

Unlike vectors of the binary type, those of the co-integrate type incorporate a replicon that is not stably maintained in *Agrobacterium*, and are therefore rescued by integration into the resident Ti plasmid (Fig. 32.7). The principle is to first engineer an intermediate plasmid. Using *in vitro* methods, a segment of DNA containing desired genes and sequences is inserted into a well-characterized *E. coli* plasmid like pBR322. The acceptor Ti plasmid that will integrate this pBR322-derived intermediate is a disarmed plasmid in which most of the central region of the T-DNA has been replaced by pBR322 sequences. The intermediate vector cannot replicate in *Agrobacterium*, but can be rescued by a single recombination event within the pBR322 sequences of the intermediate and Ti plasmids to give rise to the co-integrate plasmid in which the genes to be transmitted into the plant are now contained within the right and left borders. The relative efficiency of co-integrate systems is less than that of binary systems, but with the availability of good selection systems this has not been a serious disadvantage. With co-integrate systems, plasmid sequences outside those that are delimited by the T-DNA borders tend to be less often transmitted to the plant.

With both kinds of vector systems, the simplest way to obtain transgenic plants is to infect tissue explants, such as leaf discs, with the manipulated *Agrobacterium*, and to select for transformants in culture. *Agrobacterium*-mediated transformation has also been achieved with a variety of cell types including protoplasts, callus tissues and cells from suspension cultures. Generally, the recipient cell or tissue source employed depends largely on their capabilities for plant regeneration which varies greatly among species.

Advantages and limitations

Among the different gene transfer systems that are now available for plants, those mediated by *Agrobacterium* Ti plasmids is the best understood. The insights into the molecular biology of relevant genes and events in *Agrobacterium* that have been revealed are especially important, and it is likely that these insights can be brought to bear on fundamental questions that remain outstanding and which limit the full exploitation of this system. One can now confidently predict that this transfer system will be increasingly exploited to address fundamental problems in plant molecular biology. Although *Agrobacterium* is known to transfer genes to over a thousand plant species, a practical limitation that has to be considered is the apparent recalcitrance of some major crop species, especially the cereals. It is likely that, to overcome this limitation, a molecular understanding of factors that limit or extend host range will be needed.

Virus-mediated gene transfer

Plant viral genomes or their cDNAs can be cloned in bacterial plasmids and altered by genetic engineering. Viruses may have advantages over other vector systems. For example, host plants can be inoculated artificially without the need for invertebrates that usually act as vectors to transmit the viruses in nature. Extrachromosomal replication can also generate high copy numbers in infected plants. For these reasons, viruses are interesting candidates as vectors, but strategies to employ them effectively have yet to be developed. At this time a critical assessment of the utility of viral vectors is difficult. A substantial experimental effort is still needed to understand the complex viral life cycles in relation to the organization of their genomes and the genes they encode (see Hohn and Schell, 1987, for further readings).

Most plant viruses have genomes composed of RNA, which makes genetic engineering more complicated. Despite this, it has been shown that expression of the *cat* gene can be achieved in barley protoplasts from RNA prepared from a cDNA of brome mosaic virus (BMV). It would be advantageous to use DNA viruses, and the caulimoviruses and geminiviruses are the two groups of plant viruses with a DNA genome. Among the caulimoviruses, which have circular dsDNA genomes, the cauliflower mosaic virus (CaMV) has been most extensively explored as a vector. Replacement of a non-essential gene with the coding regions from bacterial dihydrofolate reductase or mammalian metallothionein have yielded functional expression of these proteins in turnip. Major limitations are the small size of DNA inserts that can be stably maintained when the recombinant genome is packaged as a virus particle, and the narrow host range of CaMV which is limited to the Cruciferae and some Solanaceae. The geminiviruses possess small circular ssDNA genomes. Collectively, these viruses have a much wider host range which include a number of economically important monocotyledonous species; however, their use as vectors has yet to be demonstrated.

Viral vectors have several limitations that must be considered if they are to be used for the genetic engineering of plants. (1) Viruses are generally poorly transmitted by pollen or seed to progeny; therefore this approach is not practical for the development of new crop varieties. (2) Distribution throughout a host plant may not be uniform. (3) It may not be possible to dissociate viral disease symptoms from the vectors. (4) Relatively high rates of recombination and error accumulation may generate instability. Conversely, viral vectors have the advantage that expression of specific gene products can be studied quickly after transduction without the need to regenerate plants in tissue culture which is usually the limitation with vectors that produce stable transgenic plants.

The development of *Agrobacterium*-mediated transformation technology has greatly expanded experimental opportunities for studying viruses. Complete and partial copies of viral genomes have been introduced into Ti plasmid vectors. The Ti plasmid-encoded mechanisms for transferring T-DNA into plant cells also mediate the efficient release of infectious viruses within plants if viral oligomers are used. This process can be achieved by simply inoculating plants with the bacteria. Alternatively, transgenic plants can also be produced with the same bacteria, with the virus being released from the integrated T-DNA. In both cases, virulent Ti plasmids are required to transfer the viral genomes into plants. The term 'agroinfection' has been adopted to describe these phenomena.

Agroinfection studies have been employed widely. Studies of *Agrobacterium* host range and Ti plasmid *vir* genes have been facilitated by capitalizing on virus amplification as a sensitive indicator of T-DNA transfer. The construction of transgenic plants that produce specific viral genome components from chromosomally integrated DNA permits detailed analyses by complementation and may generate new strategies for employing viral vectors. A number of agricultural applications are feasible and will be discussed in a later section.

DNA transfer without vectors

Over the years, several approaches have been used for introducing isolated DNA directly into a variety of plant cells without biological vectors. In most cases, the evidence for transformation was ambiguous. Recently, clear evidence for transformation has been obtained with protoplasts as the recipient cell type. Successful approaches include direct DNA uptake stimulated by chemical and/or electrical treatments, intranuclear microinjection of protoplasts having partially reformed cell walls, and fusion of protoplasts with liposomes that encapsulate DNA. Each of these procedures is designed to overcome the cell wall and plasma membrane barriers to DNA entry and to protect the DNA from degradation by nucleases released from the protoplasts. Since these are essentially artificial mechanisms, they are not biologically restricted to specific species. The major limitation is the prerequisite for a system of protoplast isolation and for the regeneration of the protoplasts into whole plants.

Direct DNA uptake

Direct uptake of DNA by protoplasts is the most developed mechanism for transformation with isolated DNA (Potrykus *et al.*, 1987). The uptake and integration of free or precipitated DNA is mediated by a number of factors. Often protoplast fusion treatments such as polyethylene glycol, polyvinyl alcohol and calcium combined with high pH enhance the transfer of DNA across the plasma membrane. Other factors such as high temperature treatments and electrical pulses (electroporation) facilitate this process. Generally, the concentration and type of divalent cations are important. Presumably they mediate interactions between membranes and DNA, or protect the DNA from nucleases. Some protocols present the protoplasts with a coprecipitate of DNA and calcium phosphate before uptake is induced. Other factors that affect the transformation frequencies include the use of carrier DNA and the concentrations of DNA. The condition of the protoplasts is an important factor in these experiments, as is the stage of the cell cycle. When all the factors that have been identified so far are combined, transformation frequencies may exceed 2% of the protoplasts for tobacco in the absence of selection. The optimal conditions have to be determined for each source of the protoplasts. The procedures adopted by different laboratories vary widely.

The list of protoplast systems that have been transformed by these techniques is growing rapidly. Generally, the transformation frequencies are lower than reported for tobacco. Other Solanaceae that have been used include *Nicotiana plumbaginofolia, Petunia hybrida* and *Hyoscyamus muticus*. Among the Brassiceae, *Brassica rapa* and *Brassica napus* protoplasts have been transformed using different procedures. Transformed calli have been identified after transformation of protoplasts of the Graminae species, *Lolium multiflorum, Triticum monococcum, Zea mays* and *Oryza sativa*. Clearly, the transfer of DNA directly into protoplasts using a variety of procedures for stimulating uptake is of general applicability.

The integration of the transferred DNA into the plant genome has been examined in tobacco tissues in which the uptake had been mediated by both Ti plasmids and bacterial vectors carrying selectable marker genes. In both cases, significant rearrangements of the DNA occur, presumably due to nuclease activity before it is integrated. Concatenation of smaller plasmids is commonly observed; however, the number of copies is not correlated with the level of gene expression or antibiotic resistance conveyed by the marker genes. The integration into the plant genome appears to occur randomly by non-homologous recombination mechanisms governed by the plant nucleus. Certainly the virulence genes and T-DNA borders associated with Ti plasmids (see above) do not contribute to integration when free DNA is delivered. Generally, the integrated DNA is stable and is maintained during meiosis. The location on the chromosomes appears to be random. Frequently, the marker gene is inherited as a single Mendelian factor and in a few cases *in situ* hybridization has confirmed the integration at a single locus on metaphase chromosomes. In studies with Ti plasmid DNA and calf thymus carrier DNA, it was shown that different fragments could integrate at separate locations and segregate independently. Generally, however, cotransformation, which is the transformation of non-selected DNA along with DNA carrying selectable marker genes, occurs at a relatively high frequency.

Direct DNA uptake procedures have been used to demonstrate the transformation of tobacco with total genomic DNA taken from transgenic plants in which selectable marker genes have been integrated. Selection for kanamycin resistance yielded transformants at low frequencies of 10^{-6}. These experiments demonstrate clearly the feasibility of using these techniques for transferring genes or gene families that have not been isolated from their genomes.

Recently, electroporation of protoplasts has been used to study transient expression of non-integrated DNA (Fromm and Walbot, 1987; Hauptmann *et al.*, 1987). The DNA is lost after several days in culture; however, the amount of DNA taken up, once it is optimized, is sufficient for the detection of reporter gene activity in several thousand protoplasts. Generally, the *cat* gene is employed because of the sensitivity of the methods for detecting the activity of this enzyme. Since protoplast viability, but not cell division, is a prerequisite for analyses of transient expression, the range of species that can be employed is greatly expanded over those that can be stably transformed. The approach can be used to analyze plant promotors in homologous plant systems for which regenerable protoplast technology has not been developed. This is particularly important for the cereal species.

Microinjection of DNA

Conceptually, microinjection is the most direct approach for introducing DNA into plant cells. The technology is unique in that it offers precise control over the form of the DNA that reaches the target site and the specific cells in culture that receive DNA (Miki et al., 1987). Selectable marker genes are not essential for the recovery of transformants. Preliminary experiments with regenerating alfalfa and tobacco protoplasts indicate that intranuclear microinjection yields extremely high frequencies of transformation (14%), whereas injections into the cytoplasm are very inefficient. This observation parallels those made with animal cells in culture and implies that the transfer of free DNA from the cytoplasm to the nucleus is a significant barrier in the transformation process. As with direct DNA uptake, the microinjected DNA that is integrated may be rearranged; however, the extent to which this occurs may be less severe. Presumably, delivery to the nucleus with glass capillaries protects the DNA from the extracellular nucleases and the complex chemical interactions that may extensively alter the form of the DNA before integration.

Intranuclear microinjection technologies for plant cells are in the process of being developed for a variety of cell types. Between laboratories, the technologies differ greatly depending on the systems employed for holding the protoplasts or cells, visualizing the intracellular target sites, and the equipment and methods for the identification and culture of microinjected cells. A major objective in the design of these systems is to obtain the maximum rates of injection, because only small numbers of cells (200–300) can be manipulated in an experiment. A major limitation of this technology is that it can only be applied to cells for which efficient culture conditions have been determined.

Liposome-mediated DNA transfer

The encapsulation of DNA in artificial membrane-enclosed vacuoles called liposomes, followed by transfer to protoplasts by polyethyleneglycol-induced fusion, have yielded low transformation frequencies (10^{-5}; Deshayes et al., 1985). Randomly repeated copies of plasmid DNA carrying a selectable marker gene for kanamycin resistance appeared to integrate at a single locus in transgenic tobacco plants and to segregate as a dominant genetic marker among progeny. The analyses of transformation paralleled those obtained by direct DNA uptake procedures.

Advantages and limitations

For plant species, such as tobacco, which are exceptionally well suited for manipulation in culture, a variety of artificial methods for DNA transfer yield transgenic plants. Compared with leaf disc transformation by *Agrobacterium*, these approaches are more complicated due to the tissue culture requirements for protoplast regeneration. Furthermore, the pattern of DNA integration is less predictable than T-DNA delivered by *Agrobacterium*, and the extent of rearrangement is much more severe. For species within the host range of *Agrobacterium* species, the use of Ti plasmid vectors is most practical for introducing isolated genes or specific DNA fragments into plants.

The clear advantage for the artificial delivery systems is that it bypasses the host range restrictions of *Agrobacterium*. This is particularly relevant for the Graminae species. Since they cannot be infected by *Agrobacterium*, direct DNA uptake is the most practical alternative for producing transgenic plants. With the exception of rice, most cereals cannot be regenerated from protoplasts, thereby excluding the use of direct DNA uptake or fusion with liposomes. For rye, injection of DNA into young inflorescences has proven to be a successful alternative. At specific developmental stages, the germ cells appear to be susceptible to transformation with exogenous DNA and eventually give rise to transgenic seedlings. This overcomes the restrictions imposed by the cell culture systems and the limited host range of *Agrobacterium*. Attempts to inject DNA into isolated ovules have not yet been successful. It is hoped that the adaptation of microinjection technology for isolated microspores that undergo direct embryogenesis in culture, or for small cell clumps from regenerable cell suspension cultures, will provide a more generally applicable alternative.

Microinjection offers unique experimental capabilities for studying specific molecular processes. The form of the injected molecules or complexes can be precisely controlled and targeted to specific intracellular compartments. In the animal sciences, microinjection of somatic cells is used to study such diverse processes as DNA replication, homologous recombination, or the function of specific gene products. It provides a general experimental tool for introducing molecules such as mRNA, DNA, antibodies, or complex structures such as organelles, into living cells. In the plant sciences, this kind of research is just beginning to be explored.

Applications for basic studies

A direct consequence of the development of vectors and protocols for the transfer of genes into plants is that it is enabling important questions to be addressed concerning the organization of DNA sequences that determine various aspects of plant gene expression and plant development. This has required the construction of novel second generation vectors tailored for specific purposes. In the relatively short history of their development, they have been used to delve into questions that were not otherwise easily accessible to investigation. Some examples are provided here to illustrate different applications.

A common tactic is to construct vectors from DNA that is suspected of having a role in regulation and to fuse to this a reporter gene, the expression of which can be reliably detected and from which the regulatory sequences have been deleted. Such constructs could also contain the native gene sequences whose regulation is under scrutiny. In one example, the transcriptionally active regulatory sequence found at about 1 kbp upstream from the nuclear gene (*ss*) for the small subunit of ribulose 1,5-bisphosphate carboxylase of pea was fused to the coding sequence of the bacterial gene for chloramphenicol acetyltransferase (*cat*). The vector containing this artificial *ss*:*cat* fusion was then used to dissect and assess the complexity of this upstream *cis*-acting regulatory sequence (Fluhr *et al.*, 1986). This was done by using the vector to transmit the hybrid gene to heterologous plant cells (tobacco or petunia) which could be regenerated into plants. The tissues and organelles of the transgenic plants could then be assayed for *cat* activity both with and without light induction. A similar strategy has been used to study regulatory sequences for the chlorophyll binding protein and for chalcone synthase, a key enzyme in flavonoid biosynthesis (Koulen *et al.*, 1986; Sommer and Saedler, 1986). This kind of approach is also proving useful for the analysis of organ-specific and wound-inducible genes (Sanchez-Serrano *et al.*, 1987; Thornburg *et al.*, 1987), and genes inducible by symbiotic root nodulating bacteria (Jensen *et al.*, 1986).

A second set of questions for which vector constructs are providing answers is in the studies of protein localization and the targeting of proteins for import into organelles such as chloroplasts. Such import is usually associated with the presence of a hydrophobic transit peptide at the amino-terminal end of a protein, the transit peptide being usually cleaved during import. Vectors in which the transit peptide alone or along with portions of the mature proteins have been fused to the bacterial neomycin phosphotransferase gene II are used to study aspects of targeting and translocation both *in vivo* and *in vitro* (Van den Broeck *et al.*, 1985; Smeekens *et al.*, 1987).

The T-DNA of the Ti plasmid of *Agrobacterium* can be used as an insertion element to locate plant genes by causing a mutation. It can also be used to create transcriptional and translational fusions of the types that have been so valuable in studying the molecular genetics of bacteria and of yeast (Teeri *et al.*, 1986). A particular application of the Ti plasmid exploits the normal presence, within the T-DNA, of genes for novel pathways for auxin and cytokinin biosynthesis that are not normally present in plants. For example, when mutations of appropriate phytohormone genes of the T-DNA are used, it is possible to obtain transgenic tobacco and petunia plants with altered levels of auxin in their tissues (Klee *et al.*, 1987). If the implications of these recent observations are fully realized through further experiment and predictable technology, they could have a major impact on understanding the role of these phytohormones in the control of plant development.

All the above applications involve the stable but

random integration of transferred DNA into the plant nuclear genome. Recently, techniques have also become available for studying the transient expression of transferred genes over periods in which they have not been stably integrated into the genomes of the plant. These techniques and their potential usefulness have been reviewed recently (Fromm and Walbot, 1987). In the near future, it is anticipated that these experimental capabilities will form the basis for more advanced genetic manipulations. Targeting of genes to specific regions of the chromosomes and gene replacement by homologous recombination should be feasible. Cloning of certain plant genes in plant cells is a reasonable expectation. Deliberate transformation of specific organelles other than the nucleus is challenging but feasible. These and other technological advances will have a profound effect on our understanding of basic genetic control mechanisms in plants.

Applications for agriculture

Many examples of transgenic plants that appear to be normal in all other respects have been documented. Among these the number of crop species is expanding rapidly; therefore, it is feasible to incorporate transgenic plants into breeding programs for crop improvements. The next challenge is the identification and isolation of genes that will have an impact on the quality and productivity of a particular crop. Current technological capabilities do not limit the source of these genes to plants or to natural origins. A comprehensive examination of agricultural applications is beyond the scope of this chapter; therefore, only examples that demonstrate some fundamental principles will be discussed.

Significant progress has been made in the genetic engineering of plants to make them resistant to broad-spectrum herbicides. Compared with other agronomically important characteristics, the biochemical basis for herbicide sensitivity and resistance has been well studied and the number of genes that are involved may be limited. Generally, the herbicides inhibit key biological processes or specific enzymes in metabolic pathways unique to plants (see following chapter). At least three basic genetic principles that confer resistance have been documented for nuclear genes. These are gene amplifications resulting in over-production of target enzymes, mutations that alter herbicide binding to enzymes, and expression of genes that code for enzymes which detoxify herbicides. Experimental exploitation of these principles in transgenic plants has yielded resistance to a number of herbicide classes. For instance, mutant genes from bacteria and other plants have been used to substitute for resident plant genes coding for enzymes in key metabolic pathways. By this means, resistance to the sulfonylurea herbicides and glyphosate has been achieved. Bacterial genes have been used to provide a mechanism for the detoxification of another herbicide, phosphinothricin, in transgenic plants.

Similar strategies are being examined for the protection of crops from insect damage. Preliminary analysis has been performed with a gene coding for the toxic region of the insecticidal protein of the bacterium *Bacillus thuringiensis*. Preparations of the bacterium have been used as commercial insecticides for the control of insect larvae, and a number of strains exist with selective toxicity to species of Lepidoptera, Diptera and Coleoptera. Under controlled conditions, expression of the lepidopteran protein in tobacco and tomato was toxic to feeding larvae, thereby protecting the transgenic plants from damage (Fischhoff *et al.*, 1987). Other approaches include the use of genes coding for proteins such as trypsin inhibitor that will interfere with insect digestive enzymes.

Mechanisms for the protection of transgenic plants from viral diseases are being examined. The phenomenon of viral cross-protection in which inoculation with mild strains prevents damage from more virulent strains can be mimicked by the expression of cloned viral coat protein genes in transgenic plants such as tobacco and tomato (Powell *et al.*, 1986). Studies with tobacco mosaic virus (TMV) and alfalfa mosaic virus (AMV) have shown that the appearance of disease symptoms can be significantly delayed or inhibited by this approach. A separate mechanism that suppresses disease symptoms is the expression in transgenic plants of the satellite RNA of viruses such as cucumber mosaic virus (CMV). Genetic transformation provides a tool for studying the principles responsible for these phenomena. As knowledge is acquired the strategies for acquiring viral protection will be improved.

The applications of gene transfer technology to agriculture is limited only by our understanding of the fundamental genetic principles responsible for crop quality and production. The examples offered above serve to illustrate a few of the many genetic manipulations that have been shown to be feasible. An examination of the chapters preceding and following this one reveal other areas in which genetic transformation technology will serve to acquire knowledge to eventually permit genetic manipulation for agricultural objectives. The efficiencies and sophistication of the technologies will undoubtedly improve rapidly as these experiments progress, since this field of research is still in the early stages of development.

Acknowledgments

We are grateful to colleagues and students for thoughtful comments on this chapter. Our research is supported by Agriculture Canada and the Natural Science and Engineering Research Council of Canada.

Further reading and references

Deshayes, A., Herrera-Estrella, L. and Caboche, M. (1985). Liposome-mediated transformation of tobacco mesophyll protoplasts by an *Escherichia coli* plasmid. *EMBO J.* **4**, 2731–7.

Eichholtz, D. A., Rogers, S. G., Horsch, R. B., Klee, H. J., Hayford, M., Hoffman, N. L., Braford, S. B., Fink, C., Flick, J., O'Connell, K. M. and Fraley, R. T. (1987). Expression of mouse dihydrofolate reductase gene confers methotrexate resistance in transgenic petunia plants. *Somat. Cell Molec. Genet.* **13**, 67–76.

Evans, D. A., Sharp, W. R. and Ammirato, P. V. (1986). *Handbook of Plant Cell Culture*, Vols 1–5, MacMillan Publishing Company, New York, Collier MacMillan Publishers, London.

Fischhoff, D. A., Bowdish, K. S., Perlak, F. J., Marrone, P. G., McCormick, S. M., Niedermeyer, J. G., Dean, D. A., Kusano-Kretzmer, K., Mayer, E. J., Rochester, D. E., Rogers, S. G. and Fraley, R. T. (1987). Insect tolerant transgenic tomato plants. *Bio/Technology* **5**, 807–13.

Fluhr, R., Kuhlemeier, C., Nagy, F. and Chua, N-H. (1986). Organ-specific and light-induced expression of plant genes. *Science* **232**, 1106–12.

Fromm, M. and Walbot, V. (1987). Transient expression of DNA in plant cells. In *Plant DNA Infectious Agents*, eds T. Hohn and J. Schell, Springer-Verlag, Vienna, pp. 303–10.

Gheyson, G., Dhaese, P., Van Montagu, M. and Schell, J. (1985). DNA flux across genetic barriers: The crown gall phenomenon. In *Genetic Flux in Plants*, eds B. Hohn and E. S. Dennis, Springer-Verlag, Vienna, pp. 11–47.

Hauptmann, R. M., Ozias-Akins, P., Vasil, V., Tabaeizadeh, Z., Rogers, S. G., Horsch, R. B., Vasil, I. K. Hand Fraley, R. T. (1987). Transient expression of electroporated DNA in monocotyledonous and dicotyledonous species. *Plant Cell Rep.* **6**., 265–70.

Herrera-Estrella, L., De Block, M., Messens, E., Hernalsteens, J.-P., Van Montagu, M. and Schell, J. (1983). Chimeric genes as dominant selectable markers in plant cells. *EMBO J.* **2**, 987–95.

Hille, J., Verheggen, F., Roelvink, P., Franssen, H., Van Kammen, A. and Zabel, P. (1986). Bleomycin resistance: a new dominant selectable marker for plant cell transformation. *Plant Molec. Biol.* **7**, 171–6.

Hohn, T. and Schell, J. (1987). *Plant DNA Infectious Agents*, Springer-Verlag, New York, Vienna.

Holbrook, L. A., Haffner, M. and Miki, B. L. (1986). A sensitive fluorographic method for the detection of nopaline and octopine synthase activities in *Brassica* crown gall tissues. *Biochem. Cell Biol.* **64**, 126–32.

Jefferson, R. A., Kavanagh, T. A. and Bevan, M. W. (1987). Gus fusions: β-glucoronidase as a sensitive and versatile gene fusion marker in higher plants. *EMBO J.* **6**, 3901–7.

Jensen, J. S., Mareker, K. A., Otten, L. and Schell, J. (1986). Nodule-specific expression of a chimaeric soyabean leghaemoglobin gene in transgenic *Lotus corniculatus*. *Nature* **321**, 669–74.

Kahl, G. and Schell, J. (1982). *Molecular Biology of Plant Tumors*, Academic Press, New York.

Kemp, J. D. (1982). Enzymes in octopine and nopaline metabolism. In *Molecular Biology of Plant Tumors*, eds G. Kahl and J. S. Schell, Academic Press, New York, pp. 461–74.

Klee, H., Horsch, R. B., Hinchee, M. A., Hein, M. B. and Hoffman, N. L. (1987). The effects of overproduction of two *Agrobacterium tumefaciens* T DNA auxin biosynthetic gene products in transgenic petunia plants. *Genes Devel.* **1**, 86–96.

Koukolikova-Nicola, Z., Albright, L. and Hohn, B. (1987). The mechanism of T DNA transfer from *Agrobacterium tumefaciens* to the plant cell. In *Plant*

DNA Infectious Agents, eds T. Hohn and J. Schell, Springer-Verlag, Vienna, pp. 109–48.
Koulen, H., Schell, J. and Kreuzaler, F. (1986). Light-induced expression of the chimeric chalcone synthase-NPTII gene in tobacco cells. EMBO J. 5, 1–8.
Miki, B. L. A., Reich, T. J. and Iyer, V. N. (1987). Microinjection: An experimental tool for studying and modifying plant cells. In Plant DNA Infectious Agents, eds T. Hohn and J. Schell, Springer-Verlag, Vienna, pp. 248–65.
Nester, E. W., Gordon, M. P., Arnasino, R. M. and Yanofsky, M. F. (1984). Crown gall: a molecular and physiological analysis. Ann. Rev. Plant Physiol. 35, 387–413.
Ow, D. W., Wood, K. V., Deluca, M., Dewet, J. R., Helinski, D. R. and Howell, S. H. (1986). Transient and stable expression of the firefly luciferase gene in plant cells and transgenic plants. Science 234, 856–9.
Potrykus, I., Paszkowski, J., Shillito, R. D. and Saul, M. W. (1987). Direct gene transfer to plants. In Plant DNA Infectious Agents, eds T. Hohn and J. Schell, Springer-Verlag, Vienna, pp. 229–47.
Powell Abel, P., Nelson, R. S., De, B., Hoffman, N., Rogers, S. G., Fraley, R. T. and Beachy, R. N. (1986). Delay of disease development in transgenic plants that express the tobacco mosaic virus coat protein gene. Science 232, 738–43.
Rogers, S. G. and Klee, H. (1987). Pathways to plant genetic manipulation employing Agrobacterium. In Plant DNA Infectious Agents, eds T. Hohn and J. Schell, Springer-Verlag, Vienna, pp. 179–203.
Sanchez-Serrano, J. J., Keil, M., O'Connor, A., Schell, J. and Willmitzer, L. (1987). Wound-induced expression of a potato proteinase inhibitor II gene in transgenic tobacco plants. EMBO J. 6, 303–6.
Smeekens, S., Van Steeg, H., Bauerle, C., Bettenbroek, H., Keegstra, K. and Weisbeek, P. (1987). Import into chloroplasts of a yeast mitochondrial protein directed by ferredoxin and plastocyanin transit peptides. Plant Molec. Biol. 9, 377–88.
Sommer, H. and Saedler, H. (1986). Structure of the chalcone synthase gene of Antirrhinum majus. Mol. Gen. Genet. 202, 429–34.
Teeri, T. H., Herrera-Estrella, L., Depicker, A., Van Montagu, M. and Palva, E. T. (1986). Identification of plant promoters in situ by T DNA-mediated transcriptional fusions to the npt-II gene. EMBO J. 5, 1755–60.
Thornburg, R. W., An, G., Cleveland, T. E., Johnson, R. and Ryan, C. A. (1987). Wound-inducible expression of a potato inhibitor II-chloramphenicol acetyl transferase gene fusion in transgenic plants. Proc. Natl Acad. Sci. USA 84, 744–8.
Van Den Broeck, G., Timko, M. P., Kausch, A. P., Cashmore, A. R., Van Montagu, M. and Herrera-Estrella, L. (1985). Targeting of a foreign protein to chloroplasts by fusion to the transit peptide from the small subunit of ribulose 1,5-bisphosphate carboxylase. Nature 313, 358–63.
Waldron, C., Murphy, E. B., Roberts, J. L., Gustafson, G. D., Armour, S. L. and Malcolm, S. K. (1985). Resistance to hygromycin B. A new marker for plant transformation studies. Plant Molec. Biol. 5, 103–8.
Wu, R. and Grossman, L. (1987). Recombinant DNA Part D Methods in Enzymology, Vol. 153, Academic Press, London.

33 The biochemical basis for plant improvement

C. R. Somerville

Introduction

Many of the plants used by man today have been cultivated for hundreds or thousands of years. During this time, these plants have been gradually improved by an empirical process of plant breeding which required no specific knowledge of plant biochemistry or physiology. Indeed, much plant breeding continues to be an empirical process in which large populations are simply screened for useful variation. However, our approach to plant improvement has recently undergone a major change. The recent development of techniques for stably introducing cloned genes into higher plants (Klee et al., 1987b) has created many new opportunities to undertake rational improvement of crop plants. It is now possible to transfer characters found in one species to another species. In the most extreme cases, it has been possible to transfer genes for useful characters from non-plants to plants. Thus, for instance, the introduction of a bacterial gene encoding an insecticidal protein into higher plants conferred resistance to certain insects (Vaeck et al., 1987). There are now countless opportunities to genetically modify plants in beneficial ways. However, in contrast to traditional methods of breeding, the recognition and implementation of potential applications of genetic engineering depends upon a detailed understanding of the underlying biochemistry and physiology of higher plants. Indeed, we have arrived at the point where much of the genetic technology is almost routine and the rate-limiting step is the acquisition of specific knowledge about the underlying plant biology.

The general goals of plant breeding are increased productivity, increased efficiency, improved quality and expanded uses for plants (Hallauer, 1981). Although it is not possible to describe all of the possible opportunities, a few selected examples may serve to illustrate how a detailed knowledge of plant biochemistry and physiology may lead to rational plant improvement.

Improved quality

In many cases the characteristics of the plant parts that we harvest and utilize are not ideally suited to our purposes. For instance, the fiber length in a species such as cotton or flax, which are used for textile production, may not be suitable for certain applications, or a food crop may have a noxious constituent that must be removed by a processing step before use. Although plant breeders have made substantial progress in selecting varieties with the optimal quality characteristics, there are inherent limitations to what can be accomplished by traditional methods. In many cases, knowledge of the underlying biochemistry and physiology may suggest what the limitations are and indicate ways of surmounting these limitations by conventional breeding or by genetic engineering techniques. A few examples may serve to illustrate the point.

Amino acid content of seed and forage crops

Because animals lack the ability to synthesize a number of protein amino acids, they require a balanced content of dietary amino acids in order to make efficient use of feed. Barley, wheat and maize grain provide an unbalanced diet which requires supplementation. However, if this is done they can be very effective food sources. For instance, supplementation of barley with lysine, threonine and histidine provided a diet for pigs with an almost ideal balance of the protein amino acids (Miflin et al., 1983). In other instances it is not possible to supplement the feed. Thus, for instance, wool production by New Zealand sheep, which are fed entirely on pasture grass (e.g. *Lolium*), is thought to be limited by a suboptimal content of sulfur amino acids in the leaves of *Lolium*.

The basis for the suboptimal amino acid composition of the cereal grains is the amino acid composition of the prolamin storage proteins which make up about half of the total seed protein. Attempts to alter the proportion of prolamins by conventional breeding techniques have failed. This may be due to the fact that, in most plants, the seed storage proteins are encoded by families of closely related genes which are often tightly linked on the chromosome. Thus, selection for improved amino acid composition must entail simultaneous changes in the kind or amount of many gene products – a difficult task.

The general solution to these problems is to genetically modify plants so that they accumulate more of the limiting amino acid in the tissue of interest. Because the free amino acid pools comprise a relatively small proportion of the total amino acids in a tissue, this generally means increasing the protein amino acid content. Substantial progress has been made toward identifying the genes for storage proteins (Chapter 31) and it is now possible in some cases to envision modifying these genes so that they encode proteins containing more of the limiting amino acids. This is conceptually a simple thing to do. However, answers to a number of questions must be known before this approach is practical. For example, cereal prolamins form insoluble aggregates called protein bodies inside the lumen of the rough endoplasmic reticulum. Although the storage proteins are not catalytically active, the structure of these proteins may have been strongly selected for properties related to their high-level accumulation in protein bodies. Changes must not perturb the normal protein structure, otherwise the proteins might be unstable during the long time between seed maturation and seed germination, or could prevent the protein from undergoing its normal dessication process. It is also necessary that the structure of these proteins be compatible with the proteolytic enzymes produced by the plant during germination, so peptides and amino acids can be rapidly released and transported to the developing seedling. In order to meet these requirements it should be possible to introduce genes for heterologous storage proteins from species with a suitable amino acid composition rather than restructuring endogenous storage protein genes. For example, transfer of the soybean 11S globulin to maize would result in a well-balanced amino acid composition (Larkins, 1983).

The eventual success of the approach depends upon the ability of the relevant tissue to produce increased amounts of the necessary amino acids. Detailed studies of the regulation of biosynthesis of the aspartate family of amino acids suggested that the production of amino acids by this pathway is regulated at the level of feedback inhibition. In support of this, it has been possible to increase the synthesis and accumulation of the aspartate family of amino acids by genetically modifying the allosteric properties of aspartate kinase so that it is insensitive to feedback inhibition by threonine or lysine. Lysine also regulates its own synthesis by feedback inhibition of dihydrodipicolinic acid synthase, and threonine inhibits some forms of homoserine dehydrogenase (Miflin et al., 1983). Thus, it seems probable that by increasing the demand for certain amino acids, the biosynthetic pathways will respond by increasing the rate of synthesis. The eventual production of plants with improved amino acid composition will provide a superior diet for humans unable to supplement their diet with non-plant protein, will result in more efficient use of agricultural produce for animal food and should replace the use of chemically synthesized amino acids in dietary supplements for animals.

Elimination of toxic constituents

Many plants, particularly those adapted to growth in developing countries, have toxic constituents which limit the use of the plants. Thus, for instance, castor beans (*Ricinus communis*) accumulate an industrially useful seed oil, ricinoleic acid. However, it has been largely discontinued as a crop in North America because it also contains a very toxic lectin and a poisonous alkaloid, ricinine, which prevents economical utilization of the meal and causes sensitivity in farm workers (Lord *et al.*, 1984). Another example is the pulse crop *Lathyrus sativus*, a relative of the sweetpea (*L. odoratus*), which contains a neurotoxin (β-*N*-oxalyl-α, β-diamino propionic acid). This does not, however, prevent its unfortunate use as a traditional food crop in India (Nerkar, 1976). Many of the improvements made by plant breeders have involved reduction or elimination of such constituents. A notable example is the breeding of improved oilseed rape (Canola). In older varieties, up to 48% of the fatty acids in seed triglycerides were erucic acid (22:1). Erucic acid is considered undesirable in edible oil because it affects the spreading characteristics of margarine made from the oil and, more importantly, was found to be deposited in the heart muscle of experimental animals fed with diets rich in rapeseed oil. This condition, lipidosis, was associated with the appearance of myocardial lesions. By screening a large number of rape accessions, a line was found with reduced eicosenoic and erucic acids (Boulter, 1983). The new varieties, which contain no erucic and only 1% eicosenoic (20:1) have reduced activity of the enzymes which sequentially elongate oleic acid (18:1) to eicosenoic, and eicosenoic to erucic (Fig. 33.1). Thus,

Although not desirable in edible oil, erucic acid is a valuable industrial oil and, in this context, production of rapeseed oil containing 100% erucic acid would be a valuable trait (Knauf, 1987). Rapeseed oil normally has erucic acid only in the *sn*-1 and *sn*-3 positions of triglycerides and, therefore, rapeseed oil contains at most 66% erucic Presumably, this is due to the substrate specificity of the acyltransferase which esterifies fatty acids to the *sn*-2 position of the glycerol moiety during lipid biosynthesis (Fig. 33.1). Other plants such as *Nasturtium* have erucic in all three positions of storage triglycerides. By determining the biochemical basis for this difference it may eventually be possible to move the relevant genes from *Nasturtium* into rape and, thereby, produce a rape variety with very high erucic acid content. The greater uniformity of the oil composition would increase the value of the oil and would make this vegetable oil more attractive as a starting material for chemical syntheses which

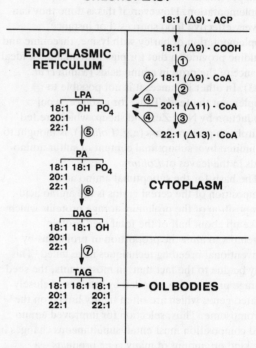

Fig. 33.1 A simplified scheme for the biosynthesis of triglycerides containing erucic acid (22:1) and eicosenoic acid (20:1). The steps leading to further desaturation of oleic acid (18:1) and the involvement of acyl exchange and biosynthesis of triglyceride from phosphatidylcholine are omitted for clarity. The enzymes indicated are: (1) acyl-CoA synthetase, (2) oleoyl-CoA elongase, (3) eicosenoyl-CoA elongase, (4) glycerol 3-phosphate acyltransferase, (5) lysophosphatidic acid acyltransferase, (6) phosphatidic acid phosphatase, (7) diacylglycerol acyltransferase. The various lipids are abbreviated as: LPA, lysophosphatidic acid; PA, phosphatidic acid; DAG, diacylglycerol; TAG, triacylglycerol. The structures of the various lipids are shown schematically by a horizontal 'E', representing the glycerol backbone, in which the range of possible acyl groups at each carbon of the glycerol moiety are listed.

would otherwise be petroleum based.

Another problem constituent of rape is a class of sulfur-containing compounds called glucosinolates (Tookey *et al.*, 1980). These compounds are present in a number of other cultivated plants and are responsible for the pungent flavors of the condiments horseradish and mustard, and contribute to the characteristic flavors of turnip, cabbage and related vegetables. In the case of rapeseed, the sulfur in these compounds poisons the nickel catalyst used in the hydrogenation reaction associated with oil processing. Also, some of the aglycones which are produced by myrosinase cleavage of glucosinolates during crushing are goitrogenic in animals. It is thought that they act by inhibiting the incorporation of iodine into precursors of thyroxine and prevent the secretion of thyroxine. In older rapeseed cultivars this severely depressed the value of the protein-rich meal left after oil extraction since it could not be used in large quantities as animal feed. Modern cultivars have been selected to contain very low levels of these compounds, although substantial research continues into the biosynthesis and function of these compounds. There is evidence that the presence of these compounds in leaves prevents predation by some insects (although they may serve as chemoattractants for other species). Since insects generally do not damage the seed, one goal of current research is to understand how to prevent expression of the glucosinolate biosynthetic pathways in seeds without affecting accumulation in vegetative tissues.

The problem of tissue specificity of gene expression has also been the focus of substantial applied research in potato (*Solanum tuberosum*). Wild accessions of this and related species of the Nightshade family contain steroid alkaloids which are poisonous to insects, microorganisms and mammals (Wink, 1988). As a result of careful breeding, the alkaloids have been reduced in tubers but were retained in leaves. Evidence for the wisdom of retaining these compounds in leaves is inferred from the results of selection for cultivars of potato resistant to the Colorado potato beetle (*Leptinotarsa decemlineata*). When such cultivars were selected they were found to have very high levels of alkaloids and were no longer suitable for human consumption because the levels were also increased in the tubers. Characterization of the genes which control alkaloid biosynthesis should permit a directed genetic modification which may overcome this problem.

In one unusual case it is the total protein content of a tissue that is deleterious. This situation occurs in the use of alfalfa as a forage crop for ruminants. In spite of the tremendous productivity of this crop, green alfalfa cannot be provided directly to ruminants as fodder in large quantities because the high protein content of the leaves causes the formation of a protein foam on the surface of the rumen contents. This acts as a gastrointestinal plug which causes the animal to bloat and frequently die. In related legumes such as trefoil, this is not as severe a problem because the presence of tannins or related compounds in the leaves causes the precipitation of protein which prevents this. Alfalfa has genes for condensed tannin production in the seed coats, but these genes are not expressed in leaves. Thus, it may be possible to genetically modify alfalfa by either introducing the genes for tannin synthesis in leaves from a related species or, possibly, to alter the regulation of the endogenous genes for tannin biosynthesis so that they are expressed in alfalfa leaves. This problem represents one of many potential applications for detailed knowledge of the mechanisms which regulate tissue-specific gene expression.

Although, as noted above, identification of natural variation and the induction of variation by chemicals or ionizing radiation (Gottschalk and Wolff, 1983) have been very useful for eliminating unwanted constituents, there are limitations to the utility of these approaches, particularly in polyploids where there may be many copies of a given gene, or in outcrossing species where it can be extremely difficult to isolate recessive mutations. Thus, there is widespread interest in the possibility that it will eventually be feasible to specifically inactivate an endogenous gene by techniques which involve introducing cloned copies of the gene into an organism. Such techniques are routine in bacteria and yeast and have recently been extended to mammalian cells. This will open up many more opportunities to eliminate unwanted metabolic pathways.

Horticultural qualities

Since many plants are valued for primarily aesthetic reasons, the biochemistry and genetics of flower color is one of the oldest and most thoroughly developed aspects of plant biochemistry. Indeed, chromatography takes its name from the techniques developed to separate plant pigments. It may not be surprising, therefore, that one of the first applications of genetic engineering in plants has been the creation of a new flower color. *Petunia hybrida* is one of the species in which the pathway of anthocyanin biosynthesis has been analyzed genetically and biochemically. In petunia, the blue and red cyanidin and delphinidin derivatives are produced as pigments, but the orange pelargonidins are not. This is due to the substrate specificity of the dihydroflavonol 4-reductase of petunia, which cannnot reduce dihydrokaempferol to leucopelargonidin (Fig. 33.2). Although maize does not have pigmented flowers in the usual sense, many maize lines accumulate anthocyanin pigments in the aleurone or in vegetative tissues. Introduction of the maize gene for dihydroquercetin 4-reductase into petunia resulted in the formation of several pelargonidin derivatives which resulted in flowers with a brick-red coloration which had never before existed in petunia (Meyer *et al.*, 1987). The importance of this example extends far beyond the domain of flower color. It represents the first example of the creation of a new metabolic pathway in a plant by interspecies transfer of genes.

A particularly interesting and unexpected outcome of some of the first genetic engineering experiments serves to illustrate the possibilities that may exist to alter the characteristics of woody species which are used primarily for lumber and which are difficult to breed by conventional approaches because of the very long times to maturity. It was recently observed that expression of one of the auxin biosynthetic genes (tryptophan monooxygenase) from the T-DNA of *Agrobacterium tumefaciens* in transgenic plants altered the morphology of the vascular tissue so that more secondary xylem and phloem cells developed than in wild-type plants of the same age (Klee *et al.*, 1987a). This is presumably due to overproduction of auxin in the vascular tissues. This observation suggests that it may eventually be possible to change the quality of wood by further elaboration of such approaches.

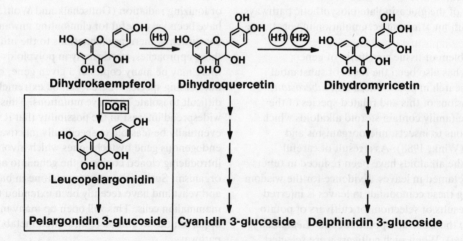

Fig. 33.2 The biosynthetic pathway for formation of a new flower color in petunia. In a wild-type petunia, dihydrokaempferol is converted by the *Ht1* gene product (flavonoid 3'-hydroxylase) to the colorless compound dihydroquercetin which can be converted by the *Hf1* and *Hf2* gene products (flavonoid 3',5'-hydroxylase) to dihydromyricetin. These compounds are, in turn, converted by a series of enzymatic reactions to the pigments cyanidin 3-glucoside and delphinidum 3-glucoside. In order to create a new color, a mutant with white petals, lacking activity for the *Ht1* and *Hf1* gene products, was transformed with the *A1* gene of maize. This gene encodes the enzyme dihydroquercetin 4-reductase (DQR) which converts dihydroquercetin into leucocyanidin, leading to the production of the novel brick-red pigment pelargonidin 3-glucoside in the petals.

Novel uses for plants

Industrial fatty acids

Modern agriculture is largely dependent on a relatively small number of plant species that are adapted to large-scale mechanized farming practices. These are used primarily to produce starch, sugar, oil, protein, fiber and animal fodder. Although some of these products, such as starch, can be chemically modified to produce an enormous number of different products, there are certain things that cannot be economically produced from available species. For instance, there are no field crops that produce short chain fatty acids such as lauric (12:0) which are used in large quantities in detergents. These compounds are obtained primarily from palm and coconut which are hand-harvested. Since it would be advantageous to be able to produce these compounds in a field crop that can be mechanically harvested, it is of interest to determine the feasibility of genetically modifying an oilseed species such as sunflower or oilseed rape so that it produces short chain fatty acids instead of primarily oleate (18:1) or linoleate (18:2) (Knauf, 1987). In principle, this might be accomplished by introducing the relevant genes from a genus such as *Cuphea*, which primarily accumulates short chain fatty acids, into an oilseed such as sunflower. However, the mechanisms by which plants such as *Cuphea* synthesize and store short chain fatty acids in the triglyceride fraction (but not the membrane fraction) has not yet been determined. Thus, the genetic engineering of a 'laurate trait' into a temperate zone crop remains speculative until more basic knowledge about the biochemical mechanisms regulating fatty acid synthesis is gained. Alternatively, it may also be possible to utilize non-plant genes to modify the pathway of lipid synthesis in plants. For instance the acyl–acyl carrier protein thioesterase II found in mammary glands generates short chain fatty acids in mammalian tissue by removing the acyl carrier protein (ACP) from acyl chains of 10,12 and 14 carbons in length. This raises the possibility that expression of a mammalian acyl–ACP thioesterase II gene in an oilseed might result in production of an oil containing significant levels of short chain fatty acids (Knauf, 1987).

In addition to chain length there are many other modifications of lipid composition which might be attractive. For instance, the presence of double bonds, hydroxyl groups or other functional groups confer useful properties on fatty acids. Hydroxylated fatty acids are used in the manufacture of jet lubricants, as solvents during the synthesis of plastics, as a component of linoleum, printers ink, dyes, cosmetics, soaps, hydraulic fluid and many other products. A major limitation to the large-scale use of modified plant oils for industrial purposes appears to be that many useful oils are produced by plants which are not suitable for domestication, and are, therefore, expensive or in short supply (Pryde *et al.*, 1981). Thus, it should be possible, by identifying the genes for the enzymes which introduce specific modifications into fatty acids and introducing them into a suitable host species, to modify the storage lipid component in the host species. Obviously, it will be necessary to know a great deal more about the physiological consequences of modifying lipid structure and also about how lipid structure is related to the underlying cell biology which supports the development of oil bodies and membranes before it is possible to undertake these modifications.

Production of secondary metabolites

It has been estimated that higher plants produce more than 100 000 secondary metabolites of which the chemical structures of more than 15 000 have reportedly been characterized (Wink, 1988). The major categories of these compounds are shown in Table 33.1. Many of the compounds are used as drugs. The magnitude of this use is indicated by the fact that in 1980, American consumers purchased about $8 billion worth of prescription drugs derived solely from higher plants (Balandrin *et al.*, 1985). Plants containing pyrethrins, rotenoids and alkaloids are also important as sources of pesticides. For example, the worldwide demand for pyrethrum flowers is about 25 000 tons annually. It is estimated that 150 million flowers are hand-harvested daily in Kenya, Tanzania and Ecuador.

Many secondary metabolites which are useful are not economically feasible to produce from plants because of the agronomic properties of the producing plant, or because the compound is present

Table 33.1 Classes of secondary metabolites of higher plants for which structures have been determined (from Wink, 1988)

Compounds	Number of structures
Monoterpenes	1000
Sesquiterpenes	1500
Diterpenes	1000
Triterpenes/steroids	800
Tetraterpenes	350
Polyketides	750
Flavonoids	1200
Phenylpropanoids	500
Amines	100
Non-protein amino acids	400
Cyanogenic glycosides	50
Glucosinolates	100

in small amounts or in a specific tissue. It may eventually be possible to produce some of these compounds by first cloning the relevant genes for biosynthetic enzymes, then transferring the genes into a suitable species. In many cases, such as in the creation of a new flower color in petunia, it may require only one or a few genes to divert an existing biosynthetic pathway toward production of a desired metabolite. Alternatively, in some cases, the amount of production of a secondary metabolite may be increased by altering the expression of the genes encoding key enzymes. This has been accomplished in several instances by selection for high level accumulation of secondary metabolites in plant cells growing in culture. Thus, for instance, shikonin, a red-colored napthoquinone produced by cell cultures of *Lithospermum erythrorhizon*, is now produced commercially from cell cultures which were repeatedly selected for high-level accumulation (Yamada and Fujita, 1983). The shikonin is used as a dye and as a topical analgesic.

Enhanced productivity

Enhanced photosynthesis

The general goal of a great deal of conventional plant breeding has been to increase the yield of some component of crop biomass per unit of land area. Because yield is a measure of the concerted action of many or all of the genes of an organism throughout the life cycle, the character is very complex and our understanding of the physiological basis of yield remains very incomplete. Attempts to approach the problem in a systematic fashion have generally focused on the identification of variation in a probable component of yield and the subsequent analysis of the effect of this variation on yield in otherwise identical material. In this respect, since plant growth is ultimately dependent upon the net amount of CO_2 fixed, there has been substantial interest in the possibility of increasing yield by identifying variants of crop species with enhanced rates of photosynthetic CO_2 fixation. Although this concept is intuitively attractive, it has not been possible, until recently, to demonstrate a strong correlation between the amount of photosynthesis and biological yield. This has been an important point because of the possibility that other factors may limit yield to such an extent that variation in photosynthetic capacity is irrelevant. However, it now appears that the difficulty in establishing a correlation was due to the way photosynthesis was measured; generally, short-term measurements of single leaf photosynthesis at a specific stage of development. Recent studies, in which the amount of photosynthesis of an entire canopy was measured throughout the growing season, demonstrated a linear relationship between net photosynthesis and yield (Christy and Porter, 1983).

A second line of evidence may be inferred from CO_2 enrichment studies in which plants were grown in partially or completely enclosed environments which permitted the CO_2 concentration to be artificially increased from about 330 $\mu l \, l^{-1}$ to as much as 1350 $\mu l \, l^{-1}$. Because net photosynthesis by C3 species in normal field conditions is generally not saturated by air levels of CO_2, the enrichment strongly stimulated net photosynthetic CO_2 fixation. Prolonged growth of several species, including wheat and soybeans, in CO_2 enriched atmospheric conditions resulted in significantly enhanced yield (Somerville and Somerville, 1985). The implication is that, if it were possible to enhance net photosynthesis by modifying the amount or kind of one or more enzymes associated with photosynthetic

CO_2 fixation, it should be possible to obtain similar increases in yield without CO_2 enrichment.

Many factors regulate photosynthetic CO_2 fixation in a network of complex interactions (Woodrow and Berry, 1988). This raises the possibility that there may be many ways to enhance net CO_2 fixation by genetic modification. However, most attention has been focused on the possibility of enhancing photosynthesis by modifying Rubisco so that it has a reduced ratio of RuBP oxygenase to RuBP carboxylase activity. This would, in principle, enhance net photosynthesis by eliminating photorespiration, which does not appear to play any essential function (Somerville and Somerville, 1985). Substantial progress has been made in determining the structure of the protein and in elucidating the molecular details of the reactions catalyzed by Rubisco (Lorimer et al., 1986). However, there is substantial uncertainty as to whether it will be possible to modify the enzyme in a desirable manner. On the one hand, Ogren and his collaborators have demonstrated that the enzyme from various species, ranging from photosynthetic bacteria to higher plants, have different ratios of RuBP carboxylase to RuBP oxygenase activity (Jordan and Ogren, 1981). They have noted that the existence of variation implies that the ratio is not immutably fixed by the inherent constraints of the chemistry of the RuBP carboxylase reaction. On the other hand, the available evidence suggests that there may be an inverse relationship between the degree of specificity of the enzyme and the rate at which it catalyzes RuBP carboxylation (Somerville, 1986). This apparent correlation, which is consistent with a general theory of enzyme reaction mechanisms, raises the possibility that it will not be possible to reduce the ratio of RuBP oxygenase to RuBP carboxylase activity without incurring an unacceptable reduction in the rate of catalysis. It is to be hoped that additional studies of the mechanistic basis of catalysis by this important enzyme will provide a clear answer to these questions.

Altered development

Many potential approaches to enhancing productivity involve modifications of plant architecture or development (Austin et al., 1986). For instance, a large proportion of soybean flowers abort early in development, limiting the seed yield. This can be prevented by local application of cytokinins to the flowers (Crosby et al., 1981). Although the molecular basis for this effect remains unknown, it may also be possible to obtain the same effect by introducing the gene for cytokinin biosynthesis from *Agrobacterium tumefaciens* into soybeans by Ti plasmid-mediated transformation. By placing the gene behind a promoter which specifically expresses the gene in developing flowers, it may be possible to mimic the effects of exogenous cytokinin application without causing deleterious side-effects on other tissue. The first steps in this direction have been taken by the expression of the cytokinin biosynthetic genes in *Petunia* (Klee et al., 1987b). There may be many other applications of this kind when more is known about the details of phytohormone biochemistry and mode of action. Furthermore, not all manipulations need involve increases in phytohormone levels. If the pathways by which the phytohormones are degraded were known, it might be possible to genetically manipulate the levels of these compounds in plant tissues by varying the expression of the degradative enzymes in a temporal or spatial fashion.

A well-known example in which alteration of phytohormone physiology has had an impact on productivity is the development of non-lodging dwarf wheats. The dwarf character is beneficial in several ways. First, it reduces the amount of photosynthate and nutrients which are utilized to produce non-harvested material (e.g. straw). Also, it prevents lodging of the wheat under conditions of nutrient availability and growing conditions which would otherwise lead to the elongation of internodes to the point that the plant cannot support vertical growth. Once lodging has taken place, the grain is susceptible to disease and predation, cannot be efficiently harvested and may be of poor quality because the grain does not dry properly. Although the precise biochemical lesion which gives rise to the dwarf phenotype is not known, it has been suggested that in the dwarf wheats the level or activity of a gibberellin receptor has been reduced (Reid, 1986).

Plant protection

An integral aspect of agricultural practices is the protection of crop species against competition by weeds and against predation by plant pathogens, insects and other pests. The prevalent solution to these problems has involved the use of large quantities of herbicides, insecticides and related agricultural chemicals. At least one widely used herbicide has recently been classified as a carcinogen and it seems possible that other agricultural chemicals will be found to have deleterious effects on some component of the ecosystem. Since it does not seem likely that such chemicals will fall from use in the forseeable future, it is important that we employ all available means to facilitate the design of safe and effective chemicals. Recent advances in plant molecular biology and biochemistry have opened up many new opportunities for improved plant protection with decreased utilization of pesticides and herbicides.

Herbicide resistance

The essential problem in designing a herbicide is to identify a chemical which kills weeds but does not harm the crop species, which may be the same genus as the weed, and is not toxic or harmful to other life forms. Because the angiosperms are relatively recently evolved from a common ancestor, it is extremely difficult to identify selective chemicals. In general, tens of thousands of compounds are screened in order to find one useful herbicide and the cost of discovery may, therefore, be in the hundreds of millions of dollars. One result of the high costs is that effective compounds are not available for many of the crop species in which the total acreage is not high enough to permit recovery of the high development costs through sales of herbicides. Because of the importance of chemical weed control in determining the economic feasibility of growing a particular crop, the lack of agrochemicals for minor crop species has reduced the diversity of agriculture in many developed countries at a time when it would otherwise be advantageous to diversify.

Recently, the discovery of the mode of action of several herbicides has opened a new approach to the development of herbicides. For example, three structurally different classes of herbicides, the sulfonylureas, the imidazolinones and glyphosate have been found to be potent inhibitors of amino acid biosynthesis in plants (Comai and Stalker, 1986). The sulfonylureas and imidazolinones inhibit acetohydroxyacid synthase, the first enzyme of the branched chain amino acid pathway. Glyphosate (N[phosphonomethyl]-glycine) is a competitive inhibitor of 5-enolpyruvylshikimic acid-3-phosphate (EPSP) synthase, an enzyme of the aromatic amino acid pathway. The compounds are highly specific for their target enzymes and are not toxic to mammals, which do not synthesize these amino acids but obtain them from nutritional sources. Certain plant species are also resistant to the herbicides because they appear to detoxify the compounds by a variety of modifications which may involve enzymes such as mixed function oxidases, amidases, decarboxylases and thiol, sugar and amino acid conjugative enzyme systems similar to those used in vertebrate liver to detoxify a wide variety of compounds. As a specific example, maize is resistant to the photosynthetic inhibitor Atrazine because it possesses an enzyme (glutathione-S-transferase) which conjugates Atrazine with glutathione, thereby rendering the herbicide non-toxic.

Mutant lines of several plant species have been isolated by selecting for resistance to these herbicides. Glyphosate-tolerant cell lines were recovered in which a gene amplification increased the amount of EPSP synthase. Resistance was also achieved by placing a cloned gene for EPSP synthase under transcriptional control of a strong promoter in transgenic plants. Mutants have also been found which have an alteration in the structural gene for the target enzyme of the herbicide, so that much higher quantities of the herbicide were required to inhibit the enzyme. Introduction of the mutant genes into other plant species resulted in the production of a herbicide-resistant enzyme which conferred resistance at the whole plant level (Haughn et al., 1988).

Mutant genes have also been identified which confer resistance to herbicides, such as the triazines, which act by inhibiting photosynthetic electron transport. The triazines are structural analogs of the mobile electron carrier, plastoquinone, and inhibit Photosystem II activity by competing with plastoquinone for binding to the reaction center polypeptide. After several decades of repeated use of triazine herbicides, resistant weed populations arose. These were found to have an amino acid substitution in the Photosystem II reaction center polypeptide so that the protein was able to transfer electrons to plastoquinone at nearly normal rates but had greatly reduced affinity for triazines. The herbicide-binding protein is encoded by the chloroplast genome. Since it has not yet been possible to reproducibly introduce genes into the chloroplast genomes of plants, genetic engineering of these herbicides may not be practical. Also, the alteration of electron transport components appears to be associated with a slight decline in yield.

The importance of developing genes which confer herbicide resistance in transgenic plants is that it reduces the stringency of the constraint that the herbicide must be selective for a specific weed. Thus, it should be possible to design herbicides which are optimized with respect to ecological and safety considerations. Because, as a general rule, genes from one plant species may function normally in many other plant species, it should also be possible to introduce the same herbicide resistance genes into a wide range of plant species. It may then be possible to use the same herbicide on many crop species which might not otherwise be feasible to grow because of the lack of suitable herbicides.

Pest resistance

Modern agriculture, horticulture and forestry depend upon the use of a wide variety of insecticides to control insect damage. Recently, several novel experiments have suggested the possibility of attractive alternate strategies for controlling insect pests. One approach has been to introduce genes for the protein toxins produced by *Bacillus thuringiensis* or *B. israelensis* and related species into plants (Vaeck *et al.*, 1987). These highly specific 'Bt-toxins' are active against Lepidoptera, Diptera and Coleoptera but have no toxic effects in other organisms. The proteins are toxic to certain insects because they inhibit an ion pump in the insect gut. Expression of a gene for a Bt-toxin in transgenic tobacco prevented feeding damage by larvae of the tobacco hornworm.

A related approach has been to exploit the antifeedant effects of proteinase inhibitors. As their metabolic target is the catalytic site of an enzyme, the ability of the insects to evolve a resistance mechanism based on mutation at this site should be minimal. Feeding trials with purified trypsin inhibitors incorporated at physiological levels into artificial diets showed these to be antimetabolic agents against a wide variety of insects which cause losses of major economic importance. Ingestion of these proteins by insects prevents the digestion of materials in the insect gut and, therefore, applied selection pressure against insects. The trypsin inhibitor of cowpea (*Vigna unguiculata*) is a small

Fig. 33.3 Effect of expression of the cowpea trypsin inhibitor gene in the leaves of transgenic tobacco plants on predation by *Heliothis virescens* larvae. The plant on the left is a control plant, and the plant on the right was transformed with a cowpea trypsin inhibitor gene. (From Hilder *et al.*, 1987; reproduced with permission.)

polypeptide of about 80 amino acids. It has recently been demonstrated that the introduction of a gene for cowpea trypsin inhibitor into transgenic tobacco provided resistance to the serious economic pest *Heliothis virescens*, the tobacco budworm (Fig. 33.3; Hilder *et al.*, 1987). These experiments demonstrated the importance of proteinase inhibitors in plant defense. The way is now open to introduce this and other proteinase inhibitors into other crop species. This could have a profound effect on the economics of growing certain crops, such as cotton, where insecticide costs are a major component of the cost of production, and should be preferable on the basis of ecological considerations. Many other polypeptides which inhibit proteases are known. In some cases these may be expected to be deleterious to the plant cell if expressed in the cytoplasm at high concentrations. However, it may be possible to target such toxic polypeptides to the cell wall, where protein turnover is thought to be low.

In principle, the general approach of producing endogenous pesticides need not be restricted to insect pests. Many plants produce secondary metabolites which prevent predation by insects, or the growth of fungal or microbial pathogens (Wink, 1988). The ability to transfer the genes for synthesis of secondary metabolites from one organism to another, and to precisely control the tissue-specific expression of these genes, will open many new opportunities in rational pest control through genetic engineering.

Concluding remarks

Plant breeders have made tremendous progress in adapting plants for human use by the patient application of relatively simple genetic methods. This 'classical' approach is now being supplemented by a variety of new techniques that have broken the species barrier which has previously limited the range of genetic material available to the breeder. Although the techniques of molecular biology have only recently begun to be applied to plant improvement, it is already apparent that there are numerous novel opportunities for rational genetic improvement of plants by manipulation of only one or a few genes. In most cases, recognition and exploitation of the opportunities depends upon detailed knowledge of the biochemical and physiological factors underlying the trait. It is to be hoped that we are on the threshold of an era in which the artificial distinctions which are commonly used to divide knowledge into compartments such as chemotaxonomy, physiology, biochemistry, genetics, pathology and plant breeding will be blurred by the combined application of knowledge from these subject areas to the rational improvement of plants.

References

Austin, R. B., Flavell, R. B., Henson, I. E. and Lowe, H. J. B. (1986). *Molecular Biology and Crop Improvement*, Cambridge University Press, London.

Balandrin, M. F., Klocke, S. A., Wurtele, E. S. and Bollinger, W. H. (1985). Natural plant chemicals: sources of industrial and medicinal materials. *Science* **228**, 1154–60.

Boulter, G. S. (1983). The history and marketing of rapeseed oil in Canada. In *High and Low Erucic Acid Rapeseed Oils*, eds J. Kramer, F. Sauer and W. Pigden, Academic Press, Toronto, pp. 62–82.

Christy, A. L. and Porter, C. A. (1983). Canopy photosynthesis and yield in soybean. In *Photosynthesis: Development and Plant Productivity*, ed. Govindjee, Academic Press, New York, pp. 499–512.

Comai, L. and Stalker, D. (1986). Mechanism of action of herbicides and their molecular manipulation. *Oxford Surv. Plant Molec. Cell Biol.* **3**, 166–95.

Crosby, K. E., Augn, L. H. and Buss, G. R. (1981). Influence of 6-benzylaminopurine on fruit-set and seed development in two soybean, *Glycine max* (L.) Merr. genotypes. *Plant Physiol.* **68**, 985–8.

Gottschalk, W. and Wolff, G. (1983). *Induced Mutations in Plant Breeding*, Springer-Verlag, Berlin.

Hallauer, A. R. (1981). Selection and breeding methods. In *Plant Breeding II*, ed. K. J. Frey, Iowa State University Press, Ames, pp. 3–57.

Haughn, G., Smith, J., Mazur, B. and Somerville, C. R. (1988). An *Arabidopsis* acetolactate synthase gene in tobacco confers resistance to sulfonylurea herbicides. *Molec. Gen. Genet.* **211**, 266–71.

Hilder, V. A., Gatehouse, A. M. R., Sheerman, S. E., Barker, R. F. and Boulter, D. (1987). A novel mechanism of insect resistance engineered into tobacco. *Nature* **330**, 160–3.

Jordan, D. B. and Ogren, W. L. (1981). Species variation in the specificity of ribulose bisphosphate carboxylase/oxygenase *Nature* **219**, 513–15.

Klee, H. J., Horsch, R. B., Hinchee, M. A., Hein,

M. B. and Hoffmann, N. L. (1987a). The effects of overproduction of two *Agrobacterium tumefaciens* T-DNA auxin biosynthetic gene products in transgenic petunia plants. *Genes Develop.* **1**, 86–96.

Klee, H., Horsch, R. B. and Rogers, S. B. (1987b). *Agrobacterium*-mediated plant transformation and its further application to plant biology. *Ann. Rev. Plant Physiol.* **38**, 467–87.

Knauf, V. C. (1987). The application of genetic engineering to oilseed crops. *Trends Biotechnol.* **5**, 40–7.

Larkins, B. A. (1983). Genetic engineering of seed storage proteins. In *Genetic Engineering of Plants: An Agricultural Perspective*, eds T. Kosuge, C. P. Meredith and A. Hollaender, Plenum Press, New York, pp. 93–118.

Lord, J. M., Lamb, F. I. and Roberts, L. M. (1984). Ricin: structure, biological activity and synthesis. *Oxford Surv. Plant Molec. Cell Biol.* **1**, 85–103.

Lorimer, G. H., Andrews, T. J., Pierce, J. and Schloss, J. V. (1986). 2′-carboxy-3-keto-D-arabinitol 1,5-bisphosphate, the six carbon intermediate of the ribulose bisphosphate carboxylase reaction. In *Ribulose bisphosphate carboxylase-oxygenase*, eds R. J. Ellis and J. C. Gray, The Royal Society, London, pp. 93–105.

Meyer, P., Heidmann I., Forkmann, G. and Saedler, H. (1987). A new petunia flower colour generated by transformation of a mutant with a maize gene. *Nature* **330**, 677–8.

Miflin, B. J., Bright, S. W. J., Rognes, S. E. and Kueh, J. S. H. (1983). Amino acids, nutrition and stress: The role of biological mutants in solving the problems of crop quality. In *Genetic Engineering of Plants: An Agricultural Perspective*, eds T. Kosuge, C. P. Meredith and A. Hollaender, Plenum Press, New York, pp. 391–414.

Nerkar, Y. S. (1976). Mutation studies in *Lathyrus sativus*. *Indian J. Genet. Plant Breed.* **36**, 223–9.

Pryde, E. H., Princen, L. H., Mukherjee, K. D. (1981). *New Sources of Fats and Oils*, American Oil Chemists Society, Champaign.

Reid, J. B. (1986). Gibberellin mutants. In *A Genetic Approach to Plant Biochemistry*, eds A. D. Blonstein and P. J. King, Springer-Verlag, New York, pp. 1–26.

Somerville, C. R. (1986). Future prospects for genetic manipulation of Rubisco, *Phil. Trans. R. Soc. London B.* **313**, 305–24.

Somerville, C. R. and Somerville, S. C. (1985). Regulation of photorespiration. In *The Biochemical Basis of Plant Breeding*, ed. C. A. Neyra, CRC Press, Boca Raton, pp. 86–131.

Tookey, H. L., Van Etten, C. H. and Daxenbichler, M. E. (1980). Glucosinolates. In *Toxic Constituents of Plant Foodstuffs*, ed. I. E. Liener, Academic Press, New York, pp. 103–41.

Vaeck, M., Reynaerts, A., Hofte, H., Jensens, S., De Beuckeleer, M., Dean, C., Zabeau, M., Van Montagu, M. and Leemans, J. (1987). Transgenic plants protected from insect attack. *Nature* **328**, 33–7.

Wink, M. (1988). Plant breeding: importance of plant secondary metabolites for protection against pathogens and herbivores. *Theor. Appl. Genet.* **75**, 225–33.

Woodrow, I. and Berry, J. (1988). Enzymatic regulation of photosynthetic CO_2 fixation in C_3 plants. *Ann. Rev. Plant Physiol.* **39**, 533–94.

Yamada, Y. and Fujita, Y. (1983). Production of useful compounds in culture. In *Handbook of Plant Cell Culture*, Vol. 1, eds D. A. Evans, W. R. Sharp, P. V. Ammirato and Y. Yamada, Macmillan, New York, pp. 717–28.

Index

*Page numbers in **bold face** refer to a major text discussion of the entry.*

abscisic acid 354
absorbance of chlorophyll 216
absorption spectra for pigment–protein complexes (figure) 214
accessory pigments 215
acetaldehyde 88
acetate 52, 113, 359
acetoacetyl-CoA 342, 355
acetoacetyl-CoA thiolase 355
acetohydroxyacid synthase 498
acetolactate synthase 52
acetyl-CoA 108, 109, 110, 114, 121, 135, 341, 355
    hydrolysis 52
    in biosynthetic reactions 121
    in cytosol 358
    oxidation 114
    source of in plastids 52, 358, (diagram) 359
acetyl-CoA:ACP transacylase 342
acetyl-CoA carboxylase 341
acetyl-CoA hydrolase 358
acetyl-CoA:malonyl-CoA transcarboxylase 342
acetyl-CoA synthetase 52, 120, 341, 358
acetylene reduction 390
acetylglucosamine 463
acetylserine 415
aconitase **109**, 182, 183, 351
aconitate 119
    content of tissues 111
actinomycete 378
acyl carrier protein (ACP) 52, **341**
    cDNA for 343
    phosphopantetheine in 341
    structure 341
    two forms of 343
acyl exchange reaction 349
acyl-ACP thioesterase 342
acyl-ACP thiotransferase II, in mammary glands 495

acyl-CoA hydrolase 120
acyl-CoA synthetase 345
acyl-CoA:lysophosphatidylcholine acyltransferase 350
acyltransferase 344
    role in fatty acid composition of oils 492
adenine nucleotides as regulators 33
adenylate kinase 33
adenylate levels, in light and dark 141
adenylate translocator 50
ADP-glucose 33, 65–69
ADP-glucose pyrophosphorylase 50, 67, 286, 287, 323
    regulation of 68–69, 318
ADP-glucose synthetase 33
AGGA sequence 18
Agrobacterium rhizogenes 476
Agrobacterium tumefaciens 475, 476
    advantages and limitations of 482
    binary transformation systems 480, (diagram) 481, 482
    chemotactic response of 476
    genes and events during transformation (diagram) 478
    infection of plants by 476
    transformation by (diagram) 477
Agrobacterium-mediated gene transfer **476–482**
alanine 116
alanine dehydrogenase (reaction) 408
alanine synthesis **412–413**
albumins 456
alcohol dehydrogenase 87, **88**, 89
    anaerobic induction of 88
    genes for 88
    isozymes of 88
    role in fermentation 88
aldolase **77**, 89, 241, 242
    anaerobically induced 77

cDNA clones of 77
classes of 77
aleurone layer 453, 462, 467
alfalfa
   genetic modification of 493
   protein content of leaves 493
allantoic acid 419
   export of 437
allantoin 332, 419
   export of 437
   synthesis (diagram) 420
allophycocyanin 214
allosteric effectors 32
allosteric enzyme 33
allosteric regulation 32
allosteric regulators 33
allosteric site 32
alternative electron transport pathway **127–129**, 424
   for energy overcharge 140
   for energy overflow 140
   for NADH oxidation 140
   heat production in 140
   inhibition of 131, 138, 139
   partitioning of electrons to **135–137**
   physiological function of 139
alternative oxidase 125, 131, 133, 139
ambiquitous enzymes 37
amino acid biosynthesis **407–419**
   anaplerotic reactions in 411
   inhibitors of 498
   requirement for carbon skeletons 424, 426, 428
amino acid families 407
amino acids
   as secondary products 407
   catabolism of 116
   functions of 407
   non-protein 418
aminooxyacetate 256
aminopeptidases 466
aminotransferases 255, **411–412**
ammonia assimilation 8, **383–386, 398–402, 408–409**
   control of 383
   energetics of **402–404**
   energy costs 402–403
   regulation of 401
   release of H^+ during 435
ammonium uptake 398, 399
amoeboid plastids 195
amphipathic lipids 339
amplification of DNA 17
amplification cascade 36

amylase 67, 70, 71, 287
amylopectin 66, 69
amyloplast 48, 60, 68, 92, 195, 196, 199, 200, 323,
   electron micrograph of 195
   transport of carbon into 48
   starch deposition in 322
amylose 66, 69
amyrin 357
amytal 130
Anabaena 376
   genes in 8
   heterocyst 7
   promoter in 8
anabolic pathways 34
anaerobic metabolism **87–90**
   enzymes of 87
   ethanol production in 87
   glycolysis during 87
anaerobic proteins **89**
anaerobic regulatory element 11
anaerobiosis 87
anaplerotic reactions **119**, (diagram) 428
anoxia-intolerant plants 89
anoxia-tolerant plants 89
antenna pigments, energy migrations in (diagram) 217
anthranilic acid 417
antimycin 127, 131–133
antiport 176, 177
antisense mRNA 7
   regulation by **6**
apoplast 434
   in sucrose transport 63
apoplastic transport 440
aquatic macrophytes **266–267**
   CO_2 compensation point in 266
Arabidopsis
   photorespiratory mutants of 255–259
   starchless mutant of 67
arabinose 5-phosphate 86
arabinose 5-phosphate ketol isomerase 86
arginase 409
arginine biosynthesis 52, **416–417**
arginosuccinate 416
arogenate dehydratase 418
arogenate dehydrogenase 418
arogenate pathway 418
aromatic amino acid biosynthesis 52, **417–418**
Arum maculatum, thermogenesis in 66, 74, 75
asaparagine, in glycosylation of proteins 463
asparaginase 413
   diurnal variation in 413

turnover of 413
asparagine 338, 413
  export of 437
  in chloroplasts 412
  increase and zinc deficiency 409
  increase under stress 409
  product of nodule 409
  synthesis 409
asparagine synthetase 409, (reaction) 408
aspartase (reaction) 408
aspartate 116, 118–120, 383
aspartate amino acid family 414, 491
aspartate aminotransferase 180, 383
aspartate dehydrogenase (reaction) 408
aspartate kinase 414
  genetic modification of 491
  isozymes of 414
  regulation of 414
aspartate semialdehyde 414
aspartate semialdehyde dehydrogenase 414
aspartate synthesis **412–413**
assimilate partitioning 434
ATP hydrolysis in mitochondria (diagram) 186
ATP synthase 131, 174, 176, 198
  gene for 147
  in chloroplast 221
  model of 133
  proteolipid channel in 221
ATP synthesis (mitochondrial) **133–134**
ATP synthesis (in chloroplast) **212–221**
ATP synthesis, *in vivo* **134**
ATP-generating pathways 34
ATP/ADP antiporter 174
atractyloside 174
atrazine 498
azaserine (AZA) 258
  structure of 400
azide 131
azido-2-nitrophenyl-4-aminobutyryl-3′-NAD (NAP^4-NAD^+) 186
Azolla 378
Azotobacter 376

Bacillus thuringiensis 487, 499
bacterial genome 149
bc_1 complex 124, 127
benzhydroxamic acid 131
benztricarboxylate 183
benzyladenine 130

binary vectors 482
biochemical unity 28
biotin 342
biotin carboxyl carrier protein 342
biotin carboxylase 342
blue light mediator 208
bongkrekic acid 174
Bradyrhizobium 376
branched chain amino acids 415
  biosynthesis of 52
branching enzyme 66, 69, 286
  multiple forms of 69
bromelain 448
Bt-toxin 499
bundle sheath cells 264, 276, 279
  decarboxylase enzymes in 276
butylmalonate 178, 180, 183

C3-C4 intermediates **265–266, 281–282**
  bundle sheath in 266
  CO_2 compensation point in 281
  glycine/serine shuttle in 281
  Kranz anatomy in 281
  leaf anatomy of 266
  NADP malic enzyme in 281
  phosphoenolpyruvate carboxylase in 281
  pyruvate phosphate dikinase in 281
C4 dicarboxylic acid pathway 274, (diagram) 275
C4 photosynthesis **274–287**
  diagram of pathway 275
  inorganic carbon pools in 281
  intercellular transport and leaf anatomy **277–279**
  intracellular transport in **282–284**
  metabolite levels in (figure) 285
  metabolite transport in (diagrams) 278, 282
  metabolites in bundle and mesophyll cells (diagram) 280
  mitochondria in 283
  movement of inorganic carbon in 281
  regulation by metabolite transport **284–287**
  regulation of starch synthesis in 286
  regulation of sucrose synthesis in 286
  symplastic transport 277, 281
  transport in bundle sheath chloroplast 283
  transport in mesophyll chloroplasts 282–283
  transport through plasmodesmata 277
  intercellular transport in **276–282**

C4 plants 254, **264–265**
  decarboxylase enzymes in 276
    Kranz anatomy in 264
    metabolite flux in **274–287**
    Rubisco activity in 265
CAAT box 11
CAAT sequence 18
carboxydismutase 224
calcium
  binding proteins 39
  channels 38
  modulated proteins 39
  pumps 38
  transport 38
  intracellular concentration 38
  role in metabolism 38–41
calcium-proton antiport 38
callose 436
calmodulin 38, 39
CAM plants 254, **265**
  chloroplast metabolite transport in 291
  diagram of pathway 275
  dicarboxylate shuttle in 290
  glycolysis in 289
  intracellular metabolite transport in **287–291**
  malate efflux from vacuole 290
  malic acid and glucan levels in (figure) 287
  malic enzyme in 265
  metabolic regulation in 288
  organelle transport in 290
  phosphate translocator in 290
  phosphoenolpyruvate carboxykinase in 265
  phosphoenolpyruvate carboxylase in 265
  properties of phosphoenolpyruvate carboxylase in 292
  regulation by metabolite transport (diagram) 291
  regulation of carbohydrate metabolism in 293
  regulation of futile metabolite cycles 292
  subtypes of 288
  titratable acid in 265
  tonoplast malate transport (diagram) 289
CaMV 8
  polyadenylation signal in 8
canavanine 419
Canola oil 492
CAP 5
cap binding protein 9
carbamoyl phosphate 416
carbamoyl phosphate synthetase 52
carbohydrate metabolism, compartmentation of **91**
carbohydrate storage in internodes 443

carbon budget in leaf 443
carbon dioxide concentrating mechanisms 264
carbon flow in amino acid biosynthesis (diagram) 410
carbon monoxide 131
carbon partitioning **442–443**
carbon transport **439–441**
  needs of 434
carbonic anhydrase 267, 269
carboxyarabinitol 1,5-bisphosphate 240
carboxyarabinitol 1-phosphate 226, 227, 230, 245
carboxyatractyloside 174, 185
carboxyl biotin 342
carboxypeptidases 466
cardiolipin (CL) 340, 348
carnitine:acetylcarnitine translocator 121
carotene 214, 353, 357, 366,
carotene biosynthesis **365–366**
carotenoids 212, 214, 215
carotenoid-protein complexes 215
casbene 367
casbene synthetase 367
  gene regulation 367
  mRNA for 367
castor beans, toxins in 491
castor endosperm 47
catabolic pathways 34
catalase, mutants of in barley 256
cauliflower mosaic virus (CaMV) **7–8**, 475, 483
  35S promoter in 475
  gene arrangement in 8
caulimoviruses 483
Calvin–Benson cycle 239
cellular redox state 35
cellulose synthesis 73, 74
chalcone synthase 486
channeling of metabolites 111
channeling of substrates 37
chaperonin 305
charge separation 217
chemical ecology 407
chemiosmotic theory 131, 137, 221
chimeric precursor genes 302
chimeric protein constructs 303
chloramphenicol 156, 205
chloramphenicol acetyltransferase (*cat*) 303, 475, 486
  gene for 23
chloramphenicol, acetylation of 475
chlorohydroxamic acid 131
chloromercuribenzoate 185
chlorophyll 212, 215
  structure of (diagram) 215

fluorescence of 217
light absorbance by (figure) 216
luminescence of 217
phosphorescence of 217
phototrap in 217
chlorophyll a/b binding protein 9
chlorophyll-protein complexes 215
chloroplast 254
    ATP synthase mutant 426
    biogenesis 207
    development **195**, 202
    dicarboxylate carrier 282
    DNA 228
    oxaloacetate carrier 282
    pyruvate carrier 282
    export of photosynthate from 311
    fatty acid biosynthesis in 341
    products of photosynthesis in 311
    protein degradation in 450
    origin of proteins in (diagram) 301
chloroplast genome 152
    differential expression in C4 plants 276
chloroplast protein uptake **301–307**
    binding of precursors 304
    chaperonin in 305
    chimeric proteins in 302
    cleavage of precursor 302
    diagram of 302
    energy requirement 302, 304, 305
    import apparatus of 303, 304
    methods of studying 302
    organelle specificity 303
    post-translational 302
    precursor protein binding 302
    precursor structure 302
    processing protease 304
    protein assembly in 305
    protein conformation in 304
    proteolytic processing of precursor 305
    receptor proteins 304
    targeting of chlorophyll binding protein 305
    targeting of lumen proteins 306
    targeting of plastocyanin (diagram) 306
    targeting within chloroplast 305
    transit peptide domains 306
    translocation across membrane 304
    translocator protein 305
    two step model for targeting 306
chloroplast structure **194–195**, (diagram) 301
    proteinaceous particles in 195
chloroplast transport **309–315**

ATP/ADP exchange 314
control of 314
dicarboxylate transporter **313**
glycerate and glycolate transport **313**
method of study 309–310
organic acid transport **313**
outer membrane permeability 310
oxaloacetate transporter **313**
phosphate translocator **311–312**
porins 310
pyruvate translocator **313**
chlororespiration 425
chorismate mutase 417
chorismic acid 417
chromoplast 195, 196, 199, 200, 358
    development **196**
    DNA in 202
cinnamic acid 380
cistron 4
cis-acting DNA sequences **5**, **10**, 22
cis-acting regulatory sequences 486
citrate 109, 119, 120, 182, 436
citrate/malate exchange 183
citrate content of tissues 111
citrate lyase 358, 360
citrate synthase 28, **109**, 182, 183, 351
citric acid cycle 134, 135, 177, 182, 186, 351, 427
    activities of enzymes in (table) 112
    activity in the light 113
    anabolic role of **118–121**, (diagram) 428
    anapleurotic reactions in **119**, 182, (diagram) 428
    catabolic role of **114–118**, (diagram) 428
    control of 121
    diagram of 107
    during photosynthesis 424
    enzymes of **107–110**
    evidence for the operation of **111–114**
    level of intermediates in 112
    maintenance of carbon flow in 422
    maintenance of (diagram) 428
    metabolism of labeled substates 113
    organization of **111**
    reactions of **107–111**
    removal of intermediate from 118, 427–429
    respiration of amino acids in **115**
    respiration of carbohydrate in **117**
    role in N assimilation (diagram) 430
    supramolecular organization of 111

citroyl-CoA 109
citrulline 416, 419
CL (cardiolipin) 340
CO_2 compensation point 225, 232, 259, 264 281
CO_2 concentrating mechanisms **264–271**
    in cyanobacteria **267–271**, (figure) 268
    in microalgae **267–271**, (figure) 268
cocoa oil 350
coenzyme-A 341
    accumulation in mitochondria 187
companion cells 440
compartmentation of metabolism 45
competitive inhibition 33
complex I 134, 186
    inhibition of 130
    iron-sulfur proteins in 126
    proton translocation in 126
coconut oil 495
complex II 134
    inhibition of 130
    iron-sulfur proteins in 126
complex III 127
    inhibition of 131
    Q-cycle in 127
    Rieske iron-sulfur protein in 127
complex IV, inhibition of 131
complex V, model of 133
condensing enzyme 342, 355
conglycinin 21, 22, 23, 463
    genes in soybean 19
consensus sequence 4
control analysis theory 29
convicilin 17, 21
cooperative photosynthesis 276
copalyl pyrophosphate 364
copper 127
cotranslational protein secretion 205
coupling site I 126
covalent modification of enzymes 34
co-translational protein transport 162
crossing over 8
cuticular lipids 339
cyanelles 199
cyanide 131
cyanoalanine 409
cyanobacteria 214, 267, 376, 378
    genes 7
Cyanophora paradoxa 199
cyano-4-hydroxycinnamate 177
cyclohexamide 156

cyclopropyl carbinyl pyrophosphate 365
cyclic electron flow 220
cystathionine 416
cystathionine lyase 416
cystathionine synthase 416
cysteine 409
cysteine synthase 415
cysteine synthesis **415–416**
    control of 415
cytochrome *550*, 208
cytochrome *a* 131
cytochrome a_3 131
cytochrome *b* 127, 131
cytochrome b_6/cytochrome *f* complex 220
cytochrome *c* 125, 127
cytochrome *c* oxidase 124, 127, 163, 165
cytochrome *c* reductase 127
cytochrome c_1 127
cytochrome P_{450} 364
cytokinesis 194
cytokinins 363, 497
    genes for biosynthesis of 497
cytoplasmic male sterility 156–158
    nuclear restorer genes 157
cytosol, pools of metabolites in 315
C-550 219

dark respiration, in light **140–141, 427–430**
DCMU 220
dehydroquinate dehydrase 52
dehydroquinate synthase 52
deoxy D-arabino heptulosonic acid 7-phosphate (DAHP) 417
deoxy-D-arabinoheptulose 7-phosphate synthase 52
de-branching enzyme 70, 71
diacyl glycerol 344, 345, 347
diaphorase 374
diazo-5-oxo-L-norvaline 258
dicarboxylate translocator 119
dicarboxylate transporter in chloroplast 259, 282
    counter exchange mechanism 313
dicarboxylate/ketoglutarate exchange 179
dichlorophenyl-dimethyl urea (DCMU) 220
dichlorophenyl-indophenol 255
digalactosyl diacylglycerol (DGDG) 340, 345
dihydroflavanol 4-reductase 494
dihydrofolate reductase 164, 483
dihydrokaempferol 494
dihydropicolinate synthase, regulation of 414
dihydropicolinic acid 414

dihydropicolinic acid synthase 491
dihydroxyacetone phosphate 77, 79, 239, 241, 242
dihydroxyacetone phosphate reductase 344
dihydroxyflavon 380
dihydroquercetin 4-reductase, gene for 494
dimethylallyl pyrophosphate 353, 361
dinitrogen reductase 391
dinitrophenol 138, 139
dipeptidases 466
diphosphoglycerate 241
disulfram 131
diterpene biosynthesis **363–365**
diterpenes 353
dithiol-disulfide interconversion of enzymes 34
dithionite 389
diurnal rhythms of enzymes activities 319, 320
DNA binding protein 5
DNA transfer
    liposome mediated **485**
    without vectors **483–486**
DNA uptake
    by electroporation 484
    by plant cells **483–485**
DNA
    amplification of 17
    endoreplication of 17
    hairpin in 6
    microinjection of **485**
DNA-dyad symmetry in 11, 12
dolichol pyrophosphate 463
dwarf wheat 497
D-enzyme 70, 72

endosymbionts 198, 199, 200, 204
eicosenoic acid, biosynthetic pathway for (diagram) 492
eicosenoic acid, in rapeseed 492
elaioplasts 196
electron acceptor, primary 218
electron transport pathway (mitochondrial) **124–128**, 134, 422, 423
    complex I 124, **125–126**
    complex II 124, **126–127**
    complex III 124, **127**
    complex IV 124, **127**
    components of **124–127**
    control by adenylates 138, 139
    cyanide resistant path **127–129**
    diagram of 125
    external NAD(P)H dehydrogenase 127
    inhibition by high ATP/ADP ratio 424

in vivo control **138–139**, (table) 138
    major substrates for 129
    regulation of **137–139**
    role in N assimilation (diagram) 430
    rotenone-insensitive dehydrogenase 125
electroporation 7, 484
elicitors 367
embryogenesis 17
    amount of DNA during 17
endoamylase 70, 71, 321
endogenous signals 13
endopeptidases 466
endoplasmic reticulum 460, 462
endoreplication of DNA 17
endosperm 16
energy charge 33, 361
energy costs of metabolism 402–403
energy transduction, efficiency of 216
enhancer core sequence 24
enhancer elements 10–12
    organ specific 23
enolase **81**, 89
    in plastids 47
    isozymes 48, 81, 92
enolpyruvylshikimate 3-phosphate synthase 52, 498
enoyl-ACP reductase 342
enzyme cascades 109
enzyme complexes 37
enzyme kinetics 31
    allosteric 33
    competitive inhibition 33
    interacting effectors 33
    mixed inhibition 33
    noncompetitive inhibition 33
enzymes
    regulation of 31–40
    activation 32
    inhibition 33
    rate-limiting 29, 30
    allosteric regulation of 32
    allosteric site of 32
    ambiquitous 37
    association of sequential enzymes 37
    calcium modulated 39
    channeling of subtrates 37
    control by pH 31
    control by substrate concentration 31
    cooperative substrate binding in 31
    covalent modification 34
    dithiol-disulfide interconversions 34

light activation of 32
micro-compartmentation of 37
multi enzyme cascades 40
multimeric 36
multisubunit 32
pacemaker 29
phosphorylation-dephosphorylation 35
polymeric 31
rate determining 36
subunit association-disassociation 36
erucic acid 350
   biosynthetic pathway for (diagram) 492
   in *Nasturtium* 492
   in rapeseed 492
erythrose 4-phosphate 81, 85, 242, 411, 417
ethanol 87, 88
ethoxyzolamide 267
ethylene synthesis 409, 416
etioplast structure **196**
exchange diffusion carriers 177
excitation energy 216
excited singlet state 215, 216, 217
excitons 216, 217
   energy of 217
   transduction of energy from 217
   transfer between photosystems 222
exogenous signals 13
exons 8
exon-intron boundary 8
exopeptidases 466
export, efficiency of 437
external NADH dehydrogenase 134
extrachromosomal element 147

farnesyl pyrophosphate 353
farnesyl pyrophosphate synthetase 362
fatty acid biosynthesis 52, 91, **340–343**
   enzymes of 341
   isozymes in 343
   location of genes for 343
   multiple genes for 343
   source of ATP and NADPH 343
   subcellular location of 341
fatty acid conversion to sucrose (diagram) 352
fatty acid desaturase 348
fatty acid distribution, eukaryotic 344, 347
fatty acid distribution, prokaryotic 344
fatty acid modification in seeds **349**
fatty acid oxidation 135, 351
fatty acids 120

in membranes (table) 340
desaturation of 348
elongation of 350
seed composition of 350
utilization of **351–352**
FCCP 138
ferredoxin 35, 220, 245, 389, 396
   A and B 218
   linked carboxylases 239
   NADP oxidoreductase 220
   thioredoxin reductase 35, 37
ferredoxin/thioredoxin **245–246**
   function of (diagram) 246
ficin 448
fix genes **381**
flavin adenine dinucleotide 126
flavodoxin 389
flavonoids 380, 411
fluorosucrose 64
folic acid in methionine biosynthesis 416
formyl-methionine 205
fraction I protein 224
Frankia 378
fructans 443
fructokinase 73
fructose 1,6-bisphosphatase 68, 241, 244, 248, 286
   effect of genetic factors 318
   effect of temperature on 318
   in amyloplasts 48
   in plastids 323
   inhibition by F2,6BP 315, 316
   inhibition by metabolites 318
fructose 1,6-bisphosphate 65, 74, 77, 241, 286
fructose 2,6-bisphosphatase 248, 286, 316, 319, 320
   regulation of 248
fructose 2,6-bisphosphate 36, 74, 75, 248, 286, 294, 319, 439
   as an indicator of photosynthate 316
   concentration of 317
   diagram of effects 316
   regulation of concentration (table) 248
   regulation of metabolism by **315–320**
   role in sucrose metabolism (diagram) 248
fructose 6-phosphate 59, 60, 74, 84, 85, 90, 241, 319
fructose 6-phosphate 2-kinase 248, 286, 316, 317, 319, 320
   diurnal rhythm in 320
   protein modification in 320
   regulation of 248
fruit
   carbon budget of 445
   carbon/nitrogen relationships in 445

nitrogen budget of 445
uptake of carbon/nitrogen into (diagram) 445
fumarase **110**
fumarate 110, 111
furanosesquiterpene 361
futile cycling 45, 244

galactolipids 339
galactosidase 4
galactosyltransferase 345
gas exchange 424, (figure) 260
geminiviruses 483
gene expression in plants 486
    development mutants of 24
    developmental regulation of 22
    effects of nutrition on 24
    environmental effects on 24
    initiation of 22
    regulation in transgenic plants 22
    regulation of 24
    temporal changes in 20, 21
    trans-acting factors in 23
    upstream regulatory sequences 486
    tissue specific 459
gene expression in transgenic plants 483
gene families 17
gene transfer **473–489**, 490
    by viruses **482–483**
    Agrobacterium-mediated **476–482**
    applications for agriculture 487
    applications for basic studies **486**
    marker genes for 474
    reporter genes for 475
    selectable markers for (table) 475
gene vectors 480–482
genes
    Anabaena 8
    cyanobacterial **7**
    eukaryotic 5, **8–13**
    expression during seed development 19
    expression of storage protein genes 22
    operators 5
    operon of 3, 4
    organellar **14**
    photosynthetic 12
    prokaryote **3–8**
    promoter **4**
    regulation of 13
    repressor 4, 5

tandem 4
transcription of 4
zein 12
genetic engineering 22, 480
    for altered development 497
    for enhanced CO_2 fixation 496
    for enhanced photosynthesis 496
    for enhanced productivity **496–497**
    for herbicide resistance **498–499**
    for new biosynthetic pathway 494
    for new flower color 494
    for novel fatty acids 495
    for pest resistance 499–500
    for secondary plant products **495–496**
    in horticulture **493–495**
    in plant protection 498–500
    plant improvement by **490–501**
    to eliminate toxins 491
    to improve quality 490
    to improve storage proteins 490
    to produce novel plants 495
genetic recombination 8
genome, organellar 14
geranyl pyrophosphate 353
geranyl pyrophosphate synthetase 362
geranylgeranyl pyrophosphate (GGPP) 353
geranylgeranyl pyrophosphate synthetase 362
gibberellane skeleton 364
gibberellic acid 453, 467
gibberellin A_{12} 364
gibberellin biosynthesis (diagram) 364
gibberellins 354, 363
    inactivation of 367
glycerolipid synthesis, on membranes (diagram) 346
glycerolipids in membranes (table) 340
gliadin 17, 457
globulin 24, 456
glucan synthesis 73
glucokinase 73
gluconeogenesis 90, 178, 114
glucono 1,5-lactone 6-phosphate 83
glucose 1-phosphate 59, 60, 65, 66, 67
    uptake by plastids 323
glucose 6-phosphate 59 , 60, 83,90
glucose 6-phosphate dehydrogenase **83**, 91, 95
    in plastids 47
    isozymes of 83
    light regulation of 83, 95
glucose 6-phosphate isomerase 59, 60, 67
glucosidase 67, 70, 71
glucosinolates

as chemoattractants 493
as insect repellants 493
biochemical pathway for 493
goitrogenic 493
in hydrogenation of oils 493
in rapeseed 492
pungent flavour in 493
glucosyltransferase 70
glucuronidase gene 475
glutamate 116, 118–120, 383
glutamate dehydrogenase **401**, 116, 259, 383, 408, 409
    anaplerotic role of 401
    in chloroplast 401
    kinetics of 401
    localization of 401
    mutations of 401
    reaction of 401
    role in ammonia assimilation 408
glutamate oxidation (diagram of) 184
glutamate semialdehyde 417
glutamate synthase (GOGAT) 183, 258, 399, **400**, 408 409
    cycle 408, 409, 411, 413
    ferredoxin dependent 258, 400
    inhibition by azaserine 400
    NAD(P)H dependent 258, 400
    reaction of 400
    types of 400
glutamate, structure of 400
glutamate-oxaloacetate transaminase 52
glutamate-pyruvate transaminase 52
glutamate/aspartate exchange 184
glutamate/dicarboxylate exchange 184
glutamate:glyoxylate aminotransferase 255, 256
    in peroxisome 259
glutamine 383
glutamine oxoglutarate aminotransferase (GOGAT)
  see glutamate synthase
glutamine synthetase (GS) 183, 258, 381, **383–386, 399–400**
    antibodies to 384
    catalytic mechanism of 400
    cDNA clones for 385
    in chloroplasts 384
    in cytosol 384
    in nodules 384
    in roots 384
    induction of (figure) 384
    inhibition by phosphinothricin 386
    isozymes of 50, 258, 384–385, 399
    methionine sulfoximine inhibition 400
    molecular weight of 384
    multigene family of 386
    other reactions of 399
    reaction of 399
glutamine, structure of 400
glutamyl 5-semialdehyde 416, 417
glutamyl phosphate, structure of 400
glutathione, oxidized 35
glutathione-S-transferase 498
glutelins 17, 456, 459, 463
glyceollin 130
glyceraldehyde 3-phosphate 77, 79, 84, 85, 90, 135, 240, 241
glyceraldehyde 3-phosphate dehydrogenase (NADP) 80, 241
    regulation by light 245
glyceraldehyde 3-phosphate dehydrogenase (NAD) **79** 87, 89
    cDNA clone for 80
    isozymes of 79
    kinetics of 80
glyceraldehyde 3-phosphate dehydrogenase (non-phosphorylating) 80
glycerate 254, 258
glycerate dehydrogenase 414
glycerate kinase 258, 265
glycerol 3-phosphate 344
glycerolipids 346
    membrane composition of (table) 339
    classes (table) 344
glycerolipid synthesis **343–349**
    acyltransferases in (diagram) 345
    location of 344
glycine 125, 129, 137, 138, 181, 254
    decarboxylation 258
    metabolism in mitochondrion 116, 117
    oxidation 185, (scheme for) 181
    synthesis, photorespiration in 413
    transporter 256
glycine decarboxylase complex 116, 256, 283
    NADH produced by 423
glycinin genes in soybean 19
glycolate 253, 254, 330, 331, 360
glycolate dehydrogenase 255
glycolate oxidase 255, 331
glycolate transporter in chloroplast 255
glycolipids 345
glycolysis 66, **77–82**, 90, 117, 129, 134, 138, 139, 177
    in CAM plants 289
    definition of 77
    diagram of pathway 78

enzymes of 77
in plastid 47, 91, 92, 322
isozymes of **92**
localization of **91**
long term regulation of 95
particulate 91
rate of 74–75
regulation of **93–95**
glycoproteins modification of 463
glycosome 91
glycosylation 334
glyoxisomes **330–331**
electron micrograph of 330
enzymes in 331
fatty acid oxidation in 330
induction during germination 331
progenitors of 335
synthesis during seed development 331
glyoxisome-peroxisome transition 332
glyoxylate 87, 255, 330
glyoxylate cycle 114, 330, 351
induction during germination 331
glyoxysome 114, 178, 201, 351
glyphosate 498
glyphosate tolerance, gene amplification in 498
Golgi apparatus 465
membrane receptors in 465
grana 194, 196
ground state of pigment 215
GS/GOGAT pathway 383, figure 374
gyrase 204

heat shock proteins in maize 90
heat-shock response 89
herbicides, detoxification of 498
Henderson-Hasselbach equation 132
herbicide binding protein, degradation of 450
herbicide resistance 487
herbicides, mode of action of 498
hesperitin 380
heterocyst of *Anabaena* 7, 8
heterogeneous nuclear RNA 8
hexokinase 64, 67, 70, 72, 73, 94
mitochondrial association of 91
mitochondrial location of 73
hexose phosphate isomerase 89
genes for isozymes 93
hexose phosphate pool **59–60**, 64, 74
hexose phosphates **59–60**, 62, 66, 68, 74, 75, 239
compartmentation of 60

in plastids 60
uptake into amyloplasts 50
hexose transporter **312–313**, 321
substrates for 313
inhibition by phloretin 313
hexose import into plastids 323
hexoses, metabolism of **72–73**
Hill coefficient 31
hnRNA 8
Hogness box 10
homocysteine 416
homoserine 414
homoserine dehydrogenase 414, 491
homoserine, in peas 414
hordein mutants 18
hordeins 17, 21
housekeeping genes 13
hydrogen cyanide 409
detoxification of 409
hydroxamic acids 131
hydroxpyruvate 180
hydroxyacyl-ACP dehydratase 342
hydroxybutyryl-ACP 342
hydroxykaurenoic acid 364
hydroxylamine 256
hydroxymethyltransferase 181
hydroxyphenylpyruvate 412, 418
hydroxypyridine methane sulfonate 255
hydroxypyruvate 255, 257, 258, 414
hydroxypyruvate reductase 331
hydroxy-3-butynoate 255
hydroxy-3-methylglutaryl-CoA 355
hydroxy-3-methylglutaryl-CoA reductase 355
in plastids 357
phosphorylation of 360
phytochrome control of 360
regulation of 360
hydroxy-3-methylglutaryl-CoA synthase 357
hyperbolic saturation kinetics 31
H^+/Pi symporter 132

imidazolinones 498
in vitro translation of mRNA 302
indole 417
inducer 4
inducer-repressor-complex 4
inhibitors of electron transport 130, 131
inhibitors of oxidative phosphorylation 130, 131
initiation codon 9
initiation enhancers 10

initiation of transcription 10
inorganic carbon, active uptake of **268–270**
inositol triphosphate 41
intergalactolipid galactosyltransferase 345
intervening sequences (introns) 8, 13, 14, 18, 228
   classes of 204
   junctions 9
invertase 63
   acid in vacuole 63, 64
   alkaline 63, 64
isocitrate 109, 114, 119, 120, 125
isocitrate dehydrogenase (NAD) **109–110**
   control by NAD^+:NADH
   oxalosuccinate as an intermediate 109
isocitrate dehydrogenase (NADP) 183
isocitrate lyase 114, 334, 351
isoflavonoids 380
isoleucine 415
isoleucine biosynthesis **414–415**
isonicotinyl hydrazide 256
isopentenoid 353
isopentenyl pyrophosphate 353
isopentenyl pyrophosphate biosynthesis **355–361**,
   diagram 356
     compartmentation of **357–358**
     precursors for 358
     regulation of **360–361**
isopentenyl pyrophosphate isomerase 361
isoprene 353
isoprenoids 120, 353
isozymes 48

kanamycin resistance 474, 484
   resistance of seedlings to (figure) 474
kaurene 361, 364
kaurene synthetase 364, 366
keto 2-carboxyarabinitol 1,5-bisphosphate
   227, 231
ketoacyl-ACP reductase 342
ketobutyrate 415
ketoglutarate 383, 411
ketoglutarate dehydrogenase 182, 184
ketoglutarate oxidation 179
ketoisocaproate 412
ketoisovalerate 415
ketomethylvalerate 414
ketosuccinamate 409
Klebsiella 376
K_m values 31

Kranz anatomy 264, 274, 276
   micrograph of 275

lactate dehydrogenase **87**, 89
   genes for 88
   hypoxically-induced 87
   induction by oxygen deficit 87
   isozymes of 87
lactic acid 87
lactose operon 4, 5
lactose permease 4
lambda virions 481
lauric acid 350, 495
leader peptide 5, 6
leaf
   carbon economy in (diagram) 444
   carbon/nitrogen relationships in **443–445**
   nitrogen economy in (diagram) 444
lectin gene 20, 23
   in soybean 19
leghemoglobin 381, 382, 392
legume symbiosis 389
legumin 17, 21, 22, 25
   box 18
   gene 20
   in peas 20
   RNA 24
leucine 412
leucine synthesis **415**
leucopelargonidin 494
leucoplasts 199
   development of **196**
   electron micrograph of 197
light absorption in photosynthesis **212–217**
   use of visible spectrum 212
light energy capture **215–217**
light harvesting chlorophyll binding protein 36
light harvesting complexes 222
   dephosphorylation of 222
   phosphorylation of 222
light harvesting pigments 215
light reactions of photosynthesis **212–219**
light-activated enzymes 244
light-deactivated enzymes 244
light-harvesting complexes 214
limit dextrins 70
linolenic acid 340
lipase 351
lipid bodies 456

lipid metabolism in seeds **349–352**
lipid synthesis
    in plastids 344
    in endoplasmic reticulum 346
    in mitochondria 348
    plastid/endoplasmic reticulum connection **347–349**
lipids
    desaturation 348
    modification of 348
    structure and function of **339–340**
lipoic acid 256
luciferase gene 475
luteolin 380
lycopene 366
lysine synthesis **414–415**
lysophosphatidic acid 344
lysosomes 465

maize
    endosperm 67
    amylose extender mutant 69
    brittle 2 mutant of 67
    shrunken mutant of 64, 67
    waxy 69
malate 110, 111, 120, 125, 129, 134, 135, 177, 178, 180, 184, 257, 436
    content of tissues 111
    flux across tonoplast 288–290
    uptake into mitochondrion 178
    storage in vacuole 287, 288
malate dehydrogenase 180–183, **110**, 111, 119, 351
    anaplerotic role in cytosol 429
    in mitochondrion 257
    in peroxisome 257
malate oxidation 181, 188, scheme for 182
malate synthase 334, 351
malate/aspartate exchange 180, 398
malate/oxaloacetate exchange 117, 180, 181
    in chloroplasts 257
    mitochondrion 257
malate/PEP ratios 429
malate/Pi antiport 178
malic enzyme (NADP) 280, 281, 264
    in C4 leaves 113
malic enzyme (NAD) 114, 120, 134, 178, 182, 186, 188
malonate 130
malonyl-CoA 341, 342
malonyl-CoA:ACP transacylase 342
maltosaccharides 70
maltose 70
maltose phosphorylase 72

maltotetraose 70
maltotriose 70
mannitol 438
marker genes 474, 481
Mehler reaction 425
membrane potential in mitochondrion 174
menthol 367
mersalyl 176, 179, 183, 185
mesophyll cells 264, 276, 279
    chloroplasts (pyruvate in) 282
metabolic control
    coarse 30
    fine 30
    types of 30
metabolic pathways, sequestration of 45
metabolic regulators 33
metabolic transducers 30
metabolism, control of 29
metabolon 38, 91, 93, 111
metallothionine 483
methionine 409, 414, 416
    synthesis **415–416**
methionine sulfoximine (MSO) 258
    structure of 400
methylguanosine cap 8
methyltriphenylphosphonium 132
mevalonate 353, 355, 357, 358
mevalonate kinase 358
    in plastids 357
mevinolin 357
Michaelis–Menten kinetics 31
micRNA 7
microalgae 267
microbodies **329–338**
    and ER 334
    and glycoproteins 334
    catalase in 329
    density of 329
    fatty acid oxidation in 329
    morphology of **329–332**
    oxidases in 329
    photorespiratory enzymes in 331
    protein import into **336–337**
    size of 329
    targeting signals 336
microbody growth and division 335
microbody matrix proteins **332–333**
    signal sequences 333
    synthesis of 332, 333
    alkaline lipase 333
microbody proliferation 334

microbody proteins, biogenesis of **332–334**
microinjection of DNA **485**
   intranuclear 485
micro-compartmentation of enzymes 37
micro-compartments 38
mitochondrial DNA 105, **147–148**
   animal 149
   circular form of 149
   function of 147
   genes encoded by 147, table 148
   human 149
   open reading frames in 147
   role in mitochondrial biogenesis 147
mitochondrial genes 147
   gene maps 155
   introns in 149, 150
   location of (diagram) 153
   plant **153–155**
   primary transcript of 150
   promoter 150
   protein 154–155
   ribosomal 153
   transcriptional units 150
   tRNA 154
mitochondrial genome 14, **147–158**
   circular 150, 152
   diagram of 151
   diversity of **148–150**
   evolution of **158**
   expression of **149–150**, 155
   fluid nature of 151
   gene order in 149
   heterogeneity of 150
   master chromosome of 151, 152
   multipartite model of 152
   organization of 149
   physical complexity of 150
   physical form of 148
   plant **150–156**
   promiscuous DNA in 152, 155
   repeated sequences in 151, 152
   size of (table) 148
   subgenomic molecules of 152
   transcription of 155–156
   tRNA genes in 149
   yeast 149
   yeast maturases
   yeast ORFs
mitochondrial matrix 111
mitochondrial matrix protease 166

mitochondrial mRNA 156
   5′ and 3′-non-coding regions 156
   lack of polyadenylation 156
   Shine-Dalgarno sequence in 156
   translation of 156
mitochondrial protein import **160–171**, 304
   alternative processes 166
   amphiphilic helix in presequence 163, diagram 163
   components of process **167**
   coordination of import 170
   cytosolic factors in 167
   electrochemical gradient in 162, 166, 170
   energy requirement for 166
   evolutionary model for 169
   heme addition in 166
   high energy compounds in 166
   import by isolated mitochondria 161
   import competent protein conformation 167
   in plants **170–171**
   intramitochondrial sorting **168–169**
   methods of study 160–161
   model of 161, figure 162
   phospholipid bilayer as a barrier 168
   presequence for 161, 163
   process **165–167**
   proteinaceous pore in 168
   proteolytic processing 165
   proton gradient in 166
   receptor for 162, 167
   site of protein synthesis 161
   stop transport domain 168
   targeting sequence 161, 162, 167, **163–165**, 170
   translocation contact sites 168
   two step processing 165
mitochondrial targeting sequences **163–165**, 170
   amino-terminal sequences of (table) 164
   conformation of precursors 165
   for ADP/ATP carrier 165
   for apocytchrome *c*
   fusion protein 164
   half-life of 165
mitochondrial
   adenine nucleotide carrier 176
   adenylate translocation **174–176**, diagram 175
   amino acid transport **183–185**
   amino acid transport, carrier proteins in 183
   amino acid transport passive diffusion in 183
   aspartate transport 184–185
   citrate carrier **182–1**

cofactor transport **185–188**
dicarboxylate carrier 177, **178–179**, 183
glutamate transport 184–185
glycine transport 185
inner membrane 174
ketoglutarate carrier **179**
metabolite exchange **173–190**
NAD$^+$ carrier 186
outer membrane 173
outer membrane as a molecular sieve 173
oxaloacetate carrier **179–182**
oxygen consumption 175
permeability of inner membrane 174
permeability of outer membrane 173
phosphate carrier **176–177, 182, 183**
phosphate translocation **176–177**, diagram 175
pyruvate carrier **177–178**
serine transport 18
translocation of citric acid cycle anions **177–183**
translocators in inner membrane 174
transport systems **173–185**
tricarboxylate carrier 182
uptake of adenylates **185–186**
uptake of coenzyme-A **187–188**
uptake of NAD$^+$ 186, diagram 187
uptake of thiamine pyrophosphate 188
mitochondrion 254, **103–105**
  adenine nucleotide concentration in 174
  adenine nucleotide content of 185
  arrangement in cell 104
  ATP synthase in 104
  bc_1 complex in 105
  biogenesis of 147
  carbon metabolism of **107–123**
  cristae of 103, 104
  cyanide resistance in 128
  cytochrome oxidase in 105
  development of 105
  endosymbiont precursor for 170
  intermembrane space of 173
  inner membrane of 104
  intermembrane space 103
  intracristal space 103
  matrix 103, 104, **105**
  micrograph of 103
  NADH dehydrogenase in 105
  nucleoids in 105
  number in cell 104
  origin of **158**
  outer membrane of 104
  oxygen uptake by (figure) 136
  plasmid-like molecules in 150
  promiscuous DNA 152
  protein synthesis in 147
  proton extrusion from 129
  size and shape 103
  state 3 and 4 138
  states of (diagram) 130
  succinic dehydrogenase in 105
  ubiquinone in 105
mixed function oxidase 364
mixed inhibition 33
monogalactosyl diacylglycerol (MGDG) 340, 345
monoterpenes 353
mRNA
  polyadenylation of 10
  turnover 13
  3′-non-coding sequences 18
  amounts of 22
  cap 8–10
  degradation of 13
  eukaryotic 8
  exons in 8
  initiation codon in 9
  initiation region 6
  introns in 8
  lariat in 9
  legumin 24
  monocistronic **9**
  polycistronic 4, **8**
  polycistronic in organelles 14
  poly(A)-tail in 8
  post-translational splicing of 8
  ribosome binding site in 6
  secondary structure loops in 6
  splicing 9
  stability of 22
  stabilization of 13
  structural features of 13
  translation 13
  transplicing in 14
  turnover of 12
multienzyme cascades 40
multienzyme complex 37, 38
multigene families 205, 228
multimeric enzymes 36
Munch pressure flow 435

NAD kinase 39
  effect of calcium on 39

NADH dehydrogenase 129
   gene 147
   internal 136
NADP$^+$:NADPH pool in chloroplast 220
NAD-formate dehydrogenase 256
nar 1 allele 375
nar 2 allele 375
nar 3 allele 375
neomycin phosphotransferase 303, 474
Nernst equation 132
neurotoxins, in pulse crops 492
nicotinamide adenine dinucleotide, structure of 126
nif gene cluster (diagram) 377
nif genes **381**
   promoter region of 378
   regulation 377
   operon 7
nigericin 176
nitrate assimilation **395–398**
   cellular location of 395
   regulation of 397
   release of OH$^-$ during 435
nitrate reductase 50, **373–376**, **395–398**
   alternate substrates for 396
   cDNA clone for 375, 376
   degradation of 397
   diurnal variation in 397
   effectors for 397
   electron donors to 396
   genes for 375, 376
   genes regulating induction 375
   genetic regulation of **374–376**
   in root cortex 439
   induction of 397
   isozymes of 396
   mechanism of 374
   model of (diagram) 374
   mRNA for 375
   mutants deficient in 374
   partial reactions of 396
   protease for 397
   regulatory role of 395
   structure and reactions of (diagram) 396
   structure of 374, 396–397
   substrate availability 397
   synthesis and degradation of 375
   turnover of 397
nitrate reduction 395
nitrate reduction to glutamate (diagram) 398
nitrate uptake and transport **395**
nitrate uptake by roots 395, 438

nitrifying bacteria 395, 398
nitrite reductase 50, 395
nitrite, concentrations of 395
nitrogen assimilation
   mitochondrial electron transport during 429
   photosynthetic electron flow in (diagram) 426
   scheme for 374
   supply of reductant for 398
nitrogen fixation **389–395**
   plant genes in 381
   symbiotic 376, **378–383**
   uptake hydrogenase (HUP) in 390
nitrogen fixing genes *(nif)* 377
nitrogen flow **409–410**
   in amino acid synthesis 413
nitrogen metabolism, molecular biology of **373–387**
nitrogen partitioning **441–442**
   role of stem in 441
   role of protein degradation in 442
nitrogen transport **438–439**
nitrogen, reallocation of 442
nitrogen, reductive assimilation of 426
nitrogenase 7, **376–383**, **389–395**
   regulation of gene for 377
   inactivation by oxygen 392
   acetylene reduction by 390
   alternative 391
   assembly of 378
   binding of nitrogen to 394
   characteristics of 376–377
   denaturation by oxygen 377
   electron allocation coefficient of 389
   genetic regulation of **377–378**
   hydrogen gas production by 389
   in heterocysts 392
   in prokaryotes 389
   inhibition by ADP 392
   inhibition by carbon monoxide 390
   inhibition by hydrogen 393
   inhibitors of 392
   measurement of 390
   measurement of total activity 390
   mechanism of (diagram) 393
   Mo-Fe protein cycle in (diagram) 394
   Mo-Fe protein of 391
   regulation of 391
   relative efficiency of 390
   repression of 378
   stoichiometry of 390
   structure of 377, 391
nod box 380

nod D gene 379
   role in regulation 379
nod gene cluster 379
nod genes 378–381
   common 379
   controlling nitrogenase 378
   mutation of 379
   products of 380
   promoter regions of 380
   regulation of 380
nodulation genes (*nod*) 378
nodulation
   effect of root exudates 380
   host range specificity 379
   plant control of 383
nodule specific genes (table) 382
   regulatory sequences of 382
nodule specific plant genes 381
nodules 331
nodulin genes, regulation of 382
nodulins (23 and 24) 381, 382
nodulins 381
noncompetitive inhibition 33
nonsense codons 19
non-cyclic electron flow 219, 221
non-photosynthetic plastids, transport in 320
nopaline synthase promoter 475
nopaline synthase 7
   detection of 475
novobiocin 204
nuclear genes for plastid proteins **206**
   in chloroplast development 207
   light activation of 207
   mutigene families 206
   phytochrome and 206
   polyadenylation signals in 206
   protein transport and assembly 207
   regulation of transcription of 206
   role of light in 208
   TATA box 206
nuclear genome, encoding plastid protein 205
nuclear protein factor 12
nucleoid in plastids 194
nucleoplasm 8
N-ethymaleimide 176
N-terminal modification of proteins 96

octopine 11
octopine synthase, promoter of 475
octulose 1,8-bisphosphate 86
oil storage bodies 351

oleic acid 340
oleoyl-ACP 342
oligomycin 141, 175
oligosaccharides, transport of 61
oligo(dT)sepharose 10
oncogenes 13
opaque mutants in maize 24, 25
operators 5
operon 3, 5
   lac 4
   nif 7
opines 476
organellar genes 14
   Shine–Dalgarno box in 14
ornithine 416
ornithine carbamoyltransferase 52
ornithine cycle (diagram) 416
o-phenanthroline 165
osmiophilic droplets 196
oxaloacetate 109, 114, 116, 119, 134, 177, 181, 182
   184, 257, 264, 351, 411
oxaloacetate carrier in chloroplast 282
oxaloacetate translocator 110
oxaloacetate
   chloroplast transporter for 257
   mitochondrial transporter for 257
beta-oxidation 114
   in mitochondria 114
oxidative phosphorylation **129–134**, 141
   inhibition of and photosynthesis 141
   proton extrusion in **131–134**
   proton transfer in 129
   states (1–4) 129, 130
   uncouplers of 131
oxoglutarate 110, 116, 120, 125, 140, 259
   into chloroplast 259
oxoglutarate dehydrogenase **110**, 119

P680 218, 220
   absorption of (figure) 218
P700 218
   absorption of (figure) 218
   structure of 218
pacemaker enzymes 29
palindrome 11, 12
palm oil 495
palmitoyl-ACP 342
papain 448
pathogens 367
PC (phosphatidylcholine) 340
PE (phosphatidylethanolamine) 340

peas, wrinkled 69
pelargonidins 494
pentose phosphate epimerase **86**
pentose phosphate isomerase **85**
   isozymes of 85
pentose phosphate pathway (oxidative) **82–87**, 90, 117, 135, 320, 322, 343, 411
   diagram of pathway 78
   enzymes of **83–86**
   F-type 86
   L-type 86
   regulation of **95**
   role of 83
pentose phosphate pathway (reductive) **239–251**, 411
   carboxylation phase 240, 244
   compartmentation of **247–248**
   diagram of 240
   light regulation of **244**
   reductive phase 240
   regeneration phase 240, 244
   regulation of **244–247**
pentose phosphate pathway, in plastids 47
pentylmalonate 178
PEPcase *see* phosphoenolpyruvate carboxylase
peptide hydrolases 466
peptidylglycans 199
peroxisome 117, 180, 181, 254–258, 329
   catalase in 330
   electron micrograph of 330
   from root nodules 331
   gluconeogenesis in 330
   in β-oxidation 114
   in leaves **330**
   morphology 335
   preforms 335
   proliferation of 335
   role in photorespiration 330
petite-mutation of yeast 147
Petunia hybrida, flower color in 494
petunia, biosynthetic pathway for flower color (diagram) 494
PG (phosphatidylglycerol) 340
pH balancing 436
   during NH_4/NO_3 uptake 435
pH stat 435, 436
phaseolin 17, 22
phaseolin gene 22, 23
phenazine methosulfate 255
phenolics 411
phenylalanine ammonia lyase 7
phenylalanine synthesis (diagram) **417–418**

phenylpyruvate 412, 417
phenylsuccinate 179
pheophytin 219, 220
phloem 435, 445
   sucrose in 438
phloem exudates, composition of (diagram) 438
phloem sap, pH of 436
phloem transport, mass flow in 435
phloretin 313
phosphatidylglycerol 345
phosphate 94
phosphate translocator 68, 236, 247, 310, **311–312**, 320, 322, 323
   control of 314
   effect of pH on 312, 314
   in chloroplasts 48
   in chromoplasts 48
   in different plants 312
   in root plastids 48
   influence of phosphate on 314
   kinetic properties 311, table 312
   light regulation of 314
   molecular properties of 312
   passive exchange in 312
   purification of 311
   substrates for 311, 312
   transport of divalent cations by 314
phosphatidic acid 344, 345, 347
phosphatidic acid phosphatase 345
phosphatidylcholine (PC) 340, 347
   acyl exchange in (diagram) 350
phosphatidylethanolamine (PE) 340, 347
phosphatidylglycerol (PG) 340
phosphatidylinositol (PI) 340
phosphatidylserine (PS) 340
phosphinothricin 386
phosphocholine 347
phosphoenolpyruvate 81, 94, 120, 129, 134, 323, 411, 417
   in CAM plants 66
phosphoenolpyruvate carboxykinase 264
   in C4 plants 113
phosphoenolpyruvate carboxylase 119, 129, 134, 181, 183, 224, 264, 280, 411, 427
   anaplerotic function of 428
   measured activity of (figure) 285
   modulation by light 284
   phosphorylation of 285, 292, 293
   properties in CAM plants 292
   regulation of **284–286**
phosphofructokinase 65, 74, 91, 94,
   in plastid 320, 322

isozymes 48, 92
  activation by F2,6BP 316
  cold lability of 322
  phosphorylation of 96
phosphoglucomutase 59, 60, 67, 89
phosphogluconate 84, 230, 246,
phosphogluconate dehydrogenase **84**, 91, 95
  allozymes of 84
  isozymes of 84, 92
  pyridoxal phosphate in 84
    in plastids 48
  regulation of 95
phosphoglucose isomerase, *Clarkia* mutants of 319
phosphoglycerate 81, 224, 231, 240, 243, 253, 254, 311, 359, 360,
  carbanion form of 231
phosphoglycerate kinase **80**, 241, 318
  cDNA clone for 80
  kinetics of 80
  role in generation of ATP 80
phosphoglycerate phosphatase 414
phosphoglyceromutase 52, **81**, 89, 92, 359
  2,3-bisphosphate dependent 81
  cofactor independent 81
  in plastids 47
  isozymes of 48, 92
phosphoglycolate 224, 243, 253, 254, 261, 264, 265
phosphoglycolate phosphatase 254
phosphohomoserine 414, 415
phosphohydroxypyruvate 414
phosphoinositides 41
phospholipid transport protein 347
phospholipids 339
phosphomevalonate 358
phosphomevalonate kinase 358
phosphopantetheine 341
phosphopentoepimerase 243
phosphoprotein phosphatase 35, 36
phosphopyruvate dehydrogenase phosphatase 108, 109
phosphoriboisomerase 243
phosphoribosyl anthranilate 417
phosphoribosyl pyrophosphate 417
phosphoribulokinase 243, 244, 246
phosphorylase 321, 322
phosphorylation-dephosphorylation of enzymes 35
phosphorylation/dephosphorylation of proteins 96
phosphoserine 414
photoassimilate, translocation of 443
photochemistry **217–219**
photoinhibition, 263

photophosphorylation 141, **220–221**
  cyclic 220
  proton build up in 220
  proton motive force in 220
photoreceptors 208
photoregulation of metabolism 40
  effect of calcium on 40
photorespiration 180, 181, 243, **253–264**, 331, 332, 360, 399, 424, 425, 427, 497
  definition of 253
  diagram of 253
  function of **262–264**
  in ammonia assimilation 409, 411
  measurement of 260
  NADH produced by 423
  organelles involved with 254
  post-illumination burst in 253, 261
  rate of 260
  regulation of 262
  salvage role for 262, 263
  to dissipate photochemical energy 263
  to protect against photo-oxidation 243
photorespiration and photosynthesis **259–260**
photorespiratory ammonia assimilation 428
photorespiratory nitrogen cycle **258–259**, 428
photosynthate 239
photosynthesis
  accessory chlorophylls in 212
  amino acid synthesis in 249
  available radiation for 212
  charge separation in 217
  cyclic electron flow in 220
  effect of N assimilation on **426–427**
  effect of respiration on **425–426**
  electron donors 218
  electron transport chain components 220
  electron transport in **219–220**
  energy traps in 217
  influence of phosphate on (figure) 314, (table) 315
  interaction of cytosol and plastid 250
  light-harvesting pigments in 212
  light-harvesting complexes of 214
  molecular biology of **198–211**
  oxygen evolution in 212
  photochemical reactions centers of 212
  phytochemistry of 217
  pigment-protein complexes in 214
  primary electron acceptor 218
  products of 239, **249–250**
  proton gradient in 220

provision of carbon skeletons by 426
quantum requirement for 219
quantum yield in C4 plants 259
reaction centers in 214
relationship of respiration to (diagrams) 423
relationship to N assimilation (diagrams) 423
secondary electron acceptor 218
state 1/state 2 transitions **221–223**
thylakoid membrane in (model) **213**
transmembrane electron flow 217
water splitting reaction in 219
photosynthetic carbon fixation 426
photosynthetic carbon oxidation cycle **253–264**
photosynthetic enzymes, coordinate regulation of **246–247**
photosynthetic gas exchange 232, 233
photosynthetic pigments **212–217**
  light energy capture by **215–217**
photosynthetically active pigments (table of) 213
photosystem I 198
  acceptor X in 220
  charge separation in 218
  ferredoxin in 220
  iron-sulfur proteins in 218, 220
  P700 in 218
  phototrap in 218, 222
  plastocyanin in 218
  reaction center of 215, 218
photosystem II 198, 208, 215
  core complex of 219
  C-550 and 219
  electron donor to 218
  oxygenic 219
  P680 in 218
  pheophytin in 218
  phototrap 218, 219, 222
  plastoquinone pool in 218, 219
  reaction center complex 215, 218
photosystems, location of in thylkoid 222
phototraps 217
phthalonate 180, 181
phycobilins 212, 213, 215
phycobiliprotein 215
phycocyanin 214
phycoerythrins 214
phylloquinone 363
phytin 456
phytoalexin 354, 361, 367, 380
phytochrome 40, 208, 450

enhancer element 206
protein kinase 41
degradation of 450
phosphorylation of 40
photoactivation of 40
phytoene 363, 365
phytol 353
phytosterols 357
PI (phosphatidylinositol) 340
piericidin 130
pigment-protein complexes 214, 215
  absorption spectra of (figure) 214
plant breeding 490
plant hormones 34
  in regulation of metabolism 41
plant productivity 262
plasma membrane receptor 41
plasmalemma 438
plasmid, helper in transformation 480
plasmodesmata 277
plastid cycle 196
plastid differentiation 52
plastid DNA
  bacterial promoter elements in 199
  copy number of 201
  cotranscription of genes in 201
  DNA inversions in 201
  gene content of 201
  genes encoded by (table) 200
  inverted repeat in 201
  localization of 201
  organization of 201
  polycistronic transcription units in 201
  rRNA transcription unit in 201
  single copy region in 201
  size of 201
plastid function **198**
  integration of **199–200**
plastid gene transcription **203–204**
  dyad symmetry in 203, 204
  prokaryotic nature of 203
  promoter for 203, 204
  RNA polymerase in 203
  supercoiled template in 203
  termination of 203
plastid genes
  prokaryotic nature of 199
  organization of 199
  transfer to nucleus 199
plastid genome 14, 198, **200–205**
  coding capacity of 200–201

coding capacity of 301
decoding of **203–205**
functions encoded by 200
gene map of (diagram) 202
polyploid nature of 201
plastid gyrase 204
plastid inheritance 193–194
   pollen mitosis 193
plastid mRNA 200
   polycistronic 199
plastid origin **198–199**
   as endosymbionts 199
plastid proteins, assembly of (diagram) 208
plastid replication 208
plastid ribosomes, number of 200
plastid RNA **204–205**
   introns in **204–205**
   primary tramscripts 204
   RNA maturases 204
   RNA processing 204
   RNase P 204
   splicing of RNA 204
   trans-spicing in 204
   tRNA 204
plastid ultrastructure **194–197**
   of chloroplast **194–196**
   of proplastids 194
plastid types **193**
   amoeboid plastids 193
   amyloplasts 193
   chloroplasts 193
   chromoplasts 193
   eoplasts 193
   etioplasts 193
   proplastids 193
   proteinoplasts 193
plastids 45–52
   biosynthesis in 48
   biosynthetic capacity of 46
   carbohydrate oxidation in 47
   DNA content of 46
   DNA in 193
   glycolysis in 47
   non-photosynthetic 46
   nucleoid in 193
   pentose phosphate pathway in 47
   plastoglobulin in 193
   protein secretion in **205**
   protein translation in **205**
   RNA polymerase in 200
plastid–cytosol interactions **309–326**

plastocyanin 215, 218, 220
plastocyanin-protein complex 215
plastoglobuli 196
plastome 93
plastoquinone-protein complex 218, 219, 220
plastoquinone 208, 215, 218, 220, 353, 357, 363, 498
plastoquinone pool 215, 219, 220, 222
pollen 156, 193
polyadenylation of mRNA 10
polyadenylation signal 8, 10, 18
polyisoprene 366
polysaccharide synthesis **74**
polysomes in protein import 161
polyterpenes 353
poly(A)polymerase 8
poly(A)-tail 8, 10
porine 173, 310
porphyrins 215
post-illumination burst 253, 261
post-translational protein transport 162
potato beetle 493
potato proteinase inhibitor II 12
potato, steriod alkaloids in 493
prenyl pyrophosphate 353
   compartmentation of 362
   formation of (diagram) 361
   regulation of 362
   synthesis of **361–362**
prenyl transferase 361, 362
prephenate 417
prephenate dehydratase 418
prephenate dehydrogenase 418
prephytoene pyrophosphate 365
presequence of mitochondrial proteins 303
presqualene pyrophosphate 366
Pribnow box 5
protein kinase 40
prolamellar body 196
prolamin gene 18
prolamins 456, 459, 463, 464
   in maize and barley 21
   suboptimal amino acid content of 491
proline 411
   accumulation in plants 116
   metabolism in mitochondrion 116
   levels under stress 417
proline synthesis **416–417**
proline oxidase
promoter 4, 5, 7, 8, 12, 18, 475
   bacterial 5
   consensus sequence in 4

efficiency of 5
eukaryotic **10**, 18
proplastids 194, 195, 196, 200, 201, 207
   electron micrograph of 195
propyl gallate 131
proteases **448–451**
   rate of turnover 453
   regulation of activity 453
   aminopeptidase 448
   ATP-dependent 450
   compartmentation of 454
   endopeptidases 448
   exopetidases 448
   in chloroplast 448
   in cytosol 448
   in vacuole 448
   zymogens of 453
protease inhibitors 453
   as insect antifeedants (figure) 499
protein bodies 16, 449, 456, 491
   breakdown during germination (diagram) 467
   crystalline 462, 463
   formation of in cereals 462–463
   formation of in legumes **461–462**, diagram 461
   structure of 457
   types of (figure) 458
protein degradation **448–455**
   effect of abnormal conformation on 452
   effect of charge on 452
   effect of molecular mass on 452
   effect of primary sequence on 452
   effect of protein conformation on 452
   regulation of **451–454**
   ubiquitin in 449
protein half-lives (table) 451
protein kinase 35, 36
   effect of calcium on 40
   calcium dependent 39
   cAMP dependent 316
protein modulase 37
protein phosphorylation, calmodulin dependent 40
protein toxins, genes for 499
protein turnover 448
   role in metabolic regulation 448
protein uptake into chloroplasts **301–397**
proteinases 466, 467
proteinoplasts 196
proteins, lifetime of 448
protein-disulfide reductase 35
proteoliposomes 176
protochlorophyll 196

protochlorophyllide 208, 210
proton extrusion sites 132
   stoichiometry of (table) 132
proton gradient 131
proton gradient in mitochondrion 174, 178
proton gradient in photosynthesis 220
proton motive force 131, 132, 133, 135, 137, 175, 220, 221
proton pumps 175
proton transport across tonoplast, ATP dependent 288
   pyrophosphate dependent 288
protoplast fusion for gene transfer 484
   by calcium 484
   by polyethylene glycol 484
   by polyvinyl alcohol 484
PS (phosphatidylserine) 340
pseudogene 19, 21, 22
purine synthesis (diagram) 419
pyrethrins 495
pyrimidines 118
pyrophosphatase 65, 69
   in plastids 65
pyrophosphate 33, 65
pyrophosphate fructose 6-phosphate
   1-phosphotransferase 36, 65, **74–75**, 248, 294, 316
pyrophosphomevalonate 357
pyrophosphomevalonate decarboxylase 358
   regulation of 361
pyrroline 5-carboxylic acid 416, 417
pyrroline 5-carboxylic acid dehydrogenase 116
pyruvate 81, 87, 88, 108, 110, 113, 114, 116, 117, 125, 129, 134, 135, 178, 411,
pyruvate decarboxylase 87, **88**, 89
   alcoholic fermentation and 88
   Hill coefficient of 88
   kinetics of 88
   thiamine pyrophosphate in 88
pyruvate dehydrogenase 36, 51, 52, **108–109**, 118, 178, 429
   in plastid 359
   kinase 108, 109
   control of 121
   dihydrolipoamide acetyltransferase in 108
   dihydrolipoamide dehydrogenase in 108
   in fatty acid biosynthesis 341
   inhibition of 109
   lipoic acid in 108
   phosphorylation/dephosphorylation of 109
   pyruvate decarboxylase in 108
   reaction of 108
   thiamine pyrophosphate in 108

pyruvate kinase **81**, 94, 134, 323, 429
   allosteric behaviour of 94
   in *S. minutum* 94
   isozymes of 48, 82, 94
   phosphorylation of 96
pyruvate oxidation 133, diagram 179
   effect of thiamine pyrophosphate on 188
pyruvate phosphate dikinase 283, 290
   activation of 282
   modulation by light 284
   in CAM plants 113
pyruvate translocator, cation ion gradient in 313
pyruvate, passive diffusion into mitochondrion 178
pyruvate/PEP ratios 429
P:O ratios 402

quantum 212
   energy of 212
quantum energy, flux of 236
quantum yield, at low light 425
Q-cycle 127, 133

r locus in peas 25
raffinose 438
rapeseed oil 492
rate-determining enzymes 36
rate-determining reactions 29
rbcL gene 14
reassimilation of ammonia 258
reducing power, competition between CO_2 and N assimilation 426
reductive carboxylic acid cycle 239
replicon in plasmid vectors 481
reporter gene 7, 23, 475
repressor 4, 5
respiration in the chloroplast 425
respiration
   during photosynthesis 422, 424
   effect of N assimilation on **427–429**
   effect of photosynthesis on **422–425**
   in supply of carbon skeletons 427
   interaction with N assimilation (figure) 428
respiratory control ratio 130
respiratory electron transport chain 257
respiratory metabolism, effectors of (table) 128
restriction fragment length polymorphism (RFLP) 17
Rhizobium 376
   symbiotic genes **378–381**, diagram 379
   legume symbiosis 378

ribose 5-phosphate 85, 242, 411, 417
ribosomal protein synthesis 6
ribosome binding, mRNA cap in 9
ribulose 5-phosphate 84, 85, 86, 240, 243
ribulose 1,5-bisphosphate 224
   regeneration of 315
   enolization of 231, 232
ribulose 1,5-bisphosphate carboxylase 93, **224–237**, 240, 243, 244, 253
   activase 225, 234, 235, 245
   activation and catalysis of (diagram) 230
   activation by light 245
   affinity for CO_2 243
   amino acid sequence of 227
   amount in chloroplast 225
   amount in leaf 233, 234
   and ribulose 1,5-bisphosphate concentration 234
   as reserve of protein 442
   assembly of 229
   availability of substrates 226
   binding protein for 225
   binding site for gases 231
   carbamylation of 230
   catalytic sites of 225
   conformation of 230
   cost of oxygenase activity 262
   deficient mutants 425
   effect of CO_2 and O_2 on 234, 245
   effect of pH on 245
   effectors of 225, 230
   energy for CO_2 and O_2 fixation (table) 263
   equilibrium of 243
   gene for large subunit 14, 225
   gene for small subunit 11, 12, 225
   genes for subunits 228
   in vivo regulation **232–234**
   inhibition of 226
   kinetic properties of (table) 233
   k_m values for substrates 254
   mechanism of catalysis (diagram) 231
   mechanism of oxygenase reaction 232
   modulation of 225
   oxygenase activity of 224, 243, 253, 259, 497
   oxygenase products 254
   properties of **224–226**
   reactions catalyzed by (diagram) 225
   regulation by carboxyarabinitol 1-phosphate 234, 235, 245
   regulation of **245**
   ribulose 1,5-bisphosphate binding site density of 427
   structure **226**

substrate concentration for 245
subunits of 225
synthesis and assembly of **227–230**, diagram 228
three dimensional structure (diagram) 226
uptake of small subunit 229
ricin D 457
ricinine 492
ricinoleic acid, as an industrial oil 491
Ricinus communis, toxins in 491
rifampicin 203
RNA maturases 204
RNA polymerase 4, 5
RNA polymerase II 10, 12, 205
  binding site for 5
  initiation of activity of 5
RNA, hairpins in 5
RNase P 204
root exudates (diagram) 380
root nodules 331, 378, 381
  differentiation of 378
  major groupings of 378
  oxygen barrier in 392
root pressure 434
rotenoids 495
rotenone 130, 132, 137, 186
  resistant pathway 424
rotenone-insensitive dehydrogenase 130, 135–137
rotenone-insensitive pathway 182, 186, 187
rotenone-sensitive dehydrogenase 135–137
rotenone-sensitive pathway 186, 187
round-seeded peas 25
rubber biosynthesis 360, 366

$S_{0.5}$ values 31
salicyl-hydroxamic acid 131
second messenger 38
secondary metabolites 411
secondary plant products (table) 496
sedoheptulose 1,7-bisphosphatase 242, 244
sedoheptulose 1,7-bisphosphate 242
sedoheptulose 7-phosphate 84, 85, 242
seed coat, transport across 440
seed drying 466
seed protein gene expression 24
seed proteins, classification of 456
seed storage proteins 16, 448
  composition of (table) 457
seed storage protein genes 17
  arrangement of 17
  expression of 24

selectable marker gene 481
senescence 449
senescence of leaves 367
sequential enzymes, association of 37
serine 117, 181, 254, 417
  metabolism in mitochondrion 116
serine acetyltransferase 415, 416
serine hydroxymethyl transferase 116
serine synthesis **413–414**
  from phosphoglycerate 414
  photorespiration in 413
serine transhydroxymethylase 256
serine:glyoxylate aminotransferase 255, 331
  in peroxisome 257
sesterterpenes 353
shikimate synthase 52
shikimate:$NADP^+$ oxidoreductase 52
shikimic acid pathway 52, 417
shikonin 496
Shine–Delgarno sequences 6, 7, 14
sieve tubes 435
  blocking of 436
  loading 440, 435
  proton co-transport in 440
  sucrose concentration in 435
  symplastic sap in 435
  unloading 440
sieve tube/companion cell complex 440
sigmoidal saturation kinetics 31
signal hypothesis 303
signal metabolite for protein kinase 35
signal molecule 38
signal peptide 303
silencer element 12
silicone oil 184
silicone oil centrifugation 309, diagram 310
singlet state 215, 216
SL (sulphoquinovosyl diacylglycerol) 340
solvent capacity of a cell 37
sorbitol 438
soyin 463
spermidine 416
spermine 416
squalene 353, 357, 363, 366
stachyose 438
starch 195
starch accumulation pathway (diagram) 323
starch biosynthesis 33, 48
starch breakdown **70–72**
starch degradation **70–72, 320–321**
  hydrolytic 321, 322

in non-photosynthetic tissues **321–322**
   influence of phosphate on (table) 321
   phosphorylytic 320, 321
   routes of (diagram) 321
starch formation from sucrose **322–324**
starch granules 66, 70
starch metabolism **65–72**
starch mobilization 322
starch phosphorylase 67, 68, 70–71, 287
starch synthase 66–69, 286
   multiple forms of 69
starch synthesis 60, **66–70**
   pathway (diagram) 311
   effect of F2,6BP on (table) 319
   influence of phosphate on (table) 315
   regulation of 318
starch accumulation 322
starch, importance of 65–66
start codon 6
start of transcription 11
state 1/state 2 transitions 221, diagram of 222
state 3 respiration 181, 182, 186
state 4 respiration
stearic acid 340
stearoyl-ACP 342
sterol esters 339
sterols 353
stomata 434
stop codon, 18 6
storage glycoproteins, structure of (diagram) 463
storage oils 349
storage protein accumulation
   effect of temperature on 465
   effects of environment on **465–466**
   effects of nutrition on 465
storage protein genes 18, 23
   cis-acting factors in 459
   expression of 18, 22
   introns in 18
   mutations in 18
   role of intron in 459
   temporal regulation of 459
   trans-acting factors in 459
   upstream regulatory elements 459
storage protein mobilization **466–468**, diagram 466
   in cereals 467
   in legumes 467
   proteases in 466
storage proteins 16, **456–469**
   accumulation of 458
   amino acid composition of 491

   composition of 456
   control by translation of synthesis 459
   cotranslational removal of signal peptide 462
   degradation of mRNA for 459
   genes for 491
   glycosylation of 462, **463–464**
   modification of oligosaccharide chains on (diagram) 464
   mRNA degradation during seed drying 466
   mRNA for 458, figure 459
   post-translational processing 460
   signal peptide of 460, 465
   soybean 21
   structure of 456
   synthesis of **457–461**
   synthesis (diagram) 460
   targeting to protein bodies **464–465**
   types 456–457
stroma of chloroplast 195, 196
suberin 339
succinate 110, 113, 114, 126, 131, 140, 178, 351, 436
succinate dehydrogenase **110, 126**
   bound FAD in 110
   iron-sulfur clusters in 110
succinate oxidation 133, 177
succinate/malate exchange 179
succinate dehydrogenase **126**
succinyl-CoA 110
succinyl-CoA ligase **110**
sucrose availability 439
sucrose breakdown **63–65**
sucrose metabolism **60–65**
sucrose phosphatase 61, 62
   allosteric regulation of 62
   inhibition by sucrose 62
sucrose phosphate synthase 61, 62, 286
   activation by glucose 6-phosphate 63
   activation by hexose phosphates 321
   allosteric effectors of 318
   covalent modification of 318
   endogenous rhythms of 319
   regulation of 63
sucrose synthase 61–64, 89, 381
sucrose synthesis **61–63, 248–249**, 286
   control of 316
   effect of phosphate on 315
   enzymes in cytosol 311
   in cytosol 62
   pathway (diagram) 311
   regulation of (diagrams) 317
sucrose to starch conversion 322–324

sucrose
 transport 63
 accumulation of 62
 role of 60
 storage in vacuole 64
 translocation of 60
 uptake into sieve tube 440
sucrose/H^+ antiport in tonoplast 62
sugar alcohols, transport of 61
sulfolipid 345
sulfonylureas 498
sulfur amino acids **415–416**
 in leaves 491
sulfur-rich hordeins 24
sulphoquinovosyl diacylglycerol (SL) 340
symplast, in sucrose transport 63
symport 176

targeting of proteins to organelles 486
TATA sequence 10, 18
termination signals 475
terminator 6
terpene 353
terpene biosynthesis **353–367**, diagram 354
 regulation by phytochrome 367
 regulation of **366–367**
terpene catabolism 367
terpenes, characteristics of (table) 354
terpenoids
 biosynthesis of **362–366**
 coupling reactions 363
 cyclization of 363
 prenyl transferase reactions of 363
 prenylation reactions 363
tetrahydrofolate 256
tetrahydropteroyl-L-glutamate 181
tetramethyl-*p*-phenylene diamine 133
tetraphenylphosphonium 132
tetrapyrroles 118
tetraterpene biosynthesis (diagram) 365
tetraterpenes 353
thermogenesis 114
thiogalactoside transacetylase 4
thioredoxin 35, 95, 245–256
 activation of enzymes by 246
thioredoxin *f*
thioredoxin *m*
thioredoxin reductase 35
threonine dehydratase 415
threonine synthase, regulation of 415

threonine synthesis **414–415**
thylakoid membrane, model of structure
 213
thylakoids 194, 195
Ti (tumor inducing) plasmid 302, 475, 476
 vir (virulence) region 476, 477
tissue culture **473–474**, diagram 473
 organogenesis in 474
 regeneration in 474
tocopherol 363
tonoplast membrane 62, 449
tonoplast proton pumps (diagram) 289
topoisomerase 204
tracheal sap 436
transaldolase **84**
 isozymes of 84
 Schiff base in 84
transaminases 52, **411–412**
 kinetic constants for 412
 pyridoxal 5-phosphate in 411
transamination reactions **411–412**
transcription 4
 attenuation of **5, 6**
 control of 5,6
 regulation of **10, 12**
 RNA polymerase II
 start of 11
 TATA box in 10
 under anaerobic conditions 11
transformation **473–489**
transformed cells, phytohormone independence of
 479
transgenic plants 22, 23, 482
 insect resistance in 487
 virus resistance·in 487
 tobacco 23
transit peptide 301, 303, 450
 amino acid composition of 303
 of small subunit of rubisco 303
transketolase **85**, 242
 as a glycoaldehyde transferase 85
 isozymes of 85
 thiamine pyrophosphate in 85
translation
 initiation of 10
 complementation interference of 7
 control by antisense RNA 6
 control of **6**
 inhibitors of 13
 initiation codon for 7
 initiation of zein storage proteins 9

initiation region for 6
stalling of 6
translocation **434–447**
transpiration 434, 445
transplicing of mRNA 14
transport of nitrate (facilitated) 438
transport of nutrients 434
transported substrates, identity of **436–438**
transposable element 7
transposable gene 7
transposon 19
trans-acting factor **5**, 11, **12**, 22, 23, 25, 382
trans-acting regulatory proteins 23
trans-splicing of RNA 204
triacylglycerol 339
triacylglycerol synthesis (diagram) **349**
   storage bodies for 351
triazines, inhibition of electron transport by 498
triose phosphate 311
triose phosphate isomerase **79**, 241, 242
   cDNA clone for 79
   equilibrium position for 312
   in plastids 48
   multigene family of 79
   multiple forms of 79
   role of 79
triose phosphate shuttle 398
triterpenes 353
trypsin inhibitor genes 20
tryptophan synthesis **417–418**
tunicamycin 464, 465
turnover of RNA 13
tyrosine 412
tyrosine synthesis (diagram) 417–418
T-DNA (transferred-DNA) 476
   border repeats of 477, 481
   function of genes in plant 479
   mechanism of insertion into nuclear genome 479
   production of phytohormones by 479
   structure in transformed plants (diagram) 479
   transfer to plant 476

ubiquinone 125, 127, 128, 131, 133, 135, 357, 363
   structure of 126
ubiquitin 450, diagram 449
   protein conjugates 450
   conservation of 450
   role in metabolic modulation 450
   role in phytochrome action 450

ubiquitin-protein lyases 450
ubisemiquinone 127
UDP-arabinose 74
UDP-galactose 74, 345
UDP-glucose 62, 64, 65, 67, 68, 73, 74
UDP-glucose pyrophosphorylase 61, 65, 68, 74
UDP-glucuronic acid 74
UDP-xylose 74
uncouplers 131, 139, 185
uptake hydrogenase (HUP) 390, 391
   expression of 390
   gene for 390
urease 457
ureide synthesis **419–420**
ureides 378
   export of 437
   in nitrogen transport 419
uricase 332, 381

vacuole
   acid hydrolases in 449
   protein storage in 449
   role in protein storage 465
   site of protein degradation 449
   sucrose in 62, 64
valine synthesis **415**
valinomycin 131, 132, 180
vectors, co-integrate type 482
verbascose 438
vibrational energy, resonance transfer of 216
vicilin 17, 21, 22, 25, 457, 460, 463
violaxanthin 366
vir region, as a regulatory unit 479
vir region, role in T-DNA transmission 479
viral genomes 482
viral vector 483
virus-mediated gene transfer **482–483**
   advantages of 482
   high copy numbers in 482
   limitations of 483

water splitting reaction 219
wax esters 339
wheat, lodging of 497
wound respiration 418
   aromatic amino acids in 418
   glycolysis in 418
   pentose phosphate pathway in 418
   phosphofructokinase in 418

pyruvate kinase in 418
wrinkled seeded peas 25

xanthine 419
xanthine dehydrogenase 419
xanthophyll 214, 353, 366
xylem 445
    solute concentration in 435
xylem exudates, composition of (diagram) 437
xylem exudates, relative composition of (diagram) 438
xylem sap, pH of 436

xylem transport 434
xylogenesis 115
xylulose 5-phosphate 85, 86, 242, 243

zein 9, 10, 17, 21, 457,
    glutamine content of 19
    structural genes 25
    synthesis 24
zein gene 12, 17, 24
    stop codon in 18
    Z-scheme (diagram) 219